Vertex Operator Algebras and the Monster

This is Volume 134 in
PURE AND APPLIED MATHEMATICS
H. Bass, A. Borel, J. Moser, and S.-T. Yau, editors
Paul A. Smith and Samuel Eilenberg, founding editors
A list of titles in this series appears at the end of this volume.

To our parents

Contents

Preface

This work grew out of our attempt to unravel the mysteries of the Monster, the most exceptional finite symmetry group in mathematics. The Monster creates a world of its own and many of the mysteries reflect the unity and diversity of this mathematical world. We began struggling with the Monster even before it was known to exist, as it was starting to reveal its true beauty. We have been able to solve some of the problems and to shed light on others, and we have added a few new ones.

The announcement of our main results was published in 1984. In writing up the exposition of the details of our work, we had to quote basic results from a new and growing mathematical subject—vertex operator representations of affine Kac-Moody algebras—including work of ours. This subject has developed into a theory of vertex operator algebras, and we decided to present a systematic introduction to this theory, in conjunction with our approach to the Monster, in book form.

At the early stages of our work we were well aware of formal connections between the Monster and string theory, a physical model of the real world not very popular among physicists at the time. In 1984, after our announcement happened to have appeared, string theory experienced a dramatic resurgence among physicists. We have now been observing deeper and deeper interactions between the mathematical world of the Monster and the physical world reflected in string theory. It has turned out that the theory of vertex operator algebras, which provides a framework for our results, is essentially the same as two-dimensional conformal quantum field theory, which provides a framework for string theory. Our methods and constructions appear to be relevant in this branch of theoretical physics, and beautiful ideas emerging in physics are being developed mathematically. In this spirit, our main theorem can be interpreted as a quantum-field-theoretic construction of the Monster and in fact as the statement that the Monster is the symmetry group of a special string theory—the first of a new type now understood as the orbifold theories. String theory may or may not blossom into a true description of fundamental physical reality, but mathematics of course does not depend on the whims of nature!

The main body of this book, written in the language of algebra, has almost no prerequisites. The book can serve as an introductory text for graduate students and researchers in mathematics and physics interested in vertex operator representations of affine algebras and the theory of vertex operator algebras and algebraic aspects of conformal field theory, or in the Golay code, the Leech lattice, the Monster and monstrous moonshine. Depending on how many initial chapters are covered, the book can be used for study programs of different lengths; for instance, the first four, seven, eight, nine or ten chapters form what might be considered essentially complete texts in themselves. Practically every chapter depends on all the preceding ones, but the reader might wish to skip many details on first reading in order to understand the flow of ideas. A few individual sections are independent of the preceding material. Our book is also a detailed research monograph which contains the complete proofs of our previously announced results.

It was far beyond the scope of the present work to include expositions of all the fields related to the Monster, such as finite group theory, modular function theory and string theory. However, since the interactions with these fields will certainly fascinate future researchers as some of them fascinated us several years ago, we have decided to include an extensive Introduction describing the interactions and their history, in particular pointing out the connections with physics. An important purpose of the Introduction is to motivate and explain the ideas developed in the text. The Introduction also states a basic conjecture about the uniqueness of our construction. This would yield the most canonical definition of the Monster as a symmetry group.

It is a pleasure to acknowledge the contributions of many people from whose insights, assistance and encouragement we have benefited greatly.

First of all, we would like to thank Richard Borcherds, whose interpretation and further development of our announced results helped us to deepen the relation between our constructions and conformal field theory. His announced results in this direction have become an integral part of this book. We are indebted to Jacques Tits for his sustained interest in our work over several years and for kindly sending us his own work related to the Monster, including unpublished mimeographed letters. His reinterpretation of Robert Griess's construction of the Monster was particularly important to us at an early stage of our research. Robert Wilson went through extensive portions of the manuscript and made many insightful comments, which we very much appreciate. We are grateful to Huang Yi-zhi for many valuable comments which are being expanded as part of a separate joint paper on the axiomatic aspects of conformal field theory. We thank Dong Chong-ying for discussions which led to the simplification of an argument. Many physicists have helped us in understanding the connections between the mathematical and physical viewpoints, and we are especially grateful to Orlando Alvarez, Daniel Friedan, Peter Goddard, Jeff Harvey, Bernard Julia, Greg Moore, David Olive, Steve Shenker, Cumrun Vafa and Edward Witten. We are glad to acknowledge stimulating discussions about the Monster and the moonshine module with John Conway, Pierre Deligne, Howard Garland, Robert Griess and John McKay. We also thank Stefano Capparelli, Byoung-Song Chwe, Jim Cogdell, Art DuPré, Cristiano Husu, Peter Ostapenko, Shari Prevost, Wan Zhe-xian, Jan Wehr, Zhou Shan-you and especially Peter Landweber and Richard Pfister for their comments and corrections. In addition, we are grateful to Adelaide Boullé, Irene Gaskill, Sandy Lefever, Lynn Lewis and Barbara Miller for their assistance with various parts of the manuscript. We deeply appreciate Armand Borel's mathematical interest in the present work and his willingness to be an editor. We are happy to acknowledge the excellent work of Academic Press in their careful and rapid publication of this book.

Finally, we thank Marina Frenkel and Lael Leslie for their very special help and encouragement.

This work was done largely at the Institute for Advanced Study, the Mathematical Sciences Research Institute and Rutgers, Stockholm and Yale Universities. All three authors gratefully acknowledge the long-term support of the National Science Foundation. In addition, I.F. thanks the Sloan Foundation, J.L. the Guggenheim Foundation and A.M. the Swedish Natural Science Research Council, the Swedish Institute and the Wenner-Gren Foundation for their generous support.

Igor Frenkel, James Lepowsky, Arne Meurman

Introduction

The power and beauty of mathematics depend to a great extent on the interrelation of general and special structures. These can influence each other in both directions. Sometimes general theories lead to distinguished examples, as often occurs with classification theorems. In other cases the understanding of a particular object gives rise to a significant general theory. It can also happen that both aspects of the theory develop in close cooperation. A characteristic feature of modern mathematics is its love of generalities, but this has not prevented the discovery of some exceptional structures which have held their own against the best examples of classical mathematics. One such modern structure, the largest sporadic finite simple group, called the **Monster**, is one of the main objects of study in this book. In spite of its name, the Monster is a remarkably beautiful mathematical entity which brings more surprises than one could have originally imagined. It is a symmetry group of a certain special structure which will be the other main object of our study, and which is an example in a general theory of **vertex operator algebras**. The latter theory was motivated by, and in fact combines, several developments in mathematics and physics.

It has been observed frequently in mathematics that apparently unrelated theories can lead to the same special objects. This often indicates a potential for new relationships yielding further development of these theories. The discovery of the Fischer–Griess Monster—also called the Friendly Giant—was one of the high points of *finite group theory*. A few years after this group was predicted to exist, and even before the group was constructed by Griess, mathematicians began to accumulate numerous astonishing facts, now collectively known as monstrous moonshine, on its relation to the theory of *modular functions*. The conceptual explanation of some of these coincidences was to emerge out of a third mathematical theory—that of *Lie algebras*. The close relation of finite groups and finite-dimensional Lie algebras had been known for several decades. But sporadic groups, which include the Monster, lay by definition beyond this relation. Approximately at the time the Monster first appeared, mathematicians noticed and began to explore the relation between modular functions and a new class of infinite-dimensional Lie algebras, known as affine Kac-Moody Lie algebras. The missing link connecting the Monster and affine Lie algebras was hinted at by similarities in their constructions—the former via automorphisms of the Griess algebra (or Griess–Norton algebra), the latter via Lie algebras of vertex operators. These coincidences and hints called forth for the present work on a natural vertex operator construction of a "moonshine module" for the Monster, equipped with a Monster-invariant action of an "affinization" of the Griess algebra. This work also includes, with the help of Borcherds' insight, the construction of a Monster-invariant vertex operator algebra structure, generated by our original structure, on the moonshine module. A single exceptional structure appearing at the intersection of several branches of mathematics has contributed to the development of a new type of algebra which may prove to be no less fundamental than associative algebras or Lie algebras.

Our motivations would not be complete, however, without our mentioning a fourth theory, arising this time from physics rather than mathematics. While physical theories, perhaps more often than their mathematical counterparts, typically start from particular structures, called "models," many of the theories that have proved the most successful in describing the fundamental laws of nature have deep conceptual bases. General relativity, quantum mechanics and Yang-Mills theories are well-known examples. Beginning around 20 years ago, physicists have been developing a new theory, known as *string theory*, which at present is the only candidate for a theory combining all the fundamental interactions. It has grown from a very special "dual resonance model" into a vast area encompassing many structures. From the early days of dual resonance models, modular functions have entered significantly. Vertex operators were first introduced in this theory in connection with describing particle interactions at a "vertex" and later, representations of affine Lie algebras by means of vertex operators have played an important role in string theory. But among the relations between string theory and the branches of mathematics that we have mentioned, the most remarkable might be the new one between string theory and the Monster. One foundation of string theory is two-dimensional conformal quantum field theory (the two dimensions corresponding to the world-sheet of the string) in the critical space–time dimension 26—or 24, in light-cone gauge. The recent formulation of two-dimensional holomorphic conformal field theory is essentially equivalent to the theory of vertex operator algebras. The vertex operator algebra associated with the Monster has exactly the critical dimension of string theory, and in fact, the Monster is precisely the symmetry group of a certain conformal field theory which may be understood as a string theory. One can perhaps assert that fundamental aspects of string theory are to a great extent unavoidable consequences of the symmetry contained in the Monster.

This book presents on the one hand an elementary, systematic and detailed introduction to a general new mathematical theory of vertex operator algebras, the algebraic counterpart of two-dimensional holomorphic conformal quantum field theory, and on the other hand, the construction of what is arguably the most exceptional mathematical structure: an infinite-dimensional $\mathbb{Z}$-graded representation of the Monster with the modular function J as the generating function of the dimensions of its homogeneous subspaces. Many exceptional as well as canonical structures, such as the Golay error-correcting code, the Leech lattice, the exceptional Lie algebra E_8, triality for the Lie algebra $\mathfrak{sl}(2)$ and the orthogonal Lie algebra $\mathfrak{o}(8)$, Heisenberg algebras, affine algebras and the Virasoro algebra, are all facets of this single mathematical object, combining coincidences beyond reasonable expectations. Its creation is the result of the extensive work of many people from several branches of mathematics and physics. We are still far from a complete understanding. What is certain, however, is that the following areas of mathematics and physics, mentioned above, play basic roles:

I. Modular functions
II. Finite groups
III. Lie algebras
IV. String theory.

We shall develop in this book the theory of vertex operator algebras as an outgrowth of special constructions of basic representations of affine Lie algebras. The other three areas will occupy a lesser part in our presentation. However, we would now like to sketch the main ideas of our work in relation to all four areas, since the connections strongly influenced the present work and, we believe, will be useful for further perspectives. Thus we recall some facts from the history of the four indicated fields of mathematics and physics, but only those facts that we view as directly related to the subject of the book. After the section on Lie algebras, we sketch the main steps and ideas of the present work, which is in fact an offspring of that theory. The reader should be aware that the main text of this book, apart from motivation, is independent of the following survey.

I. Modular Functions

The theory of modular functions arose from the theory of elliptic functions, first studied by Abel, Gauss and Jacobi in the first half of the nineteenth century. Many facts and identities are collected in Jacobi's book [M1]. Elliptic functions are functions of one complex variable which are doubly periodic, or more generally, which transform by certain factors under translation by a lattice $\mathbb{Z}\omega_1 + \mathbb{Z}\omega_2$, where $\omega_1, \omega_2 \in \mathbb{C}\backslash\{0\}$ and $\omega_1/\omega_2 \notin \mathbb{R}$. It was realized very early that the dependence of elliptic functions on their periods ω_1, ω_2, and essentially on the ratio $\tau = \omega_1/\omega_2$, is particularly interesting. The group $\mathrm{SL}(2, \mathbb{Z})$ of integral 2×2 matrices of determinant 1 acts by linear transformations on the lattice generated by the periods, inducing an action of the modular group

$$\Gamma = \mathrm{PSL}(2, \mathbb{Z}) = \mathrm{SL}(2, \mathbb{Z})/\langle \pm 1\rangle \tag{1}$$

on the upper half-plane $H = \{\tau \in \mathbb{C} \mid \operatorname{Im} \tau > 0\}$ given by:

$$g \cdot \tau = \frac{a\tau + b}{c\tau + d}, \quad \text{for} \quad g = \pm\begin{pmatrix} a & b \\ c & d \end{pmatrix} \in \Gamma, \qquad \tau \in H. \tag{2}$$

It is natural to look for functions on H invariant under Γ. But it was only in 1877 that Dedekind [M2], and later, independently, Klein [M3], constructed an example of such a function $j(\tau)$. Let us set $q = e^{2\pi i\tau}$. An explicit expression for $j(\tau)$ can be obtained from the following two functions:

$$\Delta(\tau) = \eta(\tau)^{24} = q\prod_{n=1}^{\infty}(1 - q^n)^{24}, \tag{3}$$

where $\eta(\tau)$ is the Dedekind eta-function, and the theta function

$$\Theta_L(\tau) = \sum_{\alpha \in L} q^{\langle\alpha, \alpha\rangle/2} = \left(\sum_{\alpha \in \Gamma_8} q^{\langle\alpha, \alpha\rangle/2}\right)^3, \tag{4}$$

where L is a lattice which is the orthogonal direct sum of three copies of the root lattice of E_8, here denoted Γ_8. This lattice is the unique (up to isometry) even unimodular lattice of rank 8. The functions (3) and (4) are not modular-invariant, but their ratio is:

$$j(\tau) = \frac{\Theta_L(\tau)}{\Delta(\tau)}. \tag{5}$$

This function is holomorphic on H and defines on $H/\Gamma \cup \{i\infty\}$ a meromorphic function with a simple pole at $i\infty$ and in fact a complex analytic isomorphism with the Riemann sphere $\mathbb{C} \cup \{\infty\} \simeq \mathbb{CP}^1$.

The modular invariant $j(\tau)$ is a fundamental function or Hauptmodul, i.e., the modular functions (the meromorphic modular-invariant functions on $H \cup \{i\infty\}$) comprise precisely the field of rational functions of $j(\tau)$. Up to an additive constant, $j(\tau)$ is the unique Hauptmodul on H having a simple pole at $i\infty$ with residue 1 in q. Another choice of rank-24 even unimodular lattice L in (5) affects only the constant term of $j(\tau)$. With the constant term taken to be zero—a choice not attainable from any such lattice—the modular invariant $J(\tau) = j(\tau) - 744$ (classically, J has designated $j/1728$) has the following Laurent series decomposition:

$$J(\tau) = \sum_{n=-1}^{\infty} a_n q^n = q^{-1} + 0 + 196884q + 21493760q^2 + \cdots. \tag{6}$$

The expansion coefficients of $J(\tau)$, which are all positive integers (except for the constant term), might appear unattractive. As we shall see, it took many years and an accident before their meaning was finally found.

Even before the discovery of the modular invariant $j(\tau)$ it was observed that an important characteristic, denoted by $\lambda(\tau)$, of elliptic functions with periods ω_1, ω_2, is invariant only under a certain subgroup $\Gamma(2)$ of Γ. This and other facts led Klein to the creation of the theory of congruence subgroups [M4]. He introduced a class of principal congruence subgroups $\Gamma(n)$ for any $n > 0$, and a general notion of congruence subgroup Γ' of level n, such that $\Gamma(n) \subset \Gamma' \subset \Gamma$. An important example of a congruence subgroup of level n is the group

$$\Gamma_0(n) = \left\{ g = \pm\begin{pmatrix} a & b \\ c & d \end{pmatrix} \in \Gamma \;\middle|\; c \equiv 0 \pmod{n} \right\}. \tag{7}$$

While Klein was developing his new theory, Poincaré, influenced by a paper of Fuchs [M5], launched a program to study the general class of discrete subgroups of $PSL(2, \mathbb{R}) = SL(2, \mathbb{R})/\langle \pm 1 \rangle$ and corresponding automorphic functions which he called Fuchsian [M6]. These include the groups Γ, $\Gamma(n)$ and $\Gamma_0(n)$, and the automorphic functions are analogous to the modular functions. Klein and Poincaré, using ideas of Riemann, laid the modern foundations of the theory of Fuchsian groups and their associated automorphic functions.

One of the basic results of this theory is that for any Fuchsian group Γ', a suitable compactification of H/Γ' has the structure of a compact Riemann surface. The genus g of this surface is the most obvious characteristic of Γ'. In the special case when the genus of the compactification of H/Γ' is zero the theory of automorphic functions is especially simple: The field of automorphic functions is generated by only one function, $J_{\Gamma'}(\tau)$, determined up to rational transformations and called the Hauptmodul of Γ'. In particular, in the case of the modular group Γ, the surface—the Riemann sphere—has genus zero, and the Hauptmodul of Γ is $J(\tau)$.

For a detailed account of the history of modular functions, see for example [M7].

At this point, we would like to mention some specific facts about groups corresponding to genus zero surfaces. Fricke [M8] investigated the surfaces associated with $\Gamma_0(n)$. In particular, the congruence subgroups $\Gamma_0(p)$ for p a prime provide examples of genus zero surfaces if and only if $p - 1$ is a divisor of 24. One can obtain more examples by adjoining to $\Gamma_0(n)$ the Fricke involution $w_n(\tau) = -1/n\tau$, which may of course be realized as an element of $PSL(2, \mathbb{R})$ ([M8], p. 19.1). The normalizer of $\Gamma_0(n)$ in $PSL(2, \mathbb{R})$ was fully described by Atkin and Lehner [M9]. When n is a prime p, it is just the group $\Gamma_0(p)^+$ generated by $\Gamma_0(p)$ and the Fricke involution w_p. Ogg [M10] completed Fricke's proof [M8] that for p a prime, $\Gamma_0(p)^+$ has the genus zero property if and only if

$$p = 2, 3, 5, 7, 11, 13, 17, 19, 23, 29, 31, 41, 47, 59, 71. \tag{8}$$

Certain other subgroups of the normalizer of $\Gamma_0(n)$ provide further examples of genus zero surfaces. Table 5 of [M11] gives the genus for a number of such groups.

This strange set of prime numbers might have remained one of the numerous mathematical facts which are not supposed to have any special significance. It happened however that Ogg heard a talk of Tits mentioning a certain "sporadic" finite simple group predicted—but not proved—to exist by Fischer and Griess, of order

$$2^{46}\cdot 3^{20}\cdot 5^{9}\cdot 7^{6}\cdot 11^{2}\cdot 13^{3}\cdot 17\cdot 19\cdot 23\cdot 29\cdot 31\cdot 41\cdot 47\cdot 59\cdot 71, \tag{9}$$

approximately 10^{54}. The enormous size of the group was responsible for Conway's name for it—the Monster. Impressed by the coincidence of the two sets of prime numbers, Ogg offered a bottle of Jack Daniels [M10] for its explanation. It was not realized at that time, at the beginning of 1975, that this coincidence was the tip of an iceberg.

II. Finite Groups

The discovery of the Monster was preceded by a long history of development of another branch of mathematics—the theory of finite groups, a subject originally associated with Galois. It is natural to ask for the classification of all the finite groups, yielding the enumeration of all kinds of finite symmetries, although this problem is even nowadays considered too difficult. The building blocks of an arbitrary finite group are simple groups, and the core of the problem is the classification of the finite simple groups. By the end of the nineteenth century, thanks to the work of Jordan, Dickson and others, several infinite families of simple groups were known. In addition, already in 1861, Mathieu had discovered five strange finite groups [F1]. The Mathieu groups were called "sporadic" for the first time in the book of Burnside, who noted that they "would probably repay a closer examination than they have yet received" [F2]. The pioneer of the field in our century was Brauer, who made several crucial contributions to the classification problem [F3]. Most of the finite simple groups, now called groups of Lie type or Chevalley groups, admit a uniform construction in terms of simple Lie algebras, via a systematic treatment discovered in [F4]. But it was not clear at that time how many sporadic groups besides the Mathieu groups might exist.

The modern classification race started with the work of Feit and Thompson in 1962, who proved that every nonabelian finite simple group has even order, or equivalently, contains an involution [F5]. This work made feasible the tremendous classification project led primarily by Gorenstein, resulting, after two decades of work by a large group of mathematicians, in the classification theorem (cf. [F6], [F7]). The classification of the finite simple groups was unprecedented in the history of mathematics by virtue of the length of its proof—over 10,000 pages. The result itself is no less fascinating. Besides 16 infinite families of groups of Lie type and the additional family of alternating groups on n letters, $n \geq 5$, there exist exactly 26 sporadic simple groups, each of which owes its existence to a remarkable combination of circumstances.

Five of the sporadic groups were the ones discovered by Mathieu. The largest of them, M_{24}, which contains the other four, can be realized as the symmetry group of the Golay code, an exceptional 12-dimensional subspace of a 24-dimensional vector space over the 2-element field. In 1969 Conway [F8] constructed his three simple groups from the automorphism group Co_0 of another remarkable exceptional structure—the rank 24 Leech lattice [F9]. Conway's group Co_0 involves, as quotients of subgroups, 12 sporadic simple groups, including M_{24}. In 1973 Fischer and Griess independently predicted the existence and properties of what would be the largest sporadic group—the Monster M, which would involve as subquotients either 20 or 21 of the sporadic groups. Griess, Conway and Norton then noticed that the minimal faithful representation of the Monster would have dimension at least

$$d_1 = 196883. \tag{10}$$

On the assumption that the Monster existed and had an irreducible representation of this dimension, Fischer, Livingstone and Thorne computed the full character table [F10]. Norton observed that the minimal representation would have the structure of a real commutative nonassociative algebra with an associative (in a certain sense) form. In his paper [F11], Griess explicitly constructed an appropriate algebra, exhibited enough symmetries of it, and thereby proved the existence of the Monster[1]. Griess's construction has been simplified in works of Tits [F12], [F13], [F14] and Conway [F15], [F16], and Tits has in fact proved that the Monster is the *full* automorphism group of the Griess algebra. But even in its polished version the Griess algebra does not appear as elegant as the Golay code or the Leech lattice, which have simple characterizations.

Hints that the Monster might in fact be associated with an elegant canonical structure had appeared before Griess announced his construction. We have mentioned already that Ogg, who was working in the field of modular functions, came across some coincidences of his results with the prime power orders of elements of the Monster. From the other side, McKay, who was working in finite group theory, noticed an even simpler relation between the Monster and modular functions: the near-coincidence of the minimal possible dimension of a nontrivial representation of the proposed group M and the first nontrival coefficient of $J(\tau)$:

$$a_1 = d_0 + d_1, \tag{11}$$

[1] Very recently, the Monster has been characterized through the centralizers of its involutions [F23], thus adding the final touch to the classification of the finite simple groups.

where $d_0 = 1$ and is naturally interpreted as the dimension of the trivial representation. Soon afterward, McKay and Thompson found a similar relation [F17]

$$a_2 = d_0 + d_1 + d_2, \tag{12}$$

where d_2 is the dimension of the next larger irreducible representation of M, and further simple relations of this sort. McKay and Thompson conjectured the existence of a natural infinite-dimensional representation of the Monster[2],

$$V = V_1 \oplus V_{-1} \oplus V_{-2} \oplus \cdots, \tag{13}$$

such that $\dim V_{-n} = a_n$, $n = -1, 1, 2, \ldots$ (that is, $J(\tau)$, which will be written as $J(q)$ in the main text, is what we call the *graded dimension* of the graded space V). Thompson [F18] also proposed considering, for any element $g \in M$, the modular properties of the series

$$J_g(\tau) = q^{-1} + (\operatorname{tr} g|_{V_{-1}})q + (\operatorname{tr} g|_{V_{-2}})q^2 + \cdots \tag{14}$$

(the *graded trace* of g), generalizing the case

$$J_1(\tau) = J(\tau). \tag{15}$$

Remarkable numerology concerning these graded traces, called *Thompson series*, was collected and greatly broadened by Conway and Norton in a unique paper "Monstrous moonshine" [F19]. In [F20] Conway describes as one of the most exciting moments in his life the moment when after computing several coefficients of these series using information from the character table of M, he went down to the mathematical library and found some of the series in the classical book by Jacobi [M1], with the same coefficients down to the last decimal digit!

Influenced by Ogg's observation, Thompson, Conway and Norton realized that all the series they were discovering, proceeding "experimentally" from the first few coefficients, were normalized generators of genus zero function fields arising from certain discrete subgroups of $\mathrm{PSL}(2, \mathbb{R})$. They were led to conjecture that there exists a graded representation V of the Monster with all the functions $J_g(\tau)$ having this genus-zero property. Knowing the functions $J_g(\tau)$ determines the M-module V uniquely, and the question was whether it existed, given the list of proposed functions $J_g(\tau)$,

[2] We write V_{-n} rather than V_n according to the mathematical tradition of considering highest-weight (rather than lowest-weight) modules.

one for each of the 194 conjugacy classes of M. (Actually, only 171 of the functions are distinct.) Following Thompson's strategy, Atkin, Fong and Smith (see [F21], [F22]) proved that for each n, the coefficients of q^n in the proposed Thompson series indeed define a generalized character (difference of characters) for the Monster, and using computer calculations they all but completed a proof that they define true characters.

If the conjectured V exists it should have a rich underlying structure responsible for the coincidences of monstrous moonshine. One is led therefore to the rather bizarre conclusion that the most natural representation of the Monster, a *finite* group, might be *infinite-dimensional*. This situation was never imagined before in the history of finite groups. It was not so utterly strange, however, in the theory of Lie algebras, which will be the next subject of our account.

III. Lie Algebras

The counterpart of finite group theory is the theory of continuous, or rather, differentiable, groups, initiated by the Norwegian mathematician Lie. Building the foundations of the theory, he realized in particular that a significant amount of information about a differentiable group is already contained in its algebra of "infinitesimal transformations," now called the Lie algebra of the group. Thus the unitary representations of simply-connected compact Lie groups correspond to the representations of the (complexified) Lie algebras. Lie algebras, being linear objects, are in many respects easier to study than the groups themselves.

Finite-dimensional simple Lie algebras over $\mathbb{C}$ were first classified by Killing [L1], who showed that there are four classical infinite series and found the exceptional cases. Cartan further studied the structure of complex semisimple Lie algebras [L2] and he classified their finite-dimensional irreducible representations [L3]. Developing Killing–Cartan theory, Weyl obtained his character formula for these representations, viewed as representations of compact semisimple Lie groups, as well as the complete reducibility theorem, in [L4]. In the same work he first proposed studying a certain infinite-dimensional representation of a compact semisimple Lie group, namely, the regular representation, which contains all the representations in Cartan's classification.

Another important infinite-dimensional representation of a simple Lie algebra containing all the irreducible finite-dimensional representations and directly related to the subject of this book appeared only after a new turn in the theory of Lie algebras, more than 40 years later. Around 1967, Kac [L5], Kantor [L6] and Moody [L7] introduced and began to study a (usually) infinite-dimensional generalization of finite-dimensional simple Lie algebras. (See also the discussion below of a related independent development in physics.) A few years later, Macdonald found an analogue for "affine root systems" of Weyl's identity, the special case of Weyl's character formula for the trivial one-dimensional representation [L8]. His identities pointed to a profound relation between structures associated with simple Lie algebras and the theory of modular forms. The simplest identity in Macdonald's list was already known to Jacobi and was contained in the same book [M1] that so impressed Conway. Dyson independently found some of the Macdonald identities but "missed the opportunity of discovering a deeper connection between modular forms and Lie algebras, just because the number theorist Dyson and the physicist Dyson were not speaking to each other" [L9].

The nature of Macdonald's identities was clarified when Kac in 1974 [L10] generalized finite-dimensional representation theory to the new class of Lie algebras, Kac–Moody algebras, by deriving an analogue of Weyl's character formula for the family of standard or integrable highest weight modules. In particular, these characters for the affine Kac–Moody algebras $\hat{\mathfrak{g}}$ ($\mathfrak{g}$ a finite-dimensional simple Lie algebra) and their twisted analogues—the algebras that Macdonald's affine root systems are related to—could be expressed in terms of modular functions and, it was later realized, could be given especially simple form in some cases.

In [L11] one of the authors, in collaboration with Wilson, constructed certain standard representations—the basic representations—of the simplest affine algebra $\mathfrak{sl}(2)\hat{}$ by means of some apparently new differential operators in infinitely many variables. This work was generalized in [L12] to all the basic representations of the simply-laced (equal-root-length) affine algebras and their Dynkin-diagram-induced twistings. Garland remarked that the differential operators reminded him of the "vertex operators" that physicists had been using in a theory called "dual resonance theory." The resemblance turned into a complete coincidence in the work of another author and Kac [L13] and in the independent work of Segal [L14] on what we now call the untwisted vertex operator realizations of the basic representations of the simply-laced affine algebras. The untwisted vertex operator construction of $\mathfrak{sl}(n)\hat{}$ was in fact anticipated by physicists (see the discussion below), but we remark that the papers [L11]–[L14] were independent of this. The case of $\hat{E}_8$ later proved crucial in an application to string theory, the modern version of dual resonance theory. The operators of [L11] and [L12] are now understood as examples of twisted vertex operators.

The untwisted vertex operator representations also allowed one to look in a new way at the finite-dimensional simple algebras, viewed as the subalgebras of affine algebras preserving a certain natural grading. The integral structure intrinsic to this construction had been crucial to Chevalley in his theory of finite groups of Lie type and to Steinberg [L15] in his extension of Chevalley's work. The affine analogue of the Chevalley groups was developed by Garland [L16], [L17] (see also [L18], [L19]). All the finite-dimensional irreducible representations of the finite-dimensional simply-laced simple Lie algebras are naturally contained in the untwisted vertex operator representation consisting of the direct sum of the basic modules, making it an excellent model for their study. Vertex operator representations twisted by Dynkin diagram automorphisms similarly yield all the finite-dimensional irreducible representations of the rest of the simple Lie algebras.

Many aspects of the representation theory of affine algebras—also called (extended) loop algebras—are treated in [L20], [L21].

Now, the graded dimensions of the basic representations of the simply-laced affine algebras are given by modular functions for certain Fuchsian groups [L22], [L23]. In particular, the graded dimension for the affine Lie algebra $\hat{E}_8$ is

$$j(\tau)^{1/3} = \frac{\Theta_{\Gamma_8}(\tau)}{\eta(\tau)^8} = q^{-1/3} + 248q^{2/3} + 4124q^{5/3} + \cdots, \tag{16}$$

248 being the dimension of E_8 itself—the smallest nontrivial representation of this Lie algebra.

Following an observation of McKay relating E_8 and $j(\tau)^{1/3}$ in place of the Monster and $J(\tau)$, one of the authors [L24] and Kac [L25] independently remarked the tantalizing, if at first superficial, analogy between the basic $\hat{E}_8$-module and the conjectured infinite-dimensional representation of the Monster. This was the first hint of a possible link between finite groups of Lie type and sporadic groups passing through infinite-dimensional Lie theory.

The untwisted vertex operator representation can be thought of quite generally as a correspondence from lattices to infinite-dimensional graded spaces, $L \mapsto V_L$. For $L = \Gamma_8 \oplus \Gamma_8 \oplus \Gamma_8$, the graded dimension of V_L is

$$\dim_* V_L = j(\tau) = J(\tau) + 744 \tag{17}$$

(cf. [L24]), and for the Leech lattice $L = \Lambda$,

$$\dim_* V_\Lambda = J(\tau) + 24 \tag{18}$$

(cf. [L25], [L26]). While these observations seem to suggest monstrous moonshine, the presence of the constant terms is a symptom of the unnaturality of the two spaces from this point of view. Motivated by a certain formula in [F19], Kac tried modifying the space V_Λ in [L27], but the Monster still does not seem to act naturally on the resulting space either.

The Present Work

Besides the similarity between the graded dimension of the untwisted vertex operator representation of $\hat{E}_8$ (16) and monstrous moonshine (5), (13), there appeared to us to be another one between the Lie algebra E_8 and the Griess algebra. Together these became the starting point for the present work. In our first paper [FLM1] on the subject, we constructed a twisted vertex operator representation of certain affine Lie algebras, a second generalization of [L11], or more generally, another correspondence from lattices to infinite-dimensional graded spaces, $L \mapsto V_L'$. In the special case when L is the Leech lattice Λ, this allowed us to obtain the Griess algebra, or rather, a 196884-dimensional variant $\mathfrak{B}$ of it containing a natural identity element, by means of a "cross-bracket" operation for certain general vertex operators (see below). Subsequently we realized that a model for the conjectured monstrous moonshine space (13) ought to be the $\mathbb{Z}$-graded space

$$V^\natural = V_\Lambda^+ \oplus V_\Lambda'^+ \tag{19}$$

($\natural$ for "natural"), where the symbol $+$ denotes the subspaces fixed by certain involutions of V_Λ and of V_Λ'. In fact we could verify directly that

$$\dim_* V^\natural = J(\tau); \tag{20}$$

it is important that the constant term is 0. Only later was this construction, equipped with its vertex operator structure, interpreted by physicists as the first example of a string theory on an orbifold (see the discussion of string theory below).

Another important property of $V^\natural$ was that the homogeneous component $V_{-1}^\natural$ coincided with the Griess algebra $\mathfrak{B}$, with part of the commutative nonassociative structure visible and with all of the structure canonically determined; Griess had found (his version of) $\mathfrak{B}$ after a parameter-adjustment procedure. Besides a certain manifest symmetry group of $\mathfrak{B}$, the centralizer C of a certain involution in M, Griess had constructed a highly nontrivial involution [F11], using some guesswork and a long verification. This involution could more invariantly be considered as a part of a group S_3

emphasized in [F12], [F13], and we wanted to realize this extra symmetry in a natural way from our point of view.

In order do this and thus to construct the Monster as a symmetry group of the whole infinite-dimensional space $V^\natural$, we introduced a certain permutation group $\mathcal{S}_3$ of linear automorphisms of $V^\natural$ mixing the untwisted and twisted representations, based on a new mechanism—the isomorphism of certain twisted and shifted (untwisted) vertex operator representations of $\mathfrak{sl}(2)\hat{}$. We call this rather general mechanism a principle of triality, partly because it is based on the permutations of a certain basis of $\mathfrak{sl}(2)$ and partly because for the case of the E_8 root lattice as opposed to the Leech lattice, it is reflected in the $\mathcal{S}_3$-symmetry of the Dynkin diagram D_4. A Clifford algebra analogue of the principle in this case implies a remarkable identity found by Jacobi [M1], who called it the "aequatio identica satis abstrusa":

$$\prod_{n\geq 1} (1 + q^{n-1/2})^8 - \prod_{n\geq 1} (1 - q^{n-1/2})^8 = 16q^{1/2} \prod_{n\geq 1} (1 + q^n)^8. \quad (21)$$

This formula also arose in connection with a discovery of Gliozzi, Olive and Scherk concerning the supersymmetry of what was eventually called the superstring (see below). An application to the E_8 root lattice of our approach to triality would lead to an alternate construction of the basic module for $\hat{E}_8$ exhibiting classical triality for D_4 as a consequence of the new triality. This E_8-case is sketched in [FLM3], and was a source of motivation for us, but we do not carry out the details in this book. Briefly, either of the two half-spin modules for $\hat{D}_8$ may be adjoined to the basic $\hat{D}_8$-module to form a copy of the basic $\hat{E}_8$-module, but in the Leech lattice case, the adjoining of only one of the two analogous structures leads to a nontrivial theory.

We defined the Monster as the group of linear automorphisms of $V^\natural$ generated by the group C, which acts in an obvious way, and the triality group $\mathcal{S}_3$. Using triality we also constructed a "commutative affinization" $\hat{\mathcal{B}}$ of $\mathcal{B}$ by means of vertex operators as follows: To any element $v \in V_{-1}^\natural = \mathcal{B}$ we attached a vertex operator which we now denote by $Y(v, z)$,[3] z a formal variable which can also be interpreted as a nonzero complex variable, such that $Y(v, z)$ depends linearly on v and such that

$$\lim_{z\to 0} Y(v, z) \cdot \mathbf{1} = v, \quad (22)$$

$\mathbf{1}$ denoting a distinguished element (a "vacuum vector") in $V_1^\natural$. (This

[3] The symbol Y looks like a vertex diagram in physics!

element is usually denoted $\iota(1)$ in the main text.) The components v_n, $n \in \mathbb{Z}$, of $Y(v, z) = \sum_{n \in \mathbb{Z}} v_n z^{-n-1}$ are well-defined operators on $V^\natural$, and the vertex operators can be understood as formal Laurent series in z. The components v_n, $v \in V^\natural_{-1}$, $n \in \mathbb{Z}$, provide a representation of the algebra $\hat{\mathfrak{B}}$ in the sense of the identity

$$[u_{m+1}, v_n] - [u_m, v_{n+1}] = (u \times v)_{m+n} + \tfrac{1}{2}\langle u, v\rangle m(m-1)\,\delta_{m+n,1} \quad (23)$$

for $u, v \in V^\natural_{-1}$, $m, n \in \mathbb{Z}$, where $\times$ is the commutative multiplication and $\langle\cdot,\cdot\rangle$ is the associative bilinear form on $\mathfrak{B}$. Moreover,

$$\begin{aligned} u \times v &= u_1 v, \\ \langle u, v\rangle &= u_3 v. \end{aligned} \quad (24)$$

We call the left-hand side of (23) the *cross-bracket* of the sequences of operators (u_m) and (v_n) because it is formed from two brackets which "cross." (This definition of cross-bracket is a slight variant of, and is more convenient than, the definition introduced in [FLM1] and [FLM2]. Also, the form $\langle\cdot,\cdot\rangle$ is normalized as in [FLM1] rather than [FLM2].) The representations of the Monster and of the affinization of the Griess algebra on $V^\natural$ are compatible in the following sense:

$$gY(v, z)g^{-1} = Y(g \cdot v, z) \quad (25)$$

for $g \in M$, $v \in V^\natural_{-1}$. The identity (23) for the subalgebra of $\hat{\mathfrak{B}}$ corresponding to the subspace $(V_\Lambda^+)_{-1}$ of $V^\natural_{-1}$, as well as the identity (25) for g an element of the subgroup C of the Monster and v in $(V_\Lambda^+)_{-1}$, follow directly from properties of vertex operators. The most difficult and technical part of our work is to prove the two identities (23) and (25) for the rest of the algebra and group.

As a result of our construction of $\hat{\mathfrak{B}}$, we show that $\hat{\mathfrak{B}}$ acts irreducibly on $V^\natural$. This irreducible action and formulas (24) and (25) imply the identification of our definition of the Monster with Griess's; in particular, the Monster, as we defined it, acts faithfully on $\mathfrak{B}$. We also recover Griess's result that the Monster preserves the algebra structure of $\mathfrak{B}$, and we eliminate the parameter-adjustments and the guesswork appearing in Griess's construction.

The results that we have been summarizing were announced and discussed in [FLM2] and [FLM3]; see also the account [FLM4].

In the course of our work it became clear that our affinization $\hat{\mathfrak{B}}$ of the Griess algebra should be part of a larger vertex operator algebra in the same way that the 196884-dimensional subspace $V_{-1} \simeq \mathfrak{B}$ is part of the moonshine module $V^\natural$. It was also clear to us that the general vertex operator algebra should be closed under Lie bracket, a special case of this statement having been proved before by one of the authors [L28]. But the explicit correspondence between the elements of $V^\natural$ and "general vertex operators," and more importantly, the precise commutation and other relations, were elucidated only in the announcement [L29] of Borcherds.

Motivated partly by our work on the Monster, Borcherds developed a general theory of vertex operators, the main features of which are now understood to have been implicit in string theory. Starting from the untwisted vertex operator representation V_L for the case of an even lattice L, he attached a vertex operator $Y(v, z)$ to each element $v \in V_L$, depending linearly on v, so that (22) is satisfied and so that $Y(\mathbf{1}, z)$ is the identity operator. These operators were also familiar in string theory. He then proved a number of relations for the components of vertex operators. From the algebraic structure of V_L, Borcherds axiomatized the notion of "vertex algebra" and using our announced results, he stated that the moonshine module $V^\natural$ is an example of such an algebra.

We shall modify Borcherds' definition a little, and we shall define a *vertex operator algebra* to be a $\mathbb{Z}$-graded vector space $V = \coprod_{n \in \mathbb{Z}} V_{(n)}$ such that the $V_{(n)}$ are all finite-dimensional and are 0 for n sufficiently small, equipped with a linear map $v \mapsto Y(v, z)$ from V into the vector space of formal Laurent series with coefficients in End V, and with a distinguished homogeneous element $\mathbf{1} \in V$, satisfying a number of conditions: $Y(\mathbf{1}, z) = 1$ and formula (22) should hold for $v \in V$, and the "Jacobi identity" and properties involving the Virasoro algebra, stated below, should hold. The series $Y(v, z)$ are called *vertex operators*. Here the grading is the one which will be defined by conformal weights. It is opposite to and shifted from the gradings that we have been using so far. The elements of $V_{(n)}$ are said to have *weight n*. There are many natural variants of this definition of vertex operator algebra, involving non-integral gradings, generalized Laurent series with non-integral powers of the variables, possibly infinite-dimensional spaces $V_{(n)}$, and so on, and such notions arise naturally in the course of the construction of the chief example $V^\natural$. In this work, our main business is to construct $V^\natural$, and in the process, the much easier examples V_L, and we establish a number of generalizations and analogues of Borcherds' results, including his statements themselves. We also treat a number of axiomatic aspects of vertex operator algebras.

Two major properties of a vertex operator algebra V can be expressed as follows:

$$Y(u, z_1)Y(v, z_2) \sim Y(v, z_2)Y(u, z_1) \tag{26}$$

$$Y(u, z_1)Y(v, z_2) \sim Y(Y(u, z_1 - z_2)v, z_2), \tag{27}$$

where $\sim$ is understood as an equality of complex-valued rational functions which are obtained as arbitrary suitably defined matrix coefficients of the left and right sides of (26) and (27). However, the domains of definition of the two sides are different, so that the vertex operators do not in fact commute, and they do not compose to form an associative algebra; the sense in which the vertex operator algebra is "commutative" and "associative" is only symbolic. The relations (26) and (27) turn out to be equivalent to certain families of identities among the components of vertex operators, or equivalently, to the single generating-function identity:

$$\begin{aligned} z_0^{-1}\delta\left(\frac{z_1 - z_2}{z_0}\right)Y(u, z_1)Y(v, z_2) - z_0^{-1}\delta\left(\frac{z_2 - z_1}{-z_0}\right)Y(v, z_2)Y(u, z_1) \\ = z_2^{-1}\delta\left(\frac{z_1 - z_0}{z_2}\right)Y(Y(u, z_0)v, z_2) \end{aligned} \tag{28}$$

where $\delta(z) = \sum_{n \in \mathbb{Z}} z^n$ (formally the Fourier expansion of the δ-function at $z = 1$), and where the expression $\delta[(z_1 - z_2)/z_0]$ is to be expanded as a formal power series in the second term in the numerator, z_2, and analogously for the other δ-function expressions. Each expression in (28) is to be interpreted strictly algebraically: when the expression is applied to any element of V, the coefficient of each monomial in the formal variables is a finite sum. In particular, it is assumed that for $u, v \in V$ and n sufficiently large, we have $u_n v = 0$, the components u_n of $Y(u, z)$ being defined as above: $Y(u, z) = \sum_{n \in \mathbb{Z}} u_n z^{-n-1}$. Formula (28) is very concentrated and has a great many consequences, for instance, a formula for the commutator of $Y(u, z_1)$ and $Y(v, z_2)$ [L29]. Moreover, the statement that $V^\natural$ is a vertex operator algebra and therefore satisfies (28) implies the identity (23). We call formula (28) the "Jacobi identity" for vertex operator algebras because it exhibits suggestive similarities to the Jacobi identity for Lie algebras. Borcherds has informed us that he too has found this identity, and in fact it is implicit in [L29].

We remark that while the properties (26) and (27) are more simply stated in the complex-variable approach, the property (28) is more natural in terms of the formal-variable approach. Also, the latter approach is valid over any field of characteristic zero. In fact, with suitable interpretation the two approaches are equivalent at every stage. In the text we develop the algebraic formalism, and in an appendix we discuss the complex approach.

Another important property of a vertex operator algebra V is the presence of a distinguished homogeneous element $\omega \in V$ whose vertex operator $Y(\omega, z)$ yields a representation π of the Virasoro algebra, which has basis $\{L_n \mid n \in \mathbb{Z}\} \cup \{c\}$ and commutation relations

$$[L_m, L_n] = (m - n)L_{m+n} + \tfrac{1}{12}(m^3 - m)\delta_{m+n,0}c, \qquad m, n \in \mathbb{Z}, \tag{29}$$

c central. Here $\pi(L_n) = \omega_{n+1}$ (using the component notation above), i.e.,

$$Y(\omega, z) = \sum_{n \in \mathbb{Z}} \pi(L_n) z^{-n-2}, \tag{30}$$

and the operator $\pi(c)$ is a scalar which we call the *rank* of V. In particular, rank $V_L = \operatorname{rank} L$. A homogeneous element v of V is an eigenvector for $\pi(L_0)$, and the eigenvalue agrees with the weight of v. We also have

$$\frac{d}{dz} Y(v, z) = Y(L_{-1} \cdot v, z) \quad \text{for} \quad v \in V. \tag{31}$$

This completes the definition of the term "vertex operator algebra." In such a structure, the operators given by L_{-1}, L_0, L_1 span a copy of the Lie algebra $\mathfrak{sl}(2)$ which annihilates the "vacuum vector" $\mathbf{1}$. In addition,

$$[\pi(L_{-1}), Y(v, z)] = Y(L_{-1} \cdot v, z) \quad \text{for} \quad v \in V \tag{32}$$

$$L_{-2} \cdot \mathbf{1} = \omega \tag{33}$$

$$\omega \quad \text{has weight} \quad 2. \tag{34}$$

Also, if $v \in V$ has weight m, then the operator v_n has weight $m - n - 1$.

It turns out that rank $V^\natural = 24$ and that $\frac{1}{2}\omega$ is the identity element of (our presentation of) the Griess algebra $\mathfrak{B} = V^\natural_{-1} = V_{(2)}$, the subspace of $V^\natural$ of weight 2; the relation between the grading (13) and the present grading is that the sum of the degree and the weight of a homogeneous element is (rank $V^\natural$)/24 $= 1$. The orthogonal complement of ω in $\mathfrak{B}$ consists of lowest weight vectors for the Virasoro algebra.

To establish that $V^\natural$ has the structure of a vertex operator algebra one has to define the vertex operators properly and to establish among other things the relations (26) and (27), or equivalently, (28). Naive use of Borcherds' definition of vertex operators acting on the twisted space V_L' does not provide the correct algebra. In [FLM5] we were able to modify the definition to satisfy the right relations. We remark that the answer again proved to be related to constructions originally developed in string theory.

Combining this result with the rest of our earlier work, including triality, we are able to prove the following:

(A) The $\mathbb{Z}$-graded space $V^\natural$ carries an explicitly defined vertex operator algebra structure with graded dimension $J(\tau)$ and rank 24, and which acts irreducibly on itself.
(B) The Monster acts faithfully and homogeneously on the $\mathbb{Z}$-graded space $V^\natural$, preserving the vertex operator algebra structure as in (25) (with $v \in V^\natural$) and fixing the elements $\mathbf{1}$ and ω.

A large part of this result was stated above; what we had established as of [FLM2] is now understood as the natural generating cross-bracket subalgebra of this structure, already sufficient for constructing and identifying the Monster as the automorphism group of the moonshine module equipped with the relevant vertex operator structure. Aside from the identification of the constructed group, we do not use any work of Griess, although of course this work provided valuable motivation. Combining (A) and (B) with Tits's result [F13], [F14] that the Monster is the *full* automorphism group of $\mathfrak{B}$, we find that the Monster is *precisely* the group of grading-preserving automorphisms of the vertex operator algebra $V^\natural$. In particular, this automorphism group is finite and highly nontrivial. We remark that knowledge of Griess's results or Tits's theorem does not seem to help in proving (A) or (B) or for that matter, our results in [FLM2]. In particular, it seems to be hopelessly difficult to prove *directly* that the automorphism group of $\mathfrak{B}$ acts on $V^\natural$. Weight-one vectors could be exponentiated to continuous symmetries of a vertex operator algebra, but $V^\natural$ has no such nonzero elements. Our main results are valid over any field $\mathbb{F}$ of characteristic zero, but the reader may take $\mathbb{F} = \mathbb{C}$ throughout the book if desired.

While the Griess algebra is rather complicated and special, the notion of vertex operator algebra is natural and general; even though the algebra $\mathfrak{B}$ can be written down in a short space, it is perhaps best viewed as a substructure of a certain very canonical vertex operator algebra. In fact, we conjecture that the vertex operator algebra associated with the moonshine module $V^\natural$ is characterized up to isomorphism (defined in the obvious way) by the following properties:

(a) $V^\natural$ is the only irreducible $V^\natural$-module, up to isomorphism (see below)
(b) rank $V^\natural = 24$
(c) $V_{(1)} = 0$, i.e., $V^\natural$ has no nonzero elements of weight 1.

By a *module* for a vertex operator algebra V, we mean a $\mathbb{Q}$-graded vector space $W = \coprod_{n \in \mathbb{Q}} W_{(n)}$ such that the $W_{(n)}$ are all finite-dimensional and are 0 for n sufficiently small, equipped with a linear map $v \mapsto Y(v, z)$ from V into the vector space of formal Laurent series with coefficients in End W, satisfying all the defining conditions for a vertex operator algebra that make sense. For instance, $Y(\mathbf{1}, z) = 1$, the Jacobi identity should hold and the Virasoro algebra should act on W with c acting as the scalar rank V. Property (a) would be nontrivial to show and is not discussed in this book. Our uniqueness conjecture can be viewed as the natural analogue of the known characterization, due to Niemeier and Conway (see below) of the Leech lattice as the unique (up to isometry) positive definite even lattice which:

(a) is unimodular
(b) has rank 24
(c) has no elements of square length 2.

Given the validity of our conjecture, the Monster, then, is the automorphism group of a unique structure.

The uniqueness conjecture can naturally be broken into two parts: A vertex operator algebra satisfying the three conditions above has graded dimension $J(\tau)$ (with respect to the grading considered initially, not the grading by weights), and any vertex operator algebra with graded dimension $J(\tau)$ and rank 24 is isomorphic to $V^\natural$. The analogue of the first of these two parts in the Leech lattice case—the fact that the theta function of the Leech lattice is a modular form—follows from the Poisson summation formula. One of the next steps in the Leech lattice case is to enlarge the lattice by adjoining certain orthogonal elements of square length 2, and it seems that one should adjoin analogous elements of weight 1 to a given vertex operator algebra to obtain a structure satisfying some but not all of the axioms; our construction is in fact based heavily on this sort of "scaffolding."

The series $J_g(\tau)$ [recall (14)] for our module $V^\natural$ are easy to compute for g in the subgroup C of M preserving the decomposition (19); C is the centralizer of the involution in M defining this decomposition and is the group mentioned above. For g not conjugate to an element of C, these series would be difficult to compute from our construction. It has not been verified in general that our series $J_g(\tau)$ for $g \in C$ equal those of Conway–Norton, although this is very likely true. (It has not even been checked that the two lists $J_g(\tau)$ for $g \in C$ given in [F19] define equal modular functions in general!) It should be noted that the formulas for the series $J_g(\tau)$ given in [F19] have nonzero constant terms which must be subtracted artificially; by contrast, our formulas for $g \in C$ have naturally vanishing constant terms. While the modular functions now appear naturally from our algebraic construction for at least a family of $g \in M$, a geometric explanation of the genus zero property is still missing.

Our construction is based on an element of order 2 in M. It is well known that M contains elements of orders 3, 5, 7 and 13 with analogous properties, and variants of the construction can very likely be carried out based on these primes, using general twisted vertex operators [L30], [FLM5]. The conjectured uniqueness of $V^\natural$ would then relate the different constructions.

Although vertex operator algebras are a new and fundamental mathematical object, they are not so unfamiliar in theoretical physics. It turns out that these algebras and their natural variants are essentially the same as what have recently been termed "chiral algebras" in two-dimensional conformal quantum field theory, a modern approach to, and partly an outgrowth of, string theory. The vertex operator algebra $V^\natural$ is essentially an example of a holomorphic conformal field theory, its graded dimension being modular invariant and its rank being (a multiple of) 24. We now recall some facts about string theory, not just because it played an important role in our work but also because it will most likely be valuable in explaining many of the remaining mysteries of the Monster.

IV. String Theory

The prototype of string theory, dual resonance theory, appeared at roughly the same time as the theory of Kac–Moody algebras. It originated from an observation of Veneziano [S1] who noticed that the Euler beta-function satisfies several properties required by the four-point scattering amplitude of elementary particles. In the next year Koba and Nielsen [S2] wrote the generalized Veneziano amplitude for any number of particles in a form which exhibited its holomorphic structure. Soon afterward Fubini and Veneziano [S3] introduced untwisted vertex operators as a technical tool in describing scattering amplitudes at a "vertex," and proved the factorization of n-point amplitudes. Vertex operators acting on Fock spaces—canonical modules for infinite-dimensional Heisenberg or Clifford algebras—turned out to be a very powerful tool in the dual resonance model, and as we have mentioned, these operators also proved to be crucial in constructions of representations of affine Lie algebras. Using vertex operators, Neveu and Scherk expressed the one-loop amplitude for four particles in terms of modular functions [S4]. Since then modular functions have played a major role in string theory.

Soon, however, problems began to emerge. It was realized, first by Lovelace [S5], that the theory could be mathematically consistent only in dimension 26 and only if just 24 of the dimensions were effective. In these numbers of dimensions various mathematical miracles occur—for example, the symmetry group $O(24)$ of the transversal space has an unexpected extension [S6]—a phenomenon which in retrospect is similar to our Monster construction. Further, by means of a development of vertex operator theory [S7], Brower [S8] and independently Goddard and Thorn [S9] proved the no-ghost theorem. This asserts that the physical space of the 26-dimensional theory has no vectors of negative norm and that the partition function (i.e., what we call the graded dimension) is expressed by the inverse of the 24th power of the Dedekind η-function, i.e., by the partition function of the transversal space.

The operator formalism was developed to a great extent, and the general vertex operators that were later rediscovered by Borcherds were introduced and studied by Fubini and Veneziano [S10]; see also [S11]. In their proposal of current-algebraic internal symmetry for strings, Bardakci and Halpern [S11a], independently of mathematicians, discovered affine Lie algebras in 1971, including the first case of a concrete representation—a fermionic realization of $\mathfrak{sl}(3)\hat{}$; irreducibility issues arose later in independent discoveries by mathematicians. Corrigan and Fairlie [S12] considered the twisted vertex operators later rediscovered in [L11]. Extending earlier work on fermionic strings they also introduced new vertex operators mixing the untwisted and twisted vertex operator representations. Expressions similar to theirs later occurred in our work on general twisted vertex operators. We emphasize that the pioneers of dual resonance theory discovered many important structures before mathematicians, but since they had completely different motivations and goals they did not reveal some important ideas which became apparent only in the representation-theoretic approach. The physicists did not really consider commutators of vertex operators systematically, but see [S7] and [S13], [S14]. The untwisted vertex operator construction of $\mathfrak{sl}(n)\hat{}$-representations from compactified spatial dimensions was implicit in [S15], [S16]. Bars pointed out operator-valued-distribution techniques for computing commutators at an early stage of the work in [L13]. After the general treatment by mathematicians of untwisted and twisted vertex operator representations of affine algebras and of the affinization of the Griess algebra, physicists were able to look at familiar structures from the new point of view. We single out two relevant papers on the subject from the vast literature: Goddard and Olive [S17] on the untwisted representation and Corrigan and Hollowood [S18] on the relation of the untwisted and twisted representations.

A great advantage of the physical picture of the subject is the alternative geometric approach. Already in the first period of the history of dual resonance models, it had been realized through a sequence of developments (see Nielsen [S19], Susskind [S20], Nambu [S21] and Goddard, Goldstone, Rebbi and Thorn [S22]) that the theory admitted an elegant geometric realization in terms of a quantized relativistic string. From this point of view, the more recent untwisted vertex operator construction of affine algebras could in retrospect be interpreted as the theory of a string propagating on the torus $\mathbb{R}^{\operatorname{rank} L}/L$. The geometric interpretation allowed one to use the techniques of path integrals on loop spaces, i.e., path integrals over cylinders and more general two-dimensional surfaces. The path integrals in string theory turn out to possess conformal invariance, which allowed the use of techniques of conformal mappings for their calculation [S23], [S24], [S25]. The traces of operators are given by path integrals on loops in loop spaces, i.e., over two-dimensional complex tori. This implied the manifest symmetry of these traces with respect to the modular group, and in particular, the modular invariance of the partition function (graded dimension). For a collection of reviews on the part of string theory up to this stage and for further references, see Jacob [S26]. A more recent account of the path-integral approach to the free closed bosonic string can be found in Witten [S27].

A completely new physical interpretation of the structures of string theory emerged when Yoneya [S28] and Scherk and Schwarz [S29] considered in detail that string theory, appropriately viewed, could incorporate gravity. In 1977, Gliozzi, Olive and Scherk [S30], using the classical Jacobi identity (21), obtained evidence of 10-dimensional space-time supersymmetry later explicitly constructed in [S30a]. However, it took further discoveries before a larger part of the physics community was convinced of the viability of the new interpretation. It was, incidentally, during a three-year period of quiescence of string theory, 1977–1980, that mathematicians, independently of string theory, started discovering vertex operator constructions (see the discussion of Lie algebras above).

Meanwhile, two important developments occurred in the Soviet Union. The first was Polyakov's discovery, now a basic constituent of string theory, of what could be regarded as a geometric counterpart of the no-ghost theorem [S31]. The second was an influential paper of Belavin, Polyakov and Zamolodchikov [S32] abstracting many of the features of string theory under the name "conformal quantum field theory," with emphasis on operator product expansions. In physical language, formulas (26) and (27) can be understood as "duality" in the sense of the term "dual resonance model."

In 1984 Green and Schwarz [S33] surmounted a fundamental obstruction preventing an application of strings to gravity, and string theory entered its second flourishing period. By that time physicists had at their disposal new results from the theory of infinite-dimensional Lie algebras. In particular, the untwisted vertex operator construction of the exceptional affine algebra $\hat{E}_8$ found its place in one of the most phenomenologically promising string models—the 10-dimensional $(E_8 \times E_8)\hat{}$ heterotic string [S34]. In the new period (see [S35] for a recent survey of string theory), physicists have inherited the old problem of the discrepancy between the critical dimension of string theory—26, or 10 for the superstring—and the observed dimension of space–time. The leading idea is to compactify the six unwanted dimensions in the 10-dimensional $(E_8 \times E_8)\hat{}$ heterotic string to an extremely tiny space. The first such spaces proposed were appropriate Calabi–Yau manifolds [S36]. The simplest example of a suitable compact space—a torus—lacks some important physical features. It becomes physically more sound if one instead compactifies on the quotient of a torus by the action of a discrete group [S37]. The result is usually not a manifold but rather an orbifold. It turns out that our construction of the moonshine module for the Monster gives what amounts to the first example of a theory of a string propagating on an orbifold that is not a torus (see Harvey [S38] for a discussion of "heterotic moonshine" and also [S39]). In this "toy model," which can be thought of as a light-cone formulation, the 26-dimensional bosonic string is compactified on the 24-dimensional Leech-lattice-orbifold—the torus $\mathbb{R}^{24}/\Lambda$ modulo the group $\langle \pm 1 \rangle$—to produce a theory in two-dimensional space-time. One of the main points of the present work is the explicit construction of hidden symmetries of such a theory.

The geometric approach to string theory on an orbifold was developed further in [S40] and [S41], where some of the previous results [S12] were geometrically interpreted in terms of twist operators for bosonic strings. In particular, this geometric interpretation clarifies the modular properties of the associated partition functions. However, the vertex operator algebra associated with $V^\natural$ has not so far been fully translated into purely geometric language.

Our announcement [FLM2] can in retrospect be understood as the construction of the Monster as an automorphism group of the natural conformal-weight-two generating substructure of a chiral algebra which is a holomorphic conformal field theory—the $\mathbb{Z}_2$-orbifold theory associated with the Leech lattice. We would like to emphasize that [FLM2] was contemporaneous with, and certainly independent of, Belavin–Polyakov–Zamolodchikov [S32], and preceded the 1984 resurgence of string theory

stimulated by Green–Schwarz [S33]. Rather than applying conformal field theory, we (and Borcherds) have been developing our approaches to it. Vertex operator algebras (chiral algebras) and conformal field theory are by now an established part of physics. The group-extension methods detailed in Chapter 5 of this book, used in [L30] and [FLM5] for the construction of what can in retrospect be viewed as the twisted sectors of $\mathbb{Z}_n$-orbifold theories and the associated vertex operators for the emission of untwisted states, are beginning to find physical interpretation [S42], as are aspects of what we call triality for $\hat{\mathfrak{sl}(2)}$ [S43]. We suggest that the more subtle aspects of triality and group extensions developed in [FLM2], [FLM3] and this book for the construction of highly nontrivial symmetries of conformal field theories such as $V^\natural$ may also play a role in string theory.

There have been speculations as to whether the Monster is related to the real world. Even before the new explosion of string theory, and at a time when our work was underway, Dyson wrote in [S44]: "I have a sneaking hope, a hope unsupported by any facts or any evidence, that sometime in the twenty-first century physicists will stumble upon the Monster group, built in some unsuspected way into the structure of the universe." One tends to agree with Dyson's own judgment that "this is of course only a wild speculation, almost certainly wrong." However, there could be some truth in it. (See for example [S45].) The Monster undeniably points to a string theory in the critical dimension. Examining one of the major physics problems plausibly accessible in string theory, the vanishing of the cosmological constant, Moore was led to consider partition functions symmetric with respect to Atkin–Lehner involutions [S46]. Studying the genus-zero groups associated with Ogg's list (8), he concluded that "one is thus naturally led to conjecture that the zero genus of $H/\Gamma_0(n)^+$ is connected, through the absence of appropriate weight-two forms, to the vanishing of the cosmological constant in certain theories related to the Monster." Also, the supersymmetry underlying the 10-dimensional superstring has been found in relation to our construction [S47]. The Monster does not now have anything to do with the structure of the universe. One fact, however, is undeniable: As the automorphism group of a distinguished conformal quantum field theory, the Monster is fundamentally related to one of the most spectacular chapters of modern theoretical physics—string theory.

Perspectives and Summary

Although features of a complete theory of the Monster and moonshine are now present, there are still serious open problems. First, one would like to prove the conjectures stated above. This would be analogous to the uniqueness of the Leech lattice and of the Golay code (see [PS1], [PS2]). More distant problems would be a geometric understanding of the genus-zero property and a covariant (in the sense of string theory) 26-dimensional realization of the vertex operator algebra $V^\natural$. It is clear that these two problems would be intrinsically related to string theory. For interesting and similar observations concerning the former issue, from two different viewpoints, see the appendix by Norton in [PS3] and the appendix by Witten and Moore in [S46]. Speculations in the spirit of the latter issue can be found in [PS4].

Besides the four areas of mathematics and physics discussed above, another recently discovered theory—that of elliptic genera in topology [PS5]—is also likely to be valuable for the understanding of geometric properties of the Monster. Elliptic genera have a geometric interpretation in terms of Dirac operators on infinite-dimensional vector bundles over loop spaces with Fock spaces as fibers [PS6]. In addition, the modular function $J(\tau)$ (with no constant term) appears naturally in connection with elliptic genera for manifolds of dimension 24 [PS7], the dimension relevant to the moonshine module.

There is an interesting alternative approach to the Monster, in the spirit of Fischer's original method of predicting the group's existence, based on a proposed presentation of a slight enlargement of the Monster—its wreathed square, the "Bimonster"—as a quotient of a Coxeter group associated with the incidence diagram of the projective plane over the field of three elements [PS8]. This diagram has exactly 26 vertices.[4] Also, the Monster is one of a number of finite groups that have recently been proved to be Galois groups over $\mathbb{Q}$ [PS9], [PS9a].

Other sporadic groups that are quotients of subgroups of the Monster can surely be constructed via substructures and analogues of our construction, using "twisted" vertex operator algebras related to $V^\natural$ and to the variants of $V^\natural$ mentioned above, based on primes other than 2; cf. [FLM3], [L30] and [FLM5]. It is a much more difficult but extremely important question whether any (or all?) of the rest of the sporadic groups can be realized as symmetries of natural vertex operator algebras. As for the finite simple groups of Lie type, we have remarked [FLM3] that they ought to be

[4] As the reader should be convinced by now, coincidences are not to be taken lightly, including that between the number 26 just mentioned and the critical dimension of string theory. We grant, however, that one can safely ignore the fact that the number of sporadic groups is exactly 26.

described using vertex operator ideas in the spirit of the results on $\mathbb{Z}$-forms for the universal enveloping algebras of affine Lie algebras in [L16]–[L19]. This idea is largely carried out in the announcement [L29], where these $\mathbb{Z}$-form results are given a new proof using $\mathbb{Z}$-forms for vertex operator algebras associated with Kac–Moody algebras; the finite simple groups of Lie type will then be automorphism groups of suitably twisted vertex operator algebras in finite characteristic. One can certainly hope for a uniform description of the finite simple groups as automorphism groups of certain vertex operator algebras—or conformal quantum field theories. If such a quantum field theory could somehow be attached a priori to a finite simple group, the classification of such theories, a problem of great current interest among string theorists (see [S32], [PS10]), might some day be part of a new approach to the classification of the finite simple groups. On the other hand, can the known classification of the finite simple groups help in the classification of conformal field theories?

We conclude this introduction with a brief discussion of the contents of this book. Further details and the historical attributions of the results presented in the text are contained in the introductions to the individual chapters. The main results presented in the book have been previously announced and discussed in our series of papers [FLM1–5] and in Borcherds' paper [L29]. The Monster, moonshine and the present work have been reviewed by Tits [F12], [F13], [PS11], [PS12]. See also the expositions [PS13], [PS14].

Virtually everything discussed in the main text of this book enters into the statement and/or proof of the main theorem, but we have arranged the material so that the first several chapters will be of independent interest. Since most of the theory is either only a few years old or presented here in detail for the first time, we have tried to keep the exposition elementary and almost entirely self-contained. For instance, we do not even assume knowledge of Lie algebra theory, except that we quote the Poincaré–Birkhoff–Witt theorem in Chapter 1 and the standard finite-dimensional characteristic zero representation theory of the Lie algebra $\mathfrak{sl}(2)$ in Chapter 11. We also make an effort, especially in the chapter introductions, to explain many correspondences between the mathematical language of the book and the terminology of quantum field theory.

The book may be thought of as divided into two parts—a more general one on affine Kac–Moody algebras and vertex operator algebras (Chapters 1–9) and a more special one on the actual construction of the Monster and the moonshine module and corresponding vertex operator algebra (Chapters 10–13).

In Chapter 1 we first discuss general constructions of algebras and modules. Then we introduce some of the basic objects of this book—affine Lie algebras, Heisenberg algebras and the Virasoro algebra, and graded modules.

In Chapter 2 we introduce a calculus of formal variables and we express affine Lie algebras via formal variables. It will turn out in later chapters that it is more convenient—in fact, essential—to work with generating functions rather than with individual elements of affine algebras.

In Chapters 3 and 4 we construct respectively the twisted and untwisted vertex operator representations of the simplest affine Kac–Moody algebra, $\mathfrak{sl}(2)\hat{}$. We single out this algebra for two reasons: first, the vertex operator construction is considerably simpler than for the general case, while it still shows the main features of vertex operator theory, and second, the two types of vertex operator representations of $\mathfrak{sl}(2)\hat{}$ play crucial roles in our construction of the Monster.

In order to extend the vertex operator representations to general classes of affine algebras we study central extensions of lattices and the finite groups termed "extraspecial 2-groups" and their representations in Chapter 5. These results also provide a foundation for the structure theory of simple Lie algebras presented in Chapter 6.

In Chapter 7 we generalize the untwisted and twisted vertex operator constructions of Chapters 3 and 4 to the affine algebras of types $\hat{A}_n$, $\hat{D}_n$, $\hat{E}_n$.

Chapters 8 and 9 are the culmination of the theory of vertex operator algebras and of the first part of the book. They deal respectively with the untwisted and twisted vertex operator constructions. As we have sketched above, the main idea of the construction is intrinsic to conformal quantum field theory: Any element of the representation space gives rise to a vertex operator, and these form a rich and intricate algebraic structure. In string-theoretic terminology, Chapter 8 presents various kinds of chiral algebras corresponding to two-dimensional conformal field theory on a torus. The last section of Chapter 8, which forms a kind of summary of the chapter, is devoted to the definitions of the notions of vertex operator algebra and module, and to some axiomatic properties, including "commutativity" (26) and "associativity" (27). The latter expresses the "associativity of the operator product expansion," in the language of two-dimensional conformal field theory. The twisted structure developed in Chapter 9 is a module for the algebra of Chapter 8. The simplest nontrivial untwisted and twisted vertex operators provide the vertex operator representations of affine algebras described in Chapter 7.

In Chapter 10 we start the construction of the moonshine module and of the Monster themselves. We first review the exceptional structures, namely, the Golay code and the Leech lattice, which are needed in our construction (as they were in Griess's). Then we begin the study of a particular vertex operator algebra associated with the Leech lattice and we pay special attention to a substructure which we identify with the Griess algebra. In the notation (19), the first summand V_Λ^+ is a vertex operator algebra and the second summand $V_\Lambda'^+$ is a module for it. We also compute the series $J_g(\tau)$ for $g \in C$.

To continue the construction of the vertex operator algebra associated with the moonshine module $V^\natural$, in Chapter 11 we introduce operators mixing the untwisted and twisted representations. We define and establish properties of these operators by means of a triality principle, i.e., a special action of the symmetric group $\mathbb{S}_3$, starting from the permutations of three basis elements of $\mathfrak{sl}(2)$ and using the resulting transport of structure, presented in Chapter 4, between the shifted untwisted vertex operator representations and the twisted vertex operator representations of $\mathfrak{sl}(2)\hat{}$.

In Chapter 12 we formulate the main theorem on the vertex operator algebra associated with the moonshine module $V^\natural$ and on the Monster as its group of grading-preserving automorphisms. In this chapter we define appropriate modifications of the operators introduced in Chapter 11, and we use conjugation by these operators to construct vertex operators, parametrized by the second summand $V_\Lambda'^+$ in (19), interchanging the untwisted and twisted subspaces of $V^\natural$, thus completing the definition of the desired vertex operator algebra. We define the Monster as the group of linear automorphisms of $V^\natural$ generated by the group C, which preserves the untwisted and twisted subspaces of $V^\natural$, and a modified triality group $\mathbb{S}_3$, which mixes the two subspaces. In the main theorem we collect all the final results on the moonshine module $V^\natural$, the corresponding vertex operator algebra and the Monster, and on the relations among them. In particular, the group that we call the Monster coincides with the group that Griess constructed.

Chapter 13 is devoted to the completion of the proof of the main theorem stated in Chapter 12. In order to prove that the Monster intertwines the vertex operators associated to $V^\natural$ and that these operators form a vertex operator algebra, we enlarge the triality group to a symmetry group which is an extension of a group $GL(3, \mathbb{F}_2)$ realized as a permutation group of a 7-element set. This allows us to reduce the proof to already-established properties of the first summand V_Λ^+ of $V^\natural$. The justification of this reduction requires subtle technical analysis with the use of more extensive group theory than before, as well as some group cohomology theory.

In the Appendix, working over $\mathbb{C}$, we develop the theory of vertex operator algebras from the viewpoint of elementary complex analysis, and in particular, we give another proof of the commutativity and associativity properties and of the Jacobi identity for the vertex operator algebras based on even lattices. As an alternative to the formal-variable approach of the main text, here we set up the properties of vertex operator algebras with the help of the convergence of infinite series. We also establish the relationship between the two approaches and we geometrically interpret identities involving δ-functions in terms of contour deformations.

Aside from properties of the Conway simple group Co_1 quoted in Chapters 10, 12 and 13 and group cohomology results quoted in Chapter 13, all the results in the book are developed from scratch. In [PS12], Tits has given a more invariant description of the constructions in Sections 12.1, 12.2 and 13.2 related to the "extra automorphisms" involved in building up the Monster.

The most natural way to view any group is to realize it as the symmetry group of some canonical object, which can come from algebra, geometry, physics or anything else. For a given group there is no well-defined procedure for finding such an object and there is no existence or uniqueness theorem. However, if found, the object becomes firmly tied to the group. Thus the Golay code is the object for the Mathieu group M_{24} and the Leech lattice is the one for the Conway group Co_0. Given its conjectured uniqueness, we suggest that one should think of the vertex operator algebra $V^\natural$ as such an object for the Monster.

References for the Introduction

I. Modular Functions

[M1] C. G. J. Jacobi, *Fundamenta Nova Theoriae Functionum Ellipticarum*, Königsberg, 1829, in: *Gesammelte Werke*, **1**, Chelsea, New York, 1881, 49–239.

[M2] R. Dedekind, Schreiben an Herrn Borchardt über die Theorie der elliptischen Modulfunktionen, *J. Reine Angew. Math* **83** (1877), 265–292.

[M3] F. Klein, Über die Transformation der elliptischen Funktionen und die Auflösung der Gleichungen fünften Grades, *Math. Annalen* **14** (1878/79), in: *Gesammelte mathematische Abhandlungen*, **3**, Springer, Berlin, 1923, 13–75.

[M4] F. Klein, Zur [Systematik der] Theorie der elliptischen Modulfunktionen, *Math. Annalen* **17** (1880/81), in: *Gesammelte mathematische Abhandlungen*, **3**, Springer, Berlin, 1923, 169–178.

[M5] L. Fuchs, Über eine Klasse von Funktionen mehrerer Variabeln welche durch Umkehrung der Integrale von Lösungen der linearen Differentialgleichungen mit Rationalen Coefficienten entstehen, *J. Reine Angew. Math.* **89** (1880), 151–169.

[M6] H. Poincaré, Théorie des groupes Fuchsiens, *Acta Mathematica* **1** (1882), 1-62, in: *Oeuvres*, **2**, Gauthiers-Villars, Paris, 1916, 108–168.
[M7] J. Lehner, *Discontinuous Groups and Automorphic Functions*, American Math. Soc., Providence, 1964.
[M8] R. Fricke, *Die Elliptische Funktionen und Ihre Anwendungen*, 2-ter Teil, Teubner, Leipzig, 1922.
[M9] A. O. L. Atkin and J. Lehner, Hecke operators on $\Gamma_0(m)$, *Math. Ann.* **185** (1970), 134–160.
[M10] A. P. Ogg, Automorphismes des courbes modulaires, *Séminaire Delange-Pisot-Poitou, 16e année* (1974/75), no. 7.
[M11] B. J. Birch and W. Kuyk, eds., Modular Functions of One Variable. IV, *Lecture Notes in Math.* **476**, Springer-Verlag, Berlin–Heidelberg–New York, 1975.

II. Finite Groups

[F1] E. Mathieu, Mémoire sur l'étude des functions de plusieures quantités, sur la manière de les formes et sur les substitutions qui les laissent invariables, *Crelle J.* **6** (1861), 241–323.
[F2] W. Burnside, *Theory of Groups of Finite Order*, Cambridge, 1911.
[F3] R. Brauer, On the structure of groups of finite order, in: *Proc. Internat. Congr. Math.*, **1**, Noordhoff, Groningen, North-Holland, Amsterdam, 1954, 209-217.
[F4] C. Chevalley, Sur certains groupes simples, *Tohoku Math. J.* **7** (1955), 14–66.
[F5] W. Feit and J. Thompson, Solvability of groups of odd order, *Pacific J. Math.* **13** (1963), 775-1029.
[F6] D. Gorenstein, Finite Simple Groups. *An Introduction to Their Classification*, Plenum Press, New York, 1982.
[F7] D. Gorenstein, Classifying the finite simple groups, Colloquium Lectures, Anaheim, January 1985, American Math. Soc., *Bull. American Math. Soc. (New Series)* **14** (1986), 1-98.
[F8] J. H. Conway, A group of order 8,315,553,613,086,720,000, *Bull. London Math. Soc.* **1** (1969), 79–88.
[F9] J. Leech, Notes on sphere packings, *Can. J. Math.* **19** (1967), 251-267.
[F10] B. Fischer, D. Livingstone and M. P. Thorne, *The characters of the "Monster" simple group*, Birmingham, 1978.
[F11] R. L. Griess, Jr., The Friendly Giant, *Invent. Math.* **69** (1982), 1–102.
[F12] J. Tits, Résumé de cours, *Annuaire du Collège de France*, 1982–1983, 89–102.
[F13] J. Tits, Le Monstre, Séminaire Bourbaki, exposé no. 620, 1983/84, *Astérisque* **121-122** (1985), 105-122.
[F14] J. Tits, On R. Griess' "Friendly Giant," *Invent. Math* **78** (1984), 491–499.
[F15] J. H. Conway, A simple construction for the Fischer–Griess monster group, *Invent. Math.* **79** (1985), 513-540.
[F16] J. H. Conway, The Monster group and its 196884-dimensional space, Chapter 29 of: J. H. Conway and N. J. A. Sloane, *Sphere Packings, Lattices and Groups*, Springer-Verlag, New York, 1988, 555-567.
[F17] J. G. Thompson, Some numerology between the Fischer–Griess Monster and elliptic modular functions, *Bull. London. Math. Soc.* **11** (1979), 352–353.
[F18] J. G. Thompson, Finite groups and modular functions, *Bull. London Math. Soc.* **11** (1979), 347-351.
[F19] J. H. Conway and S. P. Norton, Monstrous moonshine, *Bull. London Math. Soc.* **11** (1979), 308-339.
[F20] J. Conway, Monsters and moonshine, *Math. Intelligencer* **2** (1980), 165–172.

[F21] P. Fong, Characters arising in the Monster-modular connection, in: *The Santa Cruz Conference on Finite Groups, Proc. Symp. Pure Math., American Math. Soc.* **37** (1980), 557–559.
[F22] S. Smith, On the Head characters of the Monster simple group, in: *Finite Groups—Coming of Age, Proc. 1982 Montreal Conference*, ed. by J. McKay, *Contemporary Math.* **45** (1985), 303–313.
[F23] R. L. Griess, Jr., U. Meierfrankenfeld and Y. Segev, A uniqueness proof for the Monster, to appear.

III. Lie algebras

[L1] W. Killing, Die Zusammensetzung der stetigen endlichen Transformationsgruppen, II. *Math. Ann.* **33** (1889), 1–48.
[L2] E. Cartan, Sur la structure des groupes de transformations finis et continus, Thèse, Paris, Nony, 1894; 2ᵉ éd., Vuibert, 1933, in: *Oeuvres Complètes* I_1, Gauthiers-Villars, Paris, 1952, 137–253.
[L3] E. Cartan, Les groupes projectifs qui ne laissent invariante aucune multiplicité plane, *Bull. Soc. Math.* **41** (1913), 53–96, in: *Oeuvres Complètes* I_1, Gauthiers-Villars, Paris, 1952, 355–398.
[L4] H. Weyl, Theorie der Darstellung kontinuierlicher halbeinfacher Gruppen durch lineare Transformationen. I, *Math. Zeit.* **23** (1925), 271–309; II, *Math. Zeit.* **24** (1926), 328–376; III, *Math. Zeit.* **24** (1926), 377–395; Nachtrag, *Math. Zeit.* **24** (1926), 789–791, in: *Gesammelte Abhandlungen*, II, Springer-Verlag, Berlin–Heidelberg–New York, 1968, 543–647.
[L5] V. G. Kac, Simple irreducible graded Lie algebras of finite growth, *Izv. Akad. Nauk SSSR* **32** (1968), 1323–1367. English transl., *Math. USSR Izv.* **2** (1968), 1271–1311.
[L6] I. L. Kantor, Graded Lie algebras, *Trudy Sem. Vect. Tens. Anal., Moscow State University* **15** (1970), 227–266 (in Russian).
[L7] R. V. Moody, A new class of Lie algebras, *J. Algebra* **10** (1968), 211–230.
[L8] I. G. Macdonald, Affine root systems and Dedekind's η-function, *Invent. Math.* **15** (1972), 91–143.
[L9] F. J. Dyson, Missed opportunities, *Bull. American Math. Soc.* **78** (1972), 635–652.
[L10] V. J. Kac, Infinite-dimensional Lie algebras and Dedekind's η-function, *Funk. Anal. i Prilozhen.* **8** (1974), 77–78. English transl., *Funct. Anal. Appl.* **8** (1974), 68–70.
[L11] J. Lepowsky and R. L. Wilson, Construction of the affine Lie algebra $A_1^{(1)}$, *Commun. Math. Phys.* **62** (1978), 43–53.
[L12] V. G. Kac, D. A. Kazhdan, J. Lepowsky and R. L. Wilson, Realization of the basic representations of the Euclidean Lie algebras, *Advances in Math.* **42** (1981), 83–112.
[L13] I. B. Frenkel and V. G. Kac, Basic representations of affine Lie algebras and dual resonance models, *Invent. Math.* **62** (1980), 23–66.
[L14] G. Segal, Unitary representations of some infinite-dimensional groups, *Commun. Math. Phys.* **80** (1981), 301–342.
[L15] R. Steinberg, Lectures on Chevalley groups, lecture notes, Yale Univ. Math. Dept., New Haven, Conn., 1968.
[L16] H. Garland, The arithmetic theory of loop algebras, *J. Algebra* **53** (1978), 480–551.
[L17] H. Garland, The arithmetic theory of loop groups, *Inst. Hautes Etudes Sci. Publ. Math.* **52** (1980), 5–136.
[L18] D. Mitzman, Integral bases for affine Lie algebras and their universal enveloping algebras, *Contemporary Math.* **40**, 1985.
[L19] J. Tits, Résumé de cours, *Annuaire du Collège de France*, 1980–1981, 75–87.

[L20] V. G. Kac, *Infinite-Dimensional Lie Algebras*, 2nd ed., Cambridge Univ. Press, Cambridge, 1985.
[L21] A Pressley and G. Segal, *Loop Groups*, Oxford Univ. Press, Oxford, 1986.
[L22] A. Feingold and J. Lepowsky, The Weyl-Kac character formula and power series identities, *Advances in Math.* **29** (1978), 271–309.
[L23] V. G. Kac, Infinite-dimensional algebras, Dedekind's η-function, classical Möbius function and the very strange formula, *Advances in Math.* **30** (1978), 85–136.
[L24] J. Lepowsky, Euclidean Lie algebras and the modular function j, in: *Santa Cruz Conference on Finite Groups, 1979, Proc. Symp. Pure Math., American Math. Soc.* **37** (1980), 567–570.
[L25] V. G. Kac, An elucidation of "Infinite-dimensional . . . and the very strange formula" $E_8^{(1)}$ and the cube root of the modular invariant j, *Advances in Math.* **35** (1980), 264–273.
[L26] H. Garland, Lectures on loop algebras and the Leech lattice, Yale University, spring 1980.
[L27] V. G. Kac, A remark on the Conway–Norton conjecture about the "Monster" simple group, *Proc. Natl. Acad. Sci. USA* **77** (1980), 5048–5049.
[L28] I. B. Frenkel, Representations of Kac–Moody algebras and dual resonance models, in: *Applications of Group Theory in Physics and Mathematical Physics, Proc. 1982 Chicago Summer Seminar*, ed. by M. Flato, P. Sally and G. Zuckerman, Lectures in Applied Math., *American Math. Soc.* **21** (1985), 325–353.
[L29] R. E. Borcherds, Vertex algebras, Kac–Moody algebras, and the Monster, *Proc. Natl. Acad. Sci. USA* **83** (1986), 3068–3071.
[L30] J. Lepowsky, Calculus of twisted vertex operators, *Proc. Natl. Acad Sci. USA* **82** (1985), 8295–8299.

The Present Work

[FLM1] I. B. Frenkel, J. Lepowsky and A. Meurman, An E_8-approach to F_1, in: *Finite Groups—Coming of Age, Proc. 1982 Montreal Conference*, ed. by J. McKay, *Contemporary Math.* **45** (1985), 99–120.
[FLM2] I. B. Frenkel, J. Lepowsky and A. Meurman, A natural representation of the Fischer–Griess Monster with the modular function J as character, *Proc. Natl. Acad. Sci. USA* **81** (1984), 3256–3260.
[FLM3] I. B. Frenkel, J. Lepowsky and A. Meurman, A moonshine module for the Monster, in: *Vertex Operators in Mathematics and Physics, Proc. 1983 M.S.R.I. Conference*, ed. by J. Lepowsky, S. Mandelstam and I. M. Singer, Publ. Math. Sciences Res. Inst. #3, Springer-Verlag, New York, 1985, 231–273.
[FLM4] I. B. Frenkel, J. Lepowsky and A. Meurman, An introduction to the Monster, in: *Unified String Theories, Proc. 1985 Inst. for Theoretical Physics Workshop*, ed. by M. Green and D. Gross, World Scientific, Singapore, 1986, 533–546.
[FLM5] I. B. Frenkel, J. Lepowsky and A. Meurman, Vertex operator calculus, in: *Mathematical Aspects of String Theory, Proc. 1986 Conference, San Diego*, ed. by S.-T. Yau, World Scientific, Singapore, 1987, 150–188.

IV. String Theory

[S1] G Veneziano, Construction of a crossing-symmetric, Regge-behaved amplitude for linearly rising trajectories, *Nuovo Cim.* **57A** (1968), 190–197.

[S2] Z. Koba and H. B. Nielsen, Manifestly crossing-invariant parametrization of n-meson amplitude, *Nucl. Phys.* **B12** (1969), 517.

[S3] S. Fubini and G. Veneziano, Duality in operator formalism, *Nuovo Cim.* **67A** (1970), 29.

[S4] A. Neveu and J. Scherk, Parameter-free regularization of one-loop unitary dual diagram, *Phys. Rev.* **D1** (1970), 2355.

[S5] C. Lovelace, Pomeron form factors and dual Regge cuts, *Phys. Lett.* **34B** (1971), 500.

[S6] P. Goddard, C. Rebbi and C. B. Thorn, Lorentz covariance and the physical states in dual-resonance models, *Nuovo Cim.* **12A** (1972), 425.

[S7] E. Del Giudice, P. Di Vecchia and S. Fubini, General properties of the dual resonance model, *Ann. Phys.* **70** (1972), 378.

[S8] R. C. Brower, Spectrum-generating algebra and no-ghost theorem for the dual model, *Phys. Rev.* **D6** (1972), 1655–1662.

[S9] P. Goddard and C. B. Thorn, Compatibility of the dual Pomeron with unitarity and the absence of ghosts in the dual resonance model, *Phys. Lett.* **40B** (1972), 235.

[S10] S. Fubini and G. Veneziano, Algebraic treatment of subsidiary conditions in dual resonance models, *Ann. Phys.* **63** (1971), 12.

[S11] E. F. Corrigan and D. Olive, Fermion-meson vertices in dual theories, *Nuovo Cim.* **11A** (1972), 749–773.

[S11a] K. Bardakci and M. B. Halpern, New dual quark models, *Phys. Rev.* **D3** (1971), 2493–2506.

[S12] E. F. Corrigan and D. B. Fairlie, Off-shell states in dual resonance theory, *Nucl. Phys.* **B91** (1975), 527–545.

[S13] R. C. Brower and P. Goddard, Collinear algebra for the dual model, *Nucl. Phys.* **B40** (1972), 437–444.

[S14] R. C. Brower and P. Goddard, Physical states in the dual resonance model, in: *Proceedings of the International School of Physics "Enrico Fermi" Course LIV*, Academic Press, New York, London, 1973, 98–110.

[S15] M. B. Halpern, Quantum "solitons" which are SU(N) fermions, *Phys. Rev.* **D12** (1975), 1684–1699.

[S16] T. Banks, D. Horn and H. Neuberger, Bosonization of the SU(N) Thirring models, *Nucl. Phys.* **B108** (1976), 119.

[S17] P. Goddard and D. Olive, Algebras, lattices and strings, in: *Vertex Operators in Mathematics and Physics*, Proc. 1983 M.S.R.I. Conference, ed. by J. Lepowsky, S. Mandelstam and I. M. Singer, *Publ. Math. Sciences Res. Inst. #3*, Springer-Verlag, New York, 1985, 51–96.

[S18] E. Corrigan and T. J. Hollowood, Comments on the algebra of straight, twisted and intertwining vertex operators, *Nucl. Phys.* **B304** (1988), 77–107.

[S19] H. B. Nielsen, An almost physical interpretation of the integrand of the *n*-point Veneziano model, submitted to the 15th International Conference on High Energy Physics, Kiev, 1970.

[S20] L. Susskind, Dual-symmetric theory of hadrons. I, *Nuovo Cim.* **69A** (1970), 457–496.

[S21] Y. Nambu, Lectures at the Copenhagen symposium, 1970.

[S22] P. Goddard, J. Goldstone, C. Rebbi and C. B. Thorn, Quantum dynamics of a massless relativistic string, *Nucl. Phys.* **B56** (1973), 109.

[S23] C. S. Hsue, B. Sakita and M. A. Virasoro, Formulation of dual theory in terms of functional integrations, *Phys. Rev.* **D2** (1970), 2857.

[S24] J. L. Gervais and B. Sakita, Functional-integral approach to dual-resonance theory, *Phys. Rev.* **D4** (1971), 2291.

[S25] S. Mandelstam, Dual-resonance models, *Phys. Reports* **C13** (1974), 259. Reprinted in: Dual Theory, *Physics Reports Reprint Book Series*, **1**, ed. by M. Jacob, North-Holland, Amsterdam, American Elsevier, New York, 1974, 295–389.

[S26] Dual Theory, *Physics Reports Reprint Book Series*, **1**, ed. by M. Jacob, North-Holland, Amsterdam, American Elsevier, New York, 1974.
[S27] E. Witten, Physics and geometry, in: *Proc. Internat. Congr. Math., Berkeley, 1986*, **1**, American Math. Soc., 1987, 267–303.
[S28] T. Yoneya, Quantum gravity and the zero-slope limit of the generalized Virasoro model, *Nuovo Cim. Lett.* **8** (1973), 951–955.
[S29] J. Scherk and J. H. Schwarz, Dual models for non-hadrons, *Nucl. Phys.* **B81** (1974), 118–144.
[S30] F. Gliozzi, D. Olive and J. Scherk, Supersymmetry, supergravity theories and the dual spinor model, *Nucl. Phys.* **B122** (1977), 253–290.
[S30a] M. B. Green and J. H. Schwarz, Supersymmetrical dual string theory, *Nucl. Phys.* **B181** (1981), 502.
[S31] A. M. Polyakov, Quantum geometry of bosonic strings, *Phys. Lett.* **103B** (1981), 207–211.
[S32] A. A. Belavin, A. M. Polyakov and A. B. Zamolodchikov, Infinite conformal symmetries in two-dimensional quantum field theory, *Nucl. Phys.* **B241** (1984), 333–380.
[S33] M. B. Green and J. H. Schwarz, Anomaly cancellations in supersymmetric D = 10 gauge theory and superstring theory, *Phys. Lett.* **149B** (1984), 117.
[S34] D. J. Gross, J. A. Harvey, E. Martinec and R. Rohm, Heterotic string theory (I). The free heterotic string, *Nucl. Phys.* **B256** (1985), 253–284; (II). The interacting heterotic string, *Nucl. Phys.* **B267** (1986), 74–124.
[S35] M. B. Green, J. H. Schwarz and E. Witten, *Superstring Theory*, **1** and **2**, Cambridge University Press, Cambridge, 1987.
[S36] P. Candelas, G. Horowitz, A. Strominger and E. Witten, Vacuum configurations for superstrings, *Nucl. Phys.* **B258** (1985), 46.
[S37] L. Dixon, J. A. Harvey, C. Vafa and E. Witten, Strings on orbifolds, *Nucl. Phys.* **B261** (1985), 651; II, *Nucl. Phys.* **B274** (1986), 285.
[S38] J. A. Harvey, Twisting the heterotic string, in: *Unified String Theories, Proc. 1985 Inst. for Theoretical Physics Workshop*, ed. by M. Green and D. Gross, World Scientific, Singapore, 1986, 704–718.
[S39] W. Nahm, Quantum field theories in one and two dimensions, *Duke Math. J.* **54** (1987), 579–613.
[S40] L. Dixon, D. Friedan, E. Martinec and S. Shenker, The conformal field theory of orbifolds, *Nucl. Phys.* **B282** (1987), 13–73.
[S41] S. Hamidi and C. Vafa, Interactions on orbifolds, *Nucl. Phys.* **B279** (1987), 465–513.
[S42] K. S. Narain, M. H. Sarmadi and C. Vafa, Asymmetric orbifolds, *Nucl. Phys.* **B288** (1987), 551.
[S43] R. Dijkgraaf, E. Verlinde and H. Verlinde, $c = 1$ conformal field theories on Riemann surfaces, *Commun. Math. Phys.* **115** (1988), 649–690.
[S44] F. J. Dyson, Unfashionable pursuits, *Math. Intelligencer* **5** (1983), 47–54.
[S45] G. Chapline, Unification of gravity and elementary particle interactions in 26 dimensions?, *Phys. Lett.* **158B** (1985), 393–396.
[S46] G. Moore, Atkin–Lehner symmetry, *Nucl. Phys.* **B293** (1987), 139–188.
[S47] L. Dixon, P. Ginsparg and J. Harvey, Beauty and the Beast: Superconformal symmetry in a Monster module, to appear.

Perspectives and Summary

[PS1] H.-V. Niemeier, Definite quadratische Formen der Dimension 24 und Diskriminante 1, *J. Number Theory* **5** (1973), 142–178.

[PS2] J. Conway, A characterization of Leech's lattice, *Invent. Math.* **7** (1969), 137–142.

[PS3] G. Mason, Finite groups and modular functions (with an appendix by S. P. Norton), in: *Representations of Finite Groups, Proc. 1986 Arcata Summer Research Institute*, ed. by P. Fong, *Proc. Symp. Pure Math., American Math. Soc.* **47** (1987), 181–210.

[PS4] R. E. Borcherds, J. H. Conway, L. Queen and N. J. A. Sloane, A Monster Lie algebra?, *Advances in Math.* **53** (1984), 75–79.

[PS5] *Elliptic Curves and Modular Forms in Algebraic Topology, Proc. 1986 Princeton Conference*, ed. by P. S. Landweber, Lecture Notes in Math. **1326**, Springer-Verlag, Berlin–Heidelberg–New York, 1988.

[PS6] E. Witten, The index of the Dirac operator on loop space, in: *Elliptic Curves and Modular Forms in Algebraic Topology, Proc. 1986 Princeton Conference*, ed. by P. S. Landweber, Lecture Notes in Math. **1326**, Springer-Verlag, Berlin–Heidelberg–New York, 1988, 161–181.

[PS7] F. Hirzebruch, Introduction to elliptic genera, Arbeitstagung, Bonn, 1987.

[PS8] J. H. Conway, S. P. Norton and L. H. Soicher, The Bimonster, the group Y_{555}, and the projective plane of order 3, *Proc. "Computers in Algebra" Conference, Chicago, 1985*, to appear.

[PS9] J. Thompson, Some finite groups which appear as Gal L/K, where $K \subseteq Q(\mu_n)$, *J. Algebra* **89** (1984), 437–499.

[PS9a] J. Thompson, Some finite groups which appear as Gal L/K, where $K \subseteq Q(\mu_n)$, in: *Group Theory, Beijing 1984*, Lecture Notes in Math. **1185**, Springer-Verlag, Berlin–Heidelberg–New York, 1986, 210–230.

[PS10] D. Friedan and S. Shenker, The analytic geometry of two-dimensional conformal field theory, *Nucl. Phys.* **B281** (1987), 509–545.

[PS11] J. Tits, Résumé de cours, *Annuaire du Collège de France*, 1985–1986, 101–112.

[PS12] J. Tits, Le module du "moonshine," Séminaire Bourbaki, exposé no. 684, 1986/87.

[PS13] I. B. Frenkel, Beyond affine Lie algebras, in: *Proc. Internat. Congr. Math.*, Berkeley, 1986, **1**, American Math. Soc., 1987, 821–839.

[PS14] J. Lepowsky, Perspectives on vertex operators and the Monster, in: *Proc. 1987 Symposium on the Mathematical Heritage of Hermann Weyl, Duke Univ., Proc. Symp. Pure Math., American Math. Soc.* **48** (1988).

Notational Conventions

$\mathbb{Z}$ = set (or ring) of integers

$\mathbb{N}$ = set of integers ≥ 0

$\mathbb{Z}_+$ = set of integers > 0

$\mathbb{Q}$ = field of rational numbers

$\mathbb{R}$ = field of real numbers

$\mathbb{C}$ = field of complex numbers

$\mathbb{F}_q$ = the (finite) field with q elements

$\mathbb{F}$ = a field of characteristic zero fixed throughout this book; sometimes $\mathbb{F}$ is assumed to have special properties

$\mathbb{F}^\times$ = multiplicative group of nonzero elements of $\mathbb{F}$; similarly for any field

For a map $f: X \to Y$, $\operatorname{Im} f$ = image of f; $x \mapsto y$ means that $y = f(x)$ for $x \in X$, $X \hookrightarrow Y$ signifies that f is an injection (is one-to-one); $\operatorname{Ker} f$ = kernel of f if f is a group homomorphism

$|X|$ = cardinality of a set X

1_X = the identity operator on X

For $x, y \in X$, $\delta_{xy} = 1$ if $x = y$, 0 if $x \neq y$

For an algebraic structure A, $\mathrm{Aut}\, A$ = group of automorphisms of A; $\mathrm{End}\, A$ = algebra or ring, etc., of endomorphisms of A

$\coprod V_i$ = direct sum of vector spaces V_i

V^* = dual space of a vector space V

For a group G, $\mathrm{Cent}\, G$ = center of G; $(x, y) = xyx^{-1}y^{-1}$ for $x, y \in G$ (sometimes the notation (x, y) refers to a bilinear or hermitian form); for subgroups H, K of G, (H, K) = subgroup of G generated by the commutators (x, y) for $x \in H$, $y \in K$; $G' = (G, G)$ = commutator subgroup of G; $\langle X_1, X_2, \ldots \rangle$ = subgroup of G generated by one or more elements or subsets (or a mixture) $X_1, X_2, \ldots$ of G (very often, the notation $\langle x, y \rangle$ denotes a bilinear form, but the context should eliminate any ambiguity); for a subgroup H of G, $|G : H|$ = index of H in G; $\mathbf{N}_G(H)$ = normalizer of H in G; $\mathbf{C}_G(H)$ = centralizer of H in G; for $x, y \in G$, ${}^x y = xyx^{-1}$ and $y^x = x^{-1}yx$, the left and right conjugations; $\mathrm{Inn}\, G$ = group of inner automorphisms of G

There are four places in the text where we have found it appropriate to change our notations midsteam. In each case, the change and the expository reasons for making it will be clear. These cases are as follows:

In Section 1.7, an "obvious" grading (notion of degree) is shifted to one that is more natural.

In Sections 7.1 and 7.3, the meaning of the notation $x_\alpha(n)$ of Chapters 3 and 4 for the components of vertex operators is modified to accommodate group extension ideas (cocycles).

In Chapter 8, the approach to vertex operator calculus naturally arising in Chapters 3, 4 and 7 is changed to an approach which leads to a much more general theory, and in Sections 8.4 and 9.1, it is correspondingly convenient to change the meaning of the expressions $\alpha(z)$ and $\alpha(z)^{\pm}$ by a factor of z. Several other new objects are emphasized from this point on, but they all have new notations—for instance, the vertex operators $Y(a, z)$ as opposed to $X(a, z)$ and the normal ordering operation $\circ\!\!\circ \cdot \circ\!\!\circ$ as opposed to $: \cdot :$.

Finally, much of the important structure associated with the moonshine module can be defined by means of an extension of the Leech lattice Λ of order 2, and we denote this extension by $\hat{\Lambda}$ in Chapter 10. However, the more subtle structure and the proof of our main theorem require an extension of Λ of order 4, introduced in Chapter 12, and here it is convenient to call this new extension $\hat{\Lambda}$ and to change the notation for the extension of order 2 to $\hat{\Lambda}^{(2)}$. The same convention is used for related lattices.

Many spaces that we consider have two useful $\mathbb{Q}$-gradations—by what we call degree and by what we call weight (corresponding in physics terminology to conformal weight). These gradations are oppositely directed and are shifted from each other; that is, the sum of the degree and the weight of a homogeneous element is a nonzero constant. We adopt the convention that an operator written in the form $x(n)$ has degree n (i.e., it shifts degrees, not weights, by n). Examples that arise in the text are $\alpha(n)$, $x_\alpha(n)$, $x_a(n)$, $x_v(n)$ and $L(n)$. There should be no confusion between notations such as $\alpha(z)$ and $\alpha(n)$ because z is a formal variable, not a number.

The vertex operators denoted by the symbol X, for example, $X(a, z)$ or $X(v, z)$, are compatible with the grading by weights in that the component of z^n in such a vertex operator is an operator of weight n (and degree $-n$). For instance, $X(v, z) = \sum x_v(n)z^{-n}$. The corresponding vertex operators denoted by the symbol Y, for example, $Y(a, z)$ or $Y(v, z)$, have many other important properties, but not this simple compatibility with the weight grading.

Bibliographical citations in the main text refer to the alphabetical list at the end of the book. Together with references directly related to the main text, this bibliography includes additional material concerning vertex operator algebras and the Monster. We have chosen an unusual referencing system for the Introduction in order to separate the various areas of mathematics and physics related to this work.

1 Lie Algebras

In this chapter we introduce the Lie algebras termed affine, Heisenberg and Virasoro that will play central roles throughout this book. We also present standard constructions of important classes of modules for Heisenberg and Virasoro algebras and we discuss the notions of contravariant form and graded dimension. For completeness we have included a number of elementary concepts in the first five sections.

We start from the definitions of algebra and Lie algebra in Section 1.1 and the notion of module in Section 1.2, and we present some elementary constructions in the next two sections. We describe the notion of induced module and related concepts in Section 1.5. In this section we state one of the few results presented in this book without proof—the Poincaré–Birkhoff–Witt theorem, whose proof may be found in [Humphreys], for example. The reader familiar with basic algebra will want to refer to these first sections only for notation. For the beginner these sections will provide background for understanding the book. For extensive expositions of basic algebra we refer to [Jacobson 2, 3] and [Lang 1].

The second part of the first chapter is devoted to more specialized material immediately related to the subject of the book. In Section 1.6 we define the affine Lie algebras $\hat{\mathfrak{g}}$ and their twistings by involutions. These are examples of Kac–Moody algebras in case $\mathfrak{g}$ is finite-dimensional semisimple. See the Introduction for a discussion of the history of Kac–Moody algebras. We refer to the book [Kac 5] for a detailed exposition of the subject. Affine algebras over $\mathbb{C}$ can be realized as central extensions of loop algebras, suggesting a geometric approach to the subject explored in great detail in [Pressley–Segal].

In Section 1.7 we consider a degenerate family of affine Lie algebras—Heisenberg algebras. We study their canonical faithful irreducible representations, one for each nonzero scalar, by multiplication operators and derivations on a polynomial algebra in infinitely many variables. In physics terminology, the module is called a (bosonic) Fock space, the multiplication operators are called creation operators and the derivations are called annihilation operators. This representation, the "canonical realization of the Heisenberg communtation relations," has its roots in the study of harmonic oscillators in the early days of quantum mechanics. We also show that every representation of a Heisenberg algebra satisfying certain conditions is a direct sum of copies of the canonical irreducible representation. Here we follow [Lepowsky–Wilson 2]; see also [Kac 5]. This result is an algebraic analogue of the Stone–von Neumann theorem.

In Section 1.8 we introduce bilinear and hermitian contravariant forms for a special class of induced Lie algebra modules; see [Shapovalov]. The Virasoro algebra, closely related to affine algebras, is the subject of Section 1.9. Over $\mathbb{C}$, the Virasoro algebra admits a geometric realization as a central extension of the complexified Lie algebra of polynomial vector fields on the circle. The latter Lie algebra, called the Witt algebra, was studied in [Gelfand–Fuks], where in particular its central extensions were described. Operators realizing the essentially unique central extension in Fock spaces were first studied in [Virasoro] and by J. H. Weis (unpublished) in the context of string theory. We conclude this chapter by introducing the notion of graded dimension in Section 1.10, where the first glimpses of modular function theory start to appear. The subtle degree-shifts relating the untwisted and twisted Fock spaces and leading to graded dimensions having modular transformation properties are motivated algebraically by the Virasoro algebra.

Throughout this work, $\mathbb{F}$ *will denote a field of characteristic* 0. *All vector spaces and algebras will be over* $\mathbb{F}$ *unless another field is specified. The notation* $\mathbb{F}^{\times}$ *will denote the multiplicative group of nonzero elements of* $\mathbb{F}$.

1.1. Algebras

A *nonassociative* (= not necessarily associative) *algebra* is a vector space A equipped with a bilinear map, called *product* or *multiplication*, from $A \times A$ to A. The (nonassociative) algebra A is called *associative* if it contains an identity element 1 for multiplication, so that

$$1a = a1 = a \quad \text{for} \quad a \in A,$$

and if the associative law holds:

$$(ab)c = a(bc) \quad \text{for} \quad a, b, c \in A$$

(product being denoted by juxtaposition). The algebra A is said to be *commutative* if the commutative law holds:

$$ab = ba \quad \text{for} \quad a, b \in A.$$

For subspaces B, C of an algebra A, we write BC for the subspace of A spanned by the products bc for $b \in B$, $c \in C$. A *subalgebra* of A is a subspace B of A such that $B^2 \subset B$ and such that $1 \in B$ in the associative case. Equivalently, a subalgebra of A is a subset B of A which is an algebra under the linear and product structures induced from A.

For algebras A and B, a linear map $f: A \to B$ is a *homomorphism* if $f(ab) = f(a)f(b)$ for $a, b \in A$ and if in addition $f(1) = 1$ in the associative case. The homomorphism f is an (algebra) *isomorphism* if in addition it is a linear isomorphism. In case $A = B$, a homomorphism is called an *endomorphism* and an isomorphism is called an *automorphism*. Two algebras A and B are *isomorphic* if there is an isomorphism $f: A \to B$. In this case we sometimes write $f: A \rightrightarrows B$ or $A \simeq B$. These terms and notations are used for algebraic structures generally.

A linear endomorphism $d: A \to A$, A an algebra, is a *derivation* if

$$d(ab) = d(a)b + ad(b) \quad \text{for} \quad a, b \in A. \tag{1.1.1}$$

A *Lie algebra* is a nonassociative algebra $\mathfrak{g}$ whose product, which is conventionally denoted $[\cdot, \cdot]$ and called *bracket*, is *alternating*, i.e.,

$$[x, x] = 0 \quad \text{for} \quad x \in \mathfrak{g} \tag{1.1.2}$$

and satisfies the *Jacobi identity*

$$[x, [y, z]] + [y, [z, x]] + [z, [x, y]] = 0 \quad \text{for} \quad x, y, z \in \mathfrak{g}. \tag{1.1.3}$$

The alternating property is equivalent to the *skew-symmetry* condition

$$[x, y] = -[y, x] \quad \text{for} \quad x, y \in \mathfrak{g} \tag{1.1.4}$$

since the characteristic of $\mathbb{F}$ is not 2. Given skew-symmetry, the Jacobi identity is equivalent to the condition that $\operatorname{ad} x$ be a derivation of $\mathfrak{g}$ for all

$x \in \mathfrak{g}$. Here $\operatorname{ad} x$ ("ad" referring to "adjoint") is the linear map

$$\begin{aligned} \operatorname{ad} x\colon \mathfrak{g} &\to \mathfrak{g} \\ y &\mapsto [x, y]. \end{aligned} \tag{1.1.5}$$

A Lie algebra $\mathfrak{g}$ is *abelian* if $[\mathfrak{g}, \mathfrak{g}] = 0$. (Since the characteristic of $\mathbb{F}$ is not 2, a Lie algebra is abelian if and only if it is commutative in the general sense defined above.) Two elements x, y of a Lie algebra $\mathfrak{g}$ are said to *commute* if $[x, y] = 0$. In particular, all pairs of elements of an abelian Lie algebra commute. Every one-dimensional Lie algebra is abelian.

Let A be an associative algebra. Then a Lie algebra structure is defined on A by taking for the bracket $[x, y]$ the *commutator* $xy - yx$ of $x, y \in A$. To see that this defines a Lie algebra structure, first note that $[\cdot, \cdot]$ is alternating, and then observe that for $x \in A$, $\operatorname{ad} x\colon A \to A$ (defined with respect to the commutator) is a derivation of the associative algebra. Using this, conclude that $\operatorname{ad} x$ (and in fact any associative derivation) is a derivation of the commutator structure.

Every associative subalgebra of A becomes a Lie subalgebra of A under commutators. The associative algebra A is commutative if and only if the corresponding Lie algebra is abelian.

Take the case $A = \operatorname{End} B$, B a nonassociative algebra. Then the space of derivations of B forms a Lie subalgebra of A. That is, the commutator of two derivations is a derivation.

1.2. Modules

Let A be an associative algebra and let V be a vector space. We say that V is an *A-module* if there is a bilinear map (typically denoted with a dot)

$$\begin{aligned} A \times V &\to V \\ (a, v) &\mapsto a \cdot v \end{aligned}$$

such that

$$\begin{aligned} 1 \cdot v &= v \quad \text{for} \quad v \in V \\ (ab) \cdot v &= a \cdot (b \cdot v) \quad \text{for} \quad a, b \in A,\ v \in V. \end{aligned}$$

For $a \in A$, let $\pi(a)$ be the corresponding linear endomorphism of V, so that

$$\pi(a)v = a \cdot v \quad \text{for} \quad a \in A,\ v \in V.$$

Then the map

$$\pi: A \to \operatorname{End} V$$

is a homomorphism of associative algebras. Such a homomorphism is called a *representation* of A on V. The concepts of A-module and representation of A are equivalent. Sometimes V is called a *representation* of A.

Note that the associative algebra A has a natural representation on itself, given by the left multiplication action:

$$a \cdot b = ab \quad \text{for} \quad a, b \in A.$$

Analogously, let $\mathfrak{g}$ be a Lie algebra and let V be a vector space. Then V is called a $\mathfrak{g}$-*module* if there is a bilinear map

$$\mathfrak{g} \times V \to V$$

$$(x, v) \mapsto x \cdot v$$

such that

$$[x, y] \cdot v = x \cdot (y \cdot v) - y \cdot (x \cdot v)$$

or equivalently,

$$x \cdot (y \cdot v) = y \cdot (x \cdot v) + [x, y] \cdot v$$

for $x, y \in \mathfrak{g}$, $v \in V$. Denote by $\pi(x)$ the corresponding linear endomorphism of V:

$$\pi(x)v = x \cdot v \quad \text{for} \quad x \in \mathfrak{g},\ v \in V.$$

Then the map

$$\pi: \mathfrak{g} \to \operatorname{End} V$$

is a Lie algebra homomorphism. Such a map is called a *representation* of $\mathfrak{g}$ on V (and sometimes V is called a *representation* of $\mathfrak{g}$). The notions of $\mathfrak{g}$-module and representation of $\mathfrak{g}$ are equivalent.

The Lie algebra $\mathfrak{g}$ has a natural representation on itself—the *adjoint representation*, give by the map

$$\operatorname{ad}: \mathfrak{g} \to \operatorname{End} \mathfrak{g}$$

$$x \mapsto \operatorname{ad} x.$$

(It is easily checked that ad is a homomorphism.) Also, any vector space can be made into a *trivial* $\mathfrak{g}$-*module*, corresponding to the zero representation.

Let A be an associative algebra and let V be an A-module. Then V is also a module for A regarded as a Lie algebra, and also for any Lie subalgebra of A. In particular (taking $A = \text{End } V$), every Lie subalgebra of End V has a natural representation on V.

Let $\mathfrak{g}$ be an *associative* or *Lie* algebra, and let V be a $\mathfrak{g}$-module. For subspaces $\mathfrak{h}$ of $\mathfrak{g}$ and W of V, we denote by $\mathfrak{h} \cdot W$ the linear span of all $x \cdot w$ for $x \in \mathfrak{h}$, $w \in W$. A *submodule* of V is a subspace W of V such that $\mathfrak{g} \cdot W \subset W$, or equivalently, a subset W of V which is a $\mathfrak{g}$-module under the linear structure and $\mathfrak{g}$-module action induced from V. A subspace W of V is *invariant* (under $\mathfrak{g}$) if it is a submodule. The module V is *irreducible* or *simple* if $V \neq 0$ and if V has no proper nonzero invariant subspaces. The module V is *indecomposable* if it cannot be decomposed as a direct sum of two nonzero submodules. Clearly, an irreducible module is indecomposable. Let V and W be $\mathfrak{g}$-modules. A linear map $f: V \to W$ is called a $\mathfrak{g}$-*module homomorphism* or $\mathfrak{g}$-*module map* if $f(x \cdot v) = x \cdot f(v)$ for $x \in \mathfrak{g}$, $v \in V$. Such a map f is called a $\mathfrak{g}$-*module isomorphism* or $\mathfrak{g}$-*module equivalence* if it is a linear isomorphism. Two modules V and W are *isomorphic* or *equivalent* if there is an isomorphism $f: V \to W$, and we sometimes write $f: V \xrightarrow{\sim} W$ or $V \simeq W$ in this case.

1.3. Algebra Constructions

A subspace $\mathfrak{a}$ of a Lie algebra $\mathfrak{g}$ is called an *ideal* of $\mathfrak{g}$ if $[\mathfrak{g}, \mathfrak{a}] \subset \mathfrak{a}$. Equivalently, an ideal of $\mathfrak{g}$ is a submodule under the adjoint representation. An ideal is a subalgebra. Given an ideal $\mathfrak{a}$ of $\mathfrak{g}$, the quotient vector space $\mathfrak{g}/\mathfrak{a}$ becomes a Lie algebra, called the *quotient Lie algebra*, by means of the (well-defined) nonassociative product

$$[x + \mathfrak{a}, y + \mathfrak{a}] = [x, y] + \mathfrak{a} \quad \text{for} \quad x, y \in \mathfrak{g}.$$

The canonical map $\pi: \mathfrak{g} \to \mathfrak{g}/\mathfrak{a}$ is a homomorphism, and we have an exact sequence of Lie algebras

$$0 \to \mathfrak{a} \to \mathfrak{g} \xrightarrow{\pi} \mathfrak{g}/\mathfrak{a} \to 0.$$

A Lie algebra $\mathfrak{g}$ is said to be simple if $\mathfrak{g}$ is nonzero and has no proper nonzero ideals (equivalently, the adjoint representation is simple) and if $\dim \mathfrak{g} > 1$ (i.e., $\mathfrak{g}$ is not abelian).

A subspace I of an associative algebra A is called a *left* (resp., *right*) *ideal* of A if $AI \subset I$ (resp. $IA \subset I$), and an *ideal* (or *two-sided ideal*) if it is both a left and right ideal. Since an ideal I need not contain 1, it need not be a subalgebra. (If $1 \in I$, then in fact $I = A$.) Given an ideal I of A, the quotient vector space A/I becomes an associative algebra—the *quotient algebra*—in an obvious way.

The kernel of any homomorphism of a Lie (resp., associative) algebra $\mathfrak{g}$ into any Lie (resp., associative) algebra is an ideal. The kernel of the adjoint representation of a Lie algebra $\mathfrak{g}$ is a particularly important ideal called the *center* of $\mathfrak{g}$ and denoted Cent $\mathfrak{g}$:

$$\operatorname{Cent}\mathfrak{g} = \{x \in \mathfrak{g} \mid [x, y] = 0 \quad \text{for all} \quad y \in \mathfrak{g}\}. \tag{1.3.1}$$

Any subspace of Cent $\mathfrak{g}$ is an ideal in $\mathfrak{g}$ and is said to be a *central ideal.*

Given Lie algebras $\mathfrak{a}$ and $\mathfrak{b}$, an *extension of* $\mathfrak{a}$ *by* $\mathfrak{b}$ is a Lie algebra $\mathfrak{g}$ together with an exact sequence

$$0 \to \mathfrak{b} \to \mathfrak{g} \to \mathfrak{a} \to 0.$$

Note that $\mathfrak{b}$ is an ideal of $\mathfrak{g}$ and that $\mathfrak{g}/\mathfrak{b} \simeq \mathfrak{a}$. Sometimes $\mathfrak{g}$ itself is called an extension of $\mathfrak{a}$ by $\mathfrak{b}$. This extension is said to be *central* if $\mathfrak{b}$ is a central ideal of $\mathfrak{g}$. Two extensions $\mathfrak{g}$ and $\mathfrak{g}_1$ of $\mathfrak{a}$ by $\mathfrak{b}$ are *equivalent* if there is an isomorphism $\mathfrak{g} \rightrightarrows \mathfrak{g}_1$ making the following diagram commute:

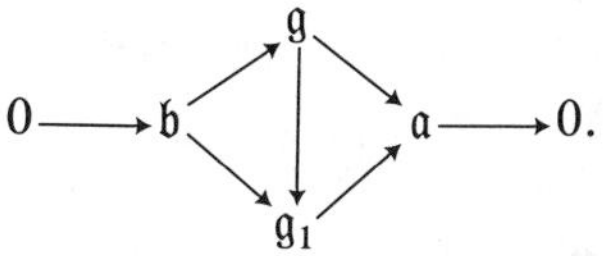

If $\mathfrak{a}$ and $\mathfrak{b}$ are ideals of a Lie algebra $\mathfrak{g}$, then $\mathfrak{a} + \mathfrak{b}$, $\mathfrak{a} \cap \mathfrak{b}$ and $[\mathfrak{a}, \mathfrak{b}]$ are ideals also, the last because of the following "derivation identity" for subspaces $\mathfrak{a}$, $\mathfrak{b}$, $\mathfrak{c}$ of $\mathfrak{g}$:

$$[\mathfrak{c}, [\mathfrak{a}, \mathfrak{b}]] \subset [[\mathfrak{c}, \mathfrak{a}], \mathfrak{b}] + [\mathfrak{a}, [\mathfrak{c}, \mathfrak{b}]],$$

which follows from the corresponding identity for elements. In particular, $[\mathfrak{g}, \mathfrak{g}]$ is an ideal of $\mathfrak{g}$, called the *commutator ideal* (or *commutator subalgebra*) and conventionally denoted $\mathfrak{g}'$:

$$\mathfrak{g}' = [\mathfrak{g}, \mathfrak{g}]. \tag{1.3.2}$$

Given two Lie algebras $\mathfrak{a}$ and $\mathfrak{b}$, their *direct product* is the Lie algebra $\mathfrak{a} \times \mathfrak{b}$ which is $\mathfrak{a} \oplus \mathfrak{b}$ as a vector space, with $\mathfrak{a}$ and $\mathfrak{b}$ retaining their original bracket structures and commuting with one another. In particular, $\mathfrak{a}$ and $\mathfrak{b}$ are ideals of $\mathfrak{a} \times \mathfrak{b}$. Of course, we can define the direct product of finitely many Lie algebras analogously.

More generally, suppose that we have a representation $\pi: \mathfrak{a} \to \operatorname{End}\mathfrak{b}$ of a Lie algebra $\mathfrak{a}$ on a Lie algebra $\mathfrak{b}$ by derivations, i.e., $\pi(x)$ is a derivation of $\mathfrak{b}$ for all $x \in \mathfrak{a}$. Then it is straightforward to check that the vector space $\mathfrak{a} \oplus \mathfrak{b}$ carries a unique Lie algebra structure such that $\mathfrak{a}$ and $\mathfrak{b}$ are subalgebras and such that $[x, y] = \pi(x)y$ for all $x \in \mathfrak{a}$, $y \in \mathfrak{b}$. Note that $\mathfrak{b}$ is an ideal, but not necessarily $\mathfrak{a}$. This Lie algebra is called the *semidirect product* of $\mathfrak{a}$ and $\mathfrak{b}$ and is denoted either $\mathfrak{a} \rtimes \mathfrak{b}$ or $\mathfrak{b} \ltimes \mathfrak{a}$. Observe that the notation picks out the ideal. A Lie algebra is a semidirect product whenever it is the vector space direct sum of a subalgebra and an ideal. Note that $\mathfrak{a} \rtimes \mathfrak{b} = \mathfrak{a} \times \mathfrak{b}$ if and only if $\pi = 0$. A semidirect product $\mathfrak{a} \rtimes \mathfrak{b}$ is an extension of $\mathfrak{a}$ by $\mathfrak{b}$. An extension of a Lie algebra $\mathfrak{a}$ by a Lie algebra $\mathfrak{b}$ is *trivial* or *split* if it is equivalent to a semidirect product $\mathfrak{a} \rtimes \mathfrak{b}$.

We mention a particular kind of semidirect product: Given a Lie algebra $\mathfrak{g}$ and a derivation d of $\mathfrak{g}$, we can form $\mathbb{F}d \rtimes \mathfrak{g}$. This procedure is called *adjoining the derivation d* to $\mathfrak{g}$.

Let S be a set. A vector space V is said to be (S-)*graded* if it is the direct sum

$$V = \coprod_{\alpha \in S} V_\alpha \tag{1.3.3}$$

of subspaces V_α ($\alpha \in S$). In this case, the elements of V_α are said to be *homogeneous of degree* α, and V_α is called the *homogeneous subspace of degree* α. For $v \in V_\alpha$ (including $v = 0$) we write

$$\deg v = \alpha. \tag{1.3.4}$$

Given another S-graded vector space W, a linear map $f\colon V \to W$ is *grading-preserving* if

$$f\colon V_\alpha \to W_\alpha \quad \text{for} \quad \alpha \in S. \tag{1.3.5}$$

If such a map f is a linear isomorphism, V and W are *graded-isomorphic*. If the set S is an abelian group, a linear map $f\colon V \to W$ is said to be *homogeneous of degree* $\beta \in S$, and we write

$$\deg f = \beta, \tag{1.3.6}$$

if

$$f\colon V_\alpha \to W_{\alpha+\beta} \quad \text{for} \quad \alpha \in S. \tag{1.3.7}$$

Note that f is grading-preserving if and only if it has degree 0. If S is an abelian group which is a subgroup of the additive group of $\mathbb{F}$, we can define the *degree operator* $d\colon V \to V$ by the condition

$$dv = \alpha v \quad \text{for} \quad v \in V_\alpha,\ \alpha \in S. \tag{1.3.8}$$

Note that a linear map $f\colon V \to W$ is grading-preserving if and only if

$$[d, f] = 0 \tag{1.3.9}$$

and is homogeneous of degree $\beta \in S$ if and only if

$$[d, f] = \beta f. \tag{1.3.10}$$

A subspace W of an S-graded vector space V is *graded* if $W = \coprod_{\alpha \in S} W_\alpha$, where $W_\alpha = W \cap V_\alpha$ for $\alpha \in S$. In this case, V/W is graded in a natural way. Given a family $(V^i)_{i \in I}$ (I any set) of S-graded vector spaces, the direct sum $X = \coprod_{i \in I} V^i$ is naturally S-graded, where we take

$$X_\alpha = \coprod_{i \in I} V_\alpha^i \quad \text{for} \quad \alpha \in S. \tag{1.3.11}$$

If S is an abelian group and if V and W are S-graded vector spaces, then $V \otimes W$ acquires a unique S-grading by the condition

$$V_\alpha \otimes W_\beta \subset (V \otimes W)_{\alpha+\beta} \quad \text{for} \quad \alpha, \beta \in S. \tag{1.3.12}$$

Using the symbol d_U for the degree operator on the space U, we have

$$d_{V \otimes W} = d_V \otimes 1 + 1 \otimes d_W. \tag{1.3.13}$$

This *tensor product grading* extends to an arbitrary finite number of tensor factors.

Now let $\mathfrak{A}$ be an abelian group and let A be a nonassociative algebra. Then A is an *$\mathfrak{A}$-graded algebra* if it is $\mathfrak{A}$-graded as a vector space, so that $A = \coprod_{\alpha \in \mathfrak{A}} A_\alpha$, and if

$$A_\alpha A_\beta \subset A_{\alpha+\beta} \quad \text{for} \quad \alpha, \beta \in \mathfrak{A}. \tag{1.3.14}$$

For A an associative algebra, it follows that $1 \in A_0$, as we observe by expanding 1 as the sum of its homogeneous components and considering the product of 1 with each of its components. Suppose that $\mathfrak{A}$ is a subgroup of the additive group of $\mathbb{F}$ and let $d: A \to A$ be the degree operator [see (1.3.8)]. Then d is a derivation of A called the *degree derivation*. If A is a Lie algebra, the extended Lie algebra $\mathbb{F}d \ltimes A$ is also $\mathfrak{A}$-graded, with d of degree 0.

1.4. Module Constructions

Fix an associative or Lie algebra $\mathfrak{g}$. Let V be a $\mathfrak{g}$-module and $U \subset V$ a submodule. Then the quotient vector space V/U becomes a $\mathfrak{g}$-module, called the *quotient module*, by means of the (well-defined) action

$$x \cdot (v + U) = x \cdot v + U \quad \text{for} \quad x \in \mathfrak{g},\ v \in V.$$

We have an exact sequence of $\mathfrak{g}$-modules

$$0 \to U \to V \to V/U \to 0.$$

Given two $\mathfrak{g}$-modules V_1 and V_2, their *direct sum* $V_1 \oplus V_2$ is the $\mathfrak{g}$-module which is $V_1 \oplus V_2$ as a vector space, with V_1 and V_2 retaining their original module structures. In particular, V_1 and V_2 are submodules of $V_1 \oplus V_2$. The direct sum of any collection $(V_i)_{i \in I}$ of $\mathfrak{g}$-modules is defined analogously and is denoted $\coprod_{i \in I} V_i$ or $\bigoplus_{i \in I} V_i$.

A $\mathfrak{g}$-module is called *completely reducible* or *semisimple* if it is a direct sum of irreducible submodules. (Here the null sum is allowed, so that the zero-dimensional module is considered completely reducible.)

Let $\mathfrak{A}$ be an abelian group and suppose that $\mathfrak{g}$ is $\mathfrak{A}$-graded. A $\mathfrak{g}$-module V is $\mathfrak{A}$-*graded* if it is $\mathfrak{A}$-graded as a vector space, so that $V = \coprod_{\alpha \in \mathfrak{A}} V_\alpha$, and if

$$\mathfrak{g}_\alpha \cdot V_\beta \subset V_{\alpha+\beta} \quad \text{for} \quad \alpha, \beta \in \mathfrak{A}, \tag{1.4.1}$$

i.e., $\mathfrak{g}_\alpha$ acts as operators of degree α [see (1.3.6), (1.3.7)]. Quotients and direct sums of $\mathfrak{A}$-graded modules are graded (as modules). In case $\mathfrak{g}$ is an ($\mathfrak{A}$-graded) Lie algebra, with $\mathfrak{A}$ a subgroup of the additive group of $\mathbb{F}$, let d be the degree derivation of $\mathfrak{g}$. Then an $\mathfrak{A}$-graded $\mathfrak{g}$-module V becomes an $\mathbb{F}d \ltimes \mathfrak{g}$-module when d is required to act as the degree operator (1.3.8) on V. Note that the symbol "d" plays two different (compatible) roles.

The grading of a graded module can be shifted in the following sense. Suppose that $\mathfrak{A}$ is a subgroup of an abelian group $\mathfrak{B}$ and that V is an $\mathfrak{A}$-graded A-module, A an $\mathfrak{A}$-graded nonassociative algebra. Let $\beta \in \mathfrak{B}$ (perhaps $\beta \in \mathfrak{A}$). Then for each $\alpha \in \mathfrak{A}$, V_α can be renamed $V_{\alpha+\beta}$, giving V the structure of a $\mathfrak{B}$-graded module with $A_\gamma = 0$ for $\gamma \in \mathfrak{B}\backslash\mathfrak{A}$ and $V_\gamma = 0$ for $\gamma \in \mathfrak{B}\backslash(\mathfrak{A} + \beta)$.

Now let $\mathfrak{g}$ be a Lie algebra. In preparation for constructing the tensor product of $\mathfrak{g}$-modules, we first note that if π_1 and π_2 are two representations of $\mathfrak{g}$ on V which commute in the sense that

$$[\pi_1(x), \pi_2(y)] = 0 \quad \text{for} \quad x, y \in \mathfrak{g},$$

then $\pi_1 + \pi_2$ is a representation of $\mathfrak{g}$ on V. Given two $\mathfrak{g}$-modules V and W, we define the *tensor product module* $V \otimes W$ to be the vector space $V \otimes W$ with the (well-defined) action of $x \in \mathfrak{g}$ determined by the condition

$$x \cdot (v \otimes w) = (x \cdot v) \otimes w + v \otimes (x \cdot w) \quad \text{for} \quad v \in V,\ w \in W.$$

This is a $\mathfrak{g}$-module action because the equations

$$x \cdot (v \otimes w) = (x \cdot v) \otimes w \qquad \text{and} \qquad x \cdot (v \otimes w) = v \otimes (x \cdot w)$$

clearly define two commuting $\mathfrak{g}$-module structures on the vector space $V \otimes W$. The tensor product of finitely many $\mathfrak{g}$-modules is defined analogously. If the tensor factors are $\mathfrak{A}$-graded modules ($\mathfrak{A}$ an abelian group), then so is the tensor product.

1.5. Induced Modules

Let B be a subalgebra of an associative algebra A and let V be a B-module. We denote by $A \otimes_B V$ the quotient of the vector space $A \otimes_{\mathbb{F}} V$ by the subspace spanned by the elements $ab \otimes v - a \otimes b \cdot v$ for $a \in A$, $b \in B$, $v \in V$, and we again write $a \otimes v$ for the image of $a \otimes v \in A \otimes_{\mathbb{F}} V$ in $A \otimes_B V$. Then $ab \otimes v = a \otimes b \cdot v$ in $A \otimes_B V$. The space $A \otimes_B V$ carries a natural A-module structure determined by the condition

$$c \cdot (a \otimes v) = ca \otimes v \quad \text{for} \quad a, c \in A,\ v \in V, \tag{1.5.1}$$

and $A \otimes_B V$ is called the A-module *induced* by the B-module V. It is sometimes denoted as follows:

$$\operatorname{Ind}_B^A V = A \otimes_B V. \tag{1.5.2}$$

There is a canonical B-module map

$$\begin{aligned} i\colon V &\to A \otimes_B V \\ v &\mapsto 1 \otimes v, \end{aligned} \tag{1.5.3}$$

and $\operatorname{Ind}_B^A V$ has the following universal property: Given any A-module W and B-module map $j\colon V \to W$, there is a unique A-module map $f\colon \operatorname{Ind}_B^A V \to W$ making the following diagram commute:

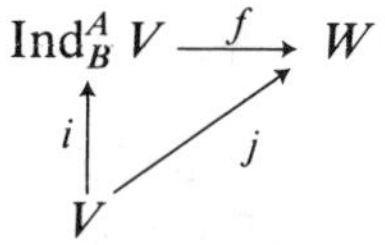

This property characterizes the A-module $\operatorname{Ind}_B^A V$ and the map i up to canonical isomorphism. In fact, if I' is another A-module with a B-module map $i'\colon V \to I'$ satisfying the same condition, then we obtain A-module maps $f\colon \operatorname{Ind}_B^A V \to I'$, $g\colon I' \to \operatorname{Ind}_B^A V$. But $g \circ f$ and the identity map both make the diagram

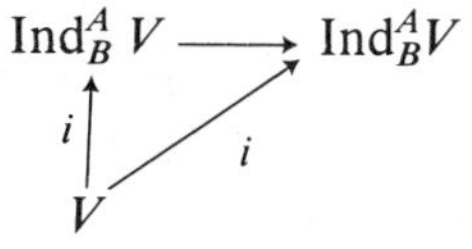

commute, so that $g \circ f$ is the identity map by the uniqueness. Similarly, $f \circ g$ is the identity on I'.

If the algebra A and subalgebra B are $\mathfrak{A}$-graded and if V is an $\mathfrak{A}$-graded B-module ($\mathfrak{A}$ an abelian group), then it is easy to see that the induced module $\operatorname{Ind}_B^A V$ is an $\mathfrak{A}$-graded A-module in a natural way.

Given a group G, we define its *group algebra* to be the associative algebra $\mathbb{F}[G]$ which is formally the set of finite linear combinations of elements of G. That is, $\mathbb{F}[G]$ has the set G as a linear basis, and multiplication in $\mathbb{F}[G]$ is simply defined by linear extension of multiplication in G. The identity element of $\mathbb{F}[G]$ is just the identity element of G.

A *representation* of the group G on a vector space V is a group homomorphism

$$\pi: G \to \operatorname{Aut} V.$$

The space V is called a *G-module* or *representation* of G, and just as for associative and Lie algebras, we often use the dot notation

$$g \cdot v = \pi(g)v \quad \text{for} \quad g \in G,\ v \in V.$$

We have

$$1 \cdot v = v$$

$$(gh) \cdot v = g \cdot (h \cdot v) \quad \text{for} \quad g, h \in G,\ v \in V.$$

We have the usual module-theoretic concepts such as irreducibility and equivalence. If $\pi(G) = 1$, π is called a *trivial* representation. Given G-modules $V_1, \ldots, V_n$, their *tensor product* is the vector space $V_1 \otimes \cdots \otimes V_n$ with G-action determined by:

$$g \cdot (v_1 \otimes \cdots \otimes v_n) = (g \cdot v_1) \otimes \cdots \otimes (g \cdot v_n) \quad \text{for} \quad g \in G,\ v_i \in V_i.$$

The group G has a natural representation on its own group algebra, given by the left multiplication action. This is called the *left regular representation* of G.

Any G-module V becomes an $\mathbb{F}[G]$-module in a canonical way—by extending the map $\pi: G \to \operatorname{Aut} V$ by linearity to an algebra homomorphism from $\mathbb{F}[G]$ to $\operatorname{End} V$. In fact, the G-modules are essentially the same as the $\mathbb{F}[G]$-modules. For example, the left regular representation of G corresponds to the left multiplication representation of $\mathbb{F}[G]$.

If the group G is an abelian group written additively, such as the group $\mathbb{Z}$, there can be confusion as to whether the symbol $a + b$ means the sum (=product) in G or the sum in $\mathbb{F}[G]$, for $a, b \in G$. For this reason we use exponential notation for the elements of G viewed as elements of $\mathbb{F}[G]$ when G is such a group: We write e^a for the element of $\mathbb{F}[G]$ corresponding to $a \in G$. In particular,

$$\begin{aligned} e^0 &= 1 \\ e^a e^b &= e^{a+b} \quad \text{for} \quad a, b \in G. \end{aligned} \tag{1.5.4}$$

Given a subgroup H of a group G and an H-module V, we define the G-module *induced* by V to be the G-module associated with the induced $\mathbb{F}[G]$-module $\mathbb{F}[G] \otimes_{\mathbb{F}[H]} V$. We sometimes write

$$\operatorname{Ind}_H^G V = \mathbb{F}[G] \otimes_{\mathbb{F}[H]} V. \tag{1.5.5}$$

There is a canonical H-module map

$$\begin{aligned} i\colon V &\to \mathbb{F}[G] \otimes_{\mathbb{F}[H]} V \\ v &\mapsto 1 \otimes v, \end{aligned} \tag{1.5.6}$$

and the induced module is characterized by the following universal property: Given any G-module W and H-module map $j\colon V \to W$, there is a unique G-module map $f\colon \operatorname{Ind}_H^G V \to W$ such that the diagram

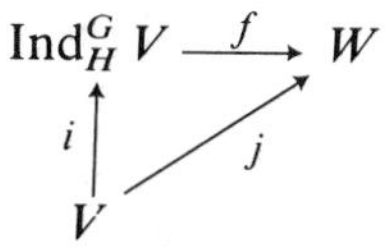

commutes. It is clear that if $X \subset G$ contains exactly one element from each of the left cosets gH of H in G, then we have a linear isomorphism

$$\operatorname{Ind}_H^G V \simeq \mathbb{F}[X] \otimes_{\mathbb{F}} V. \tag{1.5.7}$$

Here we denote by $\mathbb{F}[X]$ the linear span of X in $\mathbb{F}[G]$, even when X is not a subgroup.

In preparation for constructing the analogue for a Lie algebra of the group algebra of a group—the universal enveloping algebra—we first construct the *tensor algebra* $T(V)$ of a vector space V. For $n \geq 0$ define $T^n(V)$ to be the nth tensor power of V, i.e., the vector space

$$T^n(V) = V \otimes \cdots \otimes V \ (n \text{ times}).$$

Here it is understood that $T^0(V) = \mathbb{F}$ and that $T^1(V) = V$. Set

$$T(V) = \coprod_{n \geq 0} T^n(V) \tag{1.5.8}$$

and define an associative algebra structure on $T(V)$ by requiring that

$$(v_1 \otimes \cdots \otimes v_m)(w_1 \otimes \cdots \otimes w_n) = v_1 \otimes \cdots \otimes v_m \otimes w_1 \otimes \cdots \otimes w_n$$

in $T^{m+n}(V)$, for $v_i, w_j \in V$. Then $T(V)$ becomes a $\mathbb{Z}$-graded associative algebra with $T(V)_n = T^n(V)$ for $n \geq 0$ and $T(V)_n = 0$ for $n < 0$. It is characterized by the following universal property: Given any associative algebra A and linear map $j\colon V \to A$, there is a unique algebra map

$f: T(V) \to A$ for which the diagram

$$\begin{array}{ccc} T(V) & \xrightarrow{f} & A \\ {\scriptstyle i}\uparrow & \nearrow{\scriptstyle j} & \\ V & & \end{array}$$

commutes, where i is the inclusion of V into $T(V)$.

In the same sense that the tensor algebra is the "universal associative algebra over V," the *symmetric algebra* $S(V)$ is the universal commutative associative algebra over V. To construct it, let I be the ideal of $T(V)$ generated by all the elements $v \otimes w - w \otimes v$ for $v, w \in V$, so that I is the linear span of the products $a(v \otimes w - w \otimes v)b$ for $a, b \in T(V)$, $v, w \in V$. Form the algebra $S(V) = T(V)/I$. Since I is spanned by homogeneous elements, it is clear that $S(V)$ is a $\mathbb{Z}$-graded commutative algebra of the form

$$S(V) = \coprod_{n \geq 0} S^n(V), \tag{1.5.9}$$

where $S^n(V) = S(V)_n$, called the nth *symmetric power* of V, is the image in $S(V)$ of $T^n(V)$. We have $S^0(V) = \mathbb{F}$, $S^1(V) = V$. The algebra $S(V)$ is characterized by a universal property analogous to the one above, but for linear maps of V into *commutative* associative algebras. Given a basis $(v_j)_{j \in J}$ (J an index set) of V, $S(V)$ is naturally isomorphic to the algebra of polynomials over $\mathbb{F}$ on the generators v_j. In particular, for any total ordering $\leq$ on the set J, $S(V)$ has basis consisting of the products $v_{j_1} \cdots v_{j_n}$ for $n \geq 0$, $j_l \in J$, $j_1 \leq \cdots \leq j_n$. The space $S^n(V)$ has an obvious basis.

If V is $\mathfrak{A}$-graded ($\mathfrak{A}$ an abelian group), then $T(V)$ and $S(V)$ acquire unique algebra $\mathfrak{A}$-gradings [different from (1.5.8) and (1.5.9)] extending the grading of V.

We now turn to the *universal enveloping algebra* $U(\mathfrak{g})$ of a Lie algebra $\mathfrak{g}$. This may be constructed as the quotient associative algebra of $T(\mathfrak{g})$ by the ideal generated by the elements $x \otimes y - y \otimes x - [x, y]$ for $x, y \in \mathfrak{g}$. Clearly, $\mathbb{F}$ embeds in $U(\mathfrak{g})$; there is a canonical linear map $i: \mathfrak{g} \to U(\mathfrak{g})$ which is a homomorphism of Lie algebras; and $U(\mathfrak{g})$ is characterized by the following universal property: Given any associative algebra A and Lie algebra map $j: \mathfrak{g} \to A$, there is a unique associative algebra map $f: U(\mathfrak{g}) \to A$ making the diagram

$$\begin{array}{ccc} U(\mathfrak{g}) & \xrightarrow{f} & A \\ {\scriptstyle i}\uparrow & \nearrow{\scriptstyle j} & \\ \mathfrak{g} & & \end{array}$$

commute. [To prove this universal property, use the universal property of $T(V)$.] In particular, every $\mathfrak{g}$-module is a $U(\mathfrak{g})$-module in a natural way and conversely.

If the Lie algebra $\mathfrak{g}$ is $\mathfrak{A}$-graded ($\mathfrak{A}$ an abelian group), then $U(\mathfrak{g})$ becomes an $\mathfrak{A}$-graded algebra in a canonical way [starting from the $\mathfrak{A}$-grading of $T(\mathfrak{g})$].

If the Lie algebra $\mathfrak{g}$ is abelian then $U(\mathfrak{g})$ is just the symmetric algebra $S(\mathfrak{g})$, and in particular, the map i: $\mathfrak{g} \to U(\mathfrak{g})$ is an inclusion and we know a basis of $U(\mathfrak{g})$.

For a general Lie algebra $\mathfrak{g}$, the corresponding result is not trivial to prove, in contrast with the situation for group algebras. The Poincaré–Birkhoff–Witt theorem states the following:

The canonical map i: $\mathfrak{g} \to U(\mathfrak{g})$ is injective. Furthermore, let $(x_j)_{j \in J}$ (J a totally ordered index set) be a basis of $\mathfrak{g}$. Then the universal enveloping algebra $U(\mathfrak{g})$ has basis consisting of the ordered products $x_{j_1} \cdots x_{j_n}$ for $n \geq 0$, $j_l \in J$, $j_1 \leq \cdots \leq j_n$. (For a proof, see e.g. [Humphreys].)

Now we turn to induced Lie algebra modules. Given a subalgebra $\mathfrak{h}$ of a Lie algebra $\mathfrak{g}$ and an $\mathfrak{h}$-module V, the $\mathfrak{g}$-module *induced* by V is by definition the $\mathfrak{g}$-module corresponding to the $U(\mathfrak{g})$-module

$$\operatorname{Ind}_{\mathfrak{h}}^{\mathfrak{g}} V = U(\mathfrak{g}) \otimes_{U(\mathfrak{h})} V. \tag{1.5.10}$$

There is a canonical $\mathfrak{h}$-module map

$$\begin{aligned} i: V &\to U(\mathfrak{g}) \otimes_{U(\mathfrak{h})} V \\ v &\mapsto 1 \otimes v, \end{aligned} \tag{1.5.11}$$

and the induced module is characterized by the following universal property: For any $\mathfrak{g}$-module W and $\mathfrak{h}$-module map j: $V \to W$, there is a unique $\mathfrak{g}$-module map f: $\operatorname{Ind}_{\mathfrak{h}}^{\mathfrak{g}} V \to W$ making the diagram

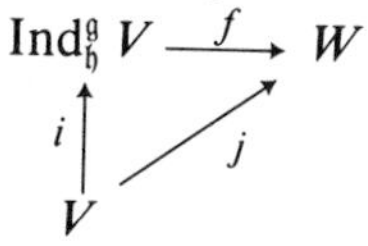

commute.

If $\mathfrak{g}$, $\mathfrak{h}$ and V are $\mathfrak{A}$-graded ($\mathfrak{A}$ an abelian group), then so is $\operatorname{Ind}_{\mathfrak{h}}^{\mathfrak{g}} V$, in a canonical way.

Suppose that $\mathfrak{k}$ and $\mathfrak{h}$ are subalgebras of $\mathfrak{g}$ such that $\mathfrak{g} = \mathfrak{k} \oplus \mathfrak{h}$ as vector spaces. Then the Poincaré–Birkhoff–Witt theorem implies that the linear map defined by

$$\begin{aligned} U(\mathfrak{k}) \otimes_{\mathbb{F}} U(\mathfrak{h}) &\to U(\mathfrak{g}) \\ x \otimes y &\mapsto xy \end{aligned} \tag{1.5.12}$$

is a linear isomorphism. (Use a basis of $\mathfrak{g}$ made up of bases of $\mathfrak{k}$ and $\mathfrak{h}$.) It follows that the linear map defined by

$$\begin{aligned} U(\mathfrak{k}) \otimes_{\mathbb{F}} V &\to U(\mathfrak{g}) \otimes_{U(\mathfrak{h})} V \\ x \otimes v &\mapsto x \otimes v \end{aligned} \tag{1.5.13}$$

is a linear isomorphism. The action of $\mathfrak{k}$ on $\mathrm{Ind}_{\mathfrak{h}}^{\mathfrak{g}} V$ carries over to the left multiplication action of $\mathfrak{k}$ on $U(\mathfrak{k}) \otimes V$, and in the case in which $\mathfrak{k}$ is an ideal of $\mathfrak{g}$, the action of $\mathfrak{h}$ carries over to the following:

$$y \cdot (x \otimes v) = [y, x] \otimes v + x \otimes y \cdot v$$

for $y \in \mathfrak{h}$, $x \in U(\mathfrak{k})$ and $v \in V$; note that $[y, x] \in U(\mathfrak{k})$ since $\operatorname{ad} y$ acts as a derivation of the associative algebra $U(\mathfrak{g})$. Even if $\mathfrak{k}$ is not an ideal, similar analysis can be used to make the action of $\mathfrak{h}$ on $U(\mathfrak{k}) \otimes V$ as explicit as we wish.

We mention an important special construction. Suppose that V is a finite-dimensional vector space with a nonsingular symmetric bilinear form $\langle \cdot, \cdot \rangle$. Let $\{v_1, \ldots, v_n\}$ be a basis of V and let $\{v_1', \ldots, v_n'\}$ be the corresponding dual basis of V, defined by:

$$\langle v_i', v_j \rangle = \delta_{ij} \quad \text{for} \quad i, j = 1, \ldots, n. \tag{1.5.14}$$

Then the element

$$\omega_0 = \sum_{i=1}^{n} v_i' \otimes v_i \in V \otimes V \tag{1.5.15}$$

is independent of the choice of basis. In fact, consider the linear isomorphism

$$i: V^* \to V$$

from the dual of V to V determined by $\langle \cdot, \cdot \rangle$, and the canonical linear isomorphism

$$j: \operatorname{End} V \to V^* \otimes V.$$

Then

$$\omega_0 = ((i \otimes 1) \circ j)(1_V), \tag{1.5.16}$$

1_V denoting the identity in End V. The canonical image

$$\omega_1 = \sum_{i=1}^{n} v_i' v_i \in S^2(V) \tag{1.5.17}$$

of ω_0 in the symmetric square of V is of course also independent of the basis. If V admits an orthonormal basis $\{e_1, \ldots, e_n\}$ (for instance, if $\mathbb{F}$ is algebraically closed), then

$$\omega_0 = \sum_{i=1}^{n} e_i \otimes e_i, \qquad \omega_1 = \sum_{i=1}^{n} e_i^2. \tag{1.5.18}$$

1.6. Affine Lie Algebras

Let $\mathfrak{g}$ be a Lie algebra and let $\langle \cdot, \cdot \rangle$ be a bilinear form on $\mathfrak{g}$—a bilinear map from $\mathfrak{g} \times \mathfrak{g}$ to $\mathbb{F}$. Then $\langle \cdot, \cdot \rangle$ is said to be *invariant* or $\mathfrak{g}$-*invariant* if

$$\langle [x, y], z \rangle + \langle y, [x, z] \rangle = 0$$

or equivalently (interchanging x and y), if

$$\langle [x, y], z \rangle = \langle x, [y, z] \rangle \tag{1.6.1}$$

for $x, y, z \in \mathfrak{g}$. Condition (1.6.1) is also called the *associativity* of $\langle \cdot, \cdot \rangle$.

Suppose that $\langle \cdot, \cdot \rangle$ is an invariant symmetric bilinear form on $\mathfrak{g}$. To the pair $(\mathfrak{g}, \langle \cdot, \cdot \rangle)$ we shall associate two (infinite-dimensional) graded Lie algebras $\hat{\mathfrak{g}}$ and $\tilde{\mathfrak{g}}$ called "affine (Lie) algebras."

Let $\mathbb{F}[t, t^{-1}]$ be the commutative associative algebra of Laurent polynomials in an indeterminate t—the algebra of finite linear combinations of integral powers of t. For a Laurent polynomial

$$f = \sum_{n \in \mathbb{Z}} a_n t^n, \qquad a_n \in \mathbb{F},$$

the sum being finite, set

$$f_0 = a_0.$$

Let d be the derivation

$$d = t\frac{d}{dt} \tag{1.6.2}$$

of $\mathbb{F}[t, t^{-1}]$. Note that $(df)_0 = 0$. Consider the vector space

$$\hat{\mathfrak{g}} = \mathfrak{g} \otimes_{\mathbb{F}} \mathbb{F}[t, t^{-1}] \oplus \mathbb{F}c, \tag{1.6.3}$$

where $\mathbb{F}c$ is a one-dimensional space. There is an (alternating) bilinear map

$$[\cdot,\cdot]: \hat{\mathfrak{g}} \times \hat{\mathfrak{g}} \to \hat{\mathfrak{g}}$$

determined by the conditions

$$\begin{aligned} &[c, \hat{\mathfrak{g}}] = [\hat{\mathfrak{g}}, c] = 0, \\ &[x \otimes f, y \otimes g] = [x, y] \otimes fg + \langle x, y\rangle (df \cdot g)_0 c \end{aligned} \tag{1.6.4}$$

for all $x, y \in \mathfrak{g}$ and $f, g \in \mathbb{F}[t, t^{-1}]$, or equivalently,

$$\begin{aligned} &[c, \hat{\mathfrak{g}}] = [\hat{\mathfrak{g}}, c] = 0, \\ &[x \otimes t^m, y \otimes t^n] = [x, y] \otimes t^{m+n} + \langle x, y\rangle m\delta_{m+n,0}c \end{aligned} \tag{1.6.5}$$

for all $x, y \in \mathfrak{g}$ and $m, n \in \mathbb{Z}$. It is easy to check using the symmetry and invariance of $\langle\cdot,\cdot\rangle$ that $\hat{\mathfrak{g}}$ is a Lie algebra, which we call the *affine* (*Lie*) *algebra* (or the *untwisted affine algebra*; cf. below) associated with $\mathfrak{g}$ and $\langle\cdot,\cdot\rangle$.

Give the space $\mathfrak{g} \otimes \mathbb{F}[t, t^{-1}]$ a Lie algebra structure by:

$$[x \otimes t^m, y \otimes t^n] = [x, y] \otimes t^{m+n} \quad \text{for} \quad x, y \in \mathfrak{g},\ m, n \in \mathbb{Z}. \tag{1.6.6}$$

Then there is an exact sequence of Lie algebras via the canonical maps

$$0 \to \mathbb{F}c \to \hat{\mathfrak{g}} \to \mathfrak{g} \otimes \mathbb{F}[t, t^{-1}] \to 0. \tag{1.6.7}$$

so that $\hat{\mathfrak{g}}$ is a central extension of the Lie algebra $\mathfrak{g} \otimes \mathbb{F}[t, t^{-1}]$.

For $x \in \mathfrak{g}$, we shall sometimes write x for the element $x \otimes t^0$ of $\mathfrak{g} \otimes \mathbb{F}[t, t^{-1}]$.

Remark 1.6.1: Suppose that $\mathfrak{g}$ is not assumed to be a Lie algebra, but only a nonassociative algebra, under $[\cdot,\cdot]$. Suppose also that the form $\langle\cdot,\cdot\rangle$ on $\mathfrak{g}$ is not assumed symmetric or invariant, but only bilinear. We can repeat the construction of the vector space $\hat{\mathfrak{g}}$ and of the nonassociative algebra structure $[\cdot,\cdot]$ on it given by (1.6.5). It is easy to see that $\hat{\mathfrak{g}}$ is a Lie algebra if and *only if* $\mathfrak{g}$ is a Lie algebra and the form $\langle\cdot,\cdot\rangle$ on $\mathfrak{g}$ is symmetric and $\mathfrak{g}$-invariant.

Let d also denote the derivation of $\hat{\mathfrak{g}}$ determined by

$$\begin{aligned} d(c) &= 0, \\ d(x \otimes f) &= x \otimes df \end{aligned} \tag{1.6.8}$$

for $x \in \mathfrak{g}$ and $f \in \mathbb{F}[t, t^{-1}]$. Form the semidirect product Lie algebra

$$\tilde{\mathfrak{g}} = \hat{\mathfrak{g}} \rtimes \mathbb{F}d, \tag{1.6.9}$$

called the *extended affine algebra* associated with $\mathfrak{g}$ and $\langle \cdot, \cdot \rangle$, or just the *affine algebra*, if no confusion is possible. We obtain a natural gradation

$$\tilde{\mathfrak{g}} = \coprod_{n \in \mathbb{Z}} \tilde{\mathfrak{g}}_n \tag{1.6.10}$$

by considering the eigenspaces

$$\tilde{\mathfrak{g}}_n = \{x \in \tilde{\mathfrak{g}} \mid [d, x] = nx\}, \quad n \in \mathbb{Z}, \tag{1.6.11}$$

of ad d. Then d is the degree derivation with respect to this grading, and

$$\tilde{\mathfrak{g}}_n = \begin{cases} \mathfrak{g} \otimes t^n & \text{for} \quad n \neq 0 \\ \mathfrak{g} \oplus \mathbb{F}c \oplus \mathbb{F}d & \text{for} \quad n = 0, \end{cases} \tag{1.6.12}$$

where we write $\mathfrak{g} \otimes t^0$ as $\mathfrak{g}$. We also have a gradation of $\hat{\mathfrak{g}}$,

$$\hat{\mathfrak{g}} = \coprod_{n \in \mathbb{Z}} \hat{\mathfrak{g}}_n \tag{1.6.13}$$

via

$$\hat{\mathfrak{g}}_n = \tilde{\mathfrak{g}}_n \cap \hat{\mathfrak{g}}. \tag{1.6.14}$$

When $\mathfrak{h}$ is a subalgebra of $\mathfrak{g}$ we shall consider $\hat{\mathfrak{h}}$ and $\tilde{\mathfrak{h}}$ as subalgebras of $\hat{\mathfrak{g}}$ and $\tilde{\mathfrak{g}}$ in the obvious way.

We shall also consider an analogue of affinization by "twisting" by an involution of $\mathfrak{g}$. An automorphism θ of a Lie algebra (or other algebraic structure) is called an *involution* if

$$\theta^2 = 1. \tag{1.6.15}$$

Let θ be an involution of $\mathfrak{g}$ which is also an isometry with respect to the form $\langle \cdot, \cdot \rangle$, i.e., which satisfies the condition

$$\langle \theta x, \theta y \rangle = \langle x, y \rangle \quad \text{for} \quad x, y \in \mathfrak{g}. \tag{1.6.16}$$

For $i \in \mathbb{Z}/2\mathbb{Z}$ set

$$\mathfrak{g}_{(i)} = \{x \in \mathfrak{g} \mid \theta x = (-1)^i x\}. \tag{1.6.17}$$

Then

$$\mathfrak{g} = \mathfrak{g}_{(0)} \oplus \mathfrak{g}_{(1)}, \tag{1.6.18}$$

$$[\mathfrak{g}_{(0)}, \mathfrak{g}_{(0)}] \subset \mathfrak{g}_{(0)}, \quad [\mathfrak{g}_{(0)}, \mathfrak{g}_{(1)}] \subset \mathfrak{g}_{(1)}, \quad [\mathfrak{g}_{(1)}, \mathfrak{g}_{(1)}] \subset \mathfrak{g}_{(0)}, \tag{1.6.19}$$

$$\langle \mathfrak{g}_{(0)}, \mathfrak{g}_{(1)} \rangle = 0. \tag{1.6.20}$$

Consider the algebra $\mathbb{F}[t^{1/2}, t^{-1/2}]$ of Laurent polynomials in an indeterminate $t^{1/2}$ whose square is t, and extend d to a derivation of $\mathbb{F}[t^{1/2}, t^{-1/2}]$ via

$$d\colon t^{n/2} \mapsto \frac{n}{2} t^{n/2}, \quad n \in \mathbb{Z}. \tag{1.6.21}$$

Form

$$\mathfrak{l} = \mathfrak{g} \otimes_{\mathbb{F}} \mathbb{F}[t^{1/2}, t^{-1/2}] \oplus \mathbb{F}c. \tag{1.6.22}$$

The formulas (1.6.4), (1.6.5) again make $\mathfrak{l}$ into a Lie algebra. Let ν be the automorphism of $\mathbb{F}[t^{1/2}, t^{-1/2}]$ such that

$$\nu\colon t^{1/2} \mapsto -t^{1/2}, \tag{1.6.23}$$

and let θ also denote the automorphism of $\mathfrak{l}$ determined by

$$\begin{aligned} \theta\colon c &\mapsto c \\ \theta\colon x \otimes f &\mapsto \theta x \otimes \nu f \end{aligned} \tag{1.6.24}$$

for $x \in \mathfrak{g}, f \in \mathbb{F}[t^{1/2}, t^{-1/2}]$. [The formula $\theta(x \otimes f) = \theta x \otimes f$ would define another automorphism of $\mathfrak{l}$.] The *twisted affine algebra* $\hat{\mathfrak{g}}[\theta]$ is then the subalgebra

$$\hat{\mathfrak{g}}[\theta] = \{x \in \mathfrak{l} \mid \theta x = x\} \tag{1.6.25}$$

of fixed points of θ in $\mathfrak{l}$. We have

$$\hat{\mathfrak{g}}[\theta] = \mathfrak{g}_{(0)} \otimes \mathbb{F}[t, t^{-1}] \oplus \mathfrak{g}_{(1)} \otimes t^{1/2}\mathbb{F}[t, t^{-1}] \oplus \mathbb{F}c. \tag{1.6.26}$$

We can again adjoin the derivation d determined by (1.6.21) as in (1.6.8) and set

$$\tilde{\mathfrak{g}}[\theta] = \hat{\mathfrak{g}}[\theta] \rtimes \mathbb{F}d, \tag{1.6.27}$$

the *extended twisted affine algebra* associated with $\mathfrak{g}$, $\langle\cdot,\cdot\rangle$ and θ. The eigenspaces of $\operatorname{ad} d$ make $\hat{\mathfrak{g}}[\theta]$ and $\tilde{\mathfrak{g}}[\theta]$ into $\frac{1}{2}\mathbb{Z}$-graded Lie algebras. The reason for considering a $\frac{1}{2}\mathbb{Z}$-grading rather that a $\mathbb{Z}$-grading will be discussed later. Note that if $\theta = 1$, then $\tilde{\mathfrak{g}}[\theta]$ degenerates to the untwisted affine algebra $\tilde{\mathfrak{g}}$.

Remark 1.6.2: The process of twisted affinization can be extended to any automorphism of finite order of $\mathfrak{g}$ which is an isometry with respect to $\langle\cdot,\cdot\rangle$, but we shall need only the case discussed.

1.7. Heisenberg Algebras

We call a Lie algebra $\mathfrak{l}$ a *Heisenberg (Lie) algebra* if

$$\operatorname{Cent} \mathfrak{l} = \mathfrak{l}' \quad \text{and} \quad \dim \operatorname{Cent} \mathfrak{l} = 1 \tag{1.7.1}$$

[recall the notations (1.3.1), (1.3.2)]. Throughout this section $\mathfrak{l}$ shall be a $\mathbb{Z}$-graded Heisenberg algebra

$$\mathfrak{l} = \coprod_{n \in \mathbb{Z}} \mathfrak{l}_n \tag{1.7.2}$$

such that

$$\dim \mathfrak{l}_n < \infty \quad \text{for} \quad n \in \mathbb{Z} \tag{1.7.3}$$

and

$$\operatorname{Cent} \mathfrak{l} = \mathfrak{l}_0 .$$

Of course, $\mathbb{Z}$ could be replaced by any isomorphic abelian group, such as $(1/N)\mathbb{Z}$ for $N \in \mathbb{Z}\backslash\{0\}$. We assume $\dim \mathfrak{l} = \infty$, although the results below hold otherwise.

Let $z \in \operatorname{Cent} \mathfrak{l}$, $z \neq 0$, so that

$$\operatorname{Cent} \mathfrak{l} = \mathbb{F}z. \tag{1.7.4}$$

Define a bilinear form $(\cdot, \cdot)$ on $\mathfrak{l}$ by:

$$[x, y] = (x, y)z \quad \text{for} \quad x, y \in \mathfrak{l}. \tag{1.7.5}$$

Then $(\cdot, \cdot)$ is an alternating form in the sense that

$$(x, x) = 0 \quad \text{for} \quad x \in \mathfrak{l}, \tag{1.7.6}$$

or equivalently,

$$(x, y) = -(y, x) \quad \text{for} \quad x, y \in \mathfrak{l}.$$

Now $(\mathbb{F}z, \mathfrak{l}) = 0$, and $(\cdot, \cdot)$ restricts to a nonsingular alternating bilinear form on $\mathfrak{l}^+ \oplus \mathfrak{l}^-$, where $\mathfrak{l}^\pm$ are the subalgebras given by:

$$\mathfrak{l}^+ = \coprod_{n > 0} \mathfrak{l}_n, \quad \mathfrak{l}^- = \coprod_{n < 0} \mathfrak{l}_n . \tag{1.7.7}$$

Clearly, $\mathfrak{l}^+$ and $\mathfrak{l}^-$ are abelian,

$$(\mathfrak{l}^+, \mathfrak{l}^+) = (\mathfrak{l}^-, \mathfrak{l}^-) = 0,$$

and the form $(\cdot, \cdot)$ is nonsingular on $\mathfrak{l}_n \oplus \mathfrak{l}_{-n}$ for each $n > 0$. Thus there are bases $(x_i)_{i \in \mathbb{Z}_+}$ of $\mathfrak{l}^+$ and $(y_i)_{i \in \mathbb{Z}_+}$ of $\mathfrak{l}^-$, consisting of homogeneous elements, such that

$$(x_i, x_j) = (y_i, y_j) = 0, \quad (x_i, y_j) = \delta_{ij}$$

for $i, j \in \mathbb{Z}_+$. (We do not assume any connection between the subscript i and $\deg x_i$ or $\deg y_i$.) In particular,

$$\begin{gathered}[z, x_i] = [z, y_i] = 0 \\ [x_i, x_j] = [y_i, y_j] = 0, \quad [x_i, y_j] = \delta_{ij} z\end{gathered} \tag{1.7.8}$$

for $i, j \in \mathbb{Z}_+$, and it follows that

$$\deg x_i + \deg y_i = 0 \quad \text{for} \quad i \in \mathbb{Z}_+.$$

The relations (1.7.8) are called the *Heisenberg commutation relations*. The subalgebras

$$\mathfrak{b} = \mathfrak{l}^+ \oplus \mathbb{F}z$$

and $\mathfrak{l}^- \oplus \mathbb{F}z$ are maximal abelian subalgebras of $\mathfrak{l}$.

We shall now construct a family of $\mathbb{Z}$-graded irreducible $\mathfrak{l}$-modules. Fix a (nonzero) scalar $k \in \mathbb{F}^\times$, and let $\mathbb{F}_k$ be the one-dimensional space $\mathbb{F}$ viewed as a $\mathbb{Z}$-graded $\mathfrak{b}$-module by:

$$z \cdot 1 = k, \quad \mathfrak{l}^+ \cdot 1 = 0, \quad \deg 1 = 0.$$

Let $M(k)$ be the $\mathbb{Z}$-graded induced $\mathfrak{l}$-module

$$M(k) = \operatorname{Ind}_{\mathfrak{b}}^{\mathfrak{l}} \mathbb{F}_k = U(\mathfrak{l}) \otimes_{U(\mathfrak{b})} \mathbb{F}_k. \tag{1.7.9}$$

Using the Poincaré–Birkhoff–Witt theorem and the fact that $U(\mathfrak{l}^-) = S(\mathfrak{l}^-)$ for the abelian Lie algebra $\mathfrak{l}^-$, we obtain a linear isomorphism

$$M(k) \simeq S(\mathfrak{l}^-). \tag{1.7.10}$$

Under the isomorphism, the grading of $M(k)$ agrees with the natural grading of $S(\mathfrak{l}^-)$ extending that of $\mathfrak{l}^-$. Viewing $S(\mathfrak{l}^-)$ as the polynomial algebra on the generators y_i, we see that z acts on $S(\mathfrak{l}^-)$ as multiplication by the scalar k, and that for $i \in \mathbb{Z}_+$, y_i acts as multiplication by the polynomial y_i. Moreover, x_i acts as the partial differentiation operator $k\partial/\partial y_i$, since both x_i and $k\partial/\partial y_i$ act as derivations of the polynomial algebra which agree on the generators y_j, $j \in \mathbb{Z}_+$. The operators provide the *canonical realization* of the *Heisenberg commutation relations* (1.7.8) associated with the nonzero scalar k. This realization shows that the $\mathfrak{l}$-module $M(k)$ is irreducible.

We say that a $\mathbb{Z}$-graded $\mathfrak{l}$-module V satisfies condition $\mathfrak{C}_k$ if

(i) z acts as multiplication by k on V and
(ii) there exists $N \in \mathbb{Z}$ such that $V_n = 0$ for $n > N$.

A nonzero vector v in an $\mathfrak{l}$-module V is called a *vacuum vector* if $\mathfrak{l}^+ \cdot v = 0$. The *vacuum space* of V is the space consisting of its vacuum vectors and 0. We denote it by Ω_V. It is easy to see that Ω_V is $\mathbb{Z}$-graded. The following results are straightforward:

Proposition 1.7.1: *Every nonzero $\mathfrak{l}$-module satisfying condition $\mathfrak{C}_k$ contains a vacuum vector.*

Proposition 1.7.2: *The module $M(k)$ satisfies condition $\mathfrak{C}_k$ and is irreducible. The vaccum space $\Omega_{M(k)}$ is one-dimensional and $\Omega_{M(k)} = \mathbb{F}(1 \otimes 1)$. In particular, every operator on $M(k)$ which commutes with the action of $\mathfrak{l}$ is a scalar multiplication operator. For any $\mathfrak{l}$-module satisfying condition $\mathfrak{C}_k$, the $\mathfrak{l}$-submodule generated by a vacuum vector is equivalent to $M(k)$. In particular, $M(k)$ is the unique (up to equivalence) irreducible module satisfying condition $\mathfrak{C}_k$.*

We now prove:

Theorem 1.7.3: *Every $\mathfrak{l}$-module satisfying condition $\mathfrak{C}_k$ is completely reducible and in particular is a direct sum of copies of $M(k)$. More precisely, for any such module V, the (well-defined) canonical linear map*

$$f: U(\mathfrak{l}) \otimes_{U(\mathfrak{b})} \Omega_V \to V$$

$$u \otimes v \mapsto u \cdot v$$

($u \in U(\mathfrak{l})$, $v \in \Omega_V$) from the induced $\mathfrak{l}$-module $\mathrm{Ind}_{\mathfrak{b}}^{\mathfrak{l}}\, \Omega_V$ to V is an $\mathfrak{l}$-module isomorphism. In particular [*see (1.5.13)*], *the linear map*

$$M(k) \otimes_{\mathbb{F}} \Omega_V = U(\mathfrak{l}^-) \otimes_{\mathbb{F}} \Omega_V \to V$$

$$u \otimes v \mapsto u \cdot v$$

($u \in U(\mathfrak{l}^-)$, $v \in \Omega_V$) defines an $\mathfrak{l}$-module isomorphism, Ω_V now being regarded as a trivial $\mathfrak{l}$-module.

Proof: We prove the second statement, the first following by choosing a basis of Ω_V. First, f is injective. In fact, let K be the kernel of f. Then K

satisfies condition $\mathfrak{C}_k$, and if $K \neq 0$ it contains a vacuum vector v (Proposition 1.7.1). Then $v \in \Omega_V$ since Ω_V is precisely the vacuum space of $\mathrm{Ind}_{\mathfrak{b}}^{\mathfrak{l}}\, \Omega_V$ (Proposition 1.7.2). But this contradicts the injectivity of f on Ω_V.

Now we show that f is surjective. Suppose instead that $V/\mathrm{Im}\, f \neq 0$. Then $V/\mathrm{Im}\, f$ is an $\mathfrak{l}$-module satisfying condition $\mathfrak{C}_k$ and so contains a vacuum vector w. Let v be a representative of w in V. Then $v \notin \mathrm{Im}\, f$, $x_i \cdot v \in \mathrm{Im}\, f$ for all $i \in \mathbb{Z}_+$ and there exists $i_0 \in \mathbb{Z}_+$ such that $x_i \cdot v = 0$ for all $i > i_0$. It is sufficient to produce $t \in \mathrm{Im}\, f$ such that $x_i \cdot t = x_i \cdot v$ for all $i \in \mathbb{Z}_+$, since $t - v$ would then be a vacuum vector in V but not in Ω_V, a contradiction.

We shall reduce the situation to the canonical realization of the Heisenberg commutation relations. Choose a basis $\{\omega_\gamma\}_{\gamma \in \Gamma}$ (Γ an index set) of Ω_V, and note that

$$\mathrm{Im}\, f = \coprod_{\gamma \in \Gamma} U(\mathfrak{l}) \otimes_{U(\mathfrak{b})} \mathbb{F}\omega_\gamma,$$

by the injectivity of f. For each $i \in \mathbb{Z}_+$ and $\gamma \in \Gamma$, let $s_{i\gamma}$ be the component of $x_i \cdot v$ in $U(\mathfrak{l}) \otimes_{U(\mathfrak{b})} \mathbb{F}\omega_\gamma$ with respect to this decomposition. Then for all $i, j \in \mathbb{Z}_+$ we have $x_i x_j \cdot v = x_j x_i \cdot v$, so that for all $\gamma \in \Gamma$,

$$x_i \cdot s_{j\gamma} = x_j \cdot s_{i\gamma}$$

and $s_{i\gamma} = 0$ for all $i > i_0$. Moreover, there is a finite subset $\Gamma_0 \subset \Gamma$ such that $s_{i\gamma} = 0$ unless $\gamma \in \Gamma_0$ and $i \leq i_0$. If we can find $t_\gamma \in U(\mathfrak{l}) \otimes_{U(\mathfrak{b})} \mathbb{F}\omega_\gamma$ such that $x_i \cdot t_\gamma = s_{i\gamma}$ for $i \in \mathbb{Z}_+$ and $\gamma \in \Gamma_0$, then we can take $t = \sum_{\gamma \in \Gamma_0} t_\gamma$ and we will be done.

Fix $\gamma \in \Gamma_0$ and identify $U(\mathfrak{l}) \otimes_{U(\mathfrak{b})} \mathbb{F}\omega_\gamma$ with the polynomial algebra on the generators y_i. Then

$$\frac{\partial}{\partial y_i} s_{j\gamma} = \frac{\partial}{\partial y_j} s_{i\gamma} \quad \text{for} \quad i, j \in \mathbb{Z}_+.$$

Recalling that $s_{i\gamma} = 0$ for $i > i_0$, we see that each $s_{j\gamma}$ lies in the polynomial algebra on the finitely many generators y_i for $i \leq i_0$. Thus there exists s in this algebra such that

$$k \frac{\partial}{\partial y_i} s = s_{i\gamma}$$

for $i \leq i_0$ and hence for all $i \in \mathbb{Z}_+$. We may therefore take $t_\gamma = s$. ∎

Certain Heisenberg algebras will play fundamental roles throughout this work, and we proceed to describe them. Let $\mathfrak{h}$ be a nonzero finite-dimensional abelian Lie algebra and let $\langle \cdot, \cdot \rangle$ be a nonsingular symmetric bilinear form on $\mathfrak{h}$.

First, we have

$$\tilde{\mathfrak{h}}' = \coprod_{\substack{n \in \mathbb{Z} \\ n \neq 0}} \mathfrak{h} \otimes t^n \oplus \mathbb{F}c, \tag{1.7.11}$$

the commutator subalgebra of $\tilde{\mathfrak{h}}$. With θ the automorphism -1 of $\mathfrak{h}$ we also consider the twisted affine algebra

$$\hat{\mathfrak{h}}[-1] = \tilde{\mathfrak{h}}[-1]' = \coprod_{n \in \mathbb{Z}+1/2} \mathfrak{h} \otimes t^n \oplus \mathbb{F}c, \tag{1.7.12}$$

the commutator subalgebra of $\tilde{\mathfrak{h}}[-1]$. We shall use the notation

$$\begin{aligned} \hat{\mathfrak{h}}_{\mathbb{Z}} &= \tilde{\mathfrak{h}}', \\ \hat{\mathfrak{h}}_{\mathbb{Z}+1/2} &= \tilde{\mathfrak{h}}[-1]'. \end{aligned} \tag{1.7.13}$$

For these algebras, the commutator formulas (1.6.5) simplify to

$$\begin{aligned} &[c, \hat{\mathfrak{h}}_{\mathbb{Z}}] = [c, \hat{\mathfrak{h}}_{\mathbb{Z}+1/2}] = 0 \\ &[x \otimes t^m, y \otimes t^n] = \langle x, y\rangle m\delta_{m+n,0}c \end{aligned} \tag{1.7.14}$$

for $x, y \in \mathfrak{h}$ and $m, n \in \mathbb{Z}\backslash\{0\}$ (resp., $\mathbb{Z} + \frac{1}{2}$). Thus $\hat{\mathfrak{h}}_{\mathbb{Z}}$ and $\hat{\mathfrak{h}}_{\mathbb{Z}+1/2}$ are Heisenberg algebras. As in Section 1.6 $\hat{\mathfrak{h}}_{\mathbb{Z}}$ is $\mathbb{Z}$-graded and $\hat{\mathfrak{h}}_{\mathbb{Z}+1/2}$ is $\frac{1}{2}\mathbb{Z}$-graded.

Let $Z = \mathbb{Z}$ or $\mathbb{Z} + \frac{1}{2}$, and take the central element z of the Heisenberg algebra $\hat{\mathfrak{h}}_Z$ to be c. Applying the discussion above with the grading group $\mathbb{Z}$ replaced by the isomorphic group $\frac{1}{2}\mathbb{Z}$ in the case $Z = \mathbb{Z} + \frac{1}{2}$, we make the identification of graded spaces

$$M(k) = S(\hat{\mathfrak{h}}_Z^-), \tag{1.7.15}$$

as in (1.7.10). The grading of the space (1.7.15) of course has the property

$$\deg 1 = 0 \quad \text{for} \quad Z = \mathbb{Z} \tag{1.7.16}$$

$$\deg 1 = 0 \quad \text{for} \quad Z = \mathbb{Z} + \tfrac{1}{2}. \tag{1.7.17}$$

Later we shall shift these two gradings in the sense of Section 1.4. The space (1.7.15) is an irreducible induced $\hat{\mathfrak{h}}_Z$-module. We make this space an (irreducible) $\mathbb{F}$-graded (in fact, $\mathbb{Z}$-, resp., $\frac{1}{2}\mathbb{Z}$-graded) $\tilde{\mathfrak{h}}$- or $\tilde{\mathfrak{h}}[-1]$-module by letting d act as the degree operator and taking $\mathfrak{h}$ to act trivially if $Z = \mathbb{Z}$; note that

$$\tilde{\mathfrak{h}} = (\tilde{\mathfrak{h}}' \rtimes \mathbb{F}d) \times \mathfrak{h}. \tag{1.7.18}$$

In both cases $S(\hat{\mathfrak{h}}_Z^-)$ is actually again an induced module—induced from its one-dimensional vacuum space, viewed as a module for the nonnegatively graded subalgebra of $\tilde{\mathfrak{h}}$ or $\tilde{\mathfrak{h}}[-1]$.

In the remainder of this work, we shall consider only the case

$$k = 1. \tag{1.7.19}$$

In particular, when $S(\hat{\mathfrak{h}}_{\bar{Z}})$ is considered as an $\tilde{\mathfrak{h}}$- or $\tilde{\mathfrak{h}}[-1]$-module, it is understood that c acts as the identity operator.

1.8. Contravariant Forms

Let $\mathfrak{g}$ be a Lie algebra and suppose that the linear map

$$\omega\colon \mathfrak{g} \to \mathfrak{g}$$

is an *anti-involution*, i.e.,

$$\omega^2 = 1, \quad \omega([x, y]) = [\omega(y), \omega(x)] \quad \text{for} \quad x, y \in \mathfrak{g}. \tag{1.8.1}$$

Then ω is of course a linear automorphism. Let

$$\mathfrak{g} = \mathfrak{n}^- \oplus \mathfrak{h} \oplus \mathfrak{n}^+ \tag{1.8.2}$$

be a *triangular decomposition* of $\mathfrak{g}$—a direct sum decomposition such that

$$[\mathfrak{h}, \mathfrak{h}] = 0, \quad [\mathfrak{n}^\pm, \mathfrak{n}^\pm] \subset \mathfrak{n}^\pm, \quad [\mathfrak{h}, \mathfrak{n}^\pm] \subset \mathfrak{n}^\pm. \tag{1.8.3}$$

Assume that

$$\omega\mathfrak{h} = \mathfrak{h}, \quad \omega\mathfrak{n}^\pm = \mathfrak{n}^\mp. \tag{1.8.4}$$

Fix a linear form $\lambda\colon \mathfrak{h} \to \mathbb{F}$ (allowing $\lambda = 0$) such that

$$\lambda(\omega h) = \lambda(h) \quad \text{for} \quad h \in \mathfrak{h} \tag{1.8.5}$$

and consider the induced module

$$M(\lambda) = \mathrm{Ind}_{\mathfrak{h}\oplus\mathfrak{n}^+}^{\mathfrak{g}} \mathbb{F}_\lambda, \tag{1.8.6}$$

where $\mathbb{F}_\lambda = \mathbb{F}v_\lambda$ (with $v_\lambda \neq 0$) is the one-dimensional $\mathfrak{h} \oplus \mathfrak{n}^+$-module such that

$$\mathfrak{n}^+ \cdot v_\lambda = 0, \quad h \cdot v_\lambda = \lambda(h)v_\lambda \quad \text{for} \quad h \in \mathfrak{h}. \tag{1.8.7}$$

The following result defines the (bilinear) *contravariant form* on $M(\lambda)$ (a hermitian contravariant form will be considered below):

Proposition 1.8.1: *There exists a unique symmetric bilinear form*

$$(\cdot, \cdot)\colon M(\lambda) \times M(\lambda) \to \mathbb{F} \tag{1.8.8}$$

such that

(i) $$(x \cdot v, w) = (v, \omega(x) \cdot w)$$

(ii) $$(v_\lambda, v_\lambda) = 1$$

for all $x \in \mathfrak{g}$ *and* $v, w \in M(\lambda)$.

Proof: We first prove the uniqueness of a form satisfying (i) and (ii). Using the construction of the universal enveloping algebra from the tensor algebra (recall Section 1.5), we see that ω extends uniquely to a linear automorphism, again denoted ω, of $U(\mathfrak{g})$ which is an *anti-involution* in the sense that

$$\omega^2 = 1, \quad \omega(xy) = \omega(y)\omega(x) \quad \text{for} \quad x, y \in U(\mathfrak{g}). \tag{1.8.9}$$

By induction on the length of a monomial in $U(\mathfrak{g})$,

$$(x \cdot v, w) = (v, \omega(x) \cdot w) \tag{1.8.10}$$

for $x \in U(\mathfrak{g})$ and $v, w \in M(\lambda)$.

Let

$$v = x \cdot v_\lambda, \quad w = y \cdot v_\lambda$$

with $x, y \in U(\mathfrak{g})$. Then

$$(v, w) = (x \cdot v_\lambda, y \cdot v_\lambda) = (v_\lambda, \omega(x)y \cdot v_\lambda).$$

By the Poincaré–Birkhoff–Witt theorem,

$$\begin{aligned} U(\mathfrak{g}) &= U(\mathfrak{n}^-) \otimes U(\mathfrak{h}) \otimes U(\mathfrak{n}^+) \\ &= (\mathbb{F} \oplus \mathfrak{n}^- U(\mathfrak{n}^-)) \otimes U(\mathfrak{h}) \otimes (\mathbb{F} \oplus U(\mathfrak{n}^+)\mathfrak{n}^+) \end{aligned}$$

in the sense of (1.5.12), so that

$$U(\mathfrak{g}) = (\mathfrak{n}^- U(\mathfrak{g}) + U(\mathfrak{g})\mathfrak{n}^+) \oplus U(\mathfrak{h}).$$

Let

$$P: U(\mathfrak{g}) \to S(\mathfrak{h}) = U(\mathfrak{h})$$

denote the corresponding projection. Since

$$\mathfrak{n}^+ \cdot v_\lambda = 0$$

and

$$(v_\lambda, \mathfrak{n}^- U(\mathfrak{g}) \cdot v_\lambda) = (\mathfrak{n}^+ \cdot v_\lambda, U(\mathfrak{g}) \cdot v_\lambda) = 0.$$

we then have

$$(v, w) = (v_\lambda, \omega(x)y \cdot v_\lambda) = (v_\lambda, P(\omega(x)y) \cdot v_\lambda).$$

Let

$$p \mapsto p(\lambda)$$

denote the algebra homomorphism

$$S(\mathfrak{h}) \to \mathbb{F}$$

determined by

$$h(\lambda) = \lambda(h)$$

for $h \in \mathfrak{h}$. Then

$$p \cdot v_\lambda = p(\lambda) v_\lambda$$

for $p \in S(\mathfrak{h})$ and we obtain

$$(v, w) = (v_\lambda, P(\omega(x)y) \cdot v_\lambda) = P(\omega(x)y)(\lambda).$$

Hence a form satisfying (i), (ii) is unique if it exists. Symmetry has not been assumed.

To prove existence, note that the left ideal

$$\mathfrak{J} = U(\mathfrak{g})\left(\mathfrak{n}^+ + \sum_{h \in \mathfrak{h}} \mathbb{F}(h - \lambda(h)1)\right)$$

is the annihilator of v_λ, so that the map

$$U(\mathfrak{g})/\mathfrak{J} \to M(\lambda)$$

$$x + \mathfrak{J} \mapsto x \cdot v_\lambda$$

is a $\mathfrak{g}$-module isomorphism. Since

$$P(xy) = P(x)y, \quad P(yx) = yP(x)$$

for $x \in U(\mathfrak{g})$, $y \in S(\mathfrak{h})$, we have

$$P(\mathfrak{J})(\lambda) = 0, \quad P(\omega(\mathfrak{J}))(\lambda) = 0.$$

Hence

$$(x \cdot v_\lambda, y \cdot v_\lambda) = P(\omega(x)y)(\lambda) \tag{1.8.11}$$

($x, y \in U(\mathfrak{g})$) is well defined, and since

$$\omega(xy)z = \omega(y)\omega(x)z,$$
$$P(\omega(1)1)(\lambda) = 1$$

for $x, y, z \in U(\mathfrak{g})$, we see that (i) and (ii) hold. We have

$$P \circ \omega = \omega \circ P,$$

and hence the form $(\cdot, \cdot)$ defined by (1.8.11) is symmetric since

$$P(\omega(y)x)(\lambda) = P(\omega(\omega(x)y))(\lambda) = \omega(P(\omega(x)y))(\lambda) = P(\omega(x)y)(\lambda)$$

for $x, y \in U(\mathfrak{g})$. The symmetry also follows from the uniqueness. ∎

Now we take $\mathfrak{g}$ to be either $\tilde{\mathfrak{h}}$ or $\tilde{\mathfrak{h}}[-1]$ in the notation of (1.7.11)–(1.7.14) (also recall Section 1.6), $\mathfrak{h}$ in (1.8.2) to be the degree-zero subalgebra and $\mathfrak{n}^{\pm}$ in (1.8.2) to be the positive and negative degree subalgebras. For the anti-involution ω we take

$$\begin{aligned} \omega(c) &= c \\ \omega(d) &= d \\ \omega(h \otimes t^n) &= h \otimes t^{-n} \quad \text{for} \quad h \in \mathfrak{h},\ n \in Z\,(= \mathbb{Z} \text{ or } \mathbb{Z} + \tfrac{1}{2}) \end{aligned} \tag{1.8.12}$$

and we assume that

$$\begin{aligned} \lambda(c) &= 1 \\ \lambda(d) &= 0 \\ \lambda(h) &= 0 \quad \text{for} \quad h \in \mathfrak{h} \quad \text{if} \quad Z = \mathbb{Z}. \end{aligned} \tag{1.8.13}$$

We conclude from Proposition 1.8.1 that there is a unique symmetric bilinear form

$$(\cdot, \cdot): S(\hat{\mathfrak{h}}_{\bar{Z}}) \times S(\hat{\mathfrak{h}}_{\bar{Z}}) \to \mathbb{F} \tag{1.8.14}$$

on the induced module $S(\hat{\mathfrak{h}}_{\bar{Z}})$ for $\tilde{\mathfrak{h}}$ or $\tilde{\mathfrak{h}}[-1]$ [recall (1.7.15)–(1.7.19)] such that

$$\begin{aligned} (d \cdot v, w) &= (v, d \cdot w) \\ ((h \otimes t^n) \cdot v, w) &= (v, (h \otimes t^{-n}) \cdot w) \\ (1, 1) &= 1 \end{aligned} \tag{1.8.15}$$

for $h \in \mathfrak{h}$, $n \in Z$ and $v, w \in S(\hat{\mathfrak{h}}_{\bar{Z}})$. Note that the first of these properties is equivalent to the condition that

$$(v, w) = 0 \text{ if } v, w \text{ are homogeneous of different degrees.} \quad (1.8.16)$$

We end this section with analogues of these results for the special case of the complex numbers. Working now over $\mathbb{C}$, we fix a square root i of -1 in $\mathbb{C}$ and write $^-$ for the conjugation map in $\mathbb{C}$.

Let $\mathfrak{g}$ be a complex Lie algebra and suppose that

$$\omega: \mathfrak{g} \to \mathfrak{g}$$

is a *conjugate-linear* map, i.e., an $\mathbb{R}$-linear map such that

$$\omega(\alpha x) = \bar{\alpha}\omega(x) \quad \text{for} \quad x \in \mathfrak{g} \quad \text{and} \quad \alpha \in \mathbb{C}, \quad (1.8.17)$$

which is an anti-involution [see (1.8.1)]. Set

$$\mathfrak{u} = \{x \in \mathfrak{g} \mid \omega x = -x\}. \quad (1.8.18)$$

Then

$$i\mathfrak{u} = \{x \in \mathfrak{g} \mid \omega x = x\}, \quad \mathfrak{u} \text{ is a real subalgebra of } \mathfrak{g} \quad (1.8.19)$$

and

$$\mathfrak{g} = \mathfrak{u} \oplus i\mathfrak{u} = \mathfrak{u} \otimes_{\mathbb{R}} \mathbb{C}. \quad (1.8.20)$$

[Conditions (1.8.19) and (1.8.20) assert that $\mathfrak{u}$ is a *real form* of $\mathfrak{g}$ and that $\mathfrak{g}$ is the *complexification* of $\mathfrak{u}$.] Fix a triangular decomposition of $\mathfrak{g}$ compatible with ω as in (1.8.2)–(1.8.4), let λ be a linear form on $\mathfrak{h}$ such that

$$\lambda(\omega h) = \overline{\lambda(h)} \quad \text{for} \quad h \in \mathfrak{h}, \quad (1.8.21)$$

and consider the induced module (1.8.6) with $\mathbb{F}$ replaced by $\mathbb{C}$. The next result constructs the *hermitian contravariant form*:

Proposition 1.8.2: *There exists a unique hermitian form (linear in the second variable)*

$$(\cdot, \cdot): M(\lambda) \times M(\lambda) \to \mathbb{C} \quad (1.8.22)$$

such that

(i) $\qquad (x \cdot v, w) = (v, \omega(x) \cdot w)$

(ii) $\qquad (v_\lambda, v_\lambda) = 1$

for all $x \in \mathfrak{g}$ and $v, w \in M(\lambda)$.

Proof: To prove the uniqueness, we first note that $\omega = -1$ on $\mathfrak{u}$ [see (1.8.18)], and we see that $\omega|_{\mathfrak{u}}$ extends uniquely to a real anti-involution of $U(\mathfrak{u})$. It follows that ω extends uniquely to a conjugate-linear anti-involution of $U(\mathfrak{g})$. The argument continues as above. ∎

We finally return to the example $\mathfrak{g} = \tilde{\mathfrak{h}}$ or $\tilde{\mathfrak{h}}[-1]$. First we take $\mathbb{F} = \mathbb{R}$ and we define the real anti-involution ω as in (1.8.12) and the real linear form λ as in (1.8.13). Then we complexify and extend ω to a conjugate-linear anti-involution and λ to a linear form satisfying (1.8.21). Proposition 1.8.2 now tells us that there is a unique hermitian form

$$(\cdot,\cdot): S(\hat{\mathfrak{h}}_{\bar{Z}}) \times S(\hat{\mathfrak{h}}_{\bar{Z}}) \to \mathbb{C} \tag{1.8.23}$$

such that

$$\begin{aligned}(d \cdot v, w) &= (v, d \cdot w)\\ ((h \otimes t^n) \cdot v, w) &= (v, (h \otimes t^{-n}) \cdot w)\\ (1, 1) &= 1\end{aligned} \tag{1.8.24}$$

for h in the *real* degree-zero subalgebra, $n \in Z$ and $v, w \in S(\hat{\mathfrak{h}}_{\bar{Z}})$. Of course, the orthogonality condition (1.8.16) holds as well.

1.9. The Virasoro Algebra

Let $p(t) \in \mathbb{F}[t, t^{-1}]$ and consider the derivation

$$T_{p(t)} = p(t)\frac{d}{dt} \tag{1.9.1}$$

of $\mathbb{F}[t, t^{-1}]$. The linear space of all derivations of $\mathbb{F}[t, t^{-1}]$ of type (1.9.1) has the structure of a Lie algebra with respect to the natural Lie bracket

$$[T_{p(t)}, T_{q(t)}] = T_{p(t)q'(t)-q(t)p'(t)} \tag{1.9.2}$$

for $p(t), q(t) \in \mathbb{F}[t, t^{-1}]$. We denote this algebra by $\mathfrak{d}$ and we choose the following basis of $\mathfrak{d}$:

$$d_n = -t^{n+1}\frac{d}{dt} = -t^n t\frac{d}{dt} \quad \text{for} \quad n \in \mathbb{Z}. \tag{1.9.3}$$

Then the commutators have the form

$$[d_m, d_n] = (m-n)d_{m+n} \quad \text{for} \quad m, n \in \mathbb{Z}. \tag{1.9.4}$$

The following statement explains the importance of $\mathfrak{d}$:

Proposition 1.9.1: *The derivations of* $\mathbb{F}[t, t^{-1}]$ *form precisely the Lie algebra* $\mathfrak{d}$.

Proof: Let $T \in \operatorname{End} \mathbb{F}[t, t^{-1}]$ be a derivation, and set

$$p(t) = T(t). \tag{1.9.5}$$

We have

$$T(1) = T(1 \cdot 1) = T(1) + T(1),$$

so that

$$T(1) = 0, \tag{1.9.6}$$

and

$$0 = T(tt^{-1}) = T(t)t^{-1} + tT(t^{-1}),$$

so that

$$T(t^{-1}) = -t^{-2}T(t). \tag{1.9.7}$$

Formulas (1.9.5)–(1.9.7) also holding for $T_{p(t)}$ in place of T, we see that $T_{p(t)}$ and T agree on all powers of t. ∎

Any three generators of $\mathfrak{d}$ of the form d_n, d_0, d_{-n}, $n \in \mathbb{Z}_+$, span a subalgebra of $\mathfrak{d}$ isomorphic to the Lie algebra $\mathfrak{sl}(2, \mathbb{F})$ of 2×2 matrices over $\mathbb{F}$ of trace 0. We shall single out the subalgebra

$$\mathfrak{p} = \mathbb{F}d_1 + \mathbb{F}d_0 + \mathbb{F}d_{-1}. \tag{1.9.8}$$

Remark 1.9.2: In Chapter 8 we shall interpret $\mathfrak{p}$ as a kind of Lie algebra of "infinitesimal projective transformations" of $\mathbb{F}[t, t^{-1}]$.

As in the case of affine Lie algebras we consider central extensions. We denote by $\mathfrak{v}$ the following one-dimensional central extension of $\mathfrak{d}$ with basis consisting of a central element c and elements L_n, $n \in \mathbb{Z}$, corresponding to the basis elements d_n, $n \in \mathbb{Z}$, of $\mathfrak{d}$: For $m, n \in \mathbb{Z}$,

$$[L_m, L_n] = (m - n)L_{m+n} + \tfrac{1}{12}(m^3 - m)\delta_{m+n,0}c. \tag{1.9.9}$$

The Lie algebra $\mathfrak{v}$ is called the *Virasoro algebra*. (It is easy to check that these relations in fact define a Lie algebra.)

Remark 1.9.3: The central term in (1.9.9) being expressible as

$$\tfrac{1}{12}(m-1)m(m+1)\delta_{m+n,0}c,$$

the central extension (1.9.9) is trivial when restricted to the subalgebra $\mathfrak{p}$ of $\mathfrak{d}$.

We can form equivalent extensions of $\mathfrak{d}$ by setting

$$L'_n = L_n + \beta_n c, \quad \beta_n \in \mathbb{F}, \quad n \in \mathbb{Z}. \tag{1.9.10}$$

Then the extension (1.9.9) is modified by the subtraction of the term

$$(m-n)\beta_{m+n}c. \tag{1.9.11}$$

The significance of the extension (1.9.9) is clear from the next result:

Proposition 1.9.4: *The extension (1.9.9) of the Lie algebra $\mathfrak{d}$ is the unique nontrivial 1-dimensional central extension up to isomorphism.*

Proof: Let $\mathfrak{v}'$ be a central extension of $\mathfrak{d}$ such that

$$\mathfrak{v}' = \mathfrak{d} \oplus \mathbb{F}c \tag{1.9.12}$$

as a vector space, and such that

$$\begin{aligned} [d_m, d_n] &= (m-n)d_{m+n} + \gamma_{m,n}c \\ [c, \mathfrak{v}'] &= 0 \end{aligned} \tag{1.9.13}$$

in $\mathfrak{v}'$, where $\gamma_{m,n} \in \mathbb{F}$. Then

$$\begin{aligned} \gamma_{m,n} + \gamma_{n,m} &= 0 \\ (m-n)\gamma_{m+n,p} + (n-p)\gamma_{n+p,m} + (p-m)\gamma_{p+m,n} &= 0 \end{aligned} \tag{1.9.14}$$

for $m, n, p \in \mathbb{Z}$. Now for $p = 0$, $m + n \neq 0$, we have

$$-(m+n)\gamma_{m,n} + (m-n)\gamma_{m+n,0} = 0,$$

so that

$$\gamma_{m,n} = \frac{m-n}{m+n}\gamma_{m+n,0}.$$

Adding $(1/n)\gamma_{n,0}c$ to d_n in (1.9.13) for $n \neq 0$, we see as in (1.9.10) and (1.9.11) that we may assume that $\gamma_{m,n} = 0$ for $m + n \neq 0$. The general solution of (1.9.14) for $m + n + p = 0$ is given by

$$\gamma_{m,-m} = \alpha m^3 + \beta m \quad \text{where} \quad \alpha, \beta \in \mathbb{F} \tag{1.9.15}$$

since this is indeed a solution, and any solution is determined by $\gamma_{1,-1}$ and $\gamma_{2,-2}$. Adding a multiple of c to d_0 permits us to change $\beta \in \mathbb{F}$ arbitrarily, and rescaling c allows us to multiply α by any nonzero scalar. We conclude that the extension (1.9.12), (1.9.13) is either equivalent to (1.9.9) or trivial. ∎

Now recall from Section 1.7 the abelian Lie algebra $\mathfrak{h}$ with its nonsingular symmetric form $\langle\cdot,\cdot\rangle$. Recall also from (1.7.15)–(1.7.19) the graded module $S(\hat{\mathfrak{h}}_Z^-)$ for $\tilde{\mathfrak{h}}$ or $\tilde{\mathfrak{h}}[-1]$ according as $Z = \mathbb{Z}$ or $\mathbb{Z} + \frac{1}{2}$. The element d acts as the degree operator and the element c acts as the identity operator. (Note: The central elements denoted c for the Lie algebras $\mathfrak{v}$ and $\hat{\mathfrak{h}}_Z$ are different and will have different normalizations as operators.) In the case $Z = \mathbb{Z}$, $\mathfrak{h} = \mathfrak{h} \otimes t^0$ acts trivially. The specific choice of the expression $\frac{1}{12}(m^3 - m)$ in (1.9.9) is best motivated by canonical representations, which we consider next, of the Virasoro algebra on the spaces $S(\hat{\mathfrak{h}}_Z^-)$.

Let $Z = \mathbb{Z}$ or $\mathbb{Z} + \frac{1}{2}$. For $h \in \mathfrak{h}$ and $n \in Z$, we shall use the notation $h(n)$ to denote the operator on $S(\hat{\mathfrak{h}}_Z^-)$ corresponding to $h \otimes t^n \in \tilde{\mathfrak{h}}$ or $\tilde{\mathfrak{h}}[-1]$:

$$h \otimes t^n \mapsto h(n). \tag{1.9.16}$$

Then we have the commutation relations

$$\begin{aligned} [g(m), h(n)] &= \langle g, h\rangle m\delta_{m+n,0} \\ [d, h(n)] &= nh(n) \end{aligned} \tag{1.9.17}$$

for $g, h \in \mathfrak{h}$ and $m, n \in Z$. The operator $h(n)$ is homogeneous of degree n [cf. (1.3.10)]. As explained in Section 1.7 the operators $h(n)$ can be realized as multiplication and partial differentiation operators on $S(\hat{\mathfrak{h}}_Z^-)$, viewed as a polynomial algebra in infinitely many variables.

Actually, for $Z = \mathbb{Z}$, we shall generalize slightly. For $\alpha \in \mathfrak{h}$, let v_α be a nonzero element of a one-dimensional $\tilde{\mathfrak{h}}$-module $\mathbb{F}v_\alpha$ on which $\mathfrak{h} \subset \tilde{\mathfrak{h}}$ acts as scalars by

$$h \cdot v_\alpha = \langle h, \alpha\rangle v_\alpha \quad \text{for} \quad h \in \mathfrak{h}, \tag{1.9.18}$$

and on which the Heisenberg algebra $\hat{\mathfrak{h}}_\mathbb{Z}$ [recall (1.7.11), (1.7.13)] acts trivially. We also let d act as multiplication by a scalar to be specified later, and we grade the one-dimensional space $\mathbb{F}v_\alpha$ accordingly. We form the tensor product $\tilde{\mathfrak{h}}$-module

$$M = S(\hat{\mathfrak{h}}_{\mathbb{Z}}^{-}) \otimes_{\mathbb{F}} \mathbb{F}v_{\alpha} \tag{1.9.19}$$

and we give it the tensor product grading. Identifying v_{α} with $1 \otimes v_{\alpha} \in M$, we find that M is the $\tilde{\mathfrak{h}}$-module induced from its one-dimensional vacuum space $\mathbb{F}v_{\alpha}$, viewed as a module for the nonnegatively graded subalgebra of $\tilde{\mathfrak{h}}$. The degree of v_{α} is left unspecified. We shall extend the notation $h(n)$ [see (1.9.16)] to M. Then the relations (1.9.17) still hold on M.

Remark 1.9.5: The module M agrees with the original module $S(\hat{\mathfrak{h}}_{\mathbb{Z}}^{-})$ in case $\alpha = 0$ and $d = 0$ on v_{α}.

For $Z = \mathbb{Z} + \frac{1}{2}$ we shall also, for convenience, use the notation

$$M = S(\hat{\mathfrak{h}}_{\mathbb{Z}+1/2}^{-}). \tag{1.9.20}$$

In order to construct a natural representation of $\mathfrak{v}$ on M, first we recall that the second tensor power $T^2(\mathfrak{h})$ contains the distinguished element denoted ω_0 in (1.5.15):

$$\omega_0 = \sum_{i=1}^{l} h_i' \otimes h_i, \tag{1.9.21}$$

where $\{h_1, \ldots, h_l\}$ is a basis of $\mathfrak{h}$ and $\{h_1', \ldots, h_l'\}$ is the corresponding dual basis with respect to $\langle \cdot, \cdot \rangle$. Let us suppose for convenience that $\mathfrak{h}$ admits an orthonormal basis, say $\{h_1, \ldots, h_l\}$, which we may always arrange by extending the field $\mathbb{F}$ if necessary. Then

$$\omega_0 = \sum_{i=1}^{l} h_i \otimes h_i \tag{1.9.22}$$

[cf. (1.5.18)]. This last expression is usually a little more convenient for computations than the equal expression (1.9.21); note that we are allowed to use (1.9.22) in place of (1.9.21) even if $\mathfrak{h}$ does not admit an orthonormal basis over the original field $\mathbb{F}$—for instance, if $\mathbb{F} = \mathbb{R}$ and the form $\langle \cdot, \cdot \rangle$ is indefinite on $\mathfrak{h}$, in which case we complexify.

Again let $Z = \mathbb{Z}$ or $\mathbb{Z} + \frac{1}{2}$. We now have the following operators on M which are canonical, i.e., independent of the choice of orthonormal basis, and well defined, because the grading of M is bounded from above and

$[g(m), h(n)] = 0$ if $m + n \neq 0$:

$$L(n) = \tfrac{1}{2} \sum_{i=1}^{\dim \mathfrak{h}} \sum_{k \in Z} h_i(n - k)h_i(k), \quad n \in \mathbb{Z}\backslash\{0\}$$
$$L(0) = \tfrac{1}{2} \sum_{i=1}^{\dim \mathfrak{h}} \sum_{k \in Z} h_i(-|k|)h_i(|k|) + \beta_0 \dim \mathfrak{h} \tag{1.9.23}$$

where

$$\beta_0 = 0 \quad \text{for} \quad Z = \mathbb{Z}$$
$$\beta_0 = \tfrac{1}{16} \quad \text{for} \quad Z = \mathbb{Z} + \tfrac{1}{2}. \tag{1.9.24}$$

Note that

$$\deg L(n) = n \quad \text{for} \quad n \in \mathbb{Z}. \tag{1.9.25}$$

We are ready to realize the Lie algebra $\mathfrak{v}$ using the infinite quadratic expressions (1.9.23):

Theorem 1.9.6: *Let $Z = \mathbb{Z}$ or $\mathbb{Z} + \frac{1}{2}$, and take M as in (1.9.19) or (1.9.20). The operators $L(n)$, $n \in \mathbb{Z}$, defined by (1.9.23) satisfy the commutation relations (1.9.9) with c replaced by the scalar multiplication operator* $\dim \mathfrak{h}$. *In particular, the correspondence*

$$L_n \mapsto L(n) \quad \text{for} \quad n \in \mathbb{Z}$$
$$c \mapsto \dim \mathfrak{h} \tag{1.9.26}$$

defines a representation of the Virasoro algebra $\mathfrak{v}$ on M.

Proof: For $h \in \mathfrak{h}$, $m \in \mathbb{Z}$ and $k \in Z$, it follows from (1.9.17) and (1.9.23) that

$$[L(m), h(k)] = -kh(k + m). \tag{1.9.27}$$

Here we use the fact that

$$\sum_{i=1}^{\dim \mathfrak{h}} \langle h_i, h \rangle h_i = h. \tag{1.9.28}$$

Therefore for $m, n \in \mathbb{Z}$ with $n \neq 0$ and $m + n \neq 0$,

$$
\begin{aligned}
[L(m), L(n)] &= [L(m), \tfrac{1}{2} \sum_{i=1}^{\dim \mathfrak{h}} \sum_{k \in Z} h_i(n-k)h_i(k)] \\
&= \tfrac{1}{2} \sum_{i=1}^{\dim \mathfrak{h}} \sum_{k \in Z} [(-n+k)h_i(m+n-k)h_i(k) \\
&\quad + (-k)h_i(n-k)h_i(m+k)] \\
&= (m-n)L(m+n). \qquad (1.9.29)
\end{aligned}
$$

In what is essentially the only remaining case, that in which $n \neq 0$ and $m + n = 0$, we express $L(-m)$ as follows:

$$
L(-m) = \tfrac{1}{2} \sum_{i=1}^{\dim \mathfrak{h}} \left(\sum_{\substack{k \in Z \\ k \leq m}} h_i(-m+k)h_i(-k) + \sum_{\substack{k \in Z \\ k > m}} h_i(-k)h_i(-m+k) \right). \qquad (1.9.30)
$$

Then by rearranging terms in the answer we obtain as in (1.9.29)

$$
[L(m), L(-m)] = 2mL(0) + \gamma_{m,-m} \qquad (1.9.31)
$$

where $\gamma_{m,-m}$ is a constant which may of course be explicitly computed by careful calculation. A relatively convenient way to determine it is to apply both sides of (1.9.31) to a vacuum vector and to use the contravariant form of Section 1.8. We shall do this. (Or one could use Proposition 1.9.4.)

Let us fix a vacuum vector v_0 in M—for example, we may take $v_0 = v_\alpha$ for $Z = \mathbb{Z}$ [see (1.9.18)] and $v_0 = 1$ for $Z = \mathbb{Z} + \frac{1}{2}$ [see (1.9.20)]. Then by Proposition 1.8.1, the $\tilde{\mathfrak{h}}$- or $\tilde{\mathfrak{h}}[-1]$-module M has a unique (bilinear) contravariant form $(\cdot, \cdot)$ normalized by the condition

$$
(v_0, v_0) = 1. \qquad (1.9.32)
$$

and determined by the properties

$$
\begin{aligned}
(h(n)v, w) &= (v, h(-n)w) \\
(d \cdot v, w) &= (v, d \cdot w)
\end{aligned} \qquad (1.9.33)
$$

for $h \in \mathfrak{h}$, $n \in Z$, $v, w \in M$ [cf. (1.8.12)–(1.8.15)]. From (1.9.23) we see also that

$$
(L(n)v, w) = (v, L(-n)w) \qquad (1.9.34)
$$

for $n \in \mathbb{Z}$, $v, w \in M$.

Another useful fact is that

$$\sum_{i=1}^{\dim \mathfrak{h}} h_i(0)^2 v_0 = \langle \alpha, \alpha \rangle v_0. \tag{1.9.35}$$

Now we compute $\gamma_{m,-m}$ in the case $m > 0$ (which is sufficient). Using (1.9.30)–(1.9.35) we have

$$\begin{aligned}
\gamma_{m,-m} &= (v_0, ([L(m), L(-m)] - 2mL(0))v_0) \\
&= (v_0, (L(m)L(-m) - 2mL(0))v_0) \\
&= (L(-m)v_0, L(-m)v_0) - 2m(v_0, L(0)v_0) \\
&= \tfrac{1}{4} \sum_{i=1}^{\dim \mathfrak{h}} \sum_{j=1}^{\dim \mathfrak{h}} \left(\sum_{\substack{k \in Z \\ 0 \le k \le m}} h_i(-m+k)h_i(-k)v_0, \sum_{\substack{l \in Z \\ 0 \le l \le m}} h_j(-m+l)h_j(-l)v_0 \right) \\
&\quad - m\left(v_0, \sum_{i=1}^{\dim \mathfrak{h}} h_i(0)^2 v_0\right) - 2m\beta_0 \dim \mathfrak{h},
\end{aligned} \tag{1.9.36}$$

the term involving $h_i(0)^2$ occurring only for $Z = \mathbb{Z}$ and equaling $-m\langle \alpha, \alpha \rangle$.

The first term in (1.9.36) equals

$$\tfrac{1}{4} \sum_{i=1}^{\dim \mathfrak{h}} \left(v_0, \left(\sum_{\substack{k \in Z \\ 0 \le k \le m}} h_i(k)h_i(m-k) \right) \left(\sum_{\substack{l \in Z \\ 0 \le l \le m}} h_i(-m+l)h_i(-l) \right) v_0 \right). \tag{1.9.37}$$

We examine the contributions from the various pairs (k, l). If $k \neq l$ and $k \neq m - l$, then the contribution of (1.9.37) is 0. If $Z = \mathbb{Z}$ and k or $l = 0$ or k or $l = m$, the total contribution is $m\langle \alpha, \alpha \rangle$, cancelling the corresponding term in (1.9.36). Assuming that $0 < k < m$ and $0 < l < m$, we see that the case $k \neq l$, $k = m - l$ gives

$$\tfrac{1}{4}(\dim \mathfrak{h})k(m-k),$$

as does the case $k = l$, $k \neq m - l$. For $k = l = m - l$ (which can occur only when m has appropriate parity), the contribution is

$$\frac{1}{2}(\dim \mathfrak{h})\left(\frac{m}{2}\right)^2 = \tfrac{1}{2}(\dim \mathfrak{h})k(m-k).$$

Thus

$$\gamma_{m,-m} = \tfrac{1}{4}\sum_{i=1}^{\dim\mathfrak{h}}\left(v_0, \left(\sum_{\substack{k\in Z\\0<k<m}} h_i(k)h_i(m-k)\right)\cdot\right.$$

$$\left.\left(\sum_{\substack{l\in Z\\0<l<m}} h_i(-m+l)h_i(-l)\right)v_0\right) - 2m\beta_0\dim\mathfrak{h}$$

$$= \tfrac{1}{2}(\dim\mathfrak{h})\left(\sum_{\substack{k\in Z\\0<k<m}} k(m-k) - 4m\beta_0\right). \tag{1.9.38}$$

But [recalling (1.9.24)]

$$1(m-1) + 2(m-2) + \cdots + (m-1)1$$
$$= (1+2+\cdots+(m-1))m - 1^2 - 2^2 - \cdots - (m-1)^2$$
$$= \tfrac{1}{6}(m^3 - m)$$

and (for $Z = \mathbb{Z} + 1/2$)

$$\tfrac{1}{2}(m-\tfrac{1}{2}) + \tfrac{3}{2}(m-\tfrac{3}{2}) + \cdots + (m-\tfrac{1}{2})\tfrac{1}{2} - 4m\beta_0$$
$$= \tfrac{1}{4}(\tfrac{1}{6})((2m)^3 - 2m) - \tfrac{1}{6}(m^3 - m) - 4m\beta_0$$
$$= \frac{1}{6}\left(m^3 + \frac{m}{2}\right) - 4m\frac{1}{16}$$
$$= \tfrac{1}{6}(m^3 - m),$$

giving the desired result. ∎

The representation of the Virasoro algebra on the space M plays an important role in the theory of vertex operator algebras as well as in the representation theory of $\mathfrak{v}$ itself. It can be considered as a motivation for the choice $\alpha = 1/12$ in the central term $\gamma_{m,-m} = \alpha m^3 + \beta m$ [recall (1.9.9) and (1.9.15)], since for this choice, c is represented as $\dim\mathfrak{h}$. As we mentioned in Remark 1.9.3, the constant $\beta = -1/12$ is fixed by the condition that the central extension be trivial on the subalgebra $\mathfrak{p}$ of $\mathfrak{d}$ [recall (1.9.8)]. These comments clarify the definition (1.9.9) of the Lie algebra $\mathfrak{v}$.

We also recall that the linear term βm in (1.9.15) can be eliminated by a choice of equivalent extension (by adding a suitable multiple of c to L_0) while the cubic term cannot be removed. Specifically, for

$$L'_n = L_n - \tfrac{1}{24}\delta_{n0}c \tag{1.9.39}$$

we have

$$[L'_m, L'_n] = (m - n)L'_{m+n} + \tfrac{1}{12}m^3\delta_{m+n,0}c. \tag{1.9.40}$$

There is an important and simple connection between the grading of the space M and $L(0)$-eigenvalues. To illustrate this, let us begin with the untwisted case $Z = \mathbb{Z}$ and $\alpha = 0$ [recall (1.9.18)–(1.9.19) and Remark 1.9.5]. From (1.9.27) with $m = 0$ we see that

$$[L(0), h(n)] = -nh(n) \quad \text{for} \quad n \in \mathbb{Z} \tag{1.9.41}$$

and by (1.9.23)–(1.9.24), we find that $L(0)$ annihilates the vacuum vector $1 \in S(\hat{\mathfrak{h}}_{\mathbb{Z}}^-) = M$:

$$L(0)1 = 0. \tag{1.9.42}$$

Since

$$\deg 1 = 0, \tag{1.9.43}$$

it follows that $L(0)$ is related as follows to the degree operator d on $S(\hat{\mathfrak{h}}_{\mathbb{Z}}^-)$:

$$L(0) = -d \quad \text{on} \quad M. \tag{1.9.44}$$

Consider next the case of arbitrary $\alpha \in \mathfrak{h}$ [see (1.9.18)–(1.9.19)]. Recall that the degree of the vacuum vector v_α of M was left unspecified. In order to preserve the elegant relation (1.9.44), we now define

$$\deg v_\alpha = -\tfrac{1}{2}\langle \alpha, \alpha \rangle. \tag{1.9.45}$$

Then (1.9.23), (1.9.35) and (1.9.41) show that (1.9.44) still holds. Note that the choice (1.9.45) is consistent with (1.9.43). When M is identified with $S(\hat{\mathfrak{h}}_{\mathbb{Z}}^-)$, we see that the definition (1.9.45) serves to shift the grading of the space $S(\hat{\mathfrak{h}}_{\mathbb{Z}}^-)$ in the sense of Section 1.4.

Now consider the twisted case $Z = \mathbb{Z} + 1/2$ and the module M given in (1.9.20). As in (1.9.41),

$$[L(0), h(n)] = -nh(n) \quad \text{for} \quad n \in \mathbb{Z} + \tfrac{1}{2}. \tag{1.9.46}$$

This time

$$L(0)1 = \tfrac{1}{16}\dim \mathfrak{h} \tag{1.9.47}$$

by (1.9.23)–(1.9.24), and so we shift the grading of the space $M = S(\hat{\mathfrak{h}}_{\mathbb{Z}+1/2}^-)$ by defining

$$\deg 1 = -\tfrac{1}{16}\dim \mathfrak{h}. \tag{1.9.48}$$

Then as above

$$L(0) = -d \quad \text{on} \quad M. \tag{1.9.49}$$

Note then that the structure and action of the Virasoro algebra suggest natural grading shifts of the spaces $S(\hat{\mathfrak{h}}_{\mathbb{Z}}^-)$ and $S(\hat{\mathfrak{h}}_{\mathbb{Z}+1/2}^-)$. But the grading shifts that we shall choose to adopt in the remainder of this work are those motivated by (1.9.39)–(1.9.40) instead. That is, *we shall insist from now on that for both $Z = \mathbb{Z}$ (with arbitrary $\alpha \in \mathfrak{h}$) and $Z = \mathbb{Z} + 1/2$ we have*

$$L'_0 = -d \quad \text{on} \quad M. \tag{1.9.50}$$

Thus for $Z = \mathbb{Z}$ and $\alpha = 0$ we have

$$\deg 1 = \tfrac{1}{24} \dim \mathfrak{h}, \tag{1.9.51}$$

for general $\alpha \in \mathfrak{h}$ we have

$$\deg v_\alpha = -\tfrac{1}{2}\langle \alpha, \alpha \rangle + \tfrac{1}{24} \dim \mathfrak{h} \tag{1.9.52}$$

and for $Z = \mathbb{Z} + 1/2$ we have

$$\deg 1 = -\tfrac{1}{48} \dim \mathfrak{h}. \tag{1.9.53}$$

Remark 1.9.7: The present degree-shifts (1.9.51)–(1.9.53) replace the earlier choice (1.7.16)–(1.7.17).

It is useful not to forget the value of the $L(0)$-eigenvalue of a homogeneous element v of M. We call this the *weight* of v and we denote it by $\mathrm{wt}\, v$. Then

$$\deg v = -\mathrm{wt}\, v + \tfrac{1}{24} \dim \mathfrak{h} \tag{1.9.54}$$

for such v. In particular,

$$\mathrm{wt}\, 1 = 0 \tag{1.9.55}$$

$$\mathrm{wt}\, v_\alpha = \tfrac{1}{2}\langle \alpha, \alpha \rangle \tag{1.9.56}$$

$$\mathrm{wt}\, 1 = \tfrac{1}{16} \dim \mathfrak{h} \tag{1.9.57}$$

in the three cases corresponding to (1.9.51)–(1.9.53).

The notions of both weight and degree will be important in the sequel. We end this section with an informal comment on their relation. In the definition (1.9.23) of the action of the Virasoro algebra on M we had to define $L(0)$ separately from the rest of the algebra. The naive formula

$$\frac{1}{2} \sum_{i=1}^{\dim \mathfrak{h}} \sum_{k \in Z} h_i(-k)h_i(k) \tag{1.9.58}$$

does not provide a well-defined operator. To make it well-defined we in fact reversed the factors of the terms $h_i(-k)h_i(k)$ for $k < 0$, formally causing the addition of the infinite expressions

$$\tfrac{1}{2}(\dim \mathfrak{h})(1 + 2 + 3 + \cdots) \quad \text{for} \quad Z = \mathbb{Z}$$

$$\tfrac{1}{2}(\dim \mathfrak{h})(\tfrac{1}{2} + \tfrac{3}{2} + \tfrac{5}{2} + \cdots) \quad \text{for} \quad Z = \mathbb{Z} + \tfrac{1}{2}.$$

The sums in parentheses can have no other interpretation than in terms of the Riemann zeta-function, namely, as

$$\zeta(-1) = -\tfrac{1}{12} \quad \text{and} \quad \tfrac{1}{2}\zeta(-1) - \zeta(-1) = \tfrac{1}{24}, \tag{1.9.59}$$

respectively. This brings us to the correct formulas (1.9.50)–(1.9.54) for the degree.

1.10. Graded Dimension

Let S be a set and let $V = \coprod_{\alpha \in S} V_\alpha$ be an S-graded vector space as in (1.3.3). We say that V *has a graded dimension* if

$$\dim V_\alpha < \infty \tag{1.10.1}$$

for all $\alpha \in S$. In this case, we call the *graded dimension* of V the formal sum

$$\dim_* V = \dim_*(V; x) = \sum_{\alpha \in S} (\dim V_\alpha)x^\alpha. \tag{1.10.2}$$

Here x is a formal variable analogous to the symbols e used in (1.5.4) and t used in (1.6.3) and (1.6.22). Formally, $\dim_* V$ (the "generating function" of the dimensions of the homogeneous subspaces of V) is an element of the abelian group $\mathbb{Z}^S$ of functions from S to $\mathbb{Z}$. In fact, $\dim_* V \in \mathbb{N}^S$, the set of functions from S to $\mathbb{N}$. The expression $\sum_{\alpha \in S} n_\alpha x^\alpha$ ($n_\alpha \in \mathbb{Z}$) is to be identified with the function which takes $\alpha \in S$ to n_α, and the addition operation in $\mathbb{Z}^S$ is pointwise addition of functions:

$$\sum_{\alpha \in S} m_\alpha x^\alpha + \sum_{\alpha \in S} n_\alpha x^\alpha = \sum_{\alpha \in S} (m_\alpha + n_\alpha)x^\alpha$$

where $m_\alpha, n_\alpha \in \mathbb{Z}$.

If W is a graded subspace of V and if V has a graded dimension, then so does V/W, and

$$\dim_*(V/W) = \dim_* V - \dim_* W. \tag{1.10.3}$$

Suppose that $(V^i)_{i \in I}$ is a family of S-graded vector spaces such that for all $\alpha \in S$,

$$\sum_{i \in I} \dim V_\alpha^i < \infty.$$

Then

$$\dim_* \coprod_{i \in I} V^i = \sum_{i \in I} \dim_* V^i, \tag{1.10.4}$$

the (possibly infinite) sum being given the obvious meaning.

Suppose that $\mathfrak{A}$ is an abelian group and that V and W are $\mathfrak{A}$-graded vector spaces such that V, W and $V \otimes W$ have graded dimensions. Then

$$\dim_*(V \otimes W) = (\dim_* V)(\dim_* W), \tag{1.10.5}$$

where the product on the right has the following natural meaning: Let

$$\sum_{\alpha \in \mathfrak{A}} m_\alpha x^\alpha, \sum_{\beta \in \mathfrak{A}} n_\beta x^\beta \in \mathbb{Z}^{\mathfrak{A}}$$

and suppose that for all $\gamma \in \mathfrak{A}$, $m_\alpha n_\beta = 0$ for all but finitely many ordered pairs $(\alpha, \beta) \in \mathfrak{A} \times \mathfrak{A}$ such that $\alpha + \beta = \gamma$. Then

$$\left(\sum m_\alpha x^\alpha\right)\left(\sum n_\beta x^\beta\right) = \sum_{\gamma \in \mathfrak{A}} \left(\sum_{\substack{\alpha, \beta \in \mathfrak{A} \\ \alpha + \beta = \gamma}} m_\alpha n_\beta \right) x^\gamma. \tag{1.10.6}$$

Formula (1.10.5) has an obvious analogue for an arbitrary finite number of tensor factors.

Let V be an $\mathfrak{A}$-graded vector space having a graded dimension. Fix $\beta \in \mathfrak{A}$. Suppose that we shift the grading of V by renaming V_α as $V_{\alpha+\beta}$ for $\alpha \in \mathfrak{A}$ (cf. Section 1.4). Then the new and old graded dimensions are related by:

$$(\dim_* V)_{\text{new}} = x^\beta (\dim_* V)_{\text{old}}. \tag{1.10.7}$$

Fix $N \in \mathbb{Z}$, $N \neq 0$. Suppose that V is a $(1/N)\mathbb{Z}$-graded vector space such that

$$\dim V_{n/N} < \infty \quad \text{for} \quad n \in \mathbb{Z} \quad \text{and} \quad V_{n/N} = 0 \quad \text{for} \quad n \leq 0. \tag{1.10.8}$$

Equipped with the corresponding grading, the symmetric algebra has graded dimension

$$\dim_* S(V) = \prod_{n \in \mathbb{Z}_+} (1 - x^{n/N})^{-\dim V_{n/N}}, \tag{1.10.9}$$

where the formal product denotes the well-defined nonnegative integral linear combination of nonnegative powers of $x^{1/N}$ obtained by formally multiplying the infinitely many geometric series

$$(1 - x^{n/N})^{-1} = \sum_{k \geq 0} x^{kn/N} \tag{1.10.10}$$

with appropriate multiplicities, or equivalently, by multiplying the binomial series

$$\begin{aligned}(1 - x^{n/N})^{-\dim V_{n/N}} &= \sum_{k \geq 0} \binom{-\dim V_{n/N}}{k} (-x^{n/N})^k \\ &= \sum_{k \geq 0} \binom{\dim V_{n/N} + k - 1}{k} x^{kn/N}.\end{aligned} \tag{1.10.11}$$

If V has a basis $(v_j)_{j \in J}$ such that

$$\deg v_j = n_j \in \frac{1}{N}\mathbb{Z} \quad \text{for} \quad j \in J, \tag{1.10.12}$$

then we can rewrite (1.10.9) as:

$$\dim_* S(V) = \prod_{j \in J} (1 - x^{n_j})^{-1}. \tag{1.10.13}$$

Note that N can be negative, in which case (1.10.8) becomes:

$$\dim V_n < \infty \quad \text{for} \quad n \in \frac{1}{N}\mathbb{Z} \quad \text{and} \quad V_n = 0 \quad \text{for} \quad n \geq 0 \tag{1.10.14}$$

and (1.10.9) becomes

$$\dim_* S(V) = \prod_{\substack{n \in (1/N)\mathbb{Z} \\ n > 0}} (1 - x^{-n})^{-\dim V_{-n}}. \tag{1.10.15}$$

We shall frequently be dealing with $(1/N)\mathbb{Z}$-graded vector spaces V with grading bounded above (i.e., $V_n = 0$ for n sufficiently large). For such V, $\dim_* V = \dim_*(V; x)$ involves powers x^n of x with n bounded above. Since we prefer to work with large positive powers of a formal variable rather than large negative powers (when there is a choice), we make the convention that when $\mathfrak{A}$ is an abelian group and V is an $\mathfrak{A}$-graded vector space which has a graded dimension, we take as our notation of graded dimension

$$\dim_* V = \dim_*(V; q^{-1}) = \sum_{\alpha \in \mathfrak{A}} (\dim V_{-\alpha}) q^{\alpha} \tag{1.10.16}$$

[cf. (1.10.2)]. That is, we formally set

$$x = q^{-1}. \tag{1.10.17}$$

Then for example (1.10.13) becomes

$$\dim_* S(V) = \prod_{j \in J} (1 - q^{-n_j})^{-1} \tag{1.10.18}$$

and (1.10.15) becomes

$$\dim_* S(V) = \prod_{n \in (1/N)\mathbb{Z}_+} (1 - q^n)^{-\dim V_{-n}}. \tag{1.10.19}$$

The graded dimensions of the spaces $S(\hat{\mathfrak{h}}_Z^-)$ for $Z = \mathbb{Z}$ and $Z = \mathbb{Z} + 1/2$ [see (1.7.13)–(1.7.15)] will be particularly important. To express these compactly, set

$$\phi(q) = \prod_{n \in \mathbb{Z}_+} (1 - q^n), \tag{1.10.20}$$

a well-defined integral linear combination of nonnegative powers of q. Also consider the related classical *Dedekind η-function*

$$\eta(q) = q^{1/24}\phi(q), \tag{1.10.21}$$

understood as an integral linear combination of positive powers of $q^{1/24}$. (Classically, q is taken as $e^{2\pi i z}$, z a complex variable in the upper half plane. In this form, $\eta(q)$ has important modular transformation properties not shared by $\phi(q)$; see for example [Knopp], [Lang 2].) From (1.10.18) and (1.10.19), we see that

$$\dim_* S(\hat{\mathfrak{h}}_{\mathbb{Z}}^-) = \frac{1}{\eta(q)^{\dim \mathfrak{h}}} \tag{1.10.22}$$

$$\dim_* S(\hat{\mathfrak{h}}_{\mathbb{Z}+1/2}^-) = \left(\frac{\eta(q)}{\eta(q^{1/2})}\right)^{\dim \mathfrak{h}}, \tag{1.10.23}$$

recalling (1.10.7) and using the grading shifts (1.9.51) and (1.9.53).

Remark 1.10.1: If we had used the original gradings of $S(\hat{\mathfrak{h}}_Z^-)$ given by (1.7.16) and (1.7.17), then the graded dimensions would have been given by expressions similar to (1.10.22) and (1.10.23) but with ϕ in place of η, and we would have lost the modular transformation properties.

2 Formal Calculus

One special feature of the representation theory of affine Lie algebras is the role of certain generating functions, which correspond in the geometric interpretation to δ-function loops, and which are known in the physics literature as quantum fields. Though these generating functions do not strictly speaking belong to the algebra, they admit an especially simple form in representations and will be used widely throughout this book. The role of such generating functions was recognized in the early works on vertex operator representations [Lepowsky–Wilson 1], [Kac–Kazhdan–Lepowsky–Wilson], [Frenkel–Kac], [Segal 1] as well as in the physics literature. The formal calculus approach was initiated in [Garland 3] as an algebraic analogue of operator techniques in physics and also in [Date–Kashiwara–Miwa], and its development was continued in [Lepowsky–Wilson 3–5], [Lepowsky–Primc 1, 2], [Meurman–Primc], [Lepowsky 4]. In this chapter we introduce the formalism, used extensively throughout the book, which allows us to work rigorously with the appropriate infinite sums. The deeper development of formal calculus will continue in Chapter 8.

In Section 2.1 we discuss elementary operations on formal series. The most important such series, which plays a fundamental role throughout this work, is the series $\delta(z)$, formally the Fourier expansion of the δ-function. The basic result about $\delta(z)$ computes its product with a Laurent polynomial.

This trivial result has useful multivariable variants, which must be formulated carefully. Section 2.2 is devoted to derivations of formal series and their relation with $\delta(z)$. In Section 2.3 we express Lie brackets for affine algebras in terms of generating functions using the language of formal variables.

2.1. Formal Series

Let V be a vector space. We denote by $V\{z\}$ the set of formal sums $\sum_{n \in \mathbb{F}} v_n z^n$ with $v_n \in V$. The sum $\sum_{n \in \mathbb{F}} v_n z^n$ can be viewed as notation for the map

$$v\colon \mathbb{F} \to V$$

$$n \mapsto v_n.$$

Note that n is not restricted to the set of integers; however, for the sums $\sum_{n \in \mathbb{F}} v_n z^n$ occurring in this work, the map v will typically be supported on the union of a finite number of cosets of $\mathbb{Z}$ in $\mathbb{F}$. When $V = \operatorname{End} W$ for a vector space W, we shall often use the notation $v(n)$ or $v(-n)$ instead of v_n.

Analogous to the symbols e of (1.5.4), t of Section 1.6 and x, q of Section 1.10, the symbol z and related symbols used below play still different roles.

The set $V\{z\}$ is a vector space under the (pointwise) operations

$$\sum_{n \in \mathbb{F}} v_n z^n + \sum_{n \in \mathbb{F}} w_n z^n = \sum_{n \in \mathbb{F}} (v_n + w_n) z^n$$

$$\alpha \sum_{n \in \mathbb{F}} v_n z^n = \sum_{n \in \mathbb{F}} \alpha v_n z^n$$

where $v_n, w_n \in V$ and $\alpha \in \mathbb{F}$. Given a linear map $\pi\colon V \to W$ of vector spaces, we shall also write

$$\pi\colon V\{z\} \to W\{z\} \tag{2.1.1}$$

for the natural extension.

The space $V\{z\}$ contains various useful subspaces:

$$V[z] = \left\{ \sum_{n \in \mathbb{N}} v_n z^n \mid v_n \in V, \quad \text{all but finitely many} \quad v_n = 0 \right\} \tag{2.1.2}$$

$$V[z, z^{-1}] = \left\{ \sum_{n \in \mathbb{Z}} v_n z^n \mid v_n \in V, \quad \text{all but finitely many} \quad v_n = 0 \right\} \tag{2.1.3}$$

$$V[[z]] = \left\{ \sum_{n \in \mathbb{N}} v_n z^n \mid v_n \in V \right\} \tag{2.1.4}$$

$$V[[z, z^{-1}]] = \left\{ \sum_{n \in \mathbb{Z}} v_n z^n \mid v_n \in V \right\}. \tag{2.1.5}$$

We have

$$V[z] = V \otimes \mathbb{F}[z], \quad V[z, z^{-1}] = V \otimes \mathbb{F}[z, z^{-1}]. \tag{2.1.6}$$

If V is an associative algebra, then so are $V[z]$, $V[z, z^{-1}]$ and $V[[z]]$—the algebras of polynomials, Laurent polynomials and formal power series, respectively, in the indeterminate z with coefficients in V.

Let $a \in \mathbb{F}^{\times}$. For

$$v(z) = \sum_{n \in \mathbb{Z}} v_n z^n \in V[z, z^{-1}]$$

set

$$v(a) = \sum_{n \in \mathbb{Z}} a^n v_n, \tag{2.1.7}$$

and for

$$v(z) = \sum_{n \in \mathbb{Z}} v_n z^n \in V[[z, z^{-1}]]$$

set

$$v(az) = \sum_{n \in \mathbb{Z}} a^n v_n z^n. \tag{2.1.8}$$

Let $(x_i)_{i \in I}$ be a family in End V (I an index set). We say that $(x_i)_{i \in I}$ is *summable* if for every $v \in V$, $x_i v = 0$ for all but a finite number of $i \in I$. In this case we denote by $\sum_{i \in I} x_i$ the operator

$$\begin{aligned} \sum_{i \in I} x_i \colon V &\to V \\ v &\mapsto \sum_{i \in I} x_i v. \end{aligned} \tag{2.1.9}$$

Of course, any finite family is summable.

Let $(X_i(z))_{i \in I}$ be a family in (End V)$\{z\}$ and set $X_i(z) = \sum_{n \in \mathbb{F}} x_i(n) z^n$. We say that $\sum_{i \in I} X_i(z)$ exists if for every $n \in \mathbb{F}$, $(x_i(n))_{i \in I}$ is summable. We then set

$$\sum_{i \in I} X_i(z) = \sum_{n \in \mathbb{F}} \left(\sum_{i \in I} x_i(n) \right) z^n. \tag{2.1.10}$$

Note that for

$$X(z) = \sum_{n \in \mathbb{F}} x(n)z^n \in (\text{End } V)\{z\},$$

the sum of the family $(x(n)z^n)_{n \in \mathbb{F}}$ exists in the new sense and equals $X(z)$.

Let $(X_i(z))_{i=1}^r$ be a finite family in $(\text{End } V)\{z\}$, with $X_i(z) = \sum_{n \in \mathbb{F}} x_i(n)z^n$. We say that $X_1(z) \cdots X_r(z)$ exists if for every $n \in \mathbb{F}$,

$$(x_1(n_1) \cdots x_r(n_r))_{\substack{n_1, \ldots, n_r \in \mathbb{F} \\ n_1 + \cdots + n_r = n}}$$

is summable. We then set

$$X_1(z) \cdots X_r(z) =$$

$$\sum_{n \in \mathbb{F}} \left(\sum_{n_1 + \cdots + n_r = n} x_1(n_1) \cdots x_r(n_r) \right) z^n \in (\text{End } V)\{z\}. \tag{2.1.11}$$

Suppose that $X_1(z) \cdots X_r(z)$ exists and that for a fixed q with $1 \le q < r$, $X_1(z) \cdots X_q(z)$ and $X_{q+1}(z) \cdots X_r(z)$ exist. Then their product exists and

$$X_1(z) \cdots X_r(z) = (X_1(z) \cdots X_q(z))(X_{q+1}(z) \cdots X_r(z)). \tag{2.1.12}$$

An example of a nonexistent product is

$$\left(\sum_{n \ge 0} z^n \right) \left(\sum_{n \le 0} z^n \right)$$

(the coefficients being viewed as scalar operators in End V or as elements of End $\mathbb{F}$). The following instructive paradox involves a nonexistent product of three series:

$$\text{“}\delta(z) = \left[\left(\sum_{n \ge 0} z^n \right) (1 - z) \right] \delta(z) = \left(\sum_{n \ge 0} z^n \right) [(1 - z)\delta(z)]$$

$$= \left(\sum_{n \ge 0} z^n \right) 0 = 0\text{”},$$

where $\delta(z) = \sum_{n \in \mathbb{Z}} z^n$, as in (2.1.22) below. Here is an example of an existent product which contains a nonexistent subproduct:

$$\left(\sum_{n \ge 0} z^n \right) \left(\sum_{n \le 0} z^n \right) 0 = 0.$$

The preceding notions generalize in the obvious way to the case of several commuting "formal variables" $z_1, z_2, \ldots, z_r$. For example,

$$V\{z_1, z_2\} = \left\{ \sum_{m,n \in \mathbb{F}} v_{mn} z_1^m z_2^n \mid v_{mn} \in V \right\}, \tag{2.1.13}$$

$$V[[z_1, z_2]] = \left\{ \sum_{m,n \in \mathbb{N}} v_{mn} z_1^m z_2^n \mid v_{mn} \in V \right\}, \tag{2.1.14}$$

$$V[[z_1, z_2^{-1}]] = \left\{ \sum_{m,n \in \mathbb{N}} v_{mn} z_1^m z_2^{-n} \mid v_{mn} \in V \right\}. \tag{2.1.15}$$

If V is an associative algebra, then so are the last two of these spaces. Given $v(z) = \sum_{n \in \mathbb{F}} v_n z^n \in V\{z\}$, we shall use notation of the following sort:

$$v(z_1/z_2) = \sum_{n \in \mathbb{F}} v_n z_1^n z_2^{-n} \in V\{z_1, z_2\}. \tag{2.1.16}$$

Let

$$\sum_{m,n \in \mathbb{F}} x(m, n) z_1^m z_2^n \in (\operatorname{End} V)\{z_1, z_2\}.$$

We say that

$$\lim_{z_1 \to z_2} \sum_{m,n \in \mathbb{F}} x(m, n) z_1^m z_2^n$$

exists if for every $n \in \mathbb{F}$, the family $(x(m, n - m))_{m \in \mathbb{F}}$ is summable. In this case we set

$$\lim_{z_1 \to z_2} \left(\sum_{m,n \in \mathbb{F}} x(m, n) z_1^m z_2^n \right) = \sum_{n \in \mathbb{F}} \left(\sum_{m \in \mathbb{F}} x(m, n - m) \right) z_2^n. \tag{2.1.17}$$

Thus the notation $\lim_{z_1 \to z_2}$ is used only as an alternative notation for $|_{z_1 = z_2}$, and we shall sometimes use some form of the latter notation. In the analogous way we define limits involving several variables, such as

$$\lim_{z_{i_1}, \ldots, z_{i_s} \to z} X(z_1, \ldots, z_r)$$

for $X(z_1, \ldots, z_r) \in (\operatorname{End} V)\{z_1, \ldots, z_r\}$, $s \leq r$ and $1 \leq i_1 < \cdots < i_s \leq r$. Note that

$$X_1(z) \cdots X_r(z) = \lim_{z_1, \ldots, z_r \to z} X_1(z_1) \cdots X_r(z_r), \tag{2.1.18}$$

both sides existing simultaneously.

In general, a limit exists or a product is defined if and only if the coefficient of every monomial in the formal variables in the indicated formal expression is summable.

There are natural "multiplication" bilinear maps

$$V[z, z^{-1}] \times \mathbb{F}\{z\} \to V\{z\} \tag{2.1.19}$$

$$V[z, z^{-1}] \times \mathbb{F}[[z, z^{-1}]] \to V[[z, z^{-1}]], \tag{2.1.20}$$

which we denote by juxtaposition. In particular (for $V = \mathbb{F}$), $F\{z\}$ is an $\mathbb{F}[z, z^{-1}]$-module and $\mathbb{F}[[z, z^{-1}]]$ is a submodule. For $v(z) \in V[z, z^{-1}]$ and $f(z) \in \mathbb{F}\{z\}$, we shall sometimes use notation such as the following:

$$v(z)f(z) = \lim_{z_1, z_2 \to z} v(z_1)f(z_2). \tag{2.1.21}$$

Define

$$\delta(z) = \sum_{n \in \mathbb{Z}} z^n \in \mathbb{F}[[z, z^{-1}]]. \tag{2.1.22}$$

Formally, this is the Laurent series expansion of the classical δ-function at $z = 1$, a fact which motivates the following fundamental properties of $\delta(z)$:

Proposition 2.1.1: *(a) Let $v(z) \in V[z, z^{-1}]$. Then in $V\{z\}$,*

$$v(z)\delta(z) = v(1)\delta(z). \tag{2.1.23}$$

More generally, for $a \in \mathbb{F}^{\times}$,

$$v(z)\delta(az) = v(a^{-1})\delta(az). \tag{2.1.24}$$

(b) Let

$$X(z_1, z_2) \in (\mathrm{End}\, V)[[z_1, z_1^{-1}, z_2, z_2^{-1}]]$$

be such that

$$\lim_{z_1 \to z_2} X(z_1, z_2) \text{ exists} \tag{2.1.25}$$

and let $a \in \mathbb{F}^{\times}$. Then in $(\mathrm{End}\, V)\{z_1, z_2\}$,

$$\begin{aligned} X(z_1, z_2)\delta(az_1/z_2) &= X(a^{-1}z_2, z_2)\delta(az_1/z_2) \\ &= X(z_1, az_1)\delta(az_1/z_2) \end{aligned} \tag{2.1.26}$$

and in particular, the indicated expressions exist.

Proof: (a) By linearity we may assume that $v(z) = v_n z^n$ ($n \in \mathbb{Z}, v_n \in V$).

In this case we have

$$\begin{aligned} v(z)\delta(az) &= (v_n z^n)\Big(\sum_{k\in\mathbb{Z}} a^k z^k\Big) = \sum_{k\in\mathbb{Z}} a^k v_n z^{k+n} \\ &= \sum_{k\in\mathbb{Z}} a^{k-n} v_n z^k = (a^{-n}v_n)\Big(\sum_{k\in\mathbb{Z}} a^k z^k\Big) \\ &= v(a^{-1})\delta(az). \end{aligned}$$

(b) Let

$$X(z_1, z_2) = \sum_{m,n\in\mathbb{Z}} x(m, n) z_1^m z_2^n.$$

Then

$$\begin{aligned} X(z_1, z_2)\delta(az_1/z_2) &= \Big(\sum_{m,n\in\mathbb{Z}} x(m, n) z_1^m z_2^n\Big)\Big(\sum_{k\in\mathbb{Z}} a^k z_1^k z_2^{-k}\Big) \\ &= \sum_{m,n,k\in\mathbb{Z}} a^k x(m, n) z_1^{m+k} z_2^{n-k} \\ &= \sum_{m,n,k\in\mathbb{Z}} a^{k-m} x(m, n) z_1^k z_2^{m+n-k} \\ &= X(a^{-1}z_2, z_2)\delta(az_1/z_2). \end{aligned}$$

all expressions existing. For the last equality, either argue similarly or rewrite $\delta(az_1/z_2)$ as $\delta(a^{-1}z_2/z_1)$ and apply the first case. ∎

Note that the substitutions $z_1 = a^{-1}z_2$ and $z_2 = az_1$ in (2.1.26) correspond formally to the substitution $az_1/z_2 = 1$; cf. (2.1.23).

Remark 2.1.2: Propositions 2.1.1(a), (b) require integral powers of z, z_1 and z_2. For instance,

$$z^{1/2}\delta(z) \neq 1^{1/2}\delta(z) \tag{2.1.27}$$

in any sense, and

$$z_1^{1/2} z_2^{1/2}\delta(z_1/z_2) \neq z_2\delta(z_1/z_2). \tag{2.1.28}$$

We shall in general omit proofs of summability of families of operators, existence of products, and so on. These will always be easy since our spaces V will typically be graded with gradations truncated from above (as in Section 1.7), and our series $\sum x(n)z^n$ ($x(n) \in \operatorname{End} V$) will typically have the property $\deg x(n) = n$ (cf. Section 1.3) or a related property.

2.2. Derivations

As in Section 2.1, let V be a vector space. Consider the endomorphism

$$\frac{d}{dz}: \sum_{n \in \mathbb{F}} v_n z^n \mapsto \sum_{n \in \mathbb{F}} n v_n z^{n-1} \tag{2.2.1}$$

of $V\{z\}$. We shall also use the prime notation of calculus. For example,

$$\delta'(z) = \frac{d}{dz}\delta(z) = \sum_{n \in \mathbb{Z}} n z^{n-1}. \tag{2.2.2}$$

For several variables $z_1, z_2, \ldots$ we shall use partial derivative notation $\partial/\partial z_1, \partial/\partial z_2, \ldots$.

The operator d/dz acts as a derivation of $\mathbb{F}[z, z^{-1}]$, and in fact, more generally,

$$\frac{d}{dz}(v(z)f(z)) = \left(\frac{d}{dz}v(z)\right)f(z) + v(z)\left(\frac{d}{dz}f(z)\right) \tag{2.2.3}$$

for all $v(z) \in V[z, z^{-1}]$, $f(z) \in \mathbb{F}\{z\}$. Analogous formulas hold for $\partial/\partial z_1$, $\partial/\partial z_2, \ldots$.

Proposition 2.2.1: *(a) Let $v(z) \in V[z, z^{-1}]$ and $a \in \mathbb{F}^{\times}$. Then in $V\{z\}$,*

$$v(z)\frac{d}{dz}(\delta(az)) = v(a^{-1})\frac{d}{dz}(\delta(az)) - v'(a^{-1})\delta(az). \tag{2.2.4}$$

(b) Let

$$X(z_1, z_2) \in (\mathrm{End}\, V)[[z_1, z_1^{-1}, z_2, z_2^{-1}]]$$

be such that

$$\lim_{z_1 \to z_2} X(z_1, z_2) \text{ exists} \tag{2.2.5}$$

and let $a \in \mathbb{F}^{\times}$. Then in $(\mathrm{End}\, V)\{z_1, z_2\}$,

$$\begin{aligned} X(z_1, z_2)&\frac{\partial}{\partial z_1}(\delta(az_1/z_2)) \\ &= X(a^{-1}z_2, z_2)\frac{\partial}{\partial z_1}(\delta(az_1/z_2)) - \left(\frac{\partial X}{\partial z_1}\right)(a^{-1}z_2, z_2)\delta(az_1/z_2), \end{aligned} \tag{2.2.6}$$

$$X(z_1, z_2)\frac{\partial}{\partial z_2}(\delta(az_1/z_2))$$

$$= X(z_1, az_1)\frac{\partial}{\partial z_2}(\delta(az_1/z_2)) - \left(\frac{\partial X}{\partial z_2}\right)(z_1, az_1)\delta(az_1/z_2), \quad (2.2.7)$$

all expressions existing.

Proof: (a) Apply d/dz to Proposition 2.1.1(a) to obtain

$$v'(z)\delta(az) + v(z)\frac{d}{dz}(\delta(az)) = v(a^{-1})\frac{d}{dz}(\delta(az)).$$

By Proposition 2.1.1(a) we also get

$$v'(z)\delta(az) = v'(a^{-1})\delta(az).$$

(b) Apply $\partial/\partial z_1$ and $\partial/\partial z_2$ to Proposition 2.1.1(b) and then use Proposition 2.1.1(b), noting that

$$\lim_{z_1 \to z_2}\left(\frac{\partial X}{\partial z_i}\right)(z_1, z_2)$$

exists for $i = 1, 2$. ∎

The same proof shows, more generally:

Proposition 2.2.2: *(a) Let $p(z) \in \mathbb{F}[z, z^{-1}]$ and consider the derivation*

$$T = T_{p(z)} = p(z)\frac{d}{dz} \quad (2.2.8)$$

of $\mathbb{F}[z, z^{-1}]$, viewed also as an endomorphism of $V[z, z^{-1}]$ and of $\mathbb{F}\{z\}$. For $v(z)$ and a as in Proposition 2.2.1(a),

$$v(z)T(\delta(az)) = v(a^{-1})T(\delta(az)) - (Tv)(a^{-1})\delta(az) \quad (2.2.9)$$

in $V\{z\}$.

(b) Let

$$p(z_1, z_2) \in \mathbb{F}[z_1, z_1^{-1}, z_2, z_2^{-1}] \quad (2.2.10)$$

and set

$$T_1 = p(z_1, z_2)\frac{\partial}{\partial z_1}, \quad T_2 = p(z_1, z_2)\frac{\partial}{\partial z_2}. \quad (2.2.11)$$

For $X(z_1, z_2)$ and a as in Proposition 2.2.1(b),

$$X(z_1, z_2)T_1(\delta(az_1/z_2))$$
$$= X(a^{-1}z_2, z_2)T_1(\delta(az_1/z_2)) - (T_1X)(a^{-1}z_2, z_2)\delta(az_1/z_2), \quad (2.2.12)$$
$$X(z_1, z_2)T_2(\delta(az_1/z_2))$$
$$= X(z_1, az_1)T_2(\delta(az_1/z_2)) - (T_2X)(z_1, az_1)\delta(az_1/z_2) \quad (2.2.13)$$

in (End V)$\{z_1, z_2\}$, *all expressions existing.*

A particularly important case of (2.2.8) is the case $p(z) = z$: Set

$$D = D_z = z\frac{d}{dz}. \quad (2.2.14)$$

As an endomorphism of $V\{z\}$, D acts as follows: For

$$v(z) = \sum_{n \in \mathbb{F}} v_n z^n,$$
$$Dv(z) = \sum_{n \in \mathbb{F}} n v_n z^n. \quad (2.2.15)$$

If V is $\mathbb{F}$-graded, with $d: V \to V$ the corresponding degree operator [see (1.3.8)], viewed also as an endomorphism of $V\{z\}$, then

$$Dv(z) = dv(z) \quad (2.2.16)$$

if and only if

$$\deg v_n = n \quad \text{for all} \quad n \in \mathbb{F} \quad (2.2.17)$$

[see (1.3.4)]. In this case, given

$$X(z) = \sum_{n \in \mathbb{F}} x(n)z^{-n} \in (\mathrm{End}\, V)\{z\}$$

(note the minus sign in the exponent), we observe similarly that

$$-DX(z) = [d, X(z)] \quad (2.2.18)$$

if and only if

$$\deg x(n) = n \quad \text{for all} \quad n \in \mathbb{F} \quad (2.2.19)$$

[see (1.3.6), (1.3.10)].

On the subspace

$$DV\{z\} = \left\{ \sum_{n \neq 0} v_n z^n \mid v_n \in V \right\} \quad (2.2.20)$$

of $V\{z\}$, the operator D is invertible. We shall denote by

$$D^{-1}: DV\{z\} \to DV\{z\} \tag{2.2.21}$$

the inverse of D on this space, so that

$$D^{-1}\left(\sum_{n\neq 0} v_n z^n\right) = \sum_{n\neq 0} \frac{v_n}{n} z^n. \tag{2.2.22}$$

Parentheses are optional in certain expressions involving D; for instance,

$$D(v(az)) = (Dv)(az) = Dv(az) \tag{2.2.23}$$

for $v(z) \in \mathbb{F}[[z, z^{-1}]]$ and $a \in \mathbb{F}^{\times}$. Similarly,

$$D_{z_1}(v(az_1/z_2)) = (Dv)(az_1/z_2). \tag{2.2.24}$$

Note also that

$$\begin{aligned} D_{z^{-1}} &= -D_z \\ D_{z_2}(v(az_1/z_2)) &= -(Dv)(az_1/z_2). \end{aligned} \tag{2.2.25}$$

Remark 2.2.3: If z were a complex variable of the form e^{τ}, then we would have

$$D_z = \frac{d}{d\tau}.$$

Formally, D_z can be interpreted as $d/d(\log z)$.

For $T = D$, Proposition 2.2.2 specializes [using (2.2.23)–(2.2.25)] to:

Proposition 2.2.4: *(a) For $v(z)$ and a as in Proposition 2.2.1(a),*

$$v(z)D\delta(az) = v(a^{-1})D\delta(az) - (Dv)(a^{-1})\delta(az).$$

(b) For $X(z_1, z_2)$ and a as in Proposition 2.2.1(b),

$$\begin{aligned} &X(z_1, z_2)(D\delta)(az_1/z_2) \\ &\quad = X(a^{-1}z_2, z_2)(D\delta)(az_1/z_2) - (D_{z_1}X)(a^{-1}z_2, z_2)\delta(az_1/z_2) \\ &\quad = X(z_1, az_1)(D\delta)(az_1/z_2) + (D_{z_2}X)(z_1, az_1)\delta(az_1/z_2). \end{aligned}$$

Remark 2.2.5: By iteration of the appropriate derivations, we can guess and derive generalizations of Propositions 2.2.1, 2.2.2 and 2.2.4 for higher derivatives, including, for instance, a formula for $v(z)T^n(\delta(az))$ for $n \geq 0$, in the notation of (2.2.9). We shall obtain such formulas by a different method in Chapter 8.

Remark 2.2.6: Recall from Proposition 1.9.1 that the derivations of $\mathbb{F}[z, z^{-1}]$ are precisely the endomorphisms of the form (2.2.8). Their Lie algebra structure is given by (1.9.4).

2.3. Affine Lie Algebras Via Formal Variables

We shall find it convenient to express the structure of an untwisted or twisted affine Lie algebra in terms of formal variables.

Given any Lie algebra $\mathfrak{l}$, there is a natural bilinear map

$$\begin{aligned} \mathfrak{l}\{z_1\} \times \mathfrak{l}\{z_2\} &\to \mathfrak{l}\{z_1, z_2\} \\ (x, y) &\mapsto [x, y] \end{aligned} \tag{2.3.1}$$

defined by:

$$\left[\sum_{m \in \mathbb{F}} x_m z_1^m, \sum_{n \in \mathbb{F}} y_n z_2^n\right] = \sum_{m,n \in \mathbb{F}} [x_m, y_n] z_1^m z_2^n \tag{2.3.2}$$

for $x_m, y_n \in \mathfrak{l}$.

Let $\mathfrak{g}$ be a Lie algebra equipped with a symmetric invariant bilinear form $\langle \cdot, \cdot \rangle$, and consider the corresponding untwisted affine algebra

$$\tilde{\mathfrak{g}} = \mathfrak{g} \otimes \mathbb{F}[t, t^{-1}] \oplus \mathbb{F}c \oplus \mathbb{F}d \tag{2.3.3}$$

(see Section 1.6). For $x \in \mathfrak{g}$, set

$$x(z) = \sum_{n \in \mathbb{Z}} (x \otimes t^n) z^{-n} \in \tilde{\mathfrak{g}}[[z, z^{-1}]]. \tag{2.3.4}$$

(Observe the minus sign in the exponent in the generating function.) Then the bracket relations of $\tilde{\mathfrak{g}}$ [see (1.6.5), (1.6.9)] can be expressed as follows:

Proposition 2.3.1: *For $x, y \in \mathfrak{g}$,*

$$[x(z_1), y(z_2)] = [x, y](z_2)\delta(z_1/z_2) - \langle x, y \rangle (D\delta)(z_1/z_2)c \tag{2.3.5}$$

$$[c, x(z)] = 0 \tag{2.3.6}$$

$$[d, x(z)] = -Dx(z) \tag{2.3.7}$$

$$[c, d] = 0. \tag{2.3.8}$$

Proof: Simply compare the coefficients of $z_1^m z_2^n$ or of z^m $(m, n \in \mathbb{Z})$ in these formulas. [The right-hand side of (2.3.5) has an obvious meaning as an element of $\tilde{\mathfrak{g}}\{z_1, z_2\}$.] ∎

There is a similar reformulation of the structure of a twisted affine algebra: Let θ be an involution of $\mathfrak{g}$ preserving $\langle \cdot, \cdot \rangle$ [see (1.6.15), (1.6.16)], so that

$$\tilde{\mathfrak{g}}[\theta] = \mathfrak{g}_{(0)} \otimes \mathbb{F}[t, t^{-1}] \oplus \mathfrak{g}_{(1)} \otimes t^{1/2}\mathbb{F}[t, t^{-1}] \oplus \mathbb{F}c \oplus \mathbb{F}d \quad (2.3.9)$$

[recall (1.6.26), (1.6.27)]. For $i \in \mathbb{Z}/2\mathbb{Z}$, denote by

$$x \mapsto x_{(i)} = \tfrac{1}{2}(x + (-1)^i \theta x) \quad (2.3.10)$$

the projection of $\mathfrak{g}$ to $\mathfrak{g}_{(i)}$ with respect to the θ-eigenspace decomposition $\mathfrak{g} = \mathfrak{g}_{(0)} \oplus \mathfrak{g}_{(1)}$ of $\mathfrak{g}$ [see (1.6.17), (1.6.18)]. For $x \in \mathfrak{g}$, also write

$$\begin{aligned} \mathfrak{g}_{(n)} &= \mathfrak{g}_{(0)}, \quad x_{(n)} = x_{(0)} \quad \text{if} \quad n \in 2\mathbb{Z} \\ \mathfrak{g}_{(n)} &= \mathfrak{g}_{(1)}, \quad x_{(n)} = x_{(1)} \quad \text{if} \quad n \in 2\mathbb{Z} + 1 \end{aligned} \quad (2.3.11)$$

and set

$$x(z) = \sum_{n \in \mathbb{Z}} (x_{(n)} \otimes t^{n/2}) z^{-n/2} \in \tilde{\mathfrak{g}}[\theta]\{z\}. \quad (2.3.12)$$

(Although the notation is the same as in (2.3.4), no confusion should arise.) Together with c and d, the coefficients of the powers of z in (2.3.12) span $\tilde{\mathfrak{g}}[\theta]$, as x ranges through $\mathfrak{g}$. Note that

$$(\theta x)(z) = \lim_{z^{1/2} \to -z^{1/2}} x(z). \quad (2.3.13)$$

The bracket structure of $\tilde{\mathfrak{g}}\,[\theta]$ is described by an analogue of Proposition 2.3.1, proved by comparing coefficients and applying the definitions:

Proposition 2.3.2: *For $x, y \in \mathfrak{g}$,*

$$\begin{aligned} [x(z_1), y(z_2)] = {} & \tfrac{1}{2} \sum_{i \in \mathbb{Z}/2\mathbb{Z}} [\theta^i x, y](z_2)\delta((-1)^i z_1^{1/2}/z_2^{1/2}) \\ & - \tfrac{1}{2} \sum_{i \in \mathbb{Z}/2\mathbb{Z}} \langle \theta^i x, y \rangle D_{z_1}\delta((-1)^i z_1^{1/2}/z_2^{1/2}) c \end{aligned} \quad (2.3.14)$$

$$[c, x(z)] = 0 \quad (2.3.15)$$

$$[d, x(z)] = -Dx(z) \quad (2.3.16)$$

$$[c, d] = 0. \quad (2.3.17)$$

For $x \in \mathfrak{g}$, it is also convenient to define

$$\begin{aligned} x^+ &= x + \theta x = 2x_{(0)} \\ x^- &= x - \theta x = 2x_{(1)}. \end{aligned} \tag{2.3.18}$$

Then for both the untwisted algebra $\tilde{\mathfrak{g}}$ and the twisted algebra $\tilde{\mathfrak{g}}[\theta]$, we have

$$\begin{aligned} x^+(z) &= x(z) + \theta x(z) \\ &= \sum_{n \in \mathbb{Z}} (x^+ \otimes t^n) z^{-n} \end{aligned} \tag{2.3.19}$$

$$\begin{aligned} x^-(z) &= x(z) - \theta x(z) \\ &= \sum_{n \in Z} (x^- \otimes t^n) z^{-n} \end{aligned} \tag{2.3.20}$$

where $Z = \mathbb{Z}$ in the untwisted case and $Z = \mathbb{Z} + 1/2$ in the twisted case. Since

$$\begin{aligned} \delta(z) &= \tfrac{1}{2}(\delta(z^{1/2}) + \delta(-z^{1/2})) \\ z^{1/2}\delta(z) &= \tfrac{1}{2}(\delta(z^{1/2}) - \delta(-z^{1/2})), \end{aligned} \tag{2.3.21}$$

the information in (2.3.5) and (2.3.14) may be equivalently expressed by the three formulas

$$[x^+(z_1), y^+(z_2)] = [x^+, y^+](z_2)\delta(z_1/z_2) - \langle x^+, y^+\rangle (D\delta)(z_1/z_2)c \tag{2.3.22}$$

$$[x^+(z_1), y^-(z_2)] = [x^+, y^-](z_2)\delta(z_1/z_2) \tag{2.3.23}$$

$$[x^-(z_1), y^-(z_2)] = \begin{cases} [x^-, y^-](z_2)\delta(z_1/z_2) \\ \quad - \langle x^-, y^-\rangle (D\delta)(z_1/z_2)c \\ \quad \text{in the untwisted case} \\ [x^-, y^-](z_2)(z_1/z_2)^{1/2}\delta(z_1/z_2) \\ \quad - \langle x^-, y^-\rangle D_{z_1}((z_1/z_2)^{1/2}\delta(z_1/z_2))c \\ \quad \text{in the twisted case,} \end{cases} \tag{2.3.24}$$

for $x, y \in \mathfrak{g}$.

3 Realizations of $\mathfrak{sl}(2)\hat{}$ by Twisted Vertex Operators

Here we study the simplest vertex operator representation of the simplest affine Kac–Moody algebra $\mathfrak{sl}(2)\hat{}$. This representation, constructed in [Lepowsky–Wilson 1], is based on a certain twisted vertex operator which turned out to have been considered by physicists studying dual resonance models (see the Introduction) and which has been interpreted in [Date–Kashiwara–Miwa] as the infinitesimal Bäcklund transformation for the Korteweg–de Vries equation in soliton theory.

In Section 3.1 we present a twisted form of the affine algebra $\mathfrak{sl}(2)\hat{}$ and we define the corresponding twisted module, a natural irreducible module for a Heisenberg algebra which is a subalgebra of $\mathfrak{sl}(2)\hat{}$. Then in Section 3.2 we motivate and introduce twisted vertex operators, expressed in terms of the annihilation and creation operators and providing a differential-operator realization of the generating function of an infinite family of $\mathfrak{sl}(2)\hat{}$-elements. In Section 3.3 we define a notion of normal ordering, which is familiar to quantum field theorists and which together with the notion of vertex operator pervades this book. In Section 3.4 we give the proof in [Garland 3] and [Date–Kashiwara–Miwa] of the vertex operator commutation relations, using formal calculus (Chapter 2). This method is an algebraic analogue of the contour integral approach discussed in the Appendix and used previously for the untwisted representation in [Frenkel–Kac]. The latter technique turned out to have appeared earlier in the physics literature (see the Introduction for a discussion and references). We complete the construction of the twisted vertex operator representation of the affine algebra $\mathfrak{sl}(2)\hat{}$ in Section 3.5.

Actually, two inequivalent irreducible $\mathfrak{sl}(2)\hat{}$-modules are constructed here. These are the ones called the basic modules—the standard, i.e., (irreducible) integrable highest weight, modules of level 1 (cf. [Kac 5]). All the standard $\mathfrak{sl}(2)\hat{}$-modules have been analogously constructed using "Z-algebras," in the same twisted gradation, in [Lepowsky–Wilson 3–5]; see also [Meurman–Primc]. The construction in this chapter will be generalized in a different direction in Chapter 7.

3.1. The Affine Lie Algebra $\mathfrak{sl}(2)\hat{}$

Set

$$\mathfrak{a} = \mathfrak{sl}(2) = \mathfrak{sl}(2, \mathbb{F}), \tag{3.1.1}$$

the Lie algebra of 2×2 matrices over $\mathbb{F}$ of trace 0. We choose the basis $\{\alpha_1, x_{\alpha_1}, x_{-\alpha_1}\}$ of $\mathfrak{a}$, where

$$\alpha_1 = \begin{bmatrix} 1 & 0 \\ 0 & -1 \end{bmatrix}, \quad x_{\alpha_1} = \begin{bmatrix} 0 & 1 \\ 0 & 0 \end{bmatrix}, \quad x_{-\alpha_1} = \begin{bmatrix} 0 & 0 \\ 1 & 0 \end{bmatrix}. \tag{3.1.2}$$

These matrices have commutation relations

$$\begin{aligned} [\alpha_1, x_{\pm\alpha_1}] &= \pm 2x_{\pm\alpha_1} = \langle \alpha_1, \pm\alpha_1 \rangle x_{\pm\alpha_1}, \\ [x_{\alpha_1}, x_{-\alpha_1}] &= \alpha_1. \end{aligned} \tag{3.1.3}$$

Here $\langle \cdot, \cdot \rangle$ denotes the nonsingular invariant symmetric bilinear form on $\mathfrak{a}$ given by

$$\langle x, y \rangle = \operatorname{tr} xy \tag{3.1.4}$$

for all $x, y \in \mathfrak{a}$, so that

$$\begin{gathered} \langle \alpha_1, \alpha_1 \rangle = 2, \qquad \langle x_{\alpha_1}, x_{-\alpha_1} \rangle = 1, \\ \langle \alpha_1, x_{\alpha_1} \rangle = \langle \alpha_1, x_{-\alpha_1} \rangle = \langle x_{\alpha_1}, x_{\alpha_1} \rangle = \langle x_{-\alpha_1}, x_{-\alpha_1} \rangle = 0. \end{gathered} \tag{3.1.5}$$

Remark 3.1.1: In the language of semisimple Lie theory (cf. e.g. [Humphreys]), $\mathbb{F}\alpha_1$ is a Cartan subalgebra of the simple Lie algebra $\mathfrak{a}$; $\pm\alpha_1$ are the roots of $\mathfrak{a}$ with respect to $\mathbb{F}\alpha_1$, which we identify with its dual $(\mathbb{F}\alpha_1)^*$

by means of the form $\langle \cdot, \cdot \rangle$; $x_{\pm\alpha_1}$ are corresponding root vectors; and $\{\alpha_1, x_{\alpha_1}, x_{-\alpha_1}\}$ is a Chevalley basis of $\mathfrak{a}$.

Let θ_1 denote the Lie algebra involution of $\mathfrak{a}$ such that

$$\theta_1 : \alpha_1 \mapsto \alpha_1, \qquad \theta_1 : x_{\alpha_1} \mapsto -x_{\alpha_1}, \qquad \theta_1 : x_{-\alpha_1} \mapsto -x_{-\alpha_1}. \tag{3.1.6}$$

Note that θ_1 preserves $\langle \cdot, \cdot \rangle$ [as in (1.6.16)]. The affine Lie algebra $\hat{\mathfrak{a}}$ (resp., $\hat{\mathfrak{a}}[\theta_1]$) has basis

$$\{c, \alpha_1 \otimes t^m, x_{\pm\alpha_1} \otimes t^n \mid m \in \mathbb{Z}, n \in \mathbb{Z} \quad (\text{resp., } n \in \mathbb{Z} + 1/2)\} \tag{3.1.7}$$

with brackets

$$[c, \hat{\mathfrak{a}}] = 0 \quad (\text{resp., } [c, \hat{\mathfrak{a}}[\theta_1]] = 0) \tag{3.1.8}$$

$$[\alpha_1 \otimes t^m, \alpha_1 \otimes t^n] = 2m\delta_{m+n,0}c \tag{3.1.9}$$

$$[\alpha_1 \otimes t^m, x_{\pm\alpha_1} \otimes t^r] = \pm 2x_{\pm\alpha_1} \otimes t^{m+r} \tag{3.1.10}$$

$$[x_{\pm\alpha_1} \otimes t^r, x_{\pm\alpha_1} \otimes t^s] = 0 \tag{3.1.11}$$

$$[x_{\alpha_1} \otimes t^r, x_{-\alpha_1} \otimes t^s] = \alpha_1 \otimes t^{r+s} + r\delta_{r+s,0}c \tag{3.1.12}$$

for $m, n \in \mathbb{Z}$, $r, s \in \mathbb{Z}$ (resp., $r, s \in \mathbb{Z} + 1/2$). [See (1.6.5), (1.6.26).]

Set

$$\mathfrak{h} = \mathbb{F}\alpha_1, \tag{3.1.13}$$

a one-dimensional subalgebra of $\mathfrak{a}$. Since $\langle \cdot, \cdot \rangle$ is nonsingular on $\mathfrak{h}$, we can form the Heisenberg algebra

$$\hat{\mathfrak{h}}_{\mathbb{Z}} = \text{span} \quad \{c, \alpha_1 \otimes t^n \mid n \in \mathbb{Z}\backslash\{0\}\} \tag{3.1.14}$$

[recall (1.7.13)], a subalgebra of both $\hat{\mathfrak{a}}$ and $\hat{\mathfrak{a}}[\theta_1]$.

We may adjoin the degree derivation d to $\hat{\mathfrak{a}}$ (resp., $\hat{\mathfrak{a}}[\theta_1]$) to form the extended affine algebra $\tilde{\mathfrak{a}}$ (resp., $\tilde{\mathfrak{a}}[\theta_1]$) [see (1.6.9), (1.6.27)].

Remark 3.1.2: The untwisted and twisted affine algebras $\tilde{\mathfrak{a}}$ and $\tilde{\mathfrak{a}}[\theta_1]$ are isomorphic (but of course not graded-isomorphic) by a map $\tilde{\mathfrak{a}} \to \tilde{\mathfrak{a}}[\theta_1]$ which has the following effect on the basis elements:

$$c \mapsto c$$

$$d \mapsto d - \tfrac{1}{4}\alpha_1 \quad (= d - \tfrac{1}{4}\alpha_1 \otimes t^0)$$

$$\alpha_1 \otimes t^n \mapsto \alpha_1 \otimes t^n + \tfrac{1}{2}\delta_{n,0}c$$

$$x_{\alpha_1} \otimes t^n \mapsto x_{\alpha_1} \otimes t^{n+1/2}$$

$$x_{-\alpha_1} \otimes t^n \mapsto x_{-\alpha_1} \otimes t^{n-1/2}$$

for $n \in \mathbb{Z}$. We shall not need this fact.

With the help of Propositions 2.3.1 and 2.3.2, we can express the structure of $\tilde{\mathfrak{a}}$ and $\tilde{\mathfrak{a}}[\theta_1]$ in terms of formal variables. Using the notation (2.3.4), (2.3.12), we observe that

$$x_{\pm\alpha_1}(z) = \sum_{n \in Z} (x_{\pm\alpha_1} \otimes t^n)z^{-n} \tag{3.1.15}$$

where $Z = \mathbb{Z}$ for $\tilde{\mathfrak{a}}$ and $Z = \mathbb{Z} + 1/2$ for $\tilde{\mathfrak{a}}[\theta_1]$,

$$\alpha_1(z) = \sum_{n \in \mathbb{Z}} (\alpha_1 \otimes t^n)z^{-n}. \tag{3.1.16}$$

It turns out to be most convenient to reformulate (3.1.10)–(3.1.12) as follows:

$$\begin{aligned}[\alpha_1 \otimes t^m, x_{\pm\alpha_1}(z)] &= \pm 2z^m x_{\pm\alpha_1}(z)\\ &= \langle \alpha_1, \pm\alpha_1 \rangle z^m x_{\pm\alpha_1}(z) \quad \text{for} \quad m \in \mathbb{Z}\end{aligned} \tag{3.1.17}$$

$$[x_{\pm\alpha_1}(z_1), x_{\pm\alpha_1}(z_2)] = 0 \tag{3.1.18}$$

for both $\tilde{\mathfrak{a}}$ and $\tilde{\mathfrak{a}}[\theta_1]$, and [cf. (2.3.24)]

$$[x_{\alpha_1}(z_1), x_{-\alpha_1}(z_2)] = \alpha_1(z_2)\delta(z_1/z_2) - (D\delta)(z_1/z_2)c \tag{3.1.19}$$

for $\tilde{\mathfrak{a}}$,

$$\begin{aligned}[x_{\alpha_1}(z_1), x_{-\alpha_1}(z_2)] = {}& \alpha_1(z_2)(z_1/z_2)^{1/2}\delta(z_1/z_2)\\ &- D_{z_1}((z_1/z_2)^{1/2}\delta(z_1/z_2))c\end{aligned} \tag{3.1.20}$$

for $\tilde{\mathfrak{a}}[\theta_1]$. Note that (3.1.17) expresses the action of the Heisenberg algebra $\hat{\mathfrak{h}}_{\mathbb{Z}}$ on the elements $x_{\pm\alpha_1}(z)$. We also have

$$[d, x_{\pm\alpha_1}(z)] = -Dx_{\pm\alpha_1}(z), \tag{3.1.21}$$

$$[d, \alpha_1(z)] = -D\alpha_1(z), \tag{3.1.22}$$

where D is the operator introduced in (2.2.14).

In addition to θ_1, we shall also consider the Lie algebra involution θ_2 of $\mathfrak{a}$ (preserving $\langle \cdot, \cdot \rangle$) such that

$$\theta_2\colon \alpha_1 \mapsto -\alpha_1, \qquad \theta_2\colon x_{\alpha_1} \mapsto x_{-\alpha_1}, \qquad \theta_2\colon x_{-\alpha_1} \mapsto x_{\alpha_1}. \tag{3.1.23}$$

As in (2.3.18), set

$$x_{\alpha_1}^{\pm} = x_{\alpha_1} \pm x_{-\alpha_1} = x_{\alpha_1} \pm \theta_2 x_{\alpha_1}, \tag{3.1.24}$$

so that in the notation of (2.3.10),

$$x_{\alpha_1}^{+} = 2(x_{\alpha_1})_{(0)}, \qquad x_{\alpha_1}^{-} = 2(x_{\alpha_1})_{(1)}. \tag{3.1.25}$$

We have

$$x_{\alpha_1}^{+} \in \mathfrak{a}_{(0)} \quad \text{and} \quad \alpha_1, x_{\alpha_1}^{-} \in \mathfrak{a}_{(1)} \tag{3.1.26}$$

[see (1.6.17)], and

$$[\alpha_1, x_{\alpha_1}^{\pm}] = 2x_{\alpha_1}^{\mp}, \qquad [x_{\alpha_1}^{+}, x_{\alpha_1}^{-}] = -2\alpha_1. \tag{3.1.27}$$

The twisted affine algebra $\hat{\mathfrak{a}}[\theta_2]$ has basis

$$\{c, \alpha_1 \otimes t^m, x_{\alpha_1}^{+} \otimes t^n, x_{\alpha_1}^{-} \otimes t^p \mid m, p \in \mathbb{Z} + \tfrac{1}{2}, n \in \mathbb{Z}\} \tag{3.1.28}$$

with brackets

$$[c, \hat{\mathfrak{a}}[\theta_2]] = 0 \tag{3.1.29}$$

$$[\alpha_1 \otimes t^m, \alpha_1 \otimes t^n] = 2m\delta_{m+n,0}c \tag{3.1.30}$$

$$[\alpha_1 \otimes t^m, x_{\alpha_1}^{+} \otimes t^r] = 2x_{\alpha_1}^{-} \otimes t^{m+r} \tag{3.1.31}$$

$$[\alpha_1 \otimes t^m, x_{\alpha_1}^{-} \otimes t^n] = 2x_{\alpha_1}^{+} \otimes t^{m+n} \tag{3.1.32}$$

$$[x_{\alpha_1}^{+} \otimes t^r, x_{\alpha_1}^{+} \otimes t^s] = 2r\delta_{r+s,0}c \tag{3.1.33}$$

$$[x_{\alpha_1}^{+} \otimes t^r, x_{\alpha_1}^{-} \otimes t^m] = -2\alpha_1 \otimes t^{r+m} \tag{3.1.34}$$

$$[x_{\alpha_1}^{-} \otimes t^m, x_{\alpha_1}^{-} \otimes t^n] = -2m\delta_{m+n,0}c \tag{3.1.35}$$

for $m, n \in \mathbb{Z} + 1/2$, $r, s \in \mathbb{Z}$. The Heinsenberg algebra

$$\hat{\mathfrak{h}}_{\mathbb{Z}+1/2} = \text{span} \quad \{c, \alpha_1 \otimes t^n \mid n \in \mathbb{Z} + \tfrac{1}{2}\} \tag{3.1.36}$$

[see (1.7.13)] is a subalgebra of $\hat{\mathfrak{a}}[\theta_2]$. The degree derivation d may be adjoined to $\hat{\mathfrak{a}}[\theta_2]$ to form the extended algebra $\tilde{\mathfrak{a}}[\theta_2]$.

We shall describe $\tilde{\mathfrak{a}}[\theta_2]$ in terms of the expressions

$$x_{\pm\alpha_1}(z) = \tfrac{1}{2} \sum_{n \in \mathbb{Z}} (x_{\alpha_1}^{+} \otimes t^n)z^{-n} \pm \tfrac{1}{2} \sum_{n \in \mathbb{Z}+1/2} (x_{\alpha_1}^{-} \otimes t^n)z^{-n}, \tag{3.1.37}$$

$$\alpha_1(z) = \sum_{n \in \mathbb{Z}+1/2} (\alpha_1 \otimes t^n)z^{-n} \tag{3.1.38}$$

[see (2.3.12), (3.1.25)]. Formulas (3.1.31)–(3.1.35) can be expressed as:

$$\begin{aligned}[\alpha_1 \otimes t^m, x_{\pm\alpha_1}(z)] &= \pm 2z^m x_{\pm\alpha_1}(z) \\ &= \langle \alpha_1, \pm\alpha_1 \rangle z^m x_{\pm\alpha_1}(z) \quad \text{for} \quad m \in \mathbb{Z} + \tfrac{1}{2}\end{aligned} \tag{3.1.39}$$

$$\begin{aligned}[x_{\alpha_1}(z_1), x_{-\alpha_1}(z_2)] &= \tfrac{1}{2}\alpha_1(z_2)\delta(z_1^{1/2}/z_2^{1/2}) \\ &\quad - \tfrac{1}{2}D_{z_1}\delta(z_1^{1/2}/z_2^{1/2})c.\end{aligned} \tag{3.1.40}$$

There are also analogues of the two formulas given by (3.1.18), but these just amount to (3.1.40) because of the relation

$$x_{-\alpha_1}(z) = \lim_{z^{1/2} \to -z^{1/2}} x_{\alpha_1}(z) \tag{3.1.41}$$

[see (2.3.13), (3.1.37)]. Proposition 2.3.2 gives the easiest proof of (3.1.40). Formula (3.1.39) exhibits the action of the Heisenberg albegra $\hat{\mathfrak{h}}_{\mathbb{Z}+1/2}$ on $x_{\pm\alpha_1}(z)$ [cf. (3.1.17)]. As in (3.1.21), (3.1.22), we have

$$[d, x_{\pm\alpha_1}(z)] = -Dx_{\pm\alpha_1}(z), \tag{3.1.42}$$

$$[d, \alpha_1(z)] = -D\alpha_1(z). \tag{3.1.43}$$

The involution θ_1 fixes α_1, while the involution θ_2 negates α_1. Nevertheless, θ_1 and θ_2 are conjugate in the following sense: Denote by σ the Lie algebra involution of $\mathfrak{a}$ (preserving $\langle \cdot, \cdot \rangle$) such that

$$\begin{aligned} \sigma&: \alpha_1 \mapsto x_{\alpha_1}^{+} \\ \sigma&: x_{\alpha_1}^{+} \mapsto \alpha_1 \\ \sigma&: x_{\alpha_1}^{-} \mapsto -x_{\alpha_1}^{-}. \end{aligned} \tag{3.1.44}$$

Then

$$\sigma\theta_2\sigma^{-1} = \theta_1, \tag{3.1.45}$$

which implies that σ induces a (grading-preserving) isomorphism

$$\sigma: \tilde{\mathfrak{a}}[\theta_2] \to \tilde{\mathfrak{a}}[\theta_1] \tag{3.1.46}$$

such that

$$\begin{aligned} \sigma&: x \otimes t^n \mapsto \sigma x \otimes t^n \\ \sigma&: c \mapsto c \\ \sigma&: d \mapsto d \end{aligned}$$

for the appropriate $x \in \mathfrak{a}$ and $n \in \frac{1}{2}\mathbb{Z}$.

We now want to construct representations of $\hat{\mathfrak{a}}$, $\hat{\mathfrak{a}}[\theta_1]$ and $\hat{\mathfrak{a}}[\theta_2]$ starting from the canonical realization of the Heisenberg commutation relations defining the Heisenberg subalgebras $\hat{\mathfrak{h}}_{\mathbb{Z}}$ (3.1.14) and $\hat{\mathfrak{h}}_{\mathbb{Z}+1/2}$ (3.1.36) (see Section 1.7). Note that

$$\hat{\mathfrak{h}}_{\mathbb{Z}} \subset \tilde{\mathfrak{h}} \subset \tilde{\mathfrak{a}}, \qquad \tilde{\mathfrak{h}} \subset \tilde{\mathfrak{a}}[\theta_1] \tag{3.1.47}$$

$$\hat{\mathfrak{h}}_{\mathbb{Z}+1/2} \subset \tilde{\mathfrak{h}}[-1] \subset \tilde{\mathfrak{a}}[\theta_2]. \tag{3.1.48}$$

Consider the $\mathbb{Q}$-graded irreducible $\tilde{\mathfrak{h}}$-module

$$M(1) = S(\hat{\mathfrak{h}}_{\mathbb{Z}}^{-}) \tag{3.1.49}$$

and the $\mathbb{Q}$-graded irreducible $\tilde{\mathfrak{h}}[-1]$-module

$$M(1) = S(\hat{\mathfrak{h}}_{\mathbb{Z}+1/2}^{-}) \tag{3.1.50}$$

as in (1.7.15), (1.7.19), (1.9.51) and (1.9.53). In both cases, c acts as the identity operator, d acts as the degree operator, and the module $M(1)$ remains irreducible under the Heisenberg algebra $\hat{\mathfrak{h}}_{\mathbb{Z}}$ or $\hat{\mathfrak{h}}_{\mathbb{Z}+1/2}$.

We face the problem of extending the modules $M(1)$ to modules for $\tilde{\mathfrak{a}}$, $\tilde{\mathfrak{a}}[\theta_1]$ and $\tilde{\mathfrak{a}}[\theta_2]$. This turns out to be easiest in the last case, since all of $\tilde{\mathfrak{a}}[\theta_2]$ can be made to act on $S(\hat{\mathfrak{h}}_{\mathbb{Z}+1/2}^{-})$. We proceed now to construct the appropriate operators. The corresponding constructions for $\tilde{\mathfrak{a}}$ and $\tilde{\mathfrak{a}}[\theta_1]$ will be carried out in Chapter 4.

Remark 3.1.3: Our choice $k = 1$ [see (1.7.19), (3.1.49), (3.1.50)] is not arbitrary. Given our normalization of the form $\langle \cdot, \cdot \rangle$ [see (3.1.4), (3.1.5)], the choice $k = 1$ is necessary for the construction of realizations of $\tilde{\mathfrak{a}}$, $\tilde{\mathfrak{a}}[\theta_1]$ and $\tilde{\mathfrak{a}}[\theta_2]$ in Chapters 3 and 4.

3.2. The Twisted Vertex Operators $X_{\mathbb{Z}+1/2}(\alpha, z)$

We could start with the one-dimensional space $\mathfrak{h} = \mathbb{F}\alpha_1$ of (3.1.13), but it is no extra effort to work more generally. Let $\mathfrak{h}$ be a nonzero finite-dimensional vector space equipped with a nonsingular symmetric bilinear form $\langle \cdot, \cdot \rangle$. Viewing $\mathfrak{h}$ as an abelian Lie algebra, consider the corresponding $\frac{1}{2}\mathbb{Z}$-graded twisted affine Lie algebra $\tilde{\mathfrak{h}}[-1]$ and its Heisenberg subalgebra

$$\hat{\mathfrak{h}}_{\mathbb{Z}+1/2} = \hat{\mathfrak{h}}[-1] \tag{3.2.1}$$

as in (1.7.12), (1.7.13). Consider also the $\mathbb{Q}$-graded irreducible $\tilde{\mathfrak{h}}[-1]$-module

$$V = M(1) = S(\hat{\mathfrak{h}}_{\mathbb{Z}+1/2}^{-}) \tag{3.2.2}$$

on which c acts as the identity operator and d acts as the degree operator [see (1.7.15), (1.7.19), (1.9.53)]. Then V remains irreducible as an $\hat{\mathfrak{h}}_{\mathbb{Z}+1/2}$-module.

For $\alpha \in \mathfrak{h}$ and $n \in \mathbb{Z} + 1/2$, we shall use the notation $\alpha(n)$ to denote the operator on V corresponding to $\alpha \otimes t^n \in \tilde{\mathfrak{h}}[-1]$, as in (1.9.16):

$$\alpha \otimes t^n \mapsto \alpha(n). \tag{3.2.3}$$

There should be no confusion with the notation $\alpha(z)$. We have the commutation relations [cf. (1.9.17)]

$$\begin{aligned} [\alpha(m), \beta(n)] &= \langle \alpha, \beta \rangle m \delta_{m+n,0} \\ [d, \alpha(m)] &= m\alpha(m) \end{aligned} \tag{3.2.4}$$

for $\alpha, \beta \in \mathfrak{h}$ and $m, n \in \mathbb{Z} + 1/2$. The operator $\alpha(m)$ is homogeneous of degree m, and the operators $\alpha(m)$ can be realized as multiplication and partial differentiation operators on V (recall Sections 1.3 and 1.7).

Take $\mathfrak{h} = \mathbb{F}\alpha_1$ as in (3.1.13). In searching for operators which realize $\tilde{\mathfrak{a}}[\theta_2]$ on V, we shall be guided by the action of $\hat{\mathfrak{h}}_{\mathbb{Z}+1/2}$ on $x_{\pm\alpha_1}(z)$, given in (3.1.39). If $\tilde{\mathfrak{a}}[\theta_2]$ is to be represented as operators on V, then there must be operators

$$x_{\pm\alpha_1}(n) \in \operatorname{End} V \quad \text{for} \quad n \in \tfrac{1}{2}\mathbb{Z} \tag{3.2.5}$$

such that if we form

$$X(\pm\alpha_1, z) = \sum_{n \in (1/2)\mathbb{Z}} x_{\pm\alpha_1}(n) z^{-n} \in (\operatorname{End} V)\{z\}, \tag{3.2.6}$$

then

$$[h(m), X(\pm\alpha_1, z)] = \langle h, \pm\alpha_1 \rangle z^m X(\pm\alpha_1, z) \tag{3.2.7}$$

for all $h \in \mathfrak{h}$ and $m \in \mathbb{Z} + 1/2$ [recall (3.2.3)]. Moreover, we must have

$$[d, X(\pm\alpha_1, z)] = -DX(\pm\alpha_1, z), \tag{3.2.8}$$

i.e.,

$$\deg x_{\pm\alpha_1}(n) = n \quad \text{for} \quad n \in \tfrac{1}{2}\mathbb{Z}, \tag{3.2.9}$$

by (3.1.42) [cf. (2.2.18), (2.2.19)].

Now return to the case of general $\mathfrak{h}$. We shall actually solve the following generalization of the system of equations (3.2.7), (3.2.8): Fix $\alpha \in \mathfrak{h}$. Consider a family of operators

$$u_\alpha(n) \in \operatorname{End} V \quad \text{for} \quad n \in \tfrac{1}{2}\mathbb{Z}, \tag{3.2.10}$$

define

$$U(\alpha, z) = \sum_{n \in (1/2)\mathbb{Z}} u_\alpha(n) z^{-n} \in (\operatorname{End} V)\{z\}, \tag{3.2.11}$$

and consider the commutation relations

$$[h(m), U(\alpha, z)] = \langle h, \alpha \rangle z^m U(\alpha, z) \tag{3.2.12}$$

for $h \in \mathfrak{h}$ and $m \in \mathbb{Z} + 1/2$, or equivalently, the relations

$$[h(m), u_\alpha(n)] = \langle h, \alpha \rangle u_\alpha(m + n) \tag{3.2.13}$$

for $h \in \mathfrak{h}$, $m \in \mathbb{Z} + 1/2$ and $n \in (1/2)\mathbb{Z}$. Consider also the relations

$$[d, U(\alpha, z)] = -DU(\alpha, z), \tag{3.2.14}$$

or equivalently,

$$\deg u_\alpha(n) = n \quad \text{for} \quad n \in \tfrac{1}{2}\mathbb{Z}. \tag{3.2.15}$$

The equations (3.2.12) suggest an exponential function, and in fact it is possible to guess such a solution. For $\alpha \in \mathfrak{h}$, define the formal series $E^+(\alpha, z)$ and $E^-(\alpha, z)$ by

$$E^\pm(\alpha, z) = \exp\left(\sum_{n \in \pm(\mathbb{N}+1/2)} \frac{\alpha(n)}{n} z^{-n} \right) \tag{3.2.16}$$
$$\in (\operatorname{End} V)[[z^{\mp 1/2}]] \subset (\operatorname{End} V)\{z\}.$$

Here exp refers to the formal power series

$$\exp x = e^x = \sum_{k \geq 0} \frac{x^k}{k!}, \tag{3.2.17}$$

the powers x^k (for $x = \sum \alpha(n)z^{-n}/n$) are to be understood in the sense of (2.1.11), and the sum is to be understood in the sense of (2.1.10). We establish the first fundamental properties of $E^\pm(\alpha, z)$:

Proposition 3.2.1: *For $\alpha \in \mathfrak{h}$, the expressions $E^\pm(\alpha, z)$ exist in the sense of (2.1.10), and we have:*

$$E^\pm(0, z) = 1 \tag{3.2.18}$$

$$E^\pm(\alpha + \beta, z) = E^\pm(\alpha, z)E^\pm(\beta, z) \tag{3.2.19}$$

$$[d, E^\pm(\alpha, z)] = -DE^\pm(\alpha, z) = \left(\sum_{n \in \pm(\mathbb{N}+1/2)} \alpha(n)z^{-n} \right) E^\pm(\alpha, z) \tag{3.2.20}$$

$$E^\pm(-\alpha, z) = \lim_{z^{1/2} \to -z^{1/2}} E^\pm(\alpha, z) \tag{3.2.21}$$

for $\alpha, \beta \in \mathfrak{h}$. *For* $h, \alpha \in \mathfrak{h}$ *and* $m \in \mathbb{Z} + 1/2$,

$$[h(m), E^{+}(\alpha, z)] = 0 \quad \text{if} \quad m > 0 \tag{3.2.22}$$

$$[h(m), E^{-}(\alpha, z)] = -\langle h, \alpha\rangle z^{m} E^{-}(\alpha, z) \quad \text{if} \quad m > 0 \tag{3.2.23}$$

$$[h(m), E^{+}(\alpha, z)] = -\langle h, \alpha\rangle z^{m} E^{+}(\alpha, z) \quad \text{if} \quad m < 0 \tag{3.2.24}$$

$$[h(m), E^{-}(\alpha, z)] = 0 \quad \text{if} \quad m < 0. \tag{3.2.25}$$

Proof: The coefficient of any given power of z in the formal expression $E^{\pm}(\alpha, z)$ is a finite linear combination of products of operators $\alpha(n)$, making the existence clear. Properties (3.2.18) and (3.2.19) follow from the corresponding simple properties of the exponential function; for (3.2.19), we use the fact that the operators $\alpha(n)$ for $n > 0$ commute with one another, as do the operators $\alpha(n)$ for $n < 0$. Property (3.2.21) is clear.

To prove (3.2.20), set

$$A = \sum_{n \in \pm(\mathbb{N}+1/2)} \frac{\alpha(n)}{n} z^{-n}$$

and observe that

$$[d, A] = \sum \alpha(n) z^{-n} = -DA.$$

Since this expression commutes with A, (3.2.20) follows from the formal principle

$$\Delta(e^{x}) = \Delta(x) e^{x} \tag{3.2.26}$$

for a derivation Δ such that $\Delta(x)$ commutes with x. Formulas (3.2.22)–(3.2.25) follow from the same principle, using (3.2.4). ∎

Formulas (3.2.22)–(3.2.25) suggest that in order to satisfy (3.2.12), we should multiply $E^{+}(-\alpha, z)$ and $E^{-}(-\alpha, z)$. However, this can be done in only one order: The product $E^{-}(-\alpha, z)E^{+}(-\alpha, z)$ exists in the sense of (2.1.11) since the coefficient of z^{-n} in $E^{\pm}(\alpha, z)$ is an operator of degree n by (3.2.20), and the grading of V is truncated from above. For later convenience, we restrict our attention to $\alpha \in \mathfrak{h}$ such that

$$\langle \alpha, \alpha \rangle \in \mathbb{Z}, \tag{3.2.27}$$

and we define

$$X(\alpha, z) = X_{\mathbb{Z}+1/2}(\alpha, z) = 2^{-\langle \alpha, \alpha\rangle} E^{-}(-\alpha, z) E^{+}(-\alpha, z) \tag{3.2.28}$$
$$\in (\operatorname{End} V)\{z\}.$$

We call this the *twisted vertex operator* associated with α. The term "twisted" refers to the fact that it is based on the twisted affine algebra $\tilde{\mathfrak{h}}[-1]$. (In Chapter 4, we shall build "untwisted" vertex operators starting from the untwisted affine algebra $\tilde{\mathfrak{h}}$.) The term "vertex operator" comes from physics.

Strictly speaking, $X(\alpha, z)$ is of course not an operator (on V), but instead is the "generating function" of an infinite family of operators $x_\alpha(n)$ ($n \in (1/2)\mathbb{Z}$) defined as the expansion coefficients of the vertex operator, by the formula

$$X(\alpha, z) = \sum_{n \in (1/2)\mathbb{Z}} x_\alpha(n) z^{-n}. \tag{3.2.29}$$

From the definitions (3.2.16), (3.2.28) and (3.2.29), it is posible to write an explicit formula for $x_\alpha(n)$ as a summable infinite linear combination of products of operators of the form $\alpha(m)$ for $m \in \mathbb{Z} + (1/2)$, but this formula is too complicated to be useful. In any case, note that $x_\alpha(n)$ is realized as a formal differential operator when the $\alpha(m)$ are realized as multiplication and partial differentiation operators. The expression (3.2.28) will be used to study as well as define the operators $x_\alpha(n)$. For instance, we have the complete solution of the equations (3.2.13), (3.2.15) in their generating function formulation (3.2.12), (3.2.14), by imitating the usual proof that the exponential function spans the space of solutions of the differential equation $f' = f$:

Proposition 3.2.2: *For $\alpha \in \mathfrak{h}$, the expression*

$$U(\alpha, z) = E^-(-\alpha, z)E^+(-\alpha, z) \in (\mathrm{End}\, V)\{z\}$$

satisfies the equations (3.2.12) and (3.2.14). Conversely, any element $W(z)$ of (End V)$\{z\}$ *which satisfies these equations is of the form*

$$W(z) = aE^-(-\alpha, z)E^+(-\alpha, z)$$

for some $a \in \mathbb{F}$.

Proof: It is clear from Proposition 3.2.1 that $U(\alpha, z)$ satisfies the equations. Suppose that $W(z) \in (\mathrm{End}\, V)\{z\}$ is any solution. We may form the product

$$Z(z) = E^-(\alpha, z)W(z)E^+(\alpha, z), \tag{3.2.30}$$

which exists in the sense of (2.1.11) because the nth coefficient of each factor is an operator of degree $-n$, and the grading of V is truncated from above. Define operators $z(n)$ on V for $n \in \mathbb{F}$ by:

$$Z(z) = \sum_{n \in \mathbb{F}} z(n)z^{-n}. \tag{3.2.31}$$

But then

$$[d, Z(z)] = -DZ(z), \tag{3.2.32}$$

$$[h(m), Z(z)] = 0 \tag{3.2.33}$$

for $h \in \mathfrak{h}$ and $m \in \mathbb{Z} + 1/2$, by Proposition 3.2.1 and the assumption on $W(z)$, so that

$$\deg z(n) = n, \tag{3.2.34}$$

$$[h(m), z(n)] = 0 \tag{3.2.35}$$

for $h \in \mathfrak{h}$, $m \in \mathbb{Z} + 1/2$ and $n \in \mathbb{F}$. By Proposition 1.7.2, the operators $z(n)$ are all scalar multiplication operators. Thus $z(n) = 0$ unless $n = 0$, and

$$Z(z) = z(0)z^0 = a \in \mathbb{F}. \tag{3.2.36}$$

Solving equation (3.2.30) for $W(z)$ [using (3.2.18) and (3.2.19)] gives the result. ∎

Thus for $\alpha \in \mathfrak{h}$ such that $\langle \alpha, \alpha \rangle \in \mathbb{Z}$, we have

$$[h(m), X(\alpha, z)] = \langle h, \alpha \rangle z^m X(\alpha, z) \tag{3.2.37}$$

$$[d, X(\alpha, z)] = -DX(\alpha, z) \tag{3.2.38}$$

for $h \in \mathfrak{h}$ and $m \in \mathbb{Z} + 1/2$, or equivalently,

$$[h(m), x_\alpha(n)] = \langle h, \alpha \rangle x_\alpha(m + n) \tag{3.2.39}$$

$$\deg x_\alpha(n) = n \tag{3.2.40}$$

for $h \in \mathfrak{h}$, $m \in \mathbb{Z} + 1/2$ and $n \in (1/2)\mathbb{Z}$. Moreover, from (3.2.21), we have

$$X(-\alpha, z) = \lim_{z^{1/2} \to -z^{1/2}} X(\alpha, z), \tag{3.2.41}$$

i.e.,

$$x_{-\alpha}(n) = \begin{cases} x_\alpha(n) & \text{if} \quad n \in \mathbb{Z} \\ -x_\alpha(n) & \text{if} \quad n \in \mathbb{Z} + \frac{1}{2}. \end{cases} \tag{3.2.42}$$

Remark 3.2.3: Proposition 3.2.2 implies that if $\tilde{\mathfrak{a}}[\theta_2]$ is representable on the space V in the case $\mathfrak{h} = \mathbb{F}\alpha_1$, then the expressions $x_{\pm\alpha_1}(z)$ must be represented by multiples of the respective vertex operators $X(\pm\alpha_1, z)$. Comparing the relations (3.1.41) and (3.2.41), we see that the respective multiples must be equal. In Section 3.4 we shall compute the commutator $[X(\alpha_1, z_1), X(-\alpha_1, z_2)]$, and we shall see that it agrees precisely with the corresponding commutator (3.1.40). This will show that $\tilde{\mathfrak{a}}[\theta_2]$ can indeed be represented on V, and in precisely two ways, where $x_{\alpha_1}(z)$ acts as either $X(\alpha_1, z)$ or as $-X(\alpha_1, z)$. This also provides the first motivation for the normalizing factor $2^{-\langle\alpha,\alpha\rangle}$ in the vertex operator.

3.3. Normal Ordering

The vertex operator $X(\alpha, z)$ and the operators $L(n)$ of (1.9.23) can be expressed using a procedure called "normal ordering." For $\alpha_1, \alpha_2 \in \mathfrak{h}$ and $n_1, n_2 \in \mathbb{Z} + 1/2$, define the *normal ordered product* $:\alpha_1(n_1)\alpha_2(n_2):$ of $\alpha_1(n_1)$ and $\alpha_2(n_2)$ to be the operator on V given by

$$:\alpha_1(n_1)\alpha_2(n_2): = \begin{cases} \alpha_1(n_1)\alpha_2(n_2) & \text{if} \quad n_1 \le n_2 \\ \alpha_2(n_2)\alpha_1(n_1) & \text{if} \quad n_1 \ge n_2. \end{cases} \tag{3.3.1}$$

(Note that the two expressions agree if $n_1 = n_2$.) Equivalently,

$$:\alpha_1(n_1)\alpha_2(n_2): = \alpha_1(n_1)\alpha_2(n_2)$$

unless $n_1 > 0$ and $n_2 < 0$, in which case $:\alpha_1(n_1)\alpha_2(n_2): = \alpha_2(n_2)\alpha_1(n_1)$. More generally, for $\alpha_1, \ldots, \alpha_k \in \mathfrak{h}$ and $n_1, \ldots, n_k \in \mathbb{Z} + 1/2$, set

$$\begin{aligned} &:\alpha_1(n_1)\alpha_2(n_2) \cdots \alpha_k(n_k): \\ &\quad = \alpha_{\pi(1)}(n_{\pi(1)})\alpha_{\pi(2)}(n_{\pi(2)}) \cdots \alpha_{\pi(k)}(n_{\pi(k)}), \end{aligned} \tag{3.3.2}$$

where π is some permutation of $\{1, \ldots, k\}$ such that $n_{\pi(1)} \le \cdots \le n_{\pi(k)}$. Note that the normal ordered product (3.3.2) is independent of the order of the "factors" $\alpha_j(n_j)$, and hence this "product" extends to a well-defined linear map (which is not an algebra map) from the symmetric algebra $S(\hat{\mathfrak{h}}_{\mathbb{Z}+1/2})$ [see (1.5.9)] into End V. This in fact is the rigorous meaning of the normal ordered product, which is *not* to be computed by first computing the ordinary product. The motivating principle behind the normal ordering procedure is that in the formal expression enclosed between colons, the operators $\alpha(n)$ for $n < 0$ are to be placed to the left of the operators $\alpha(n)$ for $n > 0$ before the multiplication is performed. Thus even when normal ordering is now extended to infinite expressions, we automatically obtain well-defined operators on V.

For $\alpha \in \mathfrak{h}$, set

$$\alpha(z) = \sum_{n \in \mathbb{Z}+1/2} \alpha(n)z^{-n},$$
$$\alpha(z)^{\pm} = \sum_{n \in \pm(\mathbb{N}+1/2)} \alpha(n)z^{-n}, \tag{3.3.3}$$

so that

$$\alpha(z) = \alpha(z)^{+} + \alpha(z)^{-}. \tag{3.3.4}$$

[There should be no confusion between the notations $\alpha(z)$ and $\alpha(n)$.] Given $\alpha_1, \ldots, \alpha_k \in \mathfrak{h}$, define recursively

$$:\alpha_1(z): = \alpha_1(z)$$
$$:\alpha_1(z)\alpha_2(z)\cdots\alpha_k(z): \tag{3.3.5}$$
$$= \alpha_k(z)^{-}:\alpha_1(z)\cdots\alpha_{k-1}(z): + :\alpha_1(z)\cdots\alpha_{k-1}(z):\alpha_k(z)^{+}$$

in $(\operatorname{End} V)\{z\}$. Note that for $\alpha \in \mathfrak{h}$ and $k \in \mathbb{N}$,

$$:\alpha(z)^k: = \sum_{l=0}^{k} \binom{k}{l} (\alpha(z)^{-})^{l}(\alpha(z)^{+})^{k-l}. \tag{3.3.6}$$

using binomial coefficient notation. The normal ordered product (3.3.5) is independent of the order of the "factors" $\alpha_j(z)$, and

$$:\alpha_1(z)\cdots\alpha_k(z):$$
$$= \sum_{n \in (1/2)\mathbb{Z}} \left(\sum_{\substack{n_1, \ldots, n_k \in \mathbb{Z}+1/2 \\ \sum n_j = n}} :\alpha_1(n_1)\cdots\alpha_k(n_k): \right) z^{-n}. \tag{3.3.7}$$

Note that we obtain a well-defined linear map

$$S(\mathfrak{h}) \to (\operatorname{End} V)\{z\} \tag{3.3.8}$$

determined by the condition

$$\alpha_1 \cdots \alpha_k \mapsto :\alpha_1(z)\cdots\alpha_k(z): \tag{3.3.9}$$

for $\alpha_1, \ldots, \alpha_k \in \mathfrak{h}$.

Recall that the inverse D^{-1} of the operator D is defined on the space $D((\operatorname{End} V)\{z\})$ [see (2.2.20)–(2.2.22)]. We have

$$D^{-1}\alpha(z) = D^{-1}\alpha(z)^{+} + D^{-1}\alpha(z)^{-} \tag{3.3.10}$$

for $\alpha \in \mathfrak{h}$. For $\alpha_1, \ldots, \alpha_k \in \mathfrak{h}$, we define

$$:D^{-1}\alpha_1(z) \cdots D^{-1}\alpha_k(z): \in (\operatorname{End} V)\{z\} \tag{3.3.11}$$

by an obvious analogue of (3.3.5), and we set

$$:e^{D^{-1}\alpha(z)}: = :\exp D^{-1}\alpha(z): = \sum_{k \geq 0} \frac{:(D^{-1}\alpha(z))^k:}{k!}, \tag{3.3.12}$$

a well-defined element of $(\operatorname{End} V)\{z\}$. [The sum exists in the sense of (2.1.10).] Then by an analogue of (3.3.6),

$$:e^{D^{-1}\alpha(z)}: = e^{D^{-1}\alpha(z)^-} e^{D^{-1}\alpha(z)^+} \tag{3.3.13}$$

[using the exponential notation (3.2.17)]. Noting that

$$E^{\pm}(\alpha, z) = e^{-D^{-1}\alpha(z)^{\pm}} \tag{3.3.14}$$

[see (3.2.16)], we can express the vertex operator (3.2.28) using the normal ordering procedure as follows: For $\alpha \in \mathfrak{h}$ such that $\langle \alpha, \alpha \rangle \in \mathbb{Z}$,

$$X(\alpha, z) = 2^{-\langle \alpha, \alpha \rangle} :e^{D^{-1}\alpha(z)}: = :X(\alpha, z):. \tag{3.3.15}$$

It is clear how to extend normal ordering to further expressions—for instance,

$$:(D\alpha_1(z))\alpha_2(z): = :\alpha_2(z) D\alpha_1(z): = D\alpha_1(z)^- \alpha_2(z) + \alpha_2(z) D\alpha_1(z)^+ \tag{3.3.16}$$

$$:\alpha_1(z) X(\alpha, z): = :X(\alpha, z)\alpha_1(z): = \alpha_1(z)^- X(\alpha, z) + X(\alpha, z)\alpha_1(z)^+ \tag{3.3.17}$$

in $(\operatorname{End} V)\{z\}$ and

$$\begin{aligned} &:X(\alpha, z_1) X(\beta, z_2): \\ &= 2^{-\langle \alpha, \alpha \rangle - \langle \beta, \beta \rangle} :e^{D^{-1}\alpha(z_1) + D^{-1}\beta(z_2)}: \\ &= 2^{-\langle \alpha, \alpha \rangle - \langle \beta, \beta \rangle} E^-(-\alpha, z_1) E^-(-\beta, z_2) E^+(-\alpha, z_1) E^+(-\beta, z_2) \end{aligned} \tag{3.3.18}$$

in $(\operatorname{End} V)\{z_1, z_2\}$, for $\alpha, \beta, \alpha_1, \alpha_2 \in \mathfrak{h}$, α and β satisfying (3.2.27). Note that

$$\lim_{z_1 \to z_2} :X(\alpha, z_1) X(\beta, z_2): = 4^{\langle \alpha, \beta \rangle} X(\alpha + \beta, z_2) \tag{3.3.19}$$

[recall the notation (2.1.17)] whenever α, β and $\alpha + \beta$ satisfy (3.2.27). It is also clear that normal ordering commutes with the (formal) application of the operator D in an obvious sense—for instance,

$$D{:}\alpha_1(z)\alpha_2(z){:} = {:}(D\alpha_1(z))\alpha_2(z){:} + {:}\alpha_1(z)D\alpha_2(z){:} \tag{3.3.20}$$

$$DX(\alpha, z) = {:}\alpha(z)X(\alpha, z){:} \tag{3.3.21}$$

$$D_{z_1}{:}X(\alpha, z_1)X(\beta, z_2){:} = {:}\alpha(z_1)X(\alpha, z_1)X(\beta, z_2){:} \tag{3.3.22}$$

$$\lim_{z_1 \to z_2} D_{z_1}{:}X(\alpha, z_1)X(\beta, z_2){:} = 4^{\langle \alpha, \beta \rangle}{:}\alpha(z_2)X(\alpha + \beta, z_2){:}. \tag{3.3.23}$$

Note also that (3.2.20) can be expressed as follows:

$$[d, E^{\pm}(\alpha, z)] = -DE^{\pm}(\alpha, z) = \alpha(z)^{\pm}E^{\pm}(\alpha, z). \tag{3.3.24}$$

Remark 3.3.1: The operators $L(n)$ introduced in (1.9.23) can be expressed using normal ordering notation as follows, in the case $Z = \mathbb{Z} + 1/2$:

$$L(n) = \tfrac{1}{2} \sum_{i=1}^{\dim \mathfrak{h}} \sum_{k \in Z} {:}h_i(n - k)h_i(k){:} + \tfrac{1}{16}\delta_{n0} \dim \mathfrak{h} \tag{3.3.25}$$

for $n \in Z$. (In Chapter 4, we shall extend normal ordering to the case $Z = \mathbb{Z}$.)

3.4. Some Commutators

We now embark on the computation of certain commutators of twisted vertex operators in order to construct representations of $\tilde{\mathfrak{a}}[\theta_2]$.

In addition to the formal exponential series notation e^x (3.2.17), we shall use the standard notation for logarithmic and binomial formal power series:

$$\log(1 + ax) = -\sum_{k \geq 1} \frac{(-a)^k}{k} x^k \tag{3.4.1}$$

$$(1 + x)^a = \sum_{k \geq 0} \binom{a}{k} x^k \tag{3.4.2}$$

for $a \in \mathbb{F}$ and suitable x. The symbol $\binom{a}{k}$ denotes the binomial coefficient

$$\binom{a}{k} = \frac{a(a - 1) \cdots (a - k + 1)}{k!} \in \mathbb{F}, \tag{3.4.3}$$

and this symbol is understood to be 0 if $k \notin \mathbb{N}$. An important special case of (3.4.2) is the geometric series

$$(1 - x)^{-1} = \sum_{k \geq 0} x^k. \tag{3.4.4}$$

These expressions, understood as elements of the formal power series algebra $\mathbb{F}[[x]]$, obey the standard rules:

$$\log(\exp x) = x$$

$$\exp(\log(1 + ax)) = 1 + ax$$

$$\log((1 + ax)(1 + bx)) = \log(1 + ax) + \log(1 + bx)$$

$$\log(1 + ax)^b = b \log(1 + ax)$$

for $a, b \in \mathbb{F}$, and so on. To justify such rules, one can for instance repeat algebraic versions of the standard calculus proofs, or one can *quote* the standard calculus rules and observe that they imply lists of algebraic identities for respective coefficients. An expression such as $((1 + x)/(1 - x))^a$ for $a \in \mathbb{F}$ can be understood either as a product of two binomial series or as $\exp(a \log((1 + x)/(1 - x)))$. Later we shall use the formula

$$\left(\frac{1 + x}{1 - x}\right)^2 = \left(1 + 2 \sum_{k \geq 1} x^k\right)^2 = 1 + 4 \sum_{k \geq 1} kx^k. \tag{3.4.5}$$

We have the following basic result:

Proposition 3.4.1: *For* $\alpha, \beta \in \mathfrak{h}$,

$$E^+(\alpha, z_1)E^-(\beta, z_2) = E^-(\beta, z_2)E^+(\alpha, z_1)\left(\frac{1 - z_2^{1/2}/z_1^{1/2}}{1 + z_2^{1/2}/z_1^{1/2}}\right)^{\langle \alpha, \beta \rangle} \tag{3.4.6}$$

in the formal power series algebra

$$(\mathrm{End}\, V)[[z_1^{-1/2}, z_2^{1/2}]] \subset (\mathrm{End}\, V)\{z_1, z_2\}.$$

In particular,

$$E^+(\alpha, z_1)E^-(\beta, z_2) = E^-(\beta, z_2)E^+(\alpha, z_1) \quad \text{if} \quad \langle \alpha, \beta \rangle = 0.$$

Proof: From the commutation relations (3.2.4), we see that

$$\left[\sum_{m \in \mathbb{N}+1/2} \frac{\alpha(m)z_1^{-m}}{m}, \sum_{n \in -(\mathbb{N}+1/2)} \frac{\beta(n)z_2^{-n}}{n}\right]$$

$$= -\langle \alpha, \beta \rangle \sum_{n \in \mathbb{N}+1/2} \frac{(z_2/z_1)^n}{n}$$

$$= \langle \alpha, \beta \rangle \log\left(\frac{1 - z_2^{1/2}/z_1^{1/2}}{1 + z_2^{1/2}/z_1^{1/2}}\right).$$

The result now follows from the formal rule

$$e^x e^y = e^y e^x e^{[x,y]} \quad \text{if} \quad [x, y] \text{ commutes with } x \text{ and } y. \tag{3.4.7}$$

This rule is easily established by the following formal argument:

$$\begin{aligned} xe^y &= e^y x + [x, e^y] \\ &= e^y x + e^y[x, y] \\ &= e^y(x + [x, y]). \end{aligned}$$

Iterating, we obtain

$$x^k e^y = e^y(x + [x, y])^k$$

for $k \geq 0$. Now divide by $k!$ and sum over k to get

$$e^x e^y = e^y e^{x+[x,y]} = e^y e^x e^{[x,y]}. \quad \blacksquare$$

Recalling (3.3.18), we now have:

Proposition 3.4.2: *Suppose that* $\alpha, \beta \in \mathfrak{h}$ *satisfy (3.2.27). Then*

$$:X(\alpha, z_1)X(\beta, z_2): \; = \; :X(\beta, z_2)X(\alpha, z_1): \tag{3.4.8}$$

$$X(\alpha, z_1)X(\beta, z_2) = :X(\alpha, z_1)X(\beta, z_2):\left(\frac{1 - z_2^{1/2}/z_1^{1/2}}{1 + z_2^{1/2}/z_1^{1/2}}\right)^{\langle\alpha,\beta\rangle}. \tag{3.4.9}$$

We can now compute the most important commutators needed to construct $\tilde{\mathfrak{a}}[\theta_2]$ by vertex operators [cf. (3.1.40)]:

Proposition 3.4.3: *For* $\alpha \in \mathfrak{h}$ *such that* $\langle\alpha, \alpha\rangle = 2$,

$$\begin{aligned} [X(\alpha, z_1), X(-\alpha, z_2)] &= \frac{1}{2}\alpha(z_2) \sum_{n \in \mathbb{Z}} (z_1/z_2)^{n/2} - \frac{1}{2} \sum_{n \in \mathbb{Z}} \frac{n}{2}(z_1/z_2)^{n/2} \\ &= \frac{1}{2}\alpha(z_2)\delta(z_1^{1/2}/z_2^{1/2}) - \frac{1}{2}D_{z_1}\delta(z_1^{1/2}/z_2^{1/2}) \end{aligned} \tag{3.4.10}$$

in (End V)$\{z_1, z_2\}$, *using the notation (2.1.22), (2.2.14) and (3.3.3). Equivalently* [*see (3.2.29)*],

$$[x_\alpha(m), x_{-\alpha}(n)] = \tfrac{1}{2}\alpha(m + n) + \tfrac{1}{2}m\delta_{m+n,0} \tag{3.4.11}$$

for all $m, n \in \frac{1}{2}\mathbb{Z}$ *where we set*

$$\alpha(n) = 0 \quad \text{for} \quad n \in \mathbb{Z}. \tag{3.4.12}$$

Proof: By (3.4.5), (3.4.8) and (3.4.9),

$$\begin{aligned}
[X(\alpha, z_1), X(-\alpha, z_2)] &= X(\alpha, z_1)X(-\alpha, z_2) - X(-\alpha, z_2)X(\alpha, z_1) \\
&= :X(\alpha, z_1)X(-\alpha, z_2): \left\{ \left(\frac{1 - z_2^{1/2}/z_1^{1/2}}{1 + z_2^{1/2}/z_1^{1/2}} \right)^{-2} - \left(\frac{1 - z_1^{1/2}/z_2^{1/2}}{1 + z_1^{1/2}/z_2^{1/2}} \right)^{-2} \right\} \\
&= 4:X(\alpha, z_1)X(-\alpha, z_2): \sum_{n \in \mathbb{Z}} n(z_2/z_1)^{n/2},
\end{aligned} \tag{3.4.13}$$

which we write using (3.3.18) as

$$= \tfrac{1}{4}:e^{D^{-1}\alpha(z_1) - D^{-1}\alpha(z_2)}:(D\delta)(z_2^{1/2}/z_1^{1/2}).$$

Note that the expression in braces in (3.4.13), being the difference of elements of the algebras $\mathbb{F}[[z_2^{1/2}/z_1^{1/2}]]$ and $\mathbb{F}[[z_1^{1/2}/z_2^{1/2}]]$, is a well-defined element of the vector space $\mathbb{F}\{z_1, z_2\}$. We shall use Proposition 2.2.4(b), but integral powers of the formal variables are needed [cf. (2.1.28)]. Thus we set

$$y_1 = z_1^{1/2}, \qquad y_2 = z_2^{1/2}, \tag{3.4.14}$$

and we obtain

$$\begin{aligned}
&[X(\alpha, z_1), X(-\alpha, z_2)] \\
&\quad = \tfrac{1}{4}:\exp((D^{-1}\alpha)(y_1^2) - (D^{-1}\alpha)(y_2^2)):(D\delta)(y_2/y_1) \\
&\quad = \tfrac{1}{4}(D\delta)(y_2/y_1) + \tfrac{1}{4}D_{y_1}:\exp(2D_{y_1}^{-1}\alpha(y_1^2) - 2D_{y_2}^{-1}\alpha(y_2^2)):\big|_{y_1 = y_2}\delta(y_1/y_2) \\
&\quad = -\tfrac{1}{4}(D\delta)(y_1/y_2) + \tfrac{1}{2}:\alpha(y_1^2)\exp(2D_{y_1}^{-1}\alpha(y_1^2) - 2D_{y_2}^{-1}\alpha(y_2^2)):\big|_{y_1 = y_2}\delta(y_1/y_2) \\
&\quad = -\tfrac{1}{4}(D\delta)(y_1/y_2) + \tfrac{1}{2}\alpha(y_2^2)\delta(y_1/y_2)
\end{aligned}$$

[cf. (3.3.22), (3.3.23)], establishing (3.4.10). Equating coefficients of $z_1^{-m}z_2^{-n}$ gives (3.4.11). ∎

Remark 3.4.4: In view of (3.2.41), (3.2.42), this result also gives $[X(\alpha, z_1), X(\alpha, z_2)]$ and $[x_\alpha(m), x_\alpha(n)]$, but there will be extra minus signs in the answers.

Remark 3.4.5: The method of proof of Proposition 3.4.3 could be extended to yield formulas for $[X(\alpha, z_1), X(\beta, z_2)]$ for all pairs $\alpha, \beta \in \mathfrak{h}$ with $\langle \alpha, \beta \rangle \in 2\mathbb{Z}$ [and with α, β satisfying (3.2.27)]. The expression in braces in (3.4.13) would be replaced by

$$\left(\frac{1 - z_2^{1/2}/z_1^{1/2}}{1 + z_2^{1/2}/z_1^{1/2}}\right)^{\langle \alpha,\beta \rangle} - \left(\frac{1 - z_1^{1/2}/z_2^{1/2}}{1 + z_1^{1/2}/z_2^{1/2}}\right)^{\langle \alpha,\beta \rangle} \tag{3.4.15}$$

and it turns out that for $\langle \alpha, \beta \rangle \in 2\mathbb{Z}$, this expression is a linear combination of expressions of the form $D^k\delta(y_1/y_2)$ for $k \in \mathbb{N}$ and y_i as in (3.4.14). In the trivial case $\langle \alpha, \beta \rangle = 0$, we find that

$$[X(\alpha, z_1), X(\beta, z_2)] = 0. \tag{3.4.16}$$

In general, the suggestion of Remark 2.2.5 could then be used to carry out the computation, but for $|\langle \alpha, \beta \rangle| > 2$, this approach becomes very complicated. A more conceptual approach will be presented later. But a more serious difficulty is the restriction $\langle \alpha, \beta \rangle \in 2\mathbb{Z}$; without this, the expression (3.4.15) is not a linear combination of derivatives of $\delta(y_1/y_2)$, and the method grinds to a halt. For instance, for $\langle \alpha, \beta \rangle = -1$, (3.4.15) becomes

$$2 \sum_{k \in \mathbb{Z}_+} (y_2/y_1)^k - 2 \sum_{k \in \mathbb{Z}_+} (y_2/y_1)^{-k}.$$

For $\langle \alpha, \beta \rangle \in \mathbb{Z}$, the situation can be corrected (and will be, later) by the introduction of certain noncommuting objects which lead to the replacement of (3.4.15) by

$$\left(\frac{1 - z_2^{1/2}/z_1^{1/2}}{1 + z_2^{1/2}/z_1^{1/2}}\right)^{\langle \alpha,\beta \rangle} - (-1)^{\langle \alpha,\beta \rangle}\left(\frac{1 - z_1^{1/2}/z_2^{1/2}}{1 + z_1^{1/2}/z_2^{1/2}}\right)^{\langle \alpha,\beta \rangle}.$$

In case $\langle \alpha, \beta \rangle = -1$, for instance, this becomes $2\delta(y_1/y_2)$. There is another way of salvaging the method for $\langle \alpha, \beta \rangle \in 2\mathbb{Z} + 1$: Compute the *anticommutator* $X(\alpha, z_1)X(\beta, z_2) + X(\beta, z_2)X(\alpha, z_1)$ instead of the commutator. For $\langle \alpha, \beta \rangle \notin \mathbb{Z}$, the situation is worse: Instead of computing commutators or anticommutators of vertex operators, we must be content to compute certain more complicated combinations. We shall not deal with this subtlety in this work; see [Lepowsky–Wilson 3–6].

In the next section, we shall summarize how Proposition 3.4.3 completes the twisted vertex operator construction of $\tilde{\mathfrak{a}}[\theta_2]$. We shall find it convenient to set

$$X^{\pm}(\alpha, z) = X(\alpha, z) \pm X(-\alpha, z) \tag{3.4.17}$$

for $\alpha \in \mathfrak{h}$ such that $\langle \alpha, \alpha \rangle \in \mathbb{Z}$, and we define operators $x_\alpha^{\pm}(n)$ for $n \in \frac{1}{2}\mathbb{Z}$ by

$$X^{\pm}(\alpha, z) = \sum_{n \in (1/2)\mathbb{Z}} x_\alpha^{\pm}(n) z^{-n}, \tag{3.4.18}$$

so that

$$x_\alpha^{\pm}(n) = x_\alpha(n) \pm x_{-\alpha}(n). \tag{3.4.19}$$

Formula (3.2.42) then gives

$$\begin{aligned} x_\alpha^{+}(n) &= 2x_\alpha(n) = 2x_{-\alpha}(n) \quad \text{for} \quad n \in \mathbb{Z}, \\ x_\alpha^{-}(n) &= 2x_\alpha(n) = -2x_{-\alpha}(n) \quad \text{for} \quad n \in \mathbb{Z} + \tfrac{1}{2}, \end{aligned} \tag{3.4.20}$$

while

$$\begin{aligned} x_\alpha^{+}(n) &= 0 \quad \text{for} \quad n \in \mathbb{Z} + \tfrac{1}{2}, \\ x_\alpha^{-}(n) &= 0 \quad \text{for} \quad n \in \mathbb{Z}. \end{aligned} \tag{3.4.21}$$

3.5. Irreducible Representations of $\mathfrak{sl}(2)\hat{}[\theta_2]$

Now we take the space $\mathfrak{h}$ of Sections 3.2–3.4 to be the one-dimensional space $\mathbb{F}\alpha_1$ of (3.1.13). Then the space V of (3.2.2) may be identified with the polynomial algebra on the generators $\alpha_1(-n)$ for $n \in \mathbb{N} + 1/2$:

$$V = \mathbb{F}[\alpha_1(-\tfrac{1}{2}), \alpha_1(-\tfrac{3}{2}), \ldots], \tag{3.5.1}$$

with the grading shifted so that

$$\deg 1 = -\tfrac{1}{48} \tag{3.5.2}$$

[recall (1.9.53)]. The algebra $\tilde{\mathfrak{h}}[-1]$ acts irreducibly on V by:

$$\begin{aligned} c &\mapsto 1, \qquad d \mapsto d \\ \alpha_1 \otimes t^n &\mapsto \alpha_1(n) \quad \text{for} \quad n \in \mathbb{Z} + \tfrac{1}{2}. \end{aligned} \tag{3.5.3}$$

Recall the elements $x_{\alpha_1}^{\pm} \in \mathfrak{a}$ defined in (3.1.24) and the operators $x_{\alpha_1}^{\pm}(n)$ of (3.4.20). As we have already explained in Remark 3.2.3, the commutation result Proposition 3.4.3 allows us to conclude:

Theorem 3.5.1: *The representation of $\tilde{\mathfrak{h}}[-1]$ on V given by (3.5.3) extends to precisely two (necessarily irreducible) representations*

$$\pi_{\pm}: \tilde{\mathfrak{a}}[\theta_2] \to \operatorname{End} V, \tag{3.5.4}$$

determined by the conditions

$$\begin{aligned}\pi_+ : x_{\alpha_1}(z) &\mapsto X(\alpha_1, z)\\ \pi_- : x_{\alpha_1}(z) &\mapsto -X(\alpha_1, z),\end{aligned} \tag{3.5.5}$$

or equivalently, the conditions

$$\begin{aligned}\pi_+ : x_{\alpha_1}^+ \otimes t^n &\mapsto x_{\alpha_1}^+(n) &&\text{for} \quad n \in \mathbb{Z}\\ \pi_+ : x_{\alpha_1}^- \otimes t^n &\mapsto x_{\alpha_1}^-(n) &&\text{for} \quad n \in \mathbb{Z} + \tfrac{1}{2}\\ \pi_- : x_{\alpha_1}^+ \otimes t^n &\mapsto -x_{\alpha_1}^+(n) &&\text{for} \quad n \in \mathbb{Z}\\ \pi_- : x_{\alpha_1}^- \otimes t^n &\mapsto -x_{\alpha_1}^-(n) &&\text{for} \quad n \in \mathbb{Z} + \tfrac{1}{2}.\end{aligned} \tag{3.5.6}$$

Remark 3.5.2: In particular, the following formal differential operators on V span a Lie algebra:

$$1, \qquad \alpha_1(-n), \qquad \frac{\partial}{\partial \alpha_1(-n)} \qquad (n \in \mathbb{N} + \tfrac{1}{2})$$

and the coefficients of z^n $(n \in (1/2)\mathbb{Z})$ in

$$\exp\left(\sum_{n \in \mathbb{N}+1/2} \frac{\alpha_1(-n)}{n} z^n\right) \exp\left(-\sum_{n \in \mathbb{N}+1/2} \frac{\partial}{\partial \alpha_1(-n)} z^{-n}\right).$$

This Lie algebra is isomorphic to $\mathfrak{sl}(2)\hat{}[\theta_2] \simeq \mathfrak{sl}(2)\hat{}$ [see Remark 3.1.2 and (3.1.46)].

Remark 3.5.3: The representations $\pi_\pm$ are inequivalent since the operators $\pi_\pm(x_{\alpha_1}^+ \otimes t^0)$ restrict to multiplication by opposite nonzero scalars on the vacuum space of V—the one-dimensional space of scalars in the realization (3.5.1) (see Proposition 1.7.2).

4 Realization of $\mathfrak{sl}(2)\hat{}$ by Untwisted Vertex Operators

This chapter largely parallels the previous one. However, the untwisted vertex operator representation of $\mathfrak{sl}(2)\hat{}$ also has new features. In particular, untwisted vertex operators contain a new factor acting trivially on the Fock space and represented on a certain group algebra. This representation is a special case of the one constructed in [Frenkel–Kac] and independently in [Segal 1]. Later it became apparent that physicists had anticipated many features of this construction in the case of the Lie algebras $\mathfrak{sl}(n)\hat{}$ (see the Introduction for a discussion and references).

In Section 4.1 we motivate and introduce untwisted vertex operators, explaining the new concept in detail. We extend the notion of normal ordering to the untwisted case in Section 4.2, and in Section 4.3 we compute the commutators of the appropriate untwisted vertex operators. In Section 4.4 we construct the untwisted vertex operator representation of $\mathfrak{sl}(2)\hat{}$ and its modification, recognized in [FLM2], arising from a shift of the relevant lattice to a coset. Finally, in Section 4.5 we show that under appropriate choices of the shifts, irreducible untwisted and twisted vertex operator modules become isomorphic as graded $\mathfrak{sl}(2)\hat{}$-modules [FLM2]. Equating the graded dimensions of the two modules gives a classical formula of Gauss involving the Dedekind η-function.

The construction in this chapter recovers the basic $\mathfrak{sl}(2)\hat{}$-modules, which have already appeared in Chapter 3. An analogous untwisted construction of all the standard $\mathfrak{sl}(2)\hat{}$-modules using "Z-algebras" has been given in [Lepowsky–Primc 1, 2]. A different generalization of the material in this chapter will be presented in Chapter 7.

4.1. The Untwisted Vertex Operators $X_{\mathbb{Z}}(\alpha, z)$

Our next goal is to construct representations of the affine Lie algebras $\tilde{\mathfrak{a}}$ and $\tilde{\mathfrak{a}}[\theta_1]$ (see Section 3.1) analogous to the representations of $\tilde{\mathfrak{a}}[\theta_2]$ based on twisted vertex operators obtained in Chapter 3. This time we start with the Heisenberg subalgebra $\hat{\mathfrak{h}}_{\mathbb{Z}}$ of $\tilde{\mathfrak{a}}$ and $\tilde{\mathfrak{a}}[\theta_1]$ [see (3.1.14)] in place of $\hat{\mathfrak{h}}_{\mathbb{Z}+1/2}$ (3.1.36).

As in Section 3.2, let $\mathfrak{h}$ be a nonzero finite-dimensional vector space and let $\langle \cdot, \cdot \rangle$ be a nonsingular symmetric bilinear form on $\mathfrak{h}$. View $\mathfrak{h}$ as an abelian Lie algebra and consider the corresponding $\mathbb{Z}$-graded untwisted affine Lie algebra $\tilde{\mathfrak{h}}$ and its Heisenberg subalgebra

$$\hat{\mathfrak{h}}_{\mathbb{Z}} = \tilde{\mathfrak{h}}' \tag{4.1.1}$$

[see (1.7.11), (1.7.13), (1.7.18)]. We have the $\mathbb{Q}$-graded $\tilde{\mathfrak{h}}$-module

$$M(1) = S(\hat{\mathfrak{h}}_{\mathbb{Z}}^{-}), \tag{4.1.2}$$

irreducible even under $\hat{\mathfrak{h}}_{\mathbb{Z}}$. On this module, c acts as 1, d acts as the degree operator and $\mathfrak{h}$ acts trivially [see (1.7.15), (1.7.19), (1.9.51)].

For $\alpha \in \mathfrak{h}$ and $n \in \mathbb{Z}$, we shall use the notation $\alpha(n)$ for the operator on $S(\hat{\mathfrak{h}}_{\mathbb{Z}}^{-})$ corresponding to $\alpha \otimes t^n \in \tilde{\mathfrak{h}}$, by analogy with (3.2.3) [see also (1.9.16)]:

$$\alpha \otimes t^n \mapsto \alpha(n). \tag{4.1.3}$$

Then [cf. (1.9.17)]

$$\begin{aligned} [\alpha(m), \beta(n)] &= \langle \alpha, \beta \rangle m \delta_{m+n,0} \\ [d, \alpha(m)] &= m\alpha(m) \end{aligned} \tag{4.1.4}$$

for $\alpha, \beta \in \mathfrak{h}$ and $m, n \in \mathbb{Z}$, and the operator $\alpha(m)$ is homogeneous of degree m. The Heisenberg algebra can be realized by multiplication and partial differentiation operators on $S(\hat{\mathfrak{h}}_{\mathbb{Z}}^{-})$ (see Section 1.7).

In the special case $\mathfrak{h} = \mathbb{F}\alpha_1$ [see (3.1.13)], we want operators whose generating functions satisfy the conditions on $x_{\pm\alpha_1}(z)$ given in (3.1.17)–(3.1.21). However, the action of $\tilde{\mathfrak{h}}$ on $S(\hat{\mathfrak{h}}_{\mathbb{Z}}^{-})$ cannot be extended to an action of $\tilde{\mathfrak{a}}$ or $\tilde{\mathfrak{a}}[\theta_1]$ on the same space. In fact, since $\mathfrak{h}$ acts trivially, condition (3.1.17) for $m = 0$ shows that $x_{\pm\alpha_1}(z)$ would have to act trivially, which is impossible, by (3.1.20). No redefinition of the action of $\mathfrak{h}$ on $S(\hat{\mathfrak{h}}_{\mathbb{Z}}^-)$ would help, since $\mathfrak{h}$ would have to act as scalar multiplication operators, by Proposition 1.7.2.

Instead, we invoke Theorem 1.7.3 as motivation to introduce (in the case of general $\mathfrak{h}$) a tensor product

$$V = S(\hat{\mathfrak{h}}_{\mathbb{Z}}^-) \otimes W \tag{4.1.5}$$

of $\tilde{\mathfrak{h}}$-modules, where W is trivial as an $\hat{\mathfrak{h}}_{\mathbb{Z}}$-module and is $\mathbb{F}$-graded, with d acting as the degree operator, and where V has the tensor product $\mathbb{F}$-grading.

Fix $\alpha \in \mathfrak{h}$. Let

$$u_\alpha(n) \in \operatorname{End} V \quad \text{for} \quad n \in \mathbb{F} \tag{4.1.6}$$

and define

$$U(\alpha, z) = \sum_{n \in \mathbb{F}} u_\alpha(n) z^{-n} \in (\operatorname{End} V)\{z\}. \tag{4.1.7}$$

Consider the following commutation action of $\tilde{\mathfrak{h}}$ on $U(\alpha, z)$:

$$[h(m), U(\alpha, z)] = \langle h, \alpha \rangle z^m U(\alpha, z) \tag{4.1.8}$$

for $h \in \mathfrak{h}$ and $m \in \mathbb{Z}$, or equivalently,

$$[h(m), u_\alpha(n)] = \langle h, \alpha \rangle u_\alpha(m + n) \tag{4.1.9}$$

for $h \in \mathfrak{h}$, $m \in \mathbb{Z}$ and $n \in \mathbb{F}$. Consider also the relations

$$[d, U(\alpha, z)] = -DU(\alpha, z), \tag{4.1.10}$$

i.e.,

$$\deg u_\alpha(n) = n \quad \text{for} \quad n \in \mathbb{F}. \tag{4.1.11}$$

We seek solutions of these equations.

Still following the pattern of Section 3.2, we define

$$\begin{aligned} E^{\pm}(\alpha, z) &= \exp\left(\sum_{n \in \pm\mathbb{Z}_+} \frac{\alpha(n)}{n} z^{-n} \right) \\ &\in (\operatorname{End} S(\hat{\mathfrak{h}}_{\mathbb{Z}}^-))[[z^{\mp 1}]] \subset (\operatorname{End} S(\hat{\mathfrak{h}}_{\mathbb{Z}}^-))\{z\} \end{aligned} \tag{4.1.12}$$

for $\alpha \in \mathfrak{h}$ [cf. (3.2.16)]. We have an obvious analogue of Proposition 3.2.1:

Proposition 4.1.1: *For $\alpha \in \mathfrak{h}$, the expressions $E^{\pm}(\alpha, z)$ exist in the sense of (2.1.10), and we have:*

$$E^{\pm}(0, z) = 1 \tag{4.1.13}$$

$$E^{\pm}(\alpha + \beta, z) = E^{\pm}(\alpha, z)E^{\pm}(\beta, z) \tag{4.1.14}$$

$$[d, E^{\pm}(\alpha, z)] = -DE^{\pm}(\alpha, z) = \left(\sum_{n \in \pm\mathbb{Z}_+} \alpha(n)z^{-n}\right)E^{\pm}(\alpha, z) \tag{4.1.15}$$

for $\alpha, \beta \in \mathfrak{h}$. For $h, \alpha \in \mathfrak{h}$,

$$[h(m), E^{+}(\alpha, z)] = 0 \quad \text{if} \quad m \in \mathbb{N} \tag{4.1.16}$$

$$[h(m), E^{-}(\alpha, z)] = -\langle h, \alpha\rangle z^{m}E^{-}(\alpha, z) \quad \text{if} \quad m \in \mathbb{Z}_+ \tag{4.1.17}$$

$$[h(m), E^{+}(\alpha, z)] = -\langle h, \alpha\rangle z^{m}E^{+}(\alpha, z) \quad \text{if} \quad m \in -\mathbb{Z}_+ \tag{4.1.18}$$

$$[h(m), E^{-}(\alpha, z)] = 0 \quad \text{if} \quad m \in -\mathbb{N}. \tag{4.1.19}$$

Now we can begin to construct expressions satisfying our equations. We embed End $S(\hat{\mathfrak{h}}_{\bar{\mathbb{Z}}})$ (resp., End W) into End V via $A \mapsto A \otimes 1$ (resp., $B \mapsto 1 \otimes B$), and we similarly embed (End $S(\hat{\mathfrak{h}}_{\bar{\mathbb{Z}}})$)$\{z\}$ and (End W)$\{z\}$ into (End V)$\{z\}$. For $\alpha \in \mathfrak{h}$, we form

$$U(\alpha, z) = E^{-}(-\alpha, z)E^{+}(-\alpha, z)Z(\alpha, z), \tag{4.1.20}$$

where

$$Z(\alpha, z) \in (\text{End } W)\{z\} \tag{4.1.21}$$

and the product (4.1.20) is of course assumed to exist in (End V)$\{z\}$. If

$$[h(0), Z(\alpha, z)] = \langle h, \alpha\rangle Z(\alpha, z) \quad \text{for} \quad h \in \mathfrak{h}, \tag{4.1.22}$$

$$[d, Z(\alpha, z)] = -DZ(\alpha, z), \tag{4.1.23}$$

then by Proposition 4.1.1, $U(\alpha, z)$ satisfies equations (4.1.8) and (4.1.10) (cf. Proposition 3.2.2).

There is a natural way to satisfy (4.1.22): Take the space W to be the group algebra

$$W = \mathbb{F}[\mathfrak{h}] = \coprod_{\alpha \in \mathfrak{h}} \mathbb{F}e^{\alpha} \tag{4.1.24}$$

of the additive group $\mathfrak{h}$ [cf. (1.5.4)]. For $\alpha \in \mathfrak{h}$, denote by

$$e^{\alpha}: \mathbb{F}[\mathfrak{h}] \to \mathbb{F}[\mathfrak{h}] \tag{4.1.25}$$

the corresponding multiplication operator, and for $h \in \mathfrak{h}$ define an operator

$$\begin{aligned} h(0)\colon \mathbb{F}[\mathfrak{h}] &\to \mathbb{F}[\mathfrak{h}] \\ e^{\alpha} &\mapsto \langle h, \alpha\rangle e^{\alpha} \end{aligned} \tag{4.1.26}$$

for $\alpha \in \mathfrak{h}$. Then for $h, \alpha \in \mathfrak{h}$,

$$[h(0), e^{\alpha}] = \langle h, \alpha\rangle e^{\alpha}, \tag{4.1.27}$$

as in (4.1.22). This suggests taking $Z(\alpha, z) = e^{\alpha}$ (which happens to be independent of z).

However, we want to be able to compute commutators of the resulting operators in order to satisfy (3.1.18)–(3.1.20), and as we shall see later, we need to modify our suggestion for $Z(\alpha, z)$ by introducing certain factors depending on z (see Remarks 4.1.2 and 4.2.1 below for further motivation): For $h \in \mathfrak{h}$, define $z^{h} \in (\operatorname{End} \mathbb{F}[\mathfrak{h}])\{z\}$ (thought of as $z^{h(0)}$) by

$$z^{h} \cdot e^{\alpha} = z^{\langle h, \alpha\rangle} e^{\alpha} \quad \text{for} \quad \alpha \in \mathfrak{h}. \tag{4.1.28}$$

(This formula expresses z^{h} as an operator from $\mathbb{F}[\mathfrak{h}]$ to $(\mathbb{F}[\mathfrak{h}])\{z\}$; its identification with an element of $(\operatorname{End} \mathbb{F}[\mathfrak{h}])\{z\}$ is clear. Its components are projection operators.) Then in $(\operatorname{End} \mathbb{F}[\mathfrak{h}])\{z\}$, for $\alpha, \beta \in \mathfrak{h}$,

$$[\alpha(0), z^{\beta}] = 0 \tag{4.1.29}$$

$$z^{\alpha} e^{\beta} = z^{\langle\alpha, \beta\rangle} e^{\beta} z^{\alpha} = e^{\beta} z^{\alpha + \langle\alpha, \beta\rangle}. \tag{4.1.30}$$

We define the (*untwisted*) *vertex operator* associated with $\alpha \in \mathfrak{h}$ by:

$$\begin{aligned} X(\alpha, z) = X_{\mathbb{Z}}(\alpha, z) &= E^{-}(-\alpha, z) E^{+}(-\alpha, z) e^{\alpha} z^{\alpha + \langle\alpha, \alpha\rangle/2} \\ &\in (\operatorname{End} V)\{z\}, \end{aligned} \tag{4.1.31}$$

where

$$V = V_{\mathfrak{h}} = S(\hat{\mathfrak{h}}_{\bar{\mathbb{Z}}}) \otimes \mathbb{F}[\mathfrak{h}]. \tag{4.1.32}$$

Here of course

$$z^{\alpha + \langle\alpha, \alpha\rangle/2} = z^{\alpha} z^{\langle\alpha, \alpha\rangle/2},$$

and the product in (4.1.31) exists in the sense of (2.1.11). By (4.1.30), we also have

$$X(\alpha, z) = E^{-}(-\alpha, z) E^{+}(-\alpha, z) z^{\alpha - \langle\alpha, \alpha\rangle/2} e^{\alpha}. \tag{4.1.33}$$

The expansion

$$X(\alpha, z) = \sum_{n \in \mathbb{F}} x_{\alpha}(n) z^{-n} \tag{4.1.34}$$

defines operators $x_\alpha(n)$ on V. We have

$$[h(m), X(\alpha, z)] = \langle h, \alpha\rangle z^m X(\alpha, z) \tag{4.1.35}$$

for $h \in \mathfrak{h}$ and $m \in \mathbb{Z}$, or equivalently,

$$[h(m), x_\alpha(n)] = \langle h, \alpha\rangle x_\alpha(m + n) \tag{4.1.36}$$

for $h \in \mathfrak{h}$, $m \in \mathbb{Z}$ and $n \in \mathbb{F}$.

In order to satisfy the degree condition (4.1.23), we must impose an appropriate $\mathbb{F}$-grading on $\mathbb{F}[\mathfrak{h}]$ (as a vector space, not an algebra): Set

$$\deg e^\alpha = -\tfrac{1}{2}\langle \alpha, \alpha\rangle \quad \text{for} \quad \alpha \in \mathfrak{h}. \tag{4.1.37}$$

Then V acquires the tensor product grading

$$V = \coprod_{n \in \mathbb{F}} V_n. \tag{4.1.38}$$

To check (4.1.23), let $\alpha, \beta \in \mathfrak{h}$. The operator e^α on $\mathbb{F}[\mathfrak{h}]$ takes the basis element e^β to $e^{\alpha+\beta}$, and by (4.1.37),

$$\deg e^{\alpha+\beta} - \deg e^\beta = -\langle \alpha, \beta\rangle - \tfrac{1}{2}\langle \alpha, \alpha\rangle.$$

Hence

$$[d, e^\alpha](e^\beta) = e^{\alpha+\beta}(-\langle \alpha, \beta\rangle - \tfrac{1}{2}\langle \alpha, \alpha\rangle)$$

so that by (4.1.25) and (4.1.26),

$$[d, e^\alpha] = e^\alpha(-\alpha(0) - \tfrac{1}{2}\langle \alpha, \alpha\rangle) \tag{4.1.39}$$

as operators on $\mathbb{F}[\mathfrak{h}]$ or $V_{\mathfrak{h}}$. On the other hand,

$$Dz^{\alpha+\langle\alpha,\alpha\rangle/2} = (\alpha(0) + \tfrac{1}{2}\langle \alpha, \alpha\rangle)z^{\alpha+\langle\alpha,\alpha\rangle/2}, \tag{4.1.40}$$

and so $e^\alpha z^{\alpha+\langle\alpha,\alpha\rangle/2}$ satisfies (4.1.23), giving

$$[d, X(\alpha, z)] = -DX(\alpha, z), \tag{4.1.41}$$

$$\deg x_\alpha(n) = n \quad \text{for} \quad n \in \mathbb{F}. \tag{4.1.42}$$

We summarize the action of $\tilde{\mathfrak{h}}$, e^α and z^α on V: For $\alpha \in \mathfrak{h}$,

$$\begin{aligned}
c &\mapsto 1\\
d &\mapsto d = d \otimes 1 + 1 \otimes d\\
\alpha = \alpha \otimes t^0 &\mapsto \alpha(0) = 1 \otimes \alpha(0)\\
\alpha \otimes t^n &\mapsto \alpha(n) = \alpha(n) \otimes 1 \quad \text{for} \quad n \in \mathbb{Z}\backslash\{0\}\\
e^\alpha &= 1 \otimes e^\alpha\\
z^\alpha &= 1 \otimes z^\alpha.
\end{aligned} \tag{4.1.43}$$

For a subset $M \subset \mathfrak{h}$, define the subspace $\mathbb{F}[M]$ of $\mathbb{F}[\mathfrak{h}]$ by

$$\mathbb{F}[M] = \coprod_{\alpha \in M} \mathbb{F}e^{\alpha} \tag{4.1.44}$$

[cf. (4.1.24)]. In case M is a subgroup of the additive group of $\mathfrak{h}$, this is the usual notation for the group algebra of M. But we shall also find this notation useful when M is not a group. Also set

$$V_M = S(\hat{\mathfrak{h}}_{\mathbb{Z}}^-) \otimes \mathbb{F}[M] \subset V_{\mathfrak{h}}. \tag{4.1.45}$$

Remark 4.1.2: We have motivated the definition of the vertex operator by means of commutation properties (some of them to be proved later) and we have motivated the definition of the grading (4.1.37) by means of the degree conditions (4.1.23), (4.1.41). We can instead motivate the grading (4.1.37) with the Virasoro algebra $\mathfrak{v}$: The space $S(\hat{\mathfrak{h}}_{\mathbb{Z}}^-) \otimes e^{\alpha}$ is of the form (1.9.19), and $\mathfrak{v}$ acts on it by the operators $L(n)$ given by (1.9.23) for $Z = \mathbb{Z}$. The action of $L(0)$ predicts (4.1.37) as in (1.9.45). [The revised value of the degree given by (1.9.52) corresponds here to the tensor product grading of the space (4.1.32).] The discussion involving (4.1.37)–(4.1.42) then motivates the factor $z^{\alpha+\langle\alpha,\alpha\rangle/2}$ in $X(\alpha, z)$.

4.2. Normal Ordering

The concept and motivating principles of normal ordering (see Section 3.3) extend naturally to the present context, with certain modifications to accommodate the second tensor factor $\mathbb{F}[\mathfrak{h}]$ of $V = V_{\mathfrak{h}}$ [see (4.1.32)]. We adopt the rules of (3.3.1) and (3.3.2), allowing the indices n_i to range now through $\mathbb{Z}$. The corresponding normal ordered products act on V. For $\alpha \in \mathfrak{h}$, set

$$\begin{aligned} \alpha(z) &= \sum_{n \in \mathbb{Z}} \alpha(n) z^{-n} \\ \alpha(z)^{\pm} &= \tfrac{1}{2}\alpha(0) + \sum_{n \in \pm\mathbb{Z}_+} \alpha(n) z^{-n} \end{aligned} \tag{4.2.1}$$

in $(\mathrm{End}\, V)\{z\}$, so that

$$\alpha(z) = \alpha(z)^+ + \alpha(z)^- \tag{4.2.2}$$

[cf. (3.3.3), (3.3.4)]. Then (3.3.5)–(3.3.9) carry over as well.

As for the normal ordering of noncommuting expressions involving e^{α} and z^{α}, we shall be motivated by the rules

$$:\alpha(z)e^{\beta}: \; = \; :e^{\beta}\alpha(z): \; = \alpha(z)^{-}e^{\beta} + e^{\beta}\alpha(z)^{+}$$
$$= (\alpha(z) - \tfrac{1}{2}\langle\alpha,\beta\rangle)e^{\beta} = e^{\beta}(\alpha(z) + \tfrac{1}{2}\langle\alpha,\beta\rangle) \tag{4.2.3}$$

and taking the constant term,

$$:\alpha(0)e^{\beta}: \; = \; :e^{\beta}\alpha(0): \; = \tfrac{1}{2}(\alpha(0)e^{\beta} + e^{\beta}\alpha(0))$$
$$= (\alpha(0) - \tfrac{1}{2}\langle\alpha,\beta\rangle)e^{\beta} = e^{\beta}(\alpha(0) + \tfrac{1}{2}\langle\alpha,\beta\rangle) \tag{4.2.4}$$

[see (4.1.27)] and correspondingly,

$$:z^{\alpha}e^{\beta}: \; = \; :e^{\beta}z^{\alpha}: \; = z^{\alpha-\langle\alpha,\beta\rangle/2}e^{\beta} = e^{\beta}z^{\alpha+\langle\alpha,\beta\rangle/2} \tag{4.2.5}$$

[see (4.1.30)] for $\alpha, \beta \in \mathfrak{h}$. The vertex operator is given by:

$$X(\alpha, z) = \; :e^{D^{-1}(\alpha(z)-\alpha(0))}e^{\alpha}z^{\alpha}: \; = \; :X(\alpha, z): \tag{4.2.6}$$

for $\alpha \in \mathfrak{h}$. (The rigorous meaning of formulas such as these should be clear.)

Remark 4.2.1: We can heuristically motivate the expression (4.1.31) for the vertex operator $X(\alpha, z)$ by analogy with (3.3.15): Take $\mathbb{F}$ to be the complex field $\mathbb{C}$. While $D^{-1}\alpha(z)$ is not defined because $\alpha(z)$ has a nonzero constant term [cf. (2.2.21)–(2.2.22)], let us think of z as a nonzero complex variable and write

$$D^{-1}\alpha(z) = -\sum_{n\neq 0} \frac{\alpha(n)}{n} z^{-n} + \alpha(0)\log z + C, \tag{4.2.7}$$

where C is a constant of integration. [Formally, $z(d/dz)$ applied to this expression gives back $\alpha(z)$.] Now $\hat{\mathfrak{h}} = \hat{\mathfrak{h}}_{\mathbb{Z}} \oplus \mathfrak{h}$ fails to be a Heisenberg algebra [see (1.7.1)] because $\mathfrak{h}$ is central. Consider the isomorphism from $\mathfrak{h}$ to its dual defined by the form $\langle\cdot,\cdot\rangle$:

$$\begin{aligned} \mathfrak{h} &\to \mathfrak{h}^* \\ \alpha &\mapsto \alpha^*, \end{aligned} \tag{4.2.8}$$

where $\alpha^*(\beta) = \langle\alpha,\beta\rangle$ for $\beta \in \mathfrak{h}$. Then $\hat{\mathfrak{h}}$ may be embedded in a Heisenberg algebra

$$\mathfrak{l} = \hat{\mathfrak{h}} \oplus \mathfrak{h}^* \tag{4.2.9}$$

where

$$[\hat{\mathfrak{h}}_{\mathbb{Z}}, \mathfrak{h}^*] = [\mathfrak{h}^*, \mathfrak{h}^*] = 0$$

$$[\alpha, \beta^*] = -[\beta^*, \alpha] = \langle \alpha, \beta \rangle c \quad \text{for} \quad \alpha, \beta \in \mathfrak{h}.$$

Identify the group algebra element $e^{\alpha} \in \mathbb{C}[\mathfrak{h}] = \mathbb{F}[\mathfrak{h}]$ [see (4.1.24)] with the exponential function $\exp \alpha^*$ on $\mathfrak{h}$. Now the Heisenberg algebra $\mathfrak{h} \oplus \mathfrak{h}^* \oplus \mathbb{C}c$ acts on the space $\mathfrak{F}$ of analytic functions on $\mathfrak{h}$ as follows: Let c act as 1, $\alpha \in \mathfrak{h}$ as the corresponding derivation [cf. (4.1.26)], and $\alpha^* \in \mathfrak{h}^*$ as the corresponding multiplication operator. Then $\mathfrak{l}$ acts on the space $S(\hat{\mathfrak{h}}_{\mathbb{Z}}^-) \otimes \mathfrak{F}$, which contains V. Using the formal rule

$$e^{x+y} = e^x e^y e^{-[x,y]/2} \tag{4.2.10}$$

if $[x, y]$ commutes with x and y, which implies (3.4.7) and which can be proved by computing $d/dt(e^{t(x+y)}e^{-tx}e^{-ty})$, we see that formally,

$$e^{\beta^* + \alpha(0) \log z} = e^{\beta} z^{\alpha + \langle \alpha, \beta \rangle /2} \tag{4.2.11}$$

for $\alpha, \beta \in \mathfrak{h}$. Hence if we choose the constant C in (4.2.7) to be

$$C = \alpha^*,$$

we find that formally,

$$X(\alpha, z) = {:}e^{D^{-1}\alpha(z)}{:} \tag{4.2.12}$$

[cf. (3.3.15) and (4.2.6)]. Note also that (4.2.11) heuristically justifies the normal ordering formula (4.2.5).

We now have the natural analogues of (3.3.16)–(3.3.24): For $\alpha, \beta \in \mathfrak{h}$ we define (for instance)

$${:}(D\alpha(z))\beta(z){:} = {:}\beta(z)D\alpha(z){:} = D\alpha(z)^- \beta(z) + \beta(z) D\alpha(z)^+ \tag{4.2.13}$$

$${:}\alpha(z)X(\beta, z){:} = {:}X(\beta, z)\alpha(z){:} = \alpha(z)^- X(\beta, z) + X(\beta, z)\alpha(z)^+ \tag{4.2.14}$$

$$\begin{aligned} &{:}X(\alpha, z_1)X(\beta, z_2){:} \\ &\quad = {:}e^{D^{-1}(\alpha(z_1) - \alpha(0)) + D^{-1}(\beta(z_2) - \beta(0))} e^{\alpha+\beta} z_1^{\alpha} z_2^{\beta}{:} \\ &\quad = {:}e^{D^{-1}(\alpha(z_1) - \alpha(0)) + D^{-1}(\beta(z_2) - \beta(0))}{:}\, e^{\alpha+\beta} z_1^{\alpha + \langle \alpha, \alpha+\beta \rangle /2} z_2^{\beta + \langle \beta, \alpha+\beta \rangle /2} \\ &\quad = E^-(-\alpha, z_1) E^-(-\beta, z_2) E^+(-\alpha, z_1) E^+(-\beta, z_2) \\ &\qquad \cdot e^{\alpha+\beta} z_1^{\alpha + \langle \alpha, \alpha+\beta \rangle /2} z_2^{\beta + \langle \beta, \alpha+\beta \rangle /2}, \end{aligned} \tag{4.2.15}$$

(4.2.15) being motivated as in (4.2.11), (4.2.12); see also (4.2.5), (4.2.6). Note the case $\beta = 0$. We have

$$\lim_{z_1 \to z_2} :X(\alpha, z_1)X(\beta, z_2): = X(\alpha + \beta, z_2) \tag{4.2.16}$$

$$D:\alpha(z)\beta(z): = :(D\alpha(z))\beta(z): + :\alpha(z)D\beta(z): \tag{4.2.17}$$

$$DX(\alpha, z) = :\alpha(z)X(\alpha, z): \tag{4.2.18}$$

$$D_{z_1}:X(\alpha, z_1)X(\beta, z_2): = :\alpha(z_1)X(\alpha, z_1)X(\beta, z_2): \tag{4.2.19}$$

$$\lim_{z_1 \to z_2} D_{z_1}:X(\alpha, z_1)X(\beta, z_2): = :\alpha(z_2)X(\alpha + \beta, z_2):. \tag{4.2.20}$$

The right-hand side of (4.2.19) is defined by analogy with (4.2.14) and using (4.2.15). For (4.2.18) and (4.2.19), we use (4.1.27). Note that (4.2.12) can serve as a mnemonic for some of these formulas. Also,

$$DE^{\pm}(\alpha, z) = -[d, E^{\pm}(\alpha, z)] = -(\alpha(z)^{\pm} - \tfrac{1}{2}\alpha(0))E^{\pm}(\alpha, z) \tag{4.2.21}$$

[cf. (4.1.15)].

Remark 4.2.2 (cf. Remark 3.3.1): The operators $L(n)$ for $n \in Z = \mathbb{Z}$ in (1.9.23) can be written as:

$$L(n) = \tfrac{1}{2} \sum_{i=1}^{\dim \mathfrak{h}} \sum_{k \in \mathbb{Z}} :h_i(n-k)h_i(k):. \tag{4.2.22}$$

4.3. Some Commutators

Following the outline of Section 3.4, we have:

Proposition 4.3.1: *For $\alpha, \beta \in \mathfrak{h}$,*

$$E^+(\alpha, z_1)E^-(\beta, z_2) = E^-(\beta, z_2)E^+(\alpha, z_1)(1 - z_2/z_1)^{\langle \alpha, \beta \rangle} \tag{4.3.1}$$

in the formal power series algebra

$$(\mathrm{End}\, V_{\mathfrak{h}})[[z_1^{-1}, z_2]] \subset (\mathrm{End}\, V_{\mathfrak{h}})\{z_1, z_2\}.$$

[Recall that $(1 - z_2/z_1)^{\langle \alpha, \beta \rangle}$ denotes the binomial series.] In particular,

$$E^+(\alpha, z_1)E^-(\beta, z_2) = E^-(\beta, z_2)E^+(\alpha, z_1) \quad \text{if} \quad \langle \alpha, \beta \rangle = 0.$$

Proof: We have

$$\left[\sum_{m\in\mathbb{Z}_+}\frac{\alpha(m)z_1^{-m}}{m}, \sum_{n\in-\mathbb{Z}_+}\frac{\beta(n)z_2^{-n}}{n}\right]$$

$$= -\langle\alpha,\beta\rangle\sum_{n\in\mathbb{Z}_+}\frac{(z_2/z_1)^n}{n}$$

$$= \langle\alpha,\beta\rangle\log(1 - z_2/z_1).$$

Now apply (3.4.7). ∎

Thus from (4.1.30) and (4.2.15) we conclude:

Proposition 4.3.2: *For* $\alpha, \beta \in \mathfrak{h}$,

$$:X(\alpha, z_1)X(\beta, z_2): = :X(\beta, z_2)X(\alpha, z_1): \tag{4.3.2}$$

$$X(\alpha, z_1)X(\beta, z_2) = :X(\alpha, z_1)X(\beta, z_2):(z_2/z_1)^{-\langle\alpha,\beta\rangle/2}(1 - z_2/z_1)^{\langle\alpha,\beta\rangle}. \tag{4.3.3}$$

Here is the main bracket computation for the construction of $\tilde{\mathfrak{a}}$ and $\tilde{\mathfrak{a}}[\theta_1]$ [recall the notation (4.1.26), (4.1.28), (4.2.1)]:

Proposition 4.3.3: *For* $\alpha, \beta \in \mathfrak{h}$,

$[X(\alpha, z_1), X(\beta, z_2)]$

$$= \begin{cases} 0 & \text{if } \langle\alpha,\beta\rangle \in 2\mathbb{N} \\ \alpha(z_2)\sum_{n\in\mathbb{Z}}(z_1/z_2)^{n+\alpha} & \\ \quad - \sum_{n\in\mathbb{Z}}(n + \alpha(0))(z_1/z_2)^{n+\alpha} & \text{if } \langle\alpha,\alpha\rangle = 2 \text{ and } \beta = -\alpha \end{cases} \tag{4.3.4}$$

in (End V)$\{z_1, z_2\}$.

Proof: By the last proposition,

$$\begin{aligned}[X(\alpha, z_1), X(\beta, z_2)] &= X(\alpha, z_1)X(\beta, z_2) - X(\beta, z_2)X(\alpha, z_1) \\ &= :X(\alpha, z_1)X(\beta, z_2):\{(z_2/z_1)^{-\langle\alpha,\beta\rangle/2}(1 - z_2/z_1)^{\langle\alpha,\beta\rangle} \\ &\quad - (z_1/z_2)^{-\langle\alpha,\beta\rangle/2}(1 - z_1/z_2)^{\langle\alpha,\beta\rangle}\}.\end{aligned} \tag{4.3.5}$$

If $\langle\alpha,\beta\rangle \in 2\mathbb{N}$, the expression in braces is 0, establishing the first assertion.

Suppose that $\langle\alpha, \alpha\rangle = 2$ and $\beta = -\alpha$. Then the expression in braces equals

$$(z_2/z_1)(1 - z_2/z_1)^{-2} - (z_1/z_2)(1 - z_1/z_2)^{-2} = \sum_{n \in \mathbb{Z}} n(z_2/z_1)^n$$
$$= (D\delta)(z_2/z_1)$$

[see (2.1.22), (2.2.14)], and (4.3.5) becomes

$$(z_2/z_1)^{-\alpha}X(z_1, z_2)(D\delta)(z_2/z_1),$$

where

$$X(z_1, z_2) = {:}e^{D^{-1}(\alpha(z_1)-\alpha(0))-D^{-1}(\alpha(z_2)-\alpha(0))}{:}$$
$$= E^-(-\alpha, z_1)E^-(\alpha, z_2)E^+(-\alpha, z_1)E^+(\alpha, z_2).$$

By (4.2.21) and Proposition 2.2.4(b) [which is applicable because $(z_2/z_1)^{-\alpha}$ has been factored out], we obtain

$$(z_2/z_1)^{-\alpha}(D\delta)(z_2/z_1) - (z_2/z_1)^{-\alpha}(-\alpha(z_2) + \alpha(0))\delta(z_2/z_1)$$
$$= \sum_{n \in \mathbb{Z}} n(z_2/z_1)^{n-\alpha} + \alpha(z_2) \sum_{n \in \mathbb{Z}} (z_2/z_1)^{n-\alpha} - \alpha(0) \sum_{n \in \mathbb{Z}} (z_2/z_1)^{n-\alpha}$$

[cf. (4.2.19), (4.2.20)], which gives (4.3.4). ∎

Remark 4.3.4: Comments exactly analogous to those in Remark 3.4.5 hold here. In particular, $[X(\alpha, z_1), X(\beta, z_2)]$ could in principle be calculated only for those $\alpha, \beta \in \mathfrak{h}$ such that $\langle\alpha, \beta\rangle \in 2\mathbb{Z}$. Later we shall introduce certain non-commuting objects which allow the computation for $\langle\alpha, \beta\rangle \in \mathbb{Z}$.

In preparation for constructing $\hat{\mathfrak{a}}$ and $\hat{\mathfrak{a}}[\theta_1]$ by vertex operators, we specialize at this point to the case $\mathfrak{h} = \mathbb{F}\alpha_1$ of (3.1.13), so that

$$S(\hat{\mathfrak{h}}_{\bar{\mathbb{Z}}}) = \mathbb{F}[\alpha_1(-1), \alpha_1(-2), \ldots], \tag{4.3.6}$$

with

$$\deg 1 = \tfrac{1}{24} \tag{4.3.7}$$

[cf. (3.5.1), (1.9.51)]. Define the additive subgroups

$$Q = \mathbb{Z}\alpha_1, \qquad P = \tfrac{1}{2}\mathbb{Z}\alpha_1, \qquad L = \tfrac{1}{4}\mathbb{Z}\alpha_1 \tag{4.3.8}$$

of $\mathfrak{h}$. We have the disjoint unions

$$P = Q \cup (Q + \tfrac{1}{2}\alpha_1)$$
$$L = P \cup (P + \tfrac{1}{4}\alpha_1). \tag{4.3.9}$$

Remark 4.3.5: We have $\langle P, Q\rangle = \mathbb{Z}$, and in fact,

$$P = \{\alpha \in \mathfrak{h} \mid \langle \alpha, Q\rangle \subset \mathbb{Z}\}.$$

In the context of Remark 3.1.1, Q is the root lattice of $\mathfrak{a}$ and P is the weight lattice.

Recall the notation $\mathbb{F}[M]$, V_M [(4.1.44), (4.1.45)] for a subset M of $\mathfrak{h}$. From (4.1.37) and (4.3.7), we see that the spaces V_Q,

$$V_P = V_Q \oplus V_{Q+\alpha_1/2} \tag{4.3.10}$$

and

$$V_L = V_Q \oplus V_{Q+\alpha_1/4} \oplus V_{Q+\alpha_1/2} \oplus V_{Q-\alpha_1/4} \tag{4.3.11}$$

are $\mathbb{Q}$-graded.

Now it follows from the definitions (4.1.31), (4.1.34) that $x_{\pm\alpha_1}(n)$ preserves each of the four spaces V_Q, $V_{Q+\alpha_1/2}$, $V_{Q\pm\alpha_1/4}$ for all $n \in \mathbb{F}$. Since

$$\langle \pm\alpha_1, P\rangle = \mathbb{Z} \tag{4.3.12}$$

$$\langle \pm\alpha_1, P + \tfrac{1}{4}\alpha_1\rangle = \mathbb{Z} + \tfrac{1}{2} \tag{4.3.13}$$

(cf. Remark 4.3.5), we see that $x_{\pm\alpha_1}(n)\big|_{V_P} = 0$ unless $n \in \mathbb{Z}$, while $x_{\pm\alpha_1}(n)\big|_{V_{P+\alpha_1/4}} = 0$ unless $n \in \mathbb{Z} + 1/2$ [recall (4.1.28)]. That is, on V_P,

$$X(\pm\alpha_1, z) = \sum_{n\in\mathbb{Z}} x_{\pm\alpha_1}(n) z^{-n} \tag{4.3.14}$$

and on $V_{P+\alpha_1/4}$,

$$X(\pm\alpha_1, z) = \sum_{n\in\mathbb{Z}+1/2} x_{\pm\alpha_1}(n) z^{-n}. \tag{4.3.15}$$

Correspondingly, the assertion of Proposition 4.3.3 simplifies as follows: Restricted to V_P,

$$[X(\alpha_1, z_1), X(-\alpha_1, z_2)] = \alpha_1(z_2)\delta(z_1/z_2) - (D\delta)(z_1/z_2), \tag{4.3.16}$$

and restricted to $V_{P+\alpha_1/4}$,

$$\begin{aligned}[X(\alpha_1, z_1), X(-\alpha_1, z_2)] &= \alpha_1(z_2)(z_1/z_2)^{1/2}\delta(z_1/z_2)\\ &\quad - D_{z_1}((z_1/z_2)^{1/2}\delta(z_1/z_2)).\end{aligned} \tag{4.3.17}$$

In terms of the component operators, Proposition 4.3.3 asserts that on V_P (resp., $V_{P+\alpha_1/4}$),

$$\begin{aligned}[x_{\pm\alpha_1}(m), x_{\pm\alpha_1}(n)] &= 0\\ [x_{\alpha_1}(m), x_{-\alpha_1}(n)] &= \alpha_1(m+n) + m\delta_{m+n,0}\end{aligned} \tag{4.3.18}$$

for $m, n \in \mathbb{Z}$ (resp., $\mathbb{Z} + 1/2$) [cf. (3.4.11)].

4.4. Irreducible Representations of $\mathfrak{sl}(2)\hat{}$ and $\mathfrak{sl}(2)\hat{}[\theta_1]$

Continuing in the setting introduced at the end of the last section, we have:

Theorem 4.4.1: *(a) The linear map*

$$\pi: \tilde{\mathfrak{a}} \to \operatorname{End} V_P \tag{4.4.1}$$

determined by

$$\pi: c \mapsto 1$$
$$\pi: d \mapsto d$$
$$\pi: \alpha_1 \otimes t^n \mapsto \alpha_1(n) \quad \text{for} \quad n \in \mathbb{Z}$$
$$\pi: x_{\pm\alpha_1} \otimes t^n \mapsto x_{\pm\alpha_1}(n) \quad \text{for} \quad n \in \mathbb{Z},$$

the last correspondence being equivalent to

$$\pi: x_{\pm\alpha_1}(z) \mapsto X(\pm\alpha_1, z)$$

[*see (3.1.15)*], *is a representation of* $\tilde{\mathfrak{a}}$ *on* V_P.
(b) The linear map

$$\pi: \tilde{\mathfrak{a}}[\theta_1] \to \operatorname{End} V_{P+\alpha_1/4} \tag{4.4.2}$$

determined by

$$\pi: c \mapsto 1$$
$$\pi: d \mapsto d$$
$$\pi: \alpha_1 \otimes t^n \mapsto \alpha_1(n) \quad \text{for} \quad n \in \mathbb{Z}$$
$$\pi: x_{\pm\alpha_1} \otimes t^n \mapsto x_{\pm\alpha_1}(n) \quad \text{for} \quad n \in \mathbb{Z} + 1/2,$$

or equivalently in the last case,

$$\pi: x_{\pm\alpha_1}(z) \mapsto X(\pm\alpha_1, z),$$

is a representation of $\tilde{\mathfrak{a}}[\theta_1]$ *on* $V_{P+\alpha_1/4}$.

Proof: Compare (3.1.17)–(3.1.21) with (4.1.35), (4.1.41), (4.3.4), (4.3.16) and (4.3.17). ∎

Note that as an $\tilde{\mathfrak{a}}$- (resp., $\tilde{\mathfrak{a}}[\theta_1]$-) module, V_P (resp., $V_{P+\alpha_1/4}$) breaks up as

$$V_P = V_Q \oplus V_{Q+\alpha_1/2}, \qquad V_{P+\alpha_1/4} = V_{Q+\alpha_1/4} \oplus V_{Q-\alpha_1/4} \tag{4.4.3}$$

[cf. (4.3.10), (4.3.11)].

Proposition 4.4.2: *The* $\hat{\mathfrak{a}}$- (resp., $\hat{\mathfrak{a}}[\theta_1]$-) *modules* V_Q, $V_{Q+\alpha_1/2}$ (resp., $V_{Q+\alpha_1/4}$, $V_{Q-\alpha_1/4}$) *are irreducible and inequivalent.*

Proof: Let W be a nonzero submodule of $V_{Q+k\alpha_1/4}$, $k = 0, 1, 2$ or 3. Since $\hat{\mathfrak{a}}$ (resp., $\hat{\mathfrak{a}}[\theta_1]$) contains the Heisenberg algebra $\hat{\mathfrak{h}}_{\mathbb{Z}}$, Theorem 1.7.3 implies that

$$W = S(\hat{\mathfrak{h}}_{\mathbb{Z}}^-) \otimes \Omega$$

for some nonzero subspace Ω of $\mathbb{F}[Q + k\alpha_1/4] = \coprod_{\alpha \in Q+k\alpha_1/4} \mathbb{F}e^{\alpha}$. Since W is invariant under $\alpha_1(0)$, for which the spaces $\mathbb{F}e^{\alpha}$ are eigenspaces with distinct eigenvalues, we must have

$$W = S(\hat{\mathfrak{h}}_{\mathbb{Z}}^-) \otimes \mathbb{F}[M] = V_M$$

for some nonempty subset M of $Q + k\alpha_1/4$. Finally, the equation

$$E^-(\alpha, z)X(\alpha, z)E^+(\alpha, z)z^{-\alpha-\langle\alpha,\alpha\rangle/2} = e^{\alpha},$$

which follows from (4.1.13), (4.1.14) and (4.1.31), shows that the $\hat{\mathfrak{a}}$- (resp., $\hat{\mathfrak{a}}[\theta_1]$-) module W is invariant under $e^{\pm\alpha_1}$, so that M must be a union of cosets of $Q = \mathbb{Z}\alpha_1$. Hence $M = Q + k\alpha_1/4$ and $W = V_{Q+k\alpha_1/4}$, proving the irreducibility.

The inequivalence follows from the fact that the modules $V_{Q+k\alpha_1/4}$ have disjoint $\alpha_1(0)$-eigenvalues. ∎

4.5. Isomorphism of Two Constructions

The twisted vertex operator module

$$V = \mathbb{F}[\alpha_1(-\tfrac{1}{2}), \alpha_1(-\tfrac{3}{2}), \ldots] \tag{4.5.1}$$

for $\tilde{\mathfrak{a}}[\theta_2]$ [see (3.5.1) and Theorem 3.5.1] and the untwisted vertex operator modules

$$V_{Q\pm\alpha_1/4} = \mathbb{F}[\alpha_1(-1), \alpha_1(-2), \ldots] \otimes \mathbb{F}[Q \pm \tfrac{1}{4}\alpha_1] \tag{4.5.2}$$

for $\tilde{\mathfrak{a}}[\theta_1]$ [see (4.3.6), Theorem 4.4.1(b) and Proposition 4.4.2] appear quite different, although certainly analogous. But since $\tilde{\mathfrak{a}}[\theta_1]$ and $\tilde{\mathfrak{a}}[\theta_2]$ are graded-isomorphic by the map σ of (3.1.46), we can establish a grading-preserving isomorphism between the spaces (4.5.1) and (4.5.2).

Remark 4.5.1: If we had not incorporated the canonical grading shifts (3.5.2) and (4.3.7) (recall Section 1.9 and Remark 1.10.1), the two spaces would have been graded-isomorphic only up to a grading shift.

Theorem 4.5.2: *There is a unique grading-preserving linear isomorphism*

$$\sigma: V = S(\hat{\mathfrak{h}}^-_{\mathbb{Z}+1/2}) \to V_{Q+\alpha_1/4} = S(\hat{\mathfrak{h}}^-_{\mathbb{Z}}) \otimes \coprod_{\alpha \in Q+\alpha_1/4} \mathbb{F}e^\alpha$$

such that

(i) $\sigma: 1 \mapsto 1 \otimes e^{\alpha_1/4}$

(ii) $\sigma \circ \pi_+(x) \circ \sigma^{-1} = \pi(\sigma x) \quad \text{for} \quad x \in \tilde{\mathfrak{a}}[\theta_2].$

Proof: We use the following standard Lie-theoretic argument. Consider the subalgebra [cf. (3.1.28)]

$$\mathfrak{b} = \coprod_{n \in \mathbb{N}+1/2} \mathbb{F}\alpha_1 \otimes t^n \oplus \coprod_{n \in \mathbb{N}} \mathbb{F}x^+_{\alpha_1} \otimes t^n \oplus \coprod_{n \in \mathbb{N}+1/2} \mathbb{F}x^-_{\alpha_1} \otimes t^n \oplus \mathbb{F}c \oplus \mathbb{F}d$$

of $\tilde{\mathfrak{a}}[\theta_2]$. Let $\mathbb{F}v_0$ be the one-dimensional $\mathfrak{b}$-module such that

$$(x \otimes t^n) \cdot v_0 = 0 \quad \text{for} \quad x \otimes t^n \in \mathfrak{b},\ n > 0$$

$$(x^+_{\alpha_1} \otimes t^0) \cdot v_0 = \tfrac{1}{2}v_0$$

$$c \cdot v_0 = v_0$$

$$d \cdot v_0 = -\tfrac{1}{48}v_0.$$

Form the induced $\tilde{\mathfrak{a}}[\theta_2]$-module

$$M = U(\tilde{\mathfrak{a}}[\theta_2]) \otimes_{U(\mathfrak{b})} \mathbb{F}v_0$$

[cf. (1.8.6)], which has a canonical $\mathbb{F}$-grading

$$M = \coprod_{n \in \mathbb{F}} M_n$$

with $1 \otimes v_0 \in M_{-1/48}$. Then M has a unique maximal proper graded submodule N since any such N is contained in the subspace $\coprod_{n<-1/48} M_n$. Make $V_{Q+\alpha_1/4}$ into an $\tilde{\mathfrak{a}}[\theta_2]$-module by means of the composition

$$\tilde{\mathfrak{a}}[\theta_2] \xrightarrow{\sigma} \tilde{\mathfrak{a}}[\theta_1] \xrightarrow{\pi} \operatorname{End} V_{Q+\alpha_1/4}.$$

Now

$$\pi_+(x^+_{\alpha_1} \otimes t^0) \cdot 1 = \tfrac{1}{2}$$

by Theorem 3.5.1, (3.4.20) and (3.2.28);

$$\pi(\sigma(x^+_{\alpha_1} \otimes t^0)) \cdot (1 \otimes e^{\alpha_1/4}) = \tfrac{1}{2}(1 \otimes e^{\alpha_1/4})$$

by (3.1.44), (3.1.46), Theorem 4.4.1(b) and (4.1.26);

$$\pi_+(c) = \pi(\sigma c) = 1;$$

and

$$\pi_+(d) \cdot 1 = -\tfrac{1}{48}, \qquad \pi(\sigma d) \cdot (1 \otimes e^{\alpha_1/4}) = -\tfrac{1}{48}(1 \otimes e^{\alpha_1/4})$$

by (3.5.2), (4.1.37) and (4.3.7). Thus the universal property of the induced module M gives unique graded $\tilde{\mathfrak{a}}[\theta_2]$-module maps

$$\begin{array}{ccc} & M & \\ {\scriptstyle\phi}\swarrow & & \searrow{\scriptstyle\psi} \\ V & & V_{Q+\alpha_1/4} \end{array}$$

such that

$$\phi: 1 \otimes v_0 \mapsto 1$$

$$\psi: 1 \otimes v_0 \mapsto 1 \otimes e^{\alpha_1/4}.$$

Since V and $V_{Q+\alpha_1/4}$ are both irreducible (see Theorem 3.5.1 and Proposition 4.4.2), we have

$$\operatorname{Ker} \phi = \operatorname{Ker} \psi = N.$$

We can now take

$$\sigma = \psi^* \circ \phi^{*-1},$$

with ϕ^*, ψ^* the associated isomorphisms

$$\begin{array}{ccc} & M/N & \\ {\scriptstyle\phi^*}\swarrow & & \searrow{\scriptstyle\psi^*} \\ V & & V_{Q+\alpha_1/4} \end{array} \qquad \blacksquare$$

Remark 4.5.3: Of course, one has a similar result using $Q - \alpha_1/4$ and π_-.

Since the spaces V and $V_{Q+\alpha_1/4}$ are graded-isomorphic, we can equate their graded dimensions (see Section 1.10). From (1.10.5), (1.10.22), (1.10.23) and (4.1.37), we obtain the following nontrivial formula, a classical result of Gauss (cf. [Andrews]):

Corollary 4.5.4: *With the notation (1.10.21) for Dedekind's η-function,*

$$\frac{\eta(q)}{\eta(q^{1/2})} = \frac{\sum_{n \in \mathbb{Z}} q^{(n+1/4)^2}}{\eta(q)}.$$

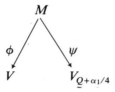

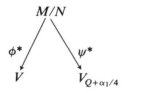

5 Central Extensions

In this chapter we study central extensions of free abelian and finite abelian groups. These extensions are required in order to modify the vertex operators which have provided representations of the affine algebra $\mathfrak{sl}(2)\hat{}$ in Chapters 3 and 4. The modification will be chosen in Chapter 7 in such a way that commutators of the appropriate vertex operators will always have a simple expression, thus giving us representations of other affine algebras. Central extensions will also be used for the structure theory of finite-dimensional simple Lie algebras in Chapter 6, and in later chapters for the study of the Monster.

We present the basic facts about central extensions, 2-cocycles and the second cohomology group in Section 5.1. In Section 5.2 we characterize equivalence classes of central extensions of finite-rank free abelian groups by means of their associated commutator maps. Then in Section 5.3 we introduce what finite group theorists term "extraspecial groups," which can be thought of as finite-group analogues of Heisenberg algebras. We characterize equivalence classes of central extensions of elementary abelian 2-groups by means of their associated "squaring maps," which are quadratic forms, and we establish the relation with the central extensions of Section 5.2. We define and describe certain classes of automorphisms of central extensions in Section 5.4. The technical Proposition 5.4.8 is not needed until Chapter 13. In Section 5.5 we construct certain uniquely determined faithful irreducible modules for a certain class of finite groups, including the extraspecial groups. These modules are finite-group analogues of Fock spaces. We also show that certain automorphisms of such groups lift to automorphisms of the modules. Much of the material in this chapter is standard; see for instance [Gorenstein 1].

5.1. 2-Cocycles

We begin with the relationship between central extensions of groups and second cohomology, restricting our attention to the situation of interest in this work.

An exact sequence of groups

$$1 \to C \to B \xrightarrow{\varphi} A \to 1$$

is called an *extension of A by C*. Sometimes the pair (B, φ) or just the group B is referred to as an extension of A by C. The extension is called a *central extension* if the image of C is contained in the center of B:

$$\operatorname{Im} C \subset \operatorname{Cent} B.$$

Two extensions (B_1, φ_1) and (B_2, φ_2) of A by C are said to be *equivalent* if there is an isomorphism $\psi: B_1 \to B_2$ such that the diagram

$$\begin{array}{ccccccccc} 1 & \longrightarrow & C & \longrightarrow & B_1 & \xrightarrow{\varphi_1} & A & \longrightarrow & 1 \\ & & \| & & \downarrow{\scriptstyle\psi} & & \| & & \\ 1 & \longrightarrow & C & \longrightarrow & B_2 & \xrightarrow{\varphi_2} & A & \longrightarrow & 1 \end{array} \tag{5.1.1}$$

is commutative.

Until further notice, A will be an abelian group written additively, s will be a positive integer, and

$$\langle \kappa \rangle = \langle \kappa \mid \kappa^s = 1 \rangle \tag{5.1.2}$$

will denote the s-element cyclic group generated by an element κ subject to the single relation $\kappa^s = 1$. We shall restrict our attention to central extensions B of A by $\langle \kappa \rangle$ such that $\langle \kappa \rangle$ is a (central) subgroup of B:

$$1 \to \langle \kappa \rangle \hookrightarrow B \xrightarrow{\varphi} A \to 1. \tag{5.1.3}$$

(We shall use the notation $\to 1$ rather than $\to 0$ at the end even though A is written additively.)

A map

$$\varepsilon_0 : A \times A \to \mathbb{Z}/s\mathbb{Z} \tag{5.1.4}$$

is called a 2-*cocycle* if

$$\begin{aligned} &\varepsilon_0(\alpha, \beta) + \varepsilon_0(\alpha + \beta, \gamma) = \varepsilon_0(\beta, \gamma) + \varepsilon_0(\alpha, \beta + \gamma) \\ &\quad \text{for} \quad \alpha, \beta, \gamma \in A \end{aligned} \tag{5.1.5}$$

and a 2-*coboundary* if

$$\varepsilon_0(\alpha, \beta) = \eta(\alpha + \beta) - \eta(\alpha) - \eta(\beta) \quad \text{for} \quad \alpha, \beta \in A \tag{5.1.6}$$

for some map $\eta : A \to \mathbb{Z}/s\mathbb{Z}$. We denote by $Z^2(A, \mathbb{Z}/s\mathbb{Z})$ the group of 2-cocycles and by $B^2(A, \mathbb{Z}/s\mathbb{Z})$ the subgroup of 2-coboundaries, and we set

$$H^2(A, \mathbb{Z}/s\mathbb{Z}) = Z^2(A, \mathbb{Z}/s\mathbb{Z})/B^2(A, \mathbb{Z}/s\mathbb{Z}), \tag{5.1.7}$$

the *second cohomology group*—the group of 2-*cohomology classes*. Two 2-cocycles in the same class are said to be *cohomologous*.

Remark 5.1.1: Any $\mathbb{Z}$-bilinear map from $A \times A$ to $\mathbb{Z}/s\mathbb{Z}$ is clearly a 2-cocycle. This provides a practical method for constructing 2-cocycles.

Consider the central extension (5.1.3), choose a section $e : A \to B$ (i.e., a map e such that $\varphi \circ e = 1$) and set

$$e_\alpha = e(\alpha) \quad \text{for} \quad \alpha \in A. \tag{5.1.8}$$

Then

$$B = \{e_\alpha \kappa^m \mid \alpha \in A, m \in \mathbb{Z}/s\mathbb{Z}\},$$

and the map ε_0 defined by

$$e_\alpha e_\beta = e_{\alpha+\beta} \kappa^{\varepsilon_0(\alpha, \beta)} \quad \text{for} \quad \alpha, \beta \in A \tag{5.1.9}$$

is a 2-cocycle. Conversely, let ε_0 be a 2-cocycle. Define a binary operation $\cdot$ on the set $B = \langle \kappa \mid \kappa^s = 1 \rangle \times A$ by

$$(\kappa^p, \alpha) \cdot (\kappa^q, \beta) = (\kappa^{p+q+\varepsilon_0(\alpha, \beta)}, \alpha + \beta).$$

Then B becomes a group with identity element

$$1 = (\kappa^{-\varepsilon_0(0,0)}, 0)$$

and we have a central extension (5.1.3) with

$$\varphi : (\kappa^p, \alpha) \mapsto \alpha, \qquad p \in \mathbb{Z}/s\mathbb{Z}, \alpha \in A,$$

where we identify κ with $\kappa = (\kappa^{1-\varepsilon_0(0,0)}, 0)$. In addition, ε_0 is the cocycle associated to the section

$$e: \alpha \mapsto (1, \alpha), \qquad \alpha \in A.$$

The following proposition is well known and easy to prove:

Proposition 5.1.2: *In the correspondences above between central extensions (5.1.3) and 2-cocycles:*

(a) different sections e determine cohomologous cocycles,
(b) central extensions with given sections inducing cohomologous cocycles are equivalent,
(c) we obtain a bijection between the set of equivalence classes of central extensions (5.1.3) and $H^2(A, \mathbb{Z}/s\mathbb{Z})$.

Remark 5.1.3: Note that the section (5.1.8) satisfies the normalization condition

$$e_0 = 1 \tag{5.1.10}$$

if and only if the corresponding cocycle ε_0 satisfies the conditions

$$\varepsilon_0(\alpha, 0) = \varepsilon_0(0, \alpha) = 0 \quad \text{for} \quad \alpha \in A. \tag{5.1.11}$$

These conditions hold in particular if ε_0 is bilinear (cf. Remark 5.1.1).

Remark 5.1.4: The results above hold even if A is nonabelian and the central subgroup is noncyclic; however, we shall always be in the above situation.

5.2. Commutator Maps

Consider a central extension

$$1 \to \langle \kappa \mid \kappa^s = 1 \rangle \hookrightarrow \hat{A} \xrightarrow{-} A \to 1. \tag{5.2.1}$$

Let $\alpha, \beta \in A$ and choose $a, b \in \hat{A}$ so that $\bar{a} = \alpha$, $\bar{b} = \beta$. Then the commutator $aba^{-1}b^{-1}$ lies in $\langle \kappa \rangle$ and depends only on the cosets of a and b modulo $\langle \kappa \rangle$, i.e., depends only on α and β. Hence the central extension (5.2.1) determines a map

$$c_0: A \times A \to \mathbb{Z}/s\mathbb{Z}, \tag{5.2.2}$$

which we shall call the *associated commutator map*, by the condition

$$aba^{-1}b^{-1} = \kappa^{c_0(\alpha,\beta)} \quad \text{for} \quad a, b \in \hat{A},\ \alpha = \bar{a},\ \beta = \bar{b}. \tag{5.2.3}$$

Let us denote group commutators by

$$(a, b) = aba^{-1}b^{-1}. \tag{5.2.4}$$

The easily verified general commutator formula

$$(ab, c) = ({}^{a}(b, c))(a, c),$$

where the left superscript a denotes left conjugation by a, simplifies for a group in which commutators are central, such as $\hat{A}$, to

$$(ab, c) = (a, c)(b, c) \quad \text{for} \quad a, b, c \in \hat{A}.$$

Similarly, we have

$$(a, bc) = (a, b)(a, c).$$

We also have $(a, a) = 1$ for $a \in \hat{A}$. These equalities translate to the following properties of c_0:

$$c_0(\alpha + \beta, \gamma) = c_0(\alpha, \gamma) + c_0(\beta, \gamma) \tag{5.2.5}$$

$$c_0(\alpha, \beta + \gamma) = c_0(\alpha, \beta) + c_0(\alpha, \gamma) \tag{5.2.6}$$

$$c_0(\alpha, \alpha) = 0, \tag{5.2.7}$$

which by polarization imply

$$c_0(\alpha, \beta) = -c_0(\beta, \alpha) \tag{5.2.8}$$

for $\alpha, \beta, \gamma \in A$. In other words, c_0 is an alternating $\mathbb{Z}$-bilinear map.

Remark 5.2.1: For a subgroup B of A, let $\hat{B}$ denote the subgroup of $\hat{A}$ which is the inverse image of B under the map $^{-}$ of (5.2.1):

$$1 \to \langle\kappa\rangle \hookrightarrow \hat{B} \to B \to 1. \tag{5.2.9}$$

Then $\hat{B}$ is abelian if and only if

$$c_0(B, B) = 0. \tag{5.2.10}$$

Let R be the radical of the form c_0—the subgroup

$$R = \{\alpha \in A \mid c_0(\alpha, A) = 0\} \tag{5.2.11}$$

of A. Then $\hat{R}$ is the center of $\hat{A}$:

$$\hat{R} = \operatorname{Cent} \hat{A}. \tag{5.2.12}$$

In particular, $\operatorname{Cent} \hat{A} = \langle \kappa \rangle$ if and only if $R = 0$.

Remark 5.2.2: If

$$1 \to \langle \kappa \mid \kappa^s = 1 \rangle \hookrightarrow \hat{A} \twoheadrightarrow A \to 1$$

is a central extension with associated commutator map c_0 and determined by the cocycle ε_0, then c_0 and ε_0 are related by the formula

$$c_0(\alpha, \beta) = \varepsilon_0(\alpha, \beta) - \varepsilon_0(\beta, \alpha), \qquad \alpha, \beta \in A. \tag{5.2.13}$$

In fact, given a section $e: \alpha \mapsto e_\alpha$ ($\alpha \in A$) corresponding to ε_0, we have

$$\begin{aligned} \kappa^{c_0(\alpha,\beta)} &= e_\alpha e_\beta (e_\beta e_\alpha)^{-1} = e_{\alpha+\beta}\kappa^{\varepsilon_0(\alpha,\beta)}(e_{\alpha+\beta}\kappa^{\varepsilon_0(\beta,\alpha)})^{-1} \\ &= \kappa^{\varepsilon_0(\alpha,\beta)-\varepsilon_0(\beta,\alpha)} \end{aligned}$$

for $\alpha, \beta \in A$.

Under a new assumption, we now establish a bijection different from that of Proposition 5.1.2(c):

Proposition 5.2.3: *Let A be a free abelian group of finite rank. The association of c_0 to the central extension $\hat{A}$ defines a bijection between the set of alternating $\mathbb{Z}$-bilinear maps*

$$c_0: A \times A \to \mathbb{Z}/s\mathbb{Z}$$

and the set of equivalence classes of central extensions

$$1 \to \langle \kappa \mid \kappa^s = 1 \rangle \to \hat{A} \twoheadrightarrow A \to 1$$

of A by $\langle \kappa \mid \kappa^s = 1 \rangle$.

Proof: Clearly, equivalent central extensions determine the same commutator map.

Let

$$c_0: A \times A \to \mathbb{Z}/s\mathbb{Z}$$

be an alternating $\mathbb{Z}$-bilinear map, and let $\{\alpha_1, \ldots, \alpha_n\}$ be a $\mathbb{Z}$-base of A. Let

$$\varepsilon_0: A \times A \to \mathbb{Z}/s\mathbb{Z}$$

be the $\mathbb{Z}$-bilinear map determined by

$$\varepsilon_0(\alpha_i, \alpha_j) = \begin{cases} c_0(\alpha_i, \alpha_j) & \text{if} \quad i > j \\ 0 & \text{if} \quad i \le j. \end{cases} \tag{5.2.14}$$

Then ε_0 is a 2-cocycle (cf. Remark 5.1.1) and

$$\varepsilon_0(\alpha, \beta) - \varepsilon_0(\beta, \alpha) = c_0(\alpha, \beta), \qquad \alpha, \beta \in A$$

since this holds for $\alpha, \beta \in \{\alpha_1, \ldots, \alpha_n\}$ and both sides are bilinear. By Proposition 5.1.2 and Remark 5.2.2, there is a central extension

$$1 \to \langle \kappa \mid \kappa^s = 1 \rangle \to \hat{A} \xrightarrow{-} A \to 1$$

with cocycle ε_0 and associated commutator map c_0.

Finally, let

$$1 \to \langle \kappa \rangle \hookrightarrow B \xrightarrow{\varphi} A \to 1$$

be any central extension with commutator map c_0. Choose $e_{\alpha_k} \in B$ so that $\varphi(e_{\alpha_k}) = \alpha_k$ for $k = 1, \ldots, n$, and define a section $e: A \to B$ by

$$e: \sum_{k=1}^{n} m_k \alpha_k \mapsto e_{\alpha_1}^{m_1} \cdots e_{\alpha_n}^{m_n} \quad \text{for} \quad m_k \in \mathbb{Z}.$$

One easily verifies that ε_0 is the cocycle associated to e and hence (B, φ) is equivalent to $(\hat{A}, {}^{-})$ by Proposition 5.1.2. ∎

5.3. Extraspecial 2-Groups

For a prime p, a *p-group* is a finite group P whose order $|P|$ satisfies

$$|P| = p^n \quad \text{for some} \quad n \in \mathbb{N}. \tag{5.3.1}$$

An *extraspecial p-group* is a p-group P such that

$$\begin{aligned} \operatorname{Cent} P &= (P, P) \simeq \mathbb{Z}/p\mathbb{Z}, \\ P/\operatorname{Cent} P &\simeq (\mathbb{Z}/p\mathbb{Z})^{n-1}, \end{aligned} \tag{5.3.2}$$

the notation (P, P) denoting the commutator subgroup of P. Equivalently (see Remark 5.2.1), an extraspecial p-group is a central extension P of the form

$$1 \to \langle \kappa \mid \kappa^p = 1 \rangle \to P \to E \to 1, \tag{5.3.3}$$

where E is an elementary abelian p-group (equivalently, the additive group of a finite-dimensional vector space over the p-element field $\mathbb{F}_p$), such that the associated commutator map

$$c_0 : E \times E \to \mathbb{Z}/p\mathbb{Z} \tag{5.3.4}$$

is nonsingular as an $\mathbb{F}_p$-bilinear vector space map.

In this work, we shall be interested in the case $p = 2$. Extraspecial 2-groups and closely related central extensions will arise as finite quotients of central extensions of free abelian groups.

We shall characterize and construct extraspecial 2-groups by an analogue of Proposition 5.2.3. For this, we shall need the notion of quadratic form over the 2-element field $\mathbb{F}_2$: Given an $\mathbb{F}_2$-vector space E, a *quadratic form* on E is a function

$$q : E \to \mathbb{F}_2 \tag{5.3.5}$$

such that the function

$$\begin{aligned} b : E \times E &\to \mathbb{F}_2 \\ (x, y) &\mapsto q(x + y) - q(x) - q(y), \end{aligned} \tag{5.3.6}$$

called the *associated form*, is a bilinear form. (The quadratic homogeneity condition $q(rx) = r^2 q(x)$ for $r \in \mathbb{F}_2$, $x \in E$ is not put into the definition because it is a consequence.) In this case, the associated form is alternating:

$$b(x, x) = 0 \quad \text{for} \quad x \in E. \tag{5.3.7}$$

The *radical* of the quadratic form q is the radical of b, and q is said to be *nonsingular* if its radical is 0. We shall use:

Proposition 5.3.1: *Let E be a finite-dimensional $\mathbb{F}_2$-vector space with a quadratic form q. There are bilinear forms*

$$\varepsilon_0 : E \times E \to \mathbb{F}_2 \tag{5.3.8}$$

whose diagonals agree with q, i.e., such that

$$q(x) = \varepsilon_0(x, x) \quad \text{for} \quad x \in E. \tag{5.3.9}$$

In particular,

$$b(x, y) = \varepsilon_0(x, y) - \varepsilon_0(y, x) \quad \text{for} \quad x, y \in E, \tag{5.3.10}$$

b being the associated bilinear form. Given an $\mathbb{F}_2$-basis $\{x_1, \ldots, x_n\}$ of E, ε_0 is uniquely determined by the values $\varepsilon_0(x_i, x_j)$ for $i < j$, which may be prescribed arbitrarily.

Proof: Given arbitrary values $\varepsilon_0(x_i, x_j)$ for $i < j$, ε_0 must satisfy the conditions

$$\varepsilon_0(x_i, x_i) = q(x_i) \quad \text{for all} \quad i$$

$$\varepsilon_0(x_j, x_i) = \varepsilon_0(x_i, x_j) + b(x_i, x_j) \quad \text{for} \quad i < j,$$

as we see by expanding $\varepsilon_0(x_i + x_j, x_i + x_j)$. These conditions uniquely determine a bilinear form ε_0, and (5.3.10) holds because both sides are bilinear and the formula holds for $x, y \in \{x_1, \ldots, x_n\}$. But then q and

$$x \mapsto \varepsilon_0(x, x) \quad \text{for} \quad x \in E$$

are two quadratic forms on E which agree on all the x_i and which have the same associated bilinear form. Hence they are equal. ∎

Remark 5.3.2: Similar considerations show that if b is an alternating bilinear form on a finite-dimensional $\mathbb{F}_2$-vector space E, then b is the associated form of some quadratic form q on E. It is easy to see that the set of all such quadratic forms is the set $\{q + \eta\}$ where η ranges through the set of linear forms on E.

Suppose that

$$1 \to \langle \kappa \mid \kappa^2 = 1 \rangle \hookrightarrow \hat{E} \twoheadrightarrow E \to 1 \tag{5.3.11}$$

is a central extension, with E an elementary abelian 2-group. Then

$$2x = 0 \quad \text{for} \quad x \in E. \tag{5.3.12}$$

In addition to the associated commutator map c_0, we have the *associated squaring map*

$$s: E \to \mathbb{Z}/2\mathbb{Z} = \mathbb{F}_2, \tag{5.3.13}$$

defined by

$$a^2 = \kappa^{s(\bar{a})} \quad \text{for} \quad a \in \hat{E}. \tag{5.3.14}$$

Note that s is well defined. Moreover, s is a quadratic form with associated bilinear form c_0:

$$c_0(x, y) = s(x + y) - s(x) - s(y) \quad \text{for} \quad x, y \in E. \tag{5.3.15}$$

The group $\hat{E}$ is an extraspecial 2-group if and only if s is nonsingular.

Proposition 5.3.3: *Let E be an elementary abelian 2-group. The association of s to the central extension $\hat{E}$ defines a bijection between the set of* *quadratic forms*

$$s: E \to \mathbb{Z}/2\mathbb{Z} = \mathbb{F}_2$$

and the set of equivalence classes of central extensions

$$1 \to \langle \kappa \mid \kappa^2 = 1 \rangle \to \hat{E} \to E \to 1,$$

and $\hat{E}$ is extraspecial if and only if s is nonsingular.

Proof: First note that equivalent central extensions determine the same squaring map. Let $s: E \to \mathbb{F}_2$ be a quadratic form. Choose an $\mathbb{F}_2$-basis $\{x_1, \ldots, x_n\}$ of E, and let

$$\varepsilon_0: E \times E \to \mathbb{F}_2$$

be the unique $\mathbb{F}_2$-bilinear map such that

$$s(x) = \varepsilon_0(x, x) \quad \text{for} \quad x \in E, \tag{5.3.16}$$

$$\varepsilon_0(x_i, x_j) = 0 \quad \text{for} \quad i < j \tag{5.3.17}$$

(see Proposition 5.3.1). Then ε_0 is a 2-cocycle (cf. Remark 5.1.1) and so there is a central extension

$$1 \to \langle \kappa \mid \kappa^2 = 1 \rangle \to \hat{E} \xrightarrow{-} E \to 1$$

with cocycle ε_0 and hence with squaring map s.

For the uniqueness, let

$$1 \to \langle \kappa \rangle \hookrightarrow B \xrightarrow{\varphi} E \to 1$$

be any central extension with squaring map s. Then the associated commutator map is the associated bilinear form of s. Fix $e_{x_k} \in B$ such that $\varphi(e_{x_k}) = x_k$ for $k = 1, \ldots, n$, and define a section $e: E \to B$ by

$$e: \sum_{k=1}^{n} m_k x_k \to e_{x_1}^{m_1} \cdots e_{x_n}^{m_n} \quad \text{for} \quad m_k \in \{0, 1\}.$$

Then using (5.3.10), (5.3.16) and (5.3.17), we see that ε_0 is the corresponding cocycle and so (B, φ) is equivalent to $(\hat{E}, {}^{-})$. ∎

It is not surprising that the central extensions $\hat{E}$ described in Proposition 5.3.3 are closely related to central extensions of free abelian groups. The following result is straightforward to prove:

Proposition 5.3.4: *Let*

$$1 \to \langle \kappa \mid \kappa^2 = 1 \rangle \hookrightarrow \hat{A} \twoheadrightarrow A \to 1$$

be a central extension of a free abelian group of finite rank and let c_0 be the associated commutator map. Then c_0 induces an alternating $\mathbb{F}_2$-bilinear form

$$c_1 : \breve{A} \times \breve{A} \to \mathbb{Z}/2\mathbb{Z} = \mathbb{F}_2, \tag{5.3.18}$$

where $\breve{A} = A/2A$, an elementary abelian 2-group. Let

$$s_1 : \breve{A} \to \mathbb{F}_2 \tag{5.3.19}$$

be a quadratic form with associated form c_1 (cf. Remark 5.3.2) and let

$$s_0 : A \to \mathbb{Z}/2\mathbb{Z} = \mathbb{F}_2 \tag{5.3.20}$$

be the pullback of s_1 to A. Then

$$K = \{a^2 \kappa^{s_0(\bar{a})} \mid a \in \hat{A}\} \tag{5.3.21}$$

is a central subgroup of $\hat{A}$ such that

$$\widehat{2A} = \langle \kappa \rangle \times K \tag{5.3.22}$$

[*using the notation of (5.2.9)*]; *the canonical map*

$$\hat{A} \to \breve{A}$$

has kernel $\widehat{2A}$; and the canonical maps

$$1 \to \langle \kappa \rangle \to \hat{A}/K \to \breve{A} \to 1 \tag{5.3.23}$$

define a central extension with squaring map s_1.

5.4. Automorphisms of Central Extensions

Let

$$1 \to \langle \kappa \mid \kappa^s = 1 \rangle \hookrightarrow \hat{A} \twoheadrightarrow A \to 1$$

be a central extension of a free abelian group A of finite rank with associated commutator map c_0. Set

$$\operatorname{Aut}(\hat{A}; \kappa) = \{g \in \operatorname{Aut} \hat{A} \mid g\kappa = \kappa\}. \tag{5.4.1}$$

Note that in case $s = 2$, we have

$$\operatorname{Aut}(\hat{A}; \kappa) = \operatorname{Aut} \hat{A} \tag{5.4.2}$$

since $\{1, \kappa\}$ is the full set of elements of finite order in $\hat{A}$. Each $g \in \operatorname{Aut}(\hat{A}; \kappa)$ induces an automorphism $\bar{g}$ of A. Define

$$\operatorname{Aut} c_0 = \{h \in \operatorname{Aut} A \mid c_0(h\alpha, h\beta) = c_0(\alpha, \beta) \quad \text{for} \quad \alpha, \beta \in A\}. \quad (5.4.3)$$

For $\lambda \in \operatorname{Hom}(A, \mathbb{Z}/s\mathbb{Z})$, define

$$\begin{aligned} \lambda^*: \hat{A} &\to \hat{A} \\ a &\mapsto a\kappa^{\lambda(\bar{a})}. \end{aligned} \quad (5.4.4)$$

Then every element of $\operatorname{Aut} c_0$ lifts to $\hat{A}$, and in fact we have:

Proposition 5.4.1: *The sequence*

$$1 \to \operatorname{Hom}(A, \mathbb{Z}/s\mathbb{Z}) \xrightarrow{*} \operatorname{Aut}(\hat{A}; \kappa) \xrightarrow{-} \operatorname{Aut} c_0 \to 1$$

is exact.

Proof: Clearly, $\lambda^* \in \operatorname{Aut}(\hat{A}; \kappa)$ and $\overline{\lambda^*} = 1$ for $\lambda \in \operatorname{Hom}(A, \mathbb{Z}/s\mathbb{Z})$, and if $g \in \operatorname{Aut}(\hat{A}; \kappa)$ with $\bar{g} = 1$, then $g = \lambda^*$ for some λ. Let $g \in \operatorname{Aut}(\hat{A}; \kappa)$. Then

$$\kappa^{c_0(\bar{g}\alpha, \bar{g}\beta)} = (ga, gb) = g(a, b) = \kappa^{c_0(\alpha, \beta)}$$

for all $a, b \in \hat{A}$, $\alpha = \bar{a}$, $\beta = \bar{b}$, so that $\bar{g} \in \operatorname{Aut} c_0$. Conversely, let $h \in \operatorname{Aut} c_0$ and consider the central extension

$$1 \longrightarrow \langle \kappa \mid \kappa^s = 1 \rangle \longrightarrow \hat{A} \xrightarrow{h \circ \bar{}} A \longrightarrow 1.$$

Since this has commutator map c_0, Proposition 5.2.3 gives an automorphism $g: \hat{A} \to \hat{A}$ such that the diagram

$$\begin{array}{ccccccccc} 1 & \longrightarrow & \langle \kappa \mid \kappa^s = 1 \rangle & \longrightarrow & \hat{A} & \xrightarrow{h \circ \bar{}} & A & \longrightarrow & 1 \\ & & \| & & \downarrow g & & \| & & \\ 1 & \longrightarrow & \langle \kappa \mid \kappa^s = 1 \rangle & \longrightarrow & \hat{A} & \xrightarrow{\bar{}} & A & \longrightarrow & 1 \end{array}$$

commutes. Then $g \in \operatorname{Aut}(\hat{A}; \kappa)$ and $\bar{g} = h$. ∎

Remark 5.4.2: If n is the rank of A, then

$$\operatorname{Hom}(A, \mathbb{Z}/s\mathbb{Z}) \simeq (\mathbb{Z}/s\mathbb{Z})^n,$$

as we see by choosing a base of A.

Liftings of the automorphism -1 of A will play a special role. Let $\theta \in \mathrm{Aut}(\hat{A}; \kappa)$ be such that $\bar{\theta} = -1$. Then for all $a \in \hat{A}$,

$$\theta(a) = \mathrm{a}^{-1}\kappa^p$$

where $p \in \mathbb{Z}$. But then

$$\theta^2(a) = \theta(a^{-1})\kappa^p = \theta(a)^{-1}\kappa^p = a,$$

so that

$$\theta^2 = 1. \tag{5.4.5}$$

[Note that Proposition 5.4.1 alone shows only that $\theta^2 = \lambda^*$ for some $\lambda \in \mathrm{Hom}(A, \mathbb{Z}/s\mathbb{Z})$.]

We now assume that $s = 2$. Then θ inverts squares:

$$\theta(a^2) = a^{-2} \quad \text{for} \quad a \in \hat{A}. \tag{5.4.6}$$

The following description of the liftings of -1 is easily proved:

Proposition 5.4.3: *In the notation of Proposition 5.3.4, the map*

$$\begin{aligned} \theta\colon \hat{A} &\to \hat{A} \\ a &\mapsto a^{-1}\kappa^{s_0(\bar{a})} \end{aligned} \tag{5.4.7}$$

lies in $\mathrm{Aut}\,\hat{A}$ *and* $\bar{\theta} = -1$. *Conversely, every element of* $\mathrm{Aut}\,\hat{A}$ *covering* -1 *arises in this way for some quadratic form* s_1 *as in (5.3.19). In particular, these forms parametrize the liftings of* -1.

Remark 5.4.4: Given the quadratic form s_1, the corresponding central subgroup K of $\hat{A}$ [see (5.3.21)] can be described using the associated automorphism θ as follows:

$$\begin{aligned} K &= \{a\theta(a^{-1}) \mid a \in \hat{A}\} = \{a^{-1}\theta(a) \mid a \in \hat{A}\} \\ &= \{\theta(a)a^{-1} \mid a \in \hat{A}\}. \end{aligned} \tag{5.4.8}$$

We have an analogue of Proposition 5.4.1 concerning automorphisms of central extensions of the type described in Proposition 5.3.3. Let

$$1 \to \langle \kappa \mid \kappa^2 = 1 \rangle \hookrightarrow \hat{E} \twoheadrightarrow E \to 1 \tag{5.4.9}$$

be a central extension of an elementary abelian 2-group, and denote by s the associated squaring map. Define

$$\mathrm{Aut}\, s = \{h \in \mathrm{Aut}\, E \mid s(hx) = s(x) \quad \text{for} \quad x \in E\}. \tag{5.4.10}$$

Then

$$\operatorname{Aut} s \subset \operatorname{Aut} c_0, \tag{5.4.11}$$

where c_0 is the associated commutator map; we are using the notation (5.4.3). We also adopt the above notations $\bar{g}$ and λ^*. The lifting theorem analogous to Proposition 5.4.1 states:

Proposition 5.4.5: *The sequence*

$$1 \to \operatorname{Hom}(E, \mathbb{Z}/2\mathbb{Z}) \xrightarrow{*} \operatorname{Aut}(\hat{E}; \kappa) \xrightarrow{-} \operatorname{Aut} s \to 1$$

is exact.

Proof: Imitate the proof of Proposition 5.4.1, using the squaring map and Proposition 5.3.3. ∎

Remark 5.4.6: By analogy with Remark 5.4.2, we have

$$\operatorname{Hom}(E, \mathbb{Z}/2\mathbb{Z}) \simeq E \simeq (\mathbb{Z}/2\mathbb{Z})^n.$$

Remark 5.4.7: If $\hat{E}$ is an extraspecial 2-group, then

$$\operatorname{Aut}(\hat{E}; \kappa) = \operatorname{Aut} \hat{E}. \tag{5.4.12}$$

[cf. (5.4.2)]. Also,

$$\operatorname{Im}(\operatorname{Hom}(E, \mathbb{Z}/2\mathbb{Z})) = \operatorname{Inn} \hat{E}, \tag{5.4.13}$$

the group of inner automorphisms of $\hat{E}$—those automorphisms of the form $a \mapsto gag^{-1}$ for $g \in \hat{E}$. There is also a *natural* isomorphism

$$\operatorname{Hom}(E, \mathbb{Z}/2\mathbb{Z}) \simeq E, \tag{5.4.14}$$

and the exact sequence in Proposition 5.4.5 can be replaced by

$$1 \to \operatorname{Inn} \hat{E} \hookrightarrow \operatorname{Aut} \hat{E} \to \operatorname{Aut} s \to 1 \tag{5.4.15}$$

or by

$$1 \to E \to \operatorname{Aut} \hat{E} \to \operatorname{Aut} s \to 1, \tag{5.4.16}$$

where the image of $x \in E$ is the automorphism

$$a \mapsto a\kappa^{c_0(x, \bar{a})} \tag{5.4.17}$$

of $\hat{E}$, c_0 being the commutator map.

Later, in Chapter 13, we shall need a more technical analogue of Proposition 5.4.1. With $\hat{A}$ and c_0 as at the beginning of this section, consider the case $s = 4$, i.e., a central extension

$$1 \to \langle \kappa \mid \kappa^4 = 1 \rangle \hookrightarrow \hat{A} \rightrightarrows A \to 1 \tag{5.4.18}$$

of a free abelian group A of finite rank. Let B be a subgroup of A containing $4A$ and such that

$$c_0(B, B) = 0. \tag{5.4.19}$$

Let K be a subgroup of $\hat{A}$ such that

$$\bar{K} = B, \qquad K \cap \langle \kappa \rangle = 1, \tag{5.4.20}$$

so that $\hat{B}$ is the abelian group

$$\hat{B} = \langle \kappa \rangle \times K, \tag{5.4.21}$$

in the notation of Remark 5.2.1. Set

$$C = \{\alpha \in A \mid 2\alpha \in B\}, \tag{5.4.22}$$

a subgroup of A containing B. Define maps s (for "square") and f (for "fourth power")

$$\begin{aligned} s&: C \to \mathbb{Z}/2\mathbb{Z}, \\ f&: A \to \mathbb{Z}/4\mathbb{Z} \end{aligned} \tag{5.4.23}$$

by the requirements

$$\begin{aligned} c^2 &\in \kappa^{s(\bar{c})} \langle \kappa^2 \rangle \times K, \\ a^4 &\in \kappa^{f(\bar{a})} K \end{aligned} \tag{5.4.24}$$

for $c \in \hat{C}$, $a \in \hat{A}$. These maps are well defined since

$$(c\kappa^m)^2 \in c^2 \langle \kappa^2 \rangle, \qquad (a\kappa^m)^4 = a^4$$

for $c, a \in \hat{A}$, $m \in \mathbb{Z}/4\mathbb{Z}$. Set

$$\operatorname{Aut}(\hat{A}; \kappa, K) = \{g \in \operatorname{Aut} \hat{A} \mid g\kappa = \kappa, gK = K\}, \tag{5.4.25}$$

and let

$$\operatorname{Aut}(A; B, c_0, s, f) \tag{5.4.26}$$

be the group of automorphisms $g \in \operatorname{Aut} A$ such that

$$\begin{aligned} gB &= B, \\ c_0(g\alpha, g\beta) &= c_0(\alpha, \beta), \\ s(g\gamma) &= s(\gamma), \\ f(g\alpha) &= f(\alpha) \end{aligned}$$

for $\alpha, \beta \in A$, $\gamma \in C$. Also, for $\lambda \in \mathrm{Hom}(A/B, \mathbb{Z}/4\mathbb{Z})$, define

$$\begin{aligned} \lambda^*: \hat{A} &\to \hat{A} \\ a &\mapsto a\kappa^{\lambda(a+B)}. \end{aligned} \tag{5.4.27}$$

Then $\lambda \mapsto \lambda^*$ defines an injective homomorphism into $\mathrm{Aut}(\hat{A}; \kappa, K)$, and λ^* induces the trivial automorphism of A.

Proposition 5.4.8: *We have an exact sequence*

$$1 \to \mathrm{Hom}(A/B, \mathbb{Z}/4\mathbb{Z}) \xrightarrow{*} \mathrm{Aut}(\hat{A}; \kappa, K) \xrightarrow{-} \mathrm{Aut}\,(A; B, c_0, s, f) \to 1.$$

Proof: It is clear that every $g \in \mathrm{Ker}(^-)$ is of the form λ^* for some $\lambda \in \mathrm{Hom}(A/B, \mathbb{Z}/4\mathbb{Z})$. Let $g \in \mathrm{Aut}(\hat{A}; \kappa, K)$. By Proposition 5.4.1, $\bar{g}$ preserves c_0. Since $gK = K$, $\bar{g}$ stabilizes $\bar{K} = B$. Let $c \in \hat{C}$. Then

$$\kappa^{s(\bar{c})} = g(\kappa^{s(\bar{c})}) \equiv g(c^2) = g(c)^2 \equiv \kappa^{s(\bar{g}\bar{c})} \operatorname{mod}(\langle \kappa^2 \rangle \times K),$$

so that $s(\bar{c}) = s(\bar{g}\bar{c})$. Let $a \in \hat{A}$. Then

$$\kappa^{f(\bar{a})} = g(\kappa^{f(\bar{a})}) \equiv g(a^4) = g(a)^4 \equiv \kappa^{f(\bar{g}\bar{a})} \bmod K,$$

so that $f(\bar{a}) = f(\bar{g}\bar{a})$, and we have shown that

$$\mathrm{Aut}(\hat{A}; \kappa, K)^- \subset \mathrm{Aut}(A; B, c_0, s, f).$$

Conversely, let $h \in \mathrm{Aut}(A; B, c_0, s, f)$. By Proposition 5.4.1, we may choose $g_0 \in \mathrm{Aut}(\hat{A}; \kappa)$ with $\bar{g}_0 = h$. Fix a base

$$\{\beta_i, \gamma_j, \delta_k \mid 1 \le i \le r, 1 \le j \le s, 1 \le k \le t\}$$

of A such that

$$B = \coprod_{i=1}^{r} 4\mathbb{Z}\beta_i \oplus \coprod_{j=1}^{s} 2\mathbb{Z}\gamma_j \oplus \coprod_{k=1}^{t} \mathbb{Z}\delta_k,$$

and choose, $b_i, c_j, d_k \in \hat{A}$ such that

$$\bar{b}_i = \beta_i, \qquad \bar{c}_j = \gamma_j, \qquad \bar{d}_k = \delta_k$$

for $1 \le i \le r$, $1 \le j \le s$, $1 \le k \le t$. Set

$$b_i' = g_0(b_i), \qquad c_j' = g_0(c_j), \qquad d_k' = g_0(d_k)$$

and let

$$c_j^2 \in \kappa^{t_j}K, \qquad c_j'^2 \in \kappa^{u_j}K,$$

$$d_k \in \kappa^{v_k}K, \qquad d_k' \in \kappa^{w_k}K$$

for $t_j, u_j, v_k, w_k \in \mathbb{Z}/4\mathbb{Z}$. Then $u_j \equiv t_j \operatorname{mod}(2\mathbb{Z}/4\mathbb{Z})$ since $s \circ h = s$. Define

$$\lambda \in \operatorname{Hom}(A, \mathbb{Z}/4\mathbb{Z})$$

by

$$\lambda(\beta_i) = 0$$

$$\lambda(\gamma_j) = \begin{cases} 0 & \text{if} \quad u_j = t_j \\ 1 + 4\mathbb{Z} & \text{if} \quad u_j = t_j + 2 + 4\mathbb{Z} \end{cases}$$

$$\lambda(\delta_k) = v_k - w_k,$$

$1 \le i \le r$, $1 \le j \le s$, $1 \le k \le t$. We have

$$\begin{aligned} K &= \langle b_i^4 \kappa^{-f(\beta_i)}, c_j^2 \kappa^{-t_j}, d_k \kappa^{-v_k} \mid 1 \le i \le r, 1 \le j \le s, 1 \le k \le t \rangle \\ &= \langle b_i'^4 \kappa^{-f(\beta_i)}, c_j'^2 \kappa^{-u_j}, d_k' \kappa^{-w_k} \mid 1 \le i \le r, 1 \le j \le s, 1 \le k \le t \rangle \end{aligned}$$

since $f \circ h = f$. Hence $g = g_0 \lambda^*$ stabilizes K, $g \in \operatorname{Aut}(\hat{A}; \kappa, K)$ and $\bar{g} = \bar{g}_0 = h$. ∎

5.5. Representations of Central Extensions

A representation $\pi: G \to \operatorname{End} V$ of a group G is said to *have a central character* if $\operatorname{Cent} G$ acts on V as scalar multiplication operators. In this case, we have a corresponding homomorphism

$$\chi: \operatorname{Cent} G \to \mathbb{F}^\times \tag{5.5.1}$$

called the *central character of* π. The homomorphisms χ as in (5.5.1) are called the *central characters of* G. A central character of an abelian group is also called a *character*. A representation π or a central character χ of G is said to be *faithful* if it is injective. An *exponent* of G is a positive integer m such that

$$g^m = 1 \quad \text{for all} \quad g \in G.$$

The following theorem generalizes the well-known result that an extra-special p-group has a unique irreducible representation with given faithful central character.

Theorem 5.5.1: *Let G be a finite group such that*

$$(G, G) \subset \operatorname{Cent} G$$

and Cent G *is a cyclic group. Suppose that the field* $\mathbb{F}$ *contains a primitive*

sth *root of unity, where s is an exponent of G. Let*

$$\chi\colon \operatorname{Cent} G \to \mathbb{F}^{\times}$$

be a faithful central character of G. Then G has a unique (up to equivalence) irreducible representation π with central character χ, and π is faithful. Let T be the corresponding G-module. If A is a maximal abelian subgroup of G and if

$$\psi\colon A \to \mathbb{F}^{\times}$$

is a homomorphism extending χ, then

$$T \simeq \operatorname{Ind}_A^G \mathbb{F}_{\psi},$$

where $\mathbb{F}_{\psi}$ is the one-dimensional A-module $\mathbb{F}$ with character ψ. Moreover,

$$\dim T = |A/\operatorname{Cent} G| = |G/A| = |G/\operatorname{Cent} G|^{1/2}.$$

In particular, $|G/\operatorname{Cent} G|$ is a perfect square.

Proof: Let

$$\operatorname{Cent} G = \langle \kappa \mid \kappa^{s_0} = 1 \rangle$$

and set

$$V = G/\operatorname{Cent} G,$$

a finite abelian group. We have a central extension

$$1 \to \operatorname{Cent} G \to G \xrightarrow{-} V \to 1.$$

Denote by

$$c_0\colon V \times V \to \mathbb{Z}/s_0\mathbb{Z}$$

the associated commutator map. Then the radical of c_0 is 0, since an element of the radical would lift to a central element of G (cf. Remark 5.2.1). In particular, V has exponent s_0.

Let A be a maximal abelian subgroup of G. Then $A \supset \operatorname{Cent} G$ and the subgroup $\bar{A} = A/\operatorname{Cent} G$ of V is maximal such that

$$c_0(\bar{A}, \bar{A}) = 0$$

[cf. (5.2.10)]. We thus have an injection

$$V/\bar{A} \hookrightarrow \bar{A}' = \operatorname{Hom}(\bar{A}, \mathbb{Z}/s_0\mathbb{Z})$$

$$v + \bar{A} \mapsto (\alpha \mapsto c_0(v, \alpha)).$$

If $\alpha \in \bar{A}$ is annihilated by the image of $V/\bar{A}$, then $\alpha = 0$. Hence

(i) $\bar{A}$ has exponent s_0
(ii) $\operatorname{Im}(V/\bar{A}) = \bar{A}'$

so that in particular

$$|G/A| = |V/\bar{A}| = |\bar{A}|. \tag{5.5.2}$$

Now G permutes by conjugation the characters of A extending χ: If $\varphi: A \to \mathbb{F}^\times$ is a character, its g-conjugate ${}^g\varphi$ for $g \in G$ is given by

$$({}^g\varphi)(a) = \varphi(g^{-1}ag) \quad \text{for} \quad a \in A.$$

Given such a character φ and $g \in G$, the condition ${}^g\varphi = \varphi$ implies that $c_0(\bar{g}, \bar{A}) = 0$ since χ is faithful. But then $g \in A$. That is, G/A acts freely on the set of such characters. Thus by (5.5.2), this set consists of only one G/A-orbit.

Now fix a character ψ of A extending χ, and let T be a G-module with central character χ. Denote by π the corresponding representation. The operators $\pi(A)$ are diagonalizable and hence simultaneously diagonalizable. Thus there exists $v \in T\backslash\{0\}$ and a character $\varphi: A \to \mathbb{F}^\times$ extending χ so that $\pi(a)v = \varphi(a)v$ for $a \in A$. By what we have proved, we may choose $g \in G$ so that $\psi = {}^g\varphi$. Then $g \cdot v$ transforms by ψ under A and hence there is a nonzero homomorphism

$$\operatorname{Ind}_A^G \mathbb{F}_\psi \to T.$$

On the other hand, as A-modules,

$$\operatorname{Ind}_A^G \mathbb{F}_\psi \simeq \coprod \mathbb{F}_\varphi,$$

where φ ranges over the characters of A extending χ. Arguments similar to the above now show that $\operatorname{Ind}_A^G \mathbb{F}_\psi$ is irreducible, and hence T is isomorphic to $\operatorname{Ind}_A^G \mathbb{F}_\psi$ if T is irreducible. The faithfulness of π follows. ∎

Remark 5.5.2: The proof shows that the conclusions all remain valid even if $\mathbb{F}$ contains only a primitive rth root of unity where r is an exponent of the maximal abelian subgroup A of G.

We shall also need:

Proposition 5.5.3: *Let G, A, χ, ψ, T and π be as in Theorem 5.5.1, and denote by $\mathbf{N}_{\operatorname{Aut} T}(\pi(G))$ the normalizer of $\pi(G)$ in* Aut T. *Then the sequence* *of canonical maps*

$$1 \longrightarrow \mathbb{F}^{\times} \longrightarrow \mathbf{N}_{\mathrm{Aut}\,T}(\pi(G)) \xrightarrow{\mathrm{int}} \{g \in \mathrm{Aut}\, G \mid g|_{\mathrm{Cent}\, G} = 1\} \longrightarrow 1 \quad (5.5.3)$$

is exact, where (identifying G with $\pi(G)$)

$$\mathrm{int}(g)(x) = gxg^{-1}. \quad (5.5.4)$$

Proof: For an extension field $\mathbb{E}$ of $\mathbb{F}$,

$$(\mathrm{Ind}_A^G \mathbb{F}_\psi) \otimes_{\mathbb{F}} \mathbb{E} \simeq \mathrm{Ind}_A^G \mathbb{E}_\psi .$$

Thus Theorem 5.5.1 implies that $\mathrm{Ind}_A^G \mathbb{F}_\psi$ is absolutely irreducible (i.e., remains irreducible upon field extension), so that the centralizer of $\pi(G)$ in Aut T is $\mathbb{F}^{\times}$. Let $g \in \mathrm{Aut}\, G$, $g|_{\mathrm{Cent}\, G} = 1$. Then the G-module T with action $\pi \circ g$ is irreducible and has central character χ, and hence is equivalent to T with action π. Thus there exists $g^* \in \mathrm{Aut}\, T$ with

$$g^* \pi(x) g^{*-1} = \pi(g(x))$$

for all $x \in G$. ∎

6 The Simple Lie Algebras A_n, D_n, E_n

In this chapter, using the results of Chapter 5, we canonically associate a Lie algebra to every positive definite even lattice. The case of indecomposable root lattices with equal root lengths leads to the simple Lie algebras of types A_n, D_n, E_6, E_7, E_8 together with a Chevalley basis with explicitly given structure constants. This construction arose from the vertex operator representations of [Frenkel–Kac], [Segal 1] that we shall describe in general in Chapter 7.

We introduce lattices and related concepts in Section 6.1. The simply-laced (equal-root-length) root lattices form an important class of examples. In Section 6.2 we construct Lie algebras from the positive definite even lattices, using certain central extensions from Chapter 5. In Section 6.3 we give a list of the simple Lie algebras obtained by this method: A_n $(n \geq 1)$, D_n $(n \geq 4)$, E_6, E_7, E_8. This is the complete list of the simply-laced simple Lie algebras with a Cartan subalgebra acting diagonally. A similar exposition, including the unequal-root-length cases B_n, C_n, F_4, G_2, has been given in [Mitzman]. While the constructions that we present might look more familiar to Lie theorists when we choose sections and cocycles for the central extensions, we increasingly often adopt a canonical section-free (cocycle-free) viewpoint as we progress through our exposition. This viewpoint will be especially valuable for the more subtle arguments in later chapters. In Section 6.4, applying the results of Chapter 5, we study a lifting of the isometry group of our lattice to a group of automorphisms of the corresponding Lie algebra. The information about simple Lie algebras needed in this book is presented in a self-contained way in this chapter. For further theory of semisimple Lie algebras, see especially [Bourbaki 1, 2], [Humphreys], [Jacobson 1].

6.1. Lattices

By a (*rational*) *lattice* of rank $n \in \mathbb{N}$ we shall mean a rank n free abelian group L provided with a rational-valued symmetric $\mathbb{Z}$-bilinear form

$$\langle \cdot, \cdot \rangle : L \times L \to \mathbb{Q}. \tag{6.1.1}$$

A lattice isomorphism is sometimes called an *isometry*. A lattice L is *nondegenerate* if its form $\langle \cdot, \cdot \rangle$ is nondegenerate in the sense that for $\alpha \in L$,

$$\langle \alpha, L \rangle = 0 \quad \text{implies} \quad \alpha = 0. \tag{6.1.2}$$

Given a lattice L, we see by choosing a $\mathbb{Z}$-base of L that $\langle L, L \rangle \subset (1/r)\mathbb{Z}$ for some $r \in \mathbb{Z}_+$:

$$\langle \cdot, \cdot \rangle : L \times L \to \frac{1}{r}\mathbb{Z}. \tag{6.1.3}$$

We canonically embed L in the $\mathbb{Q}$-vector space

$$L_{\mathbb{Q}} = L \otimes_{\mathbb{Z}} \mathbb{Q}, \tag{6.1.4}$$

which is n-dimensional since a $\mathbb{Z}$-base of L is a $\mathbb{Q}$-basis of $L_{\mathbb{Q}}$, and we extend $\langle \cdot, \cdot \rangle$ to a symmetric $\mathbb{Q}$-bilinear form

$$\langle \cdot, \cdot \rangle : L_{\mathbb{Q}} \times L_{\mathbb{Q}} \to \mathbb{Q}. \tag{6.1.5}$$

Note that every element of $L_{\mathbb{Q}}$ is of the form α/N for some $\alpha \in L$ and $N \in \mathbb{Z}\backslash\{0\}$. The lattice L is nondegenerate if and only if the form (6.1.5) is nondegenerate, and this amounts to the condition

$$\det(\langle \alpha_i, \alpha_j \rangle)_{i,j=1,\ldots,n} \neq 0 \tag{6.1.6}$$

for a $\mathbb{Z}$-base $\{\alpha_1, \ldots, \alpha_n\}$ of L. A lattice may be equivalently defined as the $\mathbb{Z}$-span of a basis of a finite-dimensional rational vector space equipped with a symmetric bilinear form.

Let L be a lattice. For $m \in \mathbb{Q}$, we set

$$L_m = \{\alpha \in L \mid \langle \alpha, \alpha \rangle = m\}. \tag{6.1.7}$$

The lattice L is said to be *even* if

$$\langle \alpha, \alpha \rangle \in 2\mathbb{Z} \quad \text{for} \quad \alpha \in L, \tag{6.1.8}$$

integral if

$$\langle \alpha, \beta \rangle \in \mathbb{Z} \quad \text{for} \quad \alpha, \beta \in L \tag{6.1.9}$$

and *positive definite* if

$$\langle \alpha, \alpha \rangle > 0 \quad \text{for} \quad \alpha \in L \backslash \{0\}, \tag{6.1.10}$$

or equivalently, for $\alpha \in L_{\mathbb{Q}} \backslash \{0\}$. The polarization formula

$$\langle \alpha, \beta \rangle = \tfrac{1}{2}(\langle \alpha + \beta, \alpha + \beta \rangle - \langle \alpha, \alpha \rangle - \langle \beta, \beta \rangle) \tag{6.1.11}$$

shows that an even lattice is integral.

The *dual* of L is the set

$$L^\circ = \{\alpha \in L_{\mathbb{Q}} \mid \langle \alpha, L \rangle \subset \mathbb{Z}\}. \tag{6.1.12}$$

This set is again a lattice if and only if L is nondegenerate, and in this case, L° has as a base the *dual base* $\{\alpha_1^*, \ldots, \alpha_n^*\}$ of a given base $\{\alpha_1, \ldots, \alpha_n\}$ of L, defined by:

$$\langle \alpha_i^*, \alpha_j \rangle = \delta_{ij} \quad \text{for} \quad i, j = 1, \ldots, n. \tag{6.1.13}$$

Note that L is integral if and only if

$$L \subset L^\circ. \tag{6.1.14}$$

The lattice L is said to be *self-dual* if

$$L = L^\circ. \tag{6.1.15}$$

This is equivalent to L being integral and *unimodular*, which means that

$$|\det(\langle \alpha_i, \alpha_j \rangle)_{ij}| = 1. \tag{6.1.16}$$

In fact, if L is integral and nondegenerate, then $(\langle \alpha_i, \alpha_j \rangle)_{i,j}$ is the matrix of the embedding map $L \hookrightarrow L^\circ$ with respect to the given base and its dual base, and the unimodularity amounts to the condition that this embedding be an isomorphism of abelian groups.

Generalizing (6.1.4) and (6.1.5), we embed L in the $\mathbb{E}$-vector space

$$L_{\mathbb{E}} = L \otimes_{\mathbb{Z}} \mathbb{E} \tag{6.1.17}$$

for any field $\mathbb{E}$ of characteristic zero, and we extend $\langle \cdot, \cdot \rangle$ to the symmetric $\mathbb{E}$-bilinear form

$$\langle \cdot, \cdot \rangle : L_{\mathbb{E}} \times L_{\mathbb{E}} \to \mathbb{E}. \tag{6.1.18}$$

Then L is positive definite if and only if the real vector space $L_{\mathbb{R}}$ is a Euclidean space. In this case,

$$|L_m| < \infty \quad \text{for} \quad m \in \mathbb{Q} \tag{6.1.19}$$

since L_m is the intersection of the discrete set L with a compact set (a sphere) in $L_{\mathbb{R}}$.

Remark 6.1.1: Using the Schwarz inequality, we observe that if the lattice L is integral and positive definite, and if $\alpha, \beta \in L_2$, then

$$\langle \alpha, \beta \rangle = 0, \pm 1 \quad \text{or} \quad \pm 2,$$

and

$$\langle \alpha, \beta \rangle = -2 \quad \text{if and only if} \quad \alpha + \beta = 0$$

$$\langle \alpha, \beta \rangle = -1 \quad \text{if and only if} \quad \alpha + \beta \in L_2$$

$$\langle \alpha, \beta \rangle \geq 0 \quad \text{if and only if} \quad \alpha + \beta \notin L_2 \cup \{0\}.$$

Let L be an even lattice. Set

$$\check{L} = L/2L, \tag{6.1.20}$$

and view the elementary abelian 2-group $\check{L}$ as a vector space over the field $\mathbb{F}_2$. Denote by

$$\begin{aligned} L &\to \check{L} \\ \alpha &\mapsto \check{\alpha} = \alpha + 2L \end{aligned} \tag{6.1.21}$$

the canonical map. Since a $\mathbb{Z}$-base of L reduces to an $\mathbb{F}_2$-basis of $\check{L}$,

$$\dim \check{L} = \operatorname{rank} L. \tag{6.1.22}$$

There is a canonical $\mathbb{Z}$-bilinear form

$$\begin{aligned} c_0 : L \times L &\to \mathbb{Z}/2\mathbb{Z} \\ (\alpha, \beta) &\mapsto \langle \alpha, \beta \rangle + 2\mathbb{Z} \end{aligned} \tag{6.1.23}$$

on L, and c_0 is alternating [cf. (5.2.5)–(5.2.8)] because L is even. The form

c_0 induces a (well-defined) alternating $\mathbb{F}_2$-bilinear map

$$\begin{aligned} c_1 : \check{L} \times \check{L} &\to \mathbb{Z}/2\mathbb{Z} = \mathbb{F}_2 \\ (\check{\alpha}, \check{\beta}) &\mapsto \langle \alpha, \beta \rangle + 2\mathbb{Z} \end{aligned} \tag{6.1.24}$$

for $\alpha, \beta \in L$. There is a canonical quadratic form q_1 on $\check{L}$ with associated bilinear form c_1 [cf. (5.3.6)]:

$$\begin{aligned} q_1 : \check{L} &\to \mathbb{Z}/2\mathbb{Z} = \mathbb{F}_2 \\ \check{\alpha} &\mapsto \tfrac{1}{2}\langle \alpha, \alpha \rangle + 2\mathbb{Z} \end{aligned} \tag{6.1.25}$$

for $\alpha \in L$. This form is well-defined: If $\check{\alpha} = \check{\beta}$, then $\beta - \alpha = \gamma \in 2L$, and

$$\begin{aligned} \tfrac{1}{2}\langle \beta, \beta \rangle &= \tfrac{1}{2}\langle \alpha, \alpha \rangle + \langle \alpha, \gamma \rangle + \tfrac{1}{2}\langle \gamma, \gamma \rangle \\ &\equiv \tfrac{1}{2}\langle \alpha, \alpha \rangle \bmod 2. \end{aligned}$$

It is clear that c_1 is the associated form. Denote by

$$\begin{aligned} q_0 : L &\to \mathbb{Z}/2\mathbb{Z} \\ \alpha &\mapsto \tfrac{1}{2}\langle \alpha, \alpha \rangle + 2\mathbb{Z} \end{aligned} \tag{6.1.26}$$

the pullback of q_1 to L. By Proposition 5.3.1, there exist $\mathbb{Z}$-bilinear forms

$$\varepsilon_0 : L \times L \to \mathbb{Z}/2\mathbb{Z} \tag{6.1.27}$$

such that

$$\varepsilon_0(\alpha, \alpha) = q_0(\alpha) = \tfrac{1}{2}\langle \alpha, \alpha \rangle + 2\mathbb{Z} \quad \text{for} \quad \alpha \in L \tag{6.1.28}$$

and consequently

$$\begin{aligned} &\varepsilon_0(\alpha, \beta) - \varepsilon_0(\beta, \alpha) = c_0(\alpha, \beta) = \langle \alpha, \beta \rangle + 2\mathbb{Z} \\ &\quad \text{for} \quad \alpha, \beta \in L. \end{aligned} \tag{6.1.29}$$

Note that q_1 (or equivalently, c_1) is nonsingular if and only if the determinant (6.1.6) is odd—in particular, if L is unimodular [see (6.1.16)]. These considerations will be important for many constructions (cf. Propositions 5.2.3, 5.3.3, 5.3.4 and 5.4.3).

Let L be a positive definite lattice. The *theta function* θ_L of L is defined to be the formal series in the variable q (cf. Section 1.10) given by:

$$\begin{aligned} \theta_L(q) &= \sum_{\alpha \in L} q^{\langle \alpha, \alpha \rangle / 2} \\ &= \sum_{m \in \mathbb{Q}} |L_m| q^{m/2} \end{aligned} \tag{6.1.30}$$

[see (6.1.7), (6.1.19)].

Remark 6.1.2: If L is even and unimodular, the theta function θ_L has important modular transformation properties under the modular group $SL(2, \mathbb{Z})$ when q is replaced by $e^{2\pi iz}$, z a complex variable in the upper half plane (see for example [Serre 1]).

6.2. A Class of Lie Algebras

Let L be a positive definite even lattice with symmetric form $\langle \cdot, \cdot \rangle$. Starting from L, we shall construct a Lie algebra in a canonical way.

Set

$$\mathfrak{h} = L_{\mathbb{F}} = L \otimes_{\mathbb{Z}} \mathbb{F} \tag{6.2.1}$$

[see (6.1.17)], a finite-dimensional $\mathbb{F}$-vector space with nonsingular symmetric form $\langle \cdot, \cdot \rangle$ [see (6.1.18)]. We view L as a subset of $\mathfrak{h}$.

Consider the alternating $\mathbb{Z}$-bilinear map

$$\begin{aligned} c_0 : L \times L &\to \mathbb{Z}/2\mathbb{Z} \\ (\alpha, \beta) &\mapsto \langle \alpha, \beta \rangle + 2\mathbb{Z} \end{aligned} \tag{6.2.2}$$

as in (6.1.23). By Proposition 5.2.3, there is a unique (up to equivalence) central extension

$$1 \to \langle \kappa \mid \kappa^2 = 1 \rangle \to \hat{L} \xrightarrow{-} L \to 1 \tag{6.2.3}$$

of L by the 2-element cyclic group $\langle \kappa \rangle$ such that

$$aba^{-1}b^{-1} = \kappa^{\langle \bar{a}, \bar{b} \rangle} \quad \text{for} \quad a, b \in \hat{L}. \tag{6.2.4}$$

Sometimes it will be convenient to choose a section

$$\begin{aligned} e : L &\to \hat{L} \\ \alpha &\mapsto e_\alpha \end{aligned} \tag{6.2.5}$$

for the central extension (6.2.3). Denote by

$$\varepsilon_0 : L \times L \to \mathbb{Z}/2\mathbb{Z} \tag{6.2.6}$$

the corresponding 2-cocycle. Then

$$e_\alpha e_\beta = e_{\alpha+\beta} \kappa^{\varepsilon_0(\alpha, \beta)} \tag{6.2.7}$$

$$\varepsilon_0(\alpha, \beta) + \varepsilon_0(\alpha + \beta, \gamma) = \varepsilon_0(\beta, \gamma) + \varepsilon_0(\alpha, \beta + \gamma) \tag{6.2.8}$$

$$\varepsilon_0(\alpha, \beta) - \varepsilon_0(\beta, \alpha) = \langle \alpha, \beta \rangle + 2\mathbb{Z} \tag{6.2.9}$$

for $\alpha, \beta, \gamma \in L$ [see (5.1.5), (5.1.9), (5.2.13)]. Conversely, given a map ε_0 as in (6.2.6) satisfying (6.2.8) and (6.2.9), there is a central extension (6.2.3) satisfying (6.2.4) and a section (6.2.5) satisfying (6.2.7). We also assume that

$$e_0 = 1, \tag{6.2.10}$$

or equivalently,

$$\varepsilon_0(\alpha, 0) = \varepsilon_0(0, \alpha) = 0 \quad \text{for} \quad \alpha \in L \tag{6.2.11}$$

(see Remark 5.1.3). Maps ε_0 with the desired properties may easily be constructed, such as the bilinear maps in (5.2.14) or (6.1.27)–(6.1.29) [cf. Remark (5.1.1)]. If the cocycle ε_0 is bilinear, then in addition to (6.2.11) we also have

$$\varepsilon_0(\alpha, \beta) = \varepsilon_0(-\alpha, \beta) = \varepsilon_0(\alpha, -\beta) \tag{6.2.12}$$

$$\varepsilon_0(\alpha, \beta) = \varepsilon_0(-\alpha, -\beta) \tag{6.2.13}$$

for $\alpha, \beta \in L$. Note that the bilinear cocycle ε_0 of (6.1.27)–(6.1.29) satisfies the condition

$$\varepsilon_0(\alpha, \alpha) = 1 + 2\mathbb{Z} \quad \text{for} \quad \alpha \in L_2. \tag{6.2.14}$$

Consider the finite set

$$\Delta = L_2 \tag{6.2.15}$$

[see (6.1.7), (6.1.19)] and write

$$\hat{\Delta} = \{a \in \hat{L} \mid \bar{a} \in \Delta\}. \tag{6.2.16}$$

Then

$$|\hat{\Delta}| = 2|\Delta|. \tag{6.2.17}$$

With the choice of the section e,

$$\hat{\Delta} = \{e_\alpha, \kappa e_\alpha \mid \alpha \in \Delta\}. \tag{6.2.18}$$

Let $\mathfrak{g}$ be the finite-dimensional vector space

$$\mathfrak{g} = \mathfrak{h} \oplus \sum_{a \in \hat{\Delta}} \mathbb{F} x_a, \tag{6.2.19}$$

where the x_a are symbols indexed by $a \in \hat{\Delta}$ and subject to the linear relations

$$x_{\kappa a} = -x_a \tag{6.2.20}$$

and no others. Then

$$\dim \mathfrak{g} = \operatorname{rank} L + |\Delta|. \tag{6.2.21}$$

In terms of the section e, we may write

$$\mathfrak{g} = \mathfrak{h} \oplus \coprod_{\alpha \in \Delta} \mathbb{F}x_{\alpha}, \tag{6.2.22}$$

where we set

$$x_{\alpha} = x_{e_{\alpha}} \quad \text{for} \quad \alpha \in \Delta. \tag{6.2.23}$$

We determine a nonassociative algebra product, designated $[\cdot, \cdot]$, on $\mathfrak{g}$ as follows:

$$[\mathfrak{h}, \mathfrak{h}] = 0 \tag{6.2.24}$$

$$[h, x_a] = -[x_a, h] = \langle h, \bar{a} \rangle x_a \tag{6.2.25}$$

$$[x_a, x_b] = \begin{cases} \bar{a} & \text{if} \quad ab = 1 \\ x_{ab} & \text{if} \quad ab \in \hat{\Delta} \\ 0 & \text{if} \quad ab \notin \hat{\Delta} \cup \{1, \kappa\} \end{cases} \tag{6.2.26}$$

for $h \in \mathfrak{h}$ and $a, b \in \hat{\Delta}$. Note that by Remark 6.1.1 and (6.2.20), the cases in (6.2.26) are exhaustive. Also note that the product is consistent with the relations (6.2.20). Taking into account (6.2.10) and (6.2.20), we see that the formulation using the section is:

$$[\mathfrak{h}, \mathfrak{h}] = 0 \tag{6.2.27}$$

$$[h, x_{\alpha}] = -[x_{\alpha}, h] = \langle h, \alpha \rangle x_{\alpha} \tag{6.2.28}$$

$$[x_{\alpha}, x_{\beta}] = \begin{cases} \varepsilon(\alpha, -\alpha)\alpha & \text{if} \quad \alpha + \beta = 0 \ (\langle \alpha, \beta \rangle = -2) \\ \varepsilon(\alpha, \beta) x_{\alpha+\beta} & \text{if} \quad \alpha + \beta \in \Delta \ (\langle \alpha, \beta \rangle = -1) \\ 0 & \text{if} \quad \alpha + \beta \notin \Delta \cup \{0\} \ (\langle \alpha, \beta \rangle \geq 0) \end{cases} \tag{6.2.29}$$

for $h \in \mathfrak{h}$ and $\alpha, \beta \in \Delta$, where we set

$$\varepsilon(\alpha, \beta) = (-1)^{\varepsilon_0(\alpha, \beta)} \quad \text{for} \quad \alpha, \beta \in L. \tag{6.2.30}$$

Note that ε is a map from $L \times L$ into the 2-element multiplicative group $\langle \pm 1 \rangle$,

$$\varepsilon : L \times L \to \langle \pm 1 \rangle, \tag{6.2.31}$$

such that

$$\varepsilon(\alpha, \beta)\varepsilon(\alpha + \beta, \gamma) = \varepsilon(\beta, \gamma)\varepsilon(\alpha, \beta + \gamma) \tag{6.2.32}$$

$$\varepsilon(\alpha, \beta)/\varepsilon(\beta, \alpha) = (-1)^{\langle \alpha, \beta \rangle} \tag{6.2.33}$$

$$\varepsilon(\alpha, 0) = \varepsilon(0, \alpha) = 1 \tag{6.2.34}$$

for $\alpha, \beta, \gamma \in L$. If ε_0 is $\mathbb{Z}$-bilinear and (6.2.14) holds, then

$$[x_\alpha, x_{-\alpha}] = -\alpha \quad \text{for} \quad \alpha \in \Delta. \tag{6.2.35}$$

Also define a nonsingular bilinear form $\langle \cdot, \cdot \rangle$ on $\mathfrak{g}$ as follows:

$$\langle \cdot, \cdot \rangle|_{\mathfrak{h}\times\mathfrak{h}} = \langle \cdot, \cdot \rangle \tag{6.2.36}$$

$$\langle \mathfrak{h}, x_a \rangle = \langle x_a, \mathfrak{h} \rangle = 0 \quad \text{for} \quad a \in \hat{\Delta} \tag{6.2.37}$$

$$\langle x_a, x_b \rangle = \begin{cases} 1 & \text{if} \quad ab = 1 \\ 0 & \text{if} \quad ab \notin \{1, \kappa\}. \end{cases} \tag{6.2.38}$$

In terms of the section, (6.2.37) is equivalent to

$$\langle \mathfrak{h}, x_\alpha \rangle = \langle x_\alpha, \mathfrak{h} \rangle = 0 \quad \text{for} \quad \alpha \in \Delta \tag{6.2.39}$$

and (6.2.38) to

$$\langle x_\alpha, x_\beta \rangle = \begin{cases} \varepsilon(\alpha, -\alpha) & \text{if} \quad \alpha + \beta = 0 \\ 0 & \text{if} \quad \alpha + \beta \neq 0 \end{cases} \tag{6.2.40}$$

for $\alpha, \beta \in \Delta$.

Theorem 6.2.1: *The nonassociative algebra $\mathfrak{g}$ is a Lie algebra and the nonsingular form $\langle \cdot, \cdot \rangle$ is symmetric and $\mathfrak{g}$-invariant.*

Proof: We use (6.2.24)–(6.2.26) rather than (6.2.27)–(6.2.29). From (6.2.4) we see that the product is alternating. (Note that $ab = \kappa ba$ if a, b $ab \in \hat{\Delta}$, since $\langle \bar{a}, \bar{b} \rangle = -1$ in this case.) To prove that $\mathfrak{g}$ is a Lie algebra, we need to verify the Jacobi identity

$$[y_1, [y_2, y_3]] + [y_2, [y_3, y_1]] + [y_3, [y_1, y_2]] = 0 \tag{6.2.41}$$

for y_1, y_2, y_3 in a spanning subset of $\mathfrak{g}$.

If at least one y_i is in $\mathfrak{h}$ and the remaining y_i's are x_a's, the identity is easy, using the alternating property. Suppose that $y_i = x_{a_i}$ ($i = 1, 2, 3$; $a_i \in \hat{\Delta}$). If $\bar{a}_1 + \bar{a}_2 + \bar{a}_3 \notin \Delta \cup \{0\}$, all three terms in (6.2.41) vanish. If

$$\bar{a}_1 + \bar{a}_2 + \bar{a}_3 = 0,$$

then since

$$\bar{a}_i + \bar{a}_j = -\bar{a}_k$$

for i, j, k distinct, the left-hand side of (6.2.41) becomes

$$\begin{aligned}&[x_{a_1}, x_{a_2a_3}] + [x_{a_2}, x_{a_3a_1}] + [x_{a_3}, x_{a_1a_2}] \\ &\quad = \chi(a_1a_2a_3)\bar{a}_1 + \chi(a_2a_3a_1)\bar{a}_2 + \chi(a_3a_1a_2)\bar{a}_3,\end{aligned} \tag{6.2.42}$$

where χ identifies the 2-element groups $\langle\kappa\rangle$ and $\langle -1\rangle$:

$$\begin{aligned} \chi\colon \langle\kappa\rangle &\tilde{\to} \langle -1\rangle \\ \kappa &\mapsto -1. \end{aligned} \tag{6.2.43}$$

Then by (6.2.4), (6.2.42) equals

$$\begin{aligned} &\chi(a_1a_2a_3)(\bar{a}_1 + (-1)^{\langle\bar{a}_1,\bar{a}_2+\bar{a}_3\rangle}\bar{a}_2 + (-1)^{\langle\bar{a}_3,\bar{a}_1+\bar{a}_2\rangle}\bar{a}_3) \\ &= \chi(a_1a_2a_3)(\bar{a}_1 + \bar{a}_2 + \bar{a}_3) \\ &= 0. \end{aligned}$$

Suppose now that

$$\bar{a}_1 + \bar{a}_2 + \bar{a}_3 \in \Delta.$$

Then

$$[x_{a_1}, [x_{a_2}, x_{a_3}]] = c_1 x_{a_1a_2a_3}$$

where

$$c_1 = \begin{cases} -\langle\bar{a}_1, \bar{a}_2\rangle & \text{if } \langle\bar{a}_2, \bar{a}_3\rangle = -2 \\ 1 & \text{if } \langle\bar{a}_2, \bar{a}_3\rangle = -1 \\ 0 & \text{if } \langle\bar{a}_2, \bar{a}_3\rangle \geq 0. \end{cases}$$

We define c_2 and c_3 analogously. Since

$$\begin{aligned} 2 &= \langle\bar{a}_1 + \bar{a}_2 + \bar{a}_3, \bar{a}_1 + \bar{a}_2 + \bar{a}_3\rangle \\ &= 6 + 2\langle\bar{a}_1, \bar{a}_2\rangle + 2\langle\bar{a}_1, \bar{a}_3\rangle + 2\langle\bar{a}_2, \bar{a}_3\rangle, \end{aligned}$$

we have

$$\langle\bar{a}_1, \bar{a}_2\rangle + \langle\bar{a}_1, \bar{a}_3\rangle + \langle\bar{a}_2, \bar{a}_3\rangle = -2. \tag{6.2.44}$$

Applying a cyclic permutation to $\bar{a}_1$, $\bar{a}_2$, $\bar{a}_3$ if necessary, we may thus assume that $\langle\bar{a}_1, \bar{a}_2\rangle \geq 0$, so that $c_3 = 0$, and (6.2.41) becomes

$$(c_1 + (-1)^{\langle\bar{a}_1,\bar{a}_2+\bar{a}_3\rangle}c_2)x_{a_1a_2a_3} = (c_1 + (-1)^{\langle\bar{a}_2,\bar{a}_3\rangle}c_2)x_{a_1a_2a_3} \tag{6.2.45}$$

using (6.2.44). If $\langle\bar{a}_1, \bar{a}_2\rangle = 0$, then

$$\langle\bar{a}_1, \bar{a}_3\rangle + \langle\bar{a}_2, \bar{a}_3\rangle = -2,$$

and $\langle\bar{a}_1, \bar{a}_3\rangle = 0, -1$ or -2. The first and third of these possibilities are the same, up to cyclic permutation of $\bar{a}_1$, $\bar{a}_2$ and $\bar{a}_3$. If $\langle\bar{a}_1, \bar{a}_2\rangle = 1$, then

$$\langle\bar{a}_1, \bar{a}_3\rangle + \langle\bar{a}_2, \bar{a}_3\rangle = -3,$$

and $\langle\bar{a}_1, \bar{a}_3\rangle = -1$ or -2. Finally, if $\langle\bar{a}_1, \bar{a}_2\rangle = 2$, then

$$\langle\bar{a}_1, \bar{a}_3\rangle = \langle\bar{a}_2, \bar{a}_3\rangle = -2.$$

In all five cases, (6.2.45) is zero, and the Jacobi identity is proved. Thus $\mathfrak{g}$ is a Lie algebra.

The symmetry of $\langle \cdot, \cdot \rangle$ follows from (6.2.4). For the $\mathfrak{g}$-invariance, we check that

$$\langle [y_1, y_2], y_3 \rangle = \langle y_1, [y_2, y_3] \rangle \tag{6.2.46}$$

for y_1, y_2, y_3 in $\mathfrak{h} \cup \{x_a \mid a \in \hat{\Delta}\}$. If at least two y_i's are in $\mathfrak{h}$, both sides are 0. If one y_i is in $\mathfrak{h}$ and the others are x_a and x_b, then both sides are 0 unless $\bar{a} + \bar{b} = 0$, in which case they are equal. Finally, if $y_i = x_{a_i}$ $(i = 1, 2, 3; a_i \in \hat{\Delta})$, then both sides of (6.2.46) are 0 unless $\bar{a}_1 + \bar{a}_2 + \bar{a}_3 = 0$, and in this case, we find, using the notation (6.2.43), that

$$\begin{aligned} \langle [x_{a_1}, x_{a_2}], x_{a_3} \rangle &= \langle x_{a_1 a_2}, x_{a_3} \rangle \\ &= \chi(a_1 a_2 a_3) \\ &= \langle x_{a_1}, x_{a_2 a_3} \rangle \\ &= \langle x_{a_1}, [x_{a_2}, x_{a_3}] \rangle . \end{aligned}$$

This proves the invariance. ∎

Remark 6.2.2: We could have given a similar (but slightly longer) proof of Theorem 6.2.1 based on the description (6.2.27)–(6.2.29) of $\mathfrak{g}$ together with the properties (6.2.8), (6.2.9) and (6.2.11) of the map $\varepsilon_0 : L \times L \to \mathbb{Z}/2\mathbb{Z}$.

Remark 6.2.3: The special case in which L has rank 1 and $\Delta = L_2 = \{\pm\alpha_1\}$ recovers the simple Lie algebra $\mathfrak{sl}(2, \mathbb{F})$ of Section 3.1 and its invariant symmetric form $\langle \cdot, \cdot \rangle$ [recall (3.1.1)–(3.1.5)]. In this case, the cocycle ε_0 can be (and was) chosen to be identically 0.

Remark 6.2.4: A completely different proof of Theorem 6.2.1, using vertex operators, will be given in Chapter 7 (cf. Remark 7.2.10).

We turn now to the untwisted affine Lie algebra $\tilde{\mathfrak{g}}$ associated with $\mathfrak{g}$ and $\langle \cdot, \cdot \rangle$:

$$\tilde{\mathfrak{g}} = \mathfrak{g} \otimes \mathbb{F}[t, t^{-1}] \oplus \mathbb{F}c \oplus \mathbb{F}d. \tag{6.2.47}$$

This algebra contains $\tilde{\mathfrak{h}}$ and its Heisenberg subalgebra

$$\hat{\mathfrak{h}}_{\mathbb{Z}} = \tilde{\mathfrak{h}}' = \coprod_{\substack{n \in \mathbb{Z} \\ n \neq 0}} \mathfrak{h} \otimes t^n \oplus \mathbb{F}c \tag{6.2.48}$$

[see (1.7.11)]. (We assume that $\mathfrak{h} \neq 0$, i.e., $L \neq 0$.) The bracket relations for $\tilde{\mathfrak{h}}$ may be expressed as follows:

$$[\alpha \otimes t^m, \beta \otimes t^n] = \langle \alpha, \beta \rangle m \delta_{m+n,0} c \tag{6.2.49}$$

$$[d, \alpha \otimes t^m] = m\alpha \otimes t^m \tag{6.2.50}$$

$$[c, \alpha \otimes t^m] = [c, d] = 0 \tag{6.2.51}$$

for $\alpha, \beta \in \mathfrak{h}$, $m, n \in \mathbb{Z}$.

It will be convenient to express the remaining linear and bracket structure of $\tilde{\mathfrak{g}}$ in terms of the elements $x_a(z)$ or $x_\alpha(z)$ of $\tilde{\mathfrak{g}}[[z, z^{-1}]]$ [see (2.3.4) and Proposition 2.3.1] as follows:

$$x_{\kappa a}(z) = -x_a(z) \tag{6.2.52}$$

$$[h \otimes t^m, x_a(z)] = \langle h, \bar{a} \rangle z^m x_a(z) \tag{6.2.53}$$

$$[x_a(z_1), x_b(z_2)] = \begin{cases} \bar{a}(z_2)\delta(z_1/z_2) - (D\delta)(z_1/z_2)c & \text{if} \quad ab = 1 \\ x_{ab}(z_2)\delta(z_1/z_2) & \text{if} \quad ab \in \hat{\Delta} \\ 0 & \text{if} \quad ab \notin \hat{\Delta} \cup \{1, \kappa\} \end{cases} \tag{6.2.54}$$

$$[d, x_a(z)] = -Dx_a(z) \tag{6.2.55}$$

$$[c, x_a(z)] = 0 \tag{6.2.56}$$

for $h \in \mathfrak{h}$, $m \in \mathbb{Z}$, $a, b \in \hat{\Delta}$, or equivalently,

$$[h \otimes t^m, x_\alpha(z)] = \langle h, \alpha \rangle z^m x_\alpha(z) \tag{6.2.57}$$

$$[x_\alpha(z_1), x_\beta(z_2)]$$
$$= \begin{cases} \varepsilon(\alpha, -\alpha)(\alpha(z_2)\delta(z_1/z_2) - (D\delta)(z_1/z_2)c) & \text{if} \quad \alpha + \beta = 0 \\ \varepsilon(\alpha, \beta)x_{\alpha+\beta}(z_2)\delta(z_1/z_2) & \text{if} \quad \alpha + \beta \in \Delta \\ 0 & \text{if} \quad \alpha + \beta \notin \Delta \cup \{0\} \end{cases} \tag{6.2.58}$$

$$[d, x_\alpha(z)] = -Dx_\alpha(z) \tag{6.2.59}$$

$$[c, x_\alpha(z)] = 0 \tag{6.2.60}$$

for $h \in \mathfrak{h}$, $m \in \mathbb{Z}$, $\alpha, \beta \in \Delta$. These formulas generalize formulas (3.1.17)–(3.1.19), which correspond to $\mathfrak{g} = \mathfrak{sl}(2, \mathbb{F})$.

6.3. The Cases A_n, D_n, E_n

We continue in the setting of the last section.

Theorem 6.3.1: *Suppose that Δ spans $\mathfrak{h}$ and that for $\beta, \gamma \in \Delta$ there exist $\alpha_1, \ldots, \alpha_m \in \Delta$ such that*

$$\beta = \alpha_1, \gamma = \alpha_m \quad \text{and} \quad \langle \alpha_i, \alpha_{i+1} \rangle \neq 0 \quad \text{for} \quad i = 1, \ldots, m-1.$$

Then the Lie algebra $\mathfrak{g}$ is simple.

Proof: Let $\mathfrak{a}$ be a nonzero ideal of $\mathfrak{g}$. Under the adjoint action of $\mathfrak{h}$ on $\mathfrak{g}$, $\mathfrak{g}$ is a semisimple $\mathfrak{h}$-module [see (6.2.27), (6.2.28)], and it is easy to see that either $\mathfrak{a} \subset \mathfrak{h}$ or $x_\alpha \in \mathfrak{a}$ for some $\alpha \in \Delta$. In the second case,

$$\alpha = \varepsilon(\alpha, -\alpha)[x_\alpha, x_{-\alpha}] \in \mathfrak{a},$$

so that $\mathfrak{a} \cap \mathfrak{h} \neq 0$ in any case. Thus by the first hypothesis there exists $\beta \in \Delta$ such that $\langle \beta, \mathfrak{a} \cap \mathfrak{h} \rangle \neq 0$. Then $x_\beta \in \mathfrak{a}$, and so $\beta \in \mathfrak{a}$, as above. Using the second hypothesis, we obtain that $\gamma \in \mathfrak{a}$ for all $\gamma \in \Delta$. Hence $x_\gamma \in \mathfrak{a}$, and we see that $\mathfrak{a} = \mathfrak{g}$. ∎

Remark 6.3.2: It is easy to see that without the assumptions of Theorem 6.3.1, $\mathfrak{g}$ is a direct product of simple Lie algebras and (if Δ does not span $\mathfrak{h}$) an (abelian) subalgebra of $\mathfrak{h}$.

Remark 6.3.3: In the language of semisimple Lie theory (cf. [Bourbaki 2], [Humphreys], [Jacobson 1]), $\mathfrak{h}$ is a Cartan subalgebra of the reductive Lie algebra $\mathfrak{g}$. Suppose that Δ spans $\mathfrak{h}$. Then $\mathfrak{g}$ is semisimple, Δ is its root system with respect to $\mathfrak{h}$, which is identified with its dual $\mathfrak{h}^*$ by means of the form $\langle \cdot, \cdot \rangle$, and the x_α for $\alpha \in \Delta$ are corresponding root vectors. Together with a base of the root system Δ, the x_α's form a Chevalley basis of $\mathfrak{g}$. The sublattice

$$Q = \mathbb{Z}\Delta = \{\textstyle\sum n_i \alpha_i \mid n_i \in \mathbb{Z}, \alpha_i \in \Delta\} \tag{6.3.1}$$

of L generated by Δ is the *root lattice* of $\mathfrak{g}$, and its dual

$$P = Q^\circ = \{\alpha \in \mathfrak{h} \mid \langle \alpha, Q \rangle \subset \mathbb{Z}\} \tag{6.3.2}$$

[see (6.1.12)] is the *weight lattice*. (Cf. Remarks 3.1.1 and 4.3.5.) The affine Lie algebra $\hat{\mathfrak{g}}$ (or $\tilde{\mathfrak{g}}$) is an example of a Kac–Moody algebra (cf. [Kac 5]). The Lie algebras $\mathfrak{g}$ constructed in Theorem 6.2.1 are precisely those finite-dimensional reductive Lie algebras containing a Cartan subalgebra acting diagonally on $\mathfrak{g}$ and such that the simple factors of $\mathfrak{g}$ are simple Lie algebras with equal root lengths. In particular, given an equal-root-length indecomposable root system Φ whose roots α are normalized so that $\langle \alpha, \alpha \rangle = 2$, the corresponding root lattice R is even, and $R_2 = \Phi$ (a fact which may be verified case-by-case; cf. Remark 6.3.4 below). The determinant (6.1.6) for R is the determinant of the Cartan matrix of Φ.

We now list examples of positive definite even lattices L such that $\Delta = L_2$ spans $\mathfrak{h}$ and is indecomposable in the sense of Theorem 6.3.1. Thus the corresponding Lie algebra $\mathfrak{g}$ is simple. In each case, L is generated by Δ, i.e.,

$$L = Q = \mathbb{Z}\Delta, \tag{6.3.3}$$

in the notation (6.3.1). The notations A_n, D_n, E_n are the standard designations of the simple Lie algebras $\mathfrak{g}$, of the lattices Q and of the root systems Δ. In each case, the subscript n designates rank $Q = \dim \mathfrak{h}$. Recall that

$$\dim \mathfrak{g} = n + |\Delta|$$

[see (6.2.21)].

Remark 6.3.4: The following is the complete list of equal-root-length indecomposable root systems (cf. [Bourbaki 1], [Humphreys], [Jacobson 1]).

For $l \geq 1$, denote by V_l an l-dimensional rational vector space equipped with a positive definite symmetric form $\langle \cdot, \cdot \rangle$ and an orthonormal basis $\{v_1, \ldots, v_l\}$.

Type A_n, $n \geq 1$:

In V_{n+1}, take

$$Q_{A_n} = \left\{ \sum_{i=1}^{n+1} m_i v_i \,\middle|\, m_i \in \mathbb{Z}, \sum m_i = 0 \right\}. \tag{6.3.4}$$

Then

$$\Delta = \{\pm(v_i - v_j) \mid 1 \leq i < j \leq n+1\},$$

$$|\Delta| = n(n+1),$$

$$\dim \mathfrak{g} = (n+1)^2 - 1.$$

The case A_1 is the case $\mathfrak{g} = \mathfrak{sl}(2, \mathbb{F})$ (see Remark 6.2.3).

Type $D_n, n \geq 3$:

In V_n, take

$$Q_{D_n} = \left\{ \sum_{i=1}^{n} m_i v_i \,\middle|\, m_i \in \mathbb{Z}, \sum m_i \in 2\mathbb{Z} \right\}. \tag{6.3.5}$$

Then

$$\Delta = \{\pm v_i \pm v_j \mid 1 \leq i < j \leq n\},$$

$$|\Delta| = 2n(n-1),$$

$$\dim \mathfrak{g} = n(2n-1).$$

The case D_3 is the same as the case A_3. The same construction for $n = 2$ gives $A_1 \times A_1$.

Type E_8:

In V_8, take

$$\begin{aligned} Q_{E_8} &= Q_{D_8} + \tfrac{1}{2}\mathbb{Z} \sum_{i=1}^{8} v_i \\ &= \left\{ \sum_{i=1}^{8} m_i v_i \,\middle|\, \text{either} \quad m_1, \ldots, m_8 \in \mathbb{Z} \quad \text{or} \right. \\ &\qquad \left. m_1, \ldots, m_8 \in \mathbb{Z} + \tfrac{1}{2};\ \sum m_i \in 2\mathbb{Z} \right\}. \end{aligned} \tag{6.3.6}$$

Then

$$\Delta = \{\pm v_i \pm v_j \mid 1 \leq i < j \leq 8\} \cup \left\{ \sum_{i=1}^{8} m_i v_i \,\middle|\, m_i = \pm\tfrac{1}{2}, \sum m_i \in 2\mathbb{Z} \right\},$$

$$|\Delta| = 240,$$

$$\dim \mathfrak{g} = 248.$$

Type E_7:

In V_8, take

$$Q_{E_7} = Q_{E_8} \cap \left\{ \sum_{i=1}^{8} m_i v_i \,\middle|\, m_i \in \tfrac{1}{2}\mathbb{Z}, m_7 + m_8 = 0 \right\}. \tag{6.3.7}$$

Then

$$\Delta = \{\pm v_i \pm v_j \mid 1 \le i < j \le 6\} \cup \{\pm(v_7 - v_8)\}$$
$$\cup \left\{ \sum_{i=1}^{6} m_i v_i \pm \tfrac{1}{2}(v_7 - v_8) \mid m_i = \pm\tfrac{1}{2}, \sum_{i=1}^{6} m_i \in 2\mathbb{Z} \right\},$$
$$|\Delta| = 126,$$
$$\dim \mathfrak{g} = 133.$$

Type E_6:

In V_8, take

$$Q_{E_6} = Q_{E_7} \cap \left\{ \sum_{i=1}^{8} m_i v_i \mid m_i \in \tfrac{1}{2}\mathbb{Z}, m_6 = m_7 \right\}. \tag{6.3.8}$$

Then

$$\Delta = \{\pm v_i \pm v_j \mid 1 \le i < j \le 5\}$$
$$\cup \left\{ \pm \left(\sum_{i=1}^{5} m_i v_i + \tfrac{1}{2}(v_6 + v_7 - v_8) \right) \mid m_i = \pm\tfrac{1}{2}, \sum_{i=1}^{5} m_i \in 2\mathbb{Z} - \tfrac{1}{2} \right\},$$
$$|\Delta| = 72,$$
$$\dim \mathfrak{g} = 78.$$

We shall be especially interested in the E_8 case, which is distinguished by unimodularity:

Proposition 6.3.5: *The root lattice Q_{E_8} of type E_8 is self-dual.*

Proof: Since the lattice is even and hence integral, it is sufficient to check unimodularity [see (6.1.16)]. But this follows from the fact that Q_{D_8} has index 2 in both Q_{E_8} and the unimodular lattice $\sum_{i=1}^{8} \mathbb{Z} v_i$. ∎

Remark 6.3.6: The lattice Q_{E_8} is the only even unimodular lattice among the root lattices of simple Lie algebras (of types $A_n, B_n, C_n, D_n, E_6, E_7, E_8, F_4, G_2$), even allowing for possible rescaling of the lattices.

Remark 6.3.7: The rank of a positive definite even unimodular lattice is divisible by 8, and Q_{E_8} is the only one (up to isometry) of rank 8. There are exactly 2 (up to isometry) of rank 16: the decomposable root lattice $Q_{E_8} \oplus Q_{E_8}$ and a certain sublattice of index 2 of the weight lattice of D_{16} containing $Q_{D_{16}}$ with index 2. There are exactly 24 of rank 24 [Niemeier]. Among these lattices there are two extreme ones: the only one generated by its elements α with $\langle \alpha, \alpha \rangle = 2$—the decomposable root lattice $Q_{E_8} \oplus Q_{E_8} \oplus Q_{E_8}$—and the only one containing no elements α with $\langle \alpha, \alpha \rangle = 2$—the Leech lattice ([Leech 2], [Conway 3], [Niemeier]). For the other 22 such lattices L, the set L_2 is a root system which spans $L_{\mathbb{Q}}$ but does not generate L.

6.4. A Group of Automorphisms of $\mathfrak{g}$

We proceed to construct a natural family of automorphisms of the Lie algebra $\mathfrak{g}$ of (6.2.19).

We define the automorphism group (or isometry group) of our positive definite even lattice L as follows:

$$\operatorname{Aut}(L; \langle \cdot, \cdot \rangle) = \{g \in \operatorname{Aut} L \mid \langle g\alpha, g\beta \rangle = \langle \alpha, \beta \rangle \quad \text{for} \quad \alpha, \beta \in L\} \tag{6.4.1}$$

where Aut L denotes the automorphism group of L as an abelian group. Since the commutator map c_0 of our central extension $\hat{L}$ [recall (6.2.2)–(6.2.4)] is the reduction of $\langle \cdot, \cdot \rangle$ mod 2, we have

$$\operatorname{Aut}(L; \langle \cdot, \cdot \rangle) \subset \operatorname{Aut} c_0, \tag{6.4.2}$$

using the notation (5.4.3). Also set

$$\begin{aligned} \operatorname{Aut}(\hat{L}; \langle \cdot, \cdot \rangle) &= \{g \in \operatorname{Aut} \hat{L} \mid \bar{g} \in \operatorname{Aut}(L; \langle \cdot, \cdot \rangle)\} \\ &= \operatorname{Aut}(\hat{L}; \kappa, \langle \cdot, \cdot \rangle) = \{g \in \operatorname{Aut} \hat{L} \mid g\kappa = \kappa, \bar{g} \in \operatorname{Aut}(L; \langle \cdot, \cdot \rangle)\} \end{aligned} \tag{6.4.3}$$

[see (5.4.1), (5.4.2)]. By Proposition 5.4.1, we have:

Proposition 6.4.1: *The sequence*

$$1 \to \operatorname{Hom}(L, \mathbb{Z}/2\mathbb{Z}) \xrightarrow{*} \operatorname{Aut}(\hat{L}; \langle \cdot, \cdot \rangle) \xrightarrow{-} \operatorname{Aut}(L; \langle \cdot, \cdot \rangle) \to 1$$

is exact.

Note also that

$$\operatorname{Hom}(L, \mathbb{Z}/2\mathbb{Z}) \simeq (\mathbb{Z}/2\mathbb{Z})^{\operatorname{rank} L} \tag{6.4.4}$$

(see Remark 5.4.2).

Define the subgroup

$$\mathrm{Aut}(\mathfrak{g}; \mathfrak{h}, \langle \cdot, \cdot \rangle) = \{g \in \mathrm{Aut}\, \mathfrak{g} \mid g\mathfrak{h} = \mathfrak{h}, \langle gx, gy \rangle = \langle x, y \rangle \quad \text{for} \quad x, y \in \mathfrak{g}\} \tag{6.4.5}$$

of the automorphism group of the Lie algebra $\mathfrak{g}$ [see (6.2.19), (6.2.36)–(6.2.38)].

Proposition 6.4.2: *We have a natural homomorphism*

$$\mathrm{Aut}(\hat{L}; \langle \cdot, \cdot \rangle) \xrightarrow{*} \mathrm{Aut}(\mathfrak{g}; \mathfrak{h}, \langle \cdot, \cdot \rangle): \tag{6.4.6}$$

For $g \in \mathrm{Aut}(\hat{L}; \langle \cdot, \cdot \rangle)$ *the corresponding automorphism* g_* of $\mathfrak{g}$ *is given by:*

$$\begin{aligned} g_* h &= \bar{g}h \quad \text{for} \quad h \in \mathfrak{h} \\ g_* x_a &= x_{ga} \quad \text{for} \quad a \in \hat{\Delta}. \end{aligned} \tag{6.4.7}$$

In the case of the root lattice $Q = \mathbb{Z}\Delta$ [*see (6.3.1)*], *the map (6.4.6) is an embedding.*

Proof: It is clear that g_* is well defined [recall (6.2.20)] and lies in $\mathrm{Aut}(\mathfrak{g}; \mathfrak{h}, \langle \cdot, \cdot \rangle)$ [see (6.2.24)–(6.2.26) and (6.2.36)–(6.2.38)]. If $L = Q$ and $g_* = 1$, then $g = 1$ since $\hat{\Delta}$ generates $\hat{Q}$. ∎

Recall from Proposition 5.4.3 that the automorphisms $\theta \in \mathrm{Aut}(\hat{L}; \langle \cdot, \cdot \rangle)$ such that $\bar{\theta} = -1$ are the involutions [cf. (5.4.5)] of the form

$$\begin{aligned} \theta: \hat{L} &\to \hat{L} \\ a &\mapsto a^{-1}\kappa^{s_0(\bar{a})} \end{aligned} \tag{6.4.8}$$

where s_0 is the pullback to L of a quadratic form s_1 on $\check{L}$ with associated bilinear form c_1 as in (6.1.24). If the section e introduced in (6.2.5) is associated with a bilinear cocycle whose diagonal agrees with s_0, then

$$e_\alpha e_{-\alpha} = \kappa^{s_0(\alpha)} \quad \text{for} \quad \alpha \in L \tag{6.4.9}$$

$$\theta e_\alpha = e_{-\alpha} \quad \text{for} \quad \alpha \in L \tag{6.4.10}$$

$$[x_\alpha, x_{-\alpha}] = (-1)^{s_0(\alpha)}\alpha \quad \text{for} \quad \alpha \in \Delta \tag{6.4.11}$$

$$\theta_* x_\alpha = x_{-\alpha} \quad \text{for} \quad \alpha \in \Delta. \tag{6.4.12}$$

Among the automorphisms (6.4.8) there is a distinguished one, determined by the map q_0 of (6.1.26):

$$\begin{aligned} \theta_0 \colon \hat{L} &\to \hat{L} \\ a &\mapsto a^{-1}\kappa^{q_0(\bar{a})} = a^{-1}\kappa^{\langle \bar{a}, \bar{a}\rangle/2}. \end{aligned} \tag{6.4.13}$$

Then

$$(\theta_0)_* x_a = -x_{a^{-1}} \quad \text{for} \quad a \in \hat{\Delta} \tag{6.4.14}$$

[see (6.2.16)].

Remark 6.4.3: In the case $\mathfrak{g} = \mathfrak{sl}(2, \mathbb{F})$ as described in Section 3.1, $(\theta_0)_*$ does not satisfy (6.4.12), but instead, the condition

$$(\theta_0)_* x_\alpha = -x_{-\alpha} \quad \text{for} \quad \alpha \in \Delta, \tag{6.4.15}$$

since $\varepsilon_0 = 0$ (recall Remark 6.2.3). Note that the two involutions θ_1 and θ_2 of $\mathfrak{g}$ defined in Section 3.1 are related by:

$$\theta_2 = (\theta_0)_* \theta_1 = \theta_1 (\theta_0)_* . \tag{6.4.16}$$

Fix a lifting $\theta \colon \hat{L} \to \hat{L}$ of -1 as in (6.4.8), and for simplicity, write θ for the corresponding automorphism θ_* of $\mathfrak{g}$, so that

$$\theta|_{\mathfrak{h}} = -1, \tag{6.4.17}$$

$$\langle \theta x, \theta y \rangle = \langle x, y \rangle \quad \text{for} \quad x, y \in \mathfrak{g}. \tag{6.4.18}$$

For $a \in \hat{\Delta}$ [see (6.2.16)] set

$$x_a^{\pm} = x_a \pm \theta x_a = x_a \pm x_{\theta a} \tag{6.4.19}$$

and for $\alpha \in \Delta$ set

$$x_\alpha^{\pm} = x_\alpha \pm \theta x_\alpha, \tag{6.4.20}$$

as in (2.3.18).

The twisted affine Lie algebra

$$\tilde{\mathfrak{g}}[\theta] = \mathfrak{g}_{(0)} \otimes \mathbb{F}[t, t^{-1}] \oplus \mathfrak{g}_{(1)} \otimes t^{1/2}\mathbb{F}[t, t^{-1}] \oplus \mathbb{F}c \oplus \mathbb{F}d \tag{6.4.21}$$

[see (2.3.9)] may be described, by analogy with (6.2.49)–(6.2.60), as follows: It contains $\tilde{\mathfrak{h}}[-1]$ and its Heisenberg subalgebra

$$\hat{\mathfrak{h}}_{\mathbb{Z}+1/2} = \hat{\mathfrak{h}}[-1] = \coprod_{n \in \mathbb{Z}+1/2} \mathfrak{h} \otimes t^n \oplus \mathbb{F}c \tag{6.4.22}$$

[see (1.7.12)]. (We assume that $\mathfrak{h} \neq 0$.) The algebra $\tilde{\mathfrak{h}}[-1]$ has bracket relations

$$[\alpha \otimes t^m, \beta \otimes t^n] = \langle \alpha, \beta \rangle m \delta_{m+n,0} c \tag{6.4.23}$$

$$[d, \alpha \otimes t^m] = m\alpha \otimes t^m \tag{6.4.24}$$

$$[c, \alpha \otimes t^m] = [c, d] = 0 \tag{6.4.25}$$

for $\alpha, \beta \in \mathfrak{h}$, $m, n \in \mathbb{Z} + 1/2$.

Now consider the elements

$$x_a(z) = \tfrac{1}{2} \sum_{n \in \mathbb{Z}} (x_a^+ \otimes t^n) z^{-n} + \tfrac{1}{2} \sum_{n \in \mathbb{Z}+1/2} (x_a^- \otimes t^n) z^{-n}, \tag{6.4.26}$$

$$x_\alpha(z) = \tfrac{1}{2} \sum_{n \in \mathbb{Z}} (x_\alpha^+ \otimes t^n) z^{-n} + \tfrac{1}{2} \sum_{n \in \mathbb{Z}+1/2} (x_\alpha^- \otimes t^n) z^{-n}, \tag{6.4.27}$$

of $\tilde{\mathfrak{g}}[\theta]\{z\}$, for $a \in \hat{\Delta}$ and $\alpha \in \Delta$ [see (2.3.12)]. Then

$$x_{\kappa a}(z) = -x_a(z) \tag{6.4.28}$$

$$(x_{\theta a})(z) = \lim_{z^{1/2} \to -z^{1/2}} x_a(z) \tag{6.4.29}$$

$$(\theta x_\alpha)(z) = \lim_{z^{1/2} \to -z^{1/2}} x_\alpha(z) \tag{6.4.30}$$

[see (6.2.20), (2.3.13)]. Using Proposition 2.3.2, we have:

$$[h \otimes t^m, x_a(z)] = \langle h, \bar{a} \rangle z^m x_a(z) \tag{6.4.31}$$

$$[x_a(z_1), x_b(z_2)] = \begin{cases} \tfrac{1}{2}\bar{a}(z_2)\delta(z_1^{1/2}/z_2^{1/2}) - \tfrac{1}{2}D_{z_1}\delta(z_1^{1/2}/z_2^{1/2})c & \text{if } ab = 1 \\ \tfrac{1}{2}x_{ab}(z_2)\delta(z_1^{1/2}/z_2^{1/2}) & \text{if } ab \in \hat{\Delta} \\ 0 & \text{if } \langle \bar{a}, \bar{b} \rangle = 0 \end{cases} \tag{6.4.32}$$

$$[d, x_a(z)] = -Dx_a(z) \tag{6.4.33}$$

$$[c, x_a(z)] = 0 \tag{6.4.34}$$

for $h \in \mathfrak{h}$, $m \in \mathbb{Z} + 1/2$, $a, b \in \hat{\Delta}$, or equivalently,

$$[h \otimes t^m, x_\alpha(z)] = \langle h, \alpha \rangle z^m x_\alpha(z) \tag{6.4.35}$$

$$[x_\alpha(z_1), x_\beta(z_2)] = \begin{cases} \tfrac{1}{2}\varepsilon(\alpha, -\alpha)(\alpha(z_2)\delta(z_1^{1/2}/z_2^{1/2}) - D_{z_1}\delta(z_1^{1/2}/z_2^{1/2})c) & \text{if } \langle \alpha, \beta \rangle = -2 \\ \tfrac{1}{2}\varepsilon(\alpha, \beta)x_{\alpha+\beta}(z_2)\delta(z_1^{1/2}/z_2^{1/2}) & \text{if } \langle \alpha, \beta \rangle = -1 \\ 0 & \text{if } \langle \alpha, \beta \rangle = 0 \end{cases} \tag{6.4.36}$$

$$[d, x_\alpha(z)] = -Dx_\alpha(z) \tag{6.4.37}$$

$$[c, x_\alpha(z)] = 0 \tag{6.4.38}$$

for $h \in \mathfrak{h}$, $m \in \mathbb{Z} + 1/2$, $\alpha, \beta \in \Delta$. Note that these formulas generalize formulas (3.1.39)–(3.1.41) for $\mathfrak{sl}(2, \mathbb{F})$.

The structure of $\mathfrak{g}$ may be described in terms of the elements $x_a^\pm$ [see (6.4.19)] as follows:

$$\mathfrak{g} = \mathfrak{h} \oplus \sum_{a \in \hat{\Delta}} \mathbb{F}x_a^+ \oplus \sum_{a \in \hat{\Delta}} \mathbb{F}x_a^- \tag{6.4.39}$$

$$\mathfrak{g}_{(0)} = \sum_{a \in \hat{\Delta}} \mathbb{F}x_a^+ \tag{6.4.40}$$

$$\mathfrak{g}_{(1)} = \mathfrak{h} \oplus \sum_{a \in \hat{\Delta}} \mathbb{F}x_a^- \tag{6.4.41}$$

$$x_{\kappa\alpha}^\pm = -x_a^\pm \tag{6.4.42}$$

$$\theta x_a^\pm = x_{\theta a}^\pm = \pm x_a^\pm \tag{6.4.43}$$

$$[h, x_a^\pm] = \langle h, \bar{a} \rangle x_a^\mp \tag{6.4.44}$$

$$[x_a^\pm, x_b^\pm] = \begin{cases} x_{ab}^+ & \text{if} \quad \langle \bar{a}, \bar{b} \rangle = -1 \\ 0 & \text{if} \quad \langle \bar{a}, \bar{b} \rangle = 0 \end{cases} \tag{6.4.45}$$

$$[x_a^+, x_b^-] = \begin{cases} 2\bar{a} & \text{if} \quad ab = 1 \\ x_{ab}^- & \text{if} \quad \langle \bar{a}, \bar{b} \rangle = -1 \\ 0 & \text{if} \quad \langle \bar{a}, \bar{b} \rangle = 0 \end{cases} \tag{6.4.46}$$

$$\langle h, x_a^\pm \rangle = \langle x_a^+, x_b^- \rangle = 0 \tag{6.4.47}$$

$$\langle x_a^\pm, x_b^\pm \rangle = \begin{cases} 2 & \text{if} \quad ab = 1 \\ 0 & \text{if} \quad \bar{a} \neq \pm\bar{b} \end{cases} \tag{6.4.48}$$

for $a, b \in \hat{\Delta}$, $h \in \mathfrak{h}$.

For both the untwisted affine algebra $\tilde{\mathfrak{g}}$ [cf. (6.2.52)–(6.2.54)] and the twisted algebra $\tilde{\mathfrak{g}}[\theta]$, we use (2.3.22)–(2.3.24) to see that the elements $x_a^\pm(z)$ satisfy the conditions

$$x_a^\pm(z) = x_a(z) \pm \theta x_a(z) = x_a(z) \pm x_{\theta a}(z) \tag{6.4.49}$$

$$x_a^+(z) = \sum_{n \in \mathbb{Z}} (x_a^+ \otimes t^n) z^{-n} \tag{6.4.50}$$

$$x_a^-(z) = \sum_{n \in Z} (x_a^- \otimes t^n) z^{-n} \tag{6.4.51}$$

where $Z = \mathbb{Z}$ in the untwisted case, $Z = \mathbb{Z} + 1/2$ in the twisted case;

$$x_{\kappa a}^{\pm}(z) = -x_a^{\pm}(z) \tag{6.4.52}$$

$$\theta x_a^{\pm}(z) = x_{\theta\alpha}^{\pm}(z) = \pm x_a^{\pm}(z) \tag{6.4.53}$$

$$[h \otimes t^m, x_a^{\pm}(z)] = \langle h, \bar{a}\rangle z^m x_a^{\mp}(z) \tag{6.4.54}$$

where $h \in \mathfrak{h}$, and $m \in \mathbb{Z}$ in the untwisted case, $m \in \mathbb{Z} + 1/2$ in the twisted case;

$$[x_a^+(z_1), x_b^+(z_2)] = \begin{cases} -2(D\delta)(z_1/z_2)c & \text{if} \quad ab = 1 \\ x_{ab}^+(z_2)\delta(z_1/z_2) & \text{if} \quad \langle \bar{a}, \bar{b}\rangle = -1 \\ 0 & \text{if} \quad \langle \bar{a}, \bar{b}\rangle = 0 \end{cases} \tag{6.4.55}$$

$$[x_a^+(z_1), x_b^-(z_2)] = \begin{cases} 2\bar{a}(z_2)\delta(z_1/z_2) & \text{if} \quad ab = 1 \\ x_{ab}^-(z_2)\delta(z_1/z_2) & \text{if} \quad \langle \bar{a}, \bar{b}\rangle = -1 \\ 0 & \text{if} \quad \langle \bar{a}, \bar{b}\rangle = 0 \end{cases} \tag{6.4.56}$$

$$[x_a^-(z_1), x_b^-(z_2)] = \begin{cases} -2(D\delta)(z_1/z_2)c & \text{if} \quad ab = 1 \\ x_{ab}^+(z_2)\delta(z_1/z_2) & \text{if} \quad \langle \bar{a}, \bar{b}\rangle = -1 \\ 0 & \text{if} \quad \langle \bar{a}, \bar{b}\rangle = 0 \end{cases} \tag{6.4.57}$$

in the untwisted case and

$$[x_a^-(z_1), x_b^-(z_2)] = \begin{cases} -2D_{z_1}((z_1/z_2)^{1/2}\delta(z_1/z_2))c & \text{if} \quad ab = 1 \\ x_{ab}^+(z_2)(z_1/z_2)^{1/2}\delta(z_1/z_2) & \text{if} \quad \langle \bar{a}, \bar{b}\rangle = -1 \\ 0 & \text{if} \quad \langle \bar{a}, \bar{b}\rangle = 0 \end{cases} \tag{6.4.58}$$

in the twisted case, for $a, b \in \hat{\Delta}$.

7 Vertex Operator Realizations of $\hat{A}_n, \hat{D}_n, \hat{E}_n$

We shall now modify the vertex operators of Chapters 3 and 4 by incorporating appropriate central extensions of the type studied in Chapter 5. By explicitly calculating commutators of vertex operators we construct representations of the affine algebras of types $\hat{A}_n$, $\hat{D}_n$ and $\hat{E}_n$ and their θ-twisted analogues, θ a suitable involution, generalizing the case of $\hat{A}_1 = \mathfrak{sl}(2)\hat{}$ studied in Chapters 3 and 4. The untwisted case of this construction is due to [Frenkel–Kac] and [Segal 1], and the twisted case was announced in [FLM1].

Starting in a general setting with a lattice that is not necessarily even or positive definite, we introduce the modified untwisted vertex operators in Section 7.1, and we obtain the untwisted vertex operator representation of affine algebras in Section 7.2. Then in Section 7.3, we define the correspondingly modified twisted vertex operators, and finally, we obtain the twisted vertex operator representation in Section 7.4. In the twisted case, we must introduce a finite-dimensional tensor factor.

The results in this chapter, together with the earlier twisted construction of [Kac–Kazhdan–Lepowsky–Wilson], have been generalized in [Kac–Peterson 4] and independently, by a different method, in [Lepowsky 4].

7.1. The Untwisted Vertex Operators $X_{\mathbb{Z}}(a, z)$

We have already defined untwisted vertex operators $X(\alpha, z) = X_{\mathbb{Z}}(\alpha, z)$ in Chapter 4 [see (4.1.31)] and we have computed some of their commutators (Proposition 4.3.3). But we encountered an intrinsic problem, explained in Remark 4.3.4, when we considered computing $[X(\alpha, z_1), X(\beta, z_2)]$ for $\langle \alpha, \beta \rangle \in 2\mathbb{Z} + 1$. Using the concept of central extension of a lattice, we shall remedy this difficulty.

We start with a lattice L with symmetric form $\langle \cdot, \cdot \rangle$, and we assume only that L is nondegenerate—not necessarily even or positive definite. As in Section 6.2 (where L was positive definite), we set

$$\mathfrak{h} = L_{\mathbb{F}} = L \otimes_{\mathbb{Z}} \mathbb{F}, \tag{7.1.1}$$

a finite-dimensional $\mathbb{F}$-vector space with nonsingular symmetric form $\langle \cdot, \cdot \rangle$, and we identify L with $L \otimes 1 \subset \mathfrak{h}$.

As in Section 4.1, we view $\mathfrak{h}$ as an abelian Lie algebra and we form the $\mathbb{Z}$-graded untwisted affine Lie algebra

$$\tilde{\mathfrak{h}} = \coprod_{n \in \mathbb{Z}} \mathfrak{h} \otimes t^n \oplus \mathbb{F}c \oplus \mathbb{F}d \tag{7.1.2}$$

and its Heisenberg subalgebra

$$\hat{\mathfrak{h}}_{\mathbb{Z}} = \tilde{\mathfrak{h}}' = \coprod_{\substack{n \in \mathbb{Z} \\ n \neq 0}} \mathfrak{h} \otimes t^n \oplus \mathbb{F}c. \tag{7.1.3}$$

Consider the usual $\mathbb{Q}$-graded $\hat{\mathfrak{h}}_{\mathbb{Z}}$-irreducible $\tilde{\mathfrak{h}}$-module

$$M(1) = S(\hat{\mathfrak{h}}_{\mathbb{Z}}^{-}), \tag{7.1.4}$$

with c acting as 1, d acting as the degree operator and $\mathfrak{h}$ acting trivially. As in (4.1.3), we use the notation $\alpha(n)$ for the action of $\alpha \otimes t^n$ on $S(\hat{\mathfrak{h}}_{\mathbb{Z}}^{-})$, for $\alpha \in \mathfrak{h}$ and $n \in \mathbb{Z}$, so that [cf. (4.1.4)]

$$\begin{aligned} [\alpha(m), \beta(n)] &= \langle \alpha, \beta \rangle m \delta_{m+n,0} \\ [d, \alpha(m)] &= m\alpha(m) \end{aligned} \tag{7.1.5}$$

for $\alpha, \beta \in \mathfrak{h}$ and $m, n \in \mathbb{Z}$.

Let $(\hat{L}, {}^-)$ be a central extension of L by a finite cyclic group

$$\langle \kappa \rangle = \langle \kappa \mid \kappa^s = 1 \rangle \tag{7.1.6}$$

of order $s > 0$:

$$1 \to \langle \kappa \rangle \to \hat{L} \xrightarrow{-} L \to 1. \tag{7.1.7}$$

Denote by

$$c_0: L \times L \to \mathbb{Z}/s\mathbb{Z} \tag{7.1.8}$$

the associated commutator map, so that

$$aba^{-1}b^{-1} = \kappa^{c_0(\bar{a},\bar{b})} \quad \text{for} \quad a, b \in \hat{L}. \tag{7.1.9}$$

Recall that c_0 characterizes the central extension up to equivalence (see Proposition 5.2.3).

Let

$$\begin{aligned} e: L &\to \hat{L} \\ \alpha &\mapsto e_\alpha \end{aligned} \tag{7.1.10}$$

be a section of $\hat{L}$ such that

$$e_0 = 1, \tag{7.1.11}$$

and denote by

$$\varepsilon_0: L \times L \to \mathbb{Z}/s\mathbb{Z} \tag{7.1.12}$$

the corresponding 2-cocycle. Then

$$\begin{aligned} e_\alpha e_\beta &= e_{\alpha+\beta}\kappa^{\varepsilon_0(\alpha,\beta)} \\ \varepsilon_0(\alpha, \beta) + \varepsilon_0(\alpha + \beta, \gamma) &= \varepsilon_0(\beta, \gamma) + \varepsilon_0(\alpha, \beta + \gamma) \\ \varepsilon_0(\alpha, \beta) - \varepsilon_0(\beta, \alpha) &= c_0(\alpha, \beta) \\ \varepsilon_0(\alpha, 0) = \varepsilon_0(0, \alpha) &= 0 \end{aligned} \tag{7.1.13}$$

for $\alpha, \beta, \gamma \in L$ [see (5.1.5), (5.1.9), (5.1.11) and (5.2.13)].

We assume that the field $\mathbb{F}$ *contains a primitive* sth *root of unity*. Fix such an element

$$\omega \in \mathbb{F}^\times \tag{7.1.14}$$

and define the faithful character (cf., Section 5.5)

$$\chi: \langle\kappa\rangle \to \mathbb{F}^\times \tag{7.1.15}$$

by the condition

$$\chi(\kappa) = \omega. \tag{7.1.16}$$

Denote by $\mathbb{F}_\chi$ the one-dimensional space $\mathbb{F}$ viewed as a $\langle\kappa\rangle$-module on which $\langle\kappa\rangle$ acts according to χ:

$$\kappa \cdot 1 = \omega, \tag{7.1.17}$$

and denote by $\mathbb{F}\{L\}$ the induced $\hat{L}$-module

$$\mathbb{F}\{L\} = \mathrm{Ind}_{\langle\kappa\rangle}^{\hat{L}} \mathbb{F}_\chi = \mathbb{F}[\hat{L}] \otimes_{\mathbb{F}[\langle\kappa\rangle]} \mathbb{F}_\chi = \mathbb{F}[\hat{L}]/(\kappa - \omega)\mathbb{F}[\hat{L}] \quad (7.1.18)$$

[see (1.5.5)]. For $a \in \hat{L}$, we set

$$\iota(a) = a \otimes 1 \in \mathbb{F}\{L\}. \quad (7.1.19)$$

These elements span $\mathbb{F}\{L\}$ and

$$\begin{aligned} a \cdot \iota(b) &= \iota(ab) \\ \kappa \cdot \iota(b) &= \iota(\kappa b) = \omega\iota(b) \end{aligned} \quad (7.1.20)$$

for $a, b \in \hat{L}$. In particular, κ acts as multiplication by ω on $\mathbb{F}\{L\}$.

Define a map

$$\begin{aligned} c: L \times L &\to \mathbb{F}^\times \\ (\alpha, \beta) &\mapsto \omega^{c_0(\alpha,\beta)}. \end{aligned} \quad (7.1.21)$$

Then as operators on $\mathbb{F}\{L\}$,

$$ab = c(\bar{a}, \bar{b})ba \quad \text{for} \quad a, b \in \hat{L}. \quad (7.1.22)$$

Our choice of section allows us to identify $\mathbb{F}\{L\}$ with the group algebra $\mathbb{F}[L]$, viewed as a vector space, by the linear isomorphism

$$\begin{aligned} \mathbb{F}[L] &\xrightarrow{\sim} \mathbb{F}\{L\} \\ e^\alpha &\mapsto \iota(e_\alpha) \quad \text{for} \quad \alpha \in L \end{aligned} \quad (7.1.23)$$

[see (1.5.7)]. It is easy to describe the corresponding action of $\hat{L}$ on $\mathbb{F}[L]$: Define

$$\begin{aligned} \varepsilon: L \times L &\to \mathbb{F}^\times \\ (\alpha, \beta) &\mapsto \omega^{\varepsilon_0(\alpha,\beta)}. \end{aligned} \quad (7.1.24)$$

Then

$$\varepsilon(\alpha, \beta)\varepsilon(\alpha + \beta, \gamma) = \varepsilon(\beta, \gamma)\varepsilon(\alpha, \beta + \gamma) \quad (7.1.25)$$

$$\varepsilon(\alpha, \beta)/\varepsilon(\beta, \alpha) = c(\alpha, \beta) \quad (7.1.26)$$

$$\varepsilon(\alpha, 0) = \varepsilon(0, \alpha) = 1 \quad (7.1.27)$$

for $\alpha, \beta, \gamma \in L$ [cf. (6.2.30)–(6.2.34)]. As operators on $\mathbb{F}\{L\}$, we have

$$e_\alpha e_\beta = \varepsilon(\alpha, \beta)e_{\alpha+\beta} \quad \text{for} \quad \alpha, \beta \in L, \quad (7.1.28)$$

and the action of $\hat{L}$ on $\mathbb{F}[L]$ is given by:

$$e_\alpha \cdot e^\beta = \varepsilon(\alpha, \beta)e^{\alpha+\beta} \tag{7.1.29}$$

$$\kappa \cdot e^\beta = \omega e^\beta \tag{7.1.30}$$

for $\alpha, \beta \in L$.

Now, under the identification (7.1.23), $\mathbb{F}\{L\}$ can be embedded into the group algebra $\mathbb{F}[\mathfrak{h}]$, and we can correspondingly transfer various constructions for Chapter 4 to $\mathbb{F}\{L\}$: Set

$$V_L = M(1) \otimes \mathbb{F}\{L\} = S(\hat{\mathfrak{h}}_{\bar{\mathbb{Z}}}) \otimes \mathbb{F}\{L\}. \tag{7.1.31}$$

[Observe that this notation is consistent with (4.1.45).] We embed End $S(\hat{\mathfrak{h}}_{\bar{\mathbb{Z}}})$ and End $\mathbb{F}\{L\}$ into End V_L, and (End $S(\hat{\mathfrak{h}}_{\bar{\mathbb{Z}}})$)$\{z\}$ and (End $\mathbb{F}\{L\}$)$\{z\}$ into (End V_L)$\{z\}$.

Regard $S(\hat{\mathfrak{h}}_{\bar{\mathbb{Z}}})$ as a trivial $\hat{L}$-module and V_L as the corresponding tensor product $\hat{L}$-module, so that $a \in \hat{L}$ acts as the operator

$$a = 1 \otimes a \in \operatorname{End} V_L. \tag{7.1.32}$$

View $\mathbb{F}\{L\}$ as a trivial $\hat{\mathfrak{h}}_{\mathbb{Z}}$-module, and for $h \in \mathfrak{h}$, define

$$\begin{aligned} h(0)\colon \mathbb{F}\{L\} &\to \mathbb{F}\{L\} \\ \iota(a) &\mapsto \langle h, \bar{a}\rangle \iota(a) \end{aligned} \tag{7.1.33}$$

for $a \in \hat{L}$; note that this is a well-defined operator. Also define

$$z^h \in (\operatorname{End} \mathbb{F}\{L\})\{z\} \tag{7.1.34}$$

[understood as $z^{h(0)}$ as in (4.1.28)] by:

$$z^h \cdot \iota(a) = z^{\langle h, \bar{a}\rangle}\iota(a). \tag{7.1.35}$$

Then

$$[h(0), a] = \langle h, \bar{a}\rangle a \tag{7.1.36}$$

$$[h_1(0), z^{h_2}] = 0 \tag{7.1.37}$$

$$z^h a = z^{\langle h, \bar{a}\rangle} a z^h = a z^{h+\langle h, \bar{a}\rangle} \tag{7.1.38}$$

for $h, h_i \in \mathfrak{h}$, $a \in \hat{L}$. Give $\mathbb{F}\{L\}$ a (well-defined) $\mathbb{F}$-gradation (actually, a $\mathbb{Q}$-gradation) by:

$$\deg \iota(a) = -\tfrac{1}{2}\langle \bar{a}, \bar{a}\rangle \quad \text{for} \quad a \in \hat{L}, \tag{7.1.39}$$

and impose the tensor product gradation on V_L:

$$V_L = \coprod_{n \in \mathbb{Q}} (V_L)_n. \tag{7.1.40}$$

$S(\hat{\mathfrak{h}}_{\bar{\mathbb{Z}}})$ having the usual (shifted) gradation [see (1.9.51)]. Let $d \in \tilde{\mathfrak{h}}$ act as the degree operator on $\mathbb{F}\{L\}$ and on V_L, and give V_L the tensor product $\tilde{\mathfrak{h}}$-module structure. Then $\tilde{\mathfrak{h}}$, $\hat{L}$ and z^h ($h \in \mathfrak{h}$) act on V_L by:

$$\begin{aligned} c &\mapsto 1 \\ d &\mapsto d = d \otimes 1 + 1 \otimes d \\ h = h \otimes t^0 &\mapsto h(0) = 1 \otimes h(0) \quad \text{for} \quad h \in \mathfrak{h} \\ h \otimes t^n &\mapsto h(n) = h(n) \otimes 1 \quad \text{for} \quad h \in \mathfrak{h},\ n \in \mathbb{Z}\backslash\{0\} \\ a &\mapsto 1 \otimes a \qquad\qquad \text{for} \quad a \in \hat{L} \\ z^h &\mapsto 1 \otimes z^h \qquad\qquad \text{for} \quad h \in \mathfrak{h}. \end{aligned} \tag{7.1.41}$$

For $\alpha \in \mathfrak{h}$, define

$$\begin{aligned} E^{\pm}(\alpha, z) &= \exp\left(\sum_{n \in \pm\mathbb{Z}_+} \frac{\alpha(n)}{n} z^{-n} \right) \\ &\in (\operatorname{End} S(\hat{\mathfrak{h}}_{\bar{\mathbb{Z}}}))[[z^{\mp 1}]] \subset (\operatorname{End} V_L)\{z\}, \end{aligned} \tag{7.1.42}$$

as in (4.1.12). Then Proposition 4.1.1 applies to these expressions. Define the (*untwisted*) *vertex operator* $X(a, z) \in (\operatorname{End} V_L)\{z\}$ associated with $a \in \hat{L}$ by:

$$\begin{aligned} X(a, z) = X_{\mathbb{Z}}(a, z) &= E^-(-\bar{a}, z)E^+(-\bar{a}, z)az^{\bar{a}+\langle \bar{a},\bar{a}\rangle/2} \\ &= {:}e^{D^{-1}(\bar{a}(z)-\bar{a}(0))}az^{\bar{a}}{:} \\ &= {:}X(a, z){:} \end{aligned} \tag{7.1.43}$$

[cf. (4.1.31) and (4.2.6); the normal-ordered expressions are explained below]. Then also [cf. (4.1.33)]

$$X(a, z) = E^-(-\bar{a}, z)E^+(-\bar{a}, z)z^{\bar{a}-\langle \bar{a},\bar{a}\rangle/2}a. \tag{7.1.44}$$

We define operators $x_a(n)$ on V_L for $n \in \mathbb{Q}$ by:

$$X(a, z) = \sum_{n \in \mathbb{Q}} x_a(n)z^{-n}. \tag{7.1.45}$$

As in Section 4.1, we have the following elementary properties:

$$[h(m), X(a, z)] = \langle h, \bar{a}\rangle z^m X(a, z) \tag{7.1.46}$$

$$[h(m), x_a(n)] = \langle h, \bar{a}\rangle x_a(m + n) \tag{7.1.47}$$

$$[d, X(a, z)] = -DX(a, z) \tag{7.1.48}$$

$$\deg x_a(n) = n \tag{7.1.49}$$

for $h \in \mathfrak{h}$, $m \in \mathbb{Z}$, $a \in \hat{L}$ and $n \in \mathbb{Q}$. Moreover,

$$\begin{aligned} X(\kappa a, z) &= \omega X(a, z) \\ x_{\kappa a}(n) &= \omega x_a(n) \end{aligned} \tag{7.1.50}$$

for $a \in \hat{L}$, $n \in \mathbb{Q}$ [see (7.1.20)].

Remark 7.1.1: The case $a = 1$ gives

$$X(1, z) = 1$$

$$x_1(n) = \delta_{n,0} \quad \text{for} \quad n \in \mathbb{Q}.$$

Note that under the identification (7.1.23), the new vertex operators are related to the old ones (4.1.31) by:

$$X(e_\alpha, z) = X(\alpha, z)\varepsilon_\alpha \quad \text{for} \quad \alpha \in L, \tag{7.1.51}$$

where

$$\begin{aligned} \varepsilon_\alpha &: \mathbb{F}[L] \to \mathbb{F}[L] \\ e^\beta &\mapsto \varepsilon(\alpha, \beta)e^\beta \quad \text{for} \quad \beta \in L \end{aligned} \tag{7.1.52}$$

[cf. (7.1.29)]. We modify the earlier definition of the operators $x_\alpha(n)$ [see (4.1.34)] by multiplying them by ε_α:

$$X(e_\alpha, z) = \sum_{n \in \mathbb{Q}} x_\alpha(n) z^{-n}. \tag{7.1.53}$$

Concepts associated with normal ordering (Section 4.2) also carry over in obvious ways. Recalling $\alpha(z)$ and $\alpha(z)^\pm$ from (4.2.1), we define:

$$\begin{aligned} {:}\alpha(z)a{:} = {:}a\alpha(z){:} &= \alpha(z)^- a + a\alpha(z)^+ \\ &= (\alpha(z) - \tfrac{1}{2}\langle \alpha, \bar{a}\rangle)a = a(\alpha(z) + \tfrac{1}{2}\langle \alpha, \bar{a}\rangle) \end{aligned} \tag{7.1.54}$$

$$\begin{aligned} {:}\alpha(0)a{:} = {:}a\alpha(0){:} &= \tfrac{1}{2}(\alpha(0)a + a\alpha(0)) \\ &= (\alpha(0) - \tfrac{1}{2}\langle \alpha, \bar{a}\rangle)a = a(\alpha(0) + \tfrac{1}{2}\langle \alpha, \bar{a}\rangle) \end{aligned} \tag{7.1.55}$$

$$:z^\alpha a: \; = \; :az^\alpha: \; = z^{\alpha - \langle \alpha, \bar{a}\rangle/2} a = a z^{\alpha + \langle \alpha, \bar{a}\rangle/2} \tag{7.1.56}$$

$$:\alpha(z)X(a, z): \; = \; :X(a, z)\alpha(z): \; = \alpha(z)^- X(a, z) + X(a, z)\alpha(z)^+ \tag{7.1.57}$$

$$\begin{aligned} &:X(a, z_1)X(b, z_2): \\ &\quad = \; :e^{D^{-1}(\bar{a}(z_1) - \bar{a}(0)) + D^{-1}(\bar{b}(z_2) - \bar{b}(0))} a b z_1^{\bar{a}} z_2^{\bar{b}}: \\ &\quad = \; :e^{D^{-1}(\bar{a}(z_1) - \bar{a}(0)) + D^{-1}(\bar{b}(z_2) - \bar{b}(0))}: ab z_1^{\bar{a} + \langle \bar{a}, \bar{a} + \bar{b}\rangle/2} z_2^{\bar{b} + \langle \bar{b}, \bar{a} + \bar{b}\rangle/2} \\ &\quad = E^-(-\bar{a}, z_1)E^-(-\bar{b}, z_2)E^+(-\bar{a}, z_1)E^+(-\bar{b}, z_2) \\ &\qquad \cdot ab z_1^{\bar{a} + \langle \bar{a}, \bar{a} + \bar{b}\rangle/2} z_2^{\bar{b} + \langle \bar{b}, \bar{a} + \bar{b}\rangle/2} \end{aligned} \tag{7.1.58}$$

for $\alpha \in \mathfrak{h}$, $a, b \in \hat{L}$. (Note the case $b = 1$.) In place of the "commutativity" relation (4.3.2), we have

$$:X(a, z_1)X(b, z_2): \; = c(\bar{a}, \bar{b}) :X(b, z_2)X(a, z_1): \tag{7.1.59}$$

for $a, b \in \hat{L}$ [see (7.1.22)], and (4.3.3) becomes

$$\begin{aligned} &X(a, z_1)X(b, z_2) \\ &\quad = \; :X(a, z_1)X(b, z_2): (z_2/z_1)^{-\langle \bar{a}, \bar{b}\rangle/2}(1 - z_2/z_1)^{\langle \bar{a}, \bar{b}\rangle}. \end{aligned} \tag{7.1.60}$$

Also,

$$\lim_{z_1 \to z_2} :X(a, z_1)X(b, z_2): \; = X(ab, z_2) \tag{7.1.61}$$

$$DX(a, z) = \; :\bar{a}(z)X(a, z): \tag{7.1.62}$$

$$D_{z_1} :X(a, z_1)X(b, z_2): \; = \; :\bar{a}(z_1)X(a, z_1)X(b, z_2): \tag{7.1.63}$$

$$\lim_{z_1 \to z_2} D_{z_1} :X(a, z_1)X(b, z_2): \; = \; :\bar{a}(z_2)X(ab, z_2): \tag{7.1.64}$$

for $a, b \in \hat{L}$.

For a subset M of L, write

$$\hat{M} = \{a \in \hat{L} \mid \bar{a} \in M\}, \tag{7.1.65}$$

and set

$$\mathbb{F}\{M\} = \operatorname{span}\{\iota(a) \mid a \in \hat{M}\} \subset \mathbb{F}\{L\}, \tag{7.1.66}$$

$$V_M = S(\hat{\mathfrak{h}}_{\mathbb{Z}}^-) \otimes \mathbb{F}\{M\} \subset V_L \tag{7.1.67}$$

[cf. (4.1.44), (4.1.45)].

Remark 7.1.2: If the lattice L is positive definite, the graded dimension of the space V_L is given by

$$\dim_* V_L = \frac{\theta_L(q)}{\eta(q)^{\operatorname{rank} L}} \tag{7.1.68}$$

[see Section 1.10 and (6.1.30); cf. Corollary 4.5.4].

Remark 7.1.3: The Virasoro algebra $\mathfrak{v}$ acts in a natural way on V_L, motivating the definition of the grading (cf. Remark 4.1.2). Also recall from (1.9.54)–(1.9.56) the notion of weight of a homogeneous element. For instance,

$$\mathrm{wt}\,\iota(a) = \tfrac{1}{2}\langle \bar{a}, \bar{a}\rangle \quad \text{for} \quad a \in \hat{L}. \tag{7.1.69}$$

7.2. Construction of $\hat{A}_n$, $\hat{D}_n$, $\hat{E}_n$

We are now prepared to compute certain commutators of the vertex operators $X(a, z)$. We begin with the following analogue and refinement of Proposition 4.3.3:

Theorem 7.2.1: *Let $a, b \in \hat{L}$ and assume that*

$$\langle \bar{a}, \bar{b}\rangle \in \mathbb{Z} \tag{7.2.1}$$

and that [*see (7.1.21)*]

$$c(\bar{a}, \bar{b}) = (-1)^{\langle \bar{a}, \bar{b}\rangle}. \tag{7.2.2}$$

[*In particular, $\langle \bar{a}, \bar{b}\rangle$ must lie in $2\mathbb{Z}$ if s* [*see (7.1.6)*] *is odd.*] *Then*

$$[X(a, z_1), X(b, z_2)]$$
$$= \begin{cases} 0 & \text{if} \quad \langle \bar{a}, \bar{b}\rangle \geq 0 \\ X(ab, z_2)(z_1/z_2)^{\bar{a} + \langle \bar{a}, \bar{a}\rangle/2}\delta(z_1/z_2) & \text{if} \quad \langle \bar{a}, \bar{b}\rangle = -1 \\ \bar{a}(z_2)(z_1/z_2)^{\bar{a}}\delta(z_1/z_2) - D_{z_1}((z_1/z_2)^{\bar{a}}\delta(z_1/z_2)) & \\ \qquad \text{if} \quad \langle \bar{a}, \bar{a}\rangle = 2 \quad \text{and} \quad b = a^{-1}. \end{cases} \tag{7.2.3}$$

Proof: We simply repeat the proof of Proposition 4.3.3, noting that the previously excluded cases $\langle \bar{a}, \bar{b}\rangle \in 2\mathbb{N} + 1$, $\langle \bar{a}, \bar{b}\rangle = -1$ can now be handled because of (7.2.2). The computation begins:

$$\begin{aligned} [X(a, z_1), X(b, z_2)] &= X(a, z_1)X(b, z_2) - X(b, z_2)X(a, z_1) \\ &= {:}X(a, z_1)X(b, z_2){:}\{(z_2/z_1)^{-\langle \bar{a}, \bar{b}\rangle/2}(1 - z_2/z_1)^{\langle \bar{a}, \bar{b}\rangle} \\ &\quad - (-1)^{\langle \bar{a}, \bar{b}\rangle}(z_1/z_2)^{-\langle \bar{a}, \bar{b}\rangle/2}(1 - z_1/z_2)^{\langle \bar{a}, \bar{b}\rangle}\} \end{aligned} \tag{7.2.4}$$

[cf. (4.3.5)], and if $\langle \bar{a}, \bar{b}\rangle \in \mathbb{N}$, the expression in braces vanishes. The last case having been covered in the earlier proof, we assume that $\langle \bar{a}, \bar{b}\rangle = -1$.

Then (7.2.4) becomes

$$
\begin{aligned}
&:X(a, z_1)X(b, z_2):(z_1/z_2)^{\langle \bar{a}, \bar{b}\rangle/2}\delta(z_1/z_2) \\
&= :e^{D^{-1}(\bar{a}(z_1)-\bar{a}(0))+D^{-1}(\bar{b}(z_2)-\bar{b}(0))}: \\
&\quad \cdot abz_1^{\bar{a}+\langle \bar{a},\bar{a}\rangle/2+\langle \bar{a},\bar{b}\rangle}z_2^{\bar{b}+\langle \bar{b},\bar{b}\rangle/2}\delta(z_1/z_2) \\
&= :e^{D^{-1}(\bar{a}(z_1)-\bar{a}(0))+D^{-1}(\bar{b}(z_2)-\bar{b}(0))}: \\
&\quad \cdot ab(z_1/z_2)^{\bar{a}+\langle \bar{a},\bar{a}\rangle/2+\langle \bar{a},\bar{b}\rangle}z_2^{(\bar{a}+\bar{b})+\langle \bar{a}+\bar{b},\bar{a}+\bar{b}\rangle/2}\delta(z_1/z_2) \\
&= X(ab, z_2)(z_1/z_2)^{\bar{a}+\langle \bar{a},\bar{a}\rangle/2}\delta(z_1/z_2),
\end{aligned}
$$

by Proposition 2.1.1(b). ∎

In terms of the choice of section and 2-cocycle, Theorem 7.2.1 asserts [taking into account (7.1.13), (7.1.26) and (7.1.50)]:

Corollary 7.2.2: *Let $\alpha, \beta \in L$ and assume that*

$$\langle \alpha, \beta\rangle \in \mathbb{Z} \tag{7.2.5}$$

and that

$$\varepsilon(\alpha, \beta)/\varepsilon(\beta, \alpha) = (-1)^{\langle \alpha,\beta\rangle}. \tag{7.2.6}$$

Then

$$
[X(e_\alpha, z_1), X(e_\beta, z_2)]
= \begin{cases}
0 & \text{if} \quad \langle \alpha, \beta\rangle \geq 0 \\
\varepsilon(\alpha, \beta)X(e_{\alpha+\beta}, z_2)(z_1/z_2)^{\alpha+\langle \alpha,\alpha\rangle/2}\delta(z_1/z_2) & \text{if} \quad \langle \alpha, \beta\rangle = -1 \\
\varepsilon(\alpha, -\alpha)(\alpha(z_2)(z_1/z_2)^{\alpha}\delta(z_1/z_2) - D_{z_1}((z_1/z_2)^{\alpha}\delta(z_1/z_2))) & \\
\qquad\qquad \text{if} \quad \langle \alpha, \alpha\rangle = 2 \quad \text{and} \quad \beta = -\alpha. &
\end{cases}
\tag{7.2.7}
$$

Suppose that

$$\theta \in \operatorname{Aut}(\hat{L}; \kappa, \langle \cdot, \cdot\rangle) \tag{7.2.8}$$

is such that

$$
\begin{aligned}
&\bar{\theta} = -1 \quad \text{on} \quad L, \\
&\theta^2 = 1,
\end{aligned}
\tag{7.2.9}
$$

using the notation (6.4.3). For $a \in \hat{L}$, set

$$X^{\pm}(a, z) = X(a, z) \pm X(\theta a, z) \tag{7.2.10}$$

and define operators $x_a^{\pm}(n)$ on V_L for $n \in \mathbb{Q}$ by

$$X^{\pm}(a, z) = \sum_{n \in \mathbb{Q}} x_a^{\pm}(n) z^{-n}, \tag{7.2.11}$$

so that

$$x_a^{\pm}(n) = x_a(n) \pm x_{\theta a}(n). \tag{7.2.12}$$

Analogously, define operators $x_\alpha^{\pm}(n)$ for $\alpha \in L$ by

$$X^{\pm}(e_\alpha, z) = \sum_{n \in \mathbb{Q}} x_\alpha^{\pm}(n) z^{-n}. \tag{7.2.13}$$

Theorem 7.2.1 and Corollary 7.2.2 imply commutation formulas for these elements. In particular, using (7.1.46)–(7.1.50), we obtain the following result, which should be compared with (6.4.52)–(6.4.58):

Corollary 7.2.3: *Let $a, b \in \hat{L}$ as in Theorem 7.2.1, and suppose that*

$$\langle \bar{a}, L \rangle \subset \tfrac{1}{2}\mathbb{Z}. \tag{7.2.14}$$

Then

$$X^{\pm}(\theta a, z) = \pm X^{\pm}(a, z) \tag{7.2.15}$$

$$[d, X^{\pm}(a, z)] = -DX^{\pm}(a, z) \tag{7.2.16}$$

$$X^{\pm}(\kappa a, z) = \omega X^{\pm}(a, z) \tag{7.2.17}$$

$$[h(m), X^{\pm}(a, z)] = \langle h, \bar{a} \rangle z^m X^{\mp}(a, z) \quad \text{for} \quad h \in \mathfrak{h} \quad \text{and} \quad m \in \mathbb{Z} \tag{7.2.18}$$

$$[X^{+}(a, z_1), X^{+}(b, z_2)] = \begin{cases} 0 & \text{if } \langle \bar{a}, \bar{b} \rangle = 0 \\ X^{+}(ab, z_2)(z_1/z_2)^{\bar{a} + \langle \bar{a}, \bar{a} \rangle/2} \delta(z_1/z_2) & \\ \qquad \text{if } \langle \bar{a}, \bar{b} \rangle = -1 & \\ -2D_{z_1}((z_1/z_2)^{\bar{a}} \delta(z_1/z_2)) & \\ \qquad \text{if } \langle \bar{a}, \bar{a} \rangle = 2 \text{ and } b = a^{-1} & \end{cases} \tag{7.2.19}$$

$$[X^+(a, z_1), X^-(b, z_2)] = \begin{cases} 0 & \text{if} \quad \langle \bar{a}, \bar{b} \rangle = 0 \\ X^-(ab, z_2)(z_1/z_2)^{\bar{a} + \langle \bar{a}, \bar{a} \rangle /2} \delta(z_1/z_2) \\ \qquad \text{if} \quad \langle \bar{a}, \bar{b} \rangle = -1 \\ 2\bar{a}(z_2)(z_1/z_2)^{\bar{a}} \delta(z_1/z_2) \\ \qquad \text{if} \quad \langle \bar{a}, \bar{a} \rangle = 2 \quad \text{and} \quad b = a^{-1} \end{cases} \tag{7.2.20}$$

$$[X^-(a, z_1), X^-(b, z_2)] = \begin{cases} 0 & \text{if} \quad \langle \bar{a}, \bar{b} \rangle = 0 \\ X^+(ab, z_2)(z_1/z_2)^{\bar{a} + \langle \bar{a}, \bar{a} \rangle /2} \delta(z_1/z_2) \\ \qquad \text{if} \quad \langle \bar{a}, \bar{b} \rangle = -1 \\ -2D_{z_1}((z_1/z_2)^{\bar{a}} \delta(z_1/z_2)) \\ \qquad \text{if} \quad \langle \bar{a}, \bar{a} \rangle = 2 \quad \text{and} \quad b = a^{-1}. \end{cases} \tag{7.2.21}$$

Now we apply Theorem 7.2.1 to the vertex operator construction of the finite-dimensional simple Lie algebras A_n, D_n, E_n and their affinizations $\hat{A}_n, \hat{D}_n, \hat{E}_n$. Take the nondegenerate lattice L to be even and take for c_0 the alternating $\mathbb{Z}$-bilinear map

$$c_0(\alpha, \beta) = \langle \alpha, \beta \rangle + 2\mathbb{Z} \tag{7.2.22}$$

as in Section 6.2, with $\hat{L}$ the corresponding central extension of L by the 2-element cyclic group $\langle \kappa \mid \kappa^2 = 1 \rangle$. Note that

$$\omega = -1 \tag{7.2.23}$$

[see (7.1.14)] and that

$$c(\alpha, \beta) = (-1)^{\langle \alpha, \beta \rangle} \quad \text{for} \quad \alpha, \beta \in L \tag{7.2.24}$$

[see (7.1.21)].

With the choice of section and 2-cocycle as in Section 6.2 or Section 7.1, we have

$$\varepsilon(\alpha, \beta) = (-1)^{\varepsilon_0(\alpha, \beta)} \quad \text{for} \quad \alpha, \beta \in L \tag{7.2.25}$$

[see (6.2.30) and (7.1.24)], with properties (6.2.32)–(6.2.34) [or (7.1.25)–(7.1.27)]. In particular,

$$\varepsilon(\alpha, \beta)/\varepsilon(\beta, \alpha) = (-1)^{\langle \alpha, \beta \rangle} \quad \text{for} \quad \alpha, \beta \in L. \tag{7.2.26}$$

Keeping in mind the evenness of L, we see that Theorem 7.2.1 and Corollary 7.2.2 take on the following simpler forms:

Corollary 7.2.4: *For $a, b \in \hat{L}$,*

$$[X(a, z_1), X(b, z_2)]$$
$$= \begin{cases} 0 & \text{if} \quad \langle \bar{a}, \bar{b} \rangle \geq 0 \\ X(ab, z_2)\delta(z_1/z_2) & \text{if} \quad \langle \bar{a}, \bar{b} \rangle = -1 \\ \bar{a}(z_2)\delta(z_1/z_2) - (D\delta)(z_1/z_2) & \text{if} \quad \langle \bar{a}, \bar{a} \rangle = 2 \quad \text{and} \quad b = a^{-1}. \end{cases} \tag{7.2.27}$$

Corollary 7.2.5: *For $\alpha, \beta \in L$,*

$$[X(e_\alpha, z_1), X(e_\beta, z_2)]$$
$$= \begin{cases} 0 \quad \text{if} \quad \langle \alpha, \beta \rangle \geq 0 \\ \varepsilon(\alpha, \beta) X(e_{\alpha+\beta}, z_2)\delta(z_1/z_2) \\ \qquad \text{if} \quad \langle \alpha, \beta \rangle = -1 \\ \varepsilon(\alpha, -\alpha)(\alpha(z_2)\delta(z_1/z_2) - (D\delta)(z_1/z_2)) \\ \qquad \text{if} \quad \langle \alpha, \alpha \rangle = 2 \quad \text{and} \quad \beta = -\alpha. \end{cases} \tag{7.2.28}$$

Now let L be positive definite as well as even. Recall the Lie algebra $\mathfrak{g}$ and invariant symmetric form $\langle \cdot, \cdot \rangle$ of Theorem 6.2.1, and recall the formal-variable description of $\tilde{\mathfrak{g}}$ in (6.2.49)–(6.2.60). If we take $a, b \in \hat{\Delta}$ in Corollary 7.2.4 [cf. (6.2.16)] or $\alpha, \beta \in \Delta$ in Corollary 7.2.5 [see also (7.1.46), (7.1.50)], we find that we have represented $\tilde{\mathfrak{g}}$ by vertex operators:

Theorem 7.2.6: *The linear map*

$$\pi: \tilde{\mathfrak{g}} \to \operatorname{End} V_L \tag{7.2.29}$$

determined by

$$\pi: c \mapsto 1$$

$$\pi: d \mapsto d$$

$$\pi: h \otimes t^n \mapsto h(n) \qquad \text{for} \quad h \in \mathfrak{h}, \quad n \in \mathbb{Z}$$

$$\pi: x_a \otimes t^n \mapsto x_a(n) \qquad \text{for} \quad a \in \hat{\Delta}, \quad n \in \mathbb{Z},$$

or equivalently in the last case,

$$\pi: x_a(z) \mapsto X(a, z) \qquad \text{for} \quad a \in \hat{\Delta}$$

or

$$\pi: x_\alpha \otimes t^n \mapsto x_\alpha(n) \qquad \text{for} \quad \alpha \in \Delta, \quad n \in \mathbb{Z}$$

[see (7.1.53)] or

$$\pi : x_\alpha(z) \mapsto X(e_\alpha, z) \quad \text{for} \quad \alpha \in \Delta,$$

is a representation of $\tilde{\mathfrak{g}}$ on V_L.

Let Q be the $\mathbb{Z}$-span of $\Delta = L_2$ in L:

$$Q = \mathbb{Z}\Delta \subset L \tag{7.2.30}$$

[cf. (6.3.1)]. Note that we are not assuming that Δ spans $\mathfrak{h}$. The proof of Proposition 4.4.2 shows:

Proposition 7.2.7: *The space V_L is irreducible as a $\hat{\mathfrak{g}}$-module if and only if $L = Q$. In general, consider the Q-coset decomposition*

$$L = \bigcup_i (Q + \lambda_i)$$

of L. Then V_L decomposes into the following direct sum of $\hat{\mathfrak{g}}$-irreducible, $\hat{\mathfrak{g}}$-inequivalent $\tilde{\mathfrak{g}}$-submodules:

$$V_L = \coprod_i V_{Q+\lambda_i},$$

using the notation (7.1.67).

Remark 7.2.8: If we take for L a rank 1 lattice Q generated by α_1 with $\langle \alpha_1, \alpha_1 \rangle = 2$, we obtain the construction of $\mathfrak{sl}(2, \mathbb{F})^{\sim}$ on V_Q included in the statement of Theorem 4.4.1(a). If more generally we take for L the root lattice of a simple Lie algebra of type A_n, D_n or E_n (see Section 6.3), then we obtain a vertex operator construction of the untwisted affine Kac–Moody algebra $\tilde{A}_n$, $\tilde{D}_n$ or $\tilde{E}_n$. The irreducible module V_Q is called the *(distinguished) basic module.*

Remark 7.2.9: It is possible to remove the evenness assumption on L and to construct irreducible vertex operator representations of the untwisted affine algebra $\tilde{\mathfrak{g}}$ on the spaces $V_{Q+\lambda_i}$, $Q + \lambda_i$ a coset of Q in L such that $\langle Q, Q + \lambda_i \rangle \subset \mathbb{Z}$. For instance, we can take L as the weight lattice P in case Δ spans $\mathfrak{h}$ [see (6.3.2)], generalizing the rest of Theorem 4.4.1(a). The resulting irreducible modules for the affine algebras $\tilde{A}_n$, $\tilde{D}_n$, $\tilde{E}_n$ are called the *basic modules*. It is also possible to "shift" these constructions to cosets M of L or Q in $\mathfrak{h}$ for which $\langle Q, M \rangle \not\subset \mathbb{Z}$, and we thereby obtain vertex operator constructions of twisted affine algebras generalizing Theorem 4.4.1(b).

Remark 7.2.10: Vertex operator commutators provide a natural proof of the fact that $\mathfrak{g}$ is a Lie algebra and $\langle \cdot, \cdot \rangle$ is an invariant symmetric form (Theorem 6.2.1), a proof completely different from the direct check in Chapter 6: Observe that the operators 1, d, $h(n)$ and $x_\alpha(n)$ for $h \in \mathfrak{h}$, $\alpha \in \Delta$ and $n \in \mathbb{Z}$ span a Lie algebra, say $\mathfrak{l}$, of operators on V_L, by (7.1.5), (7.1.47), (7.1.48) and (7.2.28). Form the vector space

$$\tilde{\mathfrak{g}} = \mathfrak{g} \otimes \mathbb{F}[t, t^{-1}] \oplus \mathbb{F}c \oplus \mathbb{F}d,$$

with $\mathfrak{g}$ the vector space defined in (6.2.22), and define a linear map

$$\pi: \tilde{\mathfrak{g}} \to \mathfrak{l}$$

as in Theorem 7.2.6. Then π is certainly injective on

$$\tilde{\mathfrak{h}} = \mathfrak{h} \otimes \mathbb{F}[t, t^{-1}] \oplus \mathbb{F}c \oplus \mathbb{F}d,$$

and the commutation action of $\pi(d)$ and $\pi(\mathfrak{h})$ on the operators $x_\alpha(n)$ shows that π is injective on $\tilde{\mathfrak{g}}$. Thus π is a linear isomorphism. But the Lie algebra structure on $\tilde{\mathfrak{g}}$ inherited from π is clearly just the affinization, in the sense of (1.6.5), of the nonassociative algebra $\mathfrak{g}$ defined in (6.2.27)–(6.2.29) with the nonsingular bilinear form $\langle \cdot, \cdot \rangle$ defined in (6.2.39)–(6.2.40). It follows from Remark 1.6.1 that $\mathfrak{g}$ is a Lie algebra and that $\langle \cdot, \cdot \rangle$ is symmetric and invariant. Also, π of course defines a representation of $\tilde{\mathfrak{g}}$ on V_L. This argument shows in fact that vertex operator commutators can be used to motivate the very definition of the Lie algebra $\mathfrak{g}$ and the invariant symmetric form $\langle \cdot, \cdot \rangle$. In particular, the exceptional simple Lie algebras E_6, E_7 and E_8 appear naturally as Lie algebras of operators built from their root lattices.

7.3. The Twisted Vertex Operators $X_{\mathbb{Z}+1/2}(a, z)$

In Section 7.1 we modified the untwisted vertex operators of Chapter 4 to permit adequate calculation of commutators. Here we modify the twisted vertex operators of Chapter 3 analogously.

As in Section 7.1, let L be a nondegenerate lattice with symmetric form $\langle \cdot, \cdot \rangle$, and set

$$\mathfrak{h} = L_{\mathbb{F}} = L \otimes_{\mathbb{Z}} \mathbb{F}. \tag{7.3.1}$$

Form the $\frac{1}{2}\mathbb{Z}$-graded twisted affine Lie algebra

$$\tilde{\mathfrak{h}}[-1] = \coprod_{n \in \mathbb{Z}+1/2} \mathfrak{h} \otimes t^n \oplus \mathbb{F}c \oplus \mathbb{F}d, \tag{7.3.2}$$

its Heisenberg subalgebra $\hat{\mathfrak{h}}_{\mathbb{Z}+1/2} = \hat{\mathfrak{h}}[-1]$ and its $\mathbb{Q}$-graded $\hat{\mathfrak{h}}_{\mathbb{Z}+1/2}$-irreducible $\tilde{\mathfrak{h}}[-1]$-module

$$M(1) = S(\hat{\mathfrak{h}}^{-}_{\mathbb{Z}+1/2}) \tag{7.3.3}$$

as in (3.2.2). Denote by $\alpha(n)$ the action of $\alpha \otimes t^n$ on $S(\hat{\mathfrak{h}}^{-}_{\mathbb{Z}+1/2})$ for $\alpha \in \mathfrak{h}$, $n \in \mathbb{Z} + 1/2$, so that

$$\begin{aligned} [\alpha(m), \beta(n)] &= \langle \alpha, \beta \rangle m \delta_{m+n,0} \\ [d, \alpha(m)] &= m\alpha(m) \end{aligned} \tag{7.3.4}$$

for $\alpha, \beta \in \mathfrak{h}$ and $m, n \in \mathbb{Z} + 1/2$.

Let $\hat{L}$ and the associated objects be as in (7.1.6)–(7.1.27)—particularly, κ, s, $^{-}$, c_0, e, ε_0, ω, c and ε. Let T be any $\hat{L}$-module such that

$$\kappa \cdot v = \omega v \quad \text{for} \quad v \in T. \tag{7.3.5}$$

Then as operators on T,

$$ab = c(\bar{a}, \bar{b})ba \qquad \text{for} \quad a, b \in \hat{L} \tag{7.3.6}$$

$$e_\alpha e_\beta = \varepsilon(\alpha, \beta) e_{\alpha+\beta} \quad \text{for} \quad \alpha, \beta \in L. \tag{7.3.7}$$

Later we shall impose additional hypotheses on T.

Form the vector space

$$V_L^T = M(1) \otimes T = S(\hat{\mathfrak{h}}^{-}_{\mathbb{Z}+1/2}) \otimes T, \tag{7.3.8}$$

and embed End $S(\hat{\mathfrak{h}}^{-}_{\mathbb{Z}+1/2})$ and End T into End V_L^T, and (End $S(\hat{\mathfrak{h}}^{-}_{\mathbb{Z}+1/2})$)$\{z\}$ and (End T)$\{z\}$ into (End V_L^T)$\{z\}$. The space V_L^T will be the replacement of the space $V = S(\hat{\mathfrak{h}}^{-}_{\mathbb{Z}+1/2})$ of Section 3.2 and the analogue of the space V_L of Section 7.1.

View $S(\hat{\mathfrak{h}}^{-}_{\mathbb{Z}+1/2})$ as a trivial $\hat{L}$-module and V_L^T as a tensor product $\hat{L}$-module. Give T the trivial grading

$$\deg T = 0 \tag{7.3.9}$$

and give V_L^T the corresponding tensor product grading [shifted as in (1.9.53)], with $d = d \otimes 1$ the degree operator. Regard T as a trivial $\tilde{\mathfrak{h}}[-1]$-module and V_L^T as the tensor product $\tilde{\mathfrak{h}}[-1]$-module. Then we have the following actions on V_L^T:

$$\begin{aligned} c &\mapsto 1 \\ d &\mapsto d = d \otimes 1 \\ h \otimes t^n &\mapsto h(n) = h(n) \otimes 1 \quad \text{for} \quad h \in \mathfrak{h},\ n \in \mathbb{Z} + \tfrac{1}{2} \\ a &\mapsto 1 \otimes a \qquad\qquad\quad\ \text{for} \quad a \in \hat{L}. \end{aligned} \tag{7.3.10}$$

Now define

$$E^{\pm}(\alpha, z) = \exp\left(\sum_{n \in \pm(\mathbb{N}+1/2)} \frac{\alpha(n)}{n} z^{-n} \right) \tag{7.3.11}$$

$$\in (\operatorname{End} S(\hat{\mathfrak{h}}^{-}_{\mathbb{Z}+1/2}))[[z^{\mp 1/2}]] \subset (\operatorname{End} V_L^T)\{z\}$$

for $\alpha \in \mathfrak{h}$, as in (3.2.16), and recall the properties given in Proposition 3.2.1. For $a \in \hat{L}$ such that

$$\langle \bar{a}, \bar{a} \rangle \in \mathbb{Z}, \tag{7.3.12}$$

define the corresponding *twisted vertex operator* $X(a, z)$ by:

$$X(a, z) = X_{\mathbb{Z}+1/2}(a, z) = 2^{-\langle \bar{a}, \bar{a} \rangle} E^{-}(-\bar{a}, z) E^{+}(-\bar{a}, z) a$$

$$= X(\bar{a}, z) a \in (\operatorname{End} V_L^T)\{z\} \tag{7.3.13}$$

[cf. (3.2.28) and (7.1.43)]. For $a \in \hat{L}$ and $\alpha \in L$ with $\langle \bar{a}, \bar{a} \rangle \in \mathbb{Z}$, $\langle \alpha, \alpha \rangle \in \mathbb{Z}$, define operators $x_a(n)$ and $x_\alpha(n)$ by:

$$X(a, z) = \sum_{n \in (1/2)\mathbb{Z}} x_a(n) z^{-n}$$

$$X(e_\alpha, z) = \sum_{n \in (1/2)\mathbb{Z}} x_\alpha(n) z^{-n}. \tag{7.3.14}$$

Note that these operators $x_\alpha(n)$ differ from the earlier operators denoted $x_\alpha(n)$ [see (3.2.29)] by the operator factor e_α [cf. (7.1.51)–(7.1.53)].

We have

$$[h(m), X(a, z)] = \langle h, \bar{a} \rangle z^m X(a, z) \tag{7.3.15}$$

$$[h(m), x_a(n)] = \langle h, \bar{a} \rangle x_a(m + n) \tag{7.3.16}$$

$$[d, X(a, z)] = -DX(a, z) \tag{7.3.17}$$

$$\deg x_a(n) = n \tag{7.3.18}$$

$$X(\kappa a, z) = \omega X(a, z) \tag{7.3.19}$$

$$x_{\kappa a}(n) = \omega x_a(n) \tag{7.3.20}$$

for all $h \in \mathfrak{h}$, $m \in \mathbb{Z} + \frac{1}{2}$ and $n \in \frac{1}{2}\mathbb{Z}$. Until we impose another condition on T, we do not yet have an analogue of (3.2.41) or (3.2.42).

Remark 7.3.1: We have

$$X(1, z) = 1$$

$$x_1(n) = \delta_{n,0} \quad \text{for} \quad n \in \tfrac{1}{2}\mathbb{Z}.$$

All the normal ordering definitions and results in Section 3.3 remain valid, with the following modifications: For $a, b \in \hat{L}$ such that a, b and ab satisfy (7.3.12) when appropriate,

$$X(a, z) = 2^{-\langle\bar{a},\bar{a}\rangle} : e^{D^{-1}\bar{a}(z)} : a = :X(a, z): \tag{7.3.21}$$

[note the absence of an operator $\bar{a}(0)$ and the presence of the operator a, which commutes with the operators $\bar{a}(n)$]

$$\begin{aligned} &:X(a, z_1)X(b, z_2): \\ &\quad = 2^{-\langle\bar{a},\bar{a}\rangle-\langle\bar{b},\bar{b}\rangle} : e^{D^{-1}\bar{a}(z_1)+D^{-1}\bar{b}(z_2)} : ab \qquad (7.3.22) \\ &\quad = 2^{-\langle\bar{a},\bar{a}\rangle-\langle\bar{b},\bar{b}\rangle} E^-(-\bar{a}, z_1)E^-(-\bar{b}, z_2)E^+(-\bar{a}, z_1)E^+(-\bar{b}, z_2)ab \end{aligned}$$

$$\lim_{z_1 \to z_2} :X(a, z_1)X(b, z_2): = 4^{\langle\bar{a},\bar{b}\rangle} X(ab, z_2) \tag{7.3.23}$$

$$DX(a, z) = :\bar{a}(z)X(a, z): \tag{7.3.24}$$

$$D_{z_1} :X(a, z_1)X(b, z_2): = :\bar{a}(z_1)X(a, z_1)X(b, z_2): \tag{7.3.25}$$

$$\lim_{z_1 \to z_2} D_{z_1} :X(a, z_1)X(b, z_2): = 4^{\langle\bar{a},\bar{b}\rangle} :\bar{a}(z_2)X(ab, z_2):. \tag{7.3.26}$$

Instead of (3.4.8) we have

$$:X(a, z_1)X(b, z_2): = c(\bar{a}, \bar{b}) :X(b, z_2)X(a, z_1): \tag{7.3.27}$$

for $a, b \in \hat{L}$ satisfying (7.3.12), and (3.4.9) becomes

$$X(a, z_1)X(b, z_2) = :X(a, z_1)X(b, z_2): \left(\frac{1 - z_2^{1/2}/z_1^{1/2}}{1 + z_2^{1/2}/z_1^{1/2}}\right)^{\langle\bar{a},\bar{b}\rangle}. \tag{7.3.28}$$

Remark 7.3.2: If T is finite-dimensional, the graded dimension of the space V_L^T is given by

$$\dim_* V_L^T = (\dim T)\left(\frac{\eta(q)}{\eta(q^{1/2})}\right)^{\operatorname{rank} L} \tag{7.3.29}$$

(see Section 1.10; cf. Corollary 4.5.4).

Remark 7.3.3: The Virasoro algebra $\mathfrak{v}$ acts in a natural way on V_L^T, motivating the definition of the grading (cf. Remark 7.1.3).

Remark 7.3.4: If $\mathbb{F}$ contains appropriate roots of 2, then we can of course remove the restriction (7.3.12).

7.4. Construction of $\hat{A}_n[\theta]$, $\hat{D}_n[\theta]$, $\hat{E}_n[\theta]$

The commutator result extrapolating Proposition 3.4.3 and Theorem 7.2.1 asserts:

Theorem 7.4.1: *Let $a, b \in \hat{L}$ and assume that*

$$\langle \bar{a}, \bar{a} \rangle, \langle \bar{b}, \bar{b} \rangle, \langle \bar{a}, \bar{b} \rangle \in \mathbb{Z} \tag{7.4.1}$$

and that

$$c(\bar{a}, \bar{b}) = (-1)^{\langle \bar{a}, \bar{b} \rangle}. \tag{7.4.2}$$

Then

$$[X(a, z_1), X(b, z_2)] = \begin{cases} 0 \quad \text{if} \quad \langle \bar{a}, \bar{b} \rangle = 0 \\ \frac{1}{2} X(ab, z_2)\delta(z_1^{1/2}/z_2^{1/2}) \\ \qquad \text{if} \quad \langle \bar{a}, \bar{b} \rangle = -1 \\ \frac{1}{2}\bar{a}(z_2)\delta(z_1^{1/2}/z_2^{1/2}) - \frac{1}{2} D_{z_1}\delta(z_1^{1/2}/z_2^{1/2}) \\ \qquad \text{if} \quad \langle \bar{a}, \bar{a} \rangle = 2 \quad \text{and} \quad b = a^{-1}. \end{cases} \tag{7.4.3}$$

Equivalently,

$$[x_a(m), x_b(n)]$$
$$= \begin{cases} 0 & \text{if} \quad \langle \bar{a}, \bar{b} \rangle = 0 \\ \frac{1}{2} x_{ab}(m+n) & \text{if} \quad \langle \bar{a}, \bar{b} \rangle = -1 \\ \frac{1}{2}\bar{a}(m+n) + \frac{1}{2} m \delta_{m+n,0} & \text{if} \quad \langle \bar{a}, \bar{a} \rangle = 2 \quad \text{and} \quad b = a^{-1} \end{cases} \tag{7.4.4}$$

for $m, n \in \frac{1}{2}\mathbb{Z}$, where we set

$$\bar{a}(n) = 0 \quad \text{for} \quad n \in \mathbb{Z}. \tag{7.4.5}$$

Proof: We have

$$[X(a, z_1), X(b, z_2)] = {:}X(a, z_1)X(b, z_2){:}\left\{\left(\frac{1 - z_2^{1/2}/z_1^{1/2}}{1 + z_2^{1/2}/z_1^{1/2}}\right)^{\langle \bar{a}, \bar{b} \rangle}\right.$$
$$\left. - (-1)^{\langle \bar{a}, \bar{b} \rangle}\left(\frac{1 - z_1^{1/2}/z_2^{1/2}}{1 + z_1^{1/2}/z_2^{1/2}}\right)^{\langle \bar{a}, \bar{b} \rangle}\right\}. \tag{7.4.6}$$

If $\langle \bar{a}, \bar{b} \rangle = 0$, this expression vanishes. The case $\langle \bar{a}, \bar{a} \rangle = 2$, $b = a^{-1}$ being treated as in the proof of Proposition 3.4.3, we assume that $\langle \bar{a}, \bar{b} \rangle = -1$.

Then (7.4.6) becomes

$$2{:}X(a, z_1)X(b, z_2){:}\delta(z_1^{1/2}/z_2^{1/2})$$
$$= \lim_{z_1^{1/2} \to z_2^{1/2}} 2{:}X(a, z_1)X(b, z_2){:}\delta(z_1^{1/2}/z_2^{1/2})$$
$$= \tfrac{1}{2}X(ab, z_2)\delta(z_1^{1/2}/z_2^{1/2}),$$

by Proposition 2.1.1(b) and (7.3.23). ∎

Remark 7.4.2: It is possible to extend this computation to the cases $\langle \bar{a}, \bar{b} \rangle = 1$ and $\langle \bar{a}, \bar{a} \rangle = 2$, $b = a$, but the results leave the realm of vertex operators. We shall soon specialize to modules T for which this difficulty is avoided.

In the spirit of Corollary 7.2.2, we reformulate Theorem 7.4.1 as follows:

Corollary 7.4.3: *Let $\alpha, \beta \in L$ and suppose that*

$$\langle \alpha, \alpha \rangle, \langle \beta, \beta \rangle, \langle \alpha, \beta \rangle \in \mathbb{Z} \tag{7.4.7}$$

and that

$$\varepsilon(\alpha, \beta)/\varepsilon(\beta, \alpha) = (-1)^{\langle \alpha, \beta \rangle}. \tag{7.4.8}$$

Then

$$[X(e_\alpha, z_1), X(e_\beta, z_2)] = \begin{cases} 0 & \text{if} \quad \langle \alpha, \beta \rangle = 0 \\ \tfrac{1}{2}\varepsilon(\alpha,\beta)X(e_{\alpha+\beta}, z_2)\delta(z_1^{1/2}/z_2^{1/2}) & \\ \qquad \text{if} \quad \langle \alpha, \beta \rangle = -1 & \\ \varepsilon(\alpha, -\alpha)(\alpha(z_2)\delta(z_1^{1/2}/z_2^{1/2}) - D_{z_1}\delta(z_1^{1/2}/z_2^{1/2})) & \\ \qquad \text{if} \quad \langle \alpha, \alpha \rangle = 2 \quad \text{and} \quad \beta = -\alpha. & \end{cases} \tag{7.4.9}$$

Now let

$$\theta \in \mathrm{Aut}(\hat{L}; \kappa, \langle \cdot, \cdot \rangle) \tag{7.4.10}$$

be such that

$$\bar{\theta} = -1 \quad \text{on} \quad L, \qquad \theta^2 = 1, \tag{7.4.11}$$

as in (7.2.8), (7.2.9). Then θ can be described by the function

$$u{:}\, \hat{L} \to \mathbb{Z}/s\mathbb{Z} \tag{7.4.12}$$

determined by the condition

$$\theta a = a^{-1}\kappa^{u(a)} \quad \text{for} \quad a \in \hat{L}. \tag{7.4.13}$$

(If $s = 2$, the function u has special properties; recall Proposition 5.4.3.) *We assume that in addition to satisfying* (7.3.5), *T is compatible with θ in the sense that for $a \in \hat{L}$,*

$$\theta a = a \tag{7.4.14}$$

as operators on T. That is,

$$\theta(a)a^{-1} = 1 \quad \text{on} \quad T. \tag{7.4.15}$$

Modules of this type will be constructed later. This assumption gives us analogues of (3.2.41) and (3.2.42): For $a \in \hat{L}$ such that $\langle \bar{a}, \bar{a} \rangle \in \mathbb{Z}$,

$$X(\theta a, z) = \lim_{z^{1/2} \to -z^{1/2}} X(a, z) \tag{7.4.16}$$

$$x_{\theta a}(n) = \begin{cases} x_a(n) & \text{if} \quad n \in \mathbb{Z} \\ -x_a(n) & \text{if} \quad n \in \mathbb{Z} + \frac{1}{2}. \end{cases} \tag{7.4.17}$$

Remark 7.4.4: We can now extend Theorem 7.4.1 (and Corollary 7.4.3) to the cases $\langle \bar{a}, \bar{b} \rangle = 1$ and $\langle \bar{a}, \bar{a} \rangle = 2$, $b = a$, where the answers may be expressed in terms of the function u of (7.4.12), (7.4.13).

By analogy with (3.4.17)–(3.4.21) and (7.2.10)–(7.2.13), define

$$X^{\pm}(a, z) = X(a, z) \pm X(\theta a, z) \tag{7.4.18}$$

for $a \in \hat{L}$ such that $\langle \bar{a}, \bar{a} \rangle \in \mathbb{Z}$, and define operators $x_a^{\pm}(n)$ for $n \in \frac{1}{2}\mathbb{Z}$ by

$$X^{\pm}(a, z) = \sum_{n \in (1/2)\mathbb{Z}} x_a^{\pm}(n) z^{-n}, \tag{7.4.19}$$

so that

$$x_a^{\pm}(n) = x_a(n) \pm x_{\theta a}(n). \tag{7.4.20}$$

Analogously, define operators $x_\alpha^{\pm}(n)$ for $\alpha \in L$ with $\langle \alpha, \alpha \rangle \in \mathbb{Z}$ by

$$X^{\pm}(e_\alpha, z) = \sum_{n \in (1/2)\mathbb{Z}} x_\alpha^{\pm}(n) z^{-n}. \tag{7.4.21}$$

Then by (7.4.17),

$$\begin{aligned} x_a^{+}(n) &= 2x_a(n) = 2x_{\theta a}(n) && \text{for} \quad n \in \mathbb{Z} \\ x_a^{-}(n) &= 2x_a(n) = -2x_{\theta a}(n) && \text{for} \quad n \in \mathbb{Z} + \tfrac{1}{2} \\ x_a^{+}(n) &= 0 && \text{for} \quad n \in \mathbb{Z} + \tfrac{1}{2} \\ x_a^{-}(n) &= 0 && \text{for} \quad n \in \mathbb{Z} \end{aligned} \tag{7.4.22}$$

and

$$\begin{aligned} x_\alpha^+(n) &= 2x_\alpha(n) && \text{for} \quad n \in \mathbb{Z} \\ x_\alpha^-(n) &= 2x_\alpha(n) && \text{for} \quad n \in \mathbb{Z} + \tfrac{1}{2} \\ x_\alpha^+(n) &= 0 && \text{for} \quad n \in \mathbb{Z} + \tfrac{1}{2} \\ x_\alpha^-(n) &= 0 && \text{for} \quad n \in \mathbb{Z}. \end{aligned} \tag{7.4.23}$$

Using (7.3.15)–(7.3.19), Theorem 7.4.1 and (7.4.16), we have:

Corollary 7.4.5: *For $a, b \in \hat{L}$ as in Theorem 7.4.1,*

$$X^\pm(\theta a, z) = \pm X^\pm(a, z) \tag{7.4.24}$$

$$[d, X^\pm(a, z)] = -DX^\pm(a, z) \tag{7.4.25}$$

$$X^\pm(\kappa a, z) = \omega X^\pm(a, z) \tag{7.4.26}$$

$$[h(m), X^\pm(a, z)] = \langle h, \bar{a}\rangle z^m X^\mp(a, z) \tag{7.4.27}$$

for $h \in \mathfrak{h}$ and $m \in \mathbb{Z} + 1/2$,

$$[X^+(a, z_1), X^+(b, z_2)]$$
$$= \begin{cases} 0 & \text{if} \quad \langle \bar{a}, \bar{b}\rangle = 0 \\ X^+(ab, z_2)\delta(z_1/z_2) & \text{if} \quad \langle \bar{a}, \bar{b}\rangle = -1 \\ -2(D\delta)(z_1/z_2) & \text{if} \quad \langle \bar{a}, \bar{a}\rangle = 2 \quad \text{and} \quad b = a^{-1} \end{cases} \tag{7.4.28}$$

$$[X^+(a, z_1), X^-(b, z_2)]$$
$$= \begin{cases} 0 & \text{if} \quad \langle \bar{a}, \bar{b}\rangle = 0 \\ X^-(ab, z_2)\delta(z_1/z_2) & \text{if} \quad \langle \bar{a}, \bar{b}\rangle = -1 \\ 2\bar{a}(z_2)\delta(z_1/z_2) & \text{if} \quad \langle \bar{a}, \bar{a}\rangle = 2 \quad \text{and} \quad b = a^{-1} \end{cases} \tag{7.4.29}$$

$$[X^-(a, z_1), X^-(b, z_2)]$$
$$= \begin{cases} 0 \quad \text{if} \quad \langle \bar{a}, \bar{b}\rangle = 0 \\ X^+(ab, z_2)(z_1/z_2)^{1/2}\delta(z_1/z_2) \\ \qquad \text{if} \quad \langle \bar{a}, \bar{b}\rangle = -1 \\ -2D_{z_1}((z_1/z_2)^{1/2}\delta(z_1/z_2)) \\ \qquad \text{if} \quad \langle \bar{a}, \bar{a}\rangle = 2 \quad \text{and} \quad b = a^{-1}. \end{cases} \tag{7.4.30}$$

This result should be compared with (6.4.52)–(6.4.58) and with Corollary 7.2.3.

Suppose now that our nondegenerate lattice L is even, with $s = 2$, $\omega = -1$ and

$$c_0(\alpha, \beta) = \langle \alpha, \beta \rangle + 2\mathbb{Z},$$
$$c(\alpha, \beta) = (-1)^{\langle \alpha, \beta \rangle} \tag{7.4.31}$$

for $\alpha, \beta \in L$, as in (7.2.22)–(7.2.24). We also have the corresponding properties of ε [see (7.2.25), (7.2.26)].

Remark 7.4.6: The hypotheses in Theorem 7.4.1 and Corollary 7.4.3 hold for all $a, b \in \hat{L}$ and $\alpha, \beta \in L$, so that (7.4.3) and (7.4.9) hold without restriction (cf. Corollaries 7.2.4 and 7.2.5).

The automorphism θ is described by the formula

$$\theta a = a^{-1}\kappa^{s_0(\bar{a})} \quad \text{for} \quad a \in \hat{L}, \tag{7.4.32}$$

with $s_0 : L \to \mathbb{Z}/2\mathbb{Z}$ as in (6.4.8). If

$$\theta e_\alpha = e_{-\alpha} \quad \text{for} \quad \alpha \in L \tag{7.4.33}$$

as in (6.4.10), then

$$e_\alpha = e_{-\alpha} \quad \text{on} \quad T \quad \text{for} \quad \alpha \in L \tag{7.4.34}$$

and

$$X(e_{-\alpha}, z) = \lim_{z^{1/2} \to -z^{1/2}} X(e_\alpha, z) \tag{7.4.35}$$

for $\alpha \in L$.

We turn to the construction of $\hat{L}$-modules T satisfying our conditions. Write

$$K = \{\theta(a)a^{-1} \mid a \in \hat{L}\}, \tag{7.4.36}$$

as in Remark 5.4.4. Then K is a central subgroup of $\hat{L}$ and

$$\bar{K} = 2L. \tag{7.4.37}$$

In particular, $\hat{L}/K$ is a finite group. More precisely [see Proposition 5.3.4 and (6.4.8)], the group $\hat{L}/K$ is a 2-group which is a central extension of the elementary abelian 2-group

$$\check{L} = L/2L: \tag{7.4.38}$$

$$1 \to \langle \kappa \rangle \xrightarrow{\iota} \hat{L}/K \to \check{L} \to 1. \tag{7.4.39}$$

Its commutator map is

$$\begin{aligned} c_1 : \check{L} \times \check{L} &\to \mathbb{Z}/2\mathbb{Z} = \mathbb{F}_2 \\ (\check{\alpha}, \check{\beta}) &\mapsto \langle \alpha, \beta \rangle + 2\mathbb{Z} \quad \text{for} \quad \alpha, \beta \in L \end{aligned} \tag{7.4.40}$$

and its squaring map is a quadratic form on $\check{L}$ with associated form c_1.

Remark 7.4.7: Our $\hat{L}$-modules T correspond precisely to the $\hat{L}/K$-modules T on which $\iota(\kappa)$ acts as multiplication by -1:

$$\iota(\kappa) \cdot v = -v \quad \text{for} \quad v \in T. \tag{7.4.41}$$

The group $\hat{L}/K$ being finite, such modules are completely reducible.

With the help of Theorem 5.5.1, we want to describe all the irreducible $\hat{L}/K$-modules on which $\iota(\kappa)$ acts as -1. Set

$$R = \{\alpha \in L \mid \langle \alpha, L \rangle \subset 2\mathbb{Z}\} \supset 2L, \tag{7.4.42}$$

let $\hat{R}$ be the pullback of R in $\hat{L}$:

$$1 \to \langle \kappa \rangle \to \hat{R} \to R \to 1 \tag{7.4.43}$$

and let $\check{R}$ be the image of R in $\check{L}$. Then $\check{R} = R/2L$ is the radical of the form c_1:

$$\check{R} = \{x \in \check{L} \mid c_1(x, \check{L}) = 0\}. \tag{7.4.44}$$

Also,

$$\hat{R} = \operatorname{Cent} \hat{L}, \tag{7.4.45}$$

$$\hat{R}/K = \operatorname{Cent}(\hat{L}/K) \tag{7.4.46}$$

and $\hat{R}/K$ is the pullback of $\check{R}$ in $\hat{L}/K$:

$$1 \to \langle \kappa \rangle \xrightarrow{\iota} \hat{R}/K \to \check{R} \to 1 \tag{7.4.47}$$

(cf. Remark 5.2.1).

Noting that the finite group $\hat{L}/K$ has exponent 4 (i.e., $g^4 = 1$ for all $g \in \hat{L}/K$), *we assume that the field* $\mathbb{F}$ *contains a primitive fourth root* i *of unity:*

$$i \in \mathbb{F}^{\times}. \tag{7.4.48}$$

Let T be any irreducible (necessarily finite-dimensional) $\hat{L}/K$-module on which $\iota(\kappa)$ acts as -1. Then the central subgroup $\hat{R}/K$ must act as multiplication by scalars, giving rise to a central character

$$\chi : \hat{R}/K \to \mathbb{F}^{\times} \tag{7.4.49}$$

such that

$$\chi(\iota(\kappa)) = -1. \tag{7.4.50}$$

There are exactly

$$|\check{R}| = |R/2L| \tag{7.4.51}$$

such central characters χ. Fix one of them. Then

$$G = (\hat{L}/K)/\mathrm{Ker}\,\chi \tag{7.4.52}$$

is a finite group with center

$$\mathrm{Cent}\, G = (\hat{R}/K)/\mathrm{Ker}\,\chi \simeq \chi(\hat{R}/K) \subset \mathbb{F}^\times, \tag{7.4.53}$$

a cyclic group of order 2 or 4; χ induces a faithful central character of G; and the hypotheses of Theorem 5.5.1 are satisfied. Thus we have:

Proposition 7.4.8: *There are exactly $|R/2L|$ central characters*

$$\chi\colon \hat{R}/K \to \mathbb{F}^\times \tag{7.4.54}$$

of $\hat{L}/K$ such that

$$\chi(\iota(\kappa)) = -1. \tag{7.4.55}$$

For each such χ, there is a unique (up to equivalence) irreducible $\hat{L}/K$-module T_χ with central character χ, and every irreducible $\hat{L}/K$-module on which $\iota(\kappa)$ acts as -1 is equivalent to one of these. We have

$$\dim T_\chi = |L/R|^{1/2} = |\check{L}/\check{R}|^{1/2}. \tag{7.4.56}$$

To construct T_χ, let Φ be any subgroup of L (necessarily containing R and $2L$) which is maximal such that the alternating form c_1 [see (7.4.40)] vanishes on $\check{\Phi}$. Then $\hat{\Phi}$ is a maximal abelian subgroup of $\hat{L}$ (cf. Remark 5.2.1). Let

$$\psi\colon \hat{\Phi}/K \to \mathbb{F}^\times \tag{7.4.57}$$

be any homomorphism extending χ and denote by $\mathbb{F}_\psi$ the one-dimensional $\hat{\Phi}$-module $\mathbb{F}$ with character ψ (pulled back to $\hat{\Phi}$). Then

$$\begin{aligned} T_\chi &\simeq \mathrm{Ind}_{\hat{\Phi}}^{\hat{L}}\, \mathbb{F}_\psi = \mathbb{F}[\hat{L}] \otimes_{\mathbb{F}[\hat{\Phi}]} \mathbb{F}_\psi \\ &\simeq \mathbb{F}[L/\Phi] \simeq \mathbb{F}[\check{L}/\check{\Phi}] \text{ (linearly).} \end{aligned} \tag{7.4.58}$$

Remark 7.4.9: Recall that the automorphism θ is determined by a map $s_0\colon L \to \mathbb{Z}/2\mathbb{Z}$ which is the pullback of a quadratic form

$$s_1\colon \check{L} \to \mathbb{Z}/2\mathbb{Z} \tag{7.4.59}$$

whose associated form is c_1, and s_1 is the squaring map of the central extension $\hat{L}/K$ [see (7.4.39)]. Suppose we can choose Φ in Proposition 7.4.8 so that s_1, not just c_1, vanishes on $\check{\Phi}$:

$$s_1(\check{\Phi}) = 0. \tag{7.4.60}$$

Then $\hat{\Phi}/K$, the pullback of $\check{\Phi}$ in the 2-group $\hat{L}/K$, has exponent 2 and must be an elementary abelian 2-group, as must its subgroup $\hat{R}/K$. It follows from Remark 5.5.2 that for such Φ, Proposition 7.4.8 remains valid even if $\mathbb{F}$ does not contain a square root of -1.

Assume now that L is positive definite, and consider the Lie algebra $\mathfrak{g}$, the invariant symmetric form $\langle \cdot, \cdot \rangle$ and the automorphism θ_* of $\mathfrak{g}$ preserving $\mathfrak{h}$ and $\langle \cdot, \cdot \rangle$ given by Proposition 6.4.2. For brevity, write

$$\theta = \theta_*, \tag{7.4.61}$$

and define $x_a^\pm$ for $a \in \hat{\Delta}$ and $x_\alpha^\pm$ for $\alpha \in \Delta$ as in (6.4.19), (6.4.20). Recall the description of the twisted affine algebra $\tilde{\mathfrak{g}}[\theta]$ by formal variables given in (6.4.23)–(6.4.38). Taking into account (7.3.4), (7.3.14), (7.3.15), (7.3.19) and (7.4.16), we obtain the following representation of $\tilde{\mathfrak{g}}[\theta]$ by twisted vertex operators:

Theorem 7.4.10: *The (well-defined) linear map*

$$\pi_T\colon \tilde{\mathfrak{g}}[\theta] \to \operatorname{End} V_L^T \tag{7.4.62}$$

determined by

$$\pi_T\colon c \mapsto 1$$

$$\pi_T\colon d \mapsto d$$

$$\pi_T\colon h \otimes t^n \mapsto h(n) \quad \text{for} \quad h \in \mathfrak{h},\ n \in \mathbb{Z} + \tfrac{1}{2}$$

$$\pi_T\colon x_a^\pm \otimes t^n \mapsto x_a^\pm(n) \quad \text{for} \quad a \in \hat{\Delta},\ n \in \mathbb{Z}\ (\text{resp., } n \in \mathbb{Z} + \tfrac{1}{2}),$$

or equivalently in the last case,

$$\pi_T\colon x_a(z) \mapsto X(a, z) \quad \text{for} \quad a \in \hat{\Delta}$$

or

$$\pi_T\colon x_\alpha^\pm \otimes t^n \mapsto x_\alpha^\pm(n) \quad \text{for} \quad \alpha \in \Delta,\ n \in \mathbb{Z}\ (\text{resp., } n \in \mathbb{Z} + \tfrac{1}{2})$$

or

$$\pi_T\colon x_\alpha(z) \mapsto X(e_\alpha, z) \quad \text{for} \quad \alpha \in \Delta,$$

is a representation of $\tilde{\mathfrak{g}}[\theta]$ *on* V_L^T.

Arguing as in the proof of Proposition 4.4.2 (cf. also Remark 3.5.3), we find [with $Q = \mathbb{Z}\Delta$ as in (7.2.30) and $\hat{Q}$ its pullback in $\hat{L}$]:

Proposition 7.4.11: *The $\tilde{\mathfrak{g}}[\theta]$-module V_L^T is irreducible under $\hat{\mathfrak{g}}[\theta]$ if and only if the $\hat{L}$-module T is irreducible under $\hat{Q}$. Given another $\hat{L}$-module T' satisfying the conditions that T satisfies, the $\tilde{\mathfrak{g}}[\theta]$-modules V_L^T and $V_L^{T'}$ are equivalent under $\tilde{\mathfrak{g}}[\theta]$ or $\hat{\mathfrak{g}}[\theta]$ if and only if T and T' are equivalent under $\hat{Q}$.*

Remark 7.4.12: If we take $L = Q$, then Proposition 7.4.8 describes the inequivalent irreducible $\hat{Q}/K$-modules T_χ and hence a corresponding family of inequivalent irreducible $\tilde{\mathfrak{g}}[\theta]$-modules (or $\hat{\mathfrak{g}}[\theta]$-modules). If $L = Q$ is the rank 1 lattice generated by α_1 with $\langle \alpha_1, \alpha_1 \rangle = 2$, then the form c_1 vanishes [see (7.4.40)], $R = Q$ and there are only two possibilities for the quadratic form s_1—either zero or the unique nonzero linear form [see (7.4.59) and Remark 5.3.2]. In Chapter 3, we chose $\varepsilon_0 = 0$ (see Remark 6.2.3), and for the automorphism θ_2 of $\mathfrak{a} = \mathfrak{sl}(2, \mathbb{F})$, we in effect chose $s_1 = 0$ [cf. (6.4.8)–(6.4.12)]. Thus, Remark 7.4.9 is applicable to $\mathfrak{a}$ and θ_2, and we find that with no restriction on $\mathbb{F}$, there are exactly two central characters χ as in Proposition 7.4.8. These correspond precisely to the two inequivalent irreducible $\tilde{\mathfrak{a}}[\theta_2]$-modules obtained in Theorem 3.5.1 and Remark 3.5.3. Note that the $\pm$ signs in formula (3.5.5) describing these two inequivalent modules are now built into the vertex operator and the (scalar) action of $\hat{Q}$ on the 1-dimensional module T.

Remark 7.4.13: If more generally we take for L the root lattice Q of a simple Lie algebra $\mathfrak{g}$ of type A_n, D_n or E_n, then we obtain representations by twisted vertex operators of the twisted affine Kac–Moody algebra $\tilde{\mathfrak{g}}[\theta] = \tilde{A}_n[\theta], \tilde{D}_n[\theta]$ or $\tilde{E}_n[\theta]$. We have $|R/2Q| = |\check{R}|$ inequivalent irreducible modules, each with $\hat{\mathfrak{h}}_{\mathbb{Z}+1/2}$-vacuum space T of dimension

$$\dim T = |Q/R|^{1/2} = |\check{Q}/\check{R}|^{1/2}. \tag{7.4.63}$$

If the determinant (6.1.6), which is the determinant of the Cartan matrix of $\mathfrak{g}$, is odd, then the bilinear form c_1 [see (7.4.40)] is nonsingular on $\check{Q}$, i.e.,

$$\check{R} = 0, \quad R = 2Q, \tag{7.4.64}$$

and $\hat{Q}/K$ is an extraspecial 2-group. In this case, we are constructing only one basic module, and

$$\dim T = 2^{(\operatorname{rank} Q)/2}. \tag{7.4.65}$$

This occurs for instance for $\mathfrak{g} = E_8$, for which

$$|\hat{Q}/K| = 2^9$$
$$\dim T = 2^4 = 16. \tag{7.4.66}$$

Remark 7.4.14: In the context of the last remark, the constructed irreducible $\tilde{\mathfrak{g}}[\theta]$-modules are among the basic modules. The algebra $\tilde{\mathfrak{g}}[\theta]$ is isomorphic to the untwisted algebra $\tilde{\mathfrak{g}}$ if and only if θ is an inner automorphism of $\mathfrak{g}$, which is the case if and only if the automorphism -1 of $\mathfrak{h}$ is in the Weyl group of $\mathfrak{g}$ (cf. [Kac 5]). This occurs for instance for $\mathfrak{g} = A_1 = \mathfrak{sl}(2, \mathbb{F})$ [cf. Remark 3.1.2 and (3.1.46)] and for $\mathfrak{g} = E_8$. In such cases, we obtain alternate realizations of the basic modules in the spirit of Theorem 4.5.2 (cf. Remark 7.2.9).

8 General Theory of Untwisted Vertex Operators

In the previous chapters our main goal was the construction of representations of affine Kac–Moody algebras. We were able to accomplish this by means of vertex operators associated with the shortest nonzero elements of a root lattice. Now we reverse the point of view and ask what kinds of algebras are generated by the vertex operators associated with all the lattice elements. This brings us to the notion of general (untwisted) vertex operators associated with arbitrary elements of the untwisted vertex operator modules studied in Chapter 4 and in Sections 7.1 and 7.2. Certain general vertex operators corresponding to a subalgebra of the Griess algebra appeared in our construction of a moonshine module for the Monster [FLM1], [FLM2]. General vertex operators are closed under Lie brackets, a fact which was known in a special case [Frenkel 4]. Moreover, general vertex operators satisfy a universal identity which we call the Jacobi identity and which is in fact analogous in deep respects to the Jacobi identity for Lie algebras. These considerations lead us to the new theory of vertex operator algebras. Such algebras were introduced in the mathematical literature in [Borcherds 3], an investigation which was partly motivated by [FLM1], [FLM2]. General vertex operators and important aspects of the corresponding algebras have also been familiar to physicists developing an algebraic formulation of two-dimensional conformal quantum field theory; see especially

[Belavin–Polyakov–Zamolodchikov] and the work of the early string theorists cited in the Introduction. In this chapter, we shall mainly consider the vertex operator algebra associated with the untwisted vertex operator representation, but in the last section of the chapter, we also formulate an axiomatic definition of vertex operator algebras and study some properties of such algebras.

We shall try to point out some of the physics terminology associated with these ideas. Vertex operator algebras, also called chiral algebras in the physics literature, are the central object in the algebraic formulation of two-dimensional conformal field theory and in special cases can be thought of as chiral (also called holomorphic) conformal field theories in their own right. Vertex operators are examples of "quantum fields." The graded dimensions of Fock spaces and of conformal field theories in general are closely related to "partition functions" in physics. The distinguished element of a vertex operator algebra that we denote **1** is called the "vacuum" or "SL(2)-invariant vacuum," the vertex operator generating the Virasoro algebra corresponds to the holomorphic part of the "stress-energy tensor" in conformal field theory, and what we call "weights" are termed "conformal weights," which are related to conformal spin and conformal dimension. The vertex operator that we denote $Y(v, z)$ is said to "create the state v" when it is applied to the vacuum **1** and when z is set equal to 0. The scalar value of the central generator of the Virasoro algebra is directly related to the "conformal anomaly." Two fundamental properties of vertex operator algebras, "commutativity" and "associativity," which together are equivalent to the Jacobi identity, correspond to two forms of "duality," in terminology going back to the dual resonance model, the precursor of modern string theory. The associativity property also yields an explicit form of the "operator product expansion," also called the "short-distance expansion," which plays an important role in quantum field theory. While the computation of the commutator of two vertex operators uses only the finitely many singular terms in the operator product expansion, the Jacobi identity (or the associativity relation) contains all the information in the full operator product expansion, expressing this information in terms of an "iteration" of vertex operators. Commutators of vertex operators are computed only when certain natural single-valuedness conditions, corresponding to "locality" in physics terminology, hold. The example of vertex operator algebras associated with untwisted vertex operator representations is related to free bosonic conformal field theory on a torus. We refer the reader to the Introduction for further discussion of string theory and references to the physics literature.

In order to present the results on vertex operator algebras we need to extend the formal calculus introduced in Chapter 2. In Section 8.1 we introduce what we call "expansions of zero," the algebraic analogues of δ-functions and their derivatives—distributions of finite support. In Section 8.2 we use our formal-variable language to discuss the exponentials of the derivations introduced in Section 2.2. This enables us to study all the derivatives of the basic formal series $\delta(z)$ at once and to generalize the results of Sections 2.1 and 2.2. From the present viewpoint, these results come from the first two terms in the exponential series expansion. The exponentials of derivations are interpreted as "global transformations" in Section 8.3, giving useful formulas in special cases. A simple but fundamental example is a formal version of Taylor's theorem, expressing the exponential of a differentiation operator as an additive change of variable, and more general cases include formal projective changes of variable. An interesting by-product of the viewpoint of this section is an immediate derivation of the classical formula for the higher derivatives of a composite function. This brief argument actually encapsulates much of the essence of this whole chapter. Proposition 8.3.12, which generalizes a case of the main result of Section 8.2 to a situation allowing non-integral powers of the fomal variables, also occurs in [Dong–Lepowsky]. Some comments at the end of Section 8.3 suggest further connections with combinatorial ideas.

In Section 8.4, we start with an arbitrary nondegenerate lattice, not necessarily even or positive definite. We recall the general setting of Section 7.1, but with some important shifts in our viewpoint and our notation, necessitated by the basic use of additive changes of variable in this chapter. Using the machinery of the last three sections, we obtain a formula, found in [Lepowsky 4] in a slightly different form, for the commutator of a pair of vertex operators associated with arbitrary lattice elements which have an integral inner product and an appropriate value for the commutator map of the central extension. The result suggests an extension, carried out in Section 8.5, of the notion of vertex operator to a notion of general vertex operator parametrized by an element of the untwisted vertex operator module. The original commutator can now be expressed in terms of an "iterate" of the two vertex operators. In Section 8.6 we extend the commutator formula to an arbitrary pair of general vertex operators satisfying natural conditions and we list some useful special cases. The general result, including the formulation of commutators in terms of iterates, is a slight generalization of a theorem of [Borcherds 3]. An exposition, with motivation, is included in [FLM5].

In Section 8.7 we focus our attention on a certain general vertex operator providing representations of the Virasoro algebra already studied in Section 1.9. In addition, we obtain commutation relations between Virasoro algebra elements and general vertex operators. These results are implicit in [Borcherds 3] and also in [Belavin–Polyakov–Zamolodchikov] and early works on string theory (see the Introduction). In particular, Propositions 8.7.7 and 8.7.9 are fundamental in the algebraic formulation of conformal (respectively, quasiconformal) two-dimensional quantum field theory; these results deal with what are called primary (respectively, quasiprimary) fields.

We motivate and derive the central identity for vertex operator algebras—the Jacobi identity—in Section 8.8. This generalizes the commutator formula (Theorem 8.6.1) for general vertex operators and incorporates an infinite family of product operations, including Lie bracket and the cross-bracket operation of [FLM1], [FLM2]. The Jacobi identity is implicit in [Borcherds 3] and in fact Borcherds has informed us that he was aware of this identity. Important consequences of this identity, such as Proposition 8.8.3 and formula (8.8.31), are stated in his paper in component form, together with the general commutator result, formula (8.6.31). Much of the material in this section, including two basic and elementary δ-function results, Propositions 8.8.5 and 8.8.15, will be reinterpreted in the Appendix in terms of contour integration. For instance, the expected "$\mathcal{S}_3$-symmetry" of the Jacobi identity explained in Remark 8.8.20 will be derived directly in the Appendix. In Section 8.9 we extract a consequence of the Jacobi identity, concerning cross-brackets and what we call commutative affinization, that will be basic in our realization of the Griess algebra as an algebra of operators. Theorem 8.9.5 was announced in [FLM1].

Finally, in Section 8.10, which is essentially self-contained, we define and discuss the notions of vertex operator algebra and module. The definitions, which could easily be varied or generalized, are motivated by the properties of general vertex operators, and they serve to summarize much of the material presented in Chapter 8. In this section we also use our algebraic language to formulate the notion of rationality of matrix coefficients—called "correlation functions" in physics—of expressions involving general vertex operators, as well as the commutativity and associativity properties. We prove that these properties are equivalent to the Jacobi identity, thus yielding an alternative definition of vertex operator algebra. We also observe natural consequences about convergent series expansions in case the field is $\mathbb{C}$ (or any complete normed field); here analytic continuation plays a basic role. In this way we explain the "associativity of the operator product expansion," in the quantum-field-theoretic terminology mentioned above. Note, however, that in spite of this phrase used in physics, vertex operator algebras are not associative algebras. The rigorous treatment of these matters is quite subtle, and throughout this book, the distinction must be clearly understood between formal Laurent series on the one hand and rational functions on the other hand. Such matters are examined in the Appendix from the viewpoint of elementary complex analysis. Axiomatic material in this chapter, including that in Section 8.10, is contained and extended in [Frenkel–Huang–Lepowsky].

8.1. Expansions of Zero

Here we lay the foundation for the deeper development of the formal calculus introduced in Chapter 2.

Consider the field $\mathbb{F}(z)$ of rational functions in the indeterminate z over $\mathbb{F}$, the field of fractions of the polynomial ring $\mathbb{F}[z]$. The elements of $\mathbb{F}(z)$ may be represented as fractions $p(z)/q(z)$, where $p(z)$, $q(z) \in \mathbb{F}[z]$ and $q(z) \neq 0$. Performing the same construction with z^{-1} in place of z, we note the identification

$$\mathbb{F}(z) = \mathbb{F}(z^{-1}). \tag{8.1.1}$$

Write $\mathbb{F}((z))$ for the field of fractions of the formal power series ring $\mathbb{F}[[z]]$. We realize this field as follows:

$$\mathbb{F}((z)) = \bigcup_{N \in \mathbb{Z}} \left\{ \sum_{j \geq N} a_j z^j \mid a_j \in \mathbb{F} \right\} \subset \mathbb{F}[[z, z^{-1}]]. \tag{8.1.2}$$

Similarly, for the field of fractions $\mathbb{F}((z^{-1}))$ of the formal power series ring $\mathbb{F}[[z^{-1}]]$, we have the realization

$$\mathbb{F}((z^{-1})) = \bigcup_{N \in \mathbb{Z}} \left\{ \sum_{j \leq N} a_j z^j \mid a_j \in \mathbb{F} \right\} \subset \mathbb{F}[[z, z^{-1}]]. \tag{8.1.3}$$

In the space $\mathbb{F}\{z\}$ (see Section 2.1),

$$\begin{aligned} \mathbb{F}[[z, z^{-1}]] &= \mathbb{F}((z)) + \mathbb{F}((z^{-1})) \\ \mathbb{F}[z, z^{-1}] &= \mathbb{F}((z)) \cap \mathbb{F}((z^{-1})) \end{aligned} \tag{8.1.4}$$

as vector spaces.

We shall often express elements of $\mathbb{F}((z))$ and $\mathbb{F}((z^{-1}))$ by means of analytic functions of z and z^{-1}, respectively. As in earlier chapters, such notations will designate the corresponding formal Taylor or Laurent expansions in z and z^{-1}, respectively. For instance, for $a \in \mathbb{F}$,

$$(1+z)^a = \sum_{n\geq 0}\binom{a}{n} z^n \in \mathbb{F}[[z]], \quad (1+z^{-1})^a = \sum_{n\geq 0}\binom{a}{n} z^{-n} \in \mathbb{F}[[z^{-1}]]$$

$$e^{az} = \exp az = \sum_{n\geq 0}\frac{a^n}{n!} z^n \in \mathbb{F}[[z]], \quad e^{az^{-1}} = \sum_{n\geq 0}\frac{a^n}{n!} z^{-n} \in \mathbb{F}[[z^{-1}]]$$

$$\log(1+az) = -\sum_{n\geq 1}\frac{(-a)^n}{n} z^n \in z\mathbb{F}[[z]], \tag{8.1.5}$$

$$\log(1+az^{-1}) = -\sum_{n\geq 1}\frac{(-a)^n}{n} z^{-n} \in z^{-1}\mathbb{F}[[z^{-1}]].$$

There are two canonical field embeddings

$$\iota_+ : \mathbb{F}(z) \hookrightarrow \mathbb{F}((z))$$
$$\iota_- : \mathbb{F}(z) = \mathbb{F}(z^{-1}) \hookrightarrow \mathbb{F}((z^{-1})). \tag{8.1.6}$$

For $f \in \mathbb{F}(z)$, $\iota_+ f$ is the expansion of f as a formal Laurent series in z, and $\iota_- f$ is its expansion as a formal Laurent series in z^{-1}.

Now we introduce a basic linear map Θ:

$$\Theta = \Theta_z : \mathbb{F}(z) \to \mathbb{F}[[z, z^{-1}]]$$
$$f \mapsto \iota_+ f - \iota_- f. \tag{8.1.7}$$

Viewing $\mathbb{F}[z, z^{-1}]$ as a subring of $\mathbb{F}(z)$, we have

$$\operatorname{Ker}\Theta = \mathbb{F}[z, z^{-1}]. \tag{8.1.8}$$

Motivated by the definition of Θ, we call the elements of the image $\operatorname{Im}\Theta$ the *expansions of zero*. For $f \in \mathbb{F}[z, z^{-1}]$ and $g \in \mathbb{F}(z)$,

$$\Theta(fg) = f\Theta(g). \tag{8.1.9}$$

Thus Θ is an $\mathbb{F}[z, z^{-1}]$-module map and $\operatorname{Im}\Theta$ is an $\mathbb{F}[z, z^{-1}]$-submodule of $\mathbb{F}[[z, z^{-1}]]$. We are interested in describing this module.

Remark 8.1.1: The most important expansion of zero is $\delta(z)$ [see (2.1.22)]. To see that $\delta(z) \in \operatorname{Im}\Theta$, observe that

$$\iota_+((1-z)^{-1}) = \sum_{n\geq 0} z^n = (1-z)^{-1},$$

$$\iota_-((1-z)^{-1}) = \iota_-(-z^{-1}(1-z^{-1})^{-1})$$
$$= -\sum_{n<0} z^n = -z^{-1}(1-z^{-1})^{-1}$$

and so

$$\Theta((1 - z)^{-1}) = \delta(z) = (1 - z)^{-1} + z^{-1}(1 - z^{-1})^{-1}. \tag{8.1.10}$$

More generally,

$$\Theta((1 - az)^{-1}) = \delta(az) = (1 - az)^{-1} + a^{-1}z^{-1}(1 - a^{-1}z^{-1})^{-1} \tag{8.1.11}$$

for $a \in \mathbb{F}^{\times}$. Of course, we are designating elements of $\mathbb{F}[[z]]$ and $\mathbb{F}[[z^{-1}]]$ by rational functions of which they are the Taylor expansions. We shall often use analogous notation to express expansions of zero.

The basic property of $\delta(z)$ is given by Proposition 2.1.1. In particular (taking $V = \mathbb{F}$ in that result),

$$f(z)\delta(az) = f(a^{-1})\delta(az) \tag{8.1.12}$$

for $f(z) \in \mathbb{F}[z, z^{-1}]$ and $a \in \mathbb{F}^{\times}$. This formula represents the first step in a description of $\operatorname{Im} \Theta$ as an $\mathbb{F}[z, z^{-1}]$-module.

It is clear that the differentiation operator d/dz (defined on $\mathbb{F}(z)$ and on $\mathbb{F}\{z\}$) commutes with Θ:

$$\Theta \circ \frac{d}{dz} = \frac{d}{dz} \circ \Theta : \mathbb{F}(z) \to \mathbb{F}[[z, z^{-1}]]. \tag{8.1.13}$$

Using standard calculus notation for the nth derivative, we have:

Proposition 8.1.2: *For $n \in \mathbb{N}$,*

$$\begin{aligned} \frac{1}{n!}\delta^{(n)}(z) &= \Theta((1 - z)^{-n-1}) \\ &= (1 - z)^{-n-1} - (-z)^{-n-1}(1 - z^{-1})^{-n-1}. \end{aligned} \tag{8.1.14}$$

Proof: We combine (8.1.10), (8.1.13) and the formula

$$\frac{1}{n!}\left(\frac{d}{dz}\right)^{n}(1 - z)^{-1} = (1 - z)^{-n-1}. \quad \blacksquare \tag{8.1.15}$$

Proposition 8.1.3: *If $\mathbb{F}$ is algebraically closed, the set*

$$\{\delta^{(n)}(az) \mid n \in \mathbb{N}, a \in \mathbb{F}^{\times}\} \tag{8.1.16}$$

is a basis of the space $\operatorname{Im} \Theta$ *of expansions of zero.*

Proof: By the partial fraction decomposition of a rational function and its uniqueness, we see that

$$\{(1 - az)^{-n-1} \mid n \geq 0, a \in \mathbb{F}^{\times}\}$$

is a basis of a linear complement of $\mathbb{F}[z, z^{-1}]$ in $\mathbb{F}(z)$. Now apply (8.1.8) and Proposition 8.1.2 with az in place of z [cf. (8.1.11)]. ∎

By Proposition 8.1.3 and its proof, we obtain the $\mathbb{F}[z, z^{-1}]$-module structure of $\mathrm{Im}\,\Theta$ in the algebraically closed case:

Proposition 8.1.4: *Suppose that $\mathbb{F}$ is algebraically closed. For $a \in \mathbb{F}^{\times}$, the linear span of*

$$\{\delta^{(n)}(az) \mid n \in \mathbb{N}\} \tag{8.1.17}$$

is an indecomposable $\mathbb{F}[z, z^{-1}]$-submodule of $\mathrm{Im}\,\Theta$, *and* $\mathrm{Im}\,\Theta$ *is the direct sum of these submodules as a ranges through $\mathbb{F}^{\times}$.*

Given $f \in \mathbb{F}[z, z^{-1}]$ and $\Theta(g) \in \mathrm{Im}\,\Theta$ $(g \in \mathbb{F}(z))$, we can in principle explicitly express $f\Theta(g) = \Theta(fg)$ [see (8.1.9)] as a linear combination of basis elements $\delta^{(n)}(az)$ using partial fractions. But we shall want explicit formulas in the spirit of (8.1.12) for expressions like $f(z)\delta^{(n)}(az)$. The expansion coefficients of this expression in terms of the elements $\delta^{(l)}(az)$ should involve the derivatives $f^{(k)}(a^{-1})$, as in Proposition 2.2.1(a). We shall find such formulas in the next section.

8.2. Exponentials of Derivations

We introduce a new formal variable w and we consider the algebra $\mathbb{F}[[w]]$ of formal power series in w. This variable will play a special new role. Expressions involving nonnegative integral powers of w and arbitrary powers of z shall be interpreted as elements of the vector space $\mathbb{F}[[z, z^{-1}]][[w]]$ or the vector space $\mathbb{F}\{z\}[[w]]$. We shall study all the $\delta^{(n)}(z)$ at once by considering

$$\sum_{n \geq 0} \frac{1}{n!} \delta^{(n)}(z) w^n = \sum_{n \geq 0} \frac{1}{n!} \left(w \frac{d}{dz}\right)^n \delta(z). \tag{8.2.1}$$

This expression may be written as $e^{w(d/dz)}\delta(z)$. According to the Lie-theoretic principle that the exponential of a derivation is an automorphism,

$e^{w(d/dz)}$ ought to act as an automorphism—actually, as a "one-parameter group of automorphisms" (parametrized by w). We now develop a context for such expressions.

Let U be a vector space. Then $(\mathrm{End}\, U)[[w]]$ is an associative algebra, and there is a natural bilinear map

$$(\mathrm{End}\, U)[[w]] \times U[[w]] \to U[[w]]$$

which we denote by juxtaposition. Let

$$X \in w(\mathrm{End}\, U)[[w]], \tag{8.2.2}$$

so that X has no constant term. Then

$$e^X = \sum_{n \geq 0} \frac{1}{n!} X^n \tag{8.2.3}$$

is a well-defined element of $(\mathrm{End}\, U)[[w]]$ and hence it acts as an endomorphism of $U[[w]]$. If X and Y are commuting elements of $w(\mathrm{End}\, U)[[w]]$, then

$$e^{X+Y} = e^X e^Y. \tag{8.2.4}$$

In fact,

$$e^{X+Y} = \sum_{n \geq 0} \frac{1}{n!} \sum_{\substack{p,q \geq 0 \\ p+q=n}} \frac{n!}{p!q!} X^p Y^q = \sum_{p,q \geq 0} \frac{1}{p!q!} X^p Y^q = e^X e^Y.$$

In particular,

$$e^X e^{-X} = 1, \tag{8.2.5}$$

so that e^X acts as a linear automorphism of $U[[w]]$.

Now for $p(z) \in \mathbb{F}[z, z^{-1}]$, the endomorphism

$$T = T_{p(z)} = p(z)\frac{d}{dz} \tag{8.2.6}$$

of $\mathbb{F}\{z\}$ acts as a derivation of $\mathbb{F}[z, z^{-1}]$ (cf. Proposition 1.9.1 and Remark 2.2.6), and in fact

$$T(f(z)g(z)) = (Tf(z))g(z) + f(z)(Tg(z)) \tag{8.2.7}$$

for all $f(z) \in \mathbb{F}[z, z^{-1}]$, $g(z) \in \mathbb{F}\{z\}$. By induction, using (8.2.7), we find that

$$T^n(f(z)g(z)) = \sum_{k=0}^{n} \binom{n}{k} (T^k f(z))(T^{n-k} g(z)) \tag{8.2.8}$$

for $n \geq 0$. Let

$$y \in w\mathbb{F}[[w]], \quad y \neq 0. \tag{8.2.9}$$

Then from (8.2.8),

$$e^{yT}(f(z)g(z)) = (e^{yT}f(z))(e^{yT}g(z)) \tag{8.2.10}$$

for $f(z) \in \mathbb{F}[z, z^{-1}]$, $g(z) \in \mathbb{F}\{z\}$. This formula takes place in $\mathbb{F}\{z\}[[w]]$.

In order to compute $f(z)\delta^{(n)}(az)$ for $f(z) \in \mathbb{F}[z, z^{-1}]$, $n \geq 0$ and $a \in \mathbb{F}^{\times}$, we examine $f(z)e^{w(d/dz)}\delta(az)$. It will be no extra effort to replace d/dz by any derivation $T = T_{p(z)}$ of $\mathbb{F}[z, z^{-1}]$ as in (8.2.6) and w by y as in (8.2.9). In $\mathbb{F}[[z, z^{-1}]][[w]]$ we have

$$\begin{aligned} f(z)e^{yT}(\delta(az)) &= e^{yT}[(e^{-yT}f(z))\delta(az)] \\ &= e^{yT}[(e^{-yT}f)(a^{-1})\delta(az)] \\ &= (e^{-yT}f)(a^{-1})e^{yT}(\delta(az)), \end{aligned} \tag{8.2.11}$$

using (8.2.10), (8.2.5) and (8.1.12).

Let V be a vector space. Recalling that

$$V[z, z^{-1}] = V \otimes \mathbb{F}[z, z^{-1}]$$

(see Section 2.1), we may tensor the expressions in (8.2.11) with an element $v \in V$ and thereby replace $f(z)$ by $v \otimes f(z) \in V[z, z^{-1}]$ in (8.2.11). Since the elements $v \otimes f(z)$ span $V[z, z^{-1}]$, we find, taking into account the expansion (8.2.3):

Proposition 8.2.1: *Let $p(z) \in \mathbb{F}[z, z^{-1}]$ and consider the derivation*

$$T = p(z)\frac{d}{dz}$$

of $\mathbb{F}[z, z^{-1}]$. Let $v(z) \in V[z, z^{-1}]$ and $a \in \mathbb{F}^{\times}$. Then

$$v(z)e^{yT}(\delta(az)) = (e^{-yT}v)(a^{-1})e^{yT}(\delta(az)). \tag{8.2.12}$$

In particular, for $n \geq 0$,

$$v(z)T^n(\delta(az)) = \sum_{k=0}^{n} (-1)^k \binom{n}{k} (T^k v)(a^{-1})T^{n-k}(\delta(az)). \tag{8.2.13}$$

Note that this result generalizes Propositions 2.1.1(a) and 2.2.2(a)—the cases $n = 0, 1$. Of course, Proposition 8.2.1 could have been guessed and derived by successive differentiation (induction on n), as in the proof of Proposition 2.2.1(a) (cf. Remark 2.2.5).

We generalize Propositions 2.1.1(b) and 2.2.2(b) as well:

Proposition 8.2.2: *Let V be a vector space,*

$$p(z_1, z_2) \in \mathbb{F}[z_1, z_1^{-1}, z_2, z_2^{-1}],$$

$a \in \mathbb{F}^\times$ *and*

$$X(z_1, z_2) \in (\mathrm{End}\, V)[[z_1, z_1^{-1}, z_2, z_2^{-1}]] \tag{8.2.14}$$

such that

$$\lim_{z_1 \to z_2} X(z_1, z_2) \text{ exists.}$$

Set

$$T_1 = p(z_1, z_2)\frac{\partial}{\partial z_1}, \quad T_2 = p(z_1, z_2)\frac{\partial}{\partial z_2}. \tag{8.2.15}$$

Then with y as in (8.2.9),

$$X(z_1, z_2)e^{yT_1}\delta(az_1/z_2) = (e^{-yT_1}X)(a^{-1}z_2, z_2)e^{yT_1}\delta(az_1/z_2), \tag{8.2.16}$$

$$X(z_1, z_2)e^{yT_2}\delta(az_1/z_2) = (e^{-yT_2}X)(z_1, az_1)e^{yT_2}\delta(az_1/z_2). \tag{8.2.17}$$

In particular, for $n \geq 0$,

$$X(z_1, z_2)T_1^n\delta(az_1/z_2)$$
$$= \sum_{k=0}^{n} (-1)^k \binom{n}{k} (T_1^k X)(a^{-1}z_2, z_2)T_1^{n-k}\delta(az_1/z_2), \tag{8.2.18}$$

$$X(z_1, z_2)T_2^n\delta(az_1/z_2)$$
$$= \sum_{k=0}^{n} (-1)^k \binom{n}{k} (T_2^k X)(z_1, az_1)T_2^{n-k}\delta(az_1/z_2), \tag{8.2.19}$$

all expressions existing.

Proof: Note that

$$\lim_{z_1 \to z_2} (T_i^k X)(z_1, z_2) \quad \text{exists for} \quad k \geq 0, \quad i = 1, 2,$$

since T_i^k is a polynomial in $\partial/\partial z_i$ with coefficients in $\mathbb{F}[z_1, z_1^{-1}, z_2, z_2^{-1}]$. It follows that all desired expressions exist. Using Proposition 2.1.1(b),

(8.2.5) and an obvious extension to two variables of (8.2.10), we have

$$\begin{aligned} X(z_1, z_2)e^{yT_1}\delta(az_1/z_2) &= e^{yT_1}[(e^{-yT_1}X(z_1, z_2))\delta(az_1/z_2)] \\ &= e^{yT_1}[(e^{-yT_1}X)(a^{-1}z_2, z_2)\delta(az_1/z_2)] \\ &= (e^{-yT_1}X)(a^{-1}z_2, z_2)e^{yT_1}\delta(az_1/z_2) \end{aligned}$$

and similarly for T_2. ∎

8.3. Projective Changes of Variable and Higher Derivatives of Composite Functions

For a derivation $T = p(z)(d/dz)$ ($p(z) \in \mathbb{F}[z, z^{-1}]$), we have studied the "one-parameter group of automorphisms" e^{yT}, where $y \in w\mathbb{F}[[w]]$, $y \neq 0$, as in (8.2.9). In the cases $T = d/dz$ and $T = z(d/dz)$, it will be useful to express the action of e^{yT} on $f(z)$ as an explicit "change of variable" in the "domain" of f—an additive or multiplicative transformation of z, respectively. At the end of this section, we shall also discuss more general transformations, including projective transformations.

Recall the notation

$$D = D_z = z\frac{d}{dz}. \tag{8.3.1}$$

Proposition 8.3.1: *Let V be a vector space and let*

$$v(z) = \sum_{n \in \mathbb{F}} v_n z^n \in V\{z\} \tag{8.3.2}$$

($v_n \in V$). Choose y as in (8.2.9). Then

$$e^{y(d/dz)}v(z) = v(z + y) = v(z(1 + y/z)) \tag{8.3.3}$$

("Taylor's theorem"). This expression is to be understood as the following (well-defined) element of $V\{z\}[[w]]$:

$$\sum_{n \in \mathbb{F}} v_n(z + y)^n = \sum_{n \in \mathbb{F}} v_n z^n(1 + y/z)^n, \tag{8.3.4}$$

each summand being expanded in nonnegative powers of y by the binomial expansion. Also,

$$e^{yD}v(z) = v(e^y z), \tag{8.3.5}$$

the series $v(e^y z)$ being understood as

$$\sum_{n \in \mathbb{F}} v_n e^{ny} z^n \in V\{z\}[[w]], \tag{8.3.6}$$

with e^{ny} expanded in nonnegative powers of y by the exponential series.

Proof: We have

$$e^{y(d/dz)}v(z) = \sum_{k \geq 0} \frac{1}{k!} y^k \left(\frac{d}{dz}\right)^k \left(\sum_{n \in \mathbb{F}} v_n z^n\right) = \sum_{k \geq 0} \frac{1}{k!} y^k \sum_{n \in \mathbb{F}} v_n k! \binom{n}{k} z^{n-k}$$

$$= \sum_{\substack{k \geq 0 \\ n \in \mathbb{F}}} v_n \binom{n}{k} z^{n-k} y^k = \sum_{n \in \mathbb{F}} v_n (z + y)^n$$

and

$$e^{yD}v(z) = \sum_{k \geq 0} \frac{1}{k!} y^k \sum_{n \in \mathbb{F}} v_n D^k z^n = \sum_{\substack{k \geq 0 \\ n \in \mathbb{F}}} v_n \frac{1}{k!} (ny)^k z^n = \sum_{n \in \mathbb{F}} v_n e^{ny} z^n. \quad \blacksquare$$

Remark 8.3.2: Formulas (8.3.3) and (8.3.5) imply

$$e^{y(d/dz)}v(z) = v(e^{y(d/dz)}z). \tag{8.3.7}$$

$$e^{yD}v(z) = v(e^{yD}z). \tag{8.3.8}$$

If $v(z) \in V[[z]]$, these formulas also follow from the "automorphism property" (8.2.10), and in fact for such $v(z)$ and for an arbitrary derivation $T = p(z)(d/dz)$ ($p(z) \in \mathbb{F}[z, z^{-1}]$),

$$e^{yT}v(z) = v(e^{yT}z). \tag{8.3.9}$$

Remark 8.3.3: We may take V in Proposition 8.3.1 to be of the form $V\{z_2\}$, so that $V\{z\}$ becomes $V\{z, z_2\}$, and we may generalize y to

$$y \in w\mathbb{F}[z_2, z_2^{-1}][[w]]. \tag{8.3.10}$$

As a result, we obtain the following two-variable analogues of (8.3.3) and (8.3.5): For $v(z, z_2) \in V\{z, z_2\}$,

$$e^{y(d/dz)}v(z, z_2) = v(z + y, z_2) \tag{8.3.11}$$

$$e^{yD}v(z, z_2) = v(e^y z, z_2). \tag{8.3.12}$$

For instance, for y as in (8.2.9),

$$e^{yz_2(\partial/\partial z)}v(z, z_2) = v(z + yz_2, z_2). \tag{8.3.13}$$

These formulas remain valid whenever their formal justification remains valid—for instance, for

$$y \in w\mathbb{F}((z_2))[[w]], \tag{8.3.14}$$

provided that

$$v(z, z_2) \in V((z_2))\{z\}. \tag{8.3.15}$$

As an interesting corollary of these methods, we can immediately derive the classical result expressing the higher derivatives of a composite function $(f \circ g)(x)$ as an explicit linear combination of products of $g^{(k)}(x)$ and $(f^{(l)} \circ g)(x)$. We express the formal content of this result as follows, for an arbitrary derivation of $\mathbb{F}[z, z^{-1}]$ in place of d/dz:

Proposition 8.3.4: *Let*

$$f(z) \in \mathbb{F}[[z]], \quad g(z_1) \in z_1\mathbb{F}[[z_1]], \tag{8.3.16}$$

so that the "composite series" $(f \circ g)(z_1)$ is a well-defined element of $\mathbb{F}[[z_1]]$. Let $p(z_1) \in \mathbb{F}[z_1, z_1^{-1}]$ and $T = p(z_1)(d/dz_1)$, and let y be as in (8.2.9). Then

$$\begin{aligned} e^{yT}(f \circ g)(z_1) &= \left[\exp\left(((e^{yT} - 1)g(z_1))\frac{d}{dz}\right)f\right](g(z_1)) \\ &= \left[\exp\left(\sum_{n \geq 1} \frac{1}{n!} T^n g(z_1) y^n \frac{d}{dz}\right)f\right](g(z_1)) \end{aligned} \tag{8.3.17}$$

in $\mathbb{F}((z_1))[[w]]$. In particular, if we define

$$p_{k,l} \in \mathbb{F}[x_1, x_2, \ldots] \tag{8.3.18}$$

for $k \geq 0$ and $0 \leq l \leq k$ by

$$\exp\left(x \sum_{n \geq 1} \frac{x_n}{n} y^n\right) = \sum_{\substack{k \geq 0 \\ 0 \leq l \leq k}} p_{k,l} x^l y^k, \tag{8.3.19}$$

then

$$\begin{aligned} &\frac{T^k}{k!}(f \circ g)(z_1) \\ &\quad = \sum_{l=0}^{k} p_{k,l}\left(T^1 g(z_1), \frac{T^2 g(z_1)}{1!}, \frac{T^3 g(z_1)}{2!}, \ldots\right)(f^{(l)} \circ g)(z_1) \end{aligned} \tag{8.3.20}$$

in $\mathbb{F}((z_1))$.

Proof: Using (8.3.9), we have

$$\begin{aligned} e^{yT}(f \circ g)(z_1) &= f(e^{yT}g(z_1)) \\ &= f\left(\sum_{n \geq 0} \frac{1}{n!} T^n g(z_1) y^n\right) \\ &= f\left(g(z_1) + \sum_{n \geq 1} \frac{1}{n!} T^n g(z_1) y^n\right) \\ &= \left[\exp\left(\sum_{n \geq 1} \frac{1}{n!} T^n g(z_1) y^n \frac{d}{dz}\right) f\right](g(z_1)). \end{aligned}$$

The last step is a consequence of (8.3.11) and (8.3.14) applied to $V = \mathbb{F}$, $z_2 = z_1$, $y = \sum (1/n!) T^n g(z_1) y^n$ and $v(z, z_2) = f(z)$. This yields an element of $\mathbb{F}((z_1))[[z, w]]$ in which $g(z_1)$ may be, and is, substituted for z. ∎

For the case $f(z) = e^z$ we get:

Corollary 8.3.5: *In the notation of Proposition 8.3.4,*

$$\begin{aligned} e^{yT} e^{g(z_1)} &= \exp((e^{yT} - 1) g(z_1)) e^{g(z_1)} \\ &= \exp\left(\sum_{n \geq 1} \frac{1}{n!} T^n g(z_1) y^n\right) e^{g(z_1)}. \end{aligned} \tag{8.3.21}$$

In particular, defining

$$p_k \in \mathbb{F}[x_1, x_2, \ldots] \tag{8.3.22}$$

for $k \geq 0$ by

$$\exp\left(\sum_{n \geq 1} \frac{x_n}{n} y^n\right) = \sum_{k \geq 0} p_k y^k, \tag{8.3.23}$$

we have

$$\frac{T^k}{k!} e^{g(z_1)} = p_k\left(Tg(z_1), \frac{T^2 g(z_1)}{1!}, \frac{T^3 g(z_1)}{2!}, \ldots\right) e^{g(z_1)}. \tag{8.3.24}$$

Remark 8.3.6: We have

$$\begin{aligned} p_0 &= 1 \\ p_1 &= x_1 \\ p_2 &= \tfrac{1}{2}(x_1^2 + x_2) \\ p_3 &= \tfrac{1}{6}(x_1^3 + 3x_1 x_2 + 2x_3). \end{aligned}$$

Remark 8.3.7: The expression (8.3.23) is reminiscent of untwisted vertex operators. This observation will be exploited later.

Remark 8.3.8: Of course, the last two results can be extended in multitudes of ways, involving more general types of series, several variables, vector and operator values, differentiable function, ...

Remark 8.3.9: The formula

$$\sum_{n \geq 1} \frac{x_n}{n} y^n = \log\left(\sum_{k \geq 0} p_k y^k \right) = \log\left(1 + \sum_{k \geq 1} p_k y^k \right), \qquad (8.3.25)$$

obtained by inverting (8.3.23), shows that every polynomial in $x_1, x_2, \ldots$ is a polynomial in $p_1, p_2, \ldots$. Thus the coefficients of the monomials in the variables $y_1, \ldots, y_l$ in the expression

$$\prod_{j=1}^{l} \exp\left(\sum_{n \geq 1} \frac{x_n}{n} y_j^n \right) = \exp\left(\sum_{j=1}^{l} \sum_{n \geq 1} \frac{x_n}{n} y_j^n \right) = \exp\left(\sum_{n \geq 1} \frac{x_n}{n} \sum_{j=1}^{l} y_j^n \right)$$

span the algebra $\mathbb{F}[x_1, x_2, \ldots]$ as l varies.

Proposition 8.3.1 can be extended as follows:

Proposition 8.3.10: *In the notation of Proposition 8.3.1,*

$$e^{yz^{n+1}(d/dz)} v(z) = v((z^{-n} - ny)^{-1/n}) = v(z(1 - nyz^n)^{-1/n}) \qquad (8.3.26)$$

for $n \in \mathbb{F}$, $n \neq 0$.

Proof: We have

$$z^{n+1} \frac{d}{dz} = -n \frac{d}{d(z^{-n})}, \qquad (8.3.27)$$

i.e.,

$$D_z = -nD_{z^{-n}} \qquad (8.3.28)$$

(recall (8.3.1)), so that by (8.3.3),

$$e^{yz^{n+1}(d/dz)} v(z) = e^{-ny(d/d(z^{-n}))} \left(\sum_{m \in \mathbb{F}} v_m (z^{-n})^{-m/n} \right) = v((z^{-n} - ny)^{-1/n}). \qquad \blacksquare$$

Remark 8.3.11: We can now justify the claim made in Remark 1.9.2 that the operators $-z^{n+1}(d/dz)$ for $n = 0, \pm 1$ can be understood as infinitesimal projective transformations: The correspondence

$$\begin{pmatrix} 0 & 1 \\ 0 & 0 \end{pmatrix} \leftrightarrow -\frac{d}{dz}, \quad \frac{1}{2}\begin{pmatrix} 1 & 0 \\ 0 & -1 \end{pmatrix} \leftrightarrow -z\frac{d}{dz}, \quad \begin{pmatrix} 0 & 0 \\ -1 & 0 \end{pmatrix} \leftrightarrow -z^2\frac{d}{dz} \tag{8.3.29}$$

defines a Lie algebra isomorphism

$$\iota: \mathfrak{sl}(2, \mathbb{F}) \xrightarrow{\sim} \mathfrak{p},$$

using the notation (1.9.8) [recall (1.9.3), (1.9.4)]. Denoting by x each of the three matrices in (8.3.29), we observe that

$$e^{-yx} = \begin{pmatrix} 1 & -y \\ 0 & 1 \end{pmatrix}, \quad \begin{pmatrix} e^{-y/2} & 0 \\ 0 & e^{y/2} \end{pmatrix}, \quad \begin{pmatrix} 1 & 0 \\ y & 1 \end{pmatrix}, \tag{8.3.30}$$

respectively, in the group G of 2×2 matrices of determinant 1 of the form $\begin{pmatrix} 1 & 0 \\ 0 & 1 \end{pmatrix} + M$, M a 2×2 matrix with coefficients in $y\mathbb{F}[[y]]$. Now G acts on $V\{z\}[[w]]$ (in the notation of Proposition 8.3.1) by the following (well-defined) projective transformations:

$$g \cdot v(z, w) = v\left(\frac{az + b}{cz + d}, w\right) \tag{8.3.31}$$

where

$$g \in G \quad \text{and} \quad g^{-1} = \begin{pmatrix} a & b \\ c & d \end{pmatrix}; \quad v(z, w) \in V\{z\}[[w]]. \tag{8.3.32}$$

For $g = e^{-yx}$, we have

$$\frac{az + b}{cz + d} = z + y, \quad e^y z, \quad \frac{z}{1 - yz}, \tag{8.3.33}$$

respectively. The point is that by Propositions 8.3.1 and 8.3.10,

$$e^{-y\iota(x)}v(z) = e^{-yx} \cdot v(z) = v\left(\frac{az + b}{cz + d}\right). \tag{8.3.34}$$

In Proposition 8.2.2, it was assumed that $X(z_1, z_2)$ involved only integral powers of z_1 and z_2. It will be important to relax this restriction in the case in which the derivation T_1 [recall (8.2.15)] is simply $\partial/\partial z_1$ and the scalar a is 1:

Proposition 8.3.12: *Let V be a vector space, $r \in \mathbb{F}$, y as in (8.2.9) and*

$$X(z_1, z_2) \in (z_1/z_2)^r(\mathrm{End}\, V)[[z_1, z_1^{-1}, z_2, z_2^{-1}]] \tag{8.3.35}$$

such that

$$\lim_{z_1 \to z_2} X(z_1, z_2) \text{ exists.}$$

Then

$$X(z_1, z_2)e^{y(\partial/\partial z_1)}\delta(z_1/z_2) = X(z_2 - y, z_2)e^{y(\partial/\partial z_1)}((z_1/z_2)^r\delta(z_1/z_2)). \tag{8.3.36}$$

Proof: We have

$$\begin{aligned}
X(z_1, z_2)e^{y(\partial/\partial z_1)}\delta\left(\frac{z_1}{z_2}\right) &= \left(\frac{z_1}{z_2}\right)^r\left(\frac{z_1}{z_2}\right)^{-r} X(z_1, z_2)e^{y(\partial/\partial z_1)}\delta\left(\frac{z_1}{z_2}\right)\\
&= \left(\frac{z_1}{z_2}\right)^r e^{y(\partial/\partial z_1)}\left(\left(\frac{z_1 - y}{z_2}\right)^{-r} X(z_1 - y, z_2)\delta\left(\frac{z_1}{z_2}\right)\right)\\
&= \left(\frac{z_1}{z_2}\right)^r e^{y(\partial/\partial z_1)}\left(\left(\frac{z_2 - y}{z_2}\right)^{-r} X(z_2 - y, z_2)\delta\left(\frac{z_1}{z_2}\right)\right)\\
&= X(z_2 - y, z_2)\left(\frac{z_1}{z_2}\right)^r e^{y(\partial/\partial z_1)}\left(\left(1 - \frac{y}{z_2}\right)^{-r}\delta\left(\frac{z_1}{z_2}\right)\right)\\
&= X(z_2 - y, z_2)\left(\frac{z_1}{z_2}\right)^r e^{y(\partial/\partial z_1)}\left(\left(1 - \frac{y}{z_1}\right)^{-r}\delta\left(\frac{z_1}{z_2}\right)\right)\\
&= X(z_2 - y, z_2)\left(\frac{z_1}{z_2}\right)^r e^{y(\partial/\partial z_1)}\left(\left(\frac{z_1}{z_1 - y}\right)^{r}\delta\left(\frac{z_1}{z_2}\right)\right)\\
&= X(z_2 - y, z_2)\left(\frac{z_1}{z_2}\right)^r \left(\frac{z_1 + y}{z_1}\right)^{r} e^{y(\partial/\partial z_1)}\delta\left(\frac{z_1}{z_2}\right)\\
&= X(z_2 - y, z_2)e^{y(\partial/\partial z_1)}\left(\left(\frac{z_1}{z_2}\right)^r\delta\left(\frac{z_1}{z_2}\right)\right). \quad \blacksquare
\end{aligned}$$

Remark 8.3.13: It is an interesting matter to take advantage of the flexibility in the variable y in (8.2.9). For instance, for such y, we have

$$\log(1 + y) \in w\mathbb{F}[[w]], \quad \log(1 + y) \neq 0, \tag{8.3.37}$$

so that $\log(1 + y)$ also satisfies the conditions of (8.2.9), and for an element

T of an associative algebra,

$$e^{(\log(1+y))T} = (1 + y)^T = \sum_{n\geq 0} \binom{T}{n} y^n, \tag{8.3.38}$$

where we use the binomial coefficient notation

$$\binom{T}{n} = \frac{T(T-1)\cdots(T-n+1)}{n!} \tag{8.3.39}$$

for $n \in \mathbb{N}$. $\left(\text{We take } \binom{T}{n} = 0 \text{ if } n \notin \mathbb{N}.\right)$ Replacing y by $\log(1+y)$ in Proposition 8.2.1, we find that for instance

$$v(z)\binom{T}{n}(\delta(az)) = \sum_{k=0}^{n} \left(\binom{-T}{k} v\right)(a^{-1}) \binom{T}{n-k}(\delta(az)), \tag{8.3.40}$$

in the notation of that result. Similarly, (8.3.5) gives

$$(1+y)^D v(z) = v((1+y)z), \tag{8.3.41}$$

and comparing this with (8.3.3) and equating the coefficients of y^n in the equation

$$(1+y)^D v(z) = v(z+yz) = \lim_{y \to yz} (e^{y(d/dz)} v)(z),$$

we get the operator equality

$$\binom{D}{n} = \frac{1}{n!} z^n \left(\frac{d}{dz}\right)^n \quad \text{for} \quad n \geq 0, \tag{8.3.42}$$

a fact which of course can also be proved by applying both sides to a monomial. Applied to $\delta(z)$, the transformation $e^{y(d/dz)}$ assumes a special form:

$$e^{y(d/dz)}\delta(z) = (1-y)^{-D-1}\delta(z). \tag{8.3.43}$$

We can also write down formulas for $\binom{T}{k}(f \circ g)(x)$ and $\binom{T}{k} e^{g(x)}$ by applying Proposition 8.3.4 and Corollary 8.3.5 to $\log(1+y)$ in place of y. Whenever we can expand e^{yT} explicitly in powers of w, we obtain analogous results. Another interesting case is $y = -\log(1-w)$:

$$e^{-(\log(1-y))T} = (1-y)^{-T} = \sum_{n\geq 0} (-1)^n \binom{-T}{n} y^n = \sum_{n \geq 0} \binom{T+n-1}{n} y^n. \tag{8.3.44}$$

8.4. Commutators of Untwisted Vertex Operators

We have set up the machinery for computing commutators of vertex operators in great generality. We work in the setting of Section 7.1. In particular, L is a nondegenerate lattice; $\mathfrak{h} = L \otimes_{\mathbb{Z}} \mathbb{F}$; $S(\hat{\mathfrak{h}}_{\mathbb{Z}}^{-})$ is viewed as the $\tilde{\mathfrak{h}}$-module $M(1)$; $(\hat{L}, {}^{-})$ is a central extension of L by $\langle \kappa \mid \kappa^s = 1 \rangle$ with commutator map c_0; χ is a faithful character of $\langle \kappa \rangle$ such that $\chi(\kappa) = \omega$; $\mathbb{F}\{L\} = \mathrm{Ind}_{\langle \kappa \rangle}^{\hat{L}} \mathbb{F}_\chi \simeq \mathbb{F}[L]$ (linearly); $V_L = S(\hat{\mathfrak{h}}_{\mathbb{Z}}^{-}) \otimes \mathbb{F}\{L\}$;

$$E^{\pm}(\alpha, z) = \exp\left(\sum_{n \in \pm\mathbb{Z}_+} \frac{\alpha(n)}{n} z^{-n} \right) = e^{-D^{-1}(\alpha(z)^{\pm} - \alpha(0)/2)} \tag{8.4.1}$$

for $\alpha \in \mathfrak{h}$; and

$$\begin{aligned} X(a, z) = X_{\mathbb{Z}}(a, z) &= E^{-}(-\bar{a}, z) E^{+}(-\bar{a}, z) a z^{\bar{a} + \langle \bar{a}, \bar{a} \rangle /2} \\ &= :e^{D^{-1}(\bar{a}(z) - \bar{a}(0))} a z^{\bar{a}}: \end{aligned} \tag{8.4.2}$$

$$X(a, z) = \sum_{n \in \mathbb{Q}} x_a(n) z^{-n} \tag{8.4.3}$$

for $a \in \hat{L}$.

Our main approach to commutators will be based on the additive transformation (8.3.3). *It turns out to be natural and convenient to emphasize a new differentiation operator, a new notion of normal ordering and a new vertex operator. In place of the operator $D = D_z$, we shall typically use d/dz.* For a vector space V, we can define an operator

$$\int = \int \cdot \, dz = \left(\frac{d}{dz} \right)^{-1} \tag{8.4.4}$$

on the space

$$\frac{d}{dz} V\{z\} = \left\{ \sum_{\substack{n \in \mathbb{F} \\ n \neq -1}} v_n z^n \mid v_n \in V \right\} \tag{8.4.5}$$

by the obvious formula

$$\int \left(\sum_{n \neq -1} v_n z^n \right) = \sum_{n \neq -1} \frac{v_n}{n+1} z^{n+1} \tag{8.4.6}$$

[cf. (2.2.20)–(2.2.22)].

We shall now make an important notation change which shall remain in force throughout the rest of this work. For $\alpha \in \mathfrak{h}$, set

$$\alpha_{\mathrm{new}}(z) = \sum_{n \in \mathbb{Z}} \alpha(n) z^{-n-1} = z^{-1} \alpha(z) \quad (= z^{-1} \alpha_{\mathrm{old}}(z)) \tag{8.4.7}$$

and by abuse of notation write

$$\alpha(z) = \alpha_{\text{new}}(z) = \sum_{n\in\mathbb{Z}} \alpha(n)z^{-n-1} = \sum_{n\in\mathbb{Z}} \alpha(n-1)z^{-n}. \tag{8.4.8}$$

This notation conflicts with the original notation $\alpha(z)$ *introduced in* (4.2.1), but no confusion should arise as the reader becomes aware of the significance of the change. Note that

$$D^{-1}(\alpha_{\text{old}}(z) - \alpha(0)) = \int (\alpha(z) - \alpha(0)z^{-1}) = \sum_{n\neq 0} \frac{\alpha(n)}{-n} z^{-n} \tag{8.4.9}$$

and we have a new way of writing an important ingredient of the vertex operator. *Also introduce the notation change* [cf. (4.2.1)]

$$\begin{aligned} \alpha(z)^+ &= \sum_{n\geq 0} \alpha(n)z^{-n-1} \\ \alpha(z)^- &= \sum_{n<0} \alpha(n)z^{-n-1}, \end{aligned} \tag{8.4.10}$$

so that

$$\alpha(z) = \alpha(z)^+ + \alpha(z)^- \tag{8.4.11}$$

[cf. (4.2.2)].

For $j \in \mathbb{N}$, note that

$$\left(\frac{d}{dz}\right)^j \alpha(z) = \left(\frac{d}{dz}\right)^j \alpha(z)^+ + \left(\frac{d}{dz}\right)^j \alpha(z)^- \tag{8.4.12}$$

and that $(d/dz)^j\alpha(z)^\pm$ involves only negative (respectively, nonnegative) powers of z.

Our new notion of normal ordering will be denoted by open colons. In place of (7.1.54)–(7.1.56), *we set*

$$\overset{\circ}{\underset{\circ}{}}\alpha(z)a\overset{\circ}{\underset{\circ}{}} = \overset{\circ}{\underset{\circ}{}}a\alpha(z)\overset{\circ}{\underset{\circ}{}} = a\alpha(z) \tag{8.4.13}$$

$$\overset{\circ}{\underset{\circ}{}}\alpha(0)a\overset{\circ}{\underset{\circ}{}} = \overset{\circ}{\underset{\circ}{}}a\alpha(0)\overset{\circ}{\underset{\circ}{}} = a\alpha(0) \tag{8.4.14}$$

$$\overset{\circ}{\underset{\circ}{}}z^\alpha a\overset{\circ}{\underset{\circ}{}} = \overset{\circ}{\underset{\circ}{}}az^\alpha\overset{\circ}{\underset{\circ}{}} = az^\alpha \tag{8.4.15}$$

for $\alpha \in \mathfrak{h}$, $a \in \hat{L}$. That is, the operators $\alpha(n)$ for all $n \in \mathbb{Z}$ *including* $n = 0$ are to be placed to the right of the operator a before the multiplication is performed, in contrast with the convention motivated in Remark 4.2.1. *For an expression built entirely from operators* $\alpha(n)$ *and series* $\alpha(z)$, *the new normal ordered product shall agree with the old one* (see the beginning of

Section 4.2). For instance,

$$
{}^{\circ}_{\circ}\left(\frac{d}{dz}\alpha_1(z)\right)\alpha_2(z){}^{\circ}_{\circ} = \left(\frac{d}{dz}\alpha_1(z)^-\right)\alpha_2(z) + \alpha_2(z)\left(\frac{d}{dz}\alpha_1(z)^+\right) \tag{8.4.16}
$$

for $\alpha_1, \alpha_2 \in \mathfrak{h}$.

Recall from Remark 7.1.3 the notion of weight of a homogeneous element of V_L. *For $a \in \hat{L}$, our new vertex operator is the normal ordered expression*

$$
\begin{aligned}
Y(a, z) = Y_{\mathbb{Z}}(a, z) &= {}^{\circ}_{\circ} e^{\int(\bar{a}(z) - \bar{a}(0)z^{-1})} a z^{\bar{a}} {}^{\circ}_{\circ} \\
&= {}^{\circ}_{\circ} e^{\int(\bar{a}(z) - \bar{a}(0)z^{-1})} {}^{\circ}_{\circ} a z^{\bar{a}} \\
&= X(a, z) z^{-\langle \bar{a}, \bar{a}\rangle/2} = X(a, z) z^{-\mathrm{wt}\,\iota(a)}
\end{aligned} \tag{8.4.17}
$$

[cf. (8.4.2)]. In the context of the motivational Remark 4.2.1, formula (4.2.12) would be replaced by the simple formula

$$
Y(\alpha, z) = {}^{\circ}_{\circ} e^{\int \alpha(z)} {}^{\circ}_{\circ}. \tag{8.4.18}
$$

Formulas (7.1.57)–(7.1.60) have the following analogues: Define

$$
{}^{\circ}_{\circ}\alpha(z) Y(a, z){}^{\circ}_{\circ} = {}^{\circ}_{\circ} Y(a, z)\alpha(z){}^{\circ}_{\circ} = \alpha(z)^- Y(a, z) + Y(a, z)\alpha(z)^+ \tag{8.4.19}
$$

$$
{}^{\circ}_{\circ} Y(a, z_1) Y(b, z_2){}^{\circ}_{\circ} = {}^{\circ}_{\circ} e^{\int(\bar{a}(z_1) - \bar{a}(0)z_1^{-1}) + \int(\bar{b}(z_2) - \bar{b}(0)z_2^{-1})} {}^{\circ}_{\circ} ab z_1^{\bar{a}} z_2^{\bar{b}} \tag{8.4.20}
$$

for $\alpha \in \mathfrak{h}$ and $a, b \in \hat{L}$. Then

$$
{}^{\circ}_{\circ} Y(a, z_1) Y(b, z_2){}^{\circ}_{\circ} = c(\bar{a}, \bar{b}) {}^{\circ}_{\circ} Y(b, z_2) Y(a, z_1){}^{\circ}_{\circ} \tag{8.4.21}
$$

$$
Y(a, z_1) Y(b, z_2) = {}^{\circ}_{\circ} Y(a, z_1) Y(b, z_2){}^{\circ}_{\circ} z_1^{\langle \bar{a}, \bar{b}\rangle} (1 - z_2/z_1)^{\langle \bar{a}, \bar{b}\rangle} \tag{8.4.22}
$$

[recall the notation c from (7.1.21)]. There are obvious analogues of formulas (7.1.61)–(7.1.64) too. For instance,

$$
\frac{d}{dz} Y(a, z) = {}^{\circ}_{\circ} \bar{a}(z) Y(a, z){}^{\circ}_{\circ}. \tag{8.4.23}
$$

Note that

$$
Y(a, z) = \sum_{n \in \mathbb{Q}} x_a(n) z^{-n - \langle \bar{a}, \bar{a}\rangle/2} = \sum_{n \in \mathbb{Q}} x_a(n - \tfrac{1}{2}\langle \bar{a}, \bar{a}\rangle) z^{-n}. \tag{8.4.24}
$$

Remark 8.4.1: Formula (8.4.22) can be written in the following more symmetrical form:

$$
Y(a, z_1) Y(b, z_2) = {}^{\circ}_{\circ} Y(a, z_1) Y(b, z_2){}^{\circ}_{\circ} (z_1 - z_2)^{\langle \bar{a}, \bar{b}\rangle}, \tag{8.4.25}
$$

where $(z_1 - z_2)^{\langle \bar{a}, \bar{b} \rangle}$ is understood as the binomial expansion in nonnegative integral powers of z_2:

$$(z_1 - z_2)^{\langle \bar{a}, \bar{b} \rangle} = \sum_{k \geq 0} (-1)^k \binom{\langle \bar{a}, \bar{b} \rangle}{k} z_1^{\langle \bar{a}, \bar{b} \rangle - k} z_2^k.$$

We proceed to compute commutators. Let $a, b \in \hat{L}$ and suppose that

$$\langle \bar{a}, L \rangle \subset \mathbb{Z} \quad \text{and} \quad \langle \bar{b}, L \rangle \subset \mathbb{Z}$$

and that

$$c(\bar{a}, \bar{b}) = (-1)^{\langle \bar{a}, \bar{b} \rangle}.$$

By analogy with (7.2.4), we have

$$\begin{aligned}
[Y(a, z_1), Y(b, z_2)] &= Y(a, z_1)Y(b, z_2) - Y(b, z_2)Y(a, z_1) \\
&= {}_{\circ}^{\circ} Y(a, z_1)Y(b, z_2) {}_{\circ}^{\circ} z_1^{\langle \bar{a}, \bar{b} \rangle} \{(1 - z_2/z_1)^{\langle \bar{a}, \bar{b} \rangle} \\
&\quad - (-z_2/z_1)^{\langle \bar{a}, \bar{b} \rangle}(1 - z_1/z_2)^{\langle \bar{a}, \bar{b} \rangle}\} \\
&= - {}_{\circ}^{\circ} Y(a, z_1)Y(b, z_2) {}_{\circ}^{\circ} (-z_2)^{\langle \bar{a}, \bar{b} \rangle} \{(1 - z_1/z_2)^{\langle \bar{a}, \bar{b} \rangle} \\
&\quad - (-z_1/z_2)^{\langle \bar{a}, \bar{b} \rangle}(1 - z_2/z_1)^{\langle \bar{a}, \bar{b} \rangle}\} \\
&= - {}_{\circ}^{\circ} Y(a, z_1)Y(b, z_2) {}_{\circ}^{\circ} (-z_2)^{\langle \bar{a}, \bar{b} \rangle} \Theta_{z_1/z_2}((1 - z_1/z_2)^{\langle \bar{a}, \bar{b} \rangle}).
\end{aligned} \tag{8.4.26}$$

[See (8.1.7) and note the distinction among formal power series in z_1/z_2, formal power series in z_2/z_1 and rational functions of z_1/z_2.] But by (8.1.8) and Proposition 8.1.2,

$$\Theta((1 - z)^{\langle \bar{a}, \bar{b} \rangle}) = \begin{cases} 0 & \text{if} \quad \langle \bar{a}, \bar{b} \rangle \geq 0 \\ \dfrac{1}{N!} \delta^{(N)}(z) & \text{if} \quad \langle \bar{a}, \bar{b} \rangle < 0 \end{cases} \tag{8.4.27}$$

where

$$N = -\langle \bar{a}, \bar{b} \rangle - 1. \tag{8.4.28}$$

Thus

$$\begin{aligned}
&-(-z_2)^{\langle \bar{a}, \bar{b} \rangle} \Theta_{z_1/z_2}((1 - z_1/z_2)^{\langle \bar{a}, \bar{b} \rangle}) \\
&\quad = \begin{cases} 0 & \text{if} \quad \langle \bar{a}, \bar{b} \rangle \geq 0 \\ z_2^{-1} \dfrac{(-1)^N}{N!} \left(\dfrac{\partial}{\partial z_1} \right)^N \delta(z_1/z_2) & \text{if} \quad \langle \bar{a}, \bar{b} \rangle < 0, \end{cases}
\end{aligned}$$

and this is the coefficient of y^N in

$$z_2^{-1}e^{-y(\partial/\partial z_1)}\delta(z_1/z_2), \tag{8.4.29}$$

for y as in (8.2.9). It follows that $[Y(a, z_1), Y(b, z_2)]$ is the coefficient of y^N in

$$z_2^{-1}\,{}^{\circ}_{\circ}Y(a, z_1)Y(b, z_2){}^{\circ}_{\circ}\,e^{-y(\partial/\partial z_1)}\delta(z_1/z_2) \tag{8.4.30}$$

(whether or not $N \geq 0$).

By the integrality assumption on $\langle \bar{a}, L\rangle$ and $\langle \bar{b}, L\rangle$, Proposition 8.2.2 applies to (8.4.30), which equals

$$z_2^{-1}\,{}^{\circ}_{\circ}Y(a, z_2 + y)Y(b, z_2){}^{\circ}_{\circ}\,e^{-y(\partial/\partial z_1)}\delta(z_1/z_2). \tag{8.4.31}$$

Here we have also used the change-of-variable formula (8.3.11). To compute (8.4.31), we use (8.3.3). From (8.4.20),

$$\begin{aligned}
&{}^{\circ}_{\circ}Y(a, z_2 + y)Y(b, z_2){}^{\circ}_{\circ}\\
&\quad = {}^{\circ}_{\circ}e^{\int(\bar{a}(z_2+y)-\bar{a}(0)(z_2+y)^{-1})d(z_2+y)+\int(\bar{b}(z_2)-\bar{b}(0)z_2^{-1})\,dz_2}{}^{\circ}_{\circ}ab(z_2 + y)^{\bar{a}}z_2^{\bar{b}}\\
&\quad = e^{\int(\bar{a}(z_2+y)^{-}d(z_2+y))}\,{}^{\circ}_{\circ}e^{\int(\bar{b}(z_2)-\bar{b}(0)z_2^{-1})\,dz_2}{}^{\circ}_{\circ}\\
&\qquad \cdot e^{\int(\bar{a}(z_2+y)^{+}-\bar{a}(0)(z_2+y)^{-1})d(z_2+y)}abz_2^{\bar{a}+\bar{b}}(1 + yz_2^{-1})^{\bar{a}}\\
&\quad = \exp\left(e^{y(d/dz_2)}\int \bar{a}(z_2)^{-}\,dz_2\right){}^{\circ}_{\circ}e^{\int(\bar{b}(z_2)-\bar{b}(0)z_2^{-1})}{}^{\circ}_{\circ}\\
&\qquad \cdot \exp\left(e^{y(d/dz_2)}\int (\bar{a}(z_2)^{+} - \bar{a}(0)z_2^{-1})\,dz_2\right)abz_2^{\overline{ab}}(1 + yz_2^{-1})^{\bar{a}}\\
&\quad = \exp\left(\sum_{n\geq 1}\frac{1}{n!}\left(\frac{d}{dz_2}\right)^{n-1}\bar{a}(z_2)^{-}y^n\right){}^{\circ}_{\circ}e^{\int((\bar{a}+\bar{b})(z_2)-(\bar{a}+\bar{b})(0)z_2^{-1})}{}^{\circ}_{\circ}\\
&\qquad \cdot \exp\left(\sum_{n\geq 1}\frac{1}{n!}\left(\frac{d}{dz_2}\right)^{n-1}(\bar{a}(z_2)^{+} - \bar{a}(0)z_2^{-1})y^n\right)abz_2^{\overline{ab}}(1 + yz_2^{-1})^{\bar{a}}\\
&\quad = \exp\left(\sum_{n\geq 1}\frac{1}{n!}\left(\frac{d}{dz_2}\right)^{n-1}\bar{a}(z_2)^{-}y^n\right)Y(ab, z_2)\\
&\qquad \cdot \exp\left(\sum_{n\geq 1}\frac{1}{n!}\left(\frac{d}{dz_2}\right)^{n-1}\bar{a}(z_2)^{+}y^n\right)\\
&\qquad \cdot \exp\left(-\sum_{n\geq 1}\frac{1}{n!}\left(\frac{d}{dz_2}\right)^{n-1}\bar{a}(0)z_2^{-1}y^n\right)(1 + yz_2^{-1})^{\bar{a}},
\end{aligned}$$

which we may write, using normal ordering notation, as

$$\begin{aligned} &{}_{\circ}^{\circ}\exp\left(\sum_{n\geq 1}\frac{1}{n!}\left(\frac{d}{dz_2}\right)^{n-1}\bar{a}(z_2)y^n\right)Y(ab, z_2)_{\circ}^{\circ} \\ &\quad\cdot\exp\left(\sum_{n\geq 1}(-1)^n\frac{1}{n}\bar{a}(0)(yz_2^{-1})^n\right)(1+yz_2^{-1})^{\bar{a}} \\ &= {}_{\circ}^{\circ}\exp\left(\sum_{n\geq 1}\frac{1}{n!}\left(\frac{d}{dz_2}\right)^{n-1}\bar{a}(z_2)y^n\right)Y(ab, z_2)_{\circ}^{\circ} \\ &\quad\cdot\exp(-\bar{a}(0)\log(1+yz_2^{-1}))(1+yz_2^{-1})^{\bar{a}} \\ &= {}_{\circ}^{\circ}\exp\left(\sum_{n\geq 1}\frac{1}{n!}\left(\frac{d}{dz_2}\right)^{n-1}\bar{a}(z_2)y^n\right)Y(ab, z_2)_{\circ}^{\circ}. \end{aligned} \tag{8.4.32}$$

This argument is nothing but an elaborate form of (8.3.21). The expressions such as $(z_2+y)^{\bar{a}}$ and $(1+yz_2^{-1})^{\bar{a}}$ have unambiguous meanings as expansions in nonnegative integral powers of y, and all the necessary relations involving these expressions are easily proved. (Also recall that $\overline{ab} = \bar{a} + \bar{b}$.)

Although the variable y has played a special role, there is no harm in replacing y by a "generic variable" z_0 whose negative and nonintegral powers are allowed. We shall use the following residue notation: For a series

$$f(z_0) = \sum_{n\in\mathbb{F}} v_n z_0^n$$

with coefficients in a vector space, we write

$$\operatorname{Res}_{z_0} f(z_0) = v_{-1}. \tag{8.4.33}$$

We have computed commutators of untwisted vertex operators in the following general form:

Theorem 8.4.2: *Let $a, b \in \hat{L}$ and suppose that*

$$\begin{aligned} &\langle\bar{a}, L\rangle \subset \mathbb{Z}, \quad \langle\bar{b}, L\rangle \subset \mathbb{Z}, \\ &c(\bar{a}, \bar{b}) = (-1)^{\langle\bar{a},\bar{b}\rangle} \end{aligned} \tag{8.4.34}$$

[*recall (7.1.21)*]. *Then*

$$\begin{aligned} &[Y(a, z_1), Y(b, z_2)] \\ &\quad = \operatorname{Res}_{z_0} z_0^{\langle\bar{a},\bar{b}\rangle} z_2^{-1}{}_{\circ}^{\circ}Y(a, z_2+z_0)Y(b, z_2)_{\circ}^{\circ}e^{-z_0(\partial/\partial z_1)}\delta(z_1/z_2). \end{aligned} \tag{8.4.35}$$

Moreover,

$$
\begin{aligned}
&{}^{\circ}_{\circ}Y(a, z_2 + z_0)Y(b, z_2){}^{\circ}_{\circ} \\
&\quad = {}^{\circ}_{\circ}\exp\left(\sum_{n \geq 1} \frac{1}{n!}\left(\frac{d}{dz_2}\right)^{n-1} \bar{a}(z_2)z_0^n\right) Y(ab, z_2){}^{\circ}_{\circ}. \qquad (8.4.36)
\end{aligned}
$$

Remark 8.4.3: A formal application of (8.4.25) shows that *in a formal sense which we shall not rigorize here,*

$$
\begin{aligned}
&[Y(a, z_1), Y(b, z_2)] \\
&\quad = \mathrm{Res}_{z_0}\, z_2^{-1}Y(a, z_2 + z_0)Y(b, z_2)e^{-z_0(\partial/\partial z_1)}\delta(z_1/z_2). \qquad (8.4.37)
\end{aligned}
$$

Note the absence of normal ordering and of the quantity $\langle \bar{a}, \bar{b} \rangle$.

Using the polynomials p_k defined in Corollary 8.3.5, we find the following more "concrete" formulation of Theorem 8.4.2:

Corollary 8.4.4: *In the notation of Theorem 8.4.2,* $[Y(a, z_1), Y(b, z_2)]$ *is the coefficient of* $z_0^{-\langle \bar{a}, \bar{b} \rangle - 1}$ *in*

$$
\begin{aligned}
&z_2^{-1}\left(\sum_{k \geq 0} {}^{\circ}_{\circ}p_k\left(\bar{a}(z_2), \bar{a}(z_2)', \frac{1}{2!}\bar{a}(z_2)'', \ldots\right)Y(ab, z_2){}^{\circ}_{\circ}z_0^k\right) \\
&\quad \cdot \sum_{l \geq 0} (-1)^l \frac{1}{l!}\left(\frac{\partial}{\partial z_1}\right)^l \delta(z_1/z_2)z_0^l \qquad (8.4.38)
\end{aligned}
$$

using prime notation for derivatives. That is,

$$
\begin{aligned}
&[Y(a, z_1), Y(b, z_2)] \\
&\quad = z_2^{-1} \sum_{k + l = -\langle \bar{a}, \bar{b} \rangle - 1} {}^{\circ}_{\circ}p_k(\bar{a}(z_2), \ldots)Y(ab, z_2){}^{\circ}_{\circ}(-1)^l \frac{1}{l!}\left(\frac{\partial}{\partial z_1}\right)^l \delta(z_1/z_2). \\
&\qquad (8.4.39)
\end{aligned}
$$

Recalling the values of p_k tabulated in Remark 8.3.6, we find that the coefficients of the first few powers of z_0 in ${}^{\circ}_{\circ}Y(a, z_2 + z_0)Y(b, z_2){}^{\circ}_{\circ}$ are:

$$
\begin{aligned}
&z_0^0 : Y(ab, z_2) \\
&z_0^1 : {}^{\circ}_{\circ}\bar{a}(z_2)Y(ab, z_2){}^{\circ}_{\circ} \\
&z_0^2 : \tfrac{1}{2}{}^{\circ}_{\circ}(\bar{a}(z_2)^2 + \bar{a}(z_2)')\, Y(ab, z_2){}^{\circ}_{\circ} \qquad (8.4.40) \\
&z_0^3 : \tfrac{1}{6}{}^{\circ}_{\circ}(\bar{a}(z_2)^3 + 3\bar{a}(z_2)\bar{a}(z_2)' + \bar{a}(z_2)'')\, Y(ab, z_2){}^{\circ}_{\circ}.
\end{aligned}
$$

Thus from (8.4.39) we see that $[Y(a, z_1), Y(b, z_2)]$, which we denote by Y, has the following values: If $\langle \bar{a}, \bar{b} \rangle \geq 0$,

$$Y = 0. \tag{8.4.41}$$

If $\langle \bar{a}, \bar{b} \rangle = -1$,

$$Y = z_2^{-1} Y(ab, z_2) \delta(z_1/z_2). \tag{8.4.42}$$

If $\langle \bar{a}, \bar{b} \rangle = -2$,

$$Y = z_2^{-1} {}^{\circ}_{\circ} \bar{a}(z_2) Y(ab, z_2) {}^{\circ}_{\circ} \delta(z_1/z_2) - z_2^{-1} Y(ab, z_2) \frac{\partial}{\partial z_1} \delta(z_1/z_2). \tag{8.4.43}$$

If $\langle \bar{a}, \bar{b} \rangle = -3$,

$$\begin{aligned} Y = {} & \tfrac{1}{2} z_2^{-1} {}^{\circ}_{\circ} (\bar{a}(z_2)^2 + \bar{a}(z_2)') Y(ab, z_2) {}^{\circ}_{\circ} \delta(z_1/z_2) \\ & - z_2^{-1} {}^{\circ}_{\circ} \bar{a}(z_2) Y(ab, z_2) {}^{\circ}_{\circ} \frac{\partial}{\partial z_1} \delta(z_1/z_2) \\ & + \tfrac{1}{2} z_2^{-1} Y(ab, z_2) \left(\frac{\partial}{\partial z_1} \right)^2 \delta(z_1/z_2). \end{aligned} \tag{8.4.44}$$

If $\langle \bar{a}, \bar{b} \rangle = -4$,

$$\begin{aligned} Y = {} & \tfrac{1}{6} z_2^{-1} {}^{\circ}_{\circ} (\bar{a}(z_2)^3 + 3\bar{a}(z_2)\bar{a}(z_2)' + \bar{a}(z_2)'') Y(ab, z_2) {}^{\circ}_{\circ} \delta(z_1/z_2) \\ & - \tfrac{1}{2} z_2^{-1} {}^{\circ}_{\circ} (\bar{a}(z_2)^2 + \bar{a}(z_2)') Y(ab, z_2) {}^{\circ}_{\circ} \frac{\partial}{\partial z_1} \delta(z_1/z_2) \\ & + \tfrac{1}{2} z_2^{-1} {}^{\circ}_{\circ} \bar{a}(z_2) Y(ab, z_2) {}^{\circ}_{\circ} \left(\frac{\partial}{\partial z_1} \right)^2 \delta(z_1/z_2) \\ & - \tfrac{1}{6} z_2^{-1} Y(ab, z_2) {}^{\circ}_{\circ} \left(\frac{\partial}{\partial z_1} \right)^3 \delta(z_1/z_2). \end{aligned} \tag{8.4.45}$$

Remark 8.4.5: A little checking using the notation-changes (8.4.7) and (8.4.17) shows that we have recovered and hence generalized Theorem 7.2.1 for the case of an integral lattice.

Remark 8.4.6: These formulas could have been derived from formula (8.2.18) (with $T_1 = \partial/\partial z_1$) for low values of n, as in the proof of Theorem 7.2.1.

8.5. General Vertex Operators

The similarity between the exponential expression in (8.4.36) and the expression for the vertex operator (cf. Remark 8.3.7) can be exploited in an interesting way: We shall define "general vertex operators" parametrized by the vector space V_L in such a way that the vertex operator $Y(a, z)$ for $a \in \hat{L}$ corresponds to $\iota(a) = a \otimes 1 \in \mathbb{F}\{L\}$ [recall (7.1.19)]. Observe in fact that the correspondence $\iota(a) \mapsto Y(a, z)$ for $a \in \hat{L}$ extends uniquely to a well-defined linear map

$$\begin{aligned} \mathbb{F}\{L\} &\to (\operatorname{End} V_L)\{z\} \\ v &\mapsto Y(v, z), \end{aligned} \tag{8.5.1}$$

by (7.1.20) and (7.1.50). [The only linear relations among the elements $\iota(a)$ are satisfied by the operators $Y(a, z)$.] Note that we are allowing the notation

$$Y(\iota(a), z) = Y(a, z) \quad \text{for} \quad a \in \hat{L}. \tag{8.5.2}$$

For instance,

$$Y(\iota(1), z) = Y(1, z) = 1. \tag{8.5.3}$$

In order to extend the map (8.5.1) to V_L, let

$$a \in \hat{L}, \quad \alpha_1, \ldots, \alpha_k \in \mathfrak{h}, \quad n_1, \ldots, n_k \in \mathbb{Z}_+$$

and set

$$\begin{aligned} v &= \alpha_1(-n_1) \cdots \alpha_k(-n_k) \otimes \iota(a) \\ &= \alpha_1(-n_1) \cdots \alpha_k(-n_k) \cdot \iota(a) \in V_L. \end{aligned} \tag{8.5.4}$$

We define

$$Y(v, z) = Y_{\mathbb{Z}}(v, z) \in (\operatorname{End} V_L)\{z\}$$

[recall the optional notation $Y_{\mathbb{Z}}$ from (8.4.17)] as follows:

$$\begin{aligned} Y(v, z) = Y_{\mathbb{Z}}(v, z) = {}_{\circ}^{\circ}&\left(\frac{1}{(n_1 - 1)!}\left(\frac{d}{dz}\right)^{n_1 - 1} \alpha_1(z)\right) \\ &\cdot \cdots \left(\frac{1}{(n_k - 1)!}\left(\frac{d}{dz}\right)^{n_k - 1} \alpha_k(z)\right) Y(a, z){}_{\circ}^{\circ}, \end{aligned} \tag{8.5.5}$$

the normal ordered product being defined in an obvious way extending the rules already given. The result is a well-defined linear map

$$\begin{aligned} V_L = S(\hat{\mathfrak{h}}_{\bar{\mathbb{Z}}}) \otimes \mathbb{F}\{L\} &\to (\operatorname{End} V_L)\{z\} \\ v &\mapsto Y(v, z) \end{aligned} \tag{8.5.6}$$

extending (8.5.1). The expressions $Y(v, z)$ are called *general (untwisted) vertex operators* (or for brevity, *vertex operators*). Note that if

$$n_1 = \cdots = n_k = 1 \quad \text{and} \quad a = 1,$$

then

$$Y(v, z) = {}^{\circ}_{\circ}\alpha_1(z) \cdots \alpha_k(z){}^{\circ}_{\circ} \tag{8.5.7}$$

[cf. (3.3.8), (3.3.9)].

It is easy to check from the definition (8.5.5) that for all $v \in V_L$,

$$\lim_{z \to 0} Y(v, z)\iota(1) = v. \tag{8.5.8}$$

That is, $Y(v, z)$ "creates the vector v." Here of course the limit has algebraic meaning because $Y(v, z)\iota(1)$ involves only nonnegative integral powers of z.

Let us replace z by z_2 in (8.5.6):

$$\begin{aligned} V_L &\to (\operatorname{End} V_L)\{z_2\} \\ v &\mapsto Y(v, z_2) \end{aligned}$$

and then extend this linear map canonically to $V_L\{z_0\}$:

$$\begin{aligned} V_L\{z_0\} &\to (\operatorname{End} V_L)\{z_0, z_2\} \\ v(z_0) &\mapsto Y(v(z_0), z_2). \end{aligned} \tag{8.5.9}$$

That is, if

$$v(z_0) = \sum_{n \in \mathbb{F}} v_n z_0^n \in V_L\{z_0\}$$

($v_n \in V_L$) then

$$Y(v(z_0), z_2) = \sum_{n \in \mathbb{F}} Y(v_n, z_2) z_0^n. \tag{8.5.10}$$

Now let $a, b \in \hat{L}$ and consider

$$Y(a, z_0)\iota(b) = \exp\left(\sum_{n \geq 1} \frac{\bar{a}(-n)}{n} z_0^n\right)\iota(ab)z_0^{\langle a, b\rangle} \tag{8.5.11}$$

in $V_L\{z_0\}$. Then we have

$$\begin{aligned} Y(Y(a, z_0)\iota(b), z_2) &= {}^{\circ}_{\circ}\exp\left(\sum_{n \geq 1} \frac{1}{n!}\left(\frac{d}{dz_2}\right)^{n-1} \bar{a}(z_2)z_0^n\right) Y(ab, z_2){}^{\circ}_{\circ} z_0^{\langle \bar{a}, \bar{b}\rangle} \\ &= {}^{\circ}_{\circ}Y(a, z_2 + z_0)Y(b, z_2){}^{\circ}_{\circ} z_0^{\langle \bar{a}, \bar{b}\rangle} \end{aligned} \tag{8.5.12}$$

[see (8.4.36)], and Theorem 8.4.2 asserts:

Corollary 8.5.1: *In the notation of Theorem 8.4.2,*

$$\begin{aligned} &[Y(a, z_1), Y(b, z_2)] \\ &\quad = \operatorname{Res}_{z_0} z_2^{-1} Y(Y(a, z_0)\iota(b), z_2) e^{-z_0(\partial/\partial z_1)} \delta(z_1/z_2). \end{aligned} \tag{8.5.13}$$

Remark 8.5.2: It is interesting to compare this corollary with Remark 8.4.3. In view of (8.5.12), a formal application of (8.4.25) gives

$$Y(Y(a, z_0)\iota(b), z_2) = Y(a, z_2 + z_0)Y(b, z_2). \tag{8.5.14}$$

This suggestive formula is not rigorous in the present setting, although at the end of Chapter 8 we shall develop an approach in which it can be interpreted rigorously.

We would also like to express Corollary 8.5.1 in terms of the component operators $v_n \in \operatorname{End} V_L$ defined by the formula

$$Y(v, z) = \sum_{n \in \mathbb{Q}} v_n z^{-n-1} = \sum_{n \in \mathbb{Q}} v_{n-1} z^{-n} \tag{8.5.15}$$

for $v \in V_L$. Note that by (8.4.24), for $a \in \hat{L}$ and $v = \iota(a)$ the operators v_n are re-indexings of the operators $x_a(n)$:

$$\iota(a)_n = x_a(n - \tfrac{1}{2}\langle \bar{a}, \bar{a}\rangle + 1) \tag{8.5.16}$$

for $n \in \mathbb{Q}$. In particular,

$$\deg \iota(a)_n = n - \tfrac{1}{2}\langle \bar{a}, \bar{a}\rangle + 1 = n - \operatorname{wt} \iota(a) + 1 \tag{8.5.17}$$

[see (7.1.49) and recall Remark 7.1.3]. We have

$$\iota(1)_n = \delta_{n,-1} \tag{8.5.18}$$

for $n \in \mathbb{Q}$ (cf. Remark 7.1.1). The case $v = \alpha(-1) = \alpha(-1) \otimes \iota(1)$ for $\alpha \in \mathfrak{h}$ gives

$$Y(\alpha(-1), z) = \alpha(z) = \sum_{n \in \mathbb{Z}} \alpha(n) z^{-n-1} \tag{8.5.19}$$

[see (8.4.8), (8.5.7)], so that

$$\begin{aligned} \alpha(-1)_n &= \alpha(n) \\ \deg \alpha(-1)_n &= n \end{aligned} \tag{8.5.20}$$

for $n \in \mathbb{Q}$. More generally, for v as in (8.5.4) and $n \in \mathbb{Q}$,

$$\deg v_n = n - n_1 - \cdots - n_k - \tfrac{1}{2}\langle \bar{a}, \bar{a} \rangle + 1. \tag{8.5.21}$$

It follows that for all homogeneous elements $v, w \in V_L$ and $n \in \mathbb{Q}$,

$$\begin{aligned} \deg v_n &= n - \operatorname{wt} v + 1 \\ \operatorname{wt}(v_n \cdot w) &= \operatorname{wt} v + \operatorname{wt} w - n - 1. \end{aligned} \tag{8.5.22}$$

From (8.5.8) and (8.5.15) we see that

$$v_{-1} \cdot \iota(1) = v \tag{8.5.23}$$

for all $v \in V_L$.

Remark 8.5.3: For $v \in V_L$,

$$v_0 = \operatorname{Res}_z Y(v, z).$$

Now in the notation of Corollary 8.5.1,

$$Y(Y(a, z_0)\iota(b), z_2) = \sum_{i,j \in \mathbb{Z}} (\iota(a)_i \cdot \iota(b))_j z_0^{-i-1} z_2^{-j-1} \tag{8.5.24}$$

and

$$e^{-z_0(\partial/\partial z_1)}\delta(z_1/z_2) = \sum_{k,l \in \mathbb{Z}} (-1)^k \binom{l}{k} z_0^k z_1^{l-k} z_2^{-l} \tag{8.5.25}$$

and so Corollary 8.5.1 can be reformulated as follows (where we note that $\iota(a)_i \cdot \iota(b) = 0$ for i sufficiently large):

Corollary 8.5.4: *In the notation of Theorem 8.4.2, let*

$$m, n \in \mathbb{Z}.$$

Then

$$[\iota(a)_m, \iota(b)_n] = \sum_{i \in \mathbb{N}} \binom{m}{i} (\iota(a)_i \cdot \iota(b))_{m+n-i}, \tag{8.5.26}$$

the sum being finite.

In particular we have:

Corollary 8.5.5: *In the same notation,*

$$[\iota(a)_0, \iota(b)_n] = (\iota(a)_0 \cdot \iota(b))_n.$$

We also want to express Corollary 8.5.1 in terms of the original operators $x_a(n)$. For every homogeneous element $v \in V_L$, set

$$\begin{aligned} X(v, z) &= X_{\mathbb{Z}}(v, z) = Y(v, z)z^{\mathrm{wt}\, v} \\ &= \sum_{n \in \mathbb{Q}} v_{n+\mathrm{wt}\, v-1} z^{-n} \end{aligned} \tag{8.5.27}$$

[cf. (8.4.17)], and define $x_v(n) \in \operatorname{End} V_L$ for $n \in \mathbb{Q}$ by the expansions

$$\begin{aligned} X(v, z) &= \sum_{n \in \mathbb{Q}} x_v(n) z^{-n} \\ Y(v, z) &= \sum_{n \in \mathbb{Q}} x_v(n) z^{-n-\mathrm{wt}\, v}. \end{aligned} \tag{8.5.28}$$

Then for $n \in \mathbb{Q}$,

$$x_v(n) = v_{n+\mathrm{wt}\, v-1} \tag{8.5.29}$$

and

$$\begin{aligned} \deg x_v(n) &= n \\ [d, X(v, z)] &= -DX(v, z) \end{aligned} \tag{8.5.30}$$

[cf. (7.1.48), (7.1.49)]. By (8.5.23) and (8.5.29),

$$x_v(-\mathrm{wt}\, v) \cdot \iota(1) = v. \tag{8.5.31}$$

Extend the correspondence $v \mapsto X(v, z)$ from homogeneous elements to arbitrary elements of V_L by linearity:

$$\begin{aligned} V_L &\to (\operatorname{End} V_L)\{z\} \\ v &\mapsto X(v, z) \end{aligned} \tag{8.5.32}$$

and use the first formula in (8.5.28) to define operators $x_v(n)$ in general. Then (8.5.30) holds for all $v \in V_L$.

For the case $v = \iota(a)$, $a \in \hat{L}$, we have

$$\begin{aligned} X(\iota(a), z) &= X(a, z) \\ x_{\iota(a)}(n) &= x_a(n) \quad \text{for} \quad n \in \mathbb{Q} \end{aligned} \tag{8.5.33}$$

For $v = \alpha(-1)$, $\alpha \in \mathfrak{h}$,

$$\begin{aligned} X(\alpha(-1), z) &= z\alpha(z) \\ x_{\alpha(-1)}(n) &= \alpha(n) \quad \text{for} \quad n \in \mathbb{Q} \end{aligned} \tag{8.5.34}$$

[recall (8.4.7), (8.4.8); $z\alpha(z)$ is the expression previously denoted $\alpha(z)$]. More generally, for v as in (8.5.4),

$$X(v, z) \\ = {}^{\circ}_{\circ}\left(\frac{z^{n_1}}{(n_1 - 1)!}\left(\frac{d}{dz}\right)^{n_1 - 1}\alpha_1(z)\right) \cdots \left(\frac{z^{n_k}}{(n_k - 1)!}\left(\frac{d}{dz}\right)^{n_k - 1}\alpha_k(z)\right)X(a, z){}^{\circ}_{\circ} \tag{8.5.35}$$

[see (8.5.5)]. But

$$\begin{aligned} \frac{z^n}{(n-1)!}\left(\frac{d}{dz}\right)^{n-1}\alpha(z) &= \frac{z^n}{(n-1)!}\left(\frac{d}{dz}\right)^{n-1}\sum_{l \in \mathbb{Z}} \alpha(l)z^{-l-1} \\ &= \sum_{l \in \mathbb{Z}} \binom{-l-1}{n-1}\alpha(l)z^{-l} \\ &= \binom{D-1}{n-1}(z\alpha(z)), \end{aligned} \tag{8.5.36}$$

using the usual binomial coefficient notation

$$\binom{T}{m} = \frac{T(T-1)\cdots(T-m+1)}{m!} \tag{8.5.37}$$

for an element T of an associative algebra. Hence (8.5.35) becomes

$$X(v, z) = {}^{\circ}_{\circ}\left(\binom{D-1}{n_1 - 1}(z\alpha_1(z))\right) \cdots \left(\binom{D-1}{n_k - 1}(z\alpha_k(z))\right)X(a, z){}^{\circ}_{\circ}. \tag{8.5.38}$$

For example,

$$X(\alpha_1(-1) \cdots a_k(-1), z) = {}^{\circ}_{\circ}(z\alpha_1(z)) \cdots (z\alpha_k(z)){}^{\circ}_{\circ} \tag{8.5.39}$$

[cf. (8.5.7)]. In this connection we generalize the notation $\alpha(n)$ [cf. (8.5.34)] by setting

$$\alpha_1 \cdots \alpha_k(n) = x_{\alpha_1(-1) \cdots \alpha_k(-1)}(n) \quad \text{for} \quad n \in \mathbb{Q}. \tag{8.5.40}$$

Here the product $\alpha_1 \cdots \alpha_k$ is of course to be viewed as an element of the kth symmetric power $S^k(\mathfrak{h})$, as in (3.3.9). We have

$${}^{\circ}_{\circ}\alpha_1(z) \cdots \alpha_k(z){}^{\circ}_{\circ} = \sum_{n \in \mathbb{Z}} \alpha_1 \cdots \alpha_k(n) z^{-n-k}. \tag{8.5.41}$$

Changing indices in Corollary 8.5.4, we find:

Corollary 8.5.6: *In the notation of Theorem 8.4.2, let*

$$m \in \mathbb{Z} - \tfrac{1}{2}\langle \bar{a}, \bar{a}\rangle, \quad n \in \mathbb{Z} - \tfrac{1}{2}\langle \bar{b}, \bar{b}\rangle.$$

Then

$$[x_a(m), x_b(n)] = \sum_{i \in \mathbb{Z}_+ - \langle \bar{a}, \bar{a}\rangle/2} \binom{m - 1 + \langle \bar{a}, \bar{a}\rangle/2}{i - 1 + \langle \bar{a}, \bar{a}\rangle/2} x_{x_a(i)\cdot \iota(b)}(m + n) \tag{8.5.42}$$

(finite sum).

To help in the use of this formula, we record that

$$X(a, z)\iota(b) = \exp\left(\sum_{n \geq 1} \frac{\bar{a}(-n)}{n} z^n\right)\iota(ab) z^{\langle \bar{a}, \bar{b}\rangle + \langle \bar{a}, \bar{a}\rangle/2}$$

$$= \sum_{k \geq 0} p_k(\bar{a}(-1), \bar{a}(-2), \ldots)\iota(ab) z^{k + \langle \bar{a}, \bar{b}\rangle + \langle \bar{a}, \bar{a}\rangle/2} \tag{8.5.43}$$

[see (8.3.23)]. Then

$$x_a(i) \cdot \iota(b) = p_{-i - \langle \bar{a}, \bar{b}\rangle - \langle \bar{a}, \bar{a}\rangle/2}(\bar{a}(-1), \ldots)\iota(ab), \tag{8.5.44}$$

$$X(x_a(i) \cdot \iota(b), z) = {}^{\circ}_{\circ} p_{-i - \langle \bar{a}, \bar{b}\rangle - \langle \bar{a}, \bar{a}\rangle/2}$$

$$\left(z\bar{a}(z), \binom{D - 1}{1}(z\bar{a}(z)), \binom{D - 1}{2}(z\bar{a}(z)), \ldots\right) X(ab, z) {}^{\circ}_{\circ}, \tag{8.5.45}$$

using (8.5.38).

For example,

$$[x_a(m), x_b(n)] = \begin{cases} 0 & \text{if } \langle \bar{a}, \bar{b}\rangle \geq 0 \\ x_{ab}(m + n) & \text{if } \langle \bar{a}, \bar{b}\rangle = -1 \\ \bar{a}(m + n) + m\delta_{m+n,0} & \text{if } \langle \bar{a}, \bar{a}\rangle = 2 \text{ and } b = a^{-1}, \end{cases} \tag{8.5.46}$$

recovering Theorem 7.2.1 in case the lattice L is integral. The results in the next section include Theorem 7.2.1 in general.

8.6. Commutators of General Vertex Operators

We shall now show that the assertions of Corollaries 8.5.1, 8.5.4 and 8.5.6 remain valid when $\iota(a)$ and $\iota(b)$ are replaced by arbitrary elements of V_L [recall that $Y(a, z) = Y(\iota(a), z)$]. In particular, the operators v_n or $x_v(n)$ span a Lie algebra whose precise structure we determine. We also relax the integrality assumption in (8.4.34).

Theorem 8.6.1: *Let $a, b \in \hat{L}$ and assume that*

$$\langle \bar{a}, \bar{b} \rangle \in \mathbb{Z} \quad \text{and} \quad c(\bar{a}, \bar{b}) = (-1)^{\langle \bar{a}, \bar{b} \rangle}. \tag{8.6.1}$$

Let $u', v' \in S(\hat{\mathfrak{h}}_{\bar{\mathbb{Z}}})$ and set

$$\begin{aligned} u &= u' \otimes \iota(a) = u' \cdot \iota(a) \in V_L \\ v &= v' \otimes \iota(b) = v' \cdot \iota(b) \in V_L. \end{aligned} \tag{8.6.2}$$

Then

$$\begin{aligned} &[Y(u, z_1), Y(v, z_2)] \\ &\quad = \operatorname{Res}_{z_0} z_2^{-1} Y(Y(u, z_0)v, z_2) e^{-z_0(\partial/\partial z_1)}((z_1/z_2)^{\bar{a}} \delta(z_1/z_2)). \end{aligned} \tag{8.6.3}$$

If L is integral and (8.6.1) holds for all $a, b \in \hat{L}$ (in which case L is also even, $c(\bar{a}, \bar{a})$ being equal to 1), then (8.6.3), with the factor $(z_1/z_2)^{\bar{a}}$ omitted, holds for all $u, v \in V_L$.

Proof: Let $k, l \geq 1$ and let

$$a_1, \ldots, a_k, b_1, \ldots, b_l \in \hat{L}$$

subject to the conditions

$$a = a_1 \cdots a_k, \quad b = b_1 \cdots b_l. \tag{8.6.4}$$

Set

$$\begin{aligned} A &= \exp\left(\sum_{i=1}^{k} \sum_{n \geq 1} \frac{\bar{a}_i(-n)}{n} w_i^n \right) \cdot \iota(a) \in V_L[[w_1, \ldots, w_k]] \\ B &= \exp\left(\sum_{j=1}^{l} \sum_{n \geq 1} \frac{\bar{b}_j(-n)}{n} x_j^n \right) \cdot \iota(b) \in V_L[[x_1, \ldots, x_l]]. \end{aligned} \tag{8.6.5}$$

Using Remark 8.3.9 we see that the coefficients in the formal power series A and B span $S(\hat{\mathfrak{h}}_{\bar{\mathbb{Z}}}) \otimes \iota(a)$ and $S(\hat{\mathfrak{h}}_{\bar{\mathbb{Z}}}) \otimes \iota(b)$, respectively, as k, l, a_i and b_j vary subject to the constraint (8.6.4). In fact, the coefficients in

$$\exp\left(\sum_{i=1}^{k-1} \sum_{n \geq 1} \frac{\bar{a}_i(-n)}{n} w_i^n \right)$$

span $S(\hat{\mathfrak{h}}_{\bar{\mathbb{Z}}})$ as k and a_i vary (let the $\bar{a}_i$ range through a fixed basis of $\mathfrak{h}$, with repetitions as necessary according to Remark 8.3.9); now multiply by

$$\exp\left(\sum_{n\geq 1}\frac{\bar{a}_k(-n)}{n}w_k^n\right)\cdot\iota(a)$$

where

$$a_k = (a_1\cdots a_{k-1})^{-1}a.$$

Hence it suffices to prove the Theorem with u and v replaced by A and B, respectively.

Recalling (8.4.20), (8.4.22) and Remark 8.4.1 we have

$$\begin{aligned} A &= \prod_{i=1}^{k}\exp\left(\sum_{n\geq 1}\frac{\bar{a}_i(-n)}{n}w_i^n\right)\cdot\iota(a_1\cdots a_k) \\ &= {}_\circ^\circ Y(a_1,w_1)\cdots Y(a_k,w_k){}_\circ^\circ\iota(1) \\ &= Y(a_1,w_1)\cdots Y(a_k,w_k)\iota(1)\prod_{1\leq i<j\leq k}(w_i-w_j)^{-\langle\bar{a}_i,\bar{a}_j\rangle}, \end{aligned} \tag{8.6.6}$$

where $(w_i - w_j)^{-\langle\bar{a}_i,\bar{a}_j\rangle}$ is understood as the binomial expansion in nonnegative integral powers of w_j, so that

$$\begin{aligned} &\prod_{1\leq i<j\leq k}(w_i-w_j)^{-\langle\bar{a}_i,\bar{a}_j\rangle} \\ &\quad = \prod_{1\leq i<j\leq k}w_i^{-\langle\bar{a}_i,\bar{a}_j\rangle}\prod_{1\leq i<j\leq k}\left(1-\frac{w_j}{w_i}\right)^{-\langle\bar{a}_i,\bar{a}_j\rangle}. \end{aligned} \tag{8.6.7}$$

(Even though A involves only nonnegative integral powers of the w_i, it is sometimes helpful to express it using arbitrary powers.) Note that without the constraint (8.6.4), the expressions (8.6.6) would have been more complicated. There are analogous expressions for B. In particular,

$$B = {}_\circ^\circ Y(b_1,x_1)\cdots Y(b_l,x_l){}_\circ^\circ\iota(1). \tag{8.6.8}$$

From (8.5.5) and (8.3.3) we see (as in the proof of Theorem 8.4.2) that

$$\begin{aligned} Y(A,z_1) &= {}_\circ^\circ\exp\left(\sum_{i=1}^{k}\sum_{n\geq 1}\frac{1}{n!}\left(\frac{d}{dz_1}\right)^{n-1}\bar{a}_i(z_1)w_i^n\right)Y(a,z_1){}_\circ^\circ \\ &= \exp\left(\sum_{i=1}^{k}\sum_{n\geq 1}\frac{1}{n!}\left(\frac{d}{dz_1}\right)^{n-1}\bar{a}_i(z_1)^- w_i^n\right)Y(a,z_1) \\ &\quad\cdot\exp\left(\sum_{i=1}^{k}\sum_{n\geq 1}\frac{1}{n!}\left(\frac{d}{dz_1}\right)^{n-1}(\bar{a}_i(z_1)^+ - \bar{a}_i(0)z_1^{-1} + \bar{a}_i(0)z_1^{-1})w_i^n\right) \end{aligned}$$

$$= \exp\left(\sum_{i=1}^{k} e^{w_i(d/dz_1)} \int \bar{a}_i(z_1)^- - \sum_{i=1}^{k} \int \bar{a}_i(z_1)^-\right) Y(a, z_1)$$

$$\cdot \exp\left(\sum_{i=1}^{k} e^{w_i(d/dz_1)} \int (\bar{a}_i(z_1)^+ - \bar{a}_i(0)z_1^{-1}) - \sum_{i=1}^{k} \int (\bar{a}_i(z_1)^+ - \bar{a}_i(0)z_1^{-1})\right)$$

$$\cdot \exp\left(\sum_{i=1}^{k} \sum_{n \geq 1} \frac{1}{n!} \left(\frac{d}{dz_1}\right)^{n-1} \bar{a}_i(0)z_1^{-1}w_i^n\right)$$

$$= \exp\left(\sum_{i=1}^{k} \int \bar{a}_i(z_1 + w_i)^- d(z_1 + w_i)\right) \exp\left(-\int \bar{a}(z_1)^-\right) Y(a, z_1)$$

$$\cdot \exp\left(-\int (\bar{a}(z_1)^+ - \bar{a}(0)z_1^{-1})\right)$$

$$\cdot \exp\left(\sum_{i=1}^{k} \int (\bar{a}_i(z_1 + w_i)^+ - \bar{a}_i(0)(z_1 + w_i)^{-1}) d(z_1 + w_i)\right)$$

$$\cdot \exp\left(\sum_{i=1}^{k} \sum_{n \geq 1} (-1)^{n-1} \frac{1}{n} \bar{a}_i(0)(z_1^{-1}w_i)^n\right)$$

$$= {}^{\circ}_{\circ} \prod_{i=1}^{k} e^{\int (\bar{a}_i(z_1+w_i) - \bar{a}_i(0)(z_1+w_i)^{-1}) d(z_1+w_i)} {}^{\circ}_{\circ} a z_1^{\bar{a}}$$

$$\cdot \exp\left(\sum_{i=1}^{k} \bar{a}_i(0) \log\left(1 + \frac{w_i}{z_1}\right)\right)$$

$$= {}^{\circ}_{\circ} Y(a_1, z_1 + w_1) \cdots Y(a_k, z_1 + w_k) {}^{\circ}_{\circ} z_1^{\bar{a}} \prod_{i=1}^{k} (z_1 + w_i)^{-\bar{a}_i} \prod_{i=1}^{k} \left(1 + \frac{w_i}{z_1}\right)^{\bar{a}_i}$$

$$= {}^{\circ}_{\circ} Y(a_1, z_1 + w_1) \cdots Y(a_k, z_1 + w_k) {}^{\circ}_{\circ}, \tag{8.6.9}$$

and similarly for $Y(B, z_2)$.

Remark 8.6.2: Note that we have a formal generalization of Remark 8.5.2: For all $a_1, \ldots, a_k \in \hat{L}$,

$$Y(Y(a_1, w_1) \cdots Y(a_k, w_k)\iota(1), z_1)$$
$$= Y(a_1, z_1 + w_1) \cdots Y(a_k, z_1 + w_k), \tag{8.6.10}$$

from (8.6.6), (8.6.9) and a formal application of (8.4.25) (for several factors). [To recover (8.5.14), set $w_k = 0$.] The hypothesis (8.6.1) is not used here. We shall interpret (8.6.10) rigorously at the end of this chapter.

Continuing with the proof, we adopt the obvious definition of the normal ordered product of $Y(A, z_1)$ and $Y(B, z_2)$ and we have

$$
{}^{\circ}_{\circ}Y(A, z_1)Y(B, z_2){}^{\circ}_{\circ} = (-1)^{\langle a, b\rangle}\,{}^{\circ}_{\circ}Y(B, z_2)Y(A, z_1){}^{\circ}_{\circ} \qquad (8.6.11)
$$

using (8.6.1) [cf. (8.4.21)]. Also, from (8.4.25) and (8.6.9),

$$
\begin{aligned}
&Y(A, z_1)Y(B, z_2)\\
&\quad = {}^{\circ}_{\circ}Y(A, z_1)Y(B, z_2){}^{\circ}_{\circ} \prod_{\substack{1\le i\le k\\ 1\le j\le l}} (z_1 - z_2 + w_i - x_j)^{\langle \bar{a}_i, \bar{b}_j\rangle},
\end{aligned} \qquad (8.6.12)
$$

where $\prod (z_1 - z_2 + w_i - x_j)^{\langle \bar{a}_i, \bar{b}_j\rangle}$ is understood as an expansion in nonnegative integral powers of z_2, w_i and x_j; it may be computed using the binomial expansion of

$$
(z_1 - (z_2 - w_i + x_j))^{\langle \bar{a}_i, \bar{b}_j\rangle}
$$

or two successive binomial expansions of

$$
((z_1 - z_2) + (w_i - x_j))^{\langle \bar{a}_i, \bar{b}_j\rangle}.
$$

Fix a monomial

$$
P = \prod_{1\le i\le k} w_i^{r_i} \prod_{1\le j\le l} x_j^{s_j} \quad (r_i, s_j \ge 0) \qquad (8.6.13)
$$

in the w_i and x_j. We may and do choose $N \ge 0$ so large that the coefficient of P and of each monomial of lower total degree than P in

$$
F_N = (z_1 - z_2)^N \prod_{\substack{1\le i\le k\\ 1\le j\le l}} (z_1 - z_2 + w_i - x_j)^{\langle \bar{a}_i, \bar{b}_j\rangle} \qquad (8.6.14)
$$

is a polynomial in $z_1 - z_2$. Let $Y_P(z_1, z_2)$ denote the coefficient of P in

$$
Y(A, z_1)Y(B, z_2)(z_1 - z_2)^N = {}^{\circ}_{\circ}Y(A, z_1)Y(B, z_2){}^{\circ}_{\circ}F_N. \qquad (8.6.15)
$$

Since

$$
\begin{aligned}
&{}^{\circ}_{\circ}Y(A, z_1)Y(B, z_2){}^{\circ}_{\circ}\\
&\quad = {}^{\circ}_{\circ}Y(a_1, z_1 + w_1)\cdots Y(a_k, z_1 + w_k)\,Y(b_1, z_2 + x_1)\cdots Y(b_l, z_2 + x_l){}^{\circ}_{\circ}\\
&\quad = {}^{\circ}_{\circ}\prod_{i=1}^{k} e^{\int(\bar{a}_i(z_1+w_i) - \bar{a}_i(0)(z_1+w_i)^{-1})d(z_1+w_i)}\\
&\qquad \cdot \prod_{j=1}^{l} e^{\int(\bar{b}_j(z_2+x_j) - \bar{b}_j(0)(z_2+x_j)^{-1})d(z_2+x_j)}\,{}^{\circ}_{\circ}\\
&\qquad \cdot ab \prod_{i=1}^{k}\left(1 + \frac{w_i}{z_1}\right)^{\bar{a}_i} \prod_{j=1}^{l}\left(1 + \frac{x_j}{z_2}\right)^{\bar{b}_j} z_1^{\bar{a}} z_2^{\bar{b}},
\end{aligned}
$$

we see that

$$Y_P(z_1, z_2) = W_P(z_1, z_2)(z_1/z_2)^{\bar{a}} z_2^{\bar{a}+\bar{b}}$$

where

$$W_P(z_1, z_2) \in (\operatorname{End} V_L)[[z_1, z_1^{-1}, z_2, z_2^{-1}]], \qquad \lim_{z_1 \to z_2} W_P(z_1, z_2) \text{ exists.} \tag{8.6.16}$$

Moreover, the coefficient of P in $Y(A, z_1)Y(B, z_2)$ is

$$Y_P(z_1, z_2)(z_1 - z_2)^{-N} = Y_P(z_1, z_2) z_1^{-N}\left(1 - \frac{z_2}{z_1}\right)^{-N}, \tag{8.6.17}$$

where $(z_1 - z_2)^{-N}$ is to be expanded in nonnegative integral powers of z_2.

Similarly, reversing the roles of A and B and of z_1 and z_2, and taking into account (8.6.11), we see that

$$\begin{aligned} Y(B, z_2)Y(A, z_1) &= {}^{\circ}_{\circ} Y(B, z_2)Y(A, z_1) {}^{\circ}_{\circ} \prod_{\substack{1 \le i \le k \\ 1 \le j \le l}} (z_2 - z_1 + x_j - w_i)^{\langle \bar{a}_i, \bar{b}_j \rangle} \\ &= {}^{\circ}_{\circ} Y(A, z_1)Y(B, z_2) {}^{\circ}_{\circ} (-1)^{\langle \bar{a}, \bar{b} \rangle} \prod_{\substack{1 \le i \le k \\ 1 \le j \le l}} (z_2 - z_1 + x_j - w_i)^{\langle \bar{a}_i, \bar{b}_j \rangle} \end{aligned} \tag{8.6.18}$$

where $\prod (z_2 - z_1 + x_j - w_i)^{\langle \bar{a}_i, \bar{b}_j \rangle}$ is now to be expanded in nonnegative integral powers of z_1, w_i and x_j. Moreover, the coefficient of P and of each monomial of lower total degree than P in

$$(z_1 - z_2)^N (-1)^{\langle \bar{a}, \bar{b} \rangle} \prod (z_2 - z_1 + x_j - w_i)^{\langle \bar{a}_i, \bar{b}_j \rangle}$$

is a polynomial in $z_1 - z_2$ which agrees with the same for F_N. Thus the coefficient of P in $Y(B, z_2)Y(A, z_1)$ is

$$Y_P(z_1, z_2)(z_1 - z_2)^{-N} = Y_P(z_1, z_2)(-z_2)^{-N}\left(1 - \frac{z_1}{z_2}\right)^{-N}, \tag{8.6.19}$$

where $(z_1 - z_2)^{-N}$ is to be expanded in nonnegative integral powers of z_1.

It follows that the coefficient of P in $[Y(A, z_1), Y(B, z_2)]$ is

$$\begin{aligned} &-Y_P(z_1, z_2)(-z_2)^{-N}\left\{\left(1 - \frac{z_1}{z_2}\right)^{-N} - \left(-\frac{z_1}{z_2}\right)^{-N}\left(1 - \frac{z_2}{z_1}\right)^{-N}\right\} \\ &\quad = -Y_P(z_1, z_2)(-z_2)^{-N} \Theta_{z_1/z_2}\left(\left(1 - \frac{z_1}{z_2}\right)^{-N}\right) \end{aligned} \tag{8.6.20}$$

[cf. (8.4.26)]. As in (8.4.27)–(8.4.30) we find that (8.6.20) is the coefficient of z_0^{N-1} in

$$z_2^{-1}Y_P(z_1, z_2)e^{-z_0(\partial/\partial z_1)}\delta(z_1/z_2). \tag{8.6.21}$$

By (8.6.16), we may apply Proposition 8.3.12, and (8.6.21) becomes

$$z_2^{-1}Y_P(z_2 + z_0, z_2)e^{-z_0(\partial/\partial z_1)}((z_1/z_2)^{\bar{a}}\delta(z_1/z_2)) \tag{8.6.22}$$

[cf. (8.4.31)]. But by (8.6.14) and (8.6.15), $Y_P(z_2 + z_0, z_2)$ is the coefficient of P in

$$\mathop{\circ}\limits^{\circ} Y(A, z_2 + z_0)Y(B, z_2)\mathop{\circ}\limits^{\circ} z_0^N \prod_{\substack{1 \le i \le k \\ 1 \le j \le l}} (z_0 + w_i - x_j)^{\langle \bar{a}_i, \bar{b}_j\rangle}, \tag{8.6.23}$$

which is understood as an expansion in nonnegative integral powers of w_i and x_j; $\mathop{\circ}\limits^{\circ} Y(A, z_2 + z_0)Y(B, z_2)\mathop{\circ}\limits^{\circ}$ is also an expansion in nonnegative integral powers of z_0. We conclude that the coefficient of P in $[Y(A, z_1), Y(B, z_2)]$ is the coefficient of $z_0^{-1}P$ in

$$z_2^{-1}\mathop{\circ}\limits^{\circ} Y(A, z_2 + z_0)Y(B, z_2)\mathop{\circ}\limits^{\circ} \cdot \prod_{\substack{1 \le i \le k \\ 1 \le j \le l}} (z_0 + w_i - x_j)^{\langle \bar{a}_i, \bar{b}_j\rangle} e^{-z_0(\partial/\partial z_1)}((z_1/z_2)^{\bar{a}}\delta(z_1/z_2)) \tag{8.6.24}$$

and therefore that $[Y(A, z_1), Y(B, z_2)]$ is the coefficient of z_0^{-1} in (8.6.24). This last assertion is independent of P and N.

On the other hand,

$$Y(A, z_0)B = \mathop{\circ}\limits^{\circ} Y(a_1, z_0 + w_1) \cdots Y(a_k, z_0 + w_k) \cdot Y(b_1, x_1) \cdots Y(b_l, x_l)\mathop{\circ}\limits^{\circ} \iota(1) \prod_{\substack{1 \le i \le k \\ 1 \le j \le l}} (z_0 + w_i - x_j)^{\langle \bar{a}_i, \bar{b}_j\rangle}, \tag{8.6.25}$$

using (8.6.6) and (8.6.9). Thus by (8.6.9) with A replaced by $Y(A, z_0)B$,

$$Y(Y(A, z_0)B, z_2) = \mathop{\circ}\limits^{\circ} Y(a_1, z_2 + z_0 + w_1) \cdots Y(a_k, z_2 + z_0 + w_k) \cdot Y(b_1, z_2 + x_1) \cdots Y(b_l, z_2 + x_l)\mathop{\circ}\limits^{\circ} \prod_{\substack{1 \le i \le k \\ 1 \le j \le l}} (z_0 + w_i - x_j)^{\langle \bar{a}_i, \bar{b}_j\rangle}. \tag{8.6.26}$$

In particular,

$$Y(Y(A, z_0)B, z_2) = \mathop{\circ}\limits^{\circ} Y(A, z_2 + z_0)Y(B, z_2)\mathop{\circ}\limits^{\circ} \prod_{\substack{1 \le i \le k \\ 1 \le j \le l}} (z_0 + w_i - x_j)^{\langle \bar{a}_i, \bar{b}_j\rangle} \tag{8.6.27}$$

where of course $(z_0 + w_i - x_j)^{\langle \bar{a}_i, \bar{b}_j\rangle}$ represents an expansion in nonnegative integral powers of w_i and x_j. This proves the Theorem. ∎

Using Remark 8.5.3 and the fact that the residue of a derivative is 0, we find:

Corollary 8.6.3: *In the notation of Theorem 8.6.1, if $\langle \bar{a}, L\rangle \subset \mathbb{Z}$, then*

$$[u_0, Y(v, z)] = Y(u_0 \cdot v, z).$$

Remark 8.6.4: A formal combination of (8.6.12) and (8.6.27) gives us [even without the assumption (8.6.1)] a second formal but suggestive generalization of Remark 8.5.2 (cf. the last remark):

$$Y(Y(u, z_0)v, z_2) = Y(u, z_2 + z_0)Y(v, z_2) \tag{8.6.28}$$

for all $u, v \in V_L$. In particular, for u and v as in Theorem 8.6.1,

$$\begin{aligned}&[Y(u, z_1),\ Y(v, z_2)]\\ &\quad = \operatorname{Res}_{z_0} z_2^{-1} Y(u, z_2 + z_0)Y(v, z_2)e^{-z_0(\partial/\partial z_1)}((z_1/z_2)^{\bar{a}}\delta(z_1/z_2)).\end{aligned} \tag{8.6.29}$$

This formally generalizes Remark 8.4.3. If we write z_1 for $z_2 + z_0$ in (8.6.28) and use the expansion (8.5.15), we obtain:

$$Y(u, z_1)\,Y(v, z_2) = Y(Y(u, z_1 - z_2)v, z_2) = \sum_{n\in\mathbb{Z}} Y(u_n \cdot v, z_2)(z_1 - z_2)^{-n-1},$$

assuming for convenience that $\langle \bar{a}, L\rangle \subset \mathbb{Z}$. This is an example of an "operator product expansion" in the sense of quantum field theory. Although not strictly rigorous in the present setting, it is very useful. For example, from the right-hand side we can read off the commutator formula (8.6.3) simply by replacing $(z_1 - z_2)^{-n-1}$ by $z_2^{-1}((-1)^n/n!)(\partial/\partial z_1)^n\delta(z_1/z_2)$ for $n \geq 0$. Here it is illuminating to observe that for $n \geq 0$,

$$(z_1 - z_2)^{-n-1} = z_2^{-1}\frac{(-1)^n}{n!}\left(\frac{\partial}{\partial z_1}\right)^n\left(\frac{z_1}{z_2} - 1\right)^{-1}$$

as rational functions [cf. (8.4.27)–(8.4.29)]. Note that only the finitely many "singular terms" in the operator product expansion contribute to the commutator. In Section 8.10, we shall interpret these and related matters in a deeper way.

Using the component operators v_n defined in (8.5.15), we have

$$Y(Y(u, z_0)v, z_2) = \sum_{i,j\in\mathbb{Q}} (u_i \cdot v)_j z_0^{-i-1} z_2^{-j-1} \tag{8.6.30}$$

for $u, v \in V_L$, as in (8.5.24). Thus Corollary 8.5.4 generalizes to:

Corollary 8.6.5: *In the notation of Theorem 8.6.1, if $\langle \bar{a}, L\rangle \subset \mathbb{Z}$ then*

$$[u_m, v_n] = \sum_{i \in \mathbb{N}} \binom{m}{i} (u_i \cdot v)_{m+n-i} \tag{8.6.31}$$

(finite sum) for $m \in \mathbb{Z}$, $n \in \mathbb{Q}$. More generally, let M be a subset of L such that

$$\langle \bar{a}, M\rangle \subset \mathbb{Z} + m_0$$

for fixed $m_0 \in \mathbb{Q}$. (If L is integral, we may take $m_0 = 0$.) For

$$m \in \mathbb{Z} - m_0, \quad n \in \mathbb{Q},$$

formula (8.6.31) holds on the space V_M [recall the notation (7.1.67)].

Hence:

Corollary 8.6.6: *In the same notation, if $\langle \bar{a}, L\rangle \subset \mathbb{Z}$, then*

$$[u_0, v_n] = (u_0 \cdot v)_n.$$

In particular, if L is integral and (8.6.1) holds for all $a, b \in \hat{L}$, then the operators u_0 for $u \in V_L$ form a Lie algebra:

$$[u_0, v_0] = (u_0 \cdot v)_0 \quad \text{for} \quad u, v \in V_L.$$

We also generalize Corollary 8.5.6, using the component operators $x_v(n)$ defined in (8.5.28). Changing indices in Corollary 8.6.5 and keeping in mind (8.5.29) and (8.5.30), we find:

Corollary 8.6.7: *In the notation of Theorem 8.6.1 and Corollary 8.6.5, suppose that the element u is homogeneous, and let*

$$m \in \mathbb{Z} - m_0 - \tfrac{1}{2}\langle \bar{a}, \bar{a}\rangle = \mathbb{Z} - m_0 - \operatorname{wt} u, \quad n \in \mathbb{Q}.$$

Then

$$[x_u(m), x_v(n)] = \sum_{i \in \mathbb{Z} - \langle \bar{a}, \bar{a}\rangle/2} \binom{m - 1 + \operatorname{wt} u}{i - 1 + \operatorname{wt} u} x_{x_u(i) \cdot v}(m + n) \tag{8.6.32}$$

on V_M (finite sum). If the lattice L is even, we take $m, i \in \mathbb{Z}$.

We list some special cases of Theorem 8.6.1 and Corollaries 8.6.5 and 8.6.7. Any of these formulas could be proved more directly (cf. Remark 8.4.6 and Section 8.7 below). We start with the case

$$u = \alpha(-1) = \alpha(-1) \cdot \iota(1) = \alpha(-1) \otimes \iota(1) \quad \text{for} \quad \alpha \in \mathfrak{h}. \tag{8.6.33}$$

Then condition (8.6.1) is vacuous since $a = 1$ and $\bar{a} = 0$. We have

$$Y(u, z_1) = \alpha(z_1) = \sum_{n \in \mathbb{Z}} \alpha(n) z_1^{-n-1} \tag{8.6.34}$$

[see (8.4.8), (8.5.19)]. Take

$$v = \iota(b) \quad \text{for} \quad b \in \hat{L}. \tag{8.6.35}$$

Then using (7.1.33) and (7.1.41), we find

$$\begin{aligned} Y(u, z_0)v = \alpha(z_0)\iota(b) &= \sum_{n \leq 0} \alpha(n) \cdot \iota(b) z_0^{-n-1} \\ &= \langle \alpha, \bar{b} \rangle \iota(b) z_0^{-1} + \sum_{n \leq -1} \alpha(n) \cdot \iota(b) z_0^{-n-1} \end{aligned}$$

and Theorem 8.6.1 asserts

$$[\alpha(z_1), Y(b, z_2)] = \langle \alpha, \bar{b} \rangle Y(b, z_2) z_2^{-1} \delta(z_1/z_2), \tag{8.6.36}$$

Corollary 8.6.5 asserts

$$[\alpha(-1)_m, \iota(b)_n] = \langle \alpha, \bar{b} \rangle \iota(b)_{m+n} \tag{8.6.37}$$

for $m \in \mathbb{Z}$, $n \in \mathbb{Q}$, and Corollary 8.6.7 asserts

$$[\alpha(m), x_b(n)] = \langle \alpha, \bar{b} \rangle x_b(m+n) \tag{8.6.38}$$

for $m \in \mathbb{Z}$, $n \in \mathbb{Q}$.

Now let

$$\begin{aligned} u &= \alpha(-1) \\ v &= \beta(-1) \end{aligned} \tag{8.6.39}$$

for $\alpha, \beta \in \mathfrak{h}$. Recalling the commutation relations (7.1.5) and the fact that

$$\alpha(0) \cdot \iota(1) = 0$$

[see (7.1.33)], we have

$$\begin{aligned} Y(u, z_0)v = \alpha(z_0) \cdot \beta(-1) &= \sum_{n \leq 1} \alpha(n) \cdot \beta(-1) z_0^{-n-1} \\ &= \alpha(1) \cdot \beta(-1) z_0^{-2} + \alpha(0) \cdot \beta(-1) z_0^{-1} + \sum_{n \leq -1} \alpha(n) \cdot \beta(-1) z_0^{-n-1} \\ &= [\alpha(1), \beta(-1)] \cdot \iota(1) z_0^{-2} + \beta(-1)\alpha(0) \cdot \iota(1) z_0^{-1} \\ &\quad + \sum_{n \leq -1} \alpha(n) \cdot \beta(-1) z_0^{-n-1} \\ &= \langle \alpha, \beta \rangle \iota(1) z_0^{-2} + \sum_{n \leq -1} \alpha(n) \cdot \beta(-1) z_0^{-n-1} \end{aligned}$$

and we find

$$[\alpha(z_1), \beta(z_2)] = -\langle\alpha, \beta\rangle z_2^{-1}\frac{\partial}{\partial z_1}\delta(z_1/z_2) \tag{8.6.40}$$

$$[\alpha(-1)_m, \beta(-1)_n] = \langle\alpha, \beta\rangle m\delta_{m+n,0} \tag{8.6.41}$$

$$[\alpha(m), \beta(n)] = \langle\alpha, \beta\rangle m\delta_{m+n,0} \tag{8.6.42}$$

for $m, n \in \mathbb{Z}$. We have (circularly!) recovered the relations (7.1.5).

Now take

$$\begin{aligned} u &= \alpha(-1) \\ v &= \beta(-1)\gamma(-1) = \beta(-1)\gamma(-1) \otimes \iota(1) \end{aligned} \tag{8.6.43}$$

for $\alpha, \beta, \gamma \in \mathfrak{h}$. Again (8.6.1) is vacuous, and we have

$$\begin{aligned} Y(u, z_0)v &= \alpha(z_0)\cdot\beta(-1)\gamma(-1) \\ &= \sum_{n\le 2}\alpha(n)\cdot\beta(-1)\gamma(-1)z_0^{-n-1} \\ &= \alpha(1)\cdot\beta(-1)\gamma(-1)z_0^{-2} + \sum_{n\le -1}\alpha(n)\cdot\beta(-1)\gamma(-1)z_0^{-n-1} \\ &= (\langle\alpha, \beta\rangle\gamma(-1) + \langle\alpha, \gamma\rangle\beta(-1))z_0^{-2} + \sum_{n\le -1}\alpha(n)\cdot\beta(-1)\gamma(-1)z_0^{-n-1}, \end{aligned}$$

so that

$$[\alpha(z_1), {}^{\circ}_{\circ}\beta(z_2)\gamma(z_2){}^{\circ}_{\circ}] = -(\langle\alpha, \beta\rangle\gamma(z_2) + \langle\alpha, \gamma\rangle\beta(z_2))z_2^{-1}\frac{\partial}{\partial z_1}\delta(z_{1/}z_2) \tag{8.6.44}$$

$$[\alpha(-1)_m, \beta(-1)\gamma(-1)_n] = m(\langle\alpha, \beta\rangle\gamma(-1) + \langle\alpha, \gamma\rangle\beta(-1))_{m+n-1} \tag{8.6.45}$$

$$[\alpha(m), \beta\gamma(n)] = m(\langle\alpha, \beta\rangle\gamma(m+n) + \langle\alpha, \gamma\rangle\beta(m+n)) \tag{8.6.46}$$

for $m, n \in \mathbb{Z}$ [recall (8.5.7) and (8.5.39)–(8.5.41)].

Remark 8.6.8: Since the powers β^2 for $\beta \in \mathfrak{h}$ span the second symmetric power $S^2(\mathfrak{h})$ in view of the usual formula

$$\beta\gamma = \tfrac{1}{2}((\beta+\gamma)^2 - \beta^2 - \gamma^2), \tag{8.6.47}$$

all the information in the last three formulas is contained in the formulas:

$$[\alpha(z_1), {}^{\circ}_{\circ}\beta(z_2)^2{}^{\circ}_{\circ}] = -2\langle\alpha,\beta\rangle\beta(z_2)z_2^{-1}\frac{\partial}{\partial z_1}\delta(z_1/z_2) \qquad (8.6.48)$$

$$[\alpha(-1)_m, \beta(-1)_n^2] = 2\langle\alpha,\beta\rangle m\beta(-1)_{m+n-1} \qquad (8.6.49)$$

$$[\alpha(m), \beta^2(n)] = 2\langle\alpha,\beta\rangle m\beta(m+n) \qquad (8.6.50)$$

for $\alpha, \beta \in \mathfrak{h}$ and $m, n \in \mathbb{Z}$.

We also describe the relations between the elements $\beta^2(m)$ and $x_b(n)$ and among the $\beta^2(m)$. Take

$$u = \beta(-1)^2 = \beta(-1)^2 \otimes \iota(1) \quad \text{for} \quad \beta \in \mathfrak{h}$$
$$v = \iota(b) \quad \text{for} \quad b \in \hat{L}. \qquad (8.6.51)$$

Then the condition (8.6.1) is still vacuous, and

$$\begin{aligned} Y(u, z_0)v &= {}^{\circ}_{\circ}\beta(z_0)^2{}^{\circ}_{\circ}\iota(b) \\ &= \beta(0)^2 \cdot \iota(b)z_0^{-2} + 2\beta(-1)\beta(0) \cdot \iota(b)z_0^{-1} + \sum_{n \geq 0} A_n z_0^n \\ &= \langle\beta,\bar{b}\rangle^2\iota(b)z_0^{-2} + 2\langle\beta,\bar{b}\rangle\beta(-1) \cdot \iota(b)z_0^{-1} + \sum_{n \geq 0} A_n z_0^n \end{aligned}$$

for suitable A_n, so that

$$\begin{aligned} [{}^{\circ}_{\circ}\beta(z_1)^2{}^{\circ}_{\circ}, Y(b, z_2)] &= 2\langle\beta,\bar{b}\rangle {}^{\circ}_{\circ}\beta(z_2)Y(b, z_2){}^{\circ}_{\circ}z_2^{-1}\delta(z_1/z_2) \\ &\quad - \langle\beta,\bar{b}\rangle^2 Y(b, z_2)z_2^{-1}\frac{\partial}{\partial z_1}\delta(z_1/z_2) \qquad (8.6.52) \end{aligned}$$

$$\begin{aligned} [\beta(-1)_m^2, \iota(b)_n] &= 2\langle\beta,\bar{b}\rangle(\beta(-1) \cdot \iota(b))_{m+n} \\ &\quad + \langle\beta,\bar{b}\rangle^2 m\iota(b)_{m+n-1} \qquad (8.6.53) \end{aligned}$$

for $m \in \mathbb{Z}$, $n \in \mathbb{Q}$,

$$\begin{aligned} [\beta^2(m), x_b(n)] &= 2\langle\beta,\bar{b}\rangle x_{\beta(-1)\cdot\iota(b)}(m+n) \\ &\quad + \langle\beta,\bar{b}\rangle^2(m+1)x_b(m+n) \qquad (8.6.54) \end{aligned}$$

for $m \in \mathbb{Z}$, $n \in \mathbb{Q}$.

Let

$$u = \beta(-1)^2$$
$$v = \gamma(-1)^2 \qquad (8.6.55)$$

for $\beta, \gamma \in \mathfrak{h}$. Then

$$\begin{aligned} Y(u, z_0)v &= \beta(1)^2 \cdot \gamma(-1)^2 z_0^{-4} + 2\beta(-1)\beta(1) \cdot \gamma(-1)^2 z_0^{-2} \\ &\quad + 2\beta(-2)\beta(1) \cdot \gamma(-1)^2 z_0^{-1} + \sum_{n \geq 0} A_n z_0^n \\ &= 2\langle \beta, \gamma \rangle^2 z_0^{-4} + 4\langle \beta, \gamma \rangle \beta(-1)\gamma(-1) z_0^{-2} \\ &\quad + 4\langle \beta, \gamma \rangle \beta(-2)\gamma(-1) z_0^{-1} + \sum_{n \geq 0} A_n z_0^n \end{aligned}$$

for suitable A_n, and so using (8.5.5) and (8.5.7) we see that

$$\begin{aligned} [{}^{\circ}_{\circ}\beta(z_1)^2{}^{\circ}_{\circ}, {}^{\circ}_{\circ}\gamma(z_2)^2{}^{\circ}_{\circ}] &= 4\langle \beta, \gamma \rangle {}^{\circ}_{\circ}\left(\frac{d}{dz_2}\beta(z_2)\right)\gamma(z_2){}^{\circ}_{\circ} z_2^{-1}\delta(z_1/z_2) \\ - 4\langle \beta, \gamma \rangle {}^{\circ}_{\circ}\beta(z_2)\gamma(z_2){}^{\circ}_{\circ} z_2^{-1} &\frac{\partial}{\partial z_1}\delta(z_1/z_2) - \frac{1}{3}\langle \beta, \gamma \rangle^2 z_2^{-1}\left(\frac{\partial}{\partial z_1}\right)^3 \delta(z_1/z_2) \end{aligned} \tag{8.6.56}$$

$$\begin{aligned} [\beta(-1)^2_m, \gamma(-1)^2_n] &= 4\langle \beta, \gamma \rangle (\beta(-2)\gamma(-1))_{m+n} \\ &+ 4\langle \beta, \gamma \rangle m(\beta(-1)\gamma(-1))_{m+n-1} + 2\langle \beta, \gamma \rangle^2 \binom{m}{3}\delta_{m+n,1} \end{aligned} \tag{8.6.57}$$

$$\begin{aligned} [\beta^2(m), \gamma^2(n)] &= 4\langle \beta, \gamma \rangle x_{\beta(-2)\gamma(-1)}(m+n) \\ &+ 4\langle \beta, \gamma \rangle (m+1)\beta\gamma(m+n) + 2\langle \beta, \gamma \rangle^2 \binom{m+1}{3}\delta_{m+n,0} \end{aligned} \tag{8.6.58}$$

(recall Remark 7.1.1); $x_{\beta(-2)\gamma(-1)}(m+n)$ is given by (8.5.28) and (8.5.38). We shall derive consequences of these relations in the next section.

Remark 8.6.9: If we allow $c(\bar{a}, \bar{b}) \neq (-1)^{\langle \bar{a}, \bar{b} \rangle}$ in (8.6.1), the results in this section clearly still hold if we insert the factor $(-1)^{\langle \bar{a}, \bar{b} \rangle} c(\bar{a}, \bar{b})$ appropriately. For instance, the left-hand side of (8.6.3) becomes

$$Y(u, z_1)Y(v, z_2) - (-1)^{\langle \bar{a}, \bar{b} \rangle} c(\bar{a}, \bar{b}) Y(v, z_2) Y(u, z_1).$$

8.7. The Virasoro Algebra Revisited

We have presented the basic properties of the Virasoro algebra $\mathfrak{v}$ in Section 1.9, and we have already used the natural action of $\mathfrak{v}$ on V_L to motivate the notions of degree and weight (recall Remarks 4.1.2 and 7.1.3). This Lie algebra has basis $\{L_n \mid n \in \mathbb{Z}\} \cup \{c\}$ and is the nonassociative algebra with structure given by

$$[L_m, L_n] = (m-n)L_{m+n} + \tfrac{1}{12}(m^3-m)\delta_{m+n,0}c \quad \text{for} \quad m, n \in \mathbb{Z}$$
$$[c, \mathfrak{v}] = [\mathfrak{v}, c] = 0. \tag{8.7.1}$$

The space spanned by $\{L_{-1}, L_0, L_1\}$ is a subalgebra isomorphic to $\mathfrak{sl}(2, \mathbb{F})$, the coefficient of c vanishing here, and this subalgebra can be understood in terms of infinitesimal projective transformations, as explained in Remark 8.3.11. The Virasoro algebra acts on V_L according to the quadratic operators $L(n)$ $(n \in \mathbb{Z})$ of (1.9.23) for $Z = \mathbb{Z}$ (see Theorem 1.9.6). We shall now reinterpret these operators in terms of the vertex operators associated with a canonical quadratic element of weight 2 in V_L.

The second symmetric power $S^2(\mathfrak{h})$ contains the distinguished element denoted ω_1 in (1.5.17). Let $\{h_1, \ldots, h_l\}$ be a basis of $\mathfrak{h}$ and let $\{h'_1, \ldots, h'_l\}$ be the corresponding dual basis of $\mathfrak{h}$ with respect to $\langle \cdot, \cdot \rangle$. Consider the canonical element

$$\omega = \tfrac{1}{2} \sum_{i=1}^{l} h'_i(-1)h_i(-1) \tag{8.7.2}$$

of weight 2 of V_L. [There should be no confusion with the notation (7.1.14).] Extending the field $\mathbb{F}$ if necessary, let us suppose for convenience that $\mathfrak{h}$ admits an orthonormal basis, say $\{h_1, \ldots, h_l\}$. Then

$$\omega = \tfrac{1}{2} \sum_{i=1}^{l} h_i(-1)^2 \tag{8.7.3}$$

[cf. (1.5.18) and (1.9.21)–(1.9.22)].

Write

$$L(z) = L_{\mathbb{Z}}(z) = Y(\omega, z) = \tfrac{1}{2} \sum_{i=1}^{l} {}_{\circ}^{\circ} h_i(z)^2 {}_{\circ}^{\circ} \tag{8.7.4}$$

and for $n \in \mathbb{Z}$ set

$$L(n) = x_\omega(n) = \tfrac{1}{2} \sum_{i=1}^{l} h_i^2(n) = \tfrac{1}{2} \sum_{i=1}^{l} \sum_{k \in \mathbb{Z}} {}_{\circ}^{\circ} h_i(n-k)h_i(k) {}_{\circ}^{\circ}, \tag{8.7.5}$$

so that

$$L(z) = \sum_{n \in \mathbb{Z}} L(n) z^{-n-2},$$
$$L(-2)\iota(1) = \omega \tag{8.7.6}$$

[recall (8.5.8), (8.5.28), (8.5.40)]. The operators $L(n)$ are the same as those in (1.9.23) for $Z = \mathbb{Z}$ (recall Remark 4.2.2).

From (8.6.56) we find that

$$[L(z_1), L(z_2)] = \left(\frac{d}{dz_2} L(z_2)\right) z_2^{-1}\delta(z_1/z_2) - 2L(z_2)z_2^{-1}\frac{\partial}{\partial z_1}\delta(z_1/z_2)$$
$$- \frac{1}{12}(\dim \mathfrak{h})z_2^{-1}\left(\frac{\partial}{\partial z_1}\right)^3 \delta(z_1/z_2). \tag{8.7.7}$$

Equating the coefficients of z_1^{-m-2} we get

$$[L(m), L(z_2)] = \left(z_2^{m+1}\frac{d}{dz_2} + 2(m+1)z_2^m\right)L(z_2)$$
$$+ \tfrac{1}{12}(m^3 - m)(\dim \mathfrak{h})z_2^{m-2} \tag{8.7.8}$$

for $m \in \mathbb{Z}$, and also equating the coefficients of z_2^{-n-2} [or using (8.6.58)] we obtain

$$[L(m), L(n)] = (m-n)L(m+n) + \tfrac{1}{12}(m^3 - m)(\dim \mathfrak{h})\delta_{m+n,0} \tag{8.7.9}$$

for $m, n \in \mathbb{Z}$. Thus from the general theory we have recovered what we had proved in Theorem 1.9.6 for $Z = \mathbb{Z}$:

Proposition 8.7.1: *The nonassociative algebra $\mathfrak{v}$ defined by (8.7.1) is a Lie algebra, and the operators $L(n)$ provide a representation of $\mathfrak{v}$ on V_L (the symbols L should not be confused!) with*

$$L_n \mapsto L(n) \quad \text{for} \quad n \in \mathbb{Z}$$
$$c \mapsto \dim \mathfrak{h}. \tag{8.7.10}$$

We next describe the commutation action of the $L(m)$ on $\alpha(z_2)$, $\mathop{\circ}\limits^{\circ}\alpha(z_2)^2\mathop{\circ}\limits^{\circ}$ and $Y(a, z_2)$: By (8.6.48) (with z_1 and z_2 reversed) and (2.2.6),

$$[L(z_1), \alpha(z_2)] = \alpha(z_1)z_1^{-1}\frac{\partial}{\partial z_2}\delta(z_1/z_2)$$
$$= -\alpha(z_1)z_2^{-1}\frac{\partial}{\partial z_1}\delta(z_1/z_2)$$
$$= \left(\frac{d}{dz_2}\alpha(z_2)\right)z_2^{-1}\delta(z_1/z_2) - \alpha(z_2)z_2^{-1}\frac{\partial}{\partial z_1}\delta(z_1/z_2) \tag{8.7.11}$$

$$[L(m), \alpha(z_2)] = \left(z_2^{m+1}\frac{d}{dz_2} + (m+1)z_2^m\right)\alpha(z_2) \tag{8.7.12}$$

$$[L(m), \alpha(n)] = -n\alpha(m+n) \tag{8.7.13}$$

for $\alpha \in \mathfrak{h}$, $m, n \in \mathbb{Z}$.

From (8.6.56),

$$[L(z_1), {}_\circ^\circ\alpha(z_2)^2{}_\circ^\circ] = \left(\frac{d}{dz_2}\,{}_\circ^\circ\alpha(z_2)^2{}_\circ^\circ\right)z_2^{-1}\delta(z_1/z_2)$$

$$- 2{}_\circ^\circ\alpha(z_2)^2{}_\circ^\circ z_2^{-1}\frac{\partial}{\partial z_1}\delta(z_1/z_2) - \frac{1}{6}\langle\alpha, \alpha\rangle z_2^{-1}\left(\frac{\partial}{\partial z_1}\right)^3\delta(z_1/z_2) \quad (8.7.14)$$

$$[L(m), {}_\circ^\circ\alpha(z_2)^2{}_\circ^\circ] = \left(z_2^{m+1}\frac{d}{dz_2} + 2(m+1)z_2^m\right){}_\circ^\circ\alpha(z_2)^2{}_\circ^\circ$$

$$+ \tfrac{1}{6}(m^3 - m)\langle\alpha, \alpha\rangle z_2^{m-2} \quad (8.7.15)$$

$$[L(m), \alpha^2(n)] = (m - n)a^2(m + n) + \tfrac{1}{6}(m^3 - m)\langle\alpha, \alpha\rangle\delta_{m+n,0} \quad (8.7.16)$$

for $\alpha \in \mathfrak{h}$, $m, n \in \mathbb{Z}$.

By (8.6.52), (8.4.23) and (8.4.24),

$$[L(z_1), Y(a, z_2)]$$

$$= {}_\circ^\circ\bar{a}(z_2)Y(a, z_2){}_\circ^\circ z_2^{-1}\delta(z_1/z_2) - \frac{1}{2}\langle\bar{a}, \bar{a}\rangle Y(a, z_2)z_2^{-1}\frac{\partial}{\partial z_1}\delta(z_1/z_2)$$

$$= \left(\frac{d}{dz_2}Y(a, z_2)\right)z_2^{-1}\delta(z_1/z_2) - \frac{1}{2}\langle\bar{a}, \bar{a}\rangle Y(a, z_2)z_2^{-1}\frac{\partial}{\partial z_1}\delta(z_1/z_2) \quad (8.7.17)$$

$$[L(m), Y(a, z_2)] = \left(z_2^{m+1}\frac{d}{dz_2} + \frac{1}{2}\langle\bar{a}, \bar{a}\rangle(m+1)z_2^m\right)Y(a, z_2) \quad (8.7.18)$$

$$[L(m), x_a(n)] = \left(\frac{m}{2}\langle\bar{a}, \bar{a}\rangle - m - n\right)x_a(m + n) \quad (8.7.19)$$

for $a \in \hat{L}$, $m \in \mathbb{Z}$ and $n \in \mathbb{Q}$. Note that if $\langle\bar{a}, \bar{a}\rangle = 2$, then

$$[L(m), x_a(n)] = -nx_a(m + n) \quad (8.7.20)$$

and if $\langle\bar{a}, \bar{a}\rangle = 4$, then

$$[L(m), x_a(n)] = (m - n)x_a(m + n). \quad (8.7.21)$$

These formulas suggest some general principles. First recall the relation, established in Section 1.9 [see especially (1.9.44), (1.9.45), (1.9.50) and (1.9.54)], between $L(0)$ and the degree operator d on V_L [recall also (7.1.39)–(7.1.41)]:

$$L(0) = -d + \tfrac{1}{24}\dim\mathfrak{h}. \quad (8.7.22)$$

Recall also that for a homogeneous element $v \in V_L$,

$$L(0)v = (\mathrm{wt}\, v)v. \tag{8.7.23}$$

There is also an important connection between $L(-1)$ and differentiation:

Proposition 8.7.2: *For all* $v \in V_L$,

$$Y(L(-1)v, z) = \frac{d}{dz} Y(v, z). \tag{8.7.24}$$

Proof: Use (8.7.13), the fact that

$$L(-1) \cdot \iota(a) = \sum_{i=1}^{l} \beta_i(-1)\beta_i(0) \cdot \iota(a) = \bar{a}(-1) \cdot \iota(a) \tag{8.7.25}$$

for $a \in \hat{L}$, and the definition (8.5.5) of general vertex operators. ∎

By iterating (8.7.24) and applying (8.3.3) we find:

Proposition 8.7.3: *For* $v \in V_L$,

$$Y(e^{z_0 L(-1)}v, z) = e^{z_0(d/dz)}Y(v, z) = Y(v, z + z_0). \tag{8.7.26}$$

We can now apply both sides to $\iota(1)$ and invoke (8.5.8) to obtain:

Proposition 8.7.4: *For* $v \in V_L$,

$$e^{zL(-1)}v = Y(v, z)\iota(1) \tag{8.7.27}$$

or equivalently,

$$v = e^{-zL(-1)}Y(v, z) \cdot \iota(1). \tag{8.7.28}$$

Now since $\mathrm{wt}\, \omega = 2$,

$$L(-1) = \omega_0 \tag{8.7.29}$$

by (8.5.29). Thus from Corollaries 8.6.3 and 8.6.6 we have:

Proposition 8.7.5: *For* $v \in V_L$,

$$[L(-1), Y(v, z)] = Y(L(-1)v, z) = \frac{d}{dz} Y(v, z) \tag{8.7.30}$$

$$[L(-1), v_n] = (L(-1)v)_n \quad \text{for} \quad n \in \mathbb{Q}. \tag{8.7.31}$$

Expressing the operator $[z_0L(-1), \cdot]$ as the difference of commuting left and right multiplication operators, we obtain by iterating (8.7.30):

Proposition 8.7.6: *For $v \in V_L$,*

$$e^{z_0L(-1)}Y(v, z)e^{-z_0L(-1)} = Y(e^{z_0L(-1)}v, z) = Y(v, z + z_0). \quad (8.7.32)$$

By Theorem 8.6.1, for $v \in V_L$,

$$\begin{aligned}[L(z_1), Y(v, z_2)] &= \mathrm{Res}_{z_0} z_2^{-1}Y(L(z_0)v, z_2)e^{-z_0(\partial/\partial z_1)}\delta(z_1/z_2) \\ &= \mathrm{Res}_{z_0} z_2^{-1} \sum_{n \geq -1} Y(L(n)v, z_2)z_0^{-n-2}e^{-z_0(\partial/\partial z_1)}\delta(z_1/z_2) \\ &= z_2^{-1}Y(L(-1)v, z_2)\delta(z_1/z_2) - z_2^{-1}Y(L(0)v, z_2)\frac{\partial}{\partial z_1}\delta(z_1/z_2) \\ &\quad + \mathrm{Res}_{z_0} z_2^{-1} \sum_{n > 0} Y(L(n)v, z_2)z_0^{-n-2}e^{-z_0(\partial/\partial z_1)}\delta(z_1/z_2). \quad (8.7.33)\end{aligned}$$

The condition for (8.7.33) to reduce to only two terms is

$$L(n)v = 0 \quad \text{for} \quad n > 0, \quad (8.7.34)$$

and in this case,

$$[L(z_1), Y(v, z_2)] = z_2^{-1}\frac{d}{dz_2}Y(v, z_2)\delta(z_1/z_2) - z_2^{-1}Y(L(0)v, z_2)\frac{\partial}{\partial z_1}\delta(z_1/z_2),$$

by (8.7.24). If also

$$L(0)v = hv \quad \text{for some} \quad h \in \mathbb{F} \quad (8.7.35)$$

(in which case $h = \mathrm{wt}\, v$), then

$$[L(z_1), Y(v, z_2)] = z_2^{-1}\frac{d}{dz_2}Y(v, z_2)\delta(z_1/z_2) - hz_2^{-1}Y(v, z_2)\frac{\partial}{\partial z_1}\delta(z_1/z_2). \quad (8.7.36)$$

A vector v which satisfies (8.7.34) and (8.7.35) is called a *lowest weight vector for* $\mathfrak{v}$. Thus we have [reformulating (8.7.36)]:

Proposition 8.7.7: *An element $v \in V_L$ is a lowest weight vector for $\mathfrak{v}$ if and only if for some $h \in \mathbb{F}$,*

$$[L(n), Y(v, z)] = \left(z^{n+1}\frac{d}{dz} + h(n + 1)z^n\right)Y(v, z) \quad \text{for} \quad n \in \mathbb{Z} \quad (8.7.37)$$

or equivalently,

$$[L(m), x_v(n)] = (hm - m - n)x_v(m + n) \quad \text{for} \quad m \in \mathbb{Z},\ n \in \mathbb{Q}. \tag{8.7.38}$$

In this case, $h = \text{wt } v$.

Remark 8.7.8: Note that $v = \alpha(-1)$ and $v = \iota(a)$ are lowest weight vectors for $\mathfrak{v}$, with $h = 1$ and $h = \frac{1}{2}\langle \bar{a}, \bar{a} \rangle$, respectively, corresponding to the commutator formulas (8.7.11)–(8.7.13) and (8.7.17)–(8.7.19). On the other hand, $\alpha(-1)^2$ does not satisfy condition (8.7.34) unless $\langle \alpha, \alpha \rangle = 0$ [cf. (8.7.14)–(8.7.16)]. Using our orthonormal basis $\{h_1, \ldots, h_l\}$, we can express the elements $v \in S^2(\mathfrak{h}(-1))$ which are lowest weight vectors (of weight 2) as follows:

$$v = \sum_{1 \le i \le j \le l} c_{ij} h_i(-1) h_j(-1) \quad \text{such that} \quad \sum_{i=1}^{l} c_{ii} = 0. \tag{8.7.39}$$

Note that the element $\omega \in S^2(\mathfrak{h}(-1))$ [see (8.7.3)] is not a lowest weight vector [cf. (8.7.7)–(8.7.9)].

Proposition 8.7.7 suggests interesting relations with the transformations discussed in Proposition 8.3.10 and Remark 8.3.11. From formula (8.7.37) we obtain:

$$[L(n), Y(v, z)] = z^{-h(n+1)} z^{n+1} \frac{d}{dz} (z^{h(n+1)} Y(v, z)). \tag{8.7.40}$$

Thus

$$[z_0 L(n), z^{h(n+1)} Y(v, z)] = z_0 z^{n+1} \frac{d}{dz} (z^{h(n+1)} Y(v, z)) \tag{8.7.41}$$

and so

$$\begin{aligned} e^{z_0 L(n)} z^{h(n+1)} Y(v, z) e^{-z_0 L(n)} &= e^{z_0 z^{n+1} d/dz} (z^{h(n+1)} Y(v, z)) \\ &= z_1^{h(n+1)} Y(v, z_1) \end{aligned} \tag{8.7.42}$$

by Propositions 8.3.1 and 8.3.10, where

$$z_1 = \begin{cases} e^{z_0} z & \text{if} \quad n = 0 \\ (z^{-n} - n z_0)^{-1/n} & \text{if} \quad n \neq 0. \end{cases} \tag{8.7.43}$$

Thus, $Y(v, z)$ is what physicists call a holomorphic primary conformal field:

$$e^{z_0 L(n)} Y(v, z) e^{-z_0 L(n)} = \left(\frac{z_1}{z}\right)^{h(n+1)} Y(v, z_1) = \left(\frac{\partial z_1}{\partial z}\right)^{h} Y(v, z_1). \quad (8.7.44)$$

The cases $n = \pm 1$ are especially interesting. We have

$$z_1 = \begin{cases} z + z_0 & \text{if} \quad n = -1 \\ \dfrac{z}{1 - z_0 z} & \text{if} \quad n = 1, \end{cases} \quad (8.7.45)$$

$$\left(\frac{z_1}{z}\right)^{n+1} = \begin{cases} 1 & \text{if} \quad n = -1 \\ e^{z_0} & \text{if} \quad n = 0 \\ (1 - z_0 z)^{-2} = (1 + z_0 z_1)^2 & \text{if} \quad n = 1. \end{cases} \quad (8.7.46)$$

An element $v \in V_L$ is said to be a *lowest weight vector for* $\mathfrak{sl}(2, \mathbb{F}) = \text{span}\{L_{-1}, L_0, L_1\}$ if

$$L(1)v = 0, \quad L(0)v = hv \quad \text{for some} \quad h \in \mathbb{F}. \quad (8.7.47)$$

Using (8.7.33) and Remark 8.3.11 we obtain the following projective transformation properties of vertex operators (cf. Proposition 8.7.6):

Proposition 8.7.9: *An element $v \in V_L$ is a lowest weight vector for $\mathfrak{sl}(2, \mathbb{F})$ if and only if for some $h \in \mathbb{F}$,*

$$[L(n), Y(v, z)] = \left(z^{n+1} \frac{d}{dz} + h(n+1)z^n\right) Y(v, z) \quad \text{for} \quad n = 0, \pm 1. \quad (8.7.48)$$

In this case, $h = \text{wt}\, v$ and

$$e^{z_0 L(n)} Y(v, z) e^{-z_0 L(n)} = (cz + d)^{-2h} Y\left(v, \frac{az + b}{cz + d}\right) = (a - cz_1)^{2h} Y(v, z_1) \quad (8.7.49)$$

for $n = 0, \pm 1$, where

$$\begin{pmatrix} a & b \\ c & d \end{pmatrix} = e^{z_0 \pi(L(n))}, \quad z_1 = \frac{az + b}{cz + d} \quad (8.7.50)$$

and

$$\pi: \text{span}\{L(-1), L(0), L(1)\} \xrightarrow{\sim} \mathfrak{sl}(2, \mathbb{F}) \quad (8.7.51)$$

is the identification given by

$$
\begin{aligned}
L(-1) &\mapsto \begin{pmatrix} 0 & 1 \\ 0 & 0 \end{pmatrix}, & e^{z_0 L(-1)} &\mapsto \begin{pmatrix} 1 & z_0 \\ 0 & 1 \end{pmatrix} \\
L(0) &\mapsto \tfrac{1}{2}\begin{pmatrix} 1 & 0 \\ 0 & -1 \end{pmatrix}, & e^{z_0 L(0)} &\mapsto \begin{pmatrix} e^{z_0/2} & 0 \\ 0 & e^{-z_0/2} \end{pmatrix} \\
L(1) &\mapsto \begin{pmatrix} 0 & 0 \\ -1 & 0 \end{pmatrix}, & e^{z_0 L(1)} &\mapsto \begin{pmatrix} 1 & 0 \\ -z_0 & 1 \end{pmatrix}
\end{aligned}
\tag{8.7.52}
$$

(cf. Remarks 1.9.2 and 8.3.11).

Remark 8.7.10: The vector $\iota(1)$ is distinguished by being the unique (up to scalar multiple) element of V_L annihilated by $\mathfrak{sl}(2, \mathbb{F})$, as we see with the help of Proposition 8.7.2.

It is sometimes useful to know how to compute commutators such as (8.6.56) [and hence (8.7.7)] directly, without the use of Theorem 8.6.1. We can proceed as follows: For $\beta, \gamma \in \mathfrak{h}$, we find from the commutation relations (7.1.5) that

$$
\begin{aligned}
[\beta(z_1)^+, \gamma(z_2)^-] &= \langle \beta, \gamma \rangle \sum_{n>0} n z_1^{-n-1} z_2^{n-1} \\
&= -\langle \beta, \gamma \rangle z_2^{-1} \frac{\partial}{\partial z_1} (1 - z_2/z_1)^{-1} \\
&= \langle \beta, \gamma \rangle (z_1 - z_2)^{-2},
\end{aligned}
\tag{8.7.53}
$$

which is to be expanded in nonnegative integral powers of z_2 [recall the definitions (8.4.10)]. Hence

$$
\begin{aligned}
{:}\beta(z_1)\gamma(z_2){:} &= \beta(z_1)\gamma(z_2) - [\beta(z_1)^+, \gamma(z_2)^-] \\
&= \beta(z_1)\gamma(z_2) - \langle \beta, \gamma \rangle (z_1 - z_2)^{-2}
\end{aligned}
\tag{8.7.54}
$$

and so

$$
{:}\gamma(z_2)^2{:} = \lim_{z_1 \to z_2} (\gamma(z_1)\gamma(z_2) - \langle \gamma, \gamma \rangle (z_1 - z_2)^{-2}). \tag{8.7.55}
$$

(Note that the limits of the individual terms on the right do not exist.) Also,

$$
[\beta(z_1), \gamma(z_2)] = -\langle \beta, \gamma \rangle z_2^{-1} \frac{\partial}{\partial z_1} \delta(z_1/z_2) \tag{8.7.56}
$$

[cf. (8.6.40)], again from (7.1.5). Thus

$$\begin{aligned}[\beta(z_1), {}^{\circ}_{\circ}\gamma(z_2)^2{}^{\circ}_{\circ}] &= \lim_{z_3 \to z_2} [\beta(z_1), \gamma(z_3)\gamma(z_2)] \\ &= \lim_{z_3 \to z_2} ([\beta(z_1), \gamma(z_3)]\gamma(z_2) + \gamma(z_3)[\beta(z_1), \gamma(z_2)]) \\ &= -\langle \beta, \gamma \rangle \lim_{z_3 \to z_2} \left(z_3^{-1} \frac{\partial}{\partial z_1} \delta(z_1/z_3)\gamma(z_2) + \gamma(z_3) z_2^{-1} \frac{\partial}{\partial z_1} \delta(z_1/z_2) \right) \\ &= -2\langle \beta, \gamma \rangle \gamma(z_2) z_2^{-1} \frac{\partial}{\partial z_1} \delta(z_1/z_2), \end{aligned} \tag{8.7.57}$$

in agreement with (8.6.44).

Finally, using (8.7.54), (8.7.55) and (8.7.57), we see that

$$\begin{aligned}&[{}^{\circ}_{\circ}\beta(z_1)^2{}^{\circ}_{\circ}, {}^{\circ}_{\circ}\gamma(z_2)^2{}^{\circ}_{\circ}] = \lim_{z_0 \to z_1} [\beta(z_0)\beta(z_1), {}^{\circ}_{\circ}\gamma(z_2)^2{}^{\circ}_{\circ}] \\ &\quad = \lim_{z_0 \to z_1} ([\beta(z_0), {}^{\circ}_{\circ}\gamma(z_2)^2{}^{\circ}_{\circ}]\beta(z_1) + \beta(z_0)[\beta(z_1), {}^{\circ}_{\circ}\gamma(z_2)^2{}^{\circ}_{\circ}]) \\ &\quad = -2\langle \beta, \gamma \rangle \lim_{z_0 \to z_1} \left(\gamma(z_2)\beta(z_1) z_2^{-1} \frac{\partial}{\partial z_0} \delta(z_0/z_2) \right. \\ &\qquad \left. + \beta(z_0)\gamma(z_2) z_2^{-1} \frac{\partial}{\partial z_1} \delta(z_1/z_2) \right) \\ &\quad = -2\langle \beta, \gamma \rangle \lim_{z_0 \to z_1} \left(({}^{\circ}_{\circ}\gamma(z_2)\beta(z_1){}^{\circ}_{\circ} + \langle \beta, \gamma \rangle (z_2 - z_1)^{-2}) z_2^{-1} \frac{\partial}{\partial z_0} \delta(z_0/z_2) \right. \\ &\qquad \left. + ({}^{\circ}_{\circ}\beta(z_0)\gamma(z_2){}^{\circ}_{\circ} + \langle \beta, \gamma \rangle (z_0 - z_2)^{-2}) z_2^{-1} \frac{\partial}{\partial z_1} \delta(z_1/z_2) \right) \\ &\quad = -4\langle \beta, \gamma \rangle {}^{\circ}_{\circ}\beta(z_1)\gamma(z_2){}^{\circ}_{\circ} z_2^{-1} \frac{\partial}{\partial z_1} \delta(z_1/z_2) \\ &\qquad - 2\langle \beta, \gamma \rangle^2 \lim_{z_0 \to z_1} [(z_2 - z_1)^{-2}((z_2 - z_0)^{-2} - (z_0 - z_2)^{-2}) \\ &\qquad + (z_0 - z_2)^{-2}((z_2 - z_1)^{-2} - (z_1 - z_2)^{-2})] \\ &\quad = -4\langle \beta, \gamma \rangle {}^{\circ}_{\circ}\beta(z_1)\gamma(z_2){}^{\circ}_{\circ} z_2^{-1} \frac{\partial}{\partial z_1} \delta(z_1/z_2) \\ &\qquad - 2\langle \beta, \gamma \rangle^2 \lim_{z_0 \to z_1} [(z_2 - z_1)^{-2}(z_2 - z_0)^{-2} - (z_0 - z_2)^{-2}(z_1 - z_2)^{-2}]\end{aligned}$$

$$= -4\langle\beta,\gamma\rangle {}^{\circ}_{\circ}\beta(z_1)\gamma(z_2){}^{\circ}_{\circ} z_2^{-1}\frac{\partial}{\partial z_1}\delta(z_1/z_2)$$

$$- 2\langle\beta,\gamma\rangle^2[(z_2 - z_1)^{-4} - (z_1 - z_2)^{-4}]$$

$$= -4\langle\beta,\gamma\rangle {}^{\circ}_{\circ}\beta(z_2)\gamma(z_2){}^{\circ}_{\circ} z_2^{-1}\frac{\partial}{\partial z_1}\delta(z_1/z_2)$$

$$+ 4\langle\beta,\gamma\rangle {}^{\circ}_{\circ}\left(\frac{d}{dz_2}\beta(z_2)\right)\gamma(z_2){}^{\circ}_{\circ} z_2^{-1}\delta(z_1/z_2)$$

$$- \tfrac{1}{3}\langle\beta,\gamma\rangle^2 z_2^{-1}\left(\frac{\partial}{\partial z_1}\right)^3\delta(z_1/z_2), \tag{8.7.58}$$

as in (8.6.56). In the last step, we have used (2.2.6). Of course, the expressions like $(z_2 - z_1)^{-2}$ are to be understood as expansions in nonnegative integral powers of the second variable.

8.8. The Jacobi Identity

As we look at the commutator formula (8.6.3), questions arise. Perhaps the first two are:

Question 8.8.1: Why is the right-hand side alternating, i.e., why does it change sign under the interchange $(u, z_1) \leftrightarrow (v, z_2)$, as it must?

Question 8.8.2: What is the significance of the expansion coefficients (with respect to z_0) of the expression

$$z_2^{-1}Y(Y(u,z_0)v, z_2)e^{-z_0(\partial/\partial z_1)}((z_1/z_2)^a\delta(z_1/z_2)) \tag{8.8.1}$$

other than Res_{z_0}?

For simplicity, we shall assume that the lattice L in Theorem 8.6.1 is even and that (8.6.1) holds for all $a, b \in \hat{L}$. (But see also the generalization at the end of this section.)

Let us recall some basic properties of the vertex operators $Y(v, z)$ for $v \in V_L$:

$$Y(\cdot, z) \quad \text{is a linear map} \quad V_L \to (\mathrm{End}\, V_L)\{z\} \tag{8.8.2}$$

[see (8.5.6)],

$$Y(\iota(1), z) = 1 \tag{8.8.3}$$

[see (8.5.3)], and for $v \in V_L$,

$$\lim_{z \to 0} Y(v, z)\iota(1) = v \tag{8.8.4}$$

[see (8.5.8)] and more generally,

$$Y(v, z)\iota(1) = e^{zL(-1)}v \tag{8.8.5}$$

(Proposition 8.7.4). Also,

$$e^{z_0 L(-1)} Y(v, z) e^{-z_0 L(-1)} = Y(e^{z_0 L(-1)}v, z) = Y(v, z + z_0) \tag{8.8.6}$$

(Proposition 8.7.6). Recall that the notation $Y(v, z + z_0)$ implies that the expression is to be expanded in nonnegative integral powers of z_0.

To answer our first question, we want to find a relation between $Y(u, z)v$ and something analogous with u and v reversed. In fact, we have the following generalization of (8.8.5):

Proposition 8.8.3: *For $u, v \in V_L$,*

$$Y(u, z)v = e^{zL(-1)}(Y(v, -z)u) \tag{8.8.7}$$

or equivalently,

$$Y(Y(u, z_0)v, z_2) = Y(Y(v, -z_0)u, z_2 + z_0). \tag{8.8.8}$$

Proof: The two statements being equivalent by (8.8.4)–(8.8.6), we prove the second, and for this it suffices to take u and v to be A and B as in (8.6.5), respectively. By (8.6.27),

$$Y(Y(A, z_0)B, z_2) = {}^{\circ}_{\circ} Y(A, z_2 + z_0) Y(B, z_2) {}^{\circ}_{\circ} \prod_{\substack{1 \le i \le k \\ 1 \le j \le l}} (z_0 + w_i - x_j)^{\langle \bar{a}_i, \bar{b}_j \rangle}.$$

On the other hand,

$$Y(Y(B, -z_0)A, z_2 + z_0) = {}^{\circ}_{\circ} Y(B, z_2) Y(A, z_2 + z_0) {}^{\circ}_{\circ} \prod_{\substack{1 \le i \le k \\ 1 \le j \le l}} (-z_0 + x_j - w_j)^{\langle \bar{a}_i, \bar{b}_j \rangle},$$

and the two expressions are equal by (8.6.11) and (8.6.4). ∎

Remark 8.8.4: From (8.6.28), we formally obtain the following suggestive "commutativity" relation:

$$Y(u, z_2 + z_0) Y(v, z_2) = Y(v, z_2) Y(u, z_2 + z_0). \tag{8.8.9}$$

To answer Question 8.8.1, we observe first that

$$e^{-z_0(\partial/\partial z_1)}\delta(z_1/z_2) = \delta\left(\frac{z_1 - z_0}{z_2}\right) \tag{8.8.10}$$

by (8.3.3), and we establish the following symmetry property of such expressions (always keeping in mind the convention of expanding as a formal power series in the second summand):

Proposition 8.8.5: *We have*

$$z_2^{-1}\delta\left(\frac{z_1 - z_0}{z_2}\right) = z_1^{-1}\delta\left(\frac{z_2 + z_0}{z_1}\right). \tag{8.8.11}$$

Proof: Expanding, we find:

$$\begin{aligned} z_2^{-1}\delta\left(\frac{z_1 - z_0}{z_2}\right) &= \sum_{n \in \mathbb{Z}} (z_1 - z_0)^n z_2^{-n-1} = \sum_n (z_1 - z_0)^{-n-1} z_2^n \\ &= \sum_{n,i} (-1)^i \binom{-n-1}{i} z_1^{-n-1-i} z_0^i z_2^n \\ &= \sum_{n,i} (-1)^i \binom{-n-1+i}{i} z_1^{-n-1} z_0^i z_2^{n-i} \\ &= \sum_{n,i} \binom{n}{i} z_1^{-n-1} z_0^i z_2^{n-i} \\ &= z_1^{-1}\delta\left(\frac{z_2 + z_0}{z_1}\right). \qquad \blacksquare \end{aligned}$$

Hence:

Proposition 8.8.6: *In the expression (8.8.1), the interchange* $(u, z_1) \leftrightarrow (v, z_2)$ *is equivalent to the change* $z_0 \leftrightarrow -z_0$.

Corollary 8.8.7: *The right-hand side in (8.6.3) is alternating. More generally, under* $(u, z_1) \leftrightarrow (v, z_2)$, *the coefficient of* z_0^{-n-1} *in (8.8.1) is alternating if n is even and symmetric if n is odd.*

Now we turn to Question 8.8.2, the last two results suggesting that the Lie bracket of $Y(u, z_1)$ with $Y(v, z_2)$ should perhaps be embedded in an infinite family of alternating or commutative products. Consideration of the proof of Theorem 8.6.1 (see the proof of the next theorem) suggests the following:

For $u, v \in V_L$ and $n \in \mathbb{N}$ define

$$[Y(u, z_1) \times_n Y(v, z_2)] = (z_1 - z_2)^n [Y(u, z_1), Y(v, z_2)] \qquad (8.8.12)$$

and more generally, for $n \in \mathbb{Z}$ define

$$[Y(u, z_1) \times_n Y(v, z_2)] = (z_1 - z_2)^n Y(u, z_1) Y(v, z_2) - (-z_2 + z_1)^n Y(v, z_2) Y(u, z_1) \qquad (8.8.13)$$

(recall the expansion convention). Then

$$[\cdot \times_0 \cdot] = [\cdot, \cdot] \quad \text{(Lie bracket)}. \qquad (8.8.14)$$

Remark 8.8.8: The product $\times_n$ is alternating if n is even and commutative if n is odd.

To understand the new products, recall the two types of expansion coefficients v_n and $x_v(n)$ from (8.5.15) and (8.5.28). For $l \in \mathbb{Z}$ and $m, n \in \mathbb{Q}$, define analogous expansion coefficients of the new products by:

$$[Y(u, z_1) \times_l Y(v, z_2)] = \sum_{m, n \in \mathbb{Q}} [u \times_l v]_{mn} z_1^{-m-1} z_2^{-n-1} \qquad (8.8.15)$$

and if u and v are homogeneous,

$$[Y(u, z_1) \times_l Y(v, z_2)] = \sum_{m, n \in \mathbb{Q}} [x_u \times_l x_v](m, n) z_1^{-m-\mathrm{wt}\, u} z_2^{-n-\mathrm{wt}\, v}$$

$$[x_u \times_l x_v](m, n) = [u \times_l v]_{m+\mathrm{wt}\, u-1, n+\mathrm{wt}\, v-1}. \qquad (8.8.16)$$

Then for instance

$$[u \times_0 v]_{mn} = [u_m, v_n] \qquad (8.8.17)$$

$$[x_u \times_0 x_v](m, n) = [x_u(m), x_v(n)]. \qquad (8.8.18)$$

Of course, we may extend the definition of $[x_u \times_l x_v](m, n)$ to all u, v by linearity.

The product $\times_1$ will be especially important. We call $[\cdot \times_1 \cdot]$ the *cross-bracket* and read it "cross" for short because

$$[u \times_1 v]_{mn} = [u_{m+1}, v_n] - [u_m, v_{n+1}] \qquad (8.8.19)$$

$$[x_u \times_1 x_v](m, n) = [x_u(m + 1), x_v(n)] - [x_u(m), x_v(n + 1)], \qquad (8.8.20)$$

each of which is made up of two brackets which "cross."

More generally, for $l \in \mathbb{Z}$,

$$[u \times_l v]_{mn} = \sum_{i \in \mathbb{N}} (-1)^i \binom{l}{i} u_{m+l-i} v_{n+i} - (-1)^l \sum_{i \in \mathbb{N}} (-1)^i \binom{l}{i} v_{n+l-i} u_{m+i} \tag{8.8.21}$$

$$\begin{aligned}[x_u \times_l x_v](m, n) = &\sum_{i \in \mathbb{N}} (-1)^i \binom{l}{i} x_u(m + l - i) x_v(n + i) \\ &- (-1)^l \sum_{i \in \mathbb{N}} (-1)^i \binom{l}{i} x_v(n + l - i) x_u(m + i).\end{aligned} \tag{8.8.22}$$

For any homogeneous elements $u, v \in V_L$, $l \in \mathbb{Z}$ and $m, n \in \mathbb{Q}$,

$$\deg[u \times_l v]_{mn} = l + m + n - \operatorname{wt} u - \operatorname{wt} v + 2 \tag{8.8.23}$$

$$\deg[x_u \times_l x_v](m, n) = l + m + n. \tag{8.8.24}$$

It is natural to form a generating function out of all the products $\times_n$: For $u, v \in V_L$ set

$$[Y(u, z_1) \times_{z_0} Y(v, z_2)] = \sum_{n \in \mathbb{Z}} [Y(u, z_1) \times_n Y(v, z_2)] z_0^{-n-1}, \tag{8.8.25}$$

so that

$$[Y(u, z_1), Y(v, z_2)] = \operatorname{Res}_{z_0} [Y(u, z_1) \times_{z_0} Y(v, z_2)] \tag{8.8.26}$$

$$[Y(u, z_1) \times_n Y(v, z_2)] = \operatorname{Res}_{z_0} z_0^n [Y(u, z_1) \times_{z_0} Y(v, z_2)] \tag{8.8.27}$$

for $n \in \mathbb{Z}$. From the definition (8.8.13) we find that

$$\begin{aligned}[Y(u, z_1) \times_{z_0} Y(v, z_2)] = &z_0^{-1} \delta\left(\frac{z_1 - z_2}{z_0}\right) Y(u, z_1) Y(v, z_2) \\ &- z_0^{-1} \delta\left(\frac{z_2 - z_1}{-z_0}\right) Y(v, z_2) Y(u, z_1).\end{aligned} \tag{8.8.28}$$

Now we have the following general result, which answers Question 8.8.2 and which connects Corollary 8.8.7 and Remark 8.8.8. We call this formula the *Jacobi identity*, since it is analogous to the Jacobi identity for Lie algebras in a number of respects; see for example Remarks 8.8.20 and 8.9.1 below.

Theorem 8.8.9: *For $u, v \in V_L$,*

$$z_0^{-1}\delta\left(\frac{z_1 - z_2}{z_0}\right)Y(u, z_1)Y(v, z_2) - z_0^{-1}\delta\left(\frac{z_2 - z_1}{-z_0}\right)Y(v, z_2)Y(u, z_1)$$

$$= z_2^{-1}\delta\left(\frac{z_1 - z_0}{z_2}\right)Y(Y(u, z_0)v, z_2). \tag{8.8.29}$$

Equivalently,

$$[Y(u, z_1) \times_n Y(v, z_2)] = \mathrm{Res}_{z_0}\, z_0^n z_2^{-1} Y(Y(u, z_0)v, z_2)e^{-z_0(\partial/\partial z_1)}\delta(z_1/z_2) \tag{8.8.30}$$

for $n \in \mathbb{Z}$.

Proof: The equivalence being clear, we prove (8.8.30). We repeat the proof of Theorem 8.6.1 except for the following changes: By (8.6.17), the coefficient of P in $(z_1 - z_2)^n\, Y(A, z_1)Y(B, z_2)$ is

$$Y_P(z_1, z_2)(z_1 - z_2)^{-(N-n)}$$

and by (8.6.19), the coefficient of P in $(-z_2 + z_1)^n Y(B, z_2)Y(A, z_1)$ is

$$Y_P(z_1, z_2)(-z_2 + z_1)^{-(N-n)}.$$

Thus by (8.6.20), the coefficient of P in $[Y(A, z_1) \times_n Y(B, z_2)]$ is

$$-Y_P(z_1, z_2)(-z_2)^{-(N-n)}\Theta_{z_1/z_2}\left(\left(1 - \frac{z_1}{z_2}\right)^{-(N-n)}\right),$$

which is the coefficient of z_0^{N-n-1} in the expression (8.6.21). Hence $[Y(A, z_1) \times_n Y(B, z_2)]$ is the coefficient of z_0^{-n-1} in (8.6.24) and the Theorem follows as in the earlier proof. ∎

Remark 8.8.10: As this argument shows, the precise significance of $\times_n$ is that it "reduces the order of the pole by n" in the proof of the commutator formula.

Remark 8.8.11: Theorem 8.6.1 is Res_{z_0} of the Jacobi identity (8.8.29). Using Proposition 8.8.5 we see that Res_{z_1} of the identity asserts that

$$Y(Y(u, z_0)v, z_2) = Y(u, z_0 + z_2)Y(v, z_2)$$

$$- \mathrm{Res}_{z_1} z_0^{-1}\delta\left(\frac{z_2 - z_1}{-z_0}\right) Y(v, z_2)Y(u, z_1); \tag{8.8.31}$$

for this, note that the first term on the left of (8.8.29) is

$$z_1^{-1}\delta\left(\frac{z_0 + z_2}{z_1}\right)Y(u, z_1)Y(v, z_2) = z_1^{-1}Y(u, z_1)Y(v, z_2)e^{z_2(\partial/\partial z_0)}\delta(z_0/z_1)$$

$$= z_1^{-1}\delta\left(\frac{z_0 + z_2}{z_1}\right)Y(u, z_0 + z_2)Y(v, z_2)$$

by a version of the usual argument (see, e.g., the proof of Proposition 8.3.12). Formula (8.8.31) is in effect a rigorous "correction" of (8.6.28)! [We shall reinterpret (8.6.28) in Section 8.10 below.]

Remark 8.8.12: A similar argument based on Res_{z_2} of the Jacobi identity shows that

$$Y(Y(u, z_0)v, z_1 - z_0) = \mathrm{Res}_{z_2} z_0^{-1}\delta\left(\frac{z_1 - z_2}{z_0}\right)Y(u, z_1)Y(v, z_2)$$

$$+ Y(v, -z_0 + z_1)Y(u, z_1). \qquad (8.8.32)$$

Remark 8.8.13: In addition to Theorem 8.6.1, (8.8.31) and (8.8.32), the Jacobi identity also implies (8.8.8) and Proposition 8.8.6: Since the left-hand side of (8.8.29) is visibly invariant under $(u, z_1, z_0) \leftrightarrow (v, z_2, -z_0)$ (cf. Remark 8.8.8), the right-hand side must be invariant, and we have

$$z_1^{-1}\delta\left(\frac{z_2 + z_0}{z_1}\right)Y(Y(u, z_0)v, z_2) = z_2^{-1}\delta\left(\frac{z_1 - z_0}{z_2}\right)Y(Y(u, z_0)v, z_2)$$

$$= z_1^{-1}\delta\left(\frac{z_2 + z_0}{z_1}\right)Y(Y(v, -z_0)u, z_1)$$

$$= z_1^{-1}\delta\left(\frac{z_2 + z_0}{z_1}\right)Y(Y(v, -z_0)u, z_2 + z_0), \qquad (8.8.33)$$

which gives (8.8.8) by taking Res_{z_1}.

Remark 8.8.14: If we apply (8.8.31) and (8.8.32) to $\iota(1)$, certain terms drop out:

$$Y(Y(u, z_0)v, z_2)\iota(1) = Y(u, z_0 + z_2)Y(v, z_2)\iota(1) \qquad (8.8.34)$$

$$Y(Y(u, z_0)v, z_1 - z_0)\iota(1) = Y(v, -z_0 + z_1)Y(u, z_1)\iota(1). \qquad (8.8.35)$$

The Jacobi identity suggests still another remarkable rewriting of the main commutator result, Theorem 8.6.1, and for that matter of the generalization (8.8.30) for $\times_n$. For this we first observe the following result, which is nothing but a distillation of the argument of (8.4.26)–(8.4.29) and which relates a power series in z_0, in z_1 and in z_2:

Proposition 8.8.15: *We have*

$$z_2^{-1}\delta\left(\frac{z_1 - z_0}{z_2}\right) = z_0^{-1}\delta\left(\frac{z_1 - z_2}{z_0}\right) - z_0^{-1}\delta\left(\frac{-z_2 + z_1}{z_0}\right). \qquad (8.8.36)$$

Proof: We can simply take $u = v = \iota(1)$ in (8.8.29) or we can argue directly: The coefficients of z_0^n for $n < 0$ are clearly 0 on the two sides. To show that the coefficients are equal for $n \geq 0$, just repeat the argument of (8.4.26)–(8.4.29), which shows that

$$\begin{aligned}(z_1 - z_2)^{-n-1} &- (-z_2 + z_1)^{-n-1} \\ &= \text{coefficient of } z_0^n \text{ in } z_2^{-1}\delta\left(\frac{z_1 - z_0}{z_2}\right)\end{aligned} \qquad (8.8.37)$$

(even for $n < 0$). ∎

Remark 8.8.16: The Jacobi identity (8.8.29) can now be rewritten:

$$\begin{aligned}&z_0^{-1}\delta\left(\frac{z_1 - z_2}{z_0}\right)Y(u, z_1)Y(v, z_2) - z_0^{-1}\delta\left(\frac{z_2 + z_1}{-z_0}\right)Y(v, z_2)Y(u, z_1)\\ &= Y\left(\left(z_0^{-1}\delta\left(\frac{z_1 - z_2}{z_0}\right)Y(u, z_0) - z_0^{-1}\delta\left(\frac{-z_2 + z_1}{z_0}\right)Y(u, z_0)\right)v, z_2\right)\\ &= Y\left(\left(z_0^{-1}\delta\left(\frac{z_1 - z_2}{z_0}\right)Y(u, z_1 - z_2)\right.\right.\\ &\qquad \left.\left. - z_0^{-1}\delta\left(\frac{-z_2 + z_1}{z_0}\right)Y(u, -z_2 + z_1)\right)v, z_2\right).\end{aligned} \qquad (8.8.38)$$

Note that these expressions indeed exist. By contrast,

$$\lim_{z_0 \to z_2} Y(Y(u, z_0)v, z_2), \quad \lim_{z_0 \to z_2} Y(Y(u, z_1 - z_0)v, z_2)$$

and

$$\lim_{z_1 \to z_2} Y(Y(u, z_1 - z_0)v, z_2)$$

do not exist for general u and v. Taking Res_{z_0} leads to an interesting alternate form of Theorem 8.6.1:

$$[Y(u, z_1), Y(v, z_2)] = Y((Y(u, z_1 - z_2) - Y(u, -z_2 + z_1))v, z_2). \qquad (8.8.39)$$

More generally, equating coefficients of z_0^{-n-1}, we find the following variant of (8.8.30):

$$\begin{aligned}
&[Y(u, z_1) \times_n Y(v, z_2)] \\
&\quad = (z_1 - z_2)^n Y(u, z_1) Y(v, z_2) - (-z_2 + z_1)^n Y(v, z_2) Y(u, z_1) \\
&\quad = Y(((z_1 - z_2)^n Y(u, z_1 - z_2) - (-z_2 + z_1)^n Y(u, -z_2 + z_1))v, z_2)
\end{aligned} \tag{8.8.40}$$

for $n \in \mathbb{Z}$. Formula (8.8.39) further illuminates formula (8.6.28). While (8.6.28) seems to suggest that the components of the operators $Y(v, z)$ form an associative algebra, this is not true. However, (8.8.39) and (8.8.40) show that these components form a Lie algebra and in fact something much more general (but not quite an associative algebra).

Just as in Corollaries 8.6.5 and 8.6.7 we can rewrite the Jacobi identity in terms of components:

Corollary 8.8.17: *For $u, v \in V_L$ and $l, m, n \in \mathbb{Z}$, we have*

$$\begin{aligned}
[u \times_l v]_{mn} &= \sum_{i \in \mathbb{N}} (-1)^i \binom{l}{i} u_{m+l-i} v_{n+i} - (-1)^l \sum_{i \in \mathbb{N}} (-1)^i \binom{l}{i} v_{n+l-i} u_{m+i} \\
&= \sum_{i \in \mathbb{N}} \binom{m}{i} (u_{l+i} \cdot v)_{m+n-i}.
\end{aligned} \tag{8.8.41}$$

Corollary 8.8.18: *In the same notation,*

$$[u \times_l v]_{1n} = (u_l \cdot v)_{n+1} + (u_{l+1} \cdot v)_n. \tag{8.8.42}$$

Corollary 8.8.19: *Suppose in addition that u is homogeneous. For*

$$m, n \in \mathbb{Z},$$

we have

$$[x_u \times_l x_v](m, n) = \sum_{i \in \mathbb{Z}} \binom{m - 1 + \operatorname{wt} u}{i - 1 + \operatorname{wt} u} x_{x_{u(l+i)} \cdot v}(l + m + n). \tag{8.8.43}$$

Remark 8.8.20: We shall next exhibit an expected "$\mathfrak{S}_3$-symmetry" property of the Jacobi identity. Let us call the assertion that (8.8.29) holds when applied to an element $w \in V_L$ "the Jacobi identity for the ordered triple (u, v, w)." Assuming this assertion we shall argue directly, using (8.8.4)–(8.8.8) [and (8.8.11)], that the identity holds for any permutation of

(u, v, w). We have already established the identity for (v, u, w) using (8.8.8) and (8.8.11), so it suffices to prove the formula for (u, w, v). First we observe that for $r, s, t \in V_L$,

$$Y(r, z_0)Y(s, z_2)t = e^{z_2 L(-1)}Y(r, z_0 - z_2)Y(t, -z_2)s \tag{8.8.44}$$

by (8.8.6) and (8.8.7), and this implies that

$$Y(Y(r, z_0)Y(s, z_2)t, z_3) = Y(Y(r, z_0 - z_2)Y(t, -z_2)s, z_3 + z_2) \tag{8.8.45}$$

by (8.8.6). Now change z_2 to $-z_2$ in (8.8.29), apply (8.8.11) to the first factor on the left, and take $Y(\cdot, z_3 + z_2)$ of both sides (applied to w) to get

$$\begin{aligned} &z_1^{-1}\delta\left(\frac{z_0 - z_2}{z_1}\right)Y(Y(u, z_0 - z_2)Y(v, -z_2)w, z_3 + z_2) \\ &\quad - z_0^{-1}\delta\left(\frac{z_2 + z_1}{z_0}\right)Y(Y(v, -z_2)Y(u, z_1)w, z_3 + z_2) \\ &\quad = -z_2^{-1}\delta\left(\frac{z_1 - z_0}{-z_2}\right)Y(Y(Y(u, z_0)v, -z_2)w, z_3 + z_2) \end{aligned} \tag{8.8.46}$$

and thus

$$\begin{aligned} &z_1^{-1}\delta\left(\frac{z_0 - z_2}{z_1}\right)Y(Y(u, z_0)Y(w, z_2)v, z_3) \\ &\quad - z_0^{-1}\delta\left(\frac{z_2 + z_1}{z_0}\right)Y(Y(Y(u, z_1)w, z_2)v, z_3) \\ &\quad = -z_2^{-1}\delta\left(\frac{z_1 - z_0}{-z_2}\right)Y(Y(w, z_2)Y(u, z_0)v, z_3) \end{aligned} \tag{8.8.47}$$

by (8.8.45) and two applications of (8.8.8). At this point, (8.8.4) and two more applications of (8.8.11) give the Jacobi identity for (u, w, v) (with z_0 and z_1 reversed), as desired. Note that (8.8.11) is used on each of the three relevant expressions, and also that the terms in the original Jabobi identity switch positions.

Remark 8.8.21: The last remark and the answer to Question 8.8.2 show that the Jacobi identity rounds out a symmetry among the variables z_0, z_1 and z_2. This was accomplished via a new type of generating function—over a family of *products*—combined with the two other types of generating function we have been using—over a family of *operators* (the vertex operator itself) and over a family of *derivatives* (such as the operator $e^{z_0(\partial/\partial z_1)}$).

At the beginning of this section we assumed that our lattice was even and that (8.6.1) held in general. Now that we know what to expect, we drop all these assumptions and work in the full generality of Theorem 8.6.1 and Remark 8.6.9. Just as in the proof of Theorem 8.8.9, we can use the proof of Theorem 8.6.1 to obtain easily a Jacobi identity more general than (8.8.29). To express the result in the form below, it is helpful to observe the following generalization of Proposition 8.8.5, proved by means of the same argument:

Proposition 8.8.22: *For $m \in \mathbb{F}$,*

$$z_2^{-1}\left(\frac{z_1 - z_0}{z_2}\right)^m \delta\left(\frac{z_1 - z_0}{z_2}\right) = z_1^{-1}\left(\frac{z_2 + z_0}{z_1}\right)^{-m} \delta\left(\frac{z_2 + z_0}{z_1}\right). \tag{8.8.48}$$

The result is:

Theorem 8.8.23: *For $u, v \in V_L$ satisfying the hypotheses of Theorem 8.6.1, or more generally, in the context of Remark 8.6.9,*

$$\begin{aligned} &z_0^{-1}\delta\left(\frac{z_1 - z_2}{z_0}\right)Y(u, z_1)Y(v, z_2) \\ &\qquad - (-1)^{\langle \bar{a}, \bar{b}\rangle}c(\bar{a}, \bar{b})z_0^{-1}\delta\left(\frac{z_2 - z_1}{-z_0}\right) Y(v, z_2)\, Y(u, z_1) \\ &= z_2^{-1}\left(\frac{z_1 - z_0}{z_2}\right)^{\bar{a}} \delta\left(\frac{z_1 - z_0}{z_2}\right)Y(Y(u, z_0)v, z_2). \end{aligned} \tag{8.8.49}$$

8.9. Cross-Brackets and Commutative Affinization

We shall be particularly interested in elements of V_L of weight 2 and their relationship with the cross-bracket $\times_1$ introduced in (8.8.12)–(8.8.20). But first we would like to use the new point of view to reformulate some of what we know concerning elements of weight 1. We assume for convenience that L is even and that (8.6.1) holds for all $a, b \in \hat{L}$. Let

$$\text{wt } u = \text{wt } v = 1. \tag{8.9.1}$$

Then by (8.5.22) and (8.5.29),

$$\begin{aligned} u_n &= x_u(n) \\ \deg u_n &= n \\ \text{wt}(u_n \cdot v) &= 1 - n \end{aligned} \tag{8.9.2}$$

for $n \in \mathbb{Z}$, so that

$$\deg u_0 = 0$$

$$\mathrm{wt}(u_0 \cdot v) = 1 \tag{8.9.3}$$

$$\deg(u_0 \cdot v)_0 = 0,$$

and $u_0 \cdot v$ defines a product on the space $(V_L)_{(\dim \mathfrak{h})/24-1}$ of elements of V_L of weight 1 [and degree $(\dim \mathfrak{h})/24 - 1$]. By Corollary 8.6.6,

$$[u_0, v_0] = (u_0 \cdot v)_0. \tag{8.9.4}$$

Remark 8.9.1: If L is (even and) positive definite, we can say more. First, $u_1 \cdot v$ is a multiple of $\iota(1)$, and therefore may be identified with a scalar. By Corollary 8.6.5 and (8.5.18),

$$[u_m, v_n] = (u_0 \cdot v)_{m+n} + (u_1 \cdot v)m\delta_{m+n,0} \tag{8.9.5}$$

for $m, n \in \mathbb{Z}$. [We can also write the operators u_m as $x_u(m)$.] Set

$$\mathfrak{g} = (V_L)_{(\dim \mathfrak{h})/24-1}, \tag{8.9.6}$$

provide $\mathfrak{g}$ with the nonassociative product

$$[u, v] = u_0 \cdot v \tag{8.9.7}$$

and the bilinear form

$$\langle u, v \rangle = u_1 \cdot v, \tag{8.9.8}$$

and form the corresponding affinization $\hat{\mathfrak{g}}$ as in (1.6.5) and Remark 1.6.1. Then $\mathfrak{g}$ is a Lie algebra and $\langle \cdot, \cdot \rangle$ is symmetric and invariant. We can see this directly: We have

$$u_0 \cdot v = -v_0 \cdot u \tag{8.9.9}$$

by (8.8.7) and the fact that

$$L(-1)\iota(1) = 0 \tag{8.9.10}$$

(see Remark 8.7.10), and the Jacobi identity (for Lie algebras!) follows from (8.9.4). Again by (8.8.7),

$$u_1 \cdot v = v_1 \cdot u, \tag{8.9.11}$$

and the formula

$$[u_0, v_1] = (u_0 \cdot v)_1 \tag{8.9.12}$$

(Corollary 8.6.6) gives the $\mathfrak{g}$-invariance. Then (8.9.5) expresses the fact that the operators u_m and the identity operator provide a representation of $\hat{\mathfrak{g}}$

on V_L. In fact of course $\mathfrak{g}$ is just the Lie algebra denoted $\mathfrak{g}$ in Theorems 6.2.1 and 7.2.6 and $\langle \cdot, \cdot \rangle$ is the corresponding form, and $\mathfrak{g}$ may be identified with the Lie algebra of operators $\{u_0\}$ on V_L (cf. Remark 7.2.10).

Now let

$$\text{wt}\, u = \text{wt}\, v = 2 \tag{8.9.13}$$

(with L not necessarily even or positive definite). Then by (8.5.22),

$$\begin{aligned} \deg u_n &= n - 1 \\ \text{wt}(u_n \cdot v) &= 3 - n \end{aligned} \tag{8.9.14}$$

for $n \in \mathbb{Z}$, so that

$$\begin{aligned} \deg u_1 &= 0 \\ \text{wt}(u_1 \cdot v) &= 2 \\ \deg(u_1 \cdot v)_1 &= 0. \end{aligned} \tag{8.9.15}$$

In particular, the formula

$$u \times v = u_1 \cdot v = x_u(0)v \tag{8.9.16}$$

defines a product on $V_{(\dim \mathfrak{h})/24-2}$.

Remark 8.9.2: The canonical quadratic element ω of (8.7.2)–(8.7.5) gives us a left identity element, by (8.7.23):

$$(\tfrac{1}{2}\omega) \times u = \tfrac{1}{2}L(0)u = u \quad \text{for} \quad u \in V_{(\dim \mathfrak{h})/24-2}.$$

Suppose again that L is positive definite. Then $u_3 \cdot v$ is a scalar multiple of $\iota(1)$, giving us a bilinear form

$$\langle u, v \rangle = u_3 \cdot v = x_u(2)v. \tag{8.9.17}$$

By Corollary 8.8.17 and (8.5.18),

$$[u \times_1 v]_{mn} = (u \times v)_{m+n} + m(u_2 \cdot v)_{m+n-1} + \tfrac{1}{2}\langle u, v \rangle m(m-1)\delta_{m+n,1} \tag{8.9.18}$$

for $m, n \in \mathbb{Z}$. Although this is more complicated than the weight-one analogue (8.9.5), we can arrange for some similarity. We first note that $\langle \cdot, \cdot \rangle$ is symmetric by (8.8.7):

$$u_3 \cdot v = v_3 \cdot u. \tag{8.9.19}$$

We would like to be able to ignore the term involving $u_2 \cdot v$ in (8.9.18). There is a fruitful way of doing this, involving both a restriction to a certain subspace of V_L and an assumption on the lattice L.

The subspace of V_L will be the subspace fixed by the involution θ introduced in (7.2.8)–(7.2.9) and (7.4.10)–(7.4.11):

$$\theta \in \mathrm{Aut}(\hat{L}; \kappa, \langle\cdot,\cdot\rangle) \tag{8.9.20}$$

such that

$$\begin{aligned} \theta &= -1 \quad \text{on} \quad L \\ \theta^2 &= 1. \end{aligned} \tag{8.9.21}$$

The action of θ on $\hat{L}$ extends naturally to a linear involution of $\mathbb{F}\{L\}$ and in fact of V_L:

$$\begin{aligned} \theta\colon V_L = S(\hat{\mathfrak{h}}_{\mathbb{Z}}^-) \otimes \mathbb{F}\{L\} &\to V_L \\ x \otimes \iota(a) &\mapsto \theta x \otimes \iota(\theta a) \end{aligned} \tag{8.9.22}$$

for $x \in S(\hat{\mathfrak{h}}_{\mathbb{Z}}^-)$ and $a \in \hat{L}$, where θ acts as on $S(\hat{\mathfrak{h}}_{\mathbb{Z}}^-)$ as the unique algebra automorphism such that

$$\theta x = -x \quad \text{for} \quad x \in \hat{\mathfrak{h}}_{\mathbb{Z}}^-. \tag{8.9.23}$$

The fact that θ is well-defined on $\mathbb{F}\{L\}$, via the formula

$$\theta(\iota(a)) = \iota(\theta a) \quad \text{for} \quad a \in \hat{L}, \tag{8.9.24}$$

follows from the definition (7.1.18) and the fact that $\theta\kappa = \kappa$. We set

$$\begin{aligned} V_L^{\theta} &= \{v \in V_L \mid \theta v = v\} \\ V_L^{-\theta} &= \{v \in V_L \mid \theta v = -v\}. \end{aligned} \tag{8.9.25}$$

From the definitions we have

$$\theta\alpha(n)\theta^{-1} = -\alpha(n) \quad \text{for} \quad \alpha \in \mathfrak{h},\ n \in \mathbb{Z} \tag{8.9.26}$$

$$\theta Y(a, z)\theta^{-1} = Y(\theta a, z) \quad \text{for} \quad a \in \hat{L} \tag{8.9.27}$$

$$\theta Y(v, z)\theta^{-1} = Y(\theta v, z) \quad \text{for} \quad v \in V_L, \tag{8.9.28}$$

and in particular,

$$\text{if} \quad v \in V_L^{\theta} \quad \text{then} \quad \theta \quad \text{commutes with} \quad Y(v, z) \tag{8.9.29}$$

and the component operators of $Y(v, z)$ preserve $V_L^{\pm\theta}$.

Remark 8.9.3: In the context of Remark 8.9.1, the content of formula (8.9.5) for u and v elements of weight 1 of V_L^θ is essentially expressed by (7.2.19), the formula for $[X^+(a, z_1), X^+(b, z_2)]$.

Now suppose that our lattice L has no elements α *such that* $\langle\alpha, \alpha\rangle = 2$:

$$L_2 = \emptyset, \tag{8.9.30}$$

in the notation of (6.1.7). Then the space of elements of weight 1 in V_L^θ is 0:

$$(V_L^\theta)_{(\dim \mathfrak{h})/24-1} = 0. \tag{8.9.31}$$

This enables us to eliminate the term involving $u_2 \cdot v$ in (8.9.18). Denote by $\mathfrak{k}$ the space of elements of V_L^θ of weight 2:

$$\mathfrak{k} = (V_L^\theta)_{(\dim \mathfrak{h})/24-2}. \tag{8.9.32}$$

Then

$$\mathfrak{k} \times \mathfrak{k} \subset \mathfrak{k} \tag{8.9.33}$$

[see (8.9.16)], and

$$[u \times_1 v]_{mn} = (u \times v)_{m+n} + \tfrac{1}{2}\langle u, v\rangle m(m-1)\delta_{m+n,1} \tag{8.9.34}$$

for $u, v \in \mathfrak{k}$, $m, n \in \mathbb{Z}$. Equivalently,

$$[x_u \times_1 x_v](m, n) = x_{u\times v}(m + n + 1) + \tfrac{1}{2}\langle u, v) m(m + 1)\delta_{m+n+1,0}. \tag{8.9.35}$$

We already know that $\langle\cdot, \cdot\rangle$ is symmetric. Using (8.9.31) together with (8.8.7), we observe also that the product $\times$ is commutative on $\mathfrak{k}$:

$$u \times v = v \times u \quad \text{for} \quad u, v \in \mathfrak{k}. \tag{8.9.36}$$

Moreover, the form $\langle\cdot, \cdot\rangle$ is *associative* with respect to the commutative product on $\mathfrak{k}$ in the sense that

$$\langle u, v \times w\rangle = \langle u \times v, w\rangle \quad \text{for} \quad u, v, w \in \mathfrak{k}. \tag{8.9.37}$$

For this, we use (8.8.19), (8.8.42) and (8.9.31), together with the fact that

$$x_s(0)\iota(1) = 0 \quad \text{for} \quad s \in V_L, \quad \text{wt}\, s > 0, \tag{8.9.38}$$

which follows from (8.8.5).

Remark 8.9.4: By the commutativity, $\mathfrak{k}$ has an identity element, namely, $\frac{1}{2}\omega$:

$$(\tfrac{1}{2}\omega) \times u = u \times (\tfrac{1}{2}\omega) = u \quad \text{for} \quad u \in \mathfrak{k}$$

(see Remark 8.9.2).

These results and Remark 8.9.1 motivate the concept of "commutative affinization": Let $\mathfrak{b}$ be a commutative nonassociative algebra with multiplication denoted $\times$, equipped with a symmetric bilinear form $\langle\cdot,\cdot\rangle$. Set

$$\hat{\mathfrak{b}} = \mathfrak{b} \otimes \mathbb{F}[t, t^{-1}] \oplus \mathbb{F}e, \tag{8.9.39}$$

where t is an indeterminate and $e \neq 0$ [cf. (1.6.3)], and provide $\hat{\mathfrak{b}}$ with the following nonassociative product $\times$, which is easily seen to be commutative:

$$u \otimes t^m \times v \otimes t^n = (u \times v) \otimes t^{m+n+1} + \tfrac{1}{2}\langle u, v\rangle m(m+1)\delta_{m+n+1,0}e$$
$$e \times \hat{\mathfrak{b}} = \hat{\mathfrak{b}} \times e = 0 \tag{8.9.40}$$

for $u, v \in \mathfrak{b}$, $m, n \in \mathbb{Z}$ [cf. (1.6.5)]. We call $\hat{\mathfrak{b}}$ the *commutative affinization* of $\mathfrak{b}$ (and $\langle\cdot,\cdot\rangle$).

The appropriate notion of representation of $\hat{\mathfrak{b}}$ is formulated as follows: Let A be an associative or Lie algebra. Given two sequences $x = (x(m))_{m \in \mathbb{Z}}$, $y = (y(n))_{n \in \mathbb{Z}}$ of elements of A, define the function

$$[x \times_1 y]: \mathbb{Z} \times \mathbb{Z} \to A$$
$$[x \times_1 y](m, n) = [x(m+1), y(n)] - [x(m), y(n+1)]. \tag{8.9.41}$$

By abuse of notation, we sometimes write

$$[x \times_1 y](m, n) = [x(m) \times_1 y(n)]. \tag{8.9.42}$$

As in (8.8.20), we call $[x \times_1 y]$ the *cross-bracket* of x and y. By a *graded* $\hat{\mathfrak{b}}$-*module* (or a *graded representation of* $\hat{\mathfrak{b}}$ *by cross-bracket*) we mean an $\mathbb{F}$-graded vector space

$$V = \coprod_{n \in \mathbb{F}} V_n$$

together with a linear map

$$\pi: \hat{\mathfrak{b}} \to \operatorname{End} V \tag{8.9.43}$$

such that $\pi(x \otimes t^m)$ is homogeneous of degree m for $x \in \mathfrak{b}$, $m \in \mathbb{Z}$ and such that

$$\pi(x \otimes t^m \times y \otimes t^n) = [\pi(x \otimes t^m) \times_1 \pi(y \otimes t^n)] \tag{8.9.44}$$

for $x, y \in \mathfrak{b}$, $m, n \in \mathbb{Z}$, where on the right-hand side we are considering the cross-bracket of the two sequences $(\pi(x \otimes t^m))_{m \in \mathbb{Z}}$ and $(\pi(y \otimes t^n))_{n \in \mathbb{Z}}$ as in (8.9.42).

We can summarize our results as follows (cf. Remark 8.9.1):

Theorem 8.9.5: *Assume that L is positive definite and that $L_2 = \emptyset$, and let $\mathfrak{k}$ be the space of elements of V_L^θ of weight 2. Then $\mathfrak{k}$ is a commutative nonassociative algebra with identity under the product $u \times v = u_1 \cdot v$, and the bilinear form $\langle u, v\rangle = u_3 \cdot v$ is symmetric and associative. Moreover, V_L is a graded $\hat{\mathfrak{k}}$-module under the action*

$$\pi: \hat{\mathfrak{k}} \to \operatorname{End} V_L \tag{8.9.45}$$

defined by

$$\begin{aligned}\pi: u \otimes t^n &\mapsto x_u(n) \quad \text{for} \quad u \in \mathfrak{k}, \quad n \in \mathbb{Z}\\ \pi: e &\mapsto 1.\end{aligned} \tag{8.9.46}$$

Remark 8.9.6: In contrast with the Lie algebra case discussed in Remark 8.9.1, the nonassociative product on $\mathfrak{k}$ is not given directly by composition of the operators $u_1 = x_u(0)$ $(u \in \mathfrak{k})$ on V_L.

Remark 8.9.7: The space $\mathfrak{k}$ has the following structure: Write

$$\hat{L}_4 = \{a \in \hat{L} \mid \bar{a} \in L_4\} \tag{8.9.47}$$

[recall the notation (6.1.7); cf. (6.2.16)] and for $a \in \hat{L}_4$ set

$$x_a^+ = \iota(a) + \iota(\theta a) = \iota(a) + \theta\iota(a) \tag{8.9.48}$$

[cf. (6.4.19)], so that [using (7.1.20)]

$$\begin{aligned}x_{\kappa a}^+ &= \omega x_a^+\\ x_{\theta a}^+ &= x_a^+.\end{aligned} \tag{8.9.49}$$

Then

$$\mathfrak{k} = S^2(\mathfrak{h}) \oplus \sum_{a \in \hat{L}_4} \mathbb{F}x_a^+ \tag{8.9.50}$$

[cf. (6.2.19) and (6.4.39)–(6.4.40)], where we make the identification

$$\mathfrak{h} = \mathfrak{h} \otimes t^{-1} \tag{8.9.51}$$

in the notation (1.7.11), (1.7.15). The linear relations among the elements

x_a^+ are such that

$$\dim(\sum \mathbb{F}x_a^+) = \tfrac{1}{2}|L_4|. \tag{8.9.52}$$

The product $\times$ on $\mathfrak{k}$ can simply be read from the coefficient of z_0^{-2} in $Y(u, z_0)v$ as computed at the end of Section 8.6, and also from (8.5.44) and Remark 8.3.6. It is given as follows:

$$g^2 \times h^2 = 4\langle g, h\rangle gh \tag{8.9.53}$$

$$g^2 \times x_a^+ = \langle g, \bar{a}\rangle^2 x_a^+ \tag{8.9.54}$$

$$x_a^+ \times x_b^+ = \begin{cases} 0 & \text{if} \quad \langle \bar{a}, \bar{b}\rangle = 0, \pm 1 \\ x_{ab}^+ & \text{if} \quad \langle \bar{a}, \bar{b}\rangle = -2 \\ \bar{a}^2 & \text{if} \quad ab = 1 \end{cases} \tag{8.9.55}$$

for $g, h \in \mathfrak{h}$, $a, b \in \hat{L}_4$ [cf.(6.4.44), (6.4.45)]; note that the possibilities in (8.9.55) are essentially exhaustive because $L_2 = \emptyset$ implies that

$$\langle \alpha, \beta\rangle = 0, \pm 1, \pm 2 \text{ or } \pm 4 \quad \text{for} \quad \alpha, \beta \in L_4, \tag{8.9.56}$$

and

$$\langle \alpha, \beta\rangle = -2 \quad \text{if and only if} \quad \alpha + \beta \in L_4 \tag{8.9.57}$$

(cf. Remark 6.1.1). Moreover,

$$(\tfrac{1}{2}\omega) \times u = u \times (\tfrac{1}{2}\omega) = u \quad \text{for} \quad u \in \mathfrak{k}. \tag{8.9.58}$$

The form $\langle \cdot, \cdot \rangle$ on $\mathfrak{k}$ is similarly read from the coefficient of z_0^{-4} in $Y(u, z_0)v$ and from (8.5.44), and is given by:

$$\langle g^2, h^2\rangle = 2\langle g, h\rangle^2 \tag{8.9.59}$$

$$\langle g^2, x_a^+\rangle = 0 \tag{8.9.60}$$

$$\langle x_a^+, x_b^+\rangle = \begin{cases} 0 & \text{if} \quad \bar{a} \neq \pm\bar{b} \\ 2 & \text{if} \quad ab = 1 \end{cases} \tag{8.9.61}$$

for g, h, a, b as above [cf. (6.4.47), (6.4.48)]. (Note the distinction between the forms denoted $\langle \cdot, \cdot \rangle$ on $\mathfrak{h}$ and on $\mathfrak{k}$.) We observe that

$$\langle g^2, \tfrac{1}{2}\omega\rangle = \tfrac{1}{2}\langle g, g\rangle \tag{8.9.62}$$

$$\langle \tfrac{1}{2}\omega, \tfrac{1}{2}\omega\rangle = \tfrac{1}{8}\dim \mathfrak{h}. \tag{8.9.63}$$

Remark 8.9.8: The formulas above show that the form $\langle \cdot, \cdot \rangle$ is nonsingular on $\mathfrak{k}$.

8.10. Vertex Operator Algebras and the Rationality, Commutativity and Associativity Properties

We now summarize much of the material presented in Chapter 8 by means of a definition of "vertex operator algebra" and a theorem exhibiting a class of them. The definition is well-motivated by the theory presented so far. There are of course many natural variants and generalizations of this notion, reflecting for example the various situations which have arisen in this chapter, but for our present purposes we have chosen this restrictive definition because the moonshine module, which will be constructed later, satisfies all these conditions. Of course, some of the properties listed below are implied by other ones. Note that the grading used here is that defined by weights, not by degrees [recall (8.7.22), (8.7.23)].

A *vertex operator algebra* is a $\mathbb{Z}$-graded vector space

$$V = \coprod_{n \in \mathbb{Z}} V_{(n)}; \quad \text{for} \quad v \in V_{(n)}, \quad n = \text{wt}\, v; \tag{8.10.1}$$

such that

$$\dim V_{(n)} < \infty \quad \text{for} \quad n \in \mathbb{Z}, \tag{8.10.2}$$

$$V_{(n)} = 0 \quad \text{for} \quad n \quad \text{sufficiently small}, \tag{8.10.3}$$

equipped with a linear map

$$\begin{aligned} V &\to (\text{End}\, V)[[z, z^{-1}]] \\ v &\mapsto Y(v, z) = \sum_{n \in \mathbb{Z}} v_n z^{-n-1} \quad (\text{where } v_n \in \text{End}\, V) \end{aligned} \tag{8.10.4}$$

and with two distinguished homogeneous vectors $\mathbf{1}, \omega \in V$, satisfying the following conditions for $u, v \in V$:

$$u_n v = 0 \quad \text{for} \quad n \quad \text{sufficiently large}; \tag{8.10.5}$$

$$Y(\mathbf{1}, z) = 1; \tag{8.10.6}$$

$$Y(v, z)\mathbf{1} \in V[[z]] \quad \text{and} \quad \lim_{z \to 0} Y(v, z)\mathbf{1} = v; \tag{8.10.7}$$

$$\begin{aligned} &z_0^{-1}\delta\left(\frac{z_1 - z_2}{z_0}\right) Y(u, z_1)Y(v, z_2) - z_0^{-1}\delta\left(\frac{z_2 - z_1}{-z_0}\right) Y(v, z_2)Y(u, z_1) \\ &= z_2^{-1}\delta\left(\frac{z_1 - z_0}{z_2}\right) Y(Y(u, z_0)v, z_2) \end{aligned} \tag{8.10.8}$$

(the Jacobi identity) where $\delta(z) = \sum_{n \in \mathbb{Z}} z^n$ and where $\delta((z_1 - z_2)/z_0)$ is to be expanded as a formal power series in the second term in the numerator,

z_2, and analogously for the other δ-function expressions; when each expression in (8.10.8) is applied to any element of V, the coefficient of each monomial in the formal variables is a finite sum;

$$[L(m), L(n)] = (m - n)L(m + n) + \tfrac{1}{12}(m^3 - m)\delta_{m+n,0}(\text{rank } V) \tag{8.10.9}$$

for $m, n \in \mathbb{Z}$, where

$$L(n) = \omega_{n+1} \quad \text{for} \quad n \in \mathbb{Z}, \quad \text{i.e.,} \quad Y(\omega, z) = \sum_{n \in \mathbb{Z}} L(n)z^{-n-2} \tag{8.10.10}$$

and

$$\text{rank } V \in \mathbb{Q}; \tag{8.10.11}$$

$$L(0)v = nv = (\text{wt } v)v \quad \text{for} \quad n \in \mathbb{Z} \quad \text{and} \quad v \in V_{(n)}; \tag{8.10.12}$$

$$\frac{d}{dz} Y(v, z) = Y(L(-1)v, z); \tag{8.10.13}$$

$$[L(-1), Y(v, z)] = Y(L(-1)v, z); \tag{8.10.14}$$

$$[L(0), Y(v, z)] = Y(L(0)v, z) + zY(L(-1)v, z); \tag{8.10.15}$$

$$L(n)\mathbf{1} = 0 \quad \text{for} \quad n \geq -1; \tag{8.10.16}$$

$$L(-2)\mathbf{1} = \omega; \tag{8.10.17}$$

$$L(0)\omega = 2\omega. \tag{8.10.18}$$

Note that (8.10.14)–(8.10.18) are consequences.

Remark 8.10.1: From (8.10.4), (8.10.12), (8.10.13) and (8.10.15), we find that if $v \in V$ is homogeneous, then v_n has weight $\text{wt } v - n - 1$ as an operator.

The series $Y(v, z)$ are called *vertex operators*. We have already established a number of axiomatic properties of vertex operator algebras, and we shall discuss more such properties below and in the Appendix. Many (but not all) of the important results so far are summarized as follows:

Theorem 8.10.2: *For a positive definite even lattice L, the space V_L based on the central extension of L by $\langle \pm 1 \rangle$ with commutator map given by* $\langle \alpha, \beta \rangle \bmod 2\mathbb{Z}$ $(\alpha, \beta \in L)$ *has the structure of a vertex operator algebra with*

$$\text{rank } V_L = \text{rank } L. \tag{8.10.19}$$

We shall construct a far more sophisticated vertex operator algebra—the moonshine module—in the succeeding chapters. Its automorphism group, in the sense of the following definition, will be the Monster.

An *automorphism* of the vertex operator algebra V is a linear automorphism g such that

$$gY(v, z)g^{-1} = Y(gv, z) \quad \text{for} \quad v \in V, \tag{8.10.20}$$

$$g\omega = \omega. \tag{8.10.21}$$

Then g is grading-preserving and

$$g\mathbf{1} = \mathbf{1}. \tag{8.10.22}$$

Isomorphism of vertex operator algebras is defined in the obvious way.

Besides the non-integral gradings, generalized Laurent series with non-integral powers of the variables, possibly infinite-dimensional spaces $V_{(n)}$, and so on, which have occurred in our treatment and which in fact will arise naturally in the course of the construction of the moonshine module and the Monster, there are other structures which might be taken into account in a definition of the term "vertex operator algebra"—for instance, a Hermitian form (at least over $\mathbb{C}$; cf. Section 12.5 below) and the associated notion of adjoint operators. The following notion of module will play an important role in the construction of the Monster:

A *module* for the vertex operator algebra V is a $\mathbb{Q}$-graded vector space

$$W = \coprod_{n \in \mathbb{Q}} W_{(n)}; \quad \text{for} \quad w \in W_{(n)}, \quad n = \text{wt } w; \tag{8.10.23}$$

such that

$$\dim W_{(n)} < \infty \quad \text{for} \quad n \in \mathbb{Q}, \tag{8.10.24}$$

$$W_{(n)} = 0 \quad \text{for} \quad n \quad \text{sufficiently small}, \tag{8.10.25}$$

equipped with a linear map

$$\begin{aligned} &V \to (\text{End } W)[[z, z^{-1}]] \\ &v \mapsto Y(v, z) = \sum_{n \in \mathbb{Z}} v_n z^{-n-1} \quad (\text{where } v_n \in \text{End } W), \end{aligned} \tag{8.10.26}$$

satisfying the following conditions for $u, v \in V$ and $w \in W$:

$$v_n w = 0 \quad \text{for} \quad n \quad \text{sufficiently large}; \tag{8.10.27}$$

$$Y(\mathbf{1}, z) = 1; \tag{8.10.28}$$

$$z_0^{-1}\delta\left(\frac{z_1 - z_2}{z_0}\right)Y(u, z_1)Y(v, z_2) - z_0^{-1}\delta\left(\frac{z_2 - z_1}{-z_0}\right)Y(v, z_2)Y(u, z_1)$$
$$= z_2^{-1}\delta\left(\frac{z_1 - z_0}{z_2}\right)Y(Y(u, z_0)v, z_2) \tag{8.10.29}$$

(the Jacobi identity); note that on the right-hand side, $Y(u, z_0)$ is the operator associated with V;

$$[L(m), L(n)] = (m - n)L(m + n) + \tfrac{1}{12}(m^3 - m)\delta_{m+n,0}(\operatorname{rank} V) \tag{8.10.30}$$

for $m, n \in \mathbb{Z}$, where

$$L(n) = \omega_{n+1} \quad \text{for} \quad n \in \mathbb{Z}, \quad \text{i.e.,} \quad Y(\omega, z) = \sum_{n \in \mathbb{Z}} L(n)z^{-n-2}; \tag{8.10.31}$$

$$L(0)w = nw = (\operatorname{wt} w)w \quad \text{for} \quad n \in \mathbb{Q} \quad \text{and} \quad w \in W_{(n)}; \tag{8.10.32}$$

$$\frac{d}{dz}Y(v, z) = Y(L(-1)v, z), \tag{8.10.33}$$

where $L(-1)$ acts on V;

$$[L(-1), Y(v, z)] = Y(L(-1)v, z); \tag{8.10.34}$$

$$[L(0), Y(v, z)] = Y(L(0)v, z) + zY(L(-1)v, z). \tag{8.10.35}$$

Of course, V is a module for itself. The notions of automorphism and isomorphism of modules are clear.

In the next chapter, we shall among other things construct modules for certain subalgebras of V_L, and the moonshine module will arise from the direct sum of a subalgebra of a suitable V_L and a certain module for it.

We shall now formulate and discuss three important properties of a vertex operator algebra—rationality, commutativity and associativity—which will also have very natural interpretations in terms of complex analysis, as we explain in the Appendix. The present approach is based on an analogue for several variables of the maps $\iota_\pm$ and Θ of Section 8.1.

Let us denote by S the set of nonzero linear polynomials in several variables:

$$S = \left\{\sum_{i=1}^{n} a_i z_i \,\middle|\, a_i \in \mathbb{F},\ a_i \text{ not all zero}\right\} \subset \mathbb{F}[z_1, \ldots, z_n]. \tag{8.10.36}$$

Consider the subring $\mathbb{F}[z_1, \ldots, z_n]_S$ of the field of rational functions $\mathbb{F}(z_1, \ldots, z_n)$ obtained by localizing (inverting) the products of elements of S. Let $(i_1 \cdots i_n)$ be a permutation of $(1 \cdots n)$. We shall recursively define maps

$$\iota_{i_1 \cdots i_n}: \mathbb{F}[z_1, \ldots, z_n]_S \to \mathbb{F}[[z_1, z_1^{-1}, \ldots, z_n, z_n^{-1}]] \tag{8.10.37}$$

as follows: For $n = 1$, ι_1 shall be the inclusion map; in this case,

$$S = \{az_1 \mid a \in \mathbb{F}^\times\}, \tag{8.10.38}$$

so that

$$\mathbb{F}[z_1]_S = \mathbb{F}[z_1, z_1^{-1}]. \tag{8.10.39}$$

Now assume that the maps $\iota_{i_1 \cdots i_{n-1}}$ are defined. To define $\iota_{i_1 \cdots i_n}$, let

$$f(z_1, \ldots, z_n) \in \mathbb{F}[z_1, \ldots, z_n]_S. \tag{8.10.40}$$

Then we can write f in the form

$$f(z_1, \ldots, z_n) = g(z_1, \ldots, z_n) \Big/ \prod_{k=1}^{r} \left(\sum_{j=2}^{n} a_{kj} z_{i_j} \right) \prod_{l=1}^{s} \left(\sum_{j=1}^{n} b_{lj} z_{i_j} \right), \tag{8.10.41}$$

where $g(z_1, \ldots, z_n) \in \mathbb{F}[z_1, \ldots, z_n]$, the denominator is nonzero, and $b_{l1} \neq 0$ for $l = 1, \ldots, s$. We can expand

$$1 \Big/ \prod_{l=1}^{s} \left(\sum_{j=1}^{n} b_{lj} z_{i_j} \right)$$

as a power series in $z_{i_2}, \ldots, z_{i_n}$ since $b_{l1} \neq 0$. Call this series $h(z_1, \ldots, z_n)$. Then for each $t \in \mathbb{Z}$ the coefficient of $z_{i_1}^t$ in $g(z_1, \ldots, z_n)h(z_1, \ldots, z_n)$ is a polynomial in $z_{i_2}, \ldots, z_{i_n}$, which we denote by $g_t(z_{i_2}, \ldots, z_{i_n})$. Using the assumed definition of

$$\iota_{i_2 \cdots i_n}\left(g_t(z_{i_2}, \ldots, z_{i_n}) \Big/ \prod_{k=1}^{r} \left(\sum_{j=2}^{n} a_{kj} z_{i_j} \right) \right),$$

we set

$$\begin{aligned} &\iota_{i_1 \cdots i_n} f(z_1, \ldots, z_n) \\ &\quad = \sum_{t \in \mathbb{Z}} \iota_{i_2 \cdots i_n}\left(g_t(z_{i_2}, \ldots, z_{i_n}) \Big/ \prod_{k=1}^{r} \left(\sum_{j=2}^{n} a_{kj} z_{i_j} \right) \right) z_{i_1}^t. \end{aligned} \tag{8.10.42}$$

For example, let $n = 2$ and suppose that $\mathbb{F}$ is algebraically closed. Then all nonzero homogeneous polynomials in two variables are inverted,

$$f(z_1, z_2) = g(z_1, z_2)/z_{i_2}^r \prod_l (b_{l1} z_{i_1} + b_{l2} z_{i_2}),$$

and $\iota_{i_1 i_2}$ is the "expansion in negative powers of z_{i_1}" or in "positive powers of z_{i_2}."

It is clear that the maps $\iota_{i_1 \cdots i_n}$ are injective.

Now let $(i_1 \cdots i_n)$ and $(j_1 \cdots j_n)$ be two permutations of $(1 \cdots n)$. By analogy with the definition of the map Θ in Section 8.1, we consider "expansions of zero" in several variables, and we set

$$\begin{aligned} \Theta_{i_1 \cdots i_n}^{j_1 \cdots j_n} \colon \mathbb{F}[z_1, \ldots, z_n]_S &\to \mathbb{F}[[z_1, z_1^{-1}, \ldots, z_n, z_n^{-1}]] \\ f &\mapsto \iota_{i_1 \cdots i_n} f - \iota_{j_1 \cdots j_n} f. \end{aligned} \tag{8.10.43}$$

For $n = 2$, we shall use the abbreviation

$$\Theta_{i_1 i_2} = \Theta_{i_1 i_2}^{i_2 i_1} \colon \mathbb{F}[z_1, z_2]_S \to \mathbb{F}[[z_1, z_1^{-1}, z_2, z_2^{-1}]]. \tag{8.10.44}$$

Then

$$\operatorname{Ker} \Theta_{i_1 i_2} = \mathbb{F}[z_1, z_1^{-1}, z_2, z_2^{-1}], \tag{8.10.45}$$

and for $f \in \mathbb{F}[z_1, z_1^{-1}, z_2, z_2^{-1}]$ and $g \in \mathbb{F}[z_1, z_2]_S$,

$$\Theta_{i_1 i_2}(fg) = f \Theta_{i_1 i_2}(g). \tag{8.10.46}$$

Moreover, if

$$\iota_{12} f = \iota_{21} g \quad \text{for} \quad f, g \in \mathbb{F}[z_1, z_2]_S,$$

then

$$f = g \in \mathbb{F}[z_1, z_1^{-1}, z_2, z_2^{-1}]. \tag{8.10.47}$$

In order to apply this formalism to vertex operator algebras, we consider "matrix coefficients" of products of vertex operators. Set

$$V' = \coprod_{n \in \mathbb{Z}} V_{(n)}^*, \tag{8.10.48}$$

the direct sum of the dual spaces of the homogeneous subspaces $V_{(n)}$ of the vertex operator algebra V, and denote by $\langle \cdot, \cdot \rangle$ the natural pairing between V' and V. Note that for $v, v_1 \in V$ and $v' \in V'$, we have

$$\langle v', Y(v_1, z_1) v \rangle \in \mathbb{F}[z_1, z_1^{-1}] = \mathbb{F}[z_1]_S. \tag{8.10.49}$$

This phenomenon generalizes to the "rationality" property of products and iterates of vertex operators, leading to the "commutativity" and "associativity" properties. Keeping in mind Remark 8.10.1, we see that the commutator formula obtained by taking Res_{z_0} of the Jacobi identity, in the form (8.8.39), easily implies:

Proposition 8.10.3: *(a)* **(rationality of products)** *For $v, v_1, v_2 \in V$ and $v' \in V'$, the formal series*

$$\langle v', Y(v_1, z_1)Y(v_2, z_2)v\rangle,$$

which involves only finitely many negative powers of z_2 and only finitely many positive powers of z_1, lies in the image of the map ι_{12}:

$$\langle v', Y(v_1, z_1)Y(v_2, z_2)v\rangle = \iota_{12}f(z_1, z_2), \tag{8.10.50}$$

where the (uniquely determined) element $f \in \mathbb{F}[z_1, z_2]_S$ is of the form

$$f(z_1, z_2) = g(z_1, z_2)/z_1^r z_2^s (z_1 - z_2)^t \tag{8.10.51}$$

for some $g \in \mathbb{F}[z_1, z_2]$ and $r, s, t \in \mathbb{Z}$.
(b) **(commutativity)** *We also have*

$$\langle v', Y(v_2, z_2)Y(v_1, z_1)v\rangle = \iota_{21}f(z_1, z_2), \tag{8.10.52}$$

that is,

$$\text{“}Y(v_1, z_1)Y(v_2, z_2) \quad \textit{agrees with} \quad Y(v_2, z_2)Y(v_1, z_1) \tag{8.10.53}$$

as operator-valued rational functions.” In particular,

$$\Theta_{12}(\iota_{12}^{-1}\langle v', Y(v_1, z_1)Y(v_2, z_2)v\rangle) = \langle v', [Y(v_1, z_2), Y(v_2, z_2)]v\rangle. \tag{8.10.54}$$

It follows that for $n \in \mathbb{Z}$,

$$\langle v', (z_1 - z_2)^n Y(v_1, z_1)Y(v_2, z_2)v\rangle = \iota_{12}(z_1 - z_2)^n f(z_1, z_2) \tag{8.10.55}$$

and that

$$\iota_{12}^{-1}\left\langle v', z_0^{-1}\delta\left(\frac{z_1 - z_2}{z_0}\right)Y(v_1, z_1)Y(v_2, z_2)v\right\rangle$$
$$= \iota_{21}^{-1}\left\langle v', z_0^{-1}\delta\left(\frac{z_2 - z_1}{-z_0}\right)Y(v_2, z_2)Y(v_1, z_1)v\right\rangle, \tag{8.10.56}$$

$$\Theta_{12}\left(\iota_{12}^{-1}\left\langle v', z_0^{-1}\delta\left(\frac{z_1 - z_2}{z_0}\right)Y(v_1, z_1)Y(v_2, z_2)v\right\rangle\right)$$
$$= \langle v', [Y(v_1, z_1) \times_{z_0} Y(v_2, z_2)]v\rangle. \tag{8.10.57}$$

Assuming these formulas, we see that the Jacobi identity asserts:

$$\Theta_{12}\left(\iota_{12}^{-1}\left\langle v', z_0^{-1}\delta\left(\frac{z_1 - z_2}{z_0}\right)Y(v_1, z_1)Y(v_2, z_2)v\right\rangle\right)$$
$$= \left\langle v', z_2^{-1}\delta\left(\frac{z_1 - z_0}{z_2}\right)Y(Y(v_1, z_0)v_2, z_2)v\right\rangle. \tag{8.10.58}$$

It is natural to look for corresponding consequences of taking Res_{z_1} of the Jacobi identity. Using Remark 8.8.11 and Propositions 8.8.5 and 8.8.15, and by analogy with Remark 8.8.16, we write the Jacobi identity:

$$z_1^{-1}\delta\left(\frac{z_2+z_0}{z_1}\right)Y(Y(u,z_0)v,z_2) - z_1^{-1}\delta\left(\frac{z_0+z_2}{z_1}\right)Y(u,z_0+z_2)Y(v,z_2)$$
$$= -z_0^{-1}\delta\left(\frac{z_2-z_1}{-z_0}\right)Y(v,z_2)Y(u,z_1)$$
$$= \left(z_1^{-1}\delta\left(\frac{z_2+z_0}{z_1}\right) - z_1^{-1}\delta\left(\frac{z_0+z_2}{z_1}\right)\right)Y(v,z_2)Y(u,z_1)$$
$$= Y(v,z_2)\left(z_1^{-1}\delta\left(\frac{z_2+z_0}{z_1}\right)Y(u,z_2+z_0) - z_1^{-1}\delta\left(\frac{z_0+z_2}{z_1}\right)Y(u,z_0+z_2)\right), \tag{8.10.59}$$

so that in particular,

$$(z_2+z_0)^n Y(Y(u,z_0)v,z_2) - (z_0+z_2)^n Y(u,z_0+z_2)Y(v,z_2)$$
$$= Y(v,z_2)((z_2+z_0)^n Y(u,z_2+z_0) - (z_0+z_2)^n Y(u,z_0+z_2)) \tag{8.10.60}$$

for $n \in \mathbb{Z}$. The case $n = 0$ (corresponding to Res_{z_1}) asserts:

$$Y(Y(u,z_0)v,z_2) - Y(u,z_0+z_2)Y(v,z_2)$$
$$= Y(v,z_2)(Y(u,z_2+z_0) - Y(u,z_0+z_2)) \tag{8.10.61}$$

[cf. (8.8.31)]. Just as in Proposition 8.10.3, we use this formula to conclude (admitting 0 into our index set):

Proposition 8.10.4: *(a)* **(rationality of iterates)** *For $v, v_1, v_2 \in V$ and $v' \in V'$, the formal series*

$$\langle v', Y(Y(v_1,z_0)v_2,z_2)v\rangle,$$

which involves only finitely many negative powers of z_0 and only finitely many positive powers of z_2, lies in the image of the map ι_{20}:

$$\langle v', Y(Y(v_1,z_0)v_2,z_2)v\rangle = \iota_{20}h(z_0,z_2), \tag{8.10.62}$$

where the (uniquely determined) element $h \in \mathbb{F}[z_0,z_2]_S$ is of the form

$$h(z_0,z_2) = k(z_0,z_2)/z_0^r z_2^s (z_0+z_2)^t \tag{8.10.63}$$

for some $k \in \mathbb{F}[z_0,z_2]$ and $r, s, t \in \mathbb{Z}$.

(b) The series

$$\langle v', Y(v_1, z_0 + z_2)Y(v_2, z_2)v\rangle,$$

which involves only finitely many negative powers of z_2 and only finitely many positive powers of z_0, lies in the image of ι_{02}, and in fact

$$\langle v', Y(v_1, z_0 + z_2)Y(v_2, z_2)v\rangle = \iota_{02} h(z_0, z_2). \tag{8.10.64}$$

That is,

$$\text{“}Y(Y(v_1, z_0)v_2, z_2) \quad \textit{agrees with} \quad Y(v_1, z_0 + z_2)Y(v_2, z_2) \tag{8.10.65}$$

as operator-valued rational functions.”

Furthermore, it is clear that for the rational function $f(z_1, z_2)$ of (8.10.51),

$$\iota_{02} f(z_0 + z_2, z_2) = (\iota_{12} f(z_1, z_2))\big|_{z_1 = z_0 + z_2}, \tag{8.10.66}$$

so that

$$h(z_0, z_2) = f(z_0 + z_2, z_2). \tag{8.10.67}$$

Thus Proposition 8.10.4 and Proposition 8.10.3(a) give:

Proposition 8.10.5 (associativity): *We have:*

$$\iota_{12}^{-1}\langle v', Y(v_1, z_1)Y(v_2, z_2)v\rangle$$
$$= (\iota_{20}^{-1}\langle v', Y(Y(v_1, z_0)v_2, z_2)v\rangle)\big|_{z_0 = z_1 - z_2}. \tag{8.10.68}$$

That is,

$$\text{“}Y(v_1, z_1)Y(v_2, z_2) \quad \textit{agrees with} \quad Y(Y(v_1, z_1 - z_2)v_2, z_2) \tag{8.10.69}$$

as operator-valued rational functions, where the right-hand expression is to be expanded as a Laurent series in $z_1 - z_2$.”

Remark 8.10.6: In two-dimensional conformal quantum field theory, the assertion (8.10.69) is called the “associativity of the operator product expansion,” and Proposition 8.10.5 precisely interprets this in terms of our formal calculus. Recall Remark 8.6.4, where these matters were discussed from an ealier point of view. If we are over a field such as $\mathbb{C}$, then the formal expansions of rational functions that we have been discussing converge in suitable domains, and the associativity of the operator product expansion can be interpreted in this way. The Appendix is devoted to this approach. As a preview, we observe the following immediate corollary of Proposition 8.10.5: Over $\mathbb{C}$, the formal series obtained by taking matrix coefficients of the two expressions in (8.10.69) converge to a common rational function in the domains

$$|z_1| > |z_2| > 0 \quad \text{and} \quad |z_2| > |z_1 - z_2| > 0,$$

respectively, and in the common domain

$$|z_1| > |z_2| > |z_1 - z_2| > 0,$$

these two series converge to the common function. Similarly, the commutativity result immediately implies: Over $\mathbb{C}$, the formal series obtained by taking matrix coefficients of the two expressions in (8.10.53) converge to a common rational function in the (disjoint) domains

$$|z_1| > |z_2| > 0 \quad \text{and} \quad |z_2| > |z_1| > 0,$$

respectively.

Using Remark 8.10.1 and Res_{z_0} and Res_{z_1} of the Jacobi identity, we have established rationality, commutativity and associativity properties of a vertex operator algebra in Propositions 8.10.3–8.10.5. Now we shall recover the Jacobi identity using Remark 8.10.1 and the three properties. From (8.10.58) and (8.8.40), it is sufficient to show that for $n \in \mathbb{Z}$,

$$\begin{aligned}&\Theta_{12}(\iota_{12}^{-1}\langle v', (z_1 - z_2)^n Y(v_1, z_1)Y(v_2, z_2)v\rangle)\\ &\quad = \langle v', Y(((z_1 - z_2)^n Y(v_1, z_1 - z_2) - (-z_2 + z_1)^n Y(v_1, -z_2 + z_1))v_2, z_2)v\rangle.\end{aligned} \tag{8.10.70}$$

But by (8.10.68), the left-hand side is

$$\begin{aligned}&\Theta_{12}((z_1 - z_2)^n(\iota_{20}^{-1}\langle v', Y(Y(v_1, z_0)v_2, z_2)v\rangle)|_{z_0 = z_1 - z_2})\\ &\quad = \Theta_{12}((\iota_{20}^{-1}\langle v', Y(z_0^n Y(v_1, z_0)v_2, z_2)v\rangle)|_{z_0 = z_1 - z_2}).\end{aligned} \tag{8.10.71}$$

We break $\langle v', Y(z_0^n Y(v_1, z_0)v_2, z_2)v\rangle$ into its "singular" and "regular" parts with respect to z_0:

$$\begin{aligned}\langle v', Y(z_0^n Y(v_1, z_0)v_2, z_2)v\rangle &= \left\langle v', Y\left(\sum_{m \geq n} (v_1)_m v_2 z_0^{-m+n-1}, z_2\right)v\right\rangle\\ &\quad + \left\langle v', Y\left(\sum_{m < n} (v_1)_m v_2 z_0^{-m+n-1}, z_2\right)v\right\rangle.\end{aligned} \tag{8.10.72}$$

The singular part, a finite sum over m and a Laurent polynomial in z_0 and z_2, clearly gives the right-hand side of (8.10.70) when we apply ι_{20}^{-1}, set $z_0 = z_1 - z_2$ and take Θ_{12}. All we have to show is that the regular part contributes nothing. But recalling (8.10.63), we see that when we apply ι_{20}^{-1} to the regular part and set $z_0 = z_1 - z_2$, we obtain a rational function of the form

$$p(z_1, z_2)/z_2^s \quad \text{for some} \quad p \in \mathbb{F}[z_1, z_2] \quad \text{and} \quad s \in \mathbb{Z};$$

there is no pole at $z_1 = z_2$ or at $z_1 = 0$. Since

$$\Theta_{12}(p(z_1, z_2)/z_2^s) = 0, \tag{8.10.73}$$

we have proved:

Proposition 8.10.7: *The Jacobi identity follows from Remark 8.10.1, the rationality of products and iterates, and commutativity and associativity.*

Many other interesting axiomatic deductions can be made; recall for instance Remarks 8.8.12, 8.8.13 and 8.8.20, as well as various other comments made in this chapter. Also, the axioms can of course be weakened. Some considerations of this sort are presented in the Appendix, where the formal calculus is also translated into the language of complex analysis. The rationality, commutativity and associativity properties and the Jacobi identity extend to several variables, using the ι and Θ maps of (8.10.37) and (8.10.43).

For instance, inductive use of the Jacobi identity gives us the rationality of products and commutativity in any number of variables: For $v_1, v_2, \ldots, v_n, v \in V$, $v' \in V'$ and any permutation $(i_1 \cdots i_n)$ of $(1 \cdots n)$, the formal series

$$\langle v', Y(v_{i_1}, z_{i_1})Y(v_{i_2}, z_{i_2}) \cdots Y(v_{i_n}, z_{i_n})v\rangle$$

lies in the image of the map $\iota_{i_1 \cdots i_n}$:

$$\langle v', Y(v_{i_1}, z_{i_1}) \cdots Y(v_{i_n}, z_{i_n})v\rangle = \iota_{i_1 \cdots i_n} f(z_1, \ldots, z_n), \tag{8.10.74}$$

where the (uniquely determined) element $f \in \mathbb{F}[z_1, \ldots, z_n]_S$ is independent of the permutation and is of the form

$$f(z_1, \ldots, z_n) = g(z_1, \ldots, z_n) \Big/ \prod_{i=1}^{n} z_i^{r_i} \prod_{j<k} (z_j - z_k)^{s_{jk}} \tag{8.10.75}$$

for some $g \in \mathbb{F}[z_1, \ldots, z_n]$ and $r_i, s_{jk} \in \mathbb{Z}$.

9 General Theory of Twisted Vertex Operators

In this chapter we present analogues of the results of Chapter 8 for the twisted vertex operators studied in Chapter 3 and in Sections 7.3 and 7.4. We extend the twisted vertex operator representation to a representation of the full algebra of general vertex operators of the previous chapter, in a general sense involving the square roots of the formal variables. The important new feature of general twisted vertex operators is that the "naive" definition of general vertex operators analogous to their definition in Chapter 8 requires a correction. This modification is achieved by composing with the exponential of a certain quadratic differential operator. A similar expression turns out to have appeared in the formula for the so-called bosonic emission vertex, later interpreted as the $\mathbb{Z}_2$-orbifold twist operator, in string theory (see the Introduction for more details on the connection with string theory and for the physics references). The resulting general vertex operators acting on the twisted space satisfy a Jacobi identity in the sense of Chapter 8. This fact will be important in the construction of the vertex operator algebra based on the moonshine module. The main results of this chapter were announced in [FLM5].

In Section 9.1, which is parallel to Section 8.4, we obtain a formula for the commutator of a pair of twisted vertex operators associated with arbitrary lattice elements satisfying natural conditions. This formula was found in [Lepowsky 4] in a slightly different form. We introduce general twisted vertex operators in Section 9.2 by analogy with the construction of Section 8.5 and parametrized by the untwisted space, but incorporating the modification mentioned above. Using the new definition we present the commutator formula in a form familiar from Chapter 8, but this time involving the square roots of the formal variables. In Section 9.3 we also extend the commutator formula to an arbitrary pair of general twisted vertex operators satisfying natural conditions, and we record some consequences and special cases. We relate the twisted construction of the Virasoro algebra and its commutation relations, from Chapter 1, with general twisted vertex operators in Section 9.4. In particular, the scalar adjustment required in the twisted case of the Virasoro algebra in Section 1.9 is now naturally determined by the exponential of the quadratic differential operator. Finally, in Section 9.5, we establish a Jacobi identity for twisted vertex operators, and from this we extract information about cross-brackets as in Section 8.9. Theorem 9.5.10 was announced in [FLM1].

The results of this chapter and of Chapter 8 are generalized to a setting based on an arbitrary lattice automorphism in [Lepowsky 6]; see [Lepowsky 4, 5] and [FLM5]. See also [Borcherds 4].

9.1. Commutators of Twisted Vertex Operators

In this section, which is analogous to Section 8.4, our aim is to obtain a very general commutator formula for twisted vertex operators, removing the restriction on $|\langle \bar{a}, \bar{b}\rangle|$ in Theorem 7.4.1.

We recall the setting of Sections 7.3 and 7.4, where the twisted vertex operators $X_{\mathbb{Z}+1/2}(a, z)$, here abbreviated $X(a, z)$ (as in Chapter 7), were introduced. We have a nondegenerate lattice L; $\mathfrak{h} = L \otimes_{\mathbb{Z}} \mathbb{F}$; $S(\hat{\mathfrak{h}}_{\mathbb{Z}+1/2}^{-})$ is the $\tilde{\mathfrak{h}}[-1]$-module $M(1)$; $(\hat{L}, {}^{-})$ is a central extension of L by $\langle \kappa \mid \kappa^s = 1\rangle$ with commutator map c_0; $c(\alpha, \beta) = \omega^{c_0(\alpha,\beta)}$ for $\alpha, \beta \in L$; $\theta \in \operatorname{Aut}(\hat{L}; \kappa, \langle\cdot,\cdot\rangle)$ such that $\bar{\theta} = -1$ on L and $\theta^2 = 1$; T is an $\hat{L}$-module such that κ acts as ω and for $a \in \hat{L}$, $\theta a = a$ as operators on T; $V_L^T = S(\hat{\mathfrak{h}}_{\mathbb{Z}+1/2}^{-}) \otimes T$;

$$E^{\pm}(\alpha, z) = \exp\left(\sum_{n \in \pm(\mathbb{N}+1/2)} \frac{\alpha(n)}{n} z^{-n}\right) = e^{-D^{-1}\alpha(z)^{\pm}} \tag{9.1.1}$$

for $\alpha \in \mathfrak{h}$; and for $a \in \hat{L}$ with

$$\langle \bar{a}, \bar{a}\rangle \in \mathbb{Z} \tag{9.1.2}$$

we have the vertex operator

$$X(a, z) = X_{\mathbb{Z}+1/2}(a, z) = 2^{-\langle \bar{a}, \bar{a} \rangle} E^{-}(-\bar{a}, z) E^{+}(-\bar{a}, z) a \quad (9.1.3)$$

$$= 2^{-\langle \bar{a}, \bar{a} \rangle} :e^{D^{-1}\bar{a}(z)}: a = 2^{-\langle \bar{a}, \bar{a} \rangle} e^{D^{-1}\bar{a}(z)^{-}} e^{D^{-1}\bar{a}(z)^{+}} a \quad (9.1.4)$$

$$= \sum_{n \in (1/2)\mathbb{Z}} x_a(n) z^{-n}, \quad (9.1.5)$$

and

$$X^{\pm}(a, z) = X(a, z) \pm X(\theta a, z) \quad (9.1.6)$$

$$= \sum_{n \in (1/2)\mathbb{Z}} x_a^{\pm}(n) z^{-n}. \quad (9.1.7)$$

(Of course, the assumption (9.1.2) is not necessary if $\mathbb{F}$ contains appropriate roots of 2, as we have commented in Remark 7.3.4.)

Just as in Section 8.4, we change notation as follows: For $\alpha \in \mathfrak{h}$, set

$$\alpha_{\text{new}}(z) = \sum_{n \in \mathbb{Z}+1/2} \alpha(n) z^{-n-1} = z^{-1}\alpha(z) \quad (= z^{-1}\alpha_{\text{old}}(z)) \quad (9.1.8)$$

and abbreviate this [as in (8.4.8)] by $\alpha(z)$:

$$\alpha(z) = \alpha_{\text{new}}(z) = \sum_{n \in \mathbb{Z}+1/2} \alpha(n) z^{-n-1} = \sum_{n \in \mathbb{Z}+1/2} \alpha(n-1) z^{-n}. \quad (9.1.9)$$

Then

$$\int \alpha(z) = \sum_{n \in \mathbb{Z}+1/2} \frac{\alpha(n)}{-n} z^{-n}. \quad (9.1.10)$$

Also introduce the notation change

$$\alpha(z)^{\pm} = \sum_{n \in \mathbb{N}+1/2} \alpha(\pm n) z^{\mp n-1} \quad (9.1.11)$$

[cf. (8.4.10)], so that

$$\alpha(z) = \alpha(z)^{+} + \alpha(z)^{-}. \quad (9.1.12)$$

We set

$$\alpha(n) = 0 \quad \text{for} \quad n \in \mathbb{Z}, \quad \text{in particular,} \quad \alpha(0) = 0, \quad (9.1.13)$$

as in (7.4.5).

Now in the twisted case there is only one natural notion of normal ordering—the one we have already used, denoted by $: \cdot :$ [recall Section 3.3 and (7.3.21)–(7.3.28)]. But to emphasize the parallels with the untwisted theory as treated in Chapter 8, *we shall replace the normal ordering notation in the twisted case by open colons:*

$$\mathring{\circ} \cdot \mathring{\circ} = : \cdot :. \quad (9.1.14)$$

In the untwisted theory, the appropriate vertex operator $Y(a, z)$ differs from the earlier one $X(a, z)$ by a factor of $z^{-\langle\bar{a},\bar{a}\rangle/2}$ [recall (8.4.17)], and it turns out that this is the correct motivation for the definition of $Y(a, z)$ in the twisted case: *For $a \in \hat{L}$ such that $\langle\bar{a},\bar{a}\rangle \in \mathbb{Z}$, define the new vertex operator by:*

$$\begin{aligned} Y(a, z) &= Y_{\mathbb{Z}+1/2}(a, z) = 2^{-\langle\bar{a},\bar{a}\rangle}\, {}^{\circ}_{\circ} e^{\int \bar{a}(z)} {}^{\circ}_{\circ}\, a z^{-\langle\bar{a},\bar{a}\rangle/2} \\ &= X(a, z) z^{-\langle\bar{a},\bar{a}\rangle/2} = X(a, z) z^{-\mathrm{wt}\,\iota(a)}. \end{aligned} \tag{9.1.15}$$

Then we have

$${}^{\circ}_{\circ}\alpha(z) Y(a, z){}^{\circ}_{\circ} = \alpha(z)^{-} Y(a, z) + Y(a, z)\alpha(z)^{+} \tag{9.1.16}$$

$${}^{\circ}_{\circ} Y(a, z_1) Y(b, z_2) {}^{\circ}_{\circ} = 2^{-\langle\bar{a},\bar{a}\rangle - \langle\bar{b},\bar{b}\rangle}\, {}^{\circ}_{\circ} e^{\int(\bar{a}(z_1)+\bar{b}(z_2))} {}^{\circ}_{\circ}\, ab z_1^{-\langle\bar{a},\bar{a}\rangle/2} z_2^{-\langle\bar{b},\bar{b}\rangle/2} \tag{9.1.17}$$

$${}^{\circ}_{\circ} Y(a, z_1) Y(b, z_2) {}^{\circ}_{\circ} = c(\bar{a}, \bar{b})\, {}^{\circ}_{\circ} Y(b, z_2) Y(a, z_1) {}^{\circ}_{\circ} \tag{9.1.18}$$

$$Y(a, z_1) Y(b, z_2) = {}^{\circ}_{\circ} Y(a, z_1) Y(b, z_2) {}^{\circ}_{\circ} \left(\frac{z_1^{1/2} - z_2^{1/2}}{z_1^{1/2} + z_2^{1/2}} \right)^{\langle\bar{a},\bar{b}\rangle} \tag{9.1.19}$$

for $\alpha \in \mathfrak{h}$ and $a, b \in \hat{L}$ satisfying the integrality condition (9.1.2). The last formula is a restatement of (7.3.28). The factor on the right is to be expanded in nonnegative integral powers of $z_2^{1/2}$, as the notation as usual suggests. We may also write (9.1.19) as follows:

$$\begin{aligned} Y(a, z_1) Y(b, z_2) &= {}^{\circ}_{\circ} Y(a, z_1) Y(b, z_2) {}^{\circ}_{\circ} \prod_{p=0,1} (z_1^{1/2} - (-1)^p z_2^{1/2})^{\langle(-1)^p \bar{a}, \bar{b}\rangle} && (9.1.20) \\ &= {}^{\circ}_{\circ} Y(a, z_1) Y(b, z_2) {}^{\circ}_{\circ} (z_1 - z_2)^{\langle\bar{a},\bar{b}\rangle} (z_1^{1/2} + z_2^{1/2})^{-2\langle\bar{a},\bar{b}\rangle}. && (9.1.21) \end{aligned}$$

Note also that

$$\frac{d}{dz} Y(a, z) = {}^{\circ}_{\circ} (\bar{a}(z) - \tfrac{1}{2}\langle\bar{a},\bar{a}\rangle z^{-1}) Y(a, z) {}^{\circ}_{\circ}, \tag{9.1.22}$$

in contrast with (8.4.23), and that

$$Y(a, z) = \sum_{n \in (1/2)\mathbb{Z}} x_a(n) z^{-n-\langle\bar{a},\bar{a}\rangle/2}. \tag{9.1.23}$$

By the properties of the involution θ and its action on the module T, we have

$$Y(\theta a, z) = \lim_{z^{1/2} \to -z^{1/2}} Y(a, z) \quad \text{if} \quad \langle\bar{a},\bar{a}\rangle \in 2\mathbb{Z}, \tag{9.1.24}$$

and more generally,

$$Y(\theta a, z) = (-1)^{\langle \bar{a}, \bar{a} \rangle} \lim_{z^{1/2} \to -z^{1/2}} Y(a, z) \quad \text{if} \quad \langle \bar{a}, \bar{a} \rangle \in \mathbb{Z} \tag{9.1.25}$$

[see (7.4.16) and (9.1.15)].

We are ready to begin computing commutators. Let $a, b \in \hat{L}$ and assume that

$$\langle \bar{a}, \bar{a} \rangle, \langle \bar{b}, \bar{b} \rangle, \langle \bar{a}, \bar{b} \rangle \in \mathbb{Z}, \tag{9.1.26}$$

$$\langle \bar{a}, \bar{b} \rangle \leq 0 \tag{9.1.27}$$

and

$$c(\bar{a}, \bar{b}) = (-1)^{\langle \bar{a}, \bar{b} \rangle}. \tag{9.1.28}$$

By (9.1.27), the last factor in (9.1.21) is a polynomial in $z_1^{1/2}$ and $z_2^{1/2}$, and we find exactly as in (8.4.26)–(8.4.30) that

$$\begin{aligned}
[Y(a, z_1), Y(b, z_2)] &= Y(a, z_1)Y(b, z_2) - Y(b, z_2)Y(a, z_1) \\
&= {}^{\circ}_{\circ} Y(a, z_1)Y(b, z_2) {}^{\circ}_{\circ} (z_1^{1/2} + z_2^{1/2})^{-2\langle \bar{a}, \bar{b} \rangle} z_1^{\langle \bar{a}, \bar{b} \rangle} \\
&\quad \cdot ((1 - z_2/z_1)^{\langle \bar{a}, \bar{b} \rangle} - (-z_2/z_1)^{\langle \bar{a}, \bar{b} \rangle}(1 - z_1/z_2)^{\langle \bar{a}, \bar{b} \rangle}) \\
&= \operatorname{Res}_{z_0} z_0^{\langle \bar{a}, \bar{b} \rangle} z_2^{-1} {}^{\circ}_{\circ} Y(a, z_1)Y(b, z_2) {}^{\circ}_{\circ} (z_1^{1/2} + z_2^{1/2})^{-2\langle \bar{a}, \bar{b} \rangle} \\
&\quad \cdot e^{-z_0(\partial/\partial z_1)} \delta(z_1/z_2).
\end{aligned} \tag{9.1.29}$$

Remark 9.1.1: The condition (9.1.27) is not a serious restriction because (9.1.24) or (9.1.25) can be combined with the present computation to find $[Y(a, z_1), Y(b, z_2)]$ if (9.1.27) does not hold.

Continuing with the computation, we would like to apply Proposition 8.2.2, but we must be careful about the fact that the exponents of z_1 and z_2 are half-integers [recall (9.1.17)]. We start by observing that

$$\delta(z) = \tfrac{1}{2}(\delta(z^{1/2}) + \delta(-z^{1/2})). \tag{9.1.30}$$

Also, the operator d/dz may be expressed as a Laurent monomial in $z^{1/2}$ times the derivative with respect to the variable $z^{1/2}$:

$$\frac{d}{dz} = \frac{1}{2} z^{-1/2} \frac{d}{d(z^{1/2})}, \tag{9.1.31}$$

as we see by formally using the chain rule or by applying both sides of (9.1.31) to a general formal series. More generally (although we shall not need this),

$$\frac{d}{dz} = \frac{1}{m} z^{1/m-1} \frac{d}{d(z^{1/m})} \tag{9.1.32}$$

for every $m \in \mathbb{F}^{\times}$, a fact which can also be expressed in the natural form

$$z \frac{d}{dz} = nz^n \frac{d}{d(z^n)} \tag{9.1.33}$$

for every $n \in \mathbb{F}^{\times}$. Actually, (9.1.31)–(9.1.33) have already arisen in the proof of Proposition 8.3.10.

We now find using Proposition 8.2.2 and (8.3.11) that

$$\begin{aligned}
&[Y(a, z_1), Y(b, z_2)] \\
&\quad = \tfrac{1}{2} \sum_{p=0,1} \operatorname{Res}_{z_0} z_0^{\langle \bar{a}, \bar{b} \rangle} z_2^{-1} {}^{\circ}_{\circ} Y(a, z_1) Y(b, z_2) {}^{\circ}_{\circ} (z_1^{1/2} + z_2^{1/2})^{-2\langle \bar{a}, \bar{b} \rangle} \\
&\qquad \cdot e^{-z_0(\partial/\partial z_1)} \delta((-1)^p z_1^{1/2}/z_2^{1/2}) \\
&\quad = \tfrac{1}{2} \sum_{p=0,1} \operatorname{Res}_{z_0} z_0^{\langle \bar{a}, \bar{b} \rangle} z_2^{-1} \lim_{z_1^{1/2} \to (-1)^p z_2^{1/2}} ({}^{\circ}_{\circ} Y(a, z_1 + z_0) Y(b, z_2) {}^{\circ}_{\circ} \\
&\qquad \cdot ((z_1 + z_0)^{1/2} + z_2^{1/2})^{-2\langle \bar{a}, \bar{b} \rangle}) e^{-z_0(\partial/\partial z_1)} \delta((-1)^p z_1^{1/2}/z_2^{1/2}).
\end{aligned} \tag{9.1.34}$$

Recalling that the exponent $-2\langle \bar{a}, \bar{b} \rangle$ is a nonnegative integer, we have

$$((z_1 + z_0)^{1/2} + z_2^{1/2})^{-2\langle \bar{a}, \bar{b} \rangle} = (z_1^{1/2}(1 + z_0/z_1)^{1/2} + z_2^{1/2})^{-2\langle \bar{a}, \bar{b} \rangle}, \tag{9.1.35}$$

and the binomial series $(1 + z_0/z_1)^{1/2}$ involves only nonnegative integral powers of z_0/z_1, with constant term 1. Thus when we replace $z_1^{1/2}$ by $-z_2^{1/2}$ in (9.1.35), the resulting formal power series in z_0 contains a factor of $z_0^{-2\langle \bar{a}, \bar{b} \rangle}$. It follows that the term corresponding to $p = 1$ in (9.1.34) contributes the residue of a formal Laurent series in z_0 in which the smallest exponent of z_0 is at least $-\langle \bar{a}, \bar{b} \rangle$. This being nonnegative, the residue is 0, and so (9.1.34) gives just

$$\begin{aligned}
[Y(a, z_1), Y(b, z_2)] &= \tfrac{1}{2} \operatorname{Res}_{z_0} z_0^{\langle \bar{a}, \bar{b} \rangle} z_2^{-1} {}^{\circ}_{\circ} Y(a, z_2 + z_0) Y(b, z_2) {}^{\circ}_{\circ} \\
&\quad \cdot ((z_2 + z_0)^{1/2} + z_2^{1/2})^{-2\langle \bar{a}, \bar{b} \rangle} e^{-z_0(\partial/\partial z_1)} \delta(z_1^{1/2}/z_2^{1/2}).
\end{aligned} \tag{9.1.36}$$

This formula is the analogue of (8.4.35). There is an interesting analogue of Remark 8.4.3:

Remark 9.1.2: Formally, (9.1.21) gives

$$\begin{aligned}
&[Y(a, z_1), Y(b, z_2)] \\
&\quad = \tfrac{1}{2} \operatorname{Res}_{z_0} z_2^{-1} Y(a, z_2 + z_0) Y(b, z_2) e^{-z_0(\partial/\partial z_1)} \delta(z_1^{1/2}/z_2^{1/2}).
\end{aligned} \tag{9.1.37}$$

We now turn to the analogue of (8.4.36). From (8.3.3), (9.1.15) and (9.1.17),

$$
\begin{aligned}
&\mathop{\circ}\limits^{\circ} Y(a, z_2 + z_0) Y(b, z_2) \mathop{\circ}\limits^{\circ} \\
&\quad = 2^{-\langle \bar{a}, \bar{a}\rangle - \langle \bar{b}, \bar{b}\rangle} \mathop{\circ}\limits^{\circ} e^{\int \bar{a}(z_2+z_0) d(z_2+z_0) + \int \bar{b}(z_2)\, dz_2} \mathop{\circ}\limits^{\circ} \\
&\qquad \cdot ab(z_2 + z_0)^{-\langle \bar{a}, \bar{a}\rangle/2} z_2^{-\langle \bar{b}, \bar{b}\rangle/2} \\
&\quad = 2^{-\langle \bar{a}, \bar{a}\rangle - \langle \bar{b}, \bar{b}\rangle} e^{\int \bar{a}(z_2+z_0)^- d(z_2+z_0)} \mathop{\circ}\limits^{\circ} e^{\int \bar{b}(z_2)} \mathop{\circ}\limits^{\circ} \\
&\qquad \cdot e^{\int \bar{a}(z_2+z_0)^+ d(z_2+z_0)} ab z_2^{-\langle \bar{a}, \bar{a}\rangle/2 - \langle \bar{b}, \bar{b}\rangle/2} (1 + z_0/z_2)^{-\langle \bar{a}, \bar{a}\rangle/2} \\
&\quad = 2^{-\langle \bar{a}+\bar{b}, \bar{a}+\bar{b}\rangle} 2^{2\langle \bar{a}, \bar{b}\rangle} \exp\left(e^{z_0 (d/dz_2)} \int \bar{a}(z_2)^-\, dz_2 \right) \mathop{\circ}\limits^{\circ} e^{\int \bar{b}(z_2)} \mathop{\circ}\limits^{\circ} \\
&\qquad \cdot \exp\left(e^{z_0 (d/dz_2)} \int \bar{a}(z_2)^+\, dz_2 \right) ab z_2^{-\langle \bar{a}+\bar{b}, \bar{a}+\bar{b}\rangle/2} \\
&\qquad \cdot z_2^{\langle \bar{a}, \bar{b}\rangle} (1 + z_0/z_2)^{-\langle \bar{a}, \bar{a}\rangle/2} \\
&\quad = 2^{-\langle \overline{ab}, \overline{ab}\rangle} 2^{2\langle \bar{a}, \bar{b}\rangle} \exp\left(\sum_{n \ge 1} \frac{1}{n!} \left(\frac{d}{dz_2}\right)^{n-1} \bar{a}(z_2)^- z_0^n \right) \mathop{\circ}\limits^{\circ} e^{\int (\bar{a}+\bar{b})(z_2)} \mathop{\circ}\limits^{\circ} \\
&\qquad \cdot \exp\left(\sum_{n \ge 1} \frac{1}{n!} \left(\frac{d}{dz_2}\right)^{n-1} \bar{a}(z_2)^+ z_0^n \right) ab z_2^{-\langle \overline{ab}, \overline{ab}\rangle/2} \\
&\qquad \cdot z_2^{\langle \bar{a}, \bar{b}\rangle} (1 + z_0/z_2)^{-\langle \bar{a}, \bar{a}\rangle/2} \\
&\quad = 2^{2\langle \bar{a}, \bar{b}\rangle} \exp\left(\sum_{n \ge 1} \frac{1}{n!} \left(\frac{d}{dz_2}\right)^{n-1} \bar{a}(z_2)^- z_0^n \right) Y(ab, z_2) \\
&\qquad \cdot \exp\left(\sum_{n \ge 1} \frac{1}{n!} \left(\frac{d}{dz_2}\right)^{n-1} \bar{a}(z_2)^+ z_0^n \right) z_2^{\langle \bar{a}, \bar{b}\rangle} (1 + z_0/z_2)^{-\langle \bar{a}, \bar{a}\rangle/2} \\
&\quad = 2^{2\langle \bar{a}, \bar{b}\rangle} \mathop{\circ}\limits^{\circ} \exp\left(\sum_{n \ge 1} \frac{1}{n!} \left(\frac{d}{dz_2}\right)^{n-1} \bar{a}(z_2) z_0^n \right) Y(ab, z_2) \mathop{\circ}\limits^{\circ} \\
&\qquad \cdot z_2^{\langle \bar{a}, \bar{b}\rangle} (1 + z_0/z_2)^{-\langle \bar{a}, \bar{a}\rangle/2}.
\end{aligned}
\tag{9.1.38}
$$

This expression is not as simple as that in (8.4.32). We shall rectify the situation below by setting up general vertex operators appropriately, but for now we summarize what we have proved (cf. Theorem 8.4.2):

Theorem 9.1.3: *Let $a, b \in \hat{L}$ and assume that*

$$\langle \bar{a}, \bar{a}\rangle, \langle \bar{b}, \bar{b}\rangle, \langle \bar{a}, \bar{b}\rangle \in \mathbb{Z}, \tag{9.1.39}$$

$$\langle \bar{a}, \bar{b}\rangle \le 0 \tag{9.1.40}$$

(recall Remark 9.1.1) and

$$c(\bar{a}, \bar{b}) = (-1)^{\langle \bar{a}, \bar{b} \rangle}. \tag{9.1.41}$$

Then

$$[Y(a, z_1), Y(b, z_2)] = \tfrac{1}{2} \operatorname{Res}_{z_0} z_0^{\langle \bar{a}, \bar{b} \rangle} z_2^{-1} {}^{\circ}_{\circ} Y(a, z_2 + z_0) Y(b, z_2) {}^{\circ}_{\circ} \cdot ((z_2 + z_0)^{1/2} + z_2^{1/2})^{-2\langle \bar{a}, \bar{b} \rangle} e^{-z_0(\partial/\partial z_1)} \delta(z_1^{1/2}/z_2^{1/2}) \tag{9.1.42}$$

and

$${}^{\circ}_{\circ} Y(a, z_2 + z_0) Y(b, z_2) {}^{\circ}_{\circ} ((z_2 + z_0)^{1/2} + z_2^{1/2})^{-2\langle \bar{a}, \bar{b} \rangle} = {}^{\circ}_{\circ} \exp\left(\sum_{n \geq 1} \frac{1}{n!} \left(\frac{d}{dz_2} \right)^{n-1} \bar{a}(z_2) z_0^n \right) Y(ab, z_2) {}^{\circ}_{\circ} \cdot \left(\frac{(1 + z_0/z_2)^{1/2} + 1}{2} \right)^{-2\langle \bar{a}, \bar{b} \rangle} (1 + z_0/z_2)^{-\langle \bar{a}, \bar{a} \rangle/2}. \tag{9.1.43}$$

Formula (9.1.43) follows immediately from (9.1.38) and does not require condition (9.1.40) or (9.1.41). Note that the factor

$$\left(\frac{(1 + z_0/z_2)^{1/2} + 1}{2} \right)^{-2\langle \bar{a}, \bar{b} \rangle} (1 + z_0/z_2)^{-\langle \bar{a}, \bar{a} \rangle/2} \tag{9.1.44}$$

is a formal power series in z_0/z_2 with constant term 1. By expanding this expression explicitly, we could write down "concrete" expressions for the commutator in the spirit of (8.4.38)–(8.4.45), but it is better to wait until we have general vertex operators available, since we will then be able to exhibit a precise parallel between the untwisted and twisted theories. The discrepancy between (9.1.22) and its simpler-looking untwisted analogue (8.4.23) will be explained in the process.

Remark 9.1.4: Theorem 9.1.3 generalizes Theorem 7.4.1 (cf. Remark 8.4.5).

Remark 9.1.5: In Section 7.4 and in the present section, we have assumed that $\theta a = a$ as operators on T for all $a \in \hat{L}$. When we investigate triality in Chapter 11, we will want to drop this assumption and to allow instead that for a particular $a \in \hat{L}$, either

$$\theta a = a \quad \text{as operators on} \quad T \tag{9.1.45}$$

or

$$\theta a = -a \quad \text{as operators on} \quad T. \tag{9.1.46}$$

The only change required in the present section is that formulas (9.1.24) and (9.1.25) gain minus signs on the right-hand sides if (9.1.46) holds.

9.2. General Twisted Vertex Operators

In the last section we have computed the commutator of the twisted vertex operators $Y(a, z_1)$ and $Y(b, z_2)$ under certain natural hypotheses on a and b. The result, Theorem 9.1.3, seems more complicated than the corresponding result in the untwisted case (Theorem 8.4.2). But there are two reasons why we might expect that a proper definition of general twisted vertex operator will reveal the untwisted and twisted theories to be entirely analogous: The vertex operator constructions of the untwisted affine algebras $\hat{A}_n$, $\hat{D}_n$, $\hat{E}_n$ and of their twisted analogues given in Sections 7.2 and 7.4 are very similar, and so are the formal versions of the commutator result given in Remarks 8.4.3 and 9.1.2. This section, which is parallel to Section 8.5, begins with an approach to the definition of general twisted vertex operator.

We shall now assume for convenience that

$$\langle \alpha, \alpha \rangle \in \mathbb{Z} \quad \text{for all} \quad \alpha \in L. \tag{9.2.1}$$

While this does not make L an integral lattice, condition (9.1.2) will always hold.

Comparison of Corollary 8.5.1 and Theorem 9.1.3 leads us to try to define general twisted vertex operators $Y(v, z) = Y_{\mathbb{Z}+1/2}(v, z)$ for $v \in V_L = S(\hat{\mathfrak{h}}_{\bar{\mathbb{Z}}}) \otimes \mathbb{F}\{L\}$ [the *un*twisted space; recall (7.1.18), (7.1.31)] having the following properties: The correspondence $v \mapsto Y(v, z)$ should be a linear map from V_L to $(\operatorname{End} V_L^T)\{z\}$ (V_L^T being the twisted space) such that when we extend the map canonically to $V_L\{z_0\}$,

$$\begin{aligned} Y(Y_{\mathbb{Z}}(a, z_0)\iota(b), z_2) = {}& \mathop{\circ}\limits^{\circ} \exp\left(\sum_{n \geq 1} \frac{1}{n!} \left(\frac{d}{dz_2}\right)^{n-1} \bar{a}(z_2) z_0^n \right) Y(ab, z_2) \mathop{\circ}\limits^{\circ} \\ & \cdot z_0^{\langle \bar{a}, \bar{b} \rangle} \left(\frac{(1 + z_0/z_2)^{1/2} + 1}{2} \right)^{-2\langle \bar{a}, \bar{b} \rangle} (1 + z_0/z_2)^{-\langle \bar{a}, \bar{a} \rangle/2} \end{aligned} \tag{9.2.2}$$

for $a, b \in \hat{L}$. In this case, we would have

$$\begin{aligned} &[Y(a, z_1), Y(b, z_2)] \\ &\quad = \tfrac{1}{2} \operatorname{Res}_{z_0} z_2^{-1} Y(Y_{\mathbb{Z}}(a, z_0)\iota(b), z_2) e^{-z_0(\partial/\partial z_1)} \delta(z_1^{1/2}/z_2^{1/2}) \end{aligned} \tag{9.2.3}$$

for a, b as in Theorem 9.1.3 [cf. (8.5.11)–(8.5.13)], and formally,

$$Y(Y_{\mathbb{Z}}(a, z_0)\iota(b), z_2) = Y(a, z_2 + z_0) Y(b, z_2) \tag{9.2.4}$$

for all $a, b \in \hat{L}$ (cf. Remark 8.5.2). Of course, Y means $Y_{\mathbb{Z}+1/2}$. [Note that we want $Y_{\mathbb{Z}}(a, z_0)$, not $Y(a, z_0)$.]

The "obvious" guess as to the right definition of $Y(v, z)$, namely, the operator denoted $Y_0(v, z)$ in (9.2.23) below [formally the same as the general untwisted vertex operator given by (8.5.5)], certainly will not work. Instead, we shall construct a linear map of exponential form,

$$\exp(\Delta_z): V_L \to V_L\{z\}, \tag{9.2.5}$$

such that

$$\exp(\Delta_z)(Y_{\mathbb{Z}}(a, z_0)\iota(b)) = Y_{\mathbb{Z}}(a, z_0)\iota(b)\left(\frac{(1 + z_0/z)^{1/2} + 1}{2}\right)^{-2\langle \bar{a}, \bar{b}\rangle}(1 + z_0/z)^{-\langle \bar{a}, \bar{a}\rangle/2} \tag{9.2.6}$$

[so that $Y_{\mathbb{Z}}(a, z_0)\iota(b)$ is an "eigenvector"], and we shall take $Y(v, z) = Y_0(\exp(\Delta_z)v, z)$. This *will* be compatible with (9.2.2). Moreover,

$$\Delta_z: V_L \to V_L\{z\} \tag{9.2.7}$$

will be realized as a quadratic expression in the operators $\alpha(n)$ for $\alpha \in \mathfrak{h}$ and $n \in \mathbb{N}$—a "quadratic differential operator."

We begin by noting that for $a, b \in \hat{L}$, $\alpha \in \mathfrak{h}$ and $n > 0$, we have

$$\alpha(n) \cdot Y_{\mathbb{Z}}(a, z_0)\iota(b) = \langle \alpha, \bar{a}\rangle z_0^n Y_{\mathbb{Z}}(a, z_0)\iota(b) \tag{9.2.8}$$

and

$$\alpha(0) \cdot Y_{\mathbb{Z}}(a, z_0)\iota(b) = \langle \alpha, \bar{a} + \bar{b}\rangle Y_{\mathbb{Z}}(a, z_0)\iota(b), \tag{9.2.9}$$

by (7.1.33), (7.1.46) and (8.4.17). Now let $\{h_1, \ldots, h_l\}$ be a basis of $\mathfrak{h}$ and let $\{h_1', \ldots, h_l'\}$ be its dual basis with respect to the form $\langle \cdot, \cdot \rangle$, as in (1.9.21) and (8.7.2). Then for $m, n > 0$,

$$\sum_{i=1}^{l} h_i(m)h_i'(n) \cdot Y_{\mathbb{Z}}(a, z_0)\iota(b) = \langle \bar{a}, \bar{a}\rangle z_0^{m+n} Y_{\mathbb{Z}}(a, z_0)\iota(b), \tag{9.2.10}$$

$$\sum_{i=1}^{l} h_i(m)h_i'(0) \cdot Y_{\mathbb{Z}}(a, z_0)\iota(b) = \langle \bar{a} + \bar{b}, \bar{a}\rangle z_0^m Y_{\mathbb{Z}}(a, z_0)\iota(b), \tag{9.2.11}$$

$$\sum_{i=1}^{l} h_i(0)h_i'(n) \cdot Y_{\mathbb{Z}}(a, z_0)\iota(b) = \langle \bar{a} + \bar{b}, \bar{a}\rangle z_0^n Y_{\mathbb{Z}}(a, z_0)\iota(b), \tag{9.2.12}$$

$$\sum_{i=1}^{l} h_i(0)h_i'(0) \cdot Y_{\mathbb{Z}}(a, z_0)\iota(b) = \langle \bar{a} + \bar{b}, \bar{a} + \bar{b}\rangle Y_{\mathbb{Z}}(a, z_0)\iota(b). \tag{9.2.13}$$

We shall take Δ_z to be of the form

$$\Delta_z = \sum_{m,n \geq 0} \sum_{i=1}^{l} c_{mn} h_i(m)h_i'(n) z^{-m-n} \tag{9.2.14}$$

with $c_{mn} \in \mathbb{F}$. This is a well-defined element of $\mathrm{End}(V_L[z^{-1}])$, that is, it preserves the space of *polynomials* in z^{-1} with coefficients in V_L. Of course, this operator can be canonically extended to the space $V_L[z^{-1}]\{z_0\}$. Assuming that

$$c_{00} = 0, \tag{9.2.15}$$

we see from (9.2.10)–(9.2.12) that

$$\begin{aligned}\Delta_z(Y_{\mathbb{Z}}(a, z_0)\iota(b))\\ &= \left(\langle \bar{a}, \bar{a}\rangle \sum_{m,n\geq 0} c_{mn}(z_0/z)^{m+n} + \langle \bar{a}, \bar{b}\rangle \sum_{n>0} (c_{n0} + c_{0n})(z_0/z)^n\right)\\ &\quad \cdot Y_{\mathbb{Z}}(a, z_0)\iota(b).\end{aligned} \tag{9.2.16}$$

Because of (9.2.15), $\exp(\Delta_z)$ is a well-defined element of $\mathrm{End}(V_L[z^{-1}])$, and from (9.2.16),

$$\begin{aligned}\exp(\Delta_z)(Y_{\mathbb{Z}}(a, z_0)\iota(b))\\ &= \exp\left(\langle \bar{a}, \bar{a}\rangle \sum_{m,n\geq 0} c_{mn}(z_0/z)^{m+n} + \langle \bar{a}, \bar{b}\rangle \sum_{n>0} (c_{n0} + c_{0n})(z_0/z)^n\right)\\ &\quad \cdot Y_{\mathbb{Z}}(a, z_0)\iota(b)\\ &= \left(\exp \sum_{m,n\geq 0} c_{mn}(z_0/z)^{m+n}\right)^{\langle \bar{a},\bar{a}\rangle} \left(\exp \sum_{n>0} (c_{n0} + c_{0n})(z_0/z)^n\right)^{\langle \bar{a},\bar{b}\rangle}\\ &\quad \cdot Y_{\mathbb{Z}}(a, z_0)\iota(b).\end{aligned} \tag{9.2.17}$$

We want (9.2.17) to match (9.2.6). It is clear that we can arrange this with the constants c_{mn} determined by the formula

$$\sum_{m,n\geq 0} c_{mn}x^m y^n = -\log\left(\frac{(1+x)^{1/2} + (1+y)^{1/2}}{2}\right). \tag{9.2.18}$$

The c_{mn} are indeed well defined, and $c_{00} = 0$. *We fix this choice of the* c_{mn} *in the expression* (9.2.14) *for* Δ_z.

Remark 9.2.1: Of course, there is certain flexibility in the choice of the constants, but (9.2.18) will prove to be correct when we consider commutators of *general* twisted vertex operators.

Before defining these operators, we introduce operators denoted $Y_0(v, z)$. As in (8.5.1)–(8.5.6), we first observe that the correspondence $\iota(a) \mapsto Y(a, z)$

for $a \in \hat{L}$ extends uniquely to a well-defined linear map

$$\begin{aligned} \mathbb{F}\{L\} &\to (\operatorname{End} V_L^T)[[z^{1/2}, z^{-1/2}]] \\ v &\mapsto Y_0(v, z), \end{aligned} \tag{9.2.19}$$

by (7.1.20) and (7.3.19). In particular, we have

$$Y_0(\iota(a), z) = Y(a, z) \quad \text{for} \quad a \in \hat{L} \tag{9.2.20}$$

and

$$Y_0(\iota(1), z) = Y(1, z) = 1. \tag{9.2.21}$$

More generally, let

$$a \in \hat{L}, \quad \alpha_1, \ldots, \alpha_k \in \mathfrak{h}, \quad n_1, \ldots, n_k \in \mathbb{Z}_+$$

and write

$$v = \alpha_1(-n_1) \cdots \alpha_k(-n_k) \cdot \iota(a) \in V_L. \tag{9.2.22}$$

We define

$$Y_0(v,z) \in (\operatorname{End} V_L^T)[[z^{1/2}, z^{-1/2}]]$$

by:

$$Y_0(v, z) = {}^{\circ}_{\circ}\left(\frac{1}{(n_1 - 1)!}\left(\frac{d}{dz}\right)^{n_1 - 1} \alpha_1(z)\right) \cdots \left(\frac{1}{(n_k - 1)!}\left(\frac{d}{dz}\right)^{n_k - 1} \alpha_k(z)\right) Y(a, z){}^{\circ}_{\circ}. \tag{9.2.23}$$

This gives us a well-defined linear map

$$\begin{aligned} V_L &\to (\operatorname{End} V_L^T)[[z^{1/2}, z^{-1/2}]] \\ v &\mapsto Y_0(v, z). \end{aligned} \tag{9.2.24}$$

Recalling [from (8.5.11), for example] that

$$Y_{\mathbb{Z}}(a, z_0)\iota(b) = \exp\left(\sum_{n \geq 1} \frac{\bar{a}(-n)}{n} z_0^n\right)\iota(ab) z_0^{\langle \bar{a}, \bar{b} \rangle} \tag{9.2.25}$$

for $a, b \in \hat{L}$, we see that

$$Y_0(Y_{\mathbb{Z}}(a, z_0)\iota(b), z) = {}^{\circ}_{\circ}\exp\left(\sum_{n \geq 1} \frac{1}{n!}\left(\frac{d}{dz}\right)^{n-1} \bar{a}(z) z_0^n\right) Y(ab, z){}^{\circ}_{\circ} z_0^{\langle \bar{a}, \bar{b} \rangle}, \tag{9.2.26}$$

where of course the map Y_0 is extended to $V_L\{z_0\}$.

We finally define the *general twisted vertex operators* $Y(v, z)$: For $v \in V_L$, we set

$$Y(v, z) = Y_{\mathbb{Z}+1/2}(v, z) = Y_0(\exp(\Delta_z)v, z), \tag{9.2.27}$$

well defined since $\exp(\Delta_z)v$ is a *polynomial* in z^{-1} with coefficients in V_L. We have a linear map

$$\begin{aligned} V_L &\to (\operatorname{End} V_L^T)[[z^{1/2}, z^{-1/2}]] \\ v &\mapsto Y(v, z). \end{aligned} \tag{9.2.28}$$

From (9.2.6) and (9.2.26) we find that we have achieved what we wanted: For $a, b \in \hat{L}$,

$$\begin{aligned} Y(Y_{\mathbb{Z}}(a, z_0)\iota(b), z) &= {}_\circ^\circ \exp\left(\sum_{n \geq 1} \frac{1}{n!}\left(\frac{d}{dz}\right)^{n-1} \bar{a}(z) z_0^n\right) Y(ab, z) {}_\circ^\circ \\ &\cdot z_0^{\langle \bar{a}, \bar{b}\rangle} \left(\frac{(1 + z_0/z)^{1/2} + 1}{2}\right)^{-2\langle \bar{a}, \bar{b}\rangle} (1 + z_0/z)^{-\langle \bar{a}, \bar{a}\rangle/2} \end{aligned} \tag{9.2.29}$$

[cf. (9.2.2)]. In particular (taking $a = 1$),

$$Y(\iota(a), z) = Y(a, z) \quad \text{for} \quad a \in \hat{L}, \tag{9.2.30}$$

$$Y(\iota(1), z) = Y(1, z) = 1, \tag{9.2.31}$$

and we have:

Corollary 9.2.2: *For a, b as in Theorem 9.1.3,*

$$[Y(a, z_1), Y(b, z_2)] = \tfrac{1}{2} \operatorname{Res}_{z_0} z_2^{-1} Y(Y_{\mathbb{Z}}(a, z_0)\iota(b), z_2) e^{-z_0(\partial/\partial z_1)} \delta(z_1^{1/2}/z_2^{1/2}). \tag{9.2.32}$$

Remark 9.2.3: Formally,

$$Y(Y_{\mathbb{Z}}(a, z_0)\iota(b), z_2) = Y(a, z_2 + z_0) Y(b, z_2) \tag{9.2.33}$$

for all $a, b \in \hat{L}$ (cf. Remark 9.1.2).

We recall from (8.9.20)–(8.9.24) the action of the involution θ [recall (7.4.10), (7.4.11)] on V_L:

$$\begin{aligned} \theta\colon V_L = S(\hat{\mathfrak{h}}_{\mathbb{Z}}^-) \otimes \mathbb{F}\{L\} &\to V_L \\ x \otimes \iota(a) &\mapsto \theta x \otimes \iota(\theta a) \end{aligned} \tag{9.2.34}$$

for $x \in S(\hat{\mathfrak{h}}_{\mathbb{Z}}^-)$ and $a \in \hat{L}$, where θ acts on $S(\hat{\mathfrak{h}}_{\mathbb{Z}}^-)$ as the unique algebra automorphism such that

$$\theta x = -x \quad \text{for} \quad x \in \hat{\mathfrak{h}}_{\mathbb{Z}}^-. \tag{9.2.35}$$

Recall that θ is indeed well defined on $\mathbb{F}\{L\}$ via the formula

$$\theta(\iota(a)) = \iota(\theta a) \quad \text{for} \quad a \in \hat{L} \tag{9.2.36}$$

by the definition (7.1.18) and the fact that $\theta\kappa = \kappa$. With this action, we have the following generalization of the action (9.1.24)–(9.1.25) of θ on the operators $Y(a, z)$:

Proposition 9.2.4: *If the lattice L is even, then for $v \in V_L$,*

$$Y(\theta v, z) = \lim_{z^{1/2} \to -z^{1/2}} Y(v, z). \tag{9.2.37}$$

More generally, for L satisfying (9.2.1) and v as in (9.2.22),

$$Y(\theta v, z) = (-1)^{\langle \bar{a}, \bar{a} \rangle} \lim_{z^{1/2} \to -z^{1/2}} Y(v, z). \tag{9.2.38}$$

Proof: From (9.1.25) and the definition (9.2.23), we see easily that

$$Y_0(\theta v, z) = (-1)^{\langle \bar{a}, \bar{a} \rangle} \lim_{z^{1/2} \to -z^{1/2}} Y_0(v, z)$$

for v as in (9.2.22). Also, applying (9.2.14) to v, we find that

$$\theta \circ \Delta_z = \Delta_z \circ \theta \tag{9.2.39}$$

and hence

$$\theta \circ \exp(\Delta_z) = \exp(\Delta_z) \circ \theta \tag{9.2.40}$$

on V_L. The result now follows from the definition (9.2.27). ∎

We would like to see what general twisted vertex operators look like in some special cases beyond (9.2.30). The expansion (9.2.18) begins as follows:

$$-\log\left(\frac{(1+x)^{1/2} + (1+y)^{1/2}}{2}\right) = -\tfrac{1}{4}(x+y) + \tfrac{3}{32}(x^2+y^2) + \tfrac{1}{16}xy + \cdots, \tag{9.2.41}$$

so that

$$\begin{aligned}\exp(\Delta_z) = 1 &- \tfrac{1}{4} \sum_{i=1}^{l} (h_i(1)h_i'(0) + h_i(0)h_i'(1))z^{-1} \\ &+ \tfrac{3}{32} \sum_{i=1}^{l} (h_i(2)h_i'(0) + h_i(0)h_i'(2))z^{-2} + \tfrac{1}{16} \sum_{i=1}^{l} h_i(1)h_i'(1)z^{-2} \\ &+ \tfrac{1}{32}\left(\sum_{i=1}^{l} (h_i(1)h_i'(0) + h_i(0)h_i'(1))\right)^2 z^{-2} + \cdots.\end{aligned} \tag{9.2.42}$$

Hence we have

$$\exp(\Delta_z)\alpha(-1) = \alpha(-1), \tag{9.2.43}$$

$$Y(\alpha(-1), z) = Y_0(\alpha(-1), z) = \alpha(z) \tag{9.2.44}$$

for $\alpha \in \mathfrak{h}$ [where $\alpha(-1)$ of course denotes $\alpha(-1) \cdot \iota(1)$, as in (8.6.33)]; more generally,

$$\exp(\Delta_z)(\alpha(-1) \cdot \iota(a)) = \alpha(-1) \cdot \iota(a) - \tfrac{1}{2}\langle \alpha, \bar{a} \rangle z^{-1}\iota(a), \tag{9.2.45}$$

$$Y(\alpha(-1) \cdot \iota(a), z) = {}^\circ_\circ(\alpha(z) - \tfrac{1}{2}\langle \alpha, \bar{a} \rangle z^{-1})Y(a, z){}^\circ_\circ \tag{9.2.46}$$

for $\alpha \in \mathfrak{h}$ and $a \in \hat{L}$ [cf. (9.1.22)];

$$\exp(\Delta_z)\alpha(-2) = \alpha(-2), \tag{9.2.47}$$

$$Y(\alpha(-2), z) = \frac{d}{dz}\alpha(z) \tag{9.2.48}$$

for $\alpha \in \mathfrak{h}$; and

$$\exp(\Delta_z)\alpha(-1)\beta(-1) = \alpha(-1)\beta(-1) + \tfrac{1}{8}\langle \alpha, \beta \rangle z^{-2}, \tag{9.2.49}$$

$$Y(\alpha(-1)\beta(-1), z) = {}^\circ_\circ\alpha(z)\beta(z){}^\circ_\circ + \tfrac{1}{8}\langle \alpha, \beta \rangle z^{-2} \tag{9.2.50}$$

for $\alpha, \beta \in \mathfrak{h}$.

Recall the canonical quadratic element ω from (8.7.2). Then by the last two formulas,

$$\exp(\Delta_z)\omega = \omega + \tfrac{1}{16}(\dim \mathfrak{h})z^{-2}, \tag{9.2.51}$$

$$Y(\omega, z) = \tfrac{1}{2} \sum_{i=1}^{l} {}^\circ_\circ h_i'(z)h_i(z){}^\circ_\circ + \tfrac{1}{16}(\dim \mathfrak{h})z^{-2} \tag{9.2.52}$$

[cf. (8.7.4)].

As in Section 8.5, we shall express our results in terms of the component operators $v_n \in \operatorname{End} V_L^T$ defined by the formula

$$Y(v, z) = \sum_{n \in \mathbb{Q}} v_n z^{-n-1} = \sum_{n \in (1/2)\mathbb{Z}} v_{n-1} z^{-n} \tag{9.2.53}$$

for $v \in V_L$. (The context should eliminate any confusion with the operators on the untwisted space V_L denoted v_n in Section 8.5.) From (9.1.23), we have

$$\iota(a)_n = x_a(n - \tfrac{1}{2}\langle \bar{a}, \bar{a} \rangle + 1) \tag{9.2.54}$$

for $a \in \hat{L}$ and $n \in \mathbb{Q}$, as in (8.5.16), and so

$$\deg \iota(a)_n = n - \operatorname{wt} \iota(a) + 1 \tag{9.2.55}$$

by (7.3.18). Note that

$$\iota(1)_n = \delta_{n,-1}. \tag{9.2.56}$$

We claim that

$$\deg v_n = n - \mathrm{wt}\, v + 1 \tag{9.2.57}$$

for every nonzero homogeneous element $v \in V_L$ and $n \in \mathbb{Q}$ [cf. (8.5.22)]. To see this, first observe that if we define operators v_n^1 by

$$Y_0(v, z) = \sum_{n \in \mathbb{Q}} v_n^1 z^{-n-1}, \tag{9.2.58}$$

then

$$\deg v_n^1 = n - \mathrm{wt}\, v + 1 \tag{9.2.59}$$

just as in (8.5.21) and (8.5.22). Now note that if A is a monomial in the operators $h_i(m)h_i'(n)z^{-m-n}$ making up Δ_z [see (9.2.14)] and if we define operators v_n^A by

$$Y_0(Av, z) = \sum_{n \in \mathbb{Q}} v_n^A z^{-n-1}, \tag{9.2.60}$$

then we still have

$$\deg v_n^A = n - \mathrm{wt}\, v + 1. \tag{9.2.61}$$

This proves the claim.

The component operator form of Corollary 9.2.2 (the commutator result) is similar to Corollary 8.5.4:

Corollary 9.2.5: *In the notation of Theorem 9.1.3, let*

$$m, n \in \tfrac{1}{2}\mathbb{Z}.$$

Then as operators on V_L^T,

$$[\iota(a)_m, \iota(b)_n] = \tfrac{1}{2} \sum_{i \in \mathbb{N}} \binom{m}{i} (\iota(a)_i \cdot \iota(b))_{m+n-i} \tag{9.2.62}$$

(finite sum). (On the right-hand side, $\iota(a)_i$ is an operator on V_L.)

Proof: As in (8.5.24) and (8.5.25),

$$Y(Y_{\mathbb{Z}}(a, z_0)\iota(b), z_2) = \sum_{\substack{i \in \mathbb{Z} \\ j \in (1/2)\mathbb{Z}}} (\iota(a)_i \cdot \iota(b))_j z_0^{-i-1} z_2^{-j-1} \tag{9.2.63}$$

and

$$e^{-z_0(\partial/\partial z_1)} \delta(z_1^{1/2}/z_2^{1/2}) = \sum_{\substack{k \in \mathbb{N} \\ l \in (1/2)\mathbb{Z}}} (-1)^k \binom{l}{k} z_0^k z_1^{l-k} z_2^{-l}. \tag{9.2.64}$$

The result follows readily, as in Corollary 8.5.3. ∎

Hence we have:

Corollary 9.2.6: *In the same notation,*

$$[\iota(a)_0, \iota(b)_n] = \tfrac{1}{2}(\iota(a)_0 \cdot \iota(b))_n.$$

Now for every homogeneous element $v \in V_L$, set

$$\begin{aligned} X(v, z) &= X_{\mathbb{Z}+1/2}(v, z) = Y(v, z)z^{\mathrm{wt}\, v} \\ &= \sum_{n \in (1/2)\mathbb{Z}} v_{n+\mathrm{wt}\, v-1} z^{-n} \end{aligned} \tag{9.2.65}$$

and define $x_v(n) \in \mathrm{End}\, V_L^T$ for $n \in \frac{1}{2}\mathbb{Z}$ by

$$\begin{aligned} X(v, z) &= \sum_{n \in (1/2)\mathbb{Z}} x_v(n) z^{-n} \\ Y(v, z) &= \sum_{n \in (1/2)\mathbb{Z}} x_v(n) z^{-n-\mathrm{wt}\, v} \end{aligned} \tag{9.2.66}$$

as in (8.5.27), (8.5.28). Then

$$x_v(n) = v_{n+\mathrm{wt}\, v-1} \tag{9.2.67}$$

and

$$\begin{aligned} \deg x_v(n) &= n \\ [d, X(v, z)] &= -DX(v, z) \end{aligned} \tag{9.2.68}$$

[cf. (7.3.17), (7.3.18)]. We define $X(v, z)$ for arbitrary elements of V_L by linearity:

$$\begin{aligned} V_L &\to (\mathrm{End}\, V_L^T)[[z^{1/2}, z^{-1/2}]] \\ v &\mapsto X(v, z). \end{aligned} \tag{9.2.69}$$

Using the first formula in (9.2.66) to define $x_v(n)$ for arbitrary $v \in V_L$, we see that (9.2.68) is valid for all such v.

We have

$$\begin{aligned} X(\iota(a), z) &= X(a, z) \\ x_{\iota(a)}(n) &= x_a(n) \end{aligned} \tag{9.2.70}$$

for $a \in \hat{L}$ and $n \in \frac{1}{2}\mathbb{Z}$, and

$$\begin{aligned} X(\alpha(-1), z) &= z\alpha(z) \\ x_{\alpha(-1)}(n) &= \alpha(n) \end{aligned} \tag{9.2.71}$$

for $\alpha \in \mathfrak{h}$ and $n \in \frac{1}{2}\mathbb{Z}$ [see (9.1.8), (9.1.9)], as in (8.5.33), (8.5.34).

Proposition 9.2.4 gives the following generalization of (7.4.16) and (7.4.17) [which we first check for v as in (9.2.22)]:

Proposition 9.2.7: *For all $v \in V_L$,*

$$X(\theta v, z) = \lim_{z^{1/2} \to -z^{1/2}} X(v, z). \tag{9.2.72}$$

In particular,

$$x_{\theta v}(n) = \begin{cases} x_v(n) & \text{if } n \in \mathbb{Z} \\ -x_v(n) & \text{if } n \in \mathbb{Z} + \tfrac{1}{2}. \end{cases} \tag{9.2.73}$$

There is an interesting formulation of $X(v, z)$ in terms of binomial coefficients, analogous to (8.5.38). For every homogeneous element $v \in V_L$, we define

$$X_0(v, z) = Y_0(v, z) z^{\operatorname{wt} v}, \tag{9.2.74}$$

and we extend to all $v \in V_L$ by linearity. Then for v homogeneous,

$$X_0(\exp(\Delta_z|_{z=1})v, z) = Y_0(\exp(\Delta_z)v, z) z^{\operatorname{wt} v}, \tag{9.2.75}$$

since for every monomial A in the operators $h_i(m)h_i'(n)z^{-m-n}$ making up Δ_z,

$$X_0(A|_{z=1}\, v, z) = Y_0(Av, z) z^{\operatorname{wt} v}. \tag{9.2.76}$$

Thus from the definitions,

$$X(v, z) = X_0(\exp(\Delta_z|_{z=1})v, z) \tag{9.2.77}$$

for v homogeneous and hence arbitrary in V_L. But we also note that for v as in (9.2.22),

$$\begin{aligned} X_0(v, z) &= {}^\circ_\circ \left(\frac{z^{n_1}}{(n_1 - 1)!} \left(\frac{d}{dz} \right)^{n_1 - 1} \alpha_1(z) \right) \cdots \left(\frac{z^{n_k}}{(n_k - 1)!} \left(\frac{d}{dz} \right)^{n_k - 1} \alpha_k(z) \right) X(a, z) {}^\circ_\circ \\ &= {}^\circ_\circ \left(\binom{D-1}{n_1 - 1} (z\alpha_1(z)) \right) \cdots \left(\binom{D-1}{n_k - 1} (z\alpha_k(z)) \right) X(a, z) {}^\circ_\circ \end{aligned} \tag{9.2.78}$$

[recall (8.5.35)–(8.5.38)], and so for arbitrary v, $X(v, z)$ can be expressed via (9.2.77) in terms of these binomial coefficients.

For example, as in (9.2.45)–(9.2.50) we see that

$$X(\alpha(-1) \cdot \iota(a), z) = {}^\circ_\circ (z\alpha(z) - \tfrac{1}{2}\langle \alpha, \bar{a} \rangle) X(a, z) {}^\circ_\circ \tag{9.2.79}$$

for $\alpha \in \mathfrak{h}$ and $a \in \hat{L}$,

$$X(\alpha(-2), z) = (D - 1)(z\alpha(z)) \tag{9.2.80}$$

for $\alpha \in \mathfrak{h}$, and

$$X(\alpha(-1)\beta(-1), z) = {}^{\circ}_{\circ}(z\alpha(z))(z\beta(z)){}^{\circ}_{\circ} + \tfrac{1}{8}\langle \alpha, \beta \rangle \tag{9.2.81}$$

for $\alpha, \beta \in \mathfrak{h}$. This last formula should be compared with (8.5.39); note the added constant. As in (8.5.40) and (8.5.41), it is convenient to define

$$\alpha\beta(n) = x_{\alpha(-1)\beta(-1)}(n) \quad \text{for} \quad n \in \tfrac{1}{2}\mathbb{Z} \text{ (actually, } n \in \mathbb{Z}\text{)}, \tag{9.2.82}$$

where the product $\alpha\beta$ takes place in $S^2(\mathfrak{h})$ [cf. (9.2.71)]; then

$${}^{\circ}_{\circ}\alpha(z)\beta(z){}^{\circ}_{\circ} + \tfrac{1}{8}\langle \alpha, \beta \rangle z^{-2} = Y(\alpha(-1)\beta(-1), z) = \sum_{n \in \mathbb{Z}} \alpha\beta(n) z^{-n-2}. \tag{9.2.83}$$

Recalling (9.2.51), we see that for the canonical quadratic element ω,

$$X(\omega, z) = \tfrac{1}{2} \sum_{i=1}^{l} {}^{\circ}_{\circ} z h_i'(z) z h_i(z) {}^{\circ}_{\circ} + \tfrac{1}{16} \dim \mathfrak{h}. \tag{9.2.84}$$

Just as in Corollary 8.5.6, we find from Corollary 9.2.5:

Corollary 9.2.8: *In the notation of Theorem 9.1.3, let*

$$m, n \in \tfrac{1}{2}\mathbb{Z}.$$

Then as operators on V_L^T,

$$[x_a(m), x_b(n)] = \frac{1}{2} \sum_{i \in \mathbb{Z}_+ - \langle \bar{a}, \bar{a} \rangle/2} \binom{m - 1 + \langle \bar{a}, \bar{a} \rangle/2}{i - 1 + \langle \bar{a}, \bar{a} \rangle/2} x_{x_a(i) \cdot \iota(b)}(m + n) \tag{9.2.85}$$

(finite sum). (On the right-hand side, $x_a(i)$ *is an operator on* V_L*.)*

With the aid of (8.5.44), (9.2.70) and (9.2.71), we see that this result gives Theorem 7.4.1 for an integral lattice, as expected:

$$[x_a(m), x_b(n)] = \begin{cases} 0 & \text{if} \quad \langle \bar{a}, \bar{b} \rangle = 0 \\ \tfrac{1}{2} x_{ab}(m + n) & \text{if} \quad \langle \bar{a}, \bar{b} \rangle = -1 \\ \tfrac{1}{2}\bar{a}(m + n) + \tfrac{1}{2} m \delta_{m+n,0} & \text{if} \quad \langle \bar{a}, \bar{a} \rangle = 2,\ b = a^{-1}. \end{cases} \tag{9.2.86}$$

We shall comment on the action of the involution θ on various structures. We have already defined an action of θ on the untwisted space V_L in (8.9.22) and (9.2.34), and we have discussed some properties in (8.9.26)–(8.9.29).

Now we define an action of θ on the twisted space V_L^T as follows:

$$\begin{aligned} \theta: V_L^T = S(\hat{\mathfrak{h}}^-_{\mathbb{Z}+1/2}) \otimes T &\to V_L^T \\ x \otimes \tau &\mapsto \theta x \otimes (-\tau) = -\theta x \otimes \tau \end{aligned} \tag{9.2.87}$$

for $x \in S(\hat{\mathfrak{h}}^-_{\mathbb{Z}+1/2})$ and $\tau \in T$, where θ acts on $S(\hat{\mathfrak{h}}^-_{\mathbb{Z}+1/2})$ as the unique algebra automorphism such that

$$\theta x = -x \quad \text{for} \quad x \in \hat{\mathfrak{h}}^-_{\mathbb{Z}+1/2}. \tag{9.2.88}$$

Then it is easy to see from the definitions (9.1.15), (9.2.23) and (9.2.34) [keeping in mind (7.4.14)] that

$$\theta\alpha(n)\theta^{-1} = -\alpha(n) \quad \text{for} \quad \alpha \in \mathfrak{h}, \quad n \in \mathbb{Z} + \tfrac{1}{2}, \tag{9.2.89}$$

$$\theta Y(a, z)\theta^{-1} = Y(\theta a, z) \quad \text{for} \quad a \in \hat{L}, \tag{9.2.90}$$

$$\theta Y_0(v, z)\theta^{-1} = Y_0(\theta v, z) \quad \text{for} \quad v \in V_L, \tag{9.2.91}$$

and it follows from (9.2.40) and (9.2.91) that

$$\theta Y(v, z)\theta^{-1} = Y(\theta v, z) \quad \text{for} \quad v \in V_L. \tag{9.2.92}$$

In particular, in the notation of (8.9.25),

$$\text{if} \quad v \in V_L^\theta \quad \text{then} \quad \theta \quad \text{commutes with} \quad Y(v, z) \tag{9.2.93}$$

and in this case, the component operators of $Y(v, z)$ preserve the ± 1-eigenspaces $(V_L^T)^{\pm\theta}$ of θ in V_L^T:

$$\begin{aligned} (V_L^T)^\theta &= \{v \in V_L^T \mid \theta v = v\} \\ (V_L^T)^{-\theta} &= \{v \in V_L^T \mid \theta v = -v\}; \end{aligned} \tag{9.2.94}$$

$$V_L^T = (V_L^T)^\theta \oplus (V_L^T)^{-\theta}. \tag{9.2.95}$$

From (9.2.87) and (9.2.88) it is clear that these two subspaces of V_L^T are distinguished by their gradings:

$$\begin{aligned} (V_L^T)^\theta &= \coprod_{n \in \mathbb{Z}+1/2} (V_L^T)_{n-(\dim \mathfrak{h})/48} \\ (V_L^T)^{-\theta} &= \coprod_{n \in \mathbb{Z}} (V_L^T)_{n-(\dim \mathfrak{h})/48} \end{aligned} \tag{9.2.96}$$

[recall (1.9.53) and (7.3.9)].

Remark 9.2.9: Continuing with the generalization discussed in Remark 9.1.5, we see that the only changes needed in the present section are the following: If the element $a \in \hat{L}$ entering into the vector v in (9.2.22) satisfies

(9.1.46), then (9.2.37), (9.2.38), (9.2.72), (9.2.73) and (9.2.90)–(9.2.92) all acquire sign changes, and (9.2.93) is replaced by:

$$\text{if} \quad v \in V_L^{-\theta} \quad \text{then} \quad \theta \quad \text{commutes with} \quad Y(v, z). \qquad (9.2.97)$$

9.3. Commutators of General Twisted Vertex Operators

The stage is now set for computing commutators in general. Following the method of Section 8.6, we shall suitably extend Corollaries 9.2.2, 9.2.5 and 9.2.8 to arbitrary elements of V_L in place of $\iota(a)$ and $\iota(b)$, thus exhibiting the span of the operators v_n or $x_v(n)$ as a Lie algebra with precisely known structure. Most important is the remarkable similarity between the Lie algebra structures in the untwisted and twisted settings. Continuing to assume (9.2.1), we start with a result which, as one easily sees, generalizes Corollary 9.2.2:

Theorem 9.3.1: *Let $a, b \in \hat{L}$ and suppose that*

$$\langle \bar{a}, \bar{b} \rangle \in \mathbb{Z} \quad \text{and} \quad c(\bar{a}, \bar{b}) = (-1)^{\langle \bar{a}, \bar{b} \rangle}. \qquad (9.3.1)$$

Let $u', v' \in S(\hat{\mathfrak{h}}_{\bar{\mathbb{Z}}})$ and set

$$\begin{aligned} u &= u' \otimes \iota(a) = u_0 \cdot \iota(a) \in V_L \\ v &= v' \otimes \iota(b) = v_0 \cdot \iota(b) \in V_L. \end{aligned} \qquad (9.3.2)$$

Then

$$[Y(u, z_1), Y(v, z_2)] = \tfrac{1}{2} \sum_{p=0,1} (-1)^{p\langle \bar{a}, \bar{a} \rangle}$$

$$\cdot \operatorname{Res}_{z_0} z_2^{-1} Y(Y_{\mathbb{Z}}(\theta^p u, z_0)v, z_2) e^{-z_0(\partial/\partial z_1)} \delta((-1)^p z_1^{1/2}/z_2^{1/2}). \qquad (9.3.3)$$

If L is even and (9.3.1) holds for all $a, b \in \hat{L}$, then

$$[Y(u, z_1), Y(v, z_2)]$$

$$= \tfrac{1}{2} \sum_{p=0,1} \operatorname{Res}_{z_0} z_2^{-1} Y(Y_{\mathbb{Z}}(\theta^p u, z_0)v, z_2) e^{-z_0(\partial/\partial z_1)} \delta((-1)^p z_1^{1/2}/z_2^{1/2}) \qquad (9.3.4)$$

for all $u, v \in V_L$.

Proof: Let $k, l \geq 1$ and let $a_1, \ldots, a_k, b_1, \ldots, b_l \in \hat{L}$ be as in (8.6.4). Define A and B as in (8.6.5). Then as we saw in the proof of Theorem 8.6.1, the coefficients in the formal power series A and B span $S(\hat{\mathfrak{h}}_{\bar{\mathbb{Z}}}) \otimes \iota(a)$ and

$S(\hat{\mathfrak{h}}_{\mathbb{Z}}^{-}) \otimes \iota(b)$, respectively, and so it suffices to prove the Theorem with u and v replaced by A and B, respectively. Recall the expressions for A and B given in (8.6.6)–(8.6.8).

Now we want the analogue of (8.6.9). Using (9.2.23) and (8.3.3), we have

$$
\begin{aligned}
Y_0(A, z) &= {\circ\atop\circ}\exp\left(\sum_{i=1}^{k} \sum_{n\geq 1} \frac{1}{n!}\left(\frac{d}{dz}\right)^{n-1} \bar{a}_i(z) w_i^n\right) Y(a, z){\circ\atop\circ} \\
&= \exp\left(\sum_{i=1}^{k} \sum_{n\geq 1} \frac{1}{n!}\left(\frac{d}{dz}\right)^{n-1} \bar{a}_i(z)^- w_i^n\right) Y(a, z) \\
&\quad \cdot \exp\left(\sum_{i=1}^{k} \sum_{n\geq 1} \frac{1}{n!}\left(\frac{d}{dz}\right)^{n-1} \bar{a}_i(z)^+ w_i^n\right) \\
&= \exp\left(\sum_{i=1}^{k} e^{w_i(d/dz)} \int \bar{a}_i(z)^- - \sum_{i=1}^{k} \int \bar{a}_i(z)^-\right) Y(a, z) \\
&\quad \cdot \exp\left(\sum_{i=1}^{k} e^{w_i(d/dz)} \int \bar{a}_i(z)^+ - \sum_{i=1}^{k} \int \bar{a}_i(z)^+\right) \\
&= \exp\left(\sum_{i=1}^{k} \int \bar{a}_i(z + w_i)^- d(z + w_i)\right) \exp\left(-\int \bar{a}(z)^-\right) Y(a, z) \\
&\quad \cdot \exp\left(-\int \bar{a}(z)^+\right) \exp\left(\sum_{i=1}^{k} \int \bar{a}_i(z + w_i)^+ d(z + w_i)\right) \\
&= {\circ\atop\circ}\prod_{i=1}^{k} e^{\int \bar{a}_i(z+w_i)d(z+w_i)}{\circ\atop\circ} 2^{-\langle \bar{a}, \bar{a}\rangle} a z^{-\langle \bar{a}, \bar{a}\rangle/2} \\
&= {\circ\atop\circ}Y(a_1, z + w_1) \cdots Y(a_k, z + w_k){\circ\atop\circ}\left(\prod_{1\leq i<j\leq k} 2^{-2\langle \bar{a}_i, \bar{a}_j\rangle} z^{-\langle \bar{a}_i, \bar{a}_j\rangle}\right) \\
&\quad \cdot \prod_{i=1}^{k} (1 + w_i/z)^{\langle \bar{a}_i, \bar{a}_i\rangle/2}.
\end{aligned} \tag{9.3.5}
$$

Just as in (9.1.38), this expression has extra factors, compared with its untwisted analogue (8.6.9). But what we really want is $Y(A, z)$, not $Y_0(A, z)$. Recalling (9.2.27), we see that we need to determine $\exp(\Delta_z)A$, and this computation will be similar to (9.2.17). From (8.6.6),

$$A = Y_{\mathbb{Z}}(a_1, w_1) \cdots Y_{\mathbb{Z}}(a_k, w_k)\iota(1) \prod_{1\leq i<j\leq k} (w_i - w_j)^{-\langle \bar{a}_i, \bar{a}_j\rangle}. \tag{9.3.6}$$

As in (9.2.8)–(9.2.13), we find that for $\alpha \in \mathfrak{h}$ and $m, n \geq 0$,

$$\alpha(n) \cdot A = \sum_{i=1}^{k} \langle \alpha, \bar{a}_i \rangle w_i^n A \tag{9.3.7}$$

$$\sum_{i=1}^{l} h_i(m) h_i'(n) \cdot A = \sum_{i,j=1}^{k} \langle \bar{a}_i, \bar{a}_j \rangle w_i^m w_j^n A \tag{9.3.8}$$

[here l is as in (9.2.10)–(9.2.14)]. Thus with Δ_z as in (9.2.14),

$$\Delta_z(A) = \sum_{i,j=1}^{k} \langle \bar{a}_i, \bar{a}_j \rangle \sum_{m,n \geq 0} c_{mn} (w_i/z)^m (w_j/z)^n A, \tag{9.3.9}$$

and if $c_{00} = 0$,

$$\exp(\Delta_z)(A) = \prod_{i,j=1}^{k} \left(\exp \sum_{m,n \geq 0} c_{mn} (w_i/z)^m (w_j/z)^n \right)^{\langle \bar{a}_i, \bar{a}_j \rangle} A \tag{9.3.10}$$

[cf. (9.2.15)–(9.2.17)].

Now take the official choice of constants c_{mn} given by (9.2.18). Then

$$\begin{aligned} \exp(\Delta_z)(A) &= \prod_{i,j=1}^{k} \left(\frac{(1 + w_i/z)^{1/2} + (1 + w_j/z)^{1/2}}{2} \right)^{-\langle \bar{a}_i, \bar{a}_j \rangle} A \\ &= \prod_{1 \leq i < j \leq k} \left(\frac{(1 + w_i/z)^{1/2} + (1 + w_j/z)^{1/2}}{2} \right)^{-2\langle \bar{a}_i, \bar{a}_j \rangle} \\ &\quad \cdot \prod_{i=1}^{k} (1 + w_i/z)^{-\langle \bar{a}_i, \bar{a}_i \rangle / 2} A, \end{aligned} \tag{9.3.11}$$

and from (9.2.27) and (9.3.5) we find that

$$\begin{aligned} Y(A, z) &= {\circ\atop\circ} Y(a_1, z + w_1) \cdots Y(a_k, z + w_k) {\circ\atop\circ} \\ &\quad \cdot \prod_{1 \leq i < j \leq k} ((z + w_i)^{1/2} + (z + w_j)^{1/2})^{-2\langle \bar{a}_i, \bar{a}_j \rangle}. \end{aligned} \tag{9.3.12}$$

Note that this last factor is to be understood as a formal power series in the w's, which is where it came from:

$$\begin{aligned} &\prod_{1 \leq i < j \leq k} ((z + w_i)^{1/2} + (z + w_j)^{1/2})^{-2\langle \bar{a}_i, \bar{a}_j \rangle} \\ &= \prod_{1 \leq i < j \leq k} 2^{-2\langle \bar{a}_i, \bar{a}_j \rangle} z^{-\langle \bar{a}_i, \bar{a}_j \rangle} \left(\frac{(1 + w_i/z)^{1/2} + (1 + w_j/z)^{1/2}}{2} \right)^{-2\langle \bar{a}_i, \bar{a}_j \rangle}. \end{aligned} \tag{9.3.13}$$

Also note that the coefficient of each monomial in the w's is a Laurent monomial in z.

We now have the analogue of Remark 8.6.2:

Remark 9.3.2: A formal application of (9.1.21) to (9.3.12) [see also (9.3.6)] yields that for all $a_1, \ldots, a_k \in \hat{L}$,

$$Y(Y_{\mathbb{Z}}(a_1, w_1) \cdots Y_{\mathbb{Z}}(a_k, w_k)\iota(1), z) = Y(a_1, z + w_1) \cdots Y(a_k, z + w_k). \tag{9.3.14}$$

This formally generalizes Remark 9.2.3. It also justifies the precise choice of constants c_{mn} in (9.2.18); recall Remark 9.2.1.

Of course, all these considerations for A also apply to B. For instance,

$$Y(B, z) = {}^{\circ}_{\circ}Y(b_1, z + x_1) \cdots Y(b_l, z + x_l){}^{\circ}_{\circ} \cdot \prod_{1 \le i < j \le l} ((z + x_i)^{1/2} + (z + x_j)^{1/2})^{-2\langle \bar{b}_i, \bar{b}_j \rangle}. \tag{9.3.15}$$

Now

$${}^{\circ}_{\circ}Y(A, z_1)Y(B, z_2){}^{\circ}_{\circ} = (-1)^{\langle \bar{a}, \bar{b} \rangle} {}^{\circ}_{\circ}Y(B, z_2)Y(A, z_1){}^{\circ}_{\circ} \tag{9.3.16}$$

from (9.3.1), (9.3.12) and (9.3.15). Also,

$$\begin{aligned} &{}^{\circ}_{\circ}Y(A, z_1)Y(B, z_2){}^{\circ}_{\circ} \\ &\quad = {}^{\circ}_{\circ}Y(a_1, z_1 + w_1) \cdots Y(a_k, z_1 + w_k)Y(b_1, z_2 + x_1) \cdots Y(b_l, z_2 + x_l){}^{\circ}_{\circ} \\ &\quad \cdot \prod_{1 \le i < j \le k} ((z_1 + w_i)^{1/2} + (z_1 + w_j)^{1/2})^{-2\langle \bar{a}_i, \bar{a}_j \rangle} \\ &\quad \cdot \prod_{1 \le i < j \le l} ((z_2 + x_i)^{1/2} + (z_2 + x_j)^{1/2})^{-2\langle \bar{b}_i, \bar{b}_j \rangle} \\ &\quad \in (\operatorname{End} V_L^T)[[z_1^{1/2}, z_1^{-1/2}, z_2^{1/2}, z_2^{-1/2}]][[w_1, \ldots, x_l]] \end{aligned} \tag{9.3.17}$$

and

$$\lim_{z_1 \to z_2} {}^{\circ}_{\circ}Y(A, z_1)Y(B, z_2){}^{\circ}_{\circ} \quad \text{exists,} \tag{9.3.18}$$

that is, if $X(z_1, z_2)$ denotes the coefficient of any fixed monomial in $w_1, \ldots, x_l$ in ${}^{\circ}_{\circ}Y(A, z_1)Y(B, z_2){}^{\circ}_{\circ}$, then

$$\lim_{z_1 \to z_2} X(z_1, z_2) \quad \text{exists.} \tag{9.3.19}$$

Moreover, from (9.1.21) we see that

$$Y(A, z_1)Y(B, z_2) = {}^{\circ}_{\circ}Y(A, z_1)Y(B, z_2){}^{\circ}_{\circ} \prod_{\substack{1 \le i \le k \\ 1 \le j \le l}} (z_1 - z_2 + w_i - x_j)^{\langle \bar{a}_i, \bar{b}_j \rangle}$$

$$\cdot \prod_{\substack{1 \le i \le k \\ 1 \le j \le l}} ((z_1 + w_i)^{1/2} + (z_2 + x_j)^{1/2})^{-2\langle \bar{a}_i, \bar{b}_j \rangle}, \quad (9.3.20)$$

where the two products over i and j are of course to be expanded in nonnegative integral powers of $(z_2 + x_j)^{1/2}$, w_i and x_j.

Now consider the factor

$$g(z_1, z_2) = ((z_1 + w_i)^{1/2} + (z_2 + x_j)^{1/2})^{-2\langle \bar{a}_i, \bar{b}_j \rangle}$$

$$= \sum_{m, n \in \mathbb{N}} g_{mn}(z_1, z_2) w_i^m x_j^n \quad (9.3.21)$$

of (9.3.20). If the exponent $-2\langle \bar{a}_i, \bar{b}_j \rangle$ $(\in \mathbb{Z})$ is negative, then

$$\lim_{z_1 \to z_2} g(z_1, z_2) \quad \text{does not exist,} \quad (9.3.22)$$

since for instance

$$\lim_{z_1 \to z_2} g_{00}(z_1, z_2) \quad \text{does not exist} \quad (9.3.23)$$

[cf. (9.3.13)!]. Thus we multiply and divide (9.3.20) by a suitable factor: Choose $M \in \mathbb{N}$ such that

$$M \ge 2\langle \bar{a}_i, \bar{b}_j \rangle \quad \text{for all} \quad i, j. \quad (9.3.24)$$

Then

$$Y(A, z_1)Y(B, z_2) = {}^{\circ}_{\circ}Y(A, z_1)Y(B, z_2){}^{\circ}_{\circ}$$

$$\cdot \prod_{\substack{1 \le i \le k \\ 1 \le j \le l}} (z_1 - z_2 + w_i - x_j)^{\langle \bar{a}_i, \bar{b}_j \rangle - M} G_M, \quad (9.3.25)$$

where

$$G_M = \prod_{\substack{1 \le i \le k \\ 1 \le j \le l}} (((z_1 + w_i)^{1/2} + (z_2 + x_j)^{1/2})^{-2\langle \bar{a}_i, \bar{b}_j \rangle + M}$$

$$\cdot ((z_1 + w_i)^{1/2} - (z_2 + x_j)^{1/2})^M), \quad (9.3.26)$$

and

$$G_M \in \mathbb{F}[z_1^{1/2}, z_1^{-1/2}, z_2^{1/2}, z_2^{-1/2}][[w_1, \ldots, x_l]], \quad (9.3.27)$$

so that

$$\lim_{z_1 \to z_2} G_M \quad \text{exists.} \quad (9.3.28)$$

If the roles of A and B and of z_1 and z_2 are switched in G_M, the result is

$$(-1)^{klM}G_M. \tag{9.3.29}$$

Fix a monomial P in $w_1, \ldots, x_l$ as in (8.6.13), and choose $N \in \mathbb{N}$ so large that the coefficient of P and of each monomial of lower total degree than P in

$$F_N = (z_1 - z_2)^N \prod_{\substack{1 \le i \le k \\ 1 \le j \le l}} (z_1 - z_2 + w_i - x_j)^{\langle \bar{a}_i, \bar{b}_j \rangle - M} \tag{9.3.30}$$

is a polynomial in $z_1 - z_2$ [cf. (8.6.14)]. Denote by $Y_P(z_1, z_2)$ the coefficient of P in

$$Y(A, z_1)Y(B, z_2)(z_1 - z_2)^N = {}^{\circ}_{\circ}Y(A, z_1)Y(B, z_2){}^{\circ}_{\circ}F_N G_M. \tag{9.3.31}$$

Then

$$Y_P(z_1, z_2) \in (\operatorname{End} V_L^T)[[z_1^{1/2}, z_1^{-1/2}, z_2^{1/2}, z_2^{-1/2}]] \tag{9.3.32}$$

and

$$\lim_{z_1 \to z_2} Y_P(z_1, z_2) \quad \text{exists.} \tag{9.3.33}$$

The coefficient of P in $Y(A, z_1)Y(B, z_2)$ is

$$Y_P(z_1, z_2)(z_1 - z_2)^{-N}, \tag{9.3.34}$$

where $(z_1 - z_2)^{-N}$ is to be expanded in nonnegative integral powers of z_2.

Similarly, reversing the roles of A and B and of z_1 and z_2, and using (9.3.16) and (9.3.29), we find that

$$Y(B, z_2)Y(A, z_1) = {}^{\circ}_{\circ}Y(A, z_1)Y(B, z_2){}^{\circ}_{\circ}(-1)^{\langle \bar{a}, \bar{b} \rangle + klM} \cdot \prod_{\substack{1 \le i \le k \\ 1 \le j \le l}} (z_2 - z_1 + x_j - w_i)^{\langle \bar{a}_i, \bar{b}_j \rangle - M} G_M,$$

where $\prod (z_2 - z_1 + x_j - w_i)^{\langle \bar{a}_i, \bar{b}_j \rangle - M}$ is now to be expanded in nonnegative integral powers of z_1, w_i and x_j [cf. (9.3.20)]. Moreover, the coefficient of P and of each monomial of lower total degree than P in

$$(z_1 - z_2)^N (-1)^{\langle \bar{a}, \bar{b} \rangle + klM} \prod (z_2 - z_1 + x_j - w_i)^{\langle \bar{a}_i, \bar{b}_j \rangle - M}$$

is a polynomial in $z_1 - z_2$ which agrees with the same for F_N [see (9.3.30)]. Hence the coefficient of P in $Y(B, z_2)Y(A, z_1)$ is

$$Y_P(z_1, z_2)(z_1 - z_2)^{-N}, \tag{9.3.35}$$

where this time the last factor is to be expanded in nonnegative integral powers of z_1 [cf. (9.3.34)].

Exactly as in (8.6.20)–(8.6.21), and using (9.1.30)–(9.1.31), Proposition 8.2.2 and (8.3.11), we see that the coefficient of P in $[Y(A, z_1), Y(B, z_2)]$ is the coefficient of z_0^{N-1} in

$$\begin{aligned} z_2^{-1}Y_P(z_1, z_2)e^{-z_0(\partial/\partial z_1)}\delta(z_1/z_2) &= \tfrac{1}{2}\sum_{p=0,1} z_2^{-1}Y_P(z_1, z_2)e^{-z_0(\partial/\partial z_1)}\delta((-1)^p z_1^{1/2}/z_2^{1/2}) \\ &= \tfrac{1}{2}\sum_{p=0,1} z_2^{-1}\left(\lim_{z_1^{1/2}\to(-1)^p z_2^{1/2}} Y_P(z_1 + z_0, z_2)\right)e^{-z_0(\partial/\partial z_1)}\delta((-1)^p z_1^{1/2}/z_2^{1/2}). \end{aligned} \tag{9.3.36}$$

We now examine the limit for $p = 0$:

$$\lim_{z_1\to z_2} Y_P(z_1 + z_0, z_2) = Y_P(z_2 + z_0, z_2), \tag{9.3.37}$$

which from (9.3.18), (9.3.26), (9.3.28), (9.3.30) and (9.3.31) is the coefficient of P in

$$\begin{aligned} &{}^{\circ}_{\circ}Y(A, z_2 + z_0)Y(B, z_2){}^{\circ}_{\circ}z_0^N \prod_{\substack{1\le i\le k\\ 1\le j\le l}} (z_0 + w_i - x_j)^{\langle \bar{a}_i, \bar{b}_j\rangle - M} \\ &\cdot \prod_{\substack{1\le i\le k\\ 1\le j\le l}} (((z_2 + z_0 + w_i)^{1/2} + (z_2 + x_j)^{1/2})^{-2\langle \bar{a}_i, \bar{b}_j\rangle + M} \\ &\cdot ((z_2 + z_0 + w_i)^{1/2} - (z_2 + x_j)^{1/2})^M). \end{aligned} \tag{9.3.38}$$

Here it is important to note that even though $\lim F_N$ does not exist, we are justified in replacing z_1 by $z_2 + z_0$ in F_N because we are looking only at the coefficient of P in (9.3.31) [cf. (8.6.23)]. Also, the factors

$$(z_2 + z_0 + w_i)^{1/2} + (z_2 + x_j)^{1/2} \quad \text{and} \quad (z_2 + z_0 + w_i)^{1/2} - (z_2 + x_j)^{1/2} \tag{9.3.39}$$

are of course to be expanded in nonnegative integral powers of z_0, w_i and x_j, and they may be raised to negative powers. Thus (9.3.38) simplifies to

$$\begin{aligned} &{}^{\circ}_{\circ}Y(A, z_2 + z_0)Y(B, z_2){}^{\circ}_{\circ}z_0^N \prod_{\substack{1\le i\le k\\ 1\le j\le l}} (z_0 + w_i - x_j)^{\langle \bar{a}_i, \bar{b}_j\rangle} \\ &\cdot \prod_{\substack{1\le i\le k\\ 1\le j\le l}} ((z_2 + z_0 + w_i)^{1/2} + (z_2 + x_j)^{1/2})^{-2\langle \bar{a}_i, \bar{b}_j\rangle}. \end{aligned} \tag{9.3.40}$$

It follows that the contribution to $[Y(A, z_1), Y(B, z_2)]$ for $p = 0$ in (9.3.36) is

$$\frac{1}{2} \operatorname{Res}_{z_0} z_2^{-1} {\circ \atop \circ} Y(A, z_2 + z_0)Y(B, z_2) {\circ \atop \circ} \prod_{\substack{1 \le i \le k \\ 1 \le j \le l}} (z_0 + w_i - x_j)^{\langle \bar{a}_i, \bar{b}_j \rangle}$$

$$\cdot \prod_{\substack{1 \le i \le k \\ 1 \le j \le l}} ((z_2 + z_0 + w_i)^{1/2} + (z_2 + x_j)^{1/2})^{-2\langle \bar{a}_i, \bar{b}_j \rangle}$$

$$\cdot e^{-z_0(\partial/\partial z_1)} \delta(z_1^{1/2}/z_2^{1/2}). \tag{9.3.41}$$

Next we determine the limit for $p = 1$ in (9.3.36). First we observe that in the notation of (9.3.31),

$$(-1)^{\langle \bar{a}, \bar{a} \rangle} \lim_{z_1^{1/2} \to -z_1^{1/2}} Y(A, z_1)Y(B, z_2)(z_1 - z_2)^N$$

$$= Y(\theta A, z_1)Y(B, z_2)(z_1 - z_2)^N \tag{9.3.42}$$

by Proposition 9.2.4 and the fact that $(z_1 - z_2)^N$ involves only integral powers of z_1. To obtain θA from A, one replaces each a_i by θa_i and each $\bar{a}_i$ by $-\bar{a}_i$ [recall (8.6.6)]. Now *without affecting the argument so far, we increase M if necessary so that*

$$M \ge -2\langle \bar{a}_i, \bar{b}_j \rangle \quad \text{for all} \quad i, j \tag{9.3.43}$$

[cf. (9.3.24)] *and then we increase N if necessary so that if we define*

$$F_N^\theta = (z_1 - z_2)^N \prod_{\substack{1 \le i \le k \\ 1 \le j \le l}} (z_1 - z_2 + w_i - x_j)^{-\langle \bar{a}_i, \bar{b}_j \rangle - M}, \tag{9.3.44}$$

then the coefficient of P and of each monomial of lower total degree than P in F_N^θ is a polynomial in $z_1 - z_2$ [cf. (9.3.30)]. Then as in (9.3.31), the expression (9.3.42) equals

$${\circ \atop \circ} Y(\theta A, z_1)Y(B, z_2) {\circ \atop \circ} F_N^\theta G_M^\theta, \tag{9.3.45}$$

where G_M^θ denotes the expression (9.3.26) with $\bar{a}_i$ replaced by $-\bar{a}_i$. Moreover, denoting by $Y_P^\theta(z_1, z_2)$ the coefficient of P in (9.3.42), we see that as in (9.3.33),

$$\lim_{z_1 \to z_2} Y_P^\theta(z_1, z_2) \quad \text{exists}, \tag{9.3.46}$$

and that this limit may be computed using (9.3.45) by the methods above.

Returning to the limit for $p = 1$ in (9.3.36), we now have

$$\lim_{z_1^{1/2} \to -z_1^{1/2}} Y_P(z_1, z_2) = (-1)^{\langle \bar{a}, \bar{a} \rangle} Y_P^\theta(z_1, z_2) \tag{9.3.47}$$

by (9.3.42), and hence

$$\begin{aligned}\lim_{z_1^{1/2}\to -z_1^{1/2}} Y_P(z_1+z_0,z_2) &= \lim_{z_1^{1/2}\to -z_1^{1/2}} Y_P(z_1(1+z_0/z_1),z_2)\\ &= (-1)^{\langle\bar{a},\bar{a}\rangle} Y_P^{\theta}(z_1(1+z_0/z_1),z_2)\\ &= (-1)^{\langle\bar{a},\bar{a}\rangle} Y_P^{\theta}(z_1+z_0,z_2) \qquad (9.3.48)\end{aligned}$$

since $1+z_0/z_1$ and its powers are unaffected by the change of sign of $z_1^{1/2}$. Thus

$$\begin{aligned}\lim_{z_1^{1/2}\to -z_2^{1/2}} Y_P(z_1+z_0,z_2) &= (-1)^{\langle\bar{a},\bar{a}\rangle} \lim_{z_1\to z_2} Y_P^{\theta}(z_1+z_0,z_2)\\ &= (-1)^{\langle\bar{a},\bar{a}\rangle} Y_P^{\theta}(z_2+z_0,z_2). \qquad (9.3.49)\end{aligned}$$

Then (9.3.38)–(9.3.41) apply with A replaced by θA, and so the contribution to $[Y(A,z_1),Y(B,z_2)]$ for $p=1$ is

$$\begin{aligned}&\tfrac{1}{2}(-1)^{\langle\bar{a},\bar{a}\rangle}\operatorname{Res}_{z_0} z_2^{-1}\,{}_{\circ}^{\circ}Y(\theta A,z_2+z_0)Y(B,z_2){}_{\circ}^{\circ}\prod_{\substack{1\le i\le k\\ 1\le j\le l}}(z_0+w_i-x_j)^{-\langle\bar{a}_i,\bar{b}_j\rangle}\\ &\quad\cdot\prod_{\substack{1\le i\le k\\ 1\le j\le l}}((z_2+z_0+w_i)^{1/2}+(z_2+x_j)^{1/2})^{2\langle\bar{a}_i,\bar{b}_j\rangle}\\ &\quad\cdot e^{-z_0(\partial/\partial z_1)}\delta(-z_1^{1/2}/z_2^{1/2}) \qquad (9.3.50)\end{aligned}$$

[note the sign changes from (9.3.41)].

All that is left is to express the commutator in terms of twisted vertex operators. First, $Y_{\mathbb{Z}}(A,z_0)B$ is given by (8.6.25). Thus by (9.3.12) with A replaced by $Y_{\mathbb{Z}}(A,z_0)B$,

$$\begin{aligned}Y(Y_{\mathbb{Z}}(A,z_0)B,z_2) &= {}_{\circ}^{\circ}Y(a_1,z_2+z_0+w_1)\cdots Y(a_k,z_2+z_0+w_k)\\ &\quad\cdot Y(b_1,z_2+x_1)\cdots Y(b_l,z_2+x_l){}_{\circ}^{\circ}\prod_{\substack{1\le i\le k\\ 1\le j\le l}}(z_0+w_i-x_j)^{\langle\bar{a}_i,\bar{b}_j\rangle}\\ &\quad\cdot\prod_{1\le i<j\le k}((z_2+z_0+w_i)^{1/2}+(z_2+z_0+w_j)^{1/2})^{-2\langle\bar{a}_i,\bar{a}_j\rangle}\\ &\quad\cdot\prod_{1\le i<j\le l}((z_2+x_i)^{1/2}+(z_2+x_j)^{1/2})^{-2\langle\bar{b}_i,\bar{b}_j\rangle}\\ &\quad\cdot\prod_{\substack{1\le i\le k\\ 1\le j\le l}}((z_2+z_0+w_i)^{1/2}+(z_2+x_j)^{1/2})^{-2\langle\bar{a}_i,\bar{b}_j\rangle}.\end{aligned}$$

Comparing with (9.3.17), we now see that

$$Y(Y_{\mathbb{Z}}(A, z_0)B, z_2) = {}^{\circ}_{\circ}Y(A, z_2 + z_0)Y(B, z_2)^{\circ}_{\circ} \prod_{\substack{1 \le i \le k \\ 1 \le j \le l}} (z_0 + w_i - x_j)^{\langle \bar{a}_i, \bar{b}_j \rangle}$$

$$\cdot \prod_{\substack{1 \le i \le k \\ 1 \le j \le l}} ((z_2 + z_0 + w_i)^{1/2} + (z_2 + x_j)^{1/2})^{-2\langle \bar{a}_i, \bar{b}_j \rangle}, \tag{9.3.51}$$

and in conjunction with (9.3.41) and (9.3.50), this proves the Theorem. ∎

As in Corollary 8.6.3, we find:

Corollary 9.3.3: *In the notation of Theorem 9.3.1,*

$$[u_0, Y(v, z)] = \tfrac{1}{2} \sum_{p=0,1} (-1)^{p\langle \bar{a}, \bar{a} \rangle} Y((\theta^p u)_0 \cdot v, z),$$

and if L is even and (9.3.1) holds for all $a, b \in \hat{L}$, then

$$[u_0, Y(v, z)] = \tfrac{1}{2} Y((u + \theta u)_0 \cdot v, z).$$

Remark 9.3.4: Formally combining (9.3.20) and (9.3.51), we obtain

$$Y(Y_{\mathbb{Z}}(u, z_0)v, z_2) = Y(u, z_2 + z_0)Y(v, z_2) \tag{9.3.52}$$

for all $u, v \in V_L$, and in particular, in the setting of Theorem 9.3.1,

$$[Y(u, z_1), Y(v, z_2)] = \tfrac{1}{2} \sum_{p=0,1} (-1)^{p\langle \bar{a}, \bar{a} \rangle}$$

$$\cdot \operatorname{Res}_{z_0} z_2^{-1} Y(\theta^p u, z_2 + z_0)Y(v, z_2) e^{-z_0(\partial/\partial z_1)} \delta((-1)^p z_1^{1/2}/z_2^{1/2}). \tag{9.3.53}$$

(Compare with Remark 8.6.4.)

As in Corollaries 8.6.5–8.6.7, 9.2.5 and 9.2.8, we can express Theorem 9.3.1 in terms of component operators as follows:

Corollary 9.3.5: *In the notation of Theorem 9.3.1, let*

$$m, n \in \tfrac{1}{2}\mathbb{Z}.$$

Then as operators on V_L^T,

$$[u_m, v_n] = \tfrac{1}{2} \sum_{p=0,1} (-1)^{p(\langle \bar{a}, \bar{a} \rangle + 2m)} \sum_{i \in \mathbb{N}} \binom{m}{i} ((\theta^p u)_i \cdot v)_{m+n-i} \tag{9.3.54}$$

(finite sum). (On the right-hand side, $(\theta^p u)_i$ is an operator on the untwisted space V_L.)

Hence:

Corollary 9.3.6: *In the same notation,*

$$[u_0, v_n] = \tfrac{1}{2} \sum_{p=0,1} (-1)^{p\langle \bar{a}, \bar{a}\rangle}((\theta^p u)_0 \cdot v)_n.$$

Corollary 9.3.7: *In the same notation, suppose that the element u is homogeneous. Then as operators on V_L^T,*

$$[x_u(m), x_v(n)] = \frac{1}{2} \sum_{p=0,1} (-1)^{2pm} \sum_{i \in \mathbb{Z} - \langle \bar{a}, \bar{a}\rangle/2} \binom{m - 1 + \operatorname{wt} u}{i - 1 + \operatorname{wt} u} x_{x_{\theta^p u}(i) \cdot v^{(m+n)}} \tag{9.3.55}$$

(finite sum). (On the right-hand side, $x_{\theta^p u}(i)$ is an operator on V_L.)

We call attention again to the similarity between these results and the corresponding results in the untwisted case—Theorem 8.6.1 and Corollaries 8.6.5–8.6.7. It is important to note that this similarity increases when we assume that the element u is fixed by the action of θ on V_L. With the help of (9.2.38), we have:

Corollary 9.3.8: *In the notation of Theorem 9.3.1 and Corollary 9.3.5, suppose that*

$$\langle \bar{a}, \bar{a}\rangle \in 2\mathbb{Z} \tag{9.3.56}$$

and set

$$w = u + \theta u. \tag{9.3.57}$$

Then

$$Y(w, z) = \sum_{n \in \mathbb{Z}} w_n z^{-n-1}, \tag{9.3.58}$$

that is, $Y(w, z)$ involves only integral powers of z, and

$$[Y(w, z_1), Y(v, z_2)] = \operatorname{Res}_{z_0} z_2^{-1} Y(Y_{\mathbb{Z}}(w, z_0)v, z_2) e^{-z_0(\partial/\partial z_1)} \delta(z_1/z_2). \tag{9.3.59}$$

Moreover, if $m \in \mathbb{Z}$, then

$$[w_m, v_n] = \sum_{i \in \mathbb{N}} \binom{m}{i} (w_i \cdot v)_{m+n-i} \tag{9.3.60}$$

and if in addition w is homogeneous,

$$[x_w(m), x_v(n)] = \sum_{i \in \mathbb{Z}} \binom{m - 1 + \operatorname{wt} w}{i - 1 + \operatorname{wt} w} x_{x_w(i) \cdot v^{(m+n)}}. \tag{9.3.61}$$

Corollary 9.3.9: *In the same notation,*

$$[w_0, Y(v, z)] = Y(w_0 \cdot v, z)$$

and

$$[w_0, v_n] = (w_0 \cdot v)_n.$$

Remark 9.3.10: The operators $X^+(a, z)$ discussed in (7.4.18)–(7.4.28) (or rather their variants $Y^+(a, z) = Y(a, z) + Y(\theta a, z)$) are examples of operators $Y(w, z)$ discussed in Corollary 9.3.8 (assuming that $\langle \bar{a}, \bar{a} \rangle \in 2\mathbb{Z}$). In particular, the commutation relation (7.4.28) follows from the corollary.

Now as at the end of Section 8.6, we write down some special cases. Let $\alpha, \beta \in \mathfrak{h}$, $b \in \hat{L}$, $m \in \mathbb{Z} + \frac{1}{2}$, $n \in \frac{1}{2}\mathbb{Z}$ and $l \in \mathbb{Z}$. Then

$$[\alpha(z_1), Y(b, z_2)] = \langle \alpha, \bar{b} \rangle Y(b, z_2)(z_1 z_2)^{-1/2} \delta(z_1/z_2), \tag{9.3.62}$$

$$[\alpha(m), x_b(n)] = \langle \alpha, \bar{b} \rangle x_b(m + n) \tag{9.3.63}$$

[cf. (7.3.15)–(7.3.16)],

$$[\alpha(z_1), \beta(z_2)] = -\langle \alpha, \beta \rangle z_2^{-1} \frac{\partial}{\partial z_1} ((z_1/z_2)^{1/2} \delta(z_1/z_2)) \tag{9.3.64}$$

$$[\alpha(m), \beta(n)] = \langle \alpha, \beta \rangle m \delta_{m+n,0} \tag{9.3.65}$$

[cf. (7.3.4)],

$$[\alpha(z_1), Y(\beta(-1)^2, z_2)] = -2\langle \alpha, \beta \rangle \beta(z_2) z_2^{-1} \frac{\partial}{\partial z_1} ((z_1/z_2)^{1/2} \delta(z_1/z_2)) \tag{9.3.66}$$

$$[\alpha(m), \beta^2(n)] = 2\langle \alpha, \beta \rangle m \beta(m + n) \tag{9.3.67}$$

[recall (9.2.50), (9.2.82)],

$$[Y(\alpha(-1)^2, z_1), Y(b, z_2)] = 2\langle \alpha, \bar{b} \rangle Y(\alpha(-1) \cdot \iota(b), z_2) z_2^{-1} \delta(z_1/z_2) - \langle \alpha, \bar{b} \rangle^2 Y(b, z_2) z_2^{-1} \frac{\partial}{\partial z_1} \delta(z_1/z_2) \tag{9.3.68}$$

$$[\alpha^2(l), x_b(n)] = 2\langle \alpha, \bar{b} \rangle x_{\alpha(-1) \cdot \iota(b)}(l + n) + \langle \alpha, \bar{b} \rangle^2 (l + 1) x_b(l + n) \tag{9.3.69}$$

[recall (9.2.46), (9.2.79)] and

$$[Y(\alpha(-1)^2, z_1), Y(\beta(-1)^2, z_2)] = 4\langle\alpha, \beta\rangle Y(\alpha(-2)\beta(-1), z_2)z_2^{-1}\delta(z_1/z_2)$$
$$- 4\langle\alpha, \beta\rangle Y(\alpha(-1)\beta(-1), z_2)z_2^{-1}\frac{\partial}{\partial z_1}\delta(z_1/z_2)$$
$$-\frac{1}{3}\langle\alpha, \beta\rangle^2 z_2^{-1}\left(\frac{\partial}{\partial z_1}\right)^3\delta(z_1/z_2) \qquad (9.3.70)$$

$$[\alpha^2(l), \beta^2(n)] = 4\langle\alpha, \beta\rangle x_{\alpha(-2)\beta(-1)}(l + n) + 4\langle\alpha, \beta\rangle(l + 1)\alpha\beta(l + n)$$
$$+ 2\langle\alpha, \beta\rangle^2\binom{l+1}{3}\delta_{l+n,0}, \qquad (9.3.71)$$

where $x_{\alpha(-2)\beta(-1)}(l + n)$ is given by (9.2.66), (9.2.77) and (9.2.78).

Remark 9.3.11: In the more general setting of Remark 9.1.5 (see also Remark 9.2.9), the only necessary changes in the present section are as follows: if the element $a \in \hat{L}$ entering into the element u in (9.3.2) satisfies (9.1.46), then the sign is to be changed in (9.3.42), (9.3.45) and (9.3.47)–(9.3.50), and correspondingly, θ is to be replaced by $-\theta$ in Theorem 9.3.1 [(9.3.3) and (9.3.4)], Corollary 9.3.3, Remark 9.3.4 (9.3.53) and Corollaries 9.3.5–9.3.9.

Remark 9.3.12: The proof of Theorem 9.3.1 shows more generally that $Y(u, z_1)Y(v, z_2) - (-1)^{\langle a,b\rangle}c(\bar{a}, \bar{b})Y(v, z_2)Y(u, z_1)$ is given by the right-hand side of (9.3.3) even if $c(\bar{a}, \bar{b}) \neq (-1)^{\langle a,b\rangle}$; similarly for (9.3.4) and the corollaries (cf. Remark 8.6.9).

9.4. The Virasoro Algebra: Twisted Construction Revisited

We recall the structure of the Virasoro algebra $\mathfrak{v}$, given by (1.9.9) or (8.7.1). In Section 1.9 we constructed $\mathfrak{v}$ in both untwisted and twisted settings, and in Section 8.7 we reinterpreted the first case in terms of general untwisted vertex operators. Here we do the same for the twisted case, having already noted in Remark 7.3.3 that $\mathfrak{v}$ acts on our space V_L^T.

In the untwisted construction, $\mathfrak{v}$ was realized using the canonical quadratic element ω of V_L defined in (8.7.2); see (8.7.4)–(8.7.10). In the present context, set

$$L(z) = L_{\mathbb{Z}+1/2}(z) = Y(\omega, z) = \tfrac{1}{2}\sum_{i=1}^{l} {}^{\circ}_{\circ}h_i'(z)h_i(z){}^{\circ}_{\circ} + \tfrac{1}{16}(\dim\mathfrak{h})z^{-2} \qquad (9.4.1)$$

[see (9.2.52)], the optional subscript $\mathbb{Z} + 1/2$ designating the twisted construction as usual. Also, for $n \in \frac{1}{2}\mathbb{Z}$ set

$$L(n) = x_\omega(n) = \tfrac{1}{2} \sum_{i=1}^{l} h_i' h_i(n)$$

$$= \tfrac{1}{2} \sum_{i=1}^{l} \sum_{k \in \mathbb{Z}+1/2} {}^{\circ}_{\circ} h_i'(n-k)h_i(k) {}^{\circ}_{\circ} + \tfrac{1}{16}\delta_{n0} \dim \mathfrak{h} \qquad (9.4.2)$$

[recall (9.2.66) and (9.2.82)]. Then

$$L(n) = 0 \quad \text{unless} \quad n \in \mathbb{Z}. \qquad (9.4.3)$$

Also,

$$L(z) = \sum_{n \in \mathbb{Z}} L(n) z^{-n-2} \qquad (9.4.4)$$

by (9.2.66). The operators $L(n)$ agree with those in (1.9.23) for $Z = \mathbb{Z} + 1/2$ (recall Remark 3.3.1).

We already know the relation between $L(0)$ and the degree operator d on the space V_L^T [recall (7.3.9), (7.3.10)]:

$$L(0) = -d + \tfrac{1}{24} \dim \mathfrak{h} \qquad (9.4.5)$$

[see (1.9.49), (1.9.50) and (1.9.54)]. For a homogeneous element $v \in V_L^T$,

$$L(0)v = (\text{wt } v)v. \qquad (9.4.6)$$

Remark 9.4.1: The scalar $\frac{1}{16} \dim \mathfrak{h}$ in (9.4.2), introduced in (1.9.24) to "make the algebra fit together," is now understood from a very general viewpoint. It comes from the action of the natural operator $\exp(\Delta_z)$ on the natural quadratic element ω [recall (9.2.51)]. In particular, the operator $\exp(\Delta_z)$ determines the correct difference in the degree-shifts (1.9.51) and (1.9.53) between the untwisted and twisted cases. Since the scalar $\frac{1}{16} \dim \mathfrak{h}$ is precisely the right one to give us the commutation relations (8.7.1) (as we shall see again below), $\exp(\Delta_z)$ can be thought of as determining these relations.

The best way to use the general theory to show that the $L(n)$ satisfy (8.7.1) is to establish some general principles first rather than to invoke (9.3.70) [cf. (8.7.7)]. By analogy with (8.7.24), we show:

Proposition 9.4.2: *For $v \in V_L$,*

$$Y(L_{\mathbb{Z}}(-1)v, z) = \frac{d}{dz} Y(v, z), \qquad (9.4.7)$$

where $L_{\mathbb{Z}}(-1)$ is the operator $L(-1)$ on V_L, not V_L^T. (We shall use the subscript $\mathbb{Z}$ to indicate operators on the untwisted space.)

Proof: First note that a special case of this already follows from (8.7.25), (9.1.22) and (9.2.46):

$$Y(L_{\mathbb{Z}}(-1)\iota(a), z) = \frac{d}{dz} Y(\iota(a), z) \quad \text{for} \quad a \in \hat{L}. \tag{9.4.8}$$

To establish (9.4.7) in general, it is most natural to refer to the following consequence of (8.7.24):

$$e^{z_0 L_{\mathbb{Z}}(-1)} v = Y_{\mathbb{Z}}(v, z_0) \cdot \iota(1) \tag{9.4.9}$$

(see Proposition 8.7.4). On the other hand, as operators on V_L^T,

$$Y(Y_{\mathbb{Z}}(v, z_0) \cdot \iota(1), z) = Y(v, z + z_0) \tag{9.4.10}$$

by (9.3.51) and the first paragraph of the proof of Theorem 9.3.1 [cf. (9.3.52)]. Hence from (8.3.3) and (9.4.9),

$$e^{z_0(d/dz)} Y(v, z) = Y(v, z + z_0) = Y(e^{z_0 L_{\mathbb{Z}}(-1)} v, z), \tag{9.4.11}$$

and (9.4.7) follows by extracting the coefficient of z_0. ∎

Using (9.4.6) and (9.4.7) it is now easy to continue the analogy with the basic results of Section 8.7. By Corollary 9.3.8, we have just as in (8.7.33) the formula

$$\begin{aligned} [L(z_1), Y(v, z_2)] &= \operatorname{Res}_{z_0} z_2^{-1} Y(L_{\mathbb{Z}}(z_0)v, z_2) e^{-z_0(\partial/\partial z_1)} \delta(z_1/z_2) \\ &= z_2^{-1} Y(L_{\mathbb{Z}}(-1)v, z_2)\delta(z_1/z_2) - z_2^{-1} Y(L_{\mathbb{Z}}(0)v, z_2) \frac{\partial}{\partial z_1} \delta(z_1/z_2) \\ &\quad + \operatorname{Res}_{z_0} z_2^{-1} \sum_{n>0} Y(L_{\mathbb{Z}}(n)v, z_2) z_0^{-n-2} e^{-z_0(\partial/\partial z_1)} \delta(z_1/z_2) \end{aligned} \tag{9.4.12}$$

for all $v \in V_L$. Taking $v = \omega$ gives us precisely the formula (8.7.7) for $[L(z_1), L(z_2)]$. Hence we also have (8.7.8) and (8.7.9). That is,

$$[L(m), L(n)] = (m - n)L(m + n) + \tfrac{1}{12}(m^3 - m)(\dim \mathfrak{h})\delta_{m+n,0} \tag{9.4.13}$$

for $m, n \in \mathbb{Z}$, and we obtain as in Proposition 8.7.1 the result of Theorem 1.9.6 for $Z = \mathbb{Z} + 1/2$:

Proposition 9.4.3: *The operators $L(n)$ provide a representation of the Virasoro algebra $\mathfrak{v}$* [see (8.7.1)] *on V_L^T with*

$$\begin{aligned} L_n &\mapsto L(n) \quad \text{for} \quad n \in \mathbb{Z} \\ c &\mapsto \dim \mathfrak{h}. \end{aligned} \tag{9.4.14}$$

To continue with other consequences of (9.4.6), (9.4.7) and (9.4.12), formulas (8.7.11)–(8.7.13) hold in the present setting. In particular,

$$[L(m), \alpha(z_2)] = \left(z_2^{m+1}\frac{d}{dz_2} + (m+1)z_2^m\right)\alpha(z_2) \quad \text{for} \quad \alpha \in \mathfrak{h}, \quad m \in \mathbb{Z}. \tag{9.4.15}$$

Also, (8.7.14)–(8.7.16) become formulas involving $Y(\alpha(-1)^2, z_2)$ in place of $\circ\!\!\!\circ\,\alpha(z_2)^2\,\circ\!\!\!\circ$, so that for example

$$[L(m), Y(\alpha(-1)^2, z_2)]$$
$$= \left(z_2^{m+1}\frac{d}{dz_2} + 2(m+1)z_2^m\right)Y(\alpha(-1)^2, z_2) + \frac{1}{6}(m^3 - m)\langle\alpha, \alpha\rangle z_2^{m-2} \tag{9.4.16}$$

for $\alpha \in \mathfrak{h}$ and $m \in \mathbb{Z}$. As in (8.7.17)–(8.7.19),

$$[L(m), Y(a, z_2)] = \left(z_2^{m+1}\frac{d}{dz_2} + \frac{1}{2}\langle\bar{a}, \bar{a}\rangle(m+1)z_2^m\right)Y(a, z_2) \tag{9.4.17}$$

for $a \in \hat{L}$ and $m \in \mathbb{Z}$.

By analogy with Proposition 8.7.5, we find (using Corollaries 9.3.3 and 9.3.6):

Proposition 9.4.4: *For $v \in V_L$,*

$$[L(-1), Y(v, z)] = Y(L_{\mathbb{Z}}(-1)v, z) = \frac{d}{dz}Y(v, z) \tag{9.4.18}$$

$$[L(-1), v_n] = (L_{\mathbb{Z}}(-1)v)_n \quad \textit{for} \quad n \in \tfrac{1}{2}\mathbb{Z}. \tag{9.4.19}$$

Just as in Proposition 8.7.6, we can iterate (9.4.18) to obtain [see also (9.4.11)]:

Proposition 9.4.5: *For $v \in V_L$,*

$$e^{z_0L(-1)}Y(v, z)e^{-z_0L(-1)} = Y(e^{z_0L_{\mathbb{Z}}(-1)}v, z) = Y(v, z + z_0). \tag{9.4.20}$$

From (9.4.7) and (9.4.12) we have the analogue of one direction of Proposition 8.7.7:

Proposition 9.4.6: *If $v \in V_L$ is a lowest weight vector for $\mathfrak{v}$ with weight h, then*

$$[L(z_1), Y(v, z_2)] = z_2^{-1}\frac{d}{dz_2}Y(v, z_2)\delta(z_1/z_2) - hz_2^{-1}Y(v, z_2)\frac{\partial}{\partial z_1}\delta(z_1/z_2), \tag{9.4.21}$$

or equivalently,

$$[L(n), Y(v, z)] = \left(z^{n+1}\frac{d}{dz} + h(n+1)z^n\right)Y(v, z) \quad \text{for} \quad n \in \mathbb{Z} \tag{9.4.22}$$

or

$$[L(m), x_v(n)] = (hm - m - n)x_v(m+n) \quad \text{for} \quad m \in \mathbb{Z}, \quad n \in \tfrac{1}{2}\mathbb{Z}. \tag{9.4.23}$$

Remark 9.4.7: The step missing in the converse argument would say that $Y(L(0)v, z) = hY(v, z)$ would imply that $L(0)v = hv$.

The results from (8.7.40) through Proposition 8.7.9 now hold, except for the converse assertion in Proposition 8.7.9. For instance: If $v \in V_L$ is a lowest weight vector for $\mathfrak{v}$ with weight h, then

$$e^{z_0 L(n)}Y(v, z)e^{-z_0 L(n)} = \left(\frac{z_1}{z}\right)^{h(n+1)}Y(v, z_1) = \left(\frac{\partial z_1}{\partial z}\right)^h Y(v, z_1) \tag{9.4.24}$$

for $n \in \mathbb{Z}$, with z_1 as in (8.7.43) [cf. (8.7.44)].

Proposition 9.4.8: *If $v \in V_L$ is a lowest weight vector for span $\{L_{-1}, L_0, L_1\}$ with weight h [see (8.7.47)], then*

$$[L(n), Y(v, z)] = \left(z^{n+1}\frac{d}{dz} + h(n+1)z^n\right)Y(v, z) \tag{9.4.25}$$

$$\begin{aligned} e^{z_0 L(n)}Y(v, z)e^{-z_0 L(n)} &= (cz + d)^{-2h}Y\left(v, \frac{az + b}{cz + d}\right) \\ &= (a - cz_1)^{2h}Y(v, z_1) \end{aligned} \tag{9.4.26}$$

for $n = 0, \pm 1$, with $\begin{pmatrix} a & b \\ c & d \end{pmatrix}$ and z_1 as in (8.7.50)–(8.7.52).

Remark 9.4.9: The more general setting of Remarks 9.1.5, 9.2.9 and 9.3.11 entails no changes in the present section.

9.5. The Jacobi Identity and Cross-Brackets: Twisted Case

Here we shall develop twisted analogues of the results of Sections 8.8 and 8.9. *For convenience, we shall assume that the lattice L is even and that (9.3.1) holds for all $a, b \in \hat{L}$.* Then in particular, (9.3.4) holds. (See also the more general assertions in Theorem 9.5.3.)

As above, we use the notation $Y(v, z)$ for the twisted operators $Y_{\mathbb{Z}+1/2}(v, z)$ and the notation $Y_{\mathbb{Z}}(v, z)$ for the untwisted operators. Recall that

$$Y(\cdot, z) \quad \text{is a linear map} \quad V_L \to (\operatorname{End} V_L^T)[[z^{1/2}, z^{-1/2}]] \tag{9.5.1}$$

[see (9.2.28)],

$$Y(\iota(1), z) = 1 \tag{9.5.2}$$

[see (9.2.31)] and

$$e^{z_0 L(-1)} Y(v, z) e^{-z_0 L(-1)} = Y(e^{z_0 L_{\mathbb{Z}}(-1)} v, z) = Y(v, z + z_0) \tag{9.5.3}$$

for $v \in V_L$ [see (9.4.20)]. If $v \in V_L^\theta$ [recall the notation (8.9.25)] then

$$Y(v, z) = \sum_{n \in \mathbb{Z}} v_n z^{-n-1}, \tag{9.5.4}$$

i.e., $Y(v, z)$ involves only integral powers of z, and if $\theta v = -v$ then

$$Y(v, z) = \sum_{n \in \mathbb{Z}+1/2} v_n z^{-n-1}, \tag{9.5.5}$$

as we see from (9.2.37) [cf. (9.3.58)].

From (8.8.7) and (9.5.3) we obtain:

Proposition 9.5.1: *For $u, v \in V_L$,*

$$Y(Y_{\mathbb{Z}}(u, z_0)v, z_2) = Y(Y_{\mathbb{Z}}(v, -z_0)u, z_2 + z_0). \tag{9.5.6}$$

Remark 9.5.2: Combining this formally with (9.3.52), we find a formal "commutativity" relation for operators on V_L^T:

$$Y(u, z_2 + z_0)Y(v, z_2) = Y(v, z_2)Y(u, z_2 + z_0) \tag{9.5.7}$$

(cf. Remark 8.8.4).

Now for $u,v \in V_L$ and $n \in \mathbb{Z}$ define the following alternating or commutative generalization of Lie bracket (the case $n = 0$) as in (8.8.13):

$$[Y(u, z_1) \times_n Y(v, z_2)]$$
$$= (z_1 - z_2)^n Y(u, z_1)Y(v, z_2) - (-z_2 + z_1)^n Y(v, z_2)Y(u, z_1). \tag{9.5.8}$$

As before we call $[\cdot \times_1 \cdot]$ the *cross-bracket*. Also define the expansion coefficients $[u \times_l v]_{mn}$ and $[x_u \times_l x_v](m, n)$ as in (8.8.15) and (8.8.16). Then (8.8.17)–(8.8.24) hold here. Also form the generating function

$$[Y(u, z_1) \times_{z_0} Y(v, z_2)] = \sum_{n \in \mathbb{Z}} [Y(u, z_1) \times_n Y(v, z_2)] z_0^{-n-1}. \tag{9.5.9}$$

Then (8.8.26)–(8.8.28) remain valid here. In particular,

$$[Y(u, z_1) \times_{z_0} Y(v, z_2)] = z_0^{-1}\delta\left(\frac{z_1 - z_2}{z_0}\right) Y(u, z_1)Y(v, z_2) - z_0^{-1}\delta\left(\frac{z_2 - z_1}{-z_0}\right) Y(v, z_2)Y(u, z_1). \tag{9.5.10}$$

We can now state the Jacobi identity for twisted operators:

Theorem 9.5.3: *For $u, v \in V_L$ (or more generally, if $\langle \bar{a}, \bar{a}\rangle \in 2\mathbb{Z}$ in the fully general setting of Theorem 9.3.1),*

$$z_0^{-1}\delta\left(\frac{z_1 - z_2}{z_0}\right) Y(u, z_1)Y(v, z_2) - z_0^{-1}\delta\left(\frac{z_2 - z_1}{-z_0}\right) Y(v, z_2)Y(u, z_1)$$
$$= \frac{1}{2} \sum_{p=0,1} z_2^{-1}\delta\left((-1)^p \frac{(z_1 - z_0)^{1/2}}{z_2^{1/2}}\right) Y(Y_{\mathbb{Z}}(\theta^p u, z_0)v, z_2). \tag{9.5.11}$$

Equivalently, for $n \in \mathbb{Z}$,

$$[Y(u, z_1) \times_n Y(v, z_2)]$$
$$= \frac{1}{2} \sum_{p=0,1} \operatorname{Res}_{z_0} z_0^n z_2^{-1} Y(Y_{\mathbb{Z}}(\theta^p u, z_0)v, z_2) e^{-z_0(\partial/\partial z_1)} \delta((-1)^p z_1^{1/2}/z_2^{1/2}) \tag{9.5.12}$$

(In the greater generality of Theorem 9.3.1 and Remarks 9.3.11 and 9.3.12, $(-1)^{p\langle \bar{a}, \bar{a}\rangle}$ is to be inserted after the summation sign; if $a \in \hat{L}$ in (9.3.2) satisfies (9.1.46), then θ is to be replaced by $-\theta$; and $(-1)^{\langle \bar{a}, \bar{b}\rangle} c(\bar{a}, \bar{b})$ is to be inserted in the second term on the left in (9.5.11).)

Proof: The equivalence is clear. To prove (9.5.12), we use the proof of Theorem 9.3.1 except for the following modifications: By (9.3.34), the coefficient of P in $(z_1 - z_2)^n Y(A, z_1)Y(B, z_2)$ is

$$Y_P(z_1, z_2)(z_1 - z_2)^{-(N-n)}$$

and similarly, the coefficient of P in $(-z_2 + z_1)^n Y(B, z_2)Y(A, z_1)$ is

$$Y_P(z_1, z_2)(-z_2 + z_1)^{-(N-n)}.$$

Thus the coefficient of P in $[Y(A, z_1) \times_n Y(B, z_2)]$ is the coefficient of z_0^{N-n-1} in (9.3.36), and it follows that $[Y(A, z_1) \times_n Y(B, z_2)]$ is the coefficient of z_0^{-n-1} instead of z_0^{-1} in the expression on the right-hand side in formula (9.3.4). ∎

If we restrict u to lie in V_L^θ, the Jacobi identity looks exactly like (8.8.29):

Corollary 9.5.4: *For $u \in V_L^\theta$ and $v \in V_L$,*

$$z_0^{-1}\delta\left(\frac{z_1 - z_2}{z_0}\right)Y(u, z_1)Y(v, z_2) - z_0^{-1}\delta\left(\frac{z_2 - z_1}{-z_0}\right)Y(v, z_2)Y(u, z_1)$$
$$= z_2^{-1}\delta\left(\frac{z_1 - z_0}{z_2}\right)Y(Y_{\mathbb{Z}}(u, z_0)v, z_2). \qquad (9.5.13)$$

Remark 9.5.5: By taking Res_{z_0} of the Jacobi identity, we of course recover formula (9.3.4). As in Remarks 8.8.11 and 8.8.12, we can take Res_{z_1} and Res_{z_2} instead. If $u \in V_L^\theta$ we find that $Y(Y_{\mathbb{Z}}(u, z_0)v, z_2)$ is given by (8.8.31) and that $Y(Y_{\mathbb{Z}}(u, z_0)v, z_1 - z_0)$ is given by (8.8.32).

Remark 9.5.6: Formula (9.5.13) can be rewritten as in Remark 8.8.16 [see (8.8.38)–(8.8.40)].

Remark 9.5.7: Formula (9.5.13) can be expressed in component forms exactly as in Corollaries 8.8.17–8.8.19.

Remark 9.5.8: The vertex operators $Y_{\mathbb{Z}}(v, z)$ and $Y_{\mathbb{Z}+1/2}(v, z)$ that we have been studying are all parametrized by $v \in V_L$. It is possible to develop a theory of operators $Y(v, z)$ for $v \in V_L^T$ by starting with (8.8.7) to define the action of such operators on V_L. We shall not pursue this direction in the present work; cf. [Frenkel–Huang–Lepowsky].

Remark 9.5.9: For $u, v \in V_L^\theta$ of weight 1, the discussion of (8.9.2)–(8.9.12) holds here, with the operators acting on V_L^T. Note that θ is an automorphism of the Lie algebra $\mathfrak{g}$ of (8.9.6) and that the form $\langle \cdot, \cdot \rangle$ of (8.9.8) is θ-invariant. Let $\mathfrak{g}_{(0)}$ denote the Lie subalgebra of θ-fixed elements of $\mathfrak{g}$:

$$\mathfrak{g}_{(0)} = \mathfrak{g} \cap V_L^\theta. \qquad (9.5.14)$$

Then V_L^T is a $\mathfrak{g}_{(0)}$-module under the action

$$u \mapsto u_0, \qquad (9.5.15)$$

and V_L^T is in fact a $(\mathfrak{g}_{(0)})\hat{}$-module, as we already knew from Theorem 7.4.10. Of course, we could also recover the full action of $\hat{\mathfrak{g}}[\theta]$ on V_L^T given in Theorem 7.4.10.

Now suppose that

$$u, v \in V_L^\theta, \quad \text{wt}\, u = \text{wt}\, v = 2. \tag{9.5.16}$$

Then the discussion at the end of Section 8.9 applies, with the operators acting on V_L^T. Suppose that L is positive definite and that $L_2 = \emptyset$, as in (8.9.30). Recall that the space $\mathfrak{k}$ consisting of the elements of V_L^θ of weight 2 (8.9.32) is a commutative nonassociative algebra under the product $u \times v = u_1 \cdot v$ and that the form $\langle u, v\rangle = u_3 \cdot v$ is symmetric and associative. As in Proposition 8.9.5 we have a graded representation of the commutative affinization $\hat{\mathfrak{k}}$ on V_L^T by cross-bracket:

Theorem 9.5.10: *The space V_L^T is a graded $\hat{\mathfrak{k}}$-module under the action*

$$\pi: \hat{\mathfrak{k}} \to \text{End}\, V_L^T \tag{9.5.17}$$

defined by

$$\begin{aligned} &\pi: u \otimes t^n \mapsto x_u(n) \quad \textit{for} \quad u \in \mathfrak{k}, \quad n \in \mathbb{Z} \\ &\pi: e \mapsto 1. \end{aligned} \tag{9.5.18}$$

Let $\mathfrak{p}$ be the space of elements of V_L^T of weight 2:

$$\mathfrak{p} = (V_L^T)_{(\dim \mathfrak{h})/24-2} \tag{9.5.19}$$

Then the identity element $\frac{1}{2}\omega$ of $\mathfrak{k}$ (recall Remarks 8.9.2 and 8.9.4) acts as the identity operator on $\mathfrak{p}$, by (9.4.6):

$$(\tfrac{1}{2}\omega)_1 \cdot w = w \quad \text{for} \quad w \in \mathfrak{p}. \tag{9.5.20}$$

Unless $\dim \mathfrak{h}$ is chosen carefully, $\mathfrak{p} = 0$ [recall (1.9.57)]. Let us assume that

$$\dim \mathfrak{h} = 24, \tag{9.5.21}$$

the case that we shall be most interested in. Then

$$\mathfrak{p} = \mathfrak{h} \otimes T, \tag{9.5.22}$$

where we make the identification

$$\mathfrak{h} = \mathfrak{h} \otimes t^{-1/2} \tag{9.5.23}$$

in the notation (1.7.12), (1.7.15) [recall (7.3.8)]. With the notation of Remark 8.9.7, we can express the action of $\mathfrak{k}$ on $\mathfrak{p}$ explicitly as follows:

$$(x_a^+)_1 \cdot (h \otimes \tau) = \tfrac{1}{8}(h - 2\langle \bar{a}, h\rangle \bar{a}) \otimes a \cdot \tau \tag{9.5.24}$$

$$(g^2)_1 \cdot (h \otimes \tau) = (\langle g, h\rangle g + \tfrac{1}{8}\langle g, g\rangle h) \otimes \tau \tag{9.5.25}$$

for $a \in \hat{L}_4$, $g, h \in \mathfrak{h}$, $\tau \in T$, using (7.3.4), (7.4.14), (9.1.15), (9.2.50) and (9.2.53).

Remark 9.5.11: Later, for the case in which L is the Leech lattice and where $\hat{L}$ and T are chosen specially, we shall make $\mathfrak{k} \oplus \mathfrak{p}$ a commutative nonassociative algebra with a nonsingular associative symmetric form—the Griess algebra. This has mostly been accomplished already. We shall also extend Theorems 8.9.5 and 9.5.10 by constructing a representation of the commutative affinization of this algebra by cross-bracket on $V_L^\theta \oplus (V_L^T)^\theta$ for a suitable choice of θ—the moonshine module [recall the notation (9.2.94)]. The Monster will act compatibly on both the algebra and the module.

10 The Moonshine Module

Here we begin the construction of the moonshine module $V^\natural$, the associated vertex operator algebra, and the Monster acting on $V^\natural$ and preserving the vertex operator algebra structure. The vertex operator algebra associated with $V^\natural$ crowns the sequence of exceptional structures starting with the Golay error-correcting code and continuing with the Leech lattice. The corresponding sequence of their automorphism groups consists of the Mathieu group M_{24}, the Conway group Co_0 and the Monster. We begin this chapter with introductions to the two exceptional structures mentioned above. For more details we refer the reader to the papers [Leech 1, 2], [Conway 1–6], [Curtis], to the book [MacWilliams–Sloane] and to the extensive collection [Conway–Sloane 5] and references therein. The history of the discoveries related to monstrous moonshine is reviewed in the Introduction.

In Section 10.1 we first introduce binary linear codes and related notions. We define and construct the Golay code, following the treatment in [Curtis], and we define the Mathieu group M_{24} as its automorphism group. Using the Golay code we construct the Leech lattice in the standard way [Leech 2] in Section 10.2, and in Theorem 10.2.5 we also present an alternative construction [Lepowsky–Meurman], [Tits 2, 3] based on three copies of the E_8 root lattice. In addition, we introduce several important lattices related to the Leech lattice, in particular, the Niemeier lattice of type

A_1^{24}, and we define the Conway groups Co_0 and Co_1. A number of properties of the Mathieu and Conway groups that we need in our construction, such as simplicity, are stated without proof; the reader can consult the cited references. This is one of the few places in this book where outside sources are needed for completeness.

We define the moonshine module $V^\natural$ in Section 10.3 by appropriately combining the vertex-operator-algebraic structures of Chapters 8 and 9 based on the Leech lattice; in particular, we take the direct sum of a canonical vertex operator algebra—the fixed subspace of an untwisted space with respect to an involution—with a module for it—the fixed subspace of a twisted space with respect to an analogous involution. We introduce the Griess algebra $\mathcal{B}$, originally discovered in [Griess 2, 3], as a part of the intrinsic structure of $V^\natural$. In Section 10.4 we define an action of a group C, a certain extension of the Conway group Co_1 by an extraspecial 2-group, on $V^\natural$ and hence on $\mathcal{B}$, intertwining the vertex operators parametrized by the untwisted subspace of $V^\natural$. This group C, well known in finite group theory (cf. [Griess 1–5]), will turn out to be the centralizer of an involution in the Monster. Finally, in Section 10.5 we define and calculate the graded characters, or equivalently, the Thompson series, for the action of C on $V^\natural$, and we briefly discuss monstrous moonshine. Again the reader should consult the Introduction for more historical details and motivation. Theorem 10.3.5 and Proposition 10.3.6, including the natural vertex-operator-algebraic definition of the Griess algebra, were found in [FLM1]. The moonshine module was constructed in [FLM2], including the determination of the Thompson series for C given in Theorem 10.5.7.

10.1. The Golay Code

Let Ω be a finite set with n elements. The power set

$$\mathcal{P}(\Omega) = \{S \mid S \subset \Omega\} \tag{10.1.1}$$

can be viewed as an $\mathbb{F}_2$-vector space under the operation $+$ of symmetric difference. By a (*binary linear*) *code* we shall understand an $\mathbb{F}_2$-subspace of $\mathcal{P}(\Omega)$. An isomorphism of codes is defined in the obvious way. The cardinality $|C|$ of an element C of a code is called the *weight* of C. A code $\mathcal{C}$ is said to be of *type* I if

$$n \in 2\mathbb{Z}, \quad |C| \in 2\mathbb{Z} \quad \text{for all} \quad C \in \mathcal{C}, \quad \text{and} \quad \Omega \in \mathcal{C}, \tag{10.1.2}$$

and $\mathcal{C}$ is said to be of *type* II if

$$n \in 4\mathbb{Z}, \quad |C| \in 4\mathbb{Z} \quad \text{for all} \quad C \in \mathcal{C}, \quad \text{and} \quad \Omega \in \mathcal{C}. \tag{10.1.3}$$

The type II codes will be seen as analogues of even lattices, and will help us construct them, while the type I codes will provide a natural setting for the study of triality in Chapter 11.

For a code $\mathcal{C}$, the *dual code* $\mathcal{C}^\circ$ is given by

$$\mathcal{C}^\circ = \{S \subset \Omega \mid |S \cap C| \in 2\mathbb{Z} \quad \text{for all} \quad C \in \mathcal{C}\}. \tag{10.1.4}$$

Thus $\mathcal{C}^\circ$ is the annihilator of $\mathcal{C}$ in $\mathcal{P}(\Omega)$ with respect to the natural nonsingular symmetric bilinear form

$$(S_1, S_2) \mapsto |S_1 \cap S_2| + 2\mathbb{Z} \tag{10.1.5}$$

on $\mathcal{P}(\Omega)$. Hence

$$\dim_{\mathbb{F}_2} \mathcal{C}^\circ = n - \dim_{\mathbb{F}_2} \mathcal{C}. \tag{10.1.6}$$

We call $\mathcal{C}$ *self-dual* if

$$\mathcal{C} = \mathcal{C}^\circ, \tag{10.1.7}$$

in which case n is even and

$$\dim_{\mathbb{F}_2} \mathcal{C} = \frac{n}{2}. \tag{10.1.8}$$

Consider the subspace

$$\mathcal{E}(\Omega) = \{S \subset \Omega \mid |S| \in 2\mathbb{Z}\}. \tag{10.1.9}$$

The map

$$\begin{aligned} q: \mathcal{E}(\Omega) &\to \mathbb{Z}/2\mathbb{Z} = \mathbb{F}_2 \\ S &\mapsto \frac{|S|}{2} + 2\mathbb{Z} \end{aligned} \tag{10.1.10}$$

is a quadratic form on $\mathcal{E}(\Omega)$ with associated bilinear form given by (10.1.5) [recall (5.3.5)–(5.3.6)]. In case $n \in 2\mathbb{Z}$, $\mathbb{F}_2\Omega$ is the radical of the form q.

A subspace of a space with a quadratic form is called *totally singular* if the form vanishes on it.

Remark 10.1.1: In case $n \in 4\mathbb{Z}$, the self-dual codes of type II correspond to the (maximal) totally singular subspaces of $\mathcal{E}(\Omega)/\mathbb{F}_2\Omega$ of dimension $(n/2) - 1$. Equivalently, the type II self-dual codes are the (maximal) totally singular subspaces of $\mathcal{E}(\Omega)$ of dimension $n/2$.

For a code $\mathcal{C}$ set

$$w(\mathcal{C}) = \sum_{C \in \mathcal{C}} q^{|C|} \in \mathbb{Z}[q], \tag{10.1.11}$$

the *weight distribution of* $\mathcal{C}$.

A code with the following properties is called a *Hamming code*:

Theorem 10.1.2: *There is a self-dual code of type II on an 8-element set* Ω.

Proof: In fact we shall exhibit two complementary 3-dimensional totally singular subspaces of $\mathcal{E}(\Omega)/\mathbb{F}_2\Omega$. We identify Ω with the projective line over the 7-element field:

$$\Omega = \mathbb{P}^1(\mathbb{F}_7) = \mathbb{F}_7 \cup \{\infty\}. \tag{10.1.12}$$

Consider the sets of squares and non-squares:

$$\begin{aligned} \mathcal{Q} &= \{x^2 \mid x \in \mathbb{F}_7\} = \{0, 1, 2, 4\} \\ \mathcal{N} &= \Omega \backslash \mathcal{Q} = \{3, 5, 6, \infty\} \end{aligned} \tag{10.1.13}$$

and define subspaces

$$\begin{aligned} \mathcal{C}_1 &= \langle \mathcal{N} + i \mid i \in \mathbb{F}_7 \rangle \\ \mathcal{C}_2 &= \langle -\mathcal{N} - i \mid i \in \mathbb{F}_7 \rangle \end{aligned} \tag{10.1.14}$$

of $\mathcal{E}(\Omega)$, $\langle \cdot \rangle$ denoting span. Then it is easy to verify that $\mathcal{C}_1$ and $\mathcal{C}_2$ have the desired properties:

$$\begin{aligned} \dim \mathcal{C}_1 &= \dim \mathcal{C}_2 = 4 \\ q(\mathcal{C}_1) &= q(\mathcal{C}_2) = 0 \\ \mathcal{C}_1 \cap \mathcal{C}_2 &= \mathbb{F}_2\Omega \\ \mathcal{C}_1 + \mathcal{C}_2 &= \mathcal{E}(\Omega). \quad \blacksquare \end{aligned} \tag{10.1.15}$$

Remark 10.1.3: It is clear that the weight distribution of the constructed Hamming codes is

$$1 + 14q^4 + q^8. \tag{10.1.16}$$

Remark 10.1.4: The Hamming code is unique up to isomorphism (see e.g. [MacWilliams–Sloane], Chapter 1, Problem 14 or Chapter 20, Problem 16; this is a simple exercise).

A code with the following properties is called a *(binary) Golay code*:

Theorem 10.1.5: *There is a self-dual code $\mathcal{C}$ of type II on a 24-element set such that $\mathcal{C}$ has no elements of weight 4.*

Proof: Let $\mathcal{C}_1, \mathcal{C}_2$ be the two Hamming codes in $\mathcal{E}(\Omega)$ constructed in the proof of Theorem 10.1.2, so that the relations (10.1.15) hold. Denote by 3Ω the disjoint union of three copies of Ω. In the 24-dimensional space $\mathcal{P}(3\Omega)$, let

$$\mathcal{C} = \langle (S, S, \emptyset), (S, \emptyset, S), (T, T, T) \mid S \in \mathcal{C}_1, \quad T \in \mathcal{C}_2 \rangle, \quad (10.1.17)$$

where the ordered triples denote the obvious unions. The space $\mathcal{C}$ is the orthogonal direct sum of three 4-dimensional totally singular subspaces of $\mathcal{E}(3\Omega)$, so that $\mathcal{C}$ is 12-dimensional and totally singular and hence is a type II self-dual code.

To show that $|C| \neq 4$ for $C \in \mathcal{C}$, suppose that

$$|S_1 + T| + |S_2 + T| + |S_3 + T| = 4 \qquad (10.1.18)$$

for $S_i \in \mathcal{C}_1$, $T \in \mathcal{C}_2$ with $S_1 + S_2 + S_3 = 0$. Then one of the summands in (10.1.18) must be 0 since they are all even. Say $S_3 + T = 0$, which implies that $T \in \mathbb{F}_2\Omega$. Thus $|S_1 + T|$ and $|S_2 + T|$ are multiples of 4, so one of them is 0, say, $S_2 + T = 0$. But then

$$S_1 = S_1 - 2T = S_1 + S_2 + S_3 = 0,$$

which implies that $|S_1 + T|$ is a multiple of 8, and we have a contradiction. ∎

Remark 10.1.6: It is not hard to see that the weight distribution of the constructed Golay code is

$$1 + 759q^8 + 2576q^{12} + 759q^{16} + q^{24} \qquad (10.1.19)$$

(see e.g. [Curtis]). Note that

$$759 = \binom{24}{5} \Big/ \binom{8}{5}. \qquad (10.1.20)$$

Remark 10.1.7: The Golay code is unique up to isomorphism (see e.g. [MacWilliams–Sloane], Chapter 20, Theorem 14).

The 759 elements of the Golay code of weight 8 are called *octads* (following [Conway 4]).

Proposition 10.1.8: *Let Ω be a 24-element set and let $\mathcal{C}$ be a Golay code in $\mathcal{E}(\Omega)$.*

(a) Every 5-element subset of Ω is included in a unique octad in $\mathcal{C}$.

(b) Let T_0 be a 4-element subset of Ω. Then T_0 lies in exactly 5 octads. These are of the form $T_0 \cup T_i$, $i = 1, \ldots, 5$, where

$$\Omega = \bigcup_{i=0}^{5} T_i \tag{10.1.21}$$

is a disjoint union of 4-element sets (called a sextet *in [Conway 4]). The union of any pair of T_i is an octad.*

Proof: It is clear that (b) follows from (a). To prove (a), first note that a 5-element set cannot be included in more than one octad, since this would give an element of $\mathcal{C}$ with positive weight less than 8. Thus distinct octads account for distinct families of $\binom{8}{5}$ 5-element sets. Now just count, using (10.1.19) and (10.1.20). ∎

Remark 10.1.9: It is easy to see from (10.1.17) that the octads generate the Golay code.

The group of automorphisms of the Golay code $\mathcal{C}$ is called the *Mathieu group* M_{24}:

$$M_{24} = \mathrm{Aut}\, \mathcal{C}. \tag{10.1.22}$$

Remark 10.1.10: The group M_{24} is a nonabelian simple group (see e.g. [Huppert]).

Remark 10.1.11 [Conway 4]: In the natural representation of M_{24} on $\mathcal{P}(\Omega)$, the complete list of submodules is:

$$0 \subset \langle \Omega \rangle \subset \mathcal{C} \subset \mathcal{E}(\Omega) \subset \mathcal{P}(\Omega). \tag{10.1.23}$$

In particular, $\mathcal{C}/\langle \Omega \rangle$ and $\mathcal{E}(\Omega)/\mathcal{C}$ are faithful irreducible modules for M_{24}.

10.2. The Leech Lattice

A self-dual code $\mathcal{C}$ of type II based on a set Ω gives rise to an even unimodular lattice as follows: Let

$$\mathfrak{h} = \coprod_{k \in \Omega} \mathbb{F}\alpha_k \tag{10.2.1}$$

be a vector space with basis $\{\alpha_k \mid k \in \Omega\}$ and provide $\mathfrak{h}$ with the symmetric bilinear form $\langle \cdot, \cdot \rangle$ such that

$$\langle \alpha_k, \alpha_l \rangle = 2\delta_{k,l} \quad \text{for} \quad k, l \in \Omega. \tag{10.2.2}$$

For $S \subset \Omega$, set

$$\alpha_S = \sum_{k \in S} \alpha_k. \tag{10.2.3}$$

Define

$$Q = \coprod_{k \in \Omega} \mathbb{Z}\alpha_k, \tag{10.2.4}$$

and for a code $\mathcal{C}$ based on Ω, define the positive definite lattice

$$L_0 = \sum_{C \in \mathcal{C}} \mathbb{Z}\tfrac{1}{2}\alpha_C + Q, \tag{10.2.5}$$

or equivalently,

$$L_0 = \left\{ \sum_{k \in \Omega} m_k \alpha_k \,\middle|\, m_k \in \tfrac{1}{2}\mathbb{Z}, \quad \{k \mid m_k \in \mathbb{Z} + \tfrac{1}{2}\} \in \mathcal{C} \right\}. \tag{10.2.6}$$

[There should be no confusion with the notation (6.1.7).] Then L_0 is even if and only if $|C| \in 4\mathbb{Z}$ for all $C \in \mathcal{C}$. Moreover, the dual lattice L_0° of L_0 [recall (6.1.12)] is the corresponding lattice based on the dual code $\mathcal{C}^\circ$ [recall (10.1.4)]:

$$L_0^\circ = \left\{ \sum_{k \in \Omega} m_k \alpha_k \,\middle|\, m_k \in \tfrac{1}{2}\mathbb{Z}, \quad \{k \mid m_k \in \mathbb{Z} + \tfrac{1}{2}\} \in \mathcal{C}^\circ \right\}. \tag{10.2.7}$$

Thus we have:

Proposition 10.2.1: *A code $\mathcal{C}$ is self-dual of type II if and only if the corresponding lattice L_0 is even self-dual, or equivalently, even unimodular.*

Now consider the following modification of the lattice L_0 associated with a code $\mathcal{C}$, which we now assume contains Ω:

$$\begin{aligned} L_0' &= \sum_{C \in \mathcal{C}} \mathbb{Z}\tfrac{1}{2}\alpha_C + \sum_{k \in \Omega} \mathbb{Z}(\tfrac{1}{4}\alpha_\Omega - \alpha_k) \\ &= \sum_{C \in \mathcal{C}} \mathbb{Z}\tfrac{1}{2}\alpha_C + \sum_{k,l} \mathbb{Z}(\alpha_k + \alpha_l) + \mathbb{Z}(\tfrac{1}{4}\alpha_\Omega - \alpha_{k_0}), \end{aligned} \tag{10.2.8}$$

where k_0 is a fixed element of Ω. Since the lattice

$$L_0 \cap L_0' = \sum_{C \in \mathcal{C}} \mathbb{Z}\tfrac{1}{2}\alpha_C + \sum_{k,l \in \Omega} \mathbb{Z}(\alpha_k + \alpha_l) \tag{10.2.9}$$

has index 2 in both L_0 and L_0', L_0' is unimodular if and only if L_0 is. A necessary condition for L_0' to be even is that

$$n = |\Omega| \in 8(2\mathbb{Z} + 1) \tag{10.2.10}$$

since

$$\langle \tfrac{1}{4}\alpha_\Omega - \alpha_k, \tfrac{1}{4}\alpha_\Omega - \alpha_k \rangle = \frac{n}{8} + 1 \tag{10.2.11}$$

for $k \in \Omega$, and we find:

Proposition 10.2.2: *If $n \in 8(2\mathbb{Z} + 1)$ and the code $\mathcal{C}$ is self-dual of type II, then the corresponding lattice L_0' is even unimodular.*

The *Leech lattice* is the even unimodular lattice

$$\Lambda = L_0' \tag{10.2.12}$$

for the case $n = 24$ and $\mathcal{C}$ the Golay code (recall Theorem 10.1.5). The lattice Λ has no "short" elements:

$$\Lambda_2 = \emptyset \tag{10.2.13}$$

[using the notation (6.1.7)]. This can be checked from (10.2.8) and the corresponding property of the Golay code, but we shall prove (10.2.13) (and in fact reconstruct Λ) instead by using another principle, which will be an analogue for lattices of Remark 10.1.1 for codes.

Let L be an even unimodular lattice of rank n and with form $\langle \cdot, \cdot \rangle$. For our special purpose now we provide L with the following rescaled form:

$$\langle \alpha, \beta \rangle_{1/2} = \tfrac{1}{2}\langle \alpha, \beta \rangle \quad \text{for} \quad \alpha, \beta \in L \tag{10.2.14}$$

and then by abuse of notation we drop the subscript $\frac{1}{2}$. With respect to the new form $\langle \cdot, \cdot \rangle$, L has the following properties:

$$\langle \alpha, \alpha \rangle \in \mathbb{Z} \quad \text{for} \quad \alpha \in L \tag{10.2.15}$$

$$|\det(\langle \alpha_i, \alpha_j \rangle)_{i,j}| = \frac{1}{2^n} \tag{10.2.16}$$

for a base $\{\alpha_1, \ldots, \alpha_n\}$ of L. As in (6.1.20)–(6.1.25) (but keeping in mind the rescaled form), set

$$\check{L} = L/2L, \tag{10.2.17}$$

an n-dimensional vector space over $\mathbb{F}_2$, and write $\alpha \mapsto \check{\alpha}$ for the canonical map. Since the original lattice L is even,

$$\begin{aligned} q_1\colon \check{L} &\to \mathbb{Z}/2\mathbb{Z} = \mathbb{F}_2 \\ \check{\alpha} &\mapsto \langle \alpha, \alpha \rangle + 2\mathbb{Z} \end{aligned} \tag{10.2.18}$$

defines a quadratic form on $\check{L}$ with associated bilinear form

$$\begin{aligned} c_1\colon \check{L} \times \check{L} &\to \mathbb{F}_2 \\ (\check{\alpha}, \check{\beta}) &\mapsto 2\langle \alpha, \beta \rangle + 2\mathbb{Z}. \end{aligned} \tag{10.2.19}$$

These forms are nonsingular since the original lattice L is unimodular. From the definitions we have:

Proposition 10.2.3: *Let M be a lattice such that*

$$2L \subset M \subset L. \tag{10.2.20}$$

The M is even unimodular with respect to the new form $\langle \cdot, \cdot \rangle$ if and only if $\check{M} = M/2L$ is a (maximal) totally singular subspace of $\check{L}$ of dimension $n/2$.

We shall apply this principle to the direct sum of three copies of the root lattice Q_{E_8}, which is an even unimodular lattice of rank 8 (recall Proposition 6.3.5). For brevity, set

$$\Gamma = Q_{E_8} \tag{10.2.21}$$

and provide Γ with the rescaled form (10.2.14) as above. Using the Hamming code (see Theorem 10.1.2) we shall first show that $\check{\Gamma}$ contains complementary 4-dimensional totally singular subspaces.

The lattice Γ is described in (6.3.6). In the notation (10.2.1)–(10.2.3),

$$\begin{aligned} \Gamma &= \sum_{k,l \in \Omega} \mathbb{Z}(\tfrac{1}{2}\alpha_k \pm \tfrac{1}{2}\alpha_l) + \mathbb{Z}\tfrac{1}{4}\alpha_\Omega \\ &= \left\{ \sum_{k \in \Omega} m_k \alpha_k \,\middle|\, \text{either} \quad m_1, \ldots, m_8 \in \tfrac{1}{2}\mathbb{Z} \quad \text{or} \right. \\ &\qquad \left. m_1, \ldots, m_8 \in \tfrac{1}{2}\mathbb{Z} + \tfrac{1}{4}; \quad \sum m_k \in \mathbb{Z} \right\}. \end{aligned} \tag{10.2.22}$$

Recall the complementary codes $\mathcal{C}_1, \mathcal{C}_2$ constructed in the proof of Theorem 10.1.2 [see (10.1.15)]. In the notations (10.2.5) and (10.2.8), consider the lattices

$$\Phi = L_0 \quad \text{for} \quad \mathcal{C} = \mathcal{C}_1 \tag{10.2.23}$$

$$\Psi = L_0' \quad \text{for} \quad \mathcal{C} = \mathcal{C}_2. \tag{10.2.24}$$

Then Φ and Ψ are even unimodular by Propositions 10.2.1 and 10.2.2. Moreover,

$$2\Gamma \subset \Phi \subset \Gamma \tag{10.2.25}$$

$$2\Gamma \subset \Psi \subset \Gamma, \tag{10.2.26}$$

and since $\mathcal{C}_1 + \mathcal{C}_2 = \mathcal{E}(\Omega)$ (10.1.15) we see that

$$\Phi + \Psi = \Gamma. \tag{10.2.27}$$

Proposition 10.2.3 thus gives:

Proposition 10.2.4: *In the notation (10.2.14)–(10.2.19), we have a decomposition*

$$\check{\Gamma} = \check{\Phi} \oplus \check{\Psi} \tag{10.2.28}$$

into complementary 4-dimensional totally singular subspaces. In particular,

$$\begin{aligned} \Phi + \Psi &= \Gamma \\ \Phi \cap \Psi &= 2\Gamma. \end{aligned} \tag{10.2.29}$$

Using the decomposition (10.2.28) we shall now reconstruct the Leech lattice by analogy with the construction of the Golay code in Theorem 10.1.5. Set

$$\Gamma^3 = \Gamma \oplus \Gamma \oplus \Gamma, \tag{10.2.30}$$

the orthogonal direct sum of three copies of Γ, equipped with the modified form (10.2.14). In Γ^3, set

$$\begin{aligned} \Lambda &= \{(\varphi, \varphi, 0) \mid \varphi \in \Phi\} \oplus \{(\varphi, 0, \varphi) \mid \varphi \in \Phi\} \oplus \{(\psi, \psi, \psi) \mid \psi \in \Psi\} \\ &= \{(\varphi_1 + \psi, \varphi_2 + \psi, \varphi_3 + \psi) \mid \varphi_i \in \Phi,\ \psi \in \Psi,\ \varphi_1 + \varphi_2 + \varphi_3 \in 2\Gamma\} \end{aligned} \tag{10.2.31}$$

(note that the sum is indeed direct). Then we have:

Theorem 10.2.5: *The lattice Λ in (10.2.31) is an even unimodular lattice such that $\Lambda_2 = \emptyset$. Moreover, it coincides with the Leech lattice defined in (10.2.12).*

Proof: Note first that

$$2\Gamma^3 \subset \Lambda \subset \Gamma^3. \tag{10.2.32}$$

Provide the space

$$(\Gamma^3)\check{} = \check{\Gamma} \oplus \check{\Gamma} \oplus \check{\Gamma} \tag{10.2.33}$$

with the quadratic form given by (10.2.18). Then

$$\check{\Lambda} = \{(x, x, 0) \mid x \in \check{\Phi}\} \oplus \{(x, 0, x) \mid x \in \check{\Phi}\} \oplus \{(y, y, y) \mid y \in \check{\Psi}\} \tag{10.2.34}$$

is the orthogonal direct sum of three 4-dimensional totally singular subspaces of $(\Gamma^3)\check{}$, so that $\check{\Lambda}$ is 12-dimensional and totally singular. Thus by Proposition 10.2.3, Λ is even unimodular. An argument exactly analogous to the one in the proof of Theorem 10.1.5 now shows that $\Lambda_2 = \emptyset$. To see that the two lattices coincide, we need only show one inclusion since both lattices are unimodular. But from the definition it is clear that the lattice of (10.2.31) is included in the earlier one. ∎

Remark 10.2.6: If we know, say from root system theory, that $\check{\Gamma}$ contains a 4-dimensional totally singular subspace (see e.g. [Bourbaki 1], p. 228), then general principles of quadratic form theory in characteristic 2 (see e.g. [Chevalley]) imply that Proposition 10.2.4 holds for some lattices Φ and Ψ. Then the straightforward argument in Theorem 10.2.5 shows that the lattice Λ in (10.2.31) is even unimodular and that $\Lambda_2 = \emptyset$. This gives a simple construction of the Leech lattice "based on Lie theory."

Remark 10.2.7: The Leech lattice is the unique positive definite even unimodular lattice Λ of rank 24 with $\Lambda_2 = \emptyset$, up to isometry (see [Niemeier], [Conway 3]), and the E_8-root lattice is the unique positive definite even unimodular lattice of rank 8 up to isometry, as we have already commented in Remark 6.3.7. Thus the lattices Φ, Ψ and Γ (with its original bilinear form) are all isometric:

$$\Gamma \simeq \Phi \simeq \Psi, \tag{10.2.35}$$

and in particular, this root lattice can be constructed either as in (6.3.6) or by the construction L_0 (10.2.6) or L_0' (10.2.8) using the Hamming code. Also, the construction (10.2.31) expresses the Leech lattice as the non-orthogonal direct sum of three rescaled copies of the E_8-root lattice.

Remark 10.2.8: We have constructed the Leech lattice as the lattice L_0' based on the Golay code. The lattice L_0 based on the Golay code is even unimodular but $(L_0)_2 \neq \emptyset$. In fact,

$$(L_0)_2 = \{\pm\alpha_i \mid i \in \Omega\}, \tag{10.2.36}$$

the root system of $\mathfrak{sl}(2, \mathbb{F})^{24} = A_1^{24}$, and the lattice Q [see (10.2.4)] is the root lattice of A_1^{24}. This lattice L_0 is called the *Niemeier lattice of type* A_1^{24}, and we sometimes write

$$L_0 = N(A_1^{24}). \tag{10.2.37}$$

Recall from (10.2.9) that $L_0 \cap L_0'$ has index two in both lattices:

$$|N(A_1^{24}) : N(A_1^{24}) \cap \Lambda| = |\Lambda : N(A_1^{24}) \cap \Lambda| = 2. \tag{10.2.38}$$

We have encountered a third even unimodular lattice of rank 24, namely, Γ^3 (10.2.30). Altogether, there are 24 even unimodular lattices of rank 24, up to isometry, called the *Niemeier lattices* (see [Niemeier]). For each such lattice L other than Λ, L_2 forms a root system which spans the vector space $L_{\mathbb{Q}}$, and [generalizing (10.2.37)] the corresponding Niemeier lattice is sometimes denoted $N(L_2)$. In particular,

$$\Gamma^3 = N(E_8^3). \tag{10.2.39}$$

Remark 10.2.9: In the setting of Proposition 10.2.3, we have

$$2M \subset 2L \subset M, \tag{10.2.40}$$

and there is clearly a reciprocity between L and M: In $M/2M$, the lattice $2L$ maps to an $n/2$-dimensional totally singular subspace with respect to the canonical quadratic form

$$\alpha + 2M \mapsto \tfrac{1}{2}\langle\alpha, \alpha\rangle + 2\mathbb{Z} \tag{10.2.41}$$

of (6.1.25). In particular,

$$2\Lambda \subset 2\Gamma^3 \subset \Lambda \tag{10.2.42}$$

[see (10.2.32)], and $\Lambda/2\Lambda$ contains a 12-dimensional totally singular subspace, namely, the image of $2\Gamma^3$ in $\Lambda/2\Lambda$.

Next we shall describe and count the shortest nonzero elements of the Leech lattice. In the notation of the beginning of this section, for $S \subset \Omega$ let

ε_S be the involution of $\mathfrak{h}$ given by:

$$\varepsilon_S : \alpha_k \mapsto \begin{cases} -\alpha_k & \text{if} \quad k \in S \\ \alpha_k & \text{if} \quad k \notin S \end{cases} \tag{10.2.43}$$

for $k \in \Omega$. It is easy to see from (10.2.6), (10.2.8), (10.2.9) and (10.2.11) that Λ_4 is composed of three types of elements:

$$\Lambda_4 = \Lambda_4^1 \cup \Lambda_4^2 \cup \Lambda_4^3 \quad \text{(disjoint)}, \tag{10.2.44}$$

where

$$\Lambda_4^1 = \{\tfrac{1}{2}\varepsilon_S \alpha_C \,|\, C \in \mathcal{C}, |C| = 8, S \subset C, |S| \in 2\mathbb{Z}\}, \tag{10.2.45}$$

$$\Lambda_4^2 = \{\pm\alpha_k \pm \alpha_l \,|\, k, l \in \Omega, k \neq l\}, \tag{10.2.46}$$

$$\Lambda_4^3 = \{\varepsilon_C(\tfrac{1}{4}\alpha_\Omega - \alpha_k) \,|\, C \in \mathcal{C}, k \in \Omega\}. \tag{10.2.47}$$

Counting, we find that

$$\begin{aligned} |\Lambda_4| &= |\Lambda_4^1| + |\Lambda_4^2| + |\Lambda_4^3| \\ &= 759 \cdot 2^7 + \binom{24}{2} \cdot 2^2 + 24 \cdot 2^{12} \\ &= 196560. \end{aligned} \tag{10.2.48}$$

Remark 10.2.10: Using Remark 10.1.9, we see from (10.2.8) that the Leech lattice is generated by Λ_4.

The analogous counts for the E_8-root lattices Φ and Ψ [see (10.2.23), (10.2.24)] are as follows:

$$\Phi_2 = \Phi_2^1 \cup \Phi_2^2 \quad \text{(disjoint)} \tag{10.2.49}$$

where

$$\Phi_2^1 = \{\tfrac{1}{2}\varepsilon_S \alpha_C \,|\, C \in \mathcal{C}_1, |C| = 4, S \subset C\}, \tag{10.2.50}$$

$$\Phi_2^2 = \{\pm\alpha_k \,|\, k \in \Omega\}, \tag{10.2.51}$$

so that

$$|\Phi_2| = |\Phi_2^1| + |\Phi_2^2| = 14 \cdot 2^4 + 16 = 240, \tag{10.2.52}$$

as expected. Also,

$$\Psi_2 = \Psi_2^1 \cup \Psi_2^2 \quad \text{(disjoint)} \tag{10.2.53}$$

where

$$\Psi_2^1 = \{\tfrac{1}{2}\varepsilon_S \alpha_C \,|\, C \in \mathcal{C}_2, |C| = 4, S \subset \mathcal{C}, |S| \in 2\mathbb{Z}\}, \tag{10.2.54}$$

$$\Psi_2^2 = \{\varepsilon_C(\tfrac{1}{4}\alpha_\Omega - \alpha_k) \,|\, C \in \mathcal{C}, k \in \Omega\}, \tag{10.2.55}$$

so that

$$|\Psi_2| = |\Psi_2^1| + |\Psi_2^2| = 14\cdot 2^3 + 8\cdot 2^4 = 240. \qquad (10.2.56)$$

The group of isometries of the Leech lattice is called the *Conway group* Co_0 (*or* $\cdot 0$):

$$\begin{aligned} \mathrm{Co}_0 = \cdot 0 &= \mathrm{Aut}(\Lambda; \langle\cdot,\cdot\rangle) \\ &= \{g \in \mathrm{Aut}\,\Lambda \mid \langle g\alpha, g\beta\rangle = \langle\alpha,\beta\rangle \quad \text{for} \quad \alpha,\beta\in\Lambda\} \end{aligned} \qquad (10.2.57)$$

[cf. (6.4.1)]. Its quotient by the central subgroup $\langle\pm 1\rangle$ is called the *Conway group* Co_1 (*or* $\cdot 1$):

$$\mathrm{Co}_1 = \cdot 1 = \mathrm{Co}_0/\langle\pm 1\rangle. \qquad (10.2.58)$$

We cite some basic facts about these groups without proof:

Remark 10.2.11 ([Conway 2], [Conway 4]): (a) The group Co_0 equals its commutator subgroup,

$$\mathrm{Co}_0 = (\mathrm{Co}_0, \mathrm{Co}_0), \qquad (10.2.59)$$

and

$$\mathrm{Cent}\,\mathrm{Co}_0 = \langle \pm 1\rangle. \qquad (10.2.60)$$

(b) The group Co_1 is a nonabelian simple group.
(c) The group Co_1 acts (faithfully and) irreducibly on $\Lambda/2\Lambda$.

Remark 10.2.12 ([Conway 2], [Conway 4]): (a) The subgroup

$$\mathrm{N}_{24} = \mathrm{Aut}(\Lambda; \langle\cdot,\cdot\rangle, L_0) \qquad (10.2.61)$$

of Co_0 which leaves L_0 invariant is given by:

$$\mathrm{N}_{24} = \{\pi\varepsilon_C \mid \pi \in \mathrm{M}_{24},\ C \in \mathcal{C}\}, \qquad (10.2.62)$$

where $\pi \in \mathrm{M}_{24}$ acts on $\mathfrak{h}$ via:

$$\pi\alpha_k = \alpha_{\pi k} \qquad (10.2.63)$$

for $k \in \Omega$, and ε_C denotes the automorphism

$$\varepsilon_C\alpha_k = \begin{cases} -\alpha_k & \text{if} \quad k \in C \\ \alpha_k & \text{if} \quad k \notin C. \end{cases} \qquad (10.2.64)$$

Thus N_{24} is an extension of M_{24} by a subgroup isomorphic to $\mathcal{C}$. It follows from Remarks 10.1.10 and 10.1.11 that N_{24} equals its commutator subgroup:

$$\mathrm{N}_{24} = (\mathrm{N}_{24}, \mathrm{N}_{24}). \qquad (10.2.65)$$

(b) Let $\{T_j \mid 0 \leq j \leq 5\}$ be a sextet in Ω as in Proposition 10.1.8(b). The linear automorphism ρ determined by:

$$\rho\alpha_k = \begin{cases} \alpha_k - \frac{1}{2}\alpha_{T_j} & \text{if} \quad k \in T_j, \quad j \neq 0 \\ -\alpha_k + \frac{1}{2}\alpha_{T_0} & \text{if} \quad k \in T_0 \end{cases} \tag{10.2.66}$$

lies in the Conway group:

$$\rho \in \mathrm{Co}_0. \tag{10.2.67}$$

In addition, N_{24} is a maximal subgroup of Co_0 and hence

$$\mathrm{Co}_0 = \langle \mathrm{N}_{24}, \rho \rangle. \tag{10.2.68}$$

(c) Set

$$\Lambda_0 = L_0 \cap \Lambda.$$

In the natural representation of N_{24} on $\Lambda_0/2\Lambda_0$,

$$\Lambda_0/2\Lambda_0 = \mathbb{F}_2\text{-span}\{g\lambda - \lambda \mid g \in \mathrm{N}_{24}, \lambda \in \Lambda_0/2\Lambda_0\}. \tag{10.2.69}$$

This follows from Remark 10.1.11, the fact that

$$\Lambda_0/(Q \cap \Lambda_0) \simeq \mathcal{C} \tag{10.2.70}$$

as modules for M_{24} and the fact that

$$\varepsilon_C \tfrac{1}{2}\alpha_D - \tfrac{1}{2}\alpha_D = -\alpha_{C \cap D}$$

for $C, D \in \mathcal{C}$; the sets $C \cap D$ generate all even subsets of Ω.

10.3. The Moonshine Module $V^\natural$ and the Griess Algebra $\mathfrak{B}$

Now that we have the Leech lattice available, we can construct our main objects of study.

First, using the Leech lattice Λ (Section 10.2), we form the untwisted space

$$V_\Lambda = S(\hat{\mathfrak{h}}_{\mathbb{Z}}^-) \otimes \mathbb{F}\{\Lambda\} \tag{10.3.1}$$

as in Chapters 7 and 8. We make the following special choices: In the notation of (7.1.6)–(7.1.9), we fix the central extension

$$1 \to \langle \kappa \rangle \to \hat{\Lambda} \xrightarrow{-} \Lambda \to 1 \tag{10.3.2}$$

where

$$\kappa^2 = 1, \quad \kappa \neq 1, \tag{10.3.3}$$

that is,

$$s = 2, \tag{10.3.4}$$

and where the commutator map is the alternating $\mathbb{Z}$-bilinear map

$$c_0(\alpha, \beta) = \langle \alpha, \beta \rangle + 2\mathbb{Z} \quad \text{for} \quad \alpha, \beta \in \Lambda, \tag{10.3.5}$$

as in (7.2.22). In the notation (7.1.14)–(7.1.21), we have

$$\omega = -1, \tag{10.3.6}$$

$$\mathbb{F}\{\Lambda\} = \mathbb{F}[\hat{\Lambda}]/(\kappa + 1)\mathbb{F}[\hat{\Lambda}] \tag{10.3.7}$$

and

$$c(\alpha, \beta) = (-1)^{\langle \alpha, \beta \rangle} \quad \text{for} \quad \alpha, \beta \in \Lambda, \tag{10.3.8}$$

so that as operators on V_Λ,

$$\kappa = -1 \tag{10.3.9}$$

and

$$ab = (-1)^{\langle \bar{a}, \bar{b} \rangle} ba \quad \text{for} \quad a, b \in \hat{\Lambda} \tag{10.3.10}$$

[recall (7.1.22), (7.1.32)].

Recall from (5.4.5) and Proposition 5.4.3 that the automorphisms of $\hat{\Lambda}$ which induce the involution -1 on Λ are automatically involutions and are parametrized by the quadratic forms on $\Lambda/2\Lambda$ with associated form induced by (10.3.5). Among these we fix the distinguished involution θ_0 determined by the canonical quadratic form q_1 given in (6.1.25):

$$\begin{aligned} q_1 \colon \Lambda/2\Lambda &\to \mathbb{F}_2 \\ \alpha + 2\Lambda &\mapsto \tfrac{1}{2}\langle \alpha, \alpha \rangle + 2\mathbb{Z}. \end{aligned} \tag{10.3.11}$$

As in (6.4.13) we have

$$\begin{aligned} \theta_0 \colon \hat{\Lambda} &\to \hat{\Lambda} \\ a &\mapsto a^{-1}\kappa^{\langle \bar{a}, \bar{a} \rangle/2}. \end{aligned} \tag{10.3.12}$$

Besides the general properties

$$\theta_0^2 = 1 \tag{10.3.13}$$

and

$$\theta_0(a^2) = a^{-2} \quad \text{for} \quad a \in \hat{\Lambda} \tag{10.3.14}$$

[recall (5.4.6)], we note that

$$\theta_0(a) = a^{-1} \quad \text{if} \quad \bar{a} \in \Lambda_4. \tag{10.3.15}$$

We also form the twisted space

$$V_\Lambda^T = S(\hat{\mathfrak{h}}_{\mathbb{Z}+1/2}^-) \otimes T \tag{10.3.16}$$

of Chapters 7 and 9. We shall fix the $\hat{\Lambda}$-module T using the considerations at the end of Section 7.4. Keeping in mind the choices above, we set

$$\begin{aligned} K &= \{\theta_0(a)a^{-1} \mid a \in \hat{\Lambda}\} \\ &= \{a^2\kappa^{\langle \bar{a}, \bar{a}\rangle/2} \mid a \in \hat{\Lambda}\} \end{aligned} \tag{10.3.17}$$

as in (7.4.36), a central subgroup of $\hat{\Lambda}$ such that

$$\bar{K} = 2\Lambda \tag{10.3.18}$$

[recall (7.4.37)]. Then $\hat{\Lambda}/K$ is a finite group which is a central extension,

$$1 \to \langle \kappa \rangle \to \hat{\Lambda}/K \to \Lambda/2\Lambda \to 1, \tag{10.3.19}$$

with commutator map induced by (10.3.5) and with squaring map the quadratic form q_1 (see Proposition 5.3.4). Since Λ is unimodular, q_1 is nonsingular, and by Proposition 5.3.3, $\hat{\Lambda}/K$ is an extraspecial 2-group with

$$|\hat{\Lambda}/K| = 2^{25}. \tag{10.3.20}$$

By Proposition 5.3.4 and (7.4.45),

$$\widehat{2\Lambda} = \langle \kappa \rangle \times K = \operatorname{Cent} \hat{\Lambda}. \tag{10.3.21}$$

Now from Remark 10.2.9 we know that $\Lambda/2\Lambda$ contains a 12-dimensional totally singular subspace with respect to q_1. Thus by Proposition 7.4.8 and Remark 7.4.9 (see also Theorem 5.5.1 and Remark 5.5.2) we conclude:

Proposition 10.3.1: *The extraspecial group $\hat{\Lambda}/K$ has a unique (up to equivalence) irreducible module T such that*

$$\kappa K \mapsto -1 \quad on \quad T. \tag{10.3.22}$$

Moreover, the corresponding representation of $\hat{\Lambda}/K$ is the unique faithful irreducible representation and

$$\dim T = 2^{12}. \tag{10.3.23}$$

To construct T, let Φ be any subgroup of Λ such that

$$\begin{gathered} 2\Lambda \subset \Phi \subset \Lambda, \\ |\Phi/2\Lambda| = 2^{12}, \\ q_1(\Phi) = 0 \end{gathered} \tag{10.3.24}$$

[*for example, the rescaled lattice* $2\Gamma^3$ *in (10.2.42)*]. *Then* $\hat{\Phi}$ *is a maximal abelian subgroup of* $\hat{\Lambda}$ *and* $\hat{\Phi}/K$ *is an elementary abelian 2-group. Let*

$$\psi\colon \hat{\Phi}/K \to \mathbb{F}^{\times} \tag{10.3.25}$$

be any homomorphism such that

$$\psi(\kappa K) = -1 \tag{10.3.26}$$

and denote by $\mathbb{F}_\psi$ *the one-dimensional* $\hat{\Phi}$*-module with the corresponding character. Then viewed as a* $\hat{\Lambda}$*-module,*

$$\begin{aligned} T &= \operatorname{Ind}_{\hat{\Phi}}^{\hat{\Lambda}} \mathbb{F}_\psi = \mathbb{F}[\hat{\Lambda}] \otimes_{\mathbb{F}[\hat{\Phi}]} \mathbb{F}_\psi \\ &\simeq \mathbb{F}[\Lambda/\Phi] \quad \textit{(linearly)}. \end{aligned} \tag{10.3.27}$$

Remark 10.3.2: This result shows further that the $\hat{\Lambda}/K$-module T has a $\mathbb{Q}$-form:

$$T \simeq (\operatorname{Ind}_{\hat{\Phi}/K}^{\hat{\Lambda}/K} \mathbb{Q}_{\psi_0}) \otimes_{\mathbb{Q}} \mathbb{F}, \tag{10.3.28}$$

where ψ_0 is any (rational-valued) character of $\hat{\Phi}/K$ such that $\psi_0(\kappa K) = -1$ (cf. the proof of Proposition 5.5.3). Thus we can choose a $\mathbb{Q}$-subspace $T_{\mathbb{Q}}$ of T, invariant under $\hat{\Lambda}/K$ and such that the canonical map

$$T_{\mathbb{Q}} \otimes_{\mathbb{Q}} \mathbb{F} \to T \tag{10.3.29}$$

is an isomorphism. It follows (by averaging a positive definite symmetric bilinear form on $T_{\mathbb{Q}}$ over the finite group $\hat{\Lambda}/K$) that T admits a nonsingular symmetric $\hat{\Lambda}/K$-invariant bilinear form, say, $\langle \cdot, \cdot \rangle$. The invariance condition asserts that

$$\langle g \cdot \tau_1, g \cdot \tau_2 \rangle = \langle \tau_1, \tau_2 \rangle \quad \text{for} \quad g \in \hat{\Lambda}/K, \quad \tau_i \in T. \tag{10.3.30}$$

By the irreducibility, such a form is unique up to normalization.

In the twisted space V_Λ^T (10.3.16), we take T to be the canonical $\hat{\Lambda}$-module described in Proposition 10.3.1. Of course for $a \in \hat{\Lambda}$ we have

$$\theta_0 a = a \quad \text{as operators on} \quad T \tag{10.3.31}$$

as in (7.4.14).

Now we can define the moonshine module—the space on which the Monster will act. Recall that θ_0 acts in a natural way on V_Λ [see (8.9.22)] and on V_Λ^T [see (9.2.87)]. We know that for $v \in V_\Lambda^{\theta_0}$ [the fixed space (8.9.25)], the component operators of both the untwisted and twisted vertex operators $Y(v, z)$ preserve the respective fixed spaces $V_\Lambda^{\theta_0}$ and $(V_\Lambda^T)^{\theta_0}$ [recall (8.9.29) and (9.2.93)–(9.2.94)]. *We define the moonshine module to be the space*

$$V^\natural = V_\Lambda^{\theta_0} \oplus (V_\Lambda^T)^{\theta_0} \tag{10.3.32}$$

($\natural$ for "natural"). For $v \in V_\Lambda$ we form the vertex operator

$$Y(v, z) = Y_{\mathbb{Z}}(v, z) \oplus Y_{\mathbb{Z}+1/2}(v, z) \tag{10.3.33}$$

acting on the larger space

$$W_\Lambda = V_\Lambda \oplus V_\Lambda^T. \tag{10.3.34}$$

Similarly, for the component operators of $Y(v, z)$ we write

$$\begin{aligned} v_n &= v_n \oplus v_n \\ x_v(n) &= x_v(n) \oplus x_v(n) \end{aligned} \tag{10.3.35}$$

on W_Λ, for $v \in V_\Lambda$, $n \in \mathbb{Q}$. Then

$$v_n \cdot V^\natural \subset V^\natural, \quad x_n(v) \cdot V^\natural \subset V^\natural \tag{10.3.36}$$

if $v \in V_\Lambda^{\theta_0}$. Of course we have special cases of (10.3.33) and (10.3.35) such as

$$\begin{aligned} \alpha(z) &= \alpha(z) \oplus \alpha(z) \\ \alpha(n) &= \alpha(n) \oplus \alpha(n) \end{aligned} \tag{10.3.37}$$

for $\alpha \in \mathfrak{h}$, $n \in \mathbb{Q}$, where at least one of the operators $\alpha(n)$ on the right is zero.

Because $\dim \mathfrak{h} = 24$ and $\Lambda_2 = \emptyset$, the space $V^\natural$ has some special structural features. First, $V^\natural$ is integrally graded, with degrees bounded above by 1:

$$V^\natural = \coprod_{\substack{n \in \mathbb{Z} \\ n \leq 1}} V_n^\natural, \tag{10.3.38}$$

as we observe using (1.9.51)–(1.9.53) and (9.2.96). Recalling (8.9.31) and the notations $\mathfrak{k}$ and $\mathfrak{p}$ from (8.9.32) and (9.5.19), we also see that

$$\begin{aligned} V_1^\natural &= \mathbb{F}\iota(1) \\ V_0^\natural &= 0 \\ V_{-1}^\natural &= \mathfrak{k} \oplus \mathfrak{p}, \end{aligned} \tag{10.3.39}$$

where in more detail,

$$\begin{aligned} \mathfrak{k} &= S^2(\mathfrak{h}) \oplus \sum_{a \in \hat{\Lambda}_4} \mathbb{F}x_a^+ \\ \mathfrak{p} &= \mathfrak{h} \otimes T \end{aligned} \tag{10.3.40}$$

by (8.9.50) and (9.5.22). We set

$$\mathcal{B} = V^{\natural}_{-1} = \mathfrak{k} \oplus \mathfrak{p}. \tag{10.3.41}$$

We have counted the elements of Λ_4 in (10.2.48), and we find that

$$\begin{aligned} \dim V^{\natural}_{1} &= 1 \\ \dim V^{\natural}_{0} &= 0 \\ \dim \mathcal{B} = \dim V^{\natural}_{-1} &= 196884, \end{aligned} \tag{10.3.42}$$

since

$$\dim \mathfrak{k} = 300 + \tfrac{1}{2}(196560) = 300 + 98280 = 98580 \tag{10.3.43}$$

[recall (8.9.52)] and

$$\dim \mathfrak{p} = 24 \cdot 2^{12} = 98304 \tag{10.3.44}$$

[from (10.3.23)].

Remark 10.3.3: The introduction of the space $V^{\natural}$ can be motivated in several ways: It admits the rich structure summarized in the next two theorems, it contains the Griess algebra in a natural way (see below) and it has graded dimension the modular function $J(q)$, with no constant term, as we shall prove later.

On $V^{\natural}$, the Jacobi identity takes the following simple form for vertex operators parametrized by $V_{\Lambda}^{\theta_0}$ [see (8.8.28), Theorem 8.8.9, (9.2.37), (9.5.10) and Corollary 9.5.4]:

Theorem 10.3.4: *For $v \in V_{\Lambda}^{\theta_0}$,*

$$Y(v, z) = \sum_{n \in \mathbb{Z}} v_n z^{-n-1} \quad on \quad V^{\natural}, \tag{10.3.45}$$

that is, $Y(v, z)$ involves only integral powers of z. For $u, v \in V_{\Lambda}^{\theta_0}$,

$$\begin{aligned} [Y(u, z_1) \times_{z_0} Y(v, z_2)] &= z_0^{-1}\delta\left(\frac{z_1 - z_2}{z_0}\right) Y(u, z_1)Y(v, z_2) \\ &\quad - z_0^{-1}\delta\left(\frac{z_2 - z_1}{-z_0}\right) Y(v, z_2)Y(u, z_1) \\ &= z_2^{-1}\delta\left(\frac{z_1 - z_0}{z_2}\right) Y(Y_{\mathbb{Z}}(u, z_0)v, z_2) \end{aligned} \tag{10.3.46}$$

on $V^{\natural}$. In particular, in the terminology of Section 8.10, $V_{\Lambda}^{\theta_0}$ is a vertex operator algebra of rank 24 and $(V_{\Lambda}^{T})^{\theta_0}$ is a $V_{\Lambda}^{\theta_0}$-module.

(Here we also use Propositions 8.7.1 and 9.4.3.)

Moreover, from Theorems 8.9.5 and 9.5.10 and Remark 8.9.8 we know the following:

Theorem 10.3.5: *The space $\mathfrak{k}$ is a commutative nonassociative algebra with identity under the product*

$$u \times v = u_1 \cdot v, \tag{10.3.47}$$

and the bilinear form

$$\langle u, v\rangle = u_3 \cdot v \tag{10.3.48}$$

is nonsingular, symmetric and associative. The space $V^\natural$ is a graded module for the commutative affinization $\hat{\mathfrak{k}}$ of $\mathfrak{k}$ under the action

$$\pi: \hat{\mathfrak{k}} \to \operatorname{End} V^\natural \tag{10.3.49}$$

defined by

$$\begin{aligned} \pi: u \otimes t^n &\mapsto x_u(n) \quad \textit{for} \quad u \in \mathfrak{k}, \quad n \in \mathbb{Z} \\ \pi: e &\mapsto 1. \end{aligned} \tag{10.3.50}$$

One of the main results in this work will be an extension of this action to a representation

$$\hat{\mathfrak{B}} \to \operatorname{End} V^\natural \tag{10.3.51}$$

of a larger commutative affinization by cross-bracket on $V^\natural$, where the space $\mathfrak{B}$ [see (10.3.41)] is given the structure of a commutative nonassociative algebra with identity and with a nonsingular symmetric associative form in the following natural way: The product, denoted $\times$, and the form, denoted $\langle\cdot,\cdot\rangle$, on $\mathfrak{B}$ extend those on $\mathfrak{k}$ [recall the definitions in Theorem 8.9.5 and the explicit formulas in (8.9.53)–(8.9.61)]. For $u \in \mathfrak{k}$ and $v \in \mathfrak{p}$ we use (10.3.47) and the commutativity of the product on $\mathfrak{k}$ as motivation to define

$$u \times v = v \times u = u_1 \cdot v, \tag{10.3.52}$$

and we use (10.3.48) and the symmetry of the form on $\mathfrak{k}$ as motivation to define

$$\langle u, v\rangle = \langle v, u\rangle = u_3 \cdot v = 0. \tag{10.3.53}$$

[The fact that $u_3 \cdot v = 0$ is clear by consideration of the gradation of the space $(V_\Lambda^T)^{\theta_0}$.] By Remark 8.9.4 and (9.5.20), $\frac{1}{2}\omega$ is an identity element of $\mathfrak{B}$. Next we define a nonsingular symmetric bilinear form $\langle\cdot,\cdot\rangle$ on

$\mathfrak{p} = \mathfrak{h} \otimes T$ by the formula

$$\langle h_1 \otimes \tau_1, h_2 \otimes \tau_2 \rangle = \tfrac{1}{2}\langle h_1, h_2 \rangle \langle \tau_1, \tau_2 \rangle \tag{10.3.54}$$

for $h_i \in \mathfrak{h}$, $\tau_i \in T$ (recall the nonsingular symmetric $\hat{\Lambda}$-invariant form $\langle \cdot, \cdot \rangle$ on T given in Remark 10.3.2). The factor $\frac{1}{2}$ in (10.3.54) will fit in naturally in Sections 12.3 and 12.5. Finally, we define a product $\times$ on $\mathfrak{p}$ so that

$$\mathfrak{p} \times \mathfrak{p} \subset \mathfrak{k} \tag{10.3.55}$$

and uniquely determined by the nonsingularity of the form on $\mathfrak{k}$ and the associativity condition

$$\langle u, v \times w \rangle = \langle u \times v, w \rangle \quad \text{for} \quad u, v \in \mathfrak{p}, \quad w \in \mathfrak{k}. \tag{10.3.56}$$

The commutativity of this product on $\mathfrak{p}$ follows from the explicit formula for it given below.

The resulting nonassociative algebra $\mathfrak{B}$ equipped with its form $\langle \cdot, \cdot \rangle$ is called the *Griess algebra*. It is actually a slight modification, with a natural identity element, of the algebra defined in [Griess 2], [Griess 3]. We have:

Proposition 10.3.6: *The Griess algebra $\mathfrak{B} = \mathfrak{k} \oplus \mathfrak{p}$ is a commutative nonassociative algebra with identity element $\frac{1}{2}\omega \in \mathfrak{k}$, and the form $\langle \cdot, \cdot \rangle$ on $\mathfrak{B}$ is nonsingular, symmetric and associative. We have*

$$\begin{aligned} &\mathfrak{k} \times \mathfrak{k} \subset \mathfrak{k} \\ &\mathfrak{k} \times \mathfrak{p} \subset \mathfrak{p} \\ &\mathfrak{p} \times \mathfrak{p} \subset \mathfrak{k}, \end{aligned} \tag{10.3.57}$$

with explicit formulas given by (8.9.53)–(8.9.55), (9.5.24)–(9.5.25) and:

$$\begin{aligned} h_1 \otimes \tau_1 \times h_2 \otimes \tau_2 = {} & \tfrac{1}{8}(2h_1h_2 + \langle h_1, h_2 \rangle \tfrac{1}{2}\omega)\langle \tau_1, \tau_2 \rangle \\ & + \tfrac{1}{128} \sum_{a \in \hat{\Lambda}_4} (\langle h_1, h_2 \rangle - 2\langle \bar{a}, h_1 \rangle \langle \bar{a}, h_2 \rangle) \langle \tau_1, a \cdot \tau_2 \rangle x_a^+ \end{aligned} \tag{10.3.58}$$

for $h_i \in \mathfrak{h}$, $\tau_i \in T$. We also have

$$\langle \mathfrak{k}, \mathfrak{p} \rangle = 0, \tag{10.3.59}$$

and explicit formulas for the form on $\mathfrak{k}$ and $\mathfrak{p}$ are given by (8.9.59)–(8.9.61) and (10.3.54). The identity element satisfies:

$$\langle \tfrac{1}{2}\omega, \tfrac{1}{2}\omega \rangle = 3. \tag{10.3.60}$$

Proof: In verifying (10.3.58) we use (10.3.15) and (10.3.31), and in checking the commutativity from this formula we also use the invariance (10.3.30). Formula (10.3.60) is a special case of (8.9.63), and the rest follows. ∎

The significance of $\mathfrak{B}$ is that Griess, who introduced this algebra, has constructed a group of automorphisms of it, preserving the form $\langle \cdot, \cdot \rangle$, and has shown this group to be a finite simple group (the Monster or Friendly Giant) [Griess 3]. In fact, Tits has shown that the Monster is the *full* automorphism group of $\mathfrak{B}$ ([Tits 4], [Tits 6]). Having reconstructed the Griess algebra using properties of vertex operators, we shall also reconstruct the Monster using properties of vertex operators, and exhibit a natural action of it on $V^\natural$.

10.4. The Group C and Its Actions on $V^\natural$ and on $\mathfrak{B}$

In a sequence of steps, we now proceed to define and establish the basic properties of a group C which will act naturally on $V^\natural$, $\mathfrak{B}$ and $\hat{\mathfrak{B}}$, and which moreover will act compatibly with the appropriate vertex operators. This group will be the centralizer of an involution in the Monster.

Starting with the central extension $\hat{\Lambda}$ (10.3.2) we first set

$$C_0 = \{g \in \operatorname{Aut} \hat{\Lambda} \mid \bar{g} \in \mathrm{Co}_0\}, \tag{10.4.1}$$

where $\bar{g}$ is the automorphism of the Leech lattice Λ induced by g and Co_0 is the isometry group of Λ (10.2.57). We know that

$$g\kappa = \kappa \tag{10.4.2}$$

automatically (5.4.2). We have met the group C_0 before, in Proposition 6.4.1. By this result (or Proposition 5.4.1), the sequence

$$1 \to \operatorname{Hom}(\Lambda, \mathbb{Z}/2\mathbb{Z}) \xrightarrow{*} C_0 \xrightarrow{-} \mathrm{Co}_0 \to 1 \tag{10.4.3}$$

is exact, where

$$\begin{aligned} \lambda^*: \hat{\Lambda} &\to \hat{\Lambda} \\ a &\mapsto a\kappa^{\lambda(\bar{a})} \end{aligned} \tag{10.4.4}$$

for $\lambda \in \operatorname{Hom}(\Lambda, \mathbb{Z}/2\mathbb{Z})$. Moreover, as in Remark 5.4.7, we have natural identifications

$$\operatorname{Hom}(\Lambda, \mathbb{Z}/2\mathbb{Z}) = \Lambda/2\Lambda = \operatorname{Inn} \hat{\Lambda}, \tag{10.4.5}$$

so that the exact sequence (10.4.3) can be written

$$1 \to \operatorname{Inn} \hat{\Lambda} \hookrightarrow C_0 \xrightarrow{-} \mathrm{Co}_0 \to 1. \tag{10.4.6}$$

Now C_0 induces a group of automorphisms of the extraspecial group $\hat{\Lambda}/K$ since C_0 preserves K [see (10.3.17) and Proposition 10.3.1], and we have a natural homomorphism

$$\varphi\colon C_0 \to \mathrm{Aut}(\hat{\Lambda}/K). \tag{10.4.7}$$

We claim that

$$\operatorname{Ker} \varphi = \langle \theta_0 \rangle \tag{10.4.8}$$

[recall (10.3.12)]. In fact, it is clear that $\theta_0 \in \operatorname{Ker} \varphi$. On the other hand,

$$\operatorname{Ker} \varphi \cap \operatorname{Inn} \hat{\Lambda} = 1 \tag{10.4.9}$$

since

$$K \cap \langle \kappa \rangle = 1. \tag{10.4.10}$$

Hence $\operatorname{Ker} \varphi$ is isomorphic to its image in Co_0 by (10.4.6). But $\overline{\operatorname{Ker} \varphi}$ acts trivially on $\Lambda/2\Lambda$ by (10.3.18), so that by the faithfulness of the action of Co_1 on $\Lambda/2\Lambda$ [Remark 10.2.11(c)],

$$\overline{\operatorname{Ker} \varphi} \subset \langle \pm 1 \rangle, \tag{10.4.11}$$

proving the claim.

Set

$$C_1 = \varphi(C_0) \subset \mathrm{Aut}(\hat{\Lambda}/K). \tag{10.4.12}$$

Then we have an exact sequence

$$1 \to \operatorname{Inn} \hat{\Lambda} \to C_1 \xrightarrow{-} \mathrm{Co}_1 \to 1, \tag{10.4.13}$$

where we continue to use the notation $^-$. Here Co_1 acts in the natural way on $\operatorname{Inn} \hat{\Lambda} = \Lambda/2\Lambda$, and it follows from Remark 10.2.11(b) and (c) that C_1 equals its commutator subgroup and has trivial center:

$$\begin{aligned} C_1 &= (C_1, C_1), \\ \operatorname{Cent} C_1 &= 1. \end{aligned} \tag{10.4.14}$$

Now from Proposition 10.3.1 we recall that $\hat{\Lambda}/K$ satisfies the hypotheses of Theorem 5.5.1 (see also Remark 5.5.2), and Proposition 5.5.3 gives the exact sequence of canonical maps

$$1 \to \mathbb{F}^\times \to \mathbf{N}_{\mathrm{Aut}\, T}(\pi(\hat{\Lambda}/K)) \xrightarrow{\mathrm{int}} \mathrm{Aut}(\hat{\Lambda}/K) \to 1, \tag{10.4.15}$$

where π denotes the (faithful) representation of $\hat{\Lambda}/K$ on T and

$$\operatorname{int}(g)(x) = gxg^{-1} \tag{10.4.16}$$

for $g \in \operatorname{Aut} T$, $x \in \hat{\Lambda}/K = \pi(\hat{\Lambda}/K)$.

Set

$$C_* = \{g \in \mathbf{N}_{\operatorname{Aut} T}(\pi(\hat{\Lambda}/K)) \mid \operatorname{int}(g) \in C_1\}, \tag{10.4.17}$$

so that we have the commutative diagram with exact rows

$$\begin{array}{ccccccccc} 1 & \to & \mathbb{F}^\times & \longrightarrow & C_* & \longrightarrow & C_1 & \longrightarrow & 1 \\ & & \| & & \downarrow & & \downarrow & & \\ 1 & \to & \mathbb{F}^\times & \to & \mathbf{N}_{\operatorname{Aut} T}(\pi(\hat{\Lambda}/K)) & \to & \operatorname{Aut}(\hat{\Lambda}/K) & \to & 1. \end{array} \tag{10.4.18}$$

Also set

$$C_T = (C_*, C_*). \tag{10.4.19}$$

We shall now show that C_T contains -1 and in fact all of $\pi(\hat{\Lambda}/K)$. Since

$$\operatorname{int}(\pi(\hat{\Lambda}/K)) = \Lambda/2\Lambda = \operatorname{Inn} \hat{\Lambda} \tag{10.4.20}$$

by Remark 5.4.7 we see that

$$\pi(\hat{\Lambda}/K) \subset C_*, \tag{10.4.21}$$

and so

$$-1 = \pi(\kappa K) \in (\pi(\hat{\Lambda}/K), \pi(\hat{\Lambda}/K)) \subset C_T. \tag{10.4.22}$$

But since Co_1 acts irreducibly on $\Lambda/2\Lambda$ [Remark 10.2.11(c)],

$$\operatorname{Inn} \hat{\Lambda} = (\operatorname{Inn} \hat{\Lambda}, C_1) \tag{10.4.23}$$

[see (10.4.13)], and it follows from (10.4.21) and (10.4.22) that

$$\pi(\hat{\Lambda}/K) = (\pi(\hat{\Lambda}/K), C_*) \subset C_T. \tag{10.4.24}$$

We claim that the sequence

$$1 \to \langle \pm 1 \rangle \hookrightarrow C_T \xrightarrow{\operatorname{int}} C_1 \to 1 \tag{10.4.25}$$

is exact. By (10.4.14), all we need to show is that

$$C_T \cap \mathbb{F}^\times = \langle \pm 1 \rangle. \tag{10.4.26}$$

To see this we use the fact that the $\hat{\Lambda}/K$-module T has a $\mathbb{Q}$-form, constructed in Remark 10.3.2: We have

$$T \simeq (\operatorname{Ind}_{\hat{\Phi}/K}^{\hat{\Lambda}/K} \mathbb{Q}_{\psi_0}) \otimes_{\mathbb{Q}} \mathbb{F}, \tag{10.4.27}$$

where ψ_0 is any (rational-valued) character of $\hat{\Phi}/K$ such that

$$\psi_0(\kappa K) = -1, \tag{10.4.28}$$

and this gives us a $\hat{\Lambda}$-invariant $\mathbb{Q}$-subspace $T_{\mathbb{Q}}$ of T such that the canonical map

$$T_{\mathbb{Q}} \otimes_{\mathbb{Q}} \mathbb{F} \to T \tag{10.4.29}$$

is an isomorphism. Let

$$C_{*,\mathbb{Q}} = C_* \cap \operatorname{Aut} T_{\mathbb{Q}}. \tag{10.4.30}$$

Proposition 5.5.3 then gives the exact sequence

$$1 \to \mathbb{Q}^\times \to C_{*,\mathbb{Q}} \to C_1 \to 1, \tag{10.4.31}$$

and so

$$C_* = C_{*,\mathbb{Q}}\mathbb{F}^\times \tag{10.4.32}$$

and

$$\begin{gathered} C_T \subset C_{*,\mathbb{Q}}, \\ C_T \cap \mathbb{F}^\times \subset C_{*,\mathbb{Q}} \cap \mathbb{F}^\times = \mathbb{Q}^\times. \end{gathered} \tag{10.4.33}$$

But since

$$\det C_T = 1, \tag{10.4.34}$$

we also have

$$C_T \cap \mathbb{F}^\times \subset \{\mu \in \mathbb{F}^\times \mid \mu^{2^{12}} = 1\}, \tag{10.4.35}$$

and this proves (10.4.26) and hence the claim.

Using (10.4.13) we have a map

$$\overline{\mathrm{int}} = {}^- \circ \mathrm{int}\colon C_T \to \mathrm{Co}_1, \tag{10.4.36}$$

and by (10.4.20) and (10.4.24),

$$\pi(\hat{\Lambda}/K) \subset \operatorname{Ker} \overline{\mathrm{int}}. \tag{10.4.37}$$

Consideration of the order of C_T [from (10.4.13) and (10.4.25)] shows that the sequence

$$1 \to \hat{\Lambda}/K \xrightarrow{\pi} C_T \xrightarrow{\overline{\mathrm{int}}} \mathrm{Co}_1 \to 1 \tag{10.4.38}$$

is exact.

Summarizing, we have an extension C_0 of Co_0 by $\Lambda/2\Lambda$ (10.4.6), an extension C_1 of Co_1 by $\Lambda/2\Lambda$ (10.4.13) and an extension C_T of Co_1 by the extraspecial group $\hat{\Lambda}/K$ (10.4.38). Now form the pullback

$$\hat{C} = \{(g, g_T) \in C_0 \times C_T \mid \varphi(g) = \overline{\mathrm{int}}(g_T)\}, \tag{10.4.39}$$

so that we have the commutative diagram of surjections

$$\begin{array}{ccc} \hat{C} & \xrightarrow{\pi_1} & C_0 \\ {\scriptstyle \pi_2}\downarrow & & \downarrow{\scriptstyle \varphi} \\ C_T & \xrightarrow{\text{int}} & C_1. \end{array} \tag{10.4.40}$$

Set

$$\begin{aligned} \hat{\theta}_0 &= (\theta_0, 1) \in \hat{C} \\ \hat{\theta} &= (1, -1) \in \hat{C}. \end{aligned} \tag{10.4.41}$$

Then

$$\begin{aligned} \operatorname{Ker} \pi_1 &= \langle \hat{\theta} \rangle \\ \operatorname{Ker} \pi_2 &= \langle \hat{\theta}_0 \rangle \end{aligned} \tag{10.4.42}$$

and

$$\operatorname{Cent} \hat{C} = \langle \hat{\theta}_0 \rangle \times \langle \hat{\theta} \rangle = \operatorname{Ker}(\varphi \circ \pi_1) \tag{10.4.43}$$

since $\operatorname{Cent} C_1 = 1$ (10.4.14) and

$$\theta_0 \in \operatorname{Cent} C_0 \tag{10.4.44}$$

from the definitions.

We are finally ready to define the group C: Set

$$C = \hat{C}/\langle \hat{\theta}_0 \hat{\theta} \rangle. \tag{10.4.45}$$

Then the diagram (10.4.40) enlarges to the commutative diagram of surjections

$$\begin{array}{ccccc} \hat{C} & & \xrightarrow{\pi_1} & & C_0 \\ & \searrow^{\pi_0} & & & \\ {\scriptstyle \pi_2}\downarrow & & C & & \downarrow{\scriptstyle \varphi} \\ & & & \searrow^{\sigma} & \\ C_T & & \xrightarrow{\text{int}} & & C_1. \end{array} \tag{10.4.46}$$

Also,

$$\operatorname{Ker} \pi_0 = \langle \hat{\theta}_0 \hat{\theta} \rangle$$

and

$$\operatorname{Cent} C = \langle z \mid z^2 = 1 \rangle = \operatorname{Ker} \sigma, \tag{10.4.47}$$

where

$$z = \pi_0(\hat{\theta}_0) = \pi_0(\hat{\theta}). \tag{10.4.48}$$

(No confusion should arise between this notation and our formal variable notation.) We have the exact sequence

$$1 \to \langle z \rangle \hookrightarrow C \xrightarrow{\sigma} C_1 \to 1. \tag{10.4.49}$$

Proceeding as in (10.4.36)–(10.4.38), we have a map

$$\bar{\sigma} = {}^{-} \circ \sigma: C \to \mathrm{Co}_1 \tag{10.4.50}$$

from (10.4.13). Moreover, there is a canonical embedding

$$\begin{aligned} \nu: \hat{\Lambda}/K &\to \hat{C} \\ gK &\mapsto (\mathrm{int}(g), \pi(g)), \end{aligned} \tag{10.4.51}$$

since

$$\varphi(\mathrm{int}(g)) = \mathrm{int}(\pi(g)). \tag{10.4.52}$$

The result is an exact sequence

$$1 \longrightarrow \hat{\Lambda}/K \xrightarrow{\pi_0 \circ \nu} C \xrightarrow{\bar{\sigma}} \mathrm{Co}_1 \longrightarrow 1, \tag{10.4.53}$$

and we have proved:

Proposition 10.4.1: *The group C is an extension of* Co_1 *by the extraspecial group $\hat{\Lambda}/K$. The nontrivial central element of $\hat{\Lambda}/K$ identifies with the nontrivial central element of C.*

Now that the group C is constructed we shall set up its canonical action on the moonshine module $V^\natural$. First we shall define an action of the larger group $\hat{C}$ on the larger space W_Λ [see (10.3.34)].

For $g \in \mathrm{Co}_0$ and $Z = \mathbb{Z}$ or $\mathbb{Z} + 1/2$ let g also denote the unique algebra automorphism

$$g: S(\hat{\mathfrak{h}}_{\bar{Z}}) \to S(\hat{\mathfrak{h}}_{\bar{Z}}) \tag{10.4.54}$$

such that g agrees with its natural action on $\hat{\mathfrak{h}}_{\bar{Z}}$. For $g \in C_0$ let g also denote the operator

$$\begin{aligned} g: \mathbb{F}\{\Lambda\} &\to \mathbb{F}\{\Lambda\} \\ \iota(a) &\mapsto \iota(ga) \end{aligned} \tag{10.4.55}$$

for $a \in \hat{\Lambda}$; note that this is well defined since $g\kappa = \kappa$. For $k = (g, g_T) \in \hat{C}$ let k also denote the operator

$$k = \bar{g} \otimes g \oplus \bar{g} \otimes g_T \tag{10.4.56}$$

on

$$W_\Lambda = S(\hat{\mathfrak{h}}_{\mathbb{Z}}^-) \otimes \mathbb{F}\{\Lambda\} \oplus S(\hat{\mathfrak{h}}_{\mathbb{Z}+1/2}^-) \otimes T. \tag{10.4.57}$$

This clearly gives a faithful representation of $\hat{C}$ on W_Λ.

Remark 10.4.2: This representation of $\hat{C}$ is faithful even on a small subspace of W_Λ, for instance, $T \oplus \mathfrak{p}$ [recall (10.3.40)].

Remark 10.4.3: The action of $\hat{C}$ on W_Λ extends the action of θ_0 already defined in (8.9.22) and (9.2.87) (for the case W_Λ), in such a way that this operator corresponds to the element $\hat{\theta}_0\hat{\theta} = (\theta_0, -1)$ of $\hat{C}$ [recall (10.4.41)]. Note that the relation $\pi(\theta_0 a) = \pi(a)$ for $a \in \hat{\Lambda}$ [(7.4.14), (10.3.31)] is a special case of the relation $\varphi(g) = \operatorname{int} h$. We have

$$\hat{C} \cdot V^\natural \subset V^\natural.$$

From the definitions of C and $V^\natural$ and the last remark, we see that C acts in a natural way on $V^\natural$: For $k = (g, g_T) \in \hat{C}$, $\pi_0(k)$ acts as the operator

$$\pi_0(k) = \bar{g} \otimes g \oplus \bar{g} \otimes g_T. \tag{10.4.58}$$

Remark 10.4.4: The action of C on $V^\natural$ is faithful, even on $\mathfrak{p} = \mathfrak{h} \otimes T$.

The decomposition (10.3.32) of $V^\natural$ is the eigenspace decomposition with respect to the central involution z in C (10.4.48), and we introduce corresponding notation:

$$\begin{aligned} V_\Lambda^{\theta_0} &= V^z = \{v \in V^\natural \mid z \cdot v = v\} \\ (V_\Lambda^T)^{\theta_0} &= V^{-z} = \{v \in V^\natural \mid z \cdot v = -v\}. \end{aligned} \tag{10.4.59}$$

Note that

$$\begin{gathered} V^\natural = V^z \oplus V^{-z} \\ C \cdot V^z \subset V^z, \quad C \cdot V^{-z} \subset V^{-z}. \end{gathered} \tag{10.4.60}$$

Remark 10.4.5: The actions of C on $V^\natural$ and of $\hat{C}$ on W_Λ preserve the homogeneous subspaces with respect to the gradings.

We now examine the relationship between the actions of $\hat{C}$ and of the vertex operators $Y(v, z)$, $v \in V_\Lambda$, on W_Λ [recall (10.3.33)–(10.3.37)]. Let

$$k = (g, g_T) \in \hat{C}, \tag{10.4.61}$$

and consider the operator k on W_Λ given by (10.4.56).

First we generalize (8.9.26) and (9.2.89) (for the case W_Λ): Since

$$[\alpha(m), \beta(n)] = \langle \alpha, \beta \rangle m \delta_{m+n,0} \tag{10.4.62}$$

for $\alpha, \beta \in \mathfrak{h}$ and $m, n \in \mathbb{Z}$ (respectively, $\mathbb{Z} + \frac{1}{2}$) on V_Λ (respectively, V_Λ^T) and since $\bar{g}$ is an isometry with respect to $\langle \cdot, \cdot \rangle$, we find easily that on W_Λ,

$$k\alpha(n)k^{-1} = (\bar{g}\alpha)(n) \tag{10.4.63}$$

for $\alpha \in \mathfrak{h}$, $n \in \frac{1}{2}\mathbb{Z}$. Equivalently,

$$k\alpha(z)k^{-1} = (\bar{g}\alpha)(z) \tag{10.4.64}$$

for $\alpha \in \mathfrak{h}$.

Next we generalize (8.9.27) and (9.2.90): From (10.4.63) and the definitions we obtain

$$kY(a, z)k^{-1} = Y(ga, z) \tag{10.4.65}$$

for $a \in \hat{\Lambda}$, where in the untwisted case we use the relation

$$g \circ a \circ g^{-1} = g(a) \quad \text{on} \quad \mathbb{F}\{\Lambda\} \tag{10.4.66}$$

and in the twisted case we use the fact that $\varphi(g) = \operatorname{int}(g_T)$.

Finally, using (10.4.64), (10.4.65) and the definitions we generalize these formulas and (8.9.28) and (9.2.92) for the case W_Λ:

Theorem 10.4.6: *For $k \in \hat{C}$ and $v \in V_\Lambda$,*

$$kY(v, z)k^{-1} = Y(kv, z) \tag{10.4.67}$$

on W_Λ.

Proof: For the twisted case we use the fact that $\bar{g}$ is an isometry with respect to $\langle \cdot, \cdot \rangle$ to verify that

$$k \circ \Delta_z = \Delta_z \circ k \tag{10.4.68}$$

on V_Λ [cf. (9.2.39)], which implies that

$$k \circ \exp(\Delta_z) = \exp(\Delta_z) \circ k \tag{10.4.69}$$

on V_Λ [cf. (9.2.40)]. ∎

In particular we see from the definitions of the actions of the Virasoro algebra on V_Λ and V_Λ^T (Sections 8.7 and 9.4) that the Virasoro algebra commutes with $\hat{C}$ on W_Λ:

Corollary 10.4.7: *For $k \in \hat{C}$,*

$$L(z)k = kL(z) \tag{10.4.70}$$

on W_Λ.

Of course these results about the action of $\hat{C}$ on W_Λ have immediate consequences about the action of C on $V^\natural$:

Corollary 10.4.8: *For $k \in C$ and $v \in V^z = V_\Lambda^{\theta_0}$,*

$$kY(v, z)k^{-1} = Y(kv, z) \tag{10.4.71}$$

on $V^\natural$. In particular, in the terminology of Section 8.10, C acts as automorphisms of the vertex operator algebra V^z and of its module V^{-z} (recall Theorem 10.3.4).

Corollary 10.4.9: *For $k \in C$,*

$$L(z)k = kL(z) \tag{10.4.72}$$

on $V^\natural$.

Remark 10.4.10: Let $k \in C$. Formula (10.4.71) is equivalent to the statement

$$kY(u, z)v = Y(ku, z)kv \tag{10.4.73}$$

for $u \in V^z$, $v \in V^\natural$. [Of course, a similar comment holds for (10.4.67).] Also,

$$k[Y(u, z_1) \times_{z_0} Y(v, z_2)]k^{-1} = [Y(ku, z_1) \times_{z_0} Y(kv, z_2)] \tag{10.4.74}$$

on $V^\natural$, for $u, v \in V^z$ [cf. (10.3.46)]. In terms of components, we have

$$\begin{aligned} k(u_n v) &= (ku)_n(kv) \\ k(x_u(n)v) &= x_{ku}(n)kv \end{aligned} \tag{10.4.75}$$

for $u \in V^z$, $v \in V^\natural$, $n \in \mathbb{Z}$, and

$$\begin{aligned} k[u \times_l v]_{mn} k^{-1} &= [ku \times_l kv]_{mn} \\ k[x_u \times_l x_v](m, n)k^{-1} &= [x_{ku} \times_l x_{kv}](m, n) \end{aligned} \tag{10.4.76}$$

on $V^\natural$, for $u, v \in V^z$, $l, m, n \in \mathbb{Z}$ [recall (8.8.15)–(8.8.16) and (9.5.8)–(9.5.9)].

The group C acts on the space $\mathfrak{B} = V^\natural_{-1}$ (10.3.41) and in fact preserves the summands $\mathfrak{k}$ and $\mathfrak{p}$. From (10.4.75) and the definition of the product $\times$ and the form $\langle\cdot,\cdot\rangle$ on $\mathfrak{B}$ given in (10.3.47)–(10.3.56), we find:

Proposition 10.4.11: *The group C acts faithfully as automorphisms of the algebra $\mathfrak{B}$ and as isometries of $\langle\cdot,\cdot\rangle$.*

Proof: In checking that the form (10.3.54) on $\mathfrak{p}$ is preserved by C, we use the fact that the form $\langle\cdot,\cdot\rangle$ on T is C_T-invariant:

$$\langle g\tau_1, g\tau_2\rangle = \langle\tau_1, \tau_2\rangle \quad \text{for} \quad g \in C_T, \quad \tau_i \in T. \tag{10.4.77}$$

This is proved as follows: As in Remark 10.3.2 or (10.4.27)–(10.4.29), T has a $\mathbb{Q}$-form and (10.4.33) shows that C_T preserves this $\mathbb{Q}$-form, which implies that C_T leaves invariant some nonsingular symmetric invariant bilinear form on T. But this form must agree with the form $\langle\cdot,\cdot\rangle$ by the uniqueness of a $\hat{\Lambda}/K$-invariant form (recall Remark 10.3.2 again), and (10.4.77) is proved. The other parts of the verification are clear. ∎

Recall from Theorem 10.3.5 that $V^\natural$ is a graded module for the commutative affinization $\hat{\mathfrak{k}}$ of $\mathfrak{k}$. We shall relate this structure to the action of C.

Given a commutative nonassociative algebra $\mathfrak{b}$ with a symmetric form and given a group G of linear automorphisms of $\mathfrak{b}$, we let G act as linear automorphisms of $\hat{\mathfrak{b}}$ by:

$$\begin{aligned} g\cdot e &= e \\ g\cdot(u\otimes t^n) &= (g\cdot u)\otimes t^n \end{aligned} \tag{10.4.78}$$

for $g \in G$, $u \in \mathfrak{b}$, $n \in \mathbb{Z}$ [recall the notation (8.9.39)–(8.9.40)]. If G acts as algebra automorphisms and isometries of $\mathfrak{b}$, then G acts as algebra automorphisms of $\hat{\mathfrak{b}}$. Suppose that V is a graded $\hat{\mathfrak{b}}$-module and that G acts as linear automorphisms of V, preserving each homogeneous subspace V_n. Then we call V a *graded* $(G, \hat{\mathfrak{b}})$*-module* if

$$gxg^{-1} = g\cdot x \tag{10.4.79}$$

as operators on V, for $g \in G$ and $x \in \hat{\mathfrak{b}}$. By (10.4.75) and Proposition 10.4.11 we have:

Theorem 10.4.12: *The space $V^\natural$ is a graded $(C, \hat{\mathfrak{k}})$-module, and C acts as automorphisms of $\hat{\mathfrak{k}}$ and in fact of $\hat{\mathfrak{B}}$.*

Later we shall enlarge C to a group M (the Monster) of automorphisms and isometries of $\mathcal{B}$ (and hence automorphisms of $\hat{\mathcal{B}}$), and we shall make $V^\natural$ a graded $(M, \hat{\mathcal{B}})$-module. Furthermore, we shall define vertex operators $Y(v, z)$ (on $V^\natural$) for all $v \in V^\natural$ and we shall extend (10.4.71) to M and $V^\natural$.

Remark 10.4.13: Under the action of C, $\mathcal{B}$ breaks into the following four invariant subspaces:

$$\mathbb{F}\omega, \quad \{u \in S^2(\mathfrak{h}) \mid \langle u, \omega\rangle = 0\}, \quad \sum_{a \in \hat{\Lambda}_4} \mathbb{F}x_a^+, \quad \mathfrak{h} \otimes T \qquad (10.4.80)$$

(recall that $\frac{1}{2}\omega$ is the identity element of $\mathcal{B}$), of dimensions

$$1,\ 299,\ 98280,\ 98304, \qquad (10.4.81)$$

respectively [see (10.3.40)–(10.3.44)]. Of course, C in fact fixes ω:

$$C \cdot \omega = \omega. \qquad (10.4.82)$$

It can be shown that each of the invariant spaces is absolutely irreducible under C; for instance, $\mathfrak{h} \otimes T$ is irreducible since T is irreducible under $\hat{\Lambda}/K$ and $\mathfrak{h}$ is irreducible under Co_0. Before the Monster was proved to exist, it was postulated to be a finite simple group containing the group C as the centralizer of the involution $z \in C$ and it was believed to have a 196883-dimensional irreducible module consisting of the direct sum of the last three C-modules listed in (10.4.80), or rather, abstract C-modules isomorphic to them. Norton had determined the existence of an invariant commutative nonassociative algebra and nonsingular associative symmetric bilinear form on this module if it and the Monster existed (cf. [Griess 3]). By constraining the possibilities for such an algebra and form on the direct sum of the C-modules, Griess was able to determine an algebra and form admitting an automorphism outside the group C. The group generated by C and this automorphism had the required properties.

10.5. The Graded Character of the C-Module $V^\natural$

In Section 1.10 we defined the notion of graded dimension of a graded vector space. Here we begin by extending that notion to that of graded character of a graded group action.

Working in the same generality as in Section 1.10, let S be a set and let $V = \coprod_{\alpha \in S} V_\alpha$ be an S-graded vector space. Assume that S has a graded dimension, i.e., that

$$\dim V_\alpha < \infty \quad \text{for} \quad \alpha \in S \tag{10.5.1}$$

[see (1.10.1)]. Recall from (1.10.2) that the graded dimension of V is the formal sum

$$\dim_* V = \dim_*(V; x) = \sum_{\alpha \in S} (\dim V_\alpha) x^\alpha. \tag{10.5.2}$$

Let G be a group and let σ be a representation of G on a finite-dimensional space U. The *character of* U (or of σ) is the function

$$\begin{aligned} \operatorname{ch} U: G &\to \mathbb{F} \\ g &\mapsto \operatorname{tr} \sigma(g). \end{aligned} \tag{10.5.3}$$

Suppose that the S-graded vector space V is a *graded G-module*, i.e., a G-module such that

$$G \cdot V_\alpha \subset V_\alpha \quad \text{for} \quad \alpha \in S. \tag{10.5.4}$$

Let us denote the corresponding representation of G by π. We define the *graded character of* V to be the formal series

$$\operatorname{ch}_* V = \operatorname{ch}_*(V; x) = \sum_{\alpha \in S} (\operatorname{ch} V_\alpha) x^\alpha \tag{10.5.5}$$

of functions from G to $\mathbb{F}$. Viewing $\operatorname{ch}_* V$ as a function on G, we write

$$\operatorname{ch}_*(g) = \operatorname{ch}_*^V(g) = \sum_{\alpha \in S} (\operatorname{tr} \pi(g)|_{V_\alpha}) x^\alpha \tag{10.5.6}$$

for $g \in G$. Note that

$$\dim_* V = \operatorname{ch}_*(1). \tag{10.5.7}$$

If W is a graded G-submodule of V, then

$$\operatorname{ch}_*(V/W) = \operatorname{ch}_* V - \operatorname{ch}_* W, \tag{10.5.8}$$

and if $(V^i)_{i \in I}$ is a family of S-graded G-modules such that for all $\alpha \in S$,

$$\sum_{i \in I} \dim V_\alpha^i < \infty,$$

then

$$\operatorname{ch}_* \coprod_{i \in I} V^i = \sum_{i \in I} \operatorname{ch}_* V^i \tag{10.5.9}$$

[cf. (1.10.3), (1.10.4)]. Suppose that $\mathfrak{A}$ is an abelian group and that V and W are $\mathfrak{A}$-graded G-modules such that V, W and $V \otimes W$ have graded dimensions. Then by the multiplicativity of characters for tensor products of finite-dimensional G-modules,

$$\mathrm{ch}_*(V \otimes W) = (\mathrm{ch}_* V)(\mathrm{ch}_* W) \tag{10.5.10}$$

[cf. (1.10.5)].

As usual we shall set $x = q^{-1}$, and we shall take

$$\mathrm{ch}_* V = \mathrm{ch}_*(V; q^{-1}) = \sum_{n \in \mathbb{F}} (\mathrm{ch}\, V_{-n}) q^n \tag{10.5.11}$$

for the graded characters of our ($\mathbb{F}$-graded) modules [cf. (1.10.16), (1.10.17)].

What we want to compute now is the graded character of the moonshine module $V^\natural$, viewed as a graded C-module. That is, we want to find a formula for

$$\mathrm{ch}_*(k) = \mathrm{ch}_*^{V^\natural}(k) = \sum_{\substack{n \in \mathbb{Z} \\ n \geq -1}} (\mathrm{tr}\, \pi(k)|_{V^\natural_{-n}}) q^n \tag{10.5.12}$$

for $k \in C$ [recall (10.3.38)]. The series (10.5.12), at least for the graded Monster-module conjectured to exist in [Thompson 5], [Conway–Norton], is called the *Thompson series for k*. At present we do not know how to compute the Thompson series $\mathrm{ch}_*(g)$ on $V^\natural$ for Monster elements g not conjugate to an element of C. (This series is defined in the obvious way, once we give $V^\natural$ the structure of a graded module for the Monster.)

Let $\bar{g}$ be an isometry of the Leech lattice of order m. (Later, we shall view $\bar{g}$ as induced by $g \in C_0$.) By extension of scalars, we shall assume whenever necessary that $\mathbb{F}$ contains a primitive mth root of unity. Let

$$\{\beta_1, \ldots, \beta_{24}\} \subset \mathfrak{h} = \Lambda \otimes_{\mathbb{Z}} \mathbb{F}$$

be a basis of eigenvectors of $\bar{g}$ with eigenvalues $\omega_1, \ldots, \omega_{24}$ (which are all mth roots of unity):

$$\bar{g}\beta_i = \omega_i \beta_i, \quad i = 1, \ldots, 24. \tag{10.5.13}$$

Since the characteristic polynomial of $\bar{g}$ has integral coefficients, all primitive kth roots of unity occur with the same multiplicity, and

$$\det(\bar{g} - x1_{\mathfrak{h}}) = \prod_{i=1}^{24} (\omega_i - x) = \prod_{k|m} f_k(x)^{n_k}, \tag{10.5.14}$$

where $1_{\mathfrak{h}}$ is the identity operator on $\mathfrak{h}$ and $f_k(x)$ is the kth cyclotomic polynomial, normalized to have constant term 1. We have

$$f_k(x) = \prod_{d|k} (1 - x^d)^{\mu(k/d)}, \tag{10.5.15}$$

where μ denotes the Möbius function,

$$\mu(n) = \begin{cases} (-1)^r & \text{if } n \text{ is a product of } r \text{ distinct primes} \\ 0 & \text{if } n \text{ has a square factor,} \end{cases} \tag{10.5.16}$$

and so

$$\det(\bar{g} - x1_{\mathfrak{h}}) = \prod_{k \mid m} (1 - x^k)^{p_k} \tag{10.5.17}$$

where the p_k are uniquely determined integers (cf. e.g. [Jacobson 2]). Note that

$$\sum_{k \mid m} kp_k = 24. \tag{10.5.18}$$

When $\bar{g}$ has characteristic polynomial given by (10.5.17), we set

$$\eta_{\bar{g}}(q) = \sum_{k \mid m} \eta(q^k)^{p_k} \tag{10.5.19}$$

[recall the Dedekind η-function $\eta(q)$ from (1.10.21)]. We now establish:

Proposition 10.5.1: *Let $\bar{g} \in \mathrm{Co}_0$ have order m and characteristic polynomial (10.5.17). Consider the action of $\bar{g}$ on $S(\hat{\mathfrak{h}}_{\bar{Z}})$, for $Z = \mathbb{Z}$ or $\mathbb{Z} + 1/2$, given by (10.4.54). Then*

$$\mathrm{ch}_*(\bar{g}) = \frac{1}{\eta_{\bar{g}}(q)} \quad \textit{if} \quad Z = \mathbb{Z} \tag{10.5.20}$$

and

$$\mathrm{ch}_*(\bar{g}) = \frac{\eta_{\bar{g}}(q)}{\eta_{\bar{g}}(q^{1/2})} \quad \textit{if} \quad Z = \mathbb{Z} + \frac{1}{2}. \tag{10.5.21}$$

Proof: As in (1.10.8)–(1.10.19), it is easy to see that

$$\mathrm{ch}_*(\bar{g}) = q^s \prod_{\substack{n \in Z \\ n > 0}} \prod_{i=1}^{24} (1 - \omega_i q^n)^{-1},$$

where

$$s = \begin{cases} -1 & \text{if } Z = \mathbb{Z} \\ \frac{1}{2} & \text{if } Z = \mathbb{Z} + \frac{1}{2} \end{cases}$$

(recall the usual grading shifts). Since

$$\det \mathrm{Co}_0 = 1 \tag{10.5.22}$$

by Remark 10.2.11(a), we have $\prod \omega_i = 1$, and substituting $x = q^n$ in

$\det(\bar{g} - x1_{\mathfrak{h}})$ we obtain

$$\prod_{i=1}^{24} (1 - \omega_i q^n) = \prod_{k|m} (1 - (q^k)^n)^{p_k}.$$

Thus

$$\mathrm{ch}_*(\bar{g}) = q^s \prod_{\substack{n \in \mathbb{Z} \\ n>0}} \prod_{k|m} (1 - (q^k)^n)^{-p_k},$$

which gives (10.5.20) and (10.5.21) by (10.5.18). ∎

Remark 10.5.2: The special case $\bar{g} = 1$ recovers the graded dimensions as computed in (1.10.22) and (1.10.23).

Now we turn to the graded character of the first summand $V_\Lambda^{\theta_0} = V^z$ in the moonshine module [recall (10.3.32), (10.4.59)]. Let $g \in C_0$ and set

$$\Lambda^g = \{\alpha \in \Lambda \mid \bar{g}\alpha = \alpha\} \tag{10.5.23}$$

[recall (10.4.1)]. For $a \in \hat{\Lambda}$ such that $\bar{a} \in \Lambda^g$, either $ga = a$ or $ga = \kappa a$. Accordingly, define $s_0(g, a) = 0$ or 1 by:

$$ga = \kappa^{s_0(g,a)} a. \tag{10.5.24}$$

We see that $s_0(g, a)$ depends only on g and the image $\bar{a} + 2\Lambda$ of a in $\Lambda/2\Lambda$, and we write

$$ga = \kappa^{s_0(g, \bar{a}+2\Lambda)} a. \tag{10.5.25}$$

Note that $ga = a$ or κa according as $\varphi(g)$ fixes or interchanges the two preimages of $\bar{a} + 2\Lambda$ in the extraspecial group $\hat{\Lambda}/K$ [recall (10.4.7)].

We have defined the theta-function of a positive definite lattice in (6.1.30). For $g \in C_0$ we now set

$$\theta_g(q) = \sum_{\substack{\alpha \in \Lambda \\ \bar{g}\alpha = \alpha}} (-1)^{s_0(g, \alpha+2\Lambda)} q^{\langle \alpha, \alpha \rangle/2}. \tag{10.5.26}$$

Viewing $V^z = V_\Lambda^{\theta_0}$ as a C_0-module as in (10.4.54)–(10.4.56), we have:

Proposition 10.5.3: *For $g \in C_0$,*

$$\mathrm{ch}_*^{V^z}(g) = \frac{1}{2}\left(\frac{\theta_g(q)}{\eta_{\bar{g}}(q)} + \frac{\theta_{\theta_0 g}(q)}{\eta_{-\bar{g}}(q)}\right), \tag{10.5.27}$$

where g acts on $V^z = V_\Lambda^{\theta_0}$ as $\bar{g} \otimes g$.

Proof: Let S^+ (respectively, S^-) denote the subspace of $S(\hat{\mathfrak{h}}_{\mathbb{Z}}^-)$ spanned by the even (respectively, odd) symmetric powers of $\hat{\mathfrak{h}}_{\mathbb{Z}}^-$. Using the action of θ_0 on $V_\Lambda = S(\hat{\mathfrak{h}}_{\mathbb{Z}}^-) \otimes \mathbb{F}\{\Lambda\}$, we see that

$$V^z = S^+ \oplus \left(S^+ \otimes \sum_{\substack{a \in \hat{\Lambda} \\ \bar{a} \neq 0}} \mathbb{F}(\iota(a) + \iota(\theta_0 a))\right)$$
$$\oplus \left(S^- \otimes \sum_{\substack{a \in \hat{\Lambda} \\ \bar{a} \neq 0}} \mathbb{F}(\iota(a) - \iota(\theta_0 a))\right). \quad (10.5.28)$$

The action of g on V^z preserves each of these three summands [using (10.4.44)]. By Proposition 10.5.1,

$$\operatorname{ch}_*^{S^\pm}(\bar{g}) = \frac{1}{2}\left(\frac{1}{\eta_{\bar{g}}(q)} \pm \frac{1}{\eta_{-\bar{g}}(q)}\right). \quad (10.5.29)$$

Decomposing the second summand in (10.5.28) as

$$\left(S^+ \otimes \sum_{\bar{a} \in \Lambda^{\bar{g}} \setminus \{0\}} \mathbb{F}(\iota(a) + \iota(\theta_0 a))\right) \oplus \left(S^+ \otimes \sum_{\bar{a} \in \Lambda^{-\bar{g}} \setminus \{0\}} \mathbb{F}(\iota(a) + \iota(\theta_0 a))\right)$$
$$\oplus \left(S^+ \otimes \sum_{\bar{a} \notin \Lambda^{\bar{g}} \cup \Lambda^{-\bar{g}}} \mathbb{F}(\iota(a) + \iota(\theta_0 a))\right),$$

we see that each of these summands is g-invariant and that $\operatorname{ch}_*(g)$ on the last is zero. We find using (10.5.10) that $\operatorname{ch}_*(g)$ on the sum of the first two is

$$\operatorname{ch}_*^{S^+}(\bar{g})(\tfrac{1}{2}(\theta_g(q) - 1) + \tfrac{1}{2}(\theta_{\theta_0 g}(q) - 1)). \quad (10.5.30)$$

Finally, the third summand in (10.5.28) is the g-invariant direct sum

$$\left(S^- \otimes \sum_{\bar{a} \in \Lambda^{\bar{g}} \setminus \{0\}} \mathbb{F}(\iota(a) - \iota(\theta_0 a))\right) \oplus \left(S^- \otimes \sum_{\bar{a} \in \Lambda^{-\bar{g}} \setminus \{0\}} \mathbb{F}(\iota(a) - \iota(\theta_0 a))\right)$$
$$\oplus \left(S^- \otimes \sum_{\bar{a} \notin \Lambda^{\bar{g}} \cup \Lambda^{-\bar{g}}} \mathbb{F}(\iota(a) - \iota(\theta_0 a))\right),$$

and we find as above that on this space, $\operatorname{ch}_*(g)$ is

$$(\operatorname{ch}_*^{S^-}(\bar{g}))(\tfrac{1}{2}(\theta_g(q) - 1) - \tfrac{1}{2}(\theta_{\theta_0 g}(q) - 1)). \quad (10.5.31)$$

Combining (10.5.29)–(10.5.31) we obtain the desired result. ∎

Remark 10.5.4: The special case $g = 1$ gives

$$\dim_* V^z = \frac{1}{2}\left(\frac{\theta_\Lambda(q)}{\eta(q)^{24}} + \frac{\eta(q)^{24}}{\eta(q^2)^{24}}\right) \quad (10.5.32)$$

[cf. (7.1.68)].

For the graded character of the second summand $V^{-z} = (V_\Lambda^T)^{\theta_0}$ of $V^\natural$, viewed as a module for the group $\hat{C}$ [recall (10.4.39) and Remark 10.4.3], we have:

Proposition 10.5.5: *For $k = (g, g_T) \in \hat{C}$,*

$$\mathrm{ch}_*^{V^{-z}}(k) = \frac{1}{2}\,\mathrm{tr}(g_T)\left(\frac{\eta_{\bar{g}}(q)}{\eta_{\bar{g}}(q^{1/2})} - \frac{\eta_{-\bar{g}}(q)}{\eta_{-\bar{g}}(q^{1/2})}\right), \tag{10.5.33}$$

where k acts on $V^{-z} = (V_\Lambda^T)^{\theta_0}$ as $\bar{g} \otimes g_T$.

Proof: Denoting by S^- the subspace of $S(\hat{\mathfrak{h}}^-_{\mathbb{Z}+1/2})$ spanned by the odd symmetric powers of $\hat{\mathfrak{h}}^-_{\mathbb{Z}+1/2}$, we see that

$$V^{-z} = S^- \otimes T, \tag{10.5.34}$$

and the result follows from (10.5.9) and Proposition 10.5.1. ∎

Remark 10.5.6: The case $k = 1$ gives

$$\dim_* V^{-z} = 2^{11}\left(\frac{\eta(q)^{24}}{\eta(q^{1/2})^{24}} - \frac{\eta(q^2)^{24}\eta(q^{1/2})^{24}}{\eta(q)^{48}}\right). \tag{10.5.35}$$

Combining Propositions 10.5.3 and 10.5.5, we have the graded character of $V^\natural$ viewed as a C-module:

Theorem 10.5.7: *Let $k = (g, g_T) \in \hat{C}$, so that the image $\pi_0(k)$ of k in C acts on $V^\natural$ as in (10.4.58). Then*

$$\mathrm{ch}_*^{V^\natural}(\pi_0(k)) = \frac{1}{2}\left(\frac{\theta_g(q)}{\eta_{\bar{g}}(q)} + \frac{\eta_{\theta_0 g}(q)}{\eta_{-\bar{g}}(q)} + \frac{\mathrm{tr}(g_T)\eta_{\bar{g}}(q)}{\eta_{\bar{g}}(q^{1/2})} + \frac{\mathrm{tr}(-g_T)\eta_{-\bar{g}}(q)}{\eta_{-\bar{g}}(q^{1/2})}\right). \tag{10.5.36}$$

Remark 10.5.8: The graded dimension of $V^\natural$ is given by the sum of (10.5.32) and (10.5.35):

$$\dim_* V^\natural = \frac{1}{2}\left(\frac{\theta_\Lambda(q)}{\eta(q)^{24}} + \frac{\eta(q)^{24}}{\eta(q^2)^{24}} + 2^{12}\frac{\eta(q)^{24}}{\eta(q^{1/2})^{24}} - 2^{12}\frac{\eta(q^2)^{24}\eta(q^{1/2})^{24}}{\eta(q)^{48}}\right). \tag{10.5.37}$$

We shall quote some fundamental facts about modular forms and modular functions from [Serre 1]. The group $\mathrm{SL}(2, \mathbb{Z})$ of integral 2×2

matrices of determinant 1 acts on the complex upper half-plane

$$H = \{z \in \mathbb{C} \mid \operatorname{Im} z > 0\} \tag{10.5.38}$$

by the formula

$$\begin{pmatrix} a & b \\ c & d \end{pmatrix} \cdot z = \frac{az + b}{cz + d} \tag{10.5.39}$$

for $\begin{pmatrix} a & b \\ c & d \end{pmatrix} \in \mathrm{SL}(2, \mathbb{Z})$, $z \in H$. The quotient

$$\mathrm{PSL}(2, \mathbb{Z}) = \mathrm{SL}(2, \mathbb{Z})/\langle \pm 1 \rangle, \tag{10.5.40}$$

which also acts on H, is called the *modular group*. The transformations (10.5.39) are called *modular transformations*.

For an integer k, a meromorphic function on H is *weakly modular of weight* $2k$ if

$$f(z) = (cz + d)^{-2k} f\left(\frac{az + b}{cz + d}\right) \tag{10.5.41}$$

for $\begin{pmatrix} a & b \\ c & d \end{pmatrix} \in \mathrm{SL}(2, \mathbb{Z})$. In this case,

$$f(z + 1) = f(z) \tag{10.5.42}$$

$$f\left(-\frac{1}{z}\right) = z^{2k} f(z) \tag{10.5.43}$$

and in fact these conditions imply (10.5.41). By (10.5.42), f can be expressed as a function, which we denote $\tilde{f}$, of

$$q = e^{2\pi i z}, \tag{10.5.44}$$

and $\tilde{f}$ is meromorphic in the punctured disk

$$\{q \in \mathbb{C} \mid 0 < |q| < 1\}. \tag{10.5.45}$$

If $\tilde{f}$ extends to a meromorphic (respectively, holomorphic) function at 0, we say that f is *meromorphic* (respectively, *holomorphic*) *at infinity*. In this case, we have a Laurent series expansion

$$f(z) = \tilde{f}(q) = \sum_{\substack{n \in \mathbb{Z} \\ n \geq N}} a_n q^n \tag{10.5.46}$$

in a neighborhood of 0, where $a_n \in \mathbb{C}$, $N \in \mathbb{Z}$ and where we may take $N = 0$ if f is holomorphic at infinity. A weakly modular function is called *modular* if it is meromorphic at infinity, and a modular function which is holomorphic everywhere, including infinity, is called a *modular form*. If such a function is zero at infinity [i.e., if $a_0 = 0$ in (10.5.46)], it is called a *cusp form*. A modular form of weight $2k$ is thus given by a series

$$f(z) = \sum_{n \in \mathbb{N}} a_n q^n = \sum_{n \in \mathbb{N}} a_n e^{2\pi i n z} \tag{10.5.47}$$

which converges for $|q| < 1$, i.e., for $z \in H$, and which satisfies (10.5.43).

Remark 10.5.9: We cite some important classical examples of modular functions: The 24th power of Dedekind's η-function,

$$\eta(q)^{24} = q \prod_{n \in \mathbb{Z}_+} (1 - q^n)^{24} = q - 24q^2 + 252q^3 + \cdots, \tag{10.5.48}$$

viewed as a function on the disk or on H, is a cusp form of weight 12. The theta function

$$\theta_L(q) = \sum_{\alpha \in L} q^{\langle \alpha, \alpha \rangle / 2} = \sum_{n \in \mathbb{N}} |L_{2n}| q^n \tag{10.5.49}$$

of a positive definite even unimodular lattice L is a modular form of weight $\frac{1}{2}$ rank L. In particular, the theta function of the Leech lattice or more generally of any of the 24 Niemeier lattices L of rank 24 (recall Remark 10.2.8) is a modular form of weight 12. Recall that $L_2 = \emptyset$ if $L = \Lambda$ and that L_2 is a root system spanning $L_{\mathbb{Q}}$ otherwise. The space of modular forms of weight 12 is two-dimensional and the space of cusp forms of weight 12 is one-dimensional, spanned by $\eta(q)^{24}$. In particular, for a Niemeier lattice L, the function

$$J(q) = \frac{\theta_L(q)}{\eta(q)^{24}} - 24 - |L_2| \tag{10.5.50}$$

is a modular function of weight 0 (and is thus invariant under the modular group) and is independent of L. The constant is chosen so that $J(q)$, whose nonzero expansion coefficients are positive integers, has no constant term:

$$J(q) = q^{-1} + 0 + 196884q + \cdots. \tag{10.5.51}$$

The modular function $J(q)$ is holomorphic in H and has a simple pole at infinity. It defines a bijection from the orbit space $H/\mathrm{PSL}(2, \mathbb{Z})$ of H under the modular group onto $\mathbb{C}$, and this bijection extends to an isomorphism of complex analytic manifolds from a natural one-point compactification of the orbit space to the Riemann sphere $S_2 = \mathbb{C} \cup \{\infty\}$. The modular functions of weight zero comprise precisely the field of rational functions $\mathbb{C}(J(q))$

of the one generator $J(q)$, corresponding to the fact that the meromorphic functions on S_2 consist precisely of the rational functions. The field of modular functions of weight 0 is correspondingly said to have *genus zero*.

The story of the discovery of monstrous moonshine is sketched in the Introduction. We mention a few points here. After McKay observed the near-coincidence between the expansion coefficient 196884 in the modular function $J(q)$ [or rather, in a conventional variant $J(q) + 744$ of $J(q)$] and the dimension 196883 of the smallest supposed nontrivial irreducible module for the Monster, McKay and Thompson postulated that each term $a_n q^n$ in the expansion (10.5.51) should be replaced by a (usually reducible) Monster-module, say V_{-n}, of dimension a_n, and that one should look for a $\mathbb{Z}$-graded Monster-module

$$V = \coprod_{n \geq -1} V_{-n} \tag{10.5.52}$$

with graded dimension $J(q)$. Thompson proposed studying the modular transformation properties of the "graded traces" of non-identity Monster elements on V, and Conway and Norton wrote down a list of 194 (the number of conjugacy classes in the Monster) normalized generators

$$J_g(q) = \sum_{n \geq -1} c_n(g) q^n \tag{10.5.53}$$

of genus zero function fields arising from certain discrete subgroups of $SL(2, \mathbb{R})/\langle \pm 1 \rangle$ [here $J_g(q)$ depends only on the conjugacy class of g in the Monster] such that the first few c_n defined characters of the Monster. They conjectured that *all* the c_n should define characters, corresponding to ch V_{-n} in the notation (10.5.52). (Actually, only 171 of these functions are distinct.) Note that we would have

$$J_1(q) = J(q). \tag{10.5.54}$$

The graded traces $J_g(q)$ are called *Thompson series*, as we have noted above. The conjecture was all but proved by Atkin, Fong, and Smith.

Conway and Norton also wrote down a second proposed list of functions $J_g(q)$ for $g \in C$, a supposed involution centralizer in the Monster, and showed that these functions did indeed define characters of C. In [Kac 4], this second list was translated into a graded C-module consisting of the direct sum of a space on which vertex operators might act (the space we call $V^{\natural}$) and a second space on which vertex operators do not seem to act. The Monster does not seem to act in any natural way on this graded C-module, and this is certainly related to the fact that the Conway–Norton formula for the graded character of this C-module has an unwanted nonzero constant term. (The Conway–Norton series $J_g(q)$ for $g \notin C$ also have this problem.) Assuming that the graded character of our moonshine module $V^\natural$ is consistent with the formulas proposed by Conway and Norton, and this is very likely, then Conway and Norton chose unnatural formulas for their Thompson series. (This is the case for both of their lists for C.) Our formula (10.5.36), which is certainly associated with a natural module, is different from either of theirs, and there is presumably a family of modular function identities equating the respective functions.

Some such identities are known to be true, and the most important one is the one which asserts that the graded dimension of $V^\natural$ [see (10.5.37)] agrees with the modular function $J(q)$:

$$\dim_* V^\natural = J(q). \tag{10.5.55}$$

We shall prove this later using (10.5.50) and triality, and we shall construct a natural action of the Monster on $V^\natural$.

11 Triality

In the previous chapter we have constructed a subalgebra of a desired vertex operator algebra structure on $V^\natural$, parametrized by the elements of the untwisted space. This subalgebra acts irreducibly on the untwisted and twisted subspaces of $V^\natural$, and in particular, does not mix them. We have also constructed a subgroup C of what will turn out to be the Monster, preserving each of the two subspaces. In this chapter we begin the construction, to be completed in the next chapter, of an involution σ on $V^\natural$ which will lie in the Monster and which will mix the untwisted and twisted subspaces. Conjugation by this involution will be used in Chapter 12 in completing the definition of a vertex operator algebra structure on $V^\natural$. In this chapter, we construct an involution σ_1, to which σ will be closely related, starting from the isomorphism between the untwisted and twisted vertex operator realizations of $\mathfrak{sl}(2)\hat{}$ obtained in Section 4.5. The involution σ_1 comes from an involution of $\mathfrak{sl}(2)\hat{}$, and in fact of the underlying finite-dimensional Lie algebra $\mathfrak{sl}(2)$. It should be considered as part of a symmetric group $\mathcal{S}_3$ permuting a standard "symmetric" basis of $\mathfrak{sl}(2)$, essentially the "vector cross product" basis. The lifting of $\mathcal{S}_3$ to corresponding automorphisms of the direct sum of an untwisted and twisted space and eventually to $V^\natural$ is what we mean by the term "triality." The notion of triality presented here also manifests itself in the classical triality of $\mathfrak{o}(8)$, as we have mentioned in the Introduction.

In Section 11.1 we introduce a setting more general than is necessary for the construction of the Monster. To a type I code, we associate a lattice and corresponding untwisted and twisted vertex operator representations. Although the Golay code is of type II, we shall need a second type I code as well, in Chapter 13. We construct the involution σ_1 by combining its actions preserving a certain subspace of the untwisted space, interchanging certain subspaces of the untwisted and twisted spaces, and preserving a certain subspace of the twisted space, in Sections 11.2, 11.3 and 11.4, respectively. We characterize σ_1 partly by its action on elements of weight 2. We prove that the conjugation action of σ_1 on the appropriate vertex operators is compatible with its action on the underlying space. In Section 11.2, elementary representation theory of the finite-dimensional Lie algebra $\mathfrak{sl}(2)$ is used, and in Sections 11.3 and 11.4 it is convenient to use "Z-operators" (cf. [Lepowsky–Wilson 4]). We summarize the description of the involution σ_1 in Section 11.5, and we point out that σ_1 is almost canonically determined by the appropriate involution of $\mathfrak{sl}(2)$. As we shall show in Chapter 12, technical formulas that Griess used in his original construction of an "extra automorphism" in the Monster are naturally motivated by the action of σ_1 on the vectors of weight 2. The results of this chapter concerning triality and its relation with the generating weight-two substructures of the vertex-operator-algebraic structures were announced in [FLM2].

11.1. The Setting

Let $\mathcal{C}$ be a type I code based on a nonempty finite set Ω [see (10.1.2)]. As in (10.2.1)–(10.2.6) and (10.2.43), let

$$\mathfrak{h} = \coprod_{k \in \Omega} \mathbb{F}\alpha_k \tag{11.1.1}$$

be a vector space with basis $\{\alpha_k \mid k \in \Omega\}$ and provide $\mathfrak{h}$ with the symmetric bilinear form $\langle \cdot, \cdot \rangle$ such that

$$\langle \alpha_k, \alpha_l \rangle = 2\delta_{k,l} \quad \text{for} \quad k, l \in \Omega. \tag{11.1.2}$$

For $S \subset \Omega$, set

$$\alpha_S = \sum_{k \in S} \alpha_k, \tag{11.1.3}$$

and let ε_S be the involution of $\mathfrak{h}$ given by

$$\varepsilon_S \colon \alpha_k \mapsto \begin{cases} -\alpha_k & \text{if} \quad k \in S \\ \alpha_k & \text{if} \quad k \notin S. \end{cases} \tag{11.1.4}$$

Set

$$Q = \coprod_{k \in \Omega} \mathbb{Z}\alpha_k, \tag{11.1.5}$$

$$L_0 = \sum_{C \in \mathcal{C}} \mathbb{Z}\tfrac{1}{2}\alpha_C + Q, \tag{11.1.6}$$

$$L_1 = L_0 + \tfrac{1}{4}\alpha_\Omega, \tag{11.1.7}$$

$$L = L_0 \cup L_1. \tag{11.1.8}$$

Then L is a lattice in $\mathfrak{h}$ since $\Omega \in \mathcal{C}$. In the notation (10.2.8),

$$L = L_0 + L_0'. \tag{11.1.9}$$

Note that L is positive definite but not integral. Since $\mathcal{C}$ is of type I, however, we know that

$$\langle \alpha, \alpha \rangle \in \mathbb{Z} \quad \text{for all} \quad \alpha \in L_0. \tag{11.1.10}$$

This will allow us to apply the results of Chapter 9 for L_0 [recall the hypothesis (9.2.1)].

Consider a central extension

$$1 \to \langle \kappa \mid \kappa^4 = 1 \rangle \hookrightarrow \hat{L} \xrightarrow{-} L \to 1 \tag{11.1.11}$$

of L by a 4-element cyclic group $\langle \kappa \rangle$ (i.e., take $s = 4$ in Chapter 5). Recall from Proposition 5.2.3 that $\hat{L}$ is specified up to equivalence by its commutator map

$$c_0 : L \times L \to \mathbb{Z}/4\mathbb{Z}. \tag{11.1.12}$$

We assume only that c_0 satisfies the condition

$$c_0(\alpha, \beta) = 2\langle \alpha, \beta \rangle + 4\mathbb{Z} \quad \text{for} \quad \alpha \in Q, \quad \beta \in L, \tag{11.1.13}$$

or equivalently, the two conditions:

$$c_0(\alpha_k, \beta) = 2\langle \alpha_k, \beta \rangle + 4\mathbb{Z} \quad \text{for} \quad k \in \Omega, \quad \beta \in L_0, \tag{11.1.14}$$

$$c_0(\alpha_k, \tfrac{1}{4}\alpha_\Omega) = 1 + 4\mathbb{Z} \quad \text{for} \quad k \in \Omega. \tag{11.1.15}$$

We shall construct such a c_0 (satisfying additional conditions) later.

We shall consider the untwisted and twisted vertex operator constructions associated with L and $\hat{L}$, as developed in Chapters 7–9. *We assume that our field $\mathbb{F}$ contains a primitive 4th root of unity i*, which we fix. Define the character

$$\begin{aligned} \chi : \langle \kappa \rangle &\to \mathbb{F}^\times \\ \kappa &\mapsto i \end{aligned} \tag{11.1.16}$$

and let $\mathbb{F}_\chi$ denote the $\langle\kappa\rangle$-module $\mathbb{F}$ affording χ. Set

$$\mathbb{F}\{L\} = \mathrm{Ind}_{\langle\kappa\rangle}^{\hat{L}}\mathbb{F}_\chi = \mathbb{F}[\hat{L}]/(\kappa - i)\mathbb{F}[\hat{L}] \tag{11.1.17}$$

and for $a \in \hat{L}$ set

$$\iota(a) = a \otimes 1 \in \mathbb{F}\{L\} \tag{11.1.18}$$

as in (7.1.18), (7.1.19). Then

$$\iota(\kappa a) = i\iota(a) \quad \text{for} \quad a \in \hat{L}. \tag{11.1.19}$$

In the notation (7.1.21),

$$c(\alpha, \beta) = i^{c_0(\alpha,\beta)} = \chi(aba^{-1}b^{-1}) \tag{11.1.20}$$

for $a, b \in \hat{L}$ with $\alpha = \bar{a}$, $\beta = \bar{b}$. We fix the untwisted space

$$V_L = S(\hat{\mathfrak{h}}_{\bar{\mathbb{Z}}}) \otimes \mathbb{F}\{L\}. \tag{11.1.21}$$

We now turn to the twisted construction. Fix elements $a_k \in \hat{L}$ for $k \in \Omega$ such that

$$\bar{a}_k = \alpha_k. \tag{11.1.22}$$

(For each k, there are four such elements.) For a subset M of L set

$$\hat{M} = \{a \in \hat{L} \mid \bar{a} \in M\}, \tag{11.1.23}$$

the inverse image of M under the map $\hat{L} \to L$ [cf. (5.2.9)]. Since by (11.1.14) a_k and a_l commute for all $k, l \in \Omega$, we see that $\hat{Q}$ is a direct product of abelian groups:

$$\hat{Q} = \langle a_k \mid k \in \Omega\rangle \times \langle\kappa\rangle, \tag{11.1.24}$$

and the group $\langle a_k \mid k \in \Omega\rangle$ is isomorphic to Q. Define the character

$$\begin{aligned} \psi\colon \hat{Q} &\to \mathbb{F}^\times \\ a_k &\mapsto 1, \quad k \in \Omega \\ \kappa &\mapsto i. \end{aligned} \tag{11.1.25}$$

As the $\hat{L}$-module T of Chapters 7 and 9 we shall take

$$\begin{aligned} T_L &= \mathrm{Ind}_{\hat{Q}}^{\hat{L}}\mathbb{F}_\psi = \mathbb{F}[\hat{L}] \otimes_{\mathbb{F}[\hat{Q}]} \mathbb{F}_\psi \\ &\simeq \mathbb{F}[L/Q] \quad \text{(linearly)}, \end{aligned} \tag{11.1.26}$$

where, as usual, $\mathbb{F}_\psi$ denotes the $\hat{Q}$-module $\mathbb{F}$ affording ψ (cf. Propositions 7.4.8 and 10.3.1). Strictly speaking, we shall view T_L as a direct sum of two $\hat{L}_0$-modules, so that the results of Chapter 9 will apply to each of these [recall (11.1.10)].

For $a \in \hat{L}$ set

$$t(a) = a \otimes 1 \in T_L, \qquad (11.1.27)$$

the twisted analogue of $\iota(a)$ (11.1.18). We then have the relations

$$t(aa_k) = t(a), \qquad (11.1.28)$$

$$t(\kappa a) = it(a) \qquad (11.1.29)$$

for $a \in \hat{L}$, $k \in \Omega$, and the $\hat{L}$-action is given by

$$a \cdot t(b) = t(ab) \qquad (11.1.30)$$

for $a, b \in \hat{L}$. We take the twisted space to be

$$V_L' = V_L^{T_L} = S(\hat{\mathfrak{h}}_{\mathbb{Z}+1/2}^-) \otimes T_L, \qquad (11.1.31)$$

the notation V_L' being an abbreviation for the usual notation $V_L^{T_L}$.

We also assume the existence of an automorphism

$$\theta \in \operatorname{Aut}(\hat{L}; \kappa, \langle \cdot, \cdot \rangle) \qquad (11.1.32)$$

such that

$$\bar{\theta} = -1 \quad \text{on} \quad L \quad \text{and} \quad \theta^2 = 1. \qquad (11.1.33)$$

However, we shall *not* assume as in (7.4.14) that $\theta a = a$ as operators on T_L for $a \in \hat{L}$ (and it is not even in general possible for this to happen), but instead, that

$$\theta a_k = a_k^{-1} \quad \text{for} \quad k \in \Omega. \qquad (11.1.34)$$

We canonically extend θ to an involution of V_L as in (9.2.34):

$$\theta \colon V_L \to V_L. \qquad (11.1.35)$$

We shall not, however, make θ act on V_L' as in (9.2.87). Chapter 9 will apply in the generality expressed in the remarks at the ends of Sections 9.1–9.4 (and in Theorem 9.5.3).

As we have already done (for the Leech lattice) in (10.3.33)–(10.3.37), we consider the untwisted and twisted constructions simultaneously and set

$$W_L = V_L \oplus V_L'. \qquad (11.1.36)$$

We shall be interested in the vertex operators $Y_{\mathbb{Z}}(v, z)$ acting on V_L and $Y_{\mathbb{Z}+1/2}(v, z)$ acting on V_L' for $v \in V_{L_0}$ (see Chapters 8 and 9; the integrality hypothesis (9.2.1) holds for L_0). For such v we form the vertex operator

$$Y(v, z) = Y_{\mathbb{Z}}(v, z) \oplus Y_{\mathbb{Z}+1/2}(v, z), \qquad (11.1.37)$$

and similarly for $X(v, z)$, acting on W_L. For the component operators of $Y(v, z)$ we similarly write

$$\begin{aligned} v_n &= v_n \oplus v_n \\ x_v(n) &= x_v(n) \oplus x_v(n) \end{aligned} \tag{11.1.38}$$

for $n \in \mathbb{Q}$. Then for example

$$\begin{aligned} \alpha(z) &= \alpha(z) \oplus \alpha(z) \\ \alpha(n) &= \alpha(n) \oplus \alpha(n) \end{aligned} \tag{11.1.39}$$

for $\alpha \in \mathfrak{h}$, $n \in \mathbb{Q}$, where at least one of the operators on the right is zero.

For a subset M of L, set

$$\begin{aligned} \mathbb{F}\{M\} &= \sum_{a \in \hat{M}} \mathbb{F}\iota(a), \\ V_M &= S(\hat{\mathfrak{h}}_{\mathbb{Z}}^-) \otimes \mathbb{F}\{M\} \subset V_L, \end{aligned} \tag{11.1.40}$$

and *for M a union of cosets of Q in L*, set

$$\begin{aligned} T_M &= \sum_{a \in \hat{M}} \mathbb{F}t(a), \\ V_M' &= S(\hat{\mathfrak{h}}_{\mathbb{Z}+1/2}^-) \otimes T_M \subset V_L'. \end{aligned} \tag{11.1.41}$$

We then have the decomposition

$$W_L = V_{L_0} \oplus V_{L_1} \oplus V_{L_0}' \oplus V_{L_1}'. \tag{11.1.42}$$

We shall sometimes use the notation

$$W_0 = V_{L_0}, \quad W_1 = V_{L_1}, \quad W_2 = V_{L_0}', \quad W_3 = V_{L_1}'. \tag{11.1.43}$$

Each W_j is invariant under the vertex operators $Y(v, z)$ for $v \in W_0$, or more precisely, under their components v_n or $x_v(n)$.

Fix $k \in \Omega$ and consider the vertex operators $Y(a_k, z)$, $Y(a_k^{-1}, z)$, $\alpha_k(z)$ and their components $x_{a_k}(n)$, $x_{a_k^{-1}}(n)$, $\alpha_k(n)$, $n \in \mathbb{Q}$, restricted to a space W_j. Note that

$$a_k^2|_{T_{L_0}} = 1, \quad \text{i.e.,} \quad a_k|_{T_{L_0}} = a_k^{-1}|_{T_{L_0}} \tag{11.1.44}$$

$$a_k^2|_{T_{L_1}} = -1, \quad \text{i.e.,} \quad a_k|_{T_{L_1}} = -a_k^{-1}|_{T_{L_1}} \tag{11.1.45}$$

since

$$\begin{aligned} c(2\alpha_k, \beta) &= 1 \quad \text{for} \quad \beta \in L_0, \\ c(2\alpha_k, \tfrac{1}{4}\alpha_\Omega) &= c(\alpha_k, \tfrac{1}{2}\alpha_\Omega) = -1 \end{aligned} \tag{11.1.46}$$

by (11.1.14) and (11.1.20). By (11.1.34) we have

$$\theta(a_k)\big|_{T_{L_0}} = a_k\big|_{T_{L_0}} \tag{11.1.47}$$

$$\theta(a_k)\big|_{T_{L_1}} = -a_k\big|_{T_{L_1}}. \tag{11.1.48}$$

Now we can apply the results of Chapters 8 and 9 to determine the commutators among the operators $x_{a_k}(n)$, $x_{a_k^{-1}}(n)$ and $\alpha_k(n)$, and between these and the operators $Y(v, z)$ for $v \in W_0 = V_{L_0}$, acting on W_j. For $j = 0$ and 1 we can use Theorem 8.6.1 and for $j = 3$ and 4, the cases (9.1.45) and (9.1.46), respectively, of the generalized form of Theorem 9.3.1 (see Remark 9.3.11). But in fact some earlier special cases of these general theorems, namely, (7.1.46), Theorem 7.2.1, (7.3.15) and Theorem 7.4.1, already imply the results below.

Recall the basis $\{\alpha_1, x_{\alpha_1}, x_{-\alpha_1}\}$ of the Lie algebra $\mathfrak{sl}(2, \mathbb{F})$ given by (3.1.2), and the canonical form (3.1.4). Consider the space

$$\mathfrak{a}_k = \mathbb{F}\iota(a_k) \oplus \mathbb{F}\alpha_k(-1) \oplus \mathbb{F}\iota(a_k^{-1}) \tag{11.1.49}$$

inside the subspace of V_{L_0} of weight 1 (recall Section 7.1), and provide $\mathfrak{a}_k$ with Lie algebra structure and form isomorphic to those of $\mathfrak{sl}(2, \mathbb{F})$ via the correspondence

$$\iota(a_k) \mapsto x_{\alpha_1}, \quad \alpha_k(-1) \mapsto \alpha_1, \quad \iota(a_k^{-1}) \mapsto x_{-\alpha_1}. \tag{11.1.50}$$

Note that these agree with the natural Lie algebra structure and form mentioned in (8.9.7)–(8.9.8):

$$\begin{aligned} [u, v] &= u_0 \cdot v = x_u(0)v \quad \text{for} \quad u, v \in \mathfrak{a}_k \\ \langle u, v\rangle &= u_1 \cdot v = x_u(1)v \quad \text{for} \quad u, v \in \mathfrak{a}_k \end{aligned} \tag{11.1.51}$$

(recall Sections 8.5 and 8.9).

Define four involutions of $\mathfrak{a}_k$ by:

$$\theta_{(0)} = 1, \tag{11.1.52}$$

$$\theta_{(1)}\colon \alpha_k(-1) \mapsto \alpha_k(-1), \quad \iota(a_k^{\pm 1}) \mapsto -\iota(a_k^{\pm 1}), \tag{11.1.53}$$

$$\theta_{(2)}\colon \alpha_k(-1) \mapsto -\alpha_k(-1), \quad \iota(a_k^{\pm 1}) \mapsto \iota(a_k^{\mp 1}), \tag{11.1.54}$$

$$\theta_{(3)}\colon \alpha_k(-1) \mapsto -\alpha_k(-1), \quad \iota(a_k^{\pm 1}) \mapsto -\iota(a_k^{\mp 1}); \tag{11.1.55}$$

cf. the notations θ_1, θ_2 of (3.1.6), (3.1.23). Note that the automorphism θ of (11.1.35) agrees with $\theta_{(2)}$:

$$\theta = \theta_{(2)} \quad \text{on} \quad \mathfrak{a}_k. \tag{11.1.56}$$

The automorphisms $\theta_{(j)}$ commute and

$$\theta_{(3)} = \theta_{(1)}\theta_{(2)} = \theta_{(2)}\theta_{(1)}; \tag{11.1.57}$$

similarly for permutations of the indices (cf. Remark 6.4.3). The involutions $\theta_{(0)}$–$\theta_{(3)}$ form an abelian group isomorphic to $\mathbb{Z}/(2) \times \mathbb{Z}/(2)$.

Now let $j = 0, 1, 2$ or 3. On W_j, the operators

$$x_{a_k^{\pm 1}}(n), \quad \alpha_k(n), \quad 1_{W_j} \quad \text{for} \quad n \in \mathbb{Q}, \tag{11.1.58}$$

via the natural correspondence

$$(\iota(a_k^{\pm 1}))(z) \mapsto X(a_k^{\pm 1}, z), \tag{11.1.59}$$

provide a representation of the affine Lie algebra

$$\hat{\mathfrak{a}}_k[\theta_{(j)}] \tag{11.1.60}$$

[recall the notations (1.6.26), (2.3.12)]. In fact, setting

$$\begin{aligned} X^{\pm}(a_k, z) &= X(a_k, z) \pm X(a_k^{-1}, z) \\ &= X(a_k, z) \pm X(\theta a_k, z) \\ &= \sum_{n \in \mathbb{Q}} x_{a_k}^{\pm}(n) z^{-n}, \end{aligned} \tag{11.1.61}$$

we have

$$\hat{\mathfrak{a}}_k[\theta_{(j)}] \xrightarrow{\sim} \sum_m \mathbb{F}\alpha_k(m) \oplus \sum_n \mathbb{F}x_{a_k}^{+}(n) \oplus \sum_p \mathbb{F}x_{a_k}^{-}(p) \oplus \mathbb{F}1_{W_j} \tag{11.1.62}$$

with

$$m, n, p \in \mathbb{Z} \quad \text{on} \quad V_{L_0} \quad (j = 0), \tag{11.1.63}$$

$$m \in \mathbb{Z}, \quad n, p \in \mathbb{Z} + \tfrac{1}{2} \quad \text{on} \quad V_{L_1} \quad (j = 1), \tag{11.1.64}$$

$$n \in \mathbb{Z}, \quad m, p \in \mathbb{Z} + \tfrac{1}{2} \quad \text{on} \quad V'_{L_0} \quad (j = 2), \tag{11.1.65}$$

$$p \in \mathbb{Z}, \quad m, n \in \mathbb{Z} + \tfrac{1}{2} \quad \text{on} \quad V'_{L_1} \quad (j = 3); \tag{11.1.66}$$

cf. Theorems 3.5.1, 4.4.1, 7.2.6 and 7.4.10.

Now form the subspace

$$\mathfrak{g} = \coprod_{k \in \Omega} \mathfrak{a}_k \tag{11.1.67}$$

of V_{L_0}. This is again a Lie algebra under the bracket given by (11.1.51), and in fact is of type $A_1^{|\Omega|}$, in the notation of Remark 6.3.4. Distinct factors A_1 are orthogonal under the form (11.1.51). (Recall that for $k \neq l \in \Omega$, $\langle \alpha_k, \alpha_l \rangle = 0$ and a_k and a_l commute.) Extending the involutions $\theta_{(0)}$–$\theta_{(3)}$ to

$\mathfrak{g}$ by their diagonal action, we note that

$$\theta = \theta_{(2)} \quad \text{on} \quad \mathfrak{g} \tag{11.1.68}$$

and that we still have

$$\langle \theta_{(0)}, \theta_{(1)}, \theta_{(2)}, \theta_{(3)} \rangle \simeq \mathbb{Z}/(2) \times \mathbb{Z}/(2). \tag{11.1.69}$$

On each W_j, the representations given by (11.1.59) and (11.1.62) extend to a representation of $\hat{\mathfrak{g}}[\theta_{(j)}]$ by means of the operators

$$x_{a_k^{\pm 1}}(n), \quad \alpha_k(n), \quad 1_{W_j} \quad \text{for} \quad k \in \Omega, \quad n \in \mathbb{Q}. \tag{11.1.70}$$

Consider the Lie algebra involution

$$\sigma_0 \colon \mathfrak{g} \mapsto \mathfrak{g} \tag{11.1.71}$$

such that

$$\sigma_0 \colon \alpha_k(-1) \mapsto \iota(a_k)^+, \quad \iota(a_k)^+ \mapsto \alpha_k(-1), \quad \iota(a_k)^- \mapsto -\iota(a_k)^- \tag{11.1.72}$$

for $k \in \Omega$, where

$$\iota(a_k)^\pm = \iota(a_k) \pm \iota(a_k^{-1}) = \iota(a_k) \pm \theta\iota(a_k) \tag{11.1.73}$$

[recall (3.1.44)]. We have

$$\sigma_0 \theta_{(2)} \sigma_0^{-1} = \theta_{(1)} \tag{11.1.74}$$

as in (3.1.45), and using (11.1.57) we see that

$$\sigma_0 \theta_{(3)} \sigma_0^{-1} = \theta_{(3)}. \tag{11.1.75}$$

Now

$$\dim T_{L_0} = \dim T_{L_1} = \dim \mathbb{F}[L_0/Q] = |L_0/Q| = |\mathcal{C}|, \tag{11.1.76}$$

and for each $j = 0, 1, 2, 3$, W_j is a direct sum of $|\mathcal{C}|$ irreducible $\hat{\mathfrak{g}}[\theta_{(j)}]$-modules. As we shall see in detail later, an application of (an $|\Omega|$-fold tensor product of) Theorem 4.5.2 shows that there exists an isomorphism

$$\sigma_1 \colon V'_{L_0} \to V_{L_1} \tag{11.1.77}$$

such that

$$\sigma_1 Y(v, z) \sigma_1^{-1} = Y(\sigma_0 v, z) \quad \text{for} \quad v \in \mathfrak{g}, \tag{11.1.78}$$

or equivalently,

$$\sigma_1(z\alpha_k(z))\sigma_1^{-1} = X^+(a_k, z), \tag{11.1.79}$$

$$\sigma_1 X^+(a_k, z)\sigma_1^{-1} = z\alpha_k(z), \tag{11.1.80}$$

$$\sigma_1 X^-(a_k, z)\sigma_1^{-1} = -X^-(a_k, z) \tag{11.1.81}$$

for $k \in \Omega$. Our goal in this chapter is to show the existence of a linear involution σ_1 of W_L with these properties, preserving V_{L_0} and V'_{L_1}, extending σ_0 on the subspace $\mathfrak{g}$, satisfying the condition

$$\sigma_1 Y(v, z)\sigma_1^{-1} = Y(\sigma_1 v, z) \quad \text{for} \quad v \in V_{L_0}, \tag{11.1.82}$$

and preserving the span of the set

$$\{\iota(b) \mid b \in \hat{L}_0, \bar{b} = \tfrac{1}{2}\varepsilon_S \alpha_C, C \in \mathcal{C}, S \subset C\} \tag{11.1.83}$$

[recall the notations ε_S, α_C from (11.1.3), (11.1.4)]. Then σ_1 will also normalize the span of the corresponding set of vertex operators

$$\{X(b, z) \mid b \in \hat{L}_0, \bar{b} = \tfrac{1}{2}\varepsilon_S \alpha_C, C \in \mathcal{C}, S \subset C\}. \tag{11.1.84}$$

11.2. Construction of $\sigma_1: V_{L_0} \to V_{L_0}$

Here we shall construct an involution

$$\sigma_1: V_{L_0} \to V_{L_0} \tag{11.2.1}$$

having the properties discussed at the end of the last section. Our strategy is to realize the involution σ_0 of $\mathfrak{g}$ as conjugation by a product of exponentials of suitable Lie algebra elements and then to invoke general principles from Chapter 8 on the commutation of such Lie algebra elements with vertex operators.

For this, we find it convenient to *assume now that the field* $\mathbb{F}$ *contains* $\sqrt{2}$. In the group SL(2, $\mathbb{F}$) of 2×2 matrices over $\mathbb{F}$ of determinant 1, set

$$\sigma_{(1)} = \frac{i}{\sqrt{2}}\begin{bmatrix} 1 & 1 \\ 1 & -1 \end{bmatrix}. \tag{11.2.2}$$

Note that

$$\sigma_{(1)}^2 = -\begin{bmatrix} 1 & 0 \\ 0 & 1 \end{bmatrix}. \tag{11.2.3}$$

Recalling the automorphism σ of $\mathfrak{sl}(2, \mathbb{F})$ given by (3.1.44), we see that for $x \in \mathfrak{sl}(2, \mathbb{F})$,

$$\sigma_{(1)} x \sigma_{(1)}^{-1} = \sigma x. \tag{11.2.4}$$

We also have

$$\sigma_{(1)} = \begin{bmatrix} 1 & 1 + \sqrt{2}i \\ 0 & 1 \end{bmatrix}\begin{bmatrix} 1 & 0 \\ i/\sqrt{2} & 1 \end{bmatrix}\begin{bmatrix} 1 & -1 + \sqrt{2}i \\ 0 & 1 \end{bmatrix}, \tag{11.2.5}$$

and each of the three factors on the right has the form of a terminating exponential series of an element of $\mathfrak{sl}(2, \mathbb{F})$.

We shall use some basic theory of representations of $\mathfrak{sl}(2)$ in characteristic zero; see for example [Humphreys]. Since $\mathfrak{g}$ acts on V_{L_0} as operators of degree 0, V_{L_0} decomposes as a direct sum of finite-dimensional irreducible modules. We define $\sigma_1: V_{L_0} \to V_{L_0}$ as the natural action of diag $\sigma_{(1)}$:

$$\sigma_1 = \prod_{k \in \Omega} \exp((1 + \sqrt{2}i)x_{a_k}(0)) \exp\left(\frac{i}{\sqrt{2}} x_{a_k^{-1}}(0)\right) \exp((-1 + \sqrt{2}i)x_{a_k}(0)); \tag{11.2.6}$$

on any given element of V_{L_0}, these exponential series terminate. But by Corollary 8.6.3,

$$[x_u(0), Y(v, z)] = Y(x_u(0)v, z) \tag{11.2.7}$$

for $u \in \mathfrak{g}$, $v \in V_{L_0}$, as operators on V_{L_0}. Hence by (11.2.6),

$$\sigma_1 Y(v, z)\sigma_1^{-1} = Y(\sigma_1 v, z). \tag{11.2.8}$$

On the other hand, if we apply σ_1 to $\mathfrak{a}_k \subset V_{L_0}$ for a fixed k and if we identify $\mathfrak{a}_k$ with $\mathfrak{sl}(2, \mathbb{F})$ as in (11.1.50), we see from (11.1.51) that σ_1 acts as conjugation by the matrix $\sigma_{(1)}$ of (11.2.5). Hence

$$\sigma_1 = \sigma_0 \quad \text{on} \quad \mathfrak{g} \tag{11.2.9}$$

by (11.2.4) and (11.1.72), and (11.1.79)–(11.1.81) (viewed as equations on V_{L_0}) follow.

We claim that

$$\sigma_1^2 = 1 \quad \text{on} \quad V_{L_0}. \tag{11.2.10}$$

In fact, by (11.2.8) and (11.2.9), σ_1^2 commutes with $Y(v, z)$ for $v \in \mathfrak{g}$, and hence with the action of $\hat{\mathfrak{h}}_{\mathbb{Z}}$, and so it suffices to show that σ_1^2 fixes $\iota(b)$ for $b \in \hat{L}_0$. Let U be the $\mathfrak{a}_k$-module generated by $\iota(b)$. Since

$$\alpha_k(0)\iota(b) = \langle \alpha_k, \bar{b}\rangle \iota(b) \tag{11.2.11}$$

$$x_{a_k}(0)\iota(b) = 0 \quad \text{if} \quad \langle \alpha_k, \bar{b}\rangle \geq 0 \tag{11.2.12}$$

$$x_{a_k^{-1}}(0)\iota(b) = 0 \quad \text{if} \quad \langle \alpha_k, \bar{b}\rangle \leq 0, \tag{11.2.13}$$

we see that U is an irreducible $\mathfrak{a}_k$-module and that

$$\dim U = |\langle \alpha_k, \bar{b}\rangle| + 1. \tag{11.2.14}$$

Thus by (11.2.3)

$$\sigma_{(1)}^2 \cdot \iota(b) = (-1)^{\langle \alpha_k, \bar{b}\rangle} \iota(b) \tag{11.2.15}$$

and it follows that

$$\sigma_1^2 \iota(b) = (-1)^{\langle \alpha_\Omega, \bar{b} \rangle} \iota(b). \tag{11.2.16}$$

But now we recall (11.1.6) and the fact that our code $\mathfrak{C}$ is of type I, and the claim is established.

Let

$$b \in \hat{L}_0 \quad \text{such that} \quad \bar{b} = \tfrac{1}{2}\varepsilon_S \alpha_C \quad \text{with} \quad C \in \mathfrak{C}, \quad S \subset C. \tag{11.2.17}$$

We want to determine the action of σ_1 on $\iota(b)$, and to show in fact that σ_1 preserves the span of such elements. Now in addition to (11.2.11)–(11.2.13) we have

$$x_{a_k}(0)\iota(b) = \iota(a_k b) \quad \text{if} \quad \langle \alpha_k, \bar{b} \rangle = -1 \tag{11.2.18}$$

$$x_{a_k^{-1}}(0)\iota(b) = \iota(a_k^{-1} b) \quad \text{if} \quad \langle \alpha_k, \bar{b} \rangle = 1 \tag{11.2.19}$$

(here b can be any element of $\hat{L}_0$). Hence setting

$$a_{R,T} = \prod_{k \in R} a_k \prod_{l \in T} a_l^{-1} \tag{11.2.20}$$

for $R, T \subset \Omega$, we see that

$$\operatorname{span}\{\iota(a_{S,T} b) \mid T \subset C\} \tag{11.2.21}$$

is a $\mathring{\mathfrak{g}}$-module of dimension $2^{|C|}$ and that the correspondence

$$\iota(a_{S,T} b) \mapsto \bigotimes_{k \in C \backslash T} \begin{bmatrix} 1 \\ 0 \end{bmatrix}_k \bigotimes_{l \in T} \begin{bmatrix} 0 \\ 1 \end{bmatrix}_l \bigotimes_{m \in \Omega \backslash C} 1_m \tag{11.2.22}$$

(using obvious notation) defines an isomorphism of $\mathfrak{g}$-modules between the module (11.2.21) and the $\mathfrak{sl}(2, \mathbb{F})^{|\Omega|}$-module $\otimes_{k \in C} (\mathbb{F}^2)_k \otimes_{l \in \Omega \backslash C} (\mathbb{F})_l$, where $\mathbb{F}^2$ (respectively, $\mathbb{F}$) denotes the natural (respectively, trivial) $\mathfrak{sl}(2, \mathbb{F})$-module. Since

$$\begin{aligned} \sigma_1 \colon & \bigotimes_{k \in C \backslash S} \begin{bmatrix} 1 \\ 0 \end{bmatrix}_k \bigotimes_{l \in S} \begin{bmatrix} 0 \\ 1 \end{bmatrix}_l \bigotimes_{m \in \Omega \backslash C} 1_m \\ & \mapsto \left(\frac{i}{\sqrt{2}} \right)^{|C|} \sum_{T \subset C} (-1)^{|S \cap T|} \bigotimes_{k \in C \backslash T} \begin{bmatrix} 1 \\ 0 \end{bmatrix}_k \bigotimes_{l \in T} \begin{bmatrix} 0 \\ 1 \end{bmatrix}_l \bigotimes_{m \in \Omega \backslash C} 1_m, \end{aligned} \tag{11.2.23}$$

by (11.2.2), we see that

$$\sigma_1(\iota(b)) = (-2)^{-|C|/2} \sum_{T \subset C} (-1)^{|S \cap T|} \iota(a_{S,T} b). \tag{11.2.24}$$

It follows from (11.2.8) that

$$\sigma_1 X(b, z)\sigma_1^{-1} = (-2)^{-|C|/2} \sum_{T \subset C} (-1)^{|S \cap T|} X(a_{S,T} b, z), \tag{11.2.25}$$

where the passage from $Y(\cdot, z)$ to $X(\cdot, z)$ is allowed since

$$\tfrac{1}{2}\langle \bar{b}, \bar{b} \rangle = \tfrac{1}{2}\langle \overline{a_{S,T} b}, \overline{a_{S,T} b} \rangle = \tfrac{1}{4}|C|. \tag{11.2.26}$$

We observe that

$$\sigma_1(\iota(1)) = \iota(1). \tag{11.2.27}$$

Also, as in the proof of Proposition 4.4.2, we see that V_{L_0} is irreducible, and in fact absolutely irreducible, under

$$\hat{\mathfrak{h}}_{\mathbb{Z}} \cup \{x_b(n) \mid n \in \mathbb{Q}, \quad b \in \hat{L}_0, \quad \bar{b} = \tfrac{1}{2}\varepsilon_S \alpha_C \quad \text{with} \quad C \in \mathcal{C}, \quad S \subset C\} \tag{11.2.28}$$

since the indicated elements $\bar{b}$ generate the lattice L_0. Hence a linear automorphism σ_1 of V_{L_0} satisfying (11.1.79), (11.2.25) and (11.2.27) is uniquely determined. Furthermore, by either the absolute irreducibility or the fact that the Virasoro algebra is built from $\hat{\mathfrak{h}}_{\mathbb{Z}}$ and contains the operator $L(0)$ (recall Remark 7.1.3 or Section 8.7), any linear automorphism of V_{L_0} satisfying only (11.1.79) and (11.2.25) is uniquely determined up to a multiplicative scalar.

Now we summarize:

Theorem 11.2.1: *There exists a linear automorphism*

$$\sigma_1 : V_{L_0} \to V_{L_0} \tag{11.2.29}$$

such that

$$\sigma_1 Y(v, z)\sigma_1^{-1} = Y(\sigma_1 v, z) \quad \text{for} \quad v \in V_{L_0}, \tag{11.2.30}$$

$$\sigma_1^2 = 1, \tag{11.2.31}$$

$$\sigma_1 \text{ is grading-preserving}, \tag{11.2.32}$$

$$\sigma_1 = \sigma_0 \quad \text{on} \quad \mathfrak{g} \subset V_{L_0}, \tag{11.2.33}$$

$$\sigma_1(\iota(b)) = (-2)^{-|C|/2} \sum_{T \subset C} (-1)^{|S \cap T|} \iota(a_{S,T} b) \tag{11.2.34}$$

for $b \in \hat{L}_0$ such that $\bar{b} = \frac{1}{2}\varepsilon_S \alpha_C$ with $C \in \mathcal{C}$, $S \subset C$, where $a_{S,T}$ is given by (11.2.20), and

$$\sigma_1(\iota(1)) = \iota(1). \tag{11.2.35}$$

Then

$$\sigma_1(z\alpha_k(z))\sigma_1^{-1} = X^+(a_k, z) \tag{11.2.36}$$

$$\sigma_1 X^+(a_k, z)\sigma_1^{-1} = z\alpha_k(z), \tag{11.2.37}$$

$$\sigma_1 X^-(a_k, z)\sigma_1^{-1} = -X^-(a_k, z) \tag{11.2.38}$$

for $k \in \Omega$ *and*

$$\sigma_1 X(b, z)\sigma_1^{-1} = (-2)^{-|C|/2} \sum_{T \subset C} (-1)^{|S \cap T|} X(a_{S,T}b, z). \tag{11.2.39}$$

The space V_{L_0} *is absolutely irreducible under*

$$\hat{\mathfrak{h}}_{\mathbb{Z}} \cup B \tag{11.2.40}$$

where B *is any subset of* $\hat{L}_0$ *such that* $\bar{B}$ *generates the lattice* L_0*, and any operator commuting with the operators (11.2.40) is a scalar. In particular, the linear automorphism* σ_1 *is uniquely determined up to a normalizing factor by (11.2.36) and (11.2.39), and is determined precisely by the further condition (11.2.35).*

Remark 11.2.2: The element $\sigma_{(1)}$ of $SL(2, \mathbb{F})$, of order 4, is one of two liftings of the automorphism σ of $\mathfrak{sl}(2, \mathbb{F})$ [recall (11.2.4)], the other being $-\sigma_{(1)}$. But both liftings give rise to the same automorphism $\sigma_1 = \operatorname{diag} \sigma_{(1)}$ of V_{L_0} [see (11.2.6)] since $|\Omega|$ is even. Thus σ_1 is canonically determined as the diagonal action on V_{L_0} of a lifting of σ.

Remark 11.2.3: Because of the type I property of $\mathcal{C}$, the numerical coefficient in (11.2.34) and (11.2.39) is rational. Thus by the irreducibility of V_{L_0} under the set (11.2.28), we see that Theorem 11.2.1 holds even if we drop our assumption that $\mathbb{F}$ contains $\sqrt{2}$.

We have described the action of σ_1 on $\mathfrak{g}$ (11.1.72) and on certain elements $\iota(b)$ (11.2.34). Now we shall compute the action on the weight-two elements $\alpha_k(-1)\alpha_l(-1) = \alpha_k(-1)\alpha_l(-1)\iota(1)$ for $k, l \in \Omega$ and related elements.

Recall from (11.1.34) and (11.1.35) the action of the involution θ, and for $v \in V_{L_0}$ set

$$v^{\pm} = v \pm \theta v, \tag{11.2.41}$$

$$X^{\pm}(v, z) = X(v^{\pm}, z) = \sum_{n \in \mathbb{Q}} x_v^{\pm}(n) z^{-n}, \tag{11.2.42}$$

$$Y^{\pm}(v, z) = Y(v^{\pm}, z), \tag{11.2.43}$$

generalizing the notations (11.1.61) and (11.1.73). As usual, we also allow the notation $X^{\pm}(a, z)$, $Y^{\pm}(a, z)$ or $x_a^{\pm}(n)$ for $a \in \hat{L}_0$.

By (11.2.30),

$$\sigma_1 Y(v, z)w = Y(\sigma_1 v, z)\sigma_1 w \tag{11.2.44}$$

for $v, w \in V_{L0}$, so that

$$\sigma_1(x_v(n)w) = x_{\sigma_1 v}(n)\sigma_1 w \tag{11.2.45}$$

for $n \in \mathbb{Q}$ [as in (10.4.75)]. Thus

$$\begin{aligned}\sigma_1(a_k(-1)\alpha_l(-1)\iota(1)) &= \sigma_1(x_{\alpha k(-1)}(-1)\alpha_l(-1)\iota(1))\\ &= x_{\sigma_1(\alpha k(-1))}(-1)\sigma_1(\alpha_l(-1)\iota(1))\\ &= x_{\iota(a_k)^+}(-1)\iota(a_l)^+\\ &= (x_{a_k}(-1) + x_{a_k^{-1}}(-1))(\iota(a_l) + \iota(a_l^{-1})).\end{aligned}$$

For $k = l$ we compute that

$$\sigma_1(\alpha_k(-1)^2) = \alpha_k(-1)^2$$

and for $k \neq l$ that

$$\sigma_1(\alpha_k(-1)\alpha_l(-1)) = \iota(a_k a_l)^+ + \iota(a_k a_l^{-1})^+.$$

Similarly, for $k \neq l$,

$$\begin{aligned}\sigma_1(\iota(a_k a_l)^+ - \iota(a_k a_l^{-1})^+) &= \sigma_1(x_{\iota(a_k)^-}(-1)\iota(a_l)^-)\\ &= -x_{-\iota(a_k)^-}(-1)\iota(a_l)^-\\ &= \iota(a_k a_l)^+ - \iota(a_k a_l^{-1})^+.\end{aligned}$$

This proves:

Corollary 11.2.4: *For $k, l \in \Omega$ with $k \neq l$ we have*

$$\sigma_1(\alpha_k(-1)^2) = \alpha_k(-1)^2 \tag{11.2.46}$$

$$\sigma_1(\alpha_k(-1)\alpha_l(-1)) = \iota(a_k a_l)^+ + \iota(a_k a_l^{-1})^+ \tag{11.2.47}$$

$$\sigma_1(\iota(a_k a_l)^+ - \iota(a_k a_l^{-1})^+) = \iota(a_k a_l)^+ - \iota(a_k a_l^{-1})^+. \tag{11.2.48}$$

In particular,

$$\sigma_1 Y(\alpha_k(-1)^2, z)\sigma_1^{-1} = Y(\alpha_k(-1)^2, z) \tag{11.2.49}$$

$$\sigma_1 Y(\alpha_k(-1)\alpha_l(-1), z)\sigma_1^{-1} = Y^+(a_k a_l, z) + Y^+(a_k a_l^{-1}, z) \tag{11.2.50}$$

$$\sigma_1(Y^+(a_k a_l, z) - Y^+(a_k a_l^{-1}, z))\sigma_1^{-1} = Y^+(a_k a_l, z) - Y^+(a_k a_l^{-1}, z) \tag{11.2.51}$$

on V_{L0}, and similarly for X in place of Y.

11.3. Construction of $\sigma_1: V'_{L_0} \to V_{L_1}$

Now we shall use the idea behind Theorem 4.5.2—which shows the isomorphism of certain untwisted and twisted vertex operator constructions of a twisted affinization of $\mathfrak{sl}(2, \mathbb{F})\hat{}$—to extend σ_1 to an isomorphism between V_{L_1} and V'_{L_0}.

We fix an element $b_0 \in \hat{L}$ with

$$\bar{b}_0 = \tfrac{1}{4}\alpha_\Omega. \tag{11.3.1}$$

As in the proof of Proposition 4.4.2, $V_{Q+\beta}$ is an irreducible $\hat{\mathfrak{g}}[\theta_{(1)}]$-module for each coset $Q + \beta$ of Q in L_1, and for each $b \in \hat{L}_0$, $S(\hat{\mathfrak{h}}^-_{\mathbb{Z}+1/2}) \otimes t(b)$ is an irreducible $\hat{\mathfrak{g}}[\theta_{(2)}]$-module. The $\hat{\mathfrak{g}}[\theta_{(1)}]$-module V_{L_1} is a direct sum of $|\mathcal{C}|$ irreducible submodules generated by highest-degree elements $\iota(b_1 b_0)$ with

$$b_1 \in \hat{L}_0, \quad \bar{b}_1 = -\tfrac{1}{2}\alpha_C \quad \text{for} \quad C \in \mathcal{C} \tag{11.3.2}$$

and the $\hat{\mathfrak{g}}[\theta_{(2)}]$-module V'_{L_0} is a direct sum of the same number of irreducible submodules generated by analogous elements $t(b_1)$ for b_1 as in (11.3.2) [cf. (11.1.76)]. From the equations [recall (11.1.14)]

$$x^+_{a_k}(0)t(b_1) = \tfrac{1}{2}t(a_k b_1) = \tfrac{1}{2}(-1)^{\langle \alpha_k, \bar{b}_1\rangle} t(b_1), \tag{11.3.3}$$

$$\alpha_k(0)\iota(b_1 b_0) = \langle \alpha_k, \overline{b_1 b_0}\rangle \iota(b_1 b_0) = \tfrac{1}{2}(-1)^{|\{k\}\cap C|}\iota(b_1 b_0), \tag{11.3.4}$$

we see as in the proof of Theorem 4.5.2 that the $|\mathcal{C}|$ $\hat{\mathfrak{g}}[\theta_{(j)}]$-modules are inequivalent for $j = 1, 2$ and that for any choice of constants

$$\lambda_C \in \mathbb{F}^\times \quad \text{for} \quad C \in \mathcal{C} \tag{11.3.5}$$

there is a unique (grading-preserving) linear isomorphism $\sigma_1: V'_{L_0} \to V_{L_1}$ such that

$$\sigma_1 Y(v, z)\sigma_1^{-1} = Y(\sigma_0 v, z) \quad \text{for} \quad v \in \mathfrak{g}, \tag{11.3.6}$$

$$\sigma_1(t(b_1)) = \lambda_C \iota(b_1 b_0) \quad \text{for} \quad b_1 \text{ as in (11.3.2)}. \tag{11.3.7}$$

Recall that (11.3.6) is equivalent to the conditions (11.1.79)–(11.1.81).

We want to choose the constants λ_C so that

$$\sigma_1 Y(v, z)\sigma_1^{-1} = Y(\sigma_1 v, z) \quad \text{for} \quad v \in V_{L_0}, \tag{11.3.8}$$

where $Y = Y_{\mathbb{Z}+1/2}$ on the left, $Y = Y_{\mathbb{Z}}$ on the right and $\sigma_1 v$ is as in Theorem 11.2.1. For $v = \iota(b_1)$ with b_1 as in (11.3.2), this amounts to the assertion of formula (11.2.39) for $S = C$. If we multiply by σ_1 on the right, apply to $t(1)$

and extract the constant term, we find that we must have

$$\sigma_1(\iota(b_1)) = (-1)^{|C|/2}\lambda_\emptyset(-1)^{|C|}\iota(b_1 b_0)$$
$$= (-1)^{|C|/2}\lambda_\emptyset \iota(b_1 b_0), \qquad (11.3.9)$$

that is,

$$\lambda_C = (-1)^{|C|/2}\lambda_\emptyset \quad \text{for} \quad C \in \mathcal{C} \qquad (11.3.10)$$

in (11.3.7). We make this choice of constants.

In order to begin proving (11.3.8), we set

$$Y'(b, z) = Y_{\mathbb{Z}}(\sigma_1(\iota(b)), z), \qquad (11.3.11)$$

$$X'(b, z) = X_{\mathbb{Z}}(\sigma_1(\iota(b)), z) = Y'(b, z)z^{\langle \bar{b}, \bar{b}\rangle/2} \qquad (11.3.12)$$

for $b \in \hat{L}_0$, acting on V_{L_1}. For $k \in \Omega$ we find by Theorem 8.6.1 and (11.2.30) that on V_{L_1},

$$\begin{aligned}[Y_{\mathbb{Z}}(\iota(a_k)^+, z_1), Y'(b, z_2)] &= [Y_{\mathbb{Z}}(\sigma_1(\alpha_k(-1)), z_1), Y_{\mathbb{Z}}(\sigma_1(\iota(b)), z_2)]\\
&= \operatorname{Res}_{z_0} z_2^{-1} Y_{\mathbb{Z}}(Y_{\mathbb{Z}}(\sigma_1(\alpha_k(-1)), z_0)\sigma_1(\iota(b)), z_2)\\
&\quad \cdot e^{-z_0(\partial/\partial z_1)}((z_1/z_2)^{1/2}\delta(z_1/z_2))\\
&= \operatorname{Res}_{z_0} z_2^{-1} Y_{\mathbb{Z}}(\sigma_1 Y_{\mathbb{Z}}(\alpha_k(-1), z_0)\iota(b), z_2)\\
&\quad \cdot e^{-z_0(\partial/\partial z_1)}((z_1/z_2)^{1/2}\delta(z_1/z_2))\\
&= Y_{\mathbb{Z}}(\sigma_1 \alpha_k(0)\iota(b), z_2) z_2^{-1}(z_1/z_2)^{1/2}\delta(z_1/z_2)\\
&= \langle \alpha_k, \bar{b}\rangle Y'(b, z_2) z_2^{-1}(z_1/z_2)^{1/2}\delta(z_1/z_2).\end{aligned} \qquad (11.3.13)$$

For $\beta \in \mathfrak{h}$ define

$$\begin{aligned}E'^{\pm}(\beta, z) &= \exp\left(\sum_{n \in \pm(\mathbb{N}+1/2)} \sum_{k \in \Omega} \frac{\langle \beta, \alpha_k\rangle}{\langle \alpha_k, \alpha_k\rangle} \frac{x^+_{a_k}(n)}{n} z^{-n}\right)\\
&= \sigma_1 \exp\left(\sum_{n \in \pm(\mathbb{N}+1/2)} \frac{\beta(n)}{n} z^{-n}\right)\sigma_1^{-1}\\
&= \sigma_1 E^{\pm}_{\mathbb{Z}+1/2}(\beta, z)\sigma_1^{-1}\end{aligned} \qquad (11.3.14)$$

acting on V_{L_1}, where we use the notation $E^{\pm}_{\mathbb{Z}+1/2}$ for the operator $E^{\pm}$ of (9.1.1), and for $b \in \hat{L}_0$ set

$$Z'(b, z) = E'^{-}(\bar{b}, z)X'(b, z)E'^{+}(\bar{b}, z). \qquad (11.3.15)$$

This operator should be thought of as an analogue of the operator (3.2.30) used in the proof of Proposition 3.2.2, and in fact we now exploit the same idea.

Consider the Heisenberg algebra

$$\mathfrak{m} = \operatorname{span}\{x_{a_k}^{+}(n), 1_{V_{L_1}} \mid k \in \Omega, n \in \mathbb{Z} + \tfrac{1}{2}\} = \sigma_1 \hat{\mathfrak{h}}_{\mathbb{Z}+1/2} \sigma_1^{-1} \quad (11.3.16)$$

acting on V_{L_1}. Then $\mathfrak{m}$ acts irreducibly on each $\hat{\mathfrak{g}}[\theta_{(1)}]$-module $V_{Q+\beta}$, so that

$$V_{L_1} = S(\mathfrak{m}^-) \otimes \sigma_1(T_{L_0}), \quad (11.3.17)$$

where

$$\mathfrak{m}^- = \operatorname{span}\{x_{a_k}^{+}(n) \mid k \in \Omega, n \in -(\mathbb{N} + \tfrac{1}{2})\} = \sigma_1 \hat{\mathfrak{h}}_{\mathbb{Z}+1/2}^{-} \sigma_1^{-1}. \quad (11.3.18)$$

By (11.3.13)–(11.3.15) and analogues of (3.2.22)–(3.2.25), we find that for $k \in \Omega$, $n \in \mathbb{Z} + \frac{1}{2}$ and $b \in \hat{L}_0$,

$$[x_{a_k}^{+}(n), X'(b, z)] = \langle \alpha_k, \bar{b} \rangle z^n X'(b, z), \quad (11.3.19)$$

$$[x_{a_k}^{+}(n), Z'(b, z)] = 0, \quad (11.3.20)$$

so that

$$[\mathfrak{m}, Z'(b, z)] = 0. \quad (11.3.21)$$

Thus since

$$[d, Z'(b, z)] = -DZ'(b, z) \quad (11.3.22)$$

as in (3.2.32), we see that

$$Z'(b, z) = z_b'(0) z^0 = z_b'(0) \quad (11.3.23)$$

where $z_b'(0)$ is an operator of degree 0 on V_{L_1}, preserving $\sigma_1(T_{L_0})$ and commuting with $\mathfrak{m}$.

Let $b \in \hat{L}_0$. Then (11.3.8) for $v = \iota(b)$ is equivalent to the assertion that

$$\sigma_1 X_{\mathbb{Z}+1/2}(b, z) \sigma_1^{-1} = X'(b, z) \quad (11.3.24)$$

or that

$$\sigma_1 \circ b \circ \sigma_1^{-1} = 2^{\langle \bar{b}, \bar{b} \rangle} z_b'(0) \quad (11.3.25)$$

[recall (9.1.3)], b of course acting as an operator of degree 0 on V_{L_0}', preserving T_{L_0} and commuting with $\hat{\mathfrak{h}}_{\mathbb{Z}+1/2}$. To prove this it is sufficient to verify that

$$\sigma_1 \circ b = 2^{\langle \bar{b}, \bar{b} \rangle} z_b'(0) \circ \sigma_1 \quad (11.3.26)$$

on T_{L_0}.

We shall now carry out this check for

$$b \in \hat{L}_0, \quad \bar{b} = \tfrac{1}{2}\alpha_C \quad \text{for} \quad C \in \mathcal{C}. \tag{11.3.27}$$

Let

$$b_1 \in \hat{L}_0, \quad \bar{b}_1 = -\tfrac{1}{2}\alpha_{C_1}, \quad C_1 \in \mathcal{C}. \tag{11.3.28}$$

Since

$$x_{a_{\emptyset,T}b}(0)\iota(b_1 b_0) \in S(\hat{\mathfrak{h}}_{\mathbb{Z}}^-) \otimes \iota(a_{\emptyset,T}bb_1 b_0) \tag{11.3.29}$$

and

$$(a_{\emptyset,T}bb_1 b_0)^- = \tfrac{1}{4}\varepsilon_{C+C_1}\alpha_\Omega + \alpha_{C\backslash C_1} - \alpha_T \tag{11.3.30}$$

we have (by degree consideration)

$$x_{a_{\emptyset,T}b}(0)\iota(b_1 b_0) = \begin{cases} \iota(a_{\emptyset,T}bb_1 b_0) & \text{if} \quad T = C\backslash C_1 \\ 0 & \text{if} \quad T \neq C\backslash C_1. \end{cases} \tag{11.3.31}$$

Thus from (11.2.34), (11.3.7), (11.3.10) and (11.3.15),

$$\begin{aligned} 2^{\langle \bar{b},\bar{b}\rangle} z'_b(0)\sigma_1(\iota(b_1)) &= 2^{|C|/2} z'_b(0)\sigma_1(\iota(b_1)) \\ &= 2^{|C|/2}(-1)^{|C_1|/2}\lambda_\emptyset(-2)^{-|C|/2} \sum_{T\subset C} x_{a_{\emptyset,T}b}(0)\iota(b_1 b_0) \\ &= (-1)^{(|C|+|C_1|)/2}\lambda_\emptyset \iota(a_{\emptyset,C\backslash C_1}bb_1 b_0) \\ &= (-1)^{|C+C_1|/2+|C\cap C_1|+|C\backslash C_1|}\lambda_\emptyset \iota(bb_1 a_{\emptyset,C\backslash C_1}b_0), \\ &= (-1)^{|C+C_1|/2+|C|}\lambda_\emptyset \iota(bb_1 a_{\emptyset,C\backslash C_1}b_0) \\ &= (-1)^{|C+C_1|/2}\lambda_\emptyset \iota(bb_1 a_{\emptyset,C\backslash C_1}b_0), \end{aligned}$$

using (11.1.14). On the other hand,

$$\begin{aligned} \sigma_1 \circ b(\iota(b_1)) = \sigma_1(\iota(bb_1)) &= \sigma_1(\iota(bb_1 a_{\emptyset,C\backslash C_1})) \\ &= (-1)^{|C+C_1|/2}\lambda_\emptyset \iota(bb_1 a_{\emptyset,C\backslash C_1}b_0), \end{aligned}$$

using (11.1.28), and we have proved (11.3.26) for b as in (11.3.27). Thus

$$\sigma_1 Y(b,z)\sigma_1^{-1} = Y(\sigma_1(\iota(b)),z) \quad \text{for} \quad b \quad \text{as in (11.3.27).} \tag{11.3.32}$$

[With a little more work, this could have been verified directly for b as in (11.2.34), but we shall obtain this result anyway by proving (11.3.8) in general.]

At this point, we have an isomorphism from V'_{L_0} to V_{L_1} and we know that (11.3.8) holds for certain elements of V_{L_0}, namely, $v \in \mathfrak{g}$ [by (11.3.6)] and

$v = \iota(b)$, b as in (11.3.27). We shall now use the main results of Chapters 8 and 9 concerning the Jacobi identity to enlarge the domain of validity of (11.3.8).

Let $v \in V_{L_0}$, $k \in \Omega$ and $u = \iota(a_k)^+ \in \mathfrak{g}$. Then $\sigma_1 u = \alpha_k(-1)$, and Theorem 8.8.23 tells us that

$$z_0^{-1}\delta\left(\frac{z_1 - z_2}{z_0}\right)Y_{\mathbb{Z}}(\sigma_1 u, z_1)Y_{\mathbb{Z}}(\sigma_1 v, z_2) - z_0^{-1}\delta\left(\frac{z_2 - z_1}{-z_0}\right)Y_{\mathbb{Z}}(\sigma_1 v, z_2)Y_{\mathbb{Z}}(\sigma_1 u, z_1) = z_2^{-1}\delta\left(\frac{z_1 - z_0}{z_2}\right)Y_{\mathbb{Z}}(Y_{\mathbb{Z}}(\sigma_1 u, z_0)\sigma_1 v, z_2) \quad (11.3.33)$$

on V_{L_1}. Also, since u is fixed by θ, (11.1.14), (11.1.47) and Theorem 9.5.3 give

$$z_0^{-1}\delta\left(\frac{z_1 - z_2}{z_0}\right)Y_{\mathbb{Z}+1/2}(u, z_1)Y_{\mathbb{Z}+1/2}(v, z_2) - z_0^{-1}\delta\left(\frac{z_2 - z_1}{-z_0}\right)Y_{\mathbb{Z}+1/2}(v, z_2)Y_{\mathbb{Z}+1/2}(u, z_1) = z_2^{-1}\delta\left(\frac{z_1 - z_0}{z_2}\right)Y_{\mathbb{Z}+1/2}(Y_{\mathbb{Z}}(u, z_0)v, z_2) \quad (11.3.34)$$

on V'_{L_0}.

Now let us suppose that (11.3.8) holds for the element v. Conjugating (11.3.34) by σ_1 and using (11.2.30), we find that

$$z_2^{-1}\delta\left(\frac{z_1 - z_0}{z_2}\right)\sigma_1 Y_{\mathbb{Z}+1/2}(Y_{\mathbb{Z}}(u, z_0)v, z_2)\sigma_1^{-1} = z_2^{-1}\delta\left(\frac{z_1 - z_0}{z_2}\right)Y_{\mathbb{Z}}(\sigma_1(Y_{\mathbb{Z}}(u, z_0)v), z_2). \quad (11.3.35)$$

Applying Proposition 8.8.5 and extracting Res_{z_1} we obtain

$$\sigma_1 Y_{\mathbb{Z}+1/2}(Y_{\mathbb{Z}}(u, z_0)v, z_2)\sigma_1^{-1} = Y_{\mathbb{Z}}(\sigma_1(Y_{\mathbb{Z}}(u, z_0)v), z_2), \quad (11.3.36)$$

and we have established that (11.3.8) holds for (each expansion coefficient of) $Y_{\mathbb{Z}}(u, z_0)v$.

Next we take u to be either $\alpha_k(-1)$ or $\iota(a_k)^-$. Then $\sigma_1 u$ is $\iota(a_k)^+$ or

$-\iota(a_k)^-$, respectively, and by (11.1.14), this time Theorem 8.8.23 gives

$$z_0^{-1}\delta\left(\frac{z_1 - z_2}{z_0}\right)Y_{\mathbb{Z}}(\sigma_1 u, z_1)Y_{\mathbb{Z}}(\sigma_1 v, z_2)$$
$$- z_0^{-1}\delta\left(\frac{z_2 - z_1}{-z_0}\right)Y_{\mathbb{Z}}(\sigma_1 v, z_2)Y_{\mathbb{Z}}(\sigma_1 u, z_1)$$
$$= z_2^{-1}\left(\frac{z_1 - z_0}{z_2}\right)^{1/2}\delta\left(\frac{z_1 - z_0}{z_2}\right)Y_{\mathbb{Z}}(Y_{\mathbb{Z}}(\sigma_1 u, z_0)\sigma_1 v, z_2) \quad (11.3.37)$$

on V_{L_1}. The elements u are now negated by θ, so that by (11.1.14) and Theorem 9.5.3,

$$z_0^{-1}\delta\left(\frac{z_1 - z_2}{z_0}\right)Y_{\mathbb{Z}+1/2}(u, z_1)Y_{\mathbb{Z}+1/2}(v, z_2)$$
$$- z_0^{-1}\delta\left(\frac{z_2 - z_1}{-z_0}\right)Y_{\mathbb{Z}+1/2}(v, z_2)Y_{\mathbb{Z}+1/2}(u, z_1)$$
$$= z_2^{-1}\left(\frac{z_1 - z_0}{z_2}\right)^{1/2}\delta\left(\frac{z_1 - z_0}{z_2}\right)Y_{\mathbb{Z}+1/2}(Y_{\mathbb{Z}}(u, z_0)v, z_2) \quad (11.3.38)$$

on V'_{L_0}. If (11.3.8) holds for v, then exactly as above we obtain

$$z_2^{-1}\left(\frac{z_1 - z_0}{z_2}\right)^{1/2}\delta\left(\frac{z_1 - z_0}{z_2}\right)\sigma_1 Y_{\mathbb{Z}+1/2}(Y_{\mathbb{Z}}(u, z_0)v, z_2)\sigma_1^{-1}$$
$$= z_2^{-1}\left(\frac{z_1 - z_0}{z_2}\right)^{1/2}\delta\left(\frac{z_1 - z_0}{z_2}\right)Y_{\mathbb{Z}}(\sigma_1(Y_{\mathbb{Z}}(u, z_0)v), z_2). \quad (11.3.39)$$

This time we invoke Proposition 8.8.22 and equate coefficients of $z_1^{-1/2}$, and we see that

$$(z_2 + z_0)^{-1/2}\sigma_1 Y_{\mathbb{Z}+1/2}(Y_{\mathbb{Z}}(u, z_0)v, z_2)\sigma_1^{-1}$$
$$= (z_2 + z_0)^{-1/2}Y_{\mathbb{Z}}(\sigma_1(Y_{\mathbb{Z}}(u, z_0)v), z_2). \quad (11.3.40)$$

By multiplying through by $(z_2 + z_0)^{1/2}$ we have again shown that (11.3.8) holds for (each expansion coefficient of) $Y_{\mathbb{Z}}(u, z_0)v$.

The conclusion is:

$$\{v \in V_{L_0} \mid (11.3.8) \text{ holds}\} \quad \text{is a } \hat{\mathfrak{g}}\text{-submodule of } V_{L_0}. \quad (11.3.41)$$

But the elements $\iota(b)$ for b as in (11.3.27), for which we know (11.3.8), clearly generate V_{L_0} as a $\hat{\mathfrak{g}}$-module, and so we have proved (11.3.8) for all $v \in V_{L_0}$. We also note that V_{L_1} is absolutely irreducible under the set (11.2.40), and that [using (11.3.3)] V'_{L_0} is absolutely irreducible under an analogous set (cf. Proposition 7.4.11); moreover, each of these sets can be used to construct the operator $L(0)$. Taking the scalar $\lambda_\emptyset$ in (11.3.10) to be 1, we summarize as follows:

Theorem 11.3.1: *There exists a linear isomorphism*

$$\sigma_1 : V'_{L_0} \to V_{L_1} \tag{11.3.42}$$

such that

$$\sigma_1 Y(v, z)\sigma_1^{-1} = Y(\sigma_1 v, z) \quad \text{for} \quad v \in V_{L_0}, \tag{11.3.43}$$

where on the right-hand side, σ_1 is as in Theorem 11.2.1,

$$\sigma_1 \text{ is grading-preserving} \tag{11.3.44}$$

and

$$\sigma_1(t(1)) = \iota(b_0), \tag{11.3.45}$$

b_0 a fixed element of $\hat{L}$ such that

$$\bar{b}_0 = \tfrac{1}{4}\alpha_\Omega. \tag{11.3.46}$$

Then in particular, the assertions of (11.2.36)–(11.2.39) and (11.2.49)–(11.2.51) hold for the present map σ_1, and

$$\sigma_1(t(b_1)) = (-1)^{|C|/2}\iota(b_1 b_0) \tag{11.3.47}$$

where

$$b_1 \in \hat{L}_0, \quad \bar{b}_1 = -\tfrac{1}{2}\alpha_C \quad \text{for} \quad C \in \mathcal{C}. \tag{11.3.48}$$

The space V_{L_1} is absolutely irreducible under

$$\hat{\mathfrak{h}}_{\mathbb{Z}} \cup B \tag{11.3.49}$$

and V'_{L_0} is absolutely irreducible under

$$\hat{\mathfrak{h}}_{\mathbb{Z}+1/2} \cup B \tag{11.3.50}$$

where B is any subset of $\hat{L}_0$ such that $\bar{B}$ generates the lattice L_0, and any operator commuting with (11.3.49) or (11.3.50) is a scalar. In particular, the linear isomorphism σ_1 is uniquely determined up to a normalizing factor by (11.2.36) and (11.2.39), and is determined precisely by the further condition (11.3.45).

We now define a reverse map

$$\sigma_1 : V_{L_1} \to V'_{L_0} \tag{11.3.51}$$

by:

$$\sigma_1 = (\sigma_1|_{V_{L_0}})^{-1}. \tag{11.3.52}$$

Then using Theorems 11.2.1 and 11.3.1 we clearly have:

Theorem 11.3.2: *The linear isomorphism (11.3.51), (11.3.52) satisfies the conditions*

$$\sigma_1 Y(v, z)\sigma_1^{-1} = Y(\sigma_1 v, z) \quad \text{for} \quad v \in V_{L_0}, \tag{11.3.53}$$

where on the right σ_1 is as in Theorem 11.2.1,

$$\sigma_1 \text{ is grading-preserving} \tag{11.3.54}$$

and

$$\sigma_1(\iota(b_0)) = t(1), \tag{11.3.55}$$

b_0 as in (11.3.46). Then the assertions of (11.2.36)–(11.2.39) and (11.2.49)–(11.2.51) hold. The linear isomorphism σ_1 is uniquely determined up to a normalizing factor by (11.2.36) and (11.2.39), and is determined precisely by (11.3.55).

Remark 11.3.3: Consider the natural representation of the group $\hat{L}_0$ on V'_{L_0} by operators of degree 0 preserving T_{L_0}. Now that (11.3.8) is known for $\iota(b)$ for all $b \in \hat{L}_0$, we see from (11.3.25) that σ_1 transports this action of $\hat{L}_0$ to an equivalent action of $\hat{L}_0$ on V_{L_1}:

$$b \mapsto 2^{\langle \bar{b}, \bar{b} \rangle} z'_b(0) \quad \text{for} \quad b \in \hat{L}_0, \tag{11.3.56}$$

where the (necessarily invertible) operators $z'_b(0)$ are given by (11.3.15) and (11.3.23).

Remark 11.3.4: Just as in Corollary 4.5.4, we can equate the graded dimensions of V_{L_1} and V'_{L_0} to obtain an identity relating a theta function and Dedekind's η-function:

$$\frac{\theta_{L_1}(q)}{\eta(q)^{|\Omega|}} = |\mathcal{C}| \frac{\eta(q)^{|\Omega|}}{\eta(q^{1/2})^{|\Omega|}}, \tag{11.3.57}$$

where

$$\theta_{L_1}(q) = \sum_{\alpha \in L_1} q^{\langle \alpha, \alpha \rangle / 2}. \tag{11.3.58}$$

Of course, this is an extension of the notation θ_L of (6.1.30) to a subset of a positive definite lattice.

11.4. Construction of $\sigma_1: V'_{L_1} \to V'_{L_1}$

To complete the construction of an involution σ_1 of the space W_L of (11.1.36), (11.1.42), we shall construct an involution of V'_{L_1}, following the general approach of Section 11.3.

Recall the element b_0 of $\hat{L}$ fixed in (11.3.1). We know that for each $b \in \hat{L}_0$, $S(\hat{\mathfrak{h}}^-_{\mathbb{Z}+1/2}) \otimes t(bb_0)$ is an irreducible $\hat{\mathfrak{g}}[\theta_{(3)}]$-module, and that V'_{L_1} is a direct sum of $|\mathfrak{C}|$ such modules, generated by highest-degree elements $t(b_1 b_0)$ spanning T_{L_1}, where b_1 is as in (11.3.2) [cf. (11.1.76)]. In fact, for $b \in \hat{L}_0$ and $k \in \Omega$,

$$x^-_{\bar{a}_k}(0)t(bb_0) = \tfrac{1}{2}t(a_k bb_0) = \tfrac{1}{2}(-1)^{\langle \alpha_k, \bar{b}\rangle} it(bb_0) \qquad (11.4.1)$$

by (9.1.3), (11.1.13), (11.1.45) and (11.1.61). Similarly,

$$x^-_{\bar{a}_k}(0)t(bb_0^{-1}) = -\tfrac{1}{2}(-1)^{\langle \alpha_k, \bar{b}\rangle} it(bb_0^{-1}). \qquad (11.4.2)$$

Thus the $|\mathfrak{C}|$ $\hat{\mathfrak{g}}[\theta_{(3)}]$-submodules are inequivalent, and since

$$\sigma_0(\iota(a_k)^-) = -\iota(a_k)^- \qquad (11.4.3)$$

and

$$\sigma_0 \theta_{(3)} \sigma_0^{-1} = \theta_{(3)} \qquad (11.4.4)$$

[recall (11.1.72), (11.1.75)], we see that for any choice of constants

$$\mu_C \in \mathbb{F}^\times \quad \text{for} \quad C \in \mathfrak{C} \qquad (11.4.5)$$

there is a unique (grading-preserving) linear automorphism σ_1 of V'_{L_1} such that

$$\sigma_1 Y(v, z)\sigma_1^{-1} = Y(\sigma_0 v, z) \quad \text{for} \quad v \in \mathfrak{g}, \qquad (11.4.6)$$

$$\sigma_1(t(b_1 b_0)) = \mu_C t(b_1 b_0^{-1}) \quad \text{for} \quad b_1 \quad \text{as in (11.3.2).} \qquad (11.4.7)$$

As usual, (11.4.6) is equivalent to the conditions (11.1.79)–(11.1.81).

Again our goal is the formula

$$\sigma_1 Y(v, z)\sigma_1^{-1} = Y(\sigma_1 v, z) \quad \text{for} \quad v \in V_{L_0} \qquad (11.4.8)$$

with $\sigma_1 v$ as in Theorem 11.2.1, and we use this for $v = \iota(b_1)$, b_1 as in (11.3.2), to motivate a choice of the constants μ_C. As in (11.3.9), we apply (11.2.39) for $S = C$ to $\sigma_1(t(b_0))$ to obtain

$$\begin{aligned}
\sigma_1(t(b_1 b_0)) &= \mu_\emptyset(-2)^{-|C|/2} \sum_{T \subset C} (-1)^{|T|} t(a_{C,T} b_1 b_0^{-1}) \\
&= \mu_\emptyset(-2)^{-|C|/2} \sum_{T \subset C} (-1)^{|T|}(-1)^{|C \setminus T|} i^{-|C \setminus T|} t(b_1 b_0^{-1}) \\
&= \mu_\emptyset 2^{-|C|/2} \left(\sum_{T \subset C} i^{|T|} \right) t(b_1 b_0^{-1}) \\
&= \mu_\emptyset 2^{-|C|/2} (1 + i)^{|C|} t(b_1 b_0^{-1}) \\
&= \mu_\emptyset 2^{-|C|/2} (2i)^{|C|/2} t(b_1 b_0^{-1}) \\
&= \mu_\emptyset i^{|C|/2} t(b_1 b_0^{-1}),
\end{aligned} \tag{11.4.9}$$

using (11.1.13), and this gives

$$\mu_C = i^{|C|/2} \mu_\emptyset \quad \text{for} \quad C \in \mathcal{C}. \tag{11.4.10}$$

We choose the μ_C in this way.

As a consequence we have more generally

$$\sigma_1(t(b b_0)) = \mu_\emptyset(-1)^{|S|} i^{|C|/2} t(b b_0^{-1}) \tag{11.4.11}$$

for

$$b \in \hat{L}_0 \quad \text{with} \quad \bar{b} = \tfrac{1}{2}\varepsilon_S \alpha_C \quad \text{for} \quad C \in \mathcal{C}, \quad S \subset C. \tag{11.4.12}$$

We also want $\sigma_1^2 = 1$, and this will restrict $\mu_\emptyset$ as follows: For b_1 as in (11.3.2), we have by (11.4.9), (11.1.16) and (11.1.29)

$$\begin{aligned}
\sigma_1^2(t(b_1 b_0)) &= \mu_\emptyset i^{|C|/2} \sigma_1(t(b_1 b_0^{-1})) \\
&= \mu_\emptyset i^{|C|/2 - |C|} \sigma_1(t(b_1 b_0^{-2} a_{C,\emptyset} b_0)) \\
&= \mu_\emptyset^2 i^{|C|/2 + |\Omega \setminus C|/2 - |C|} t(b_1 b_0^{-2} a_{C,\emptyset} b_0^{-1}) \\
&= \mu_\emptyset^2 i^{|\Omega|/2 - 2|C|} t(b_1 b_0^{-3}) \\
&= \mu_\emptyset^2 i^{|\Omega|/2} \chi(b_0^{-4} a_{\Omega,\emptyset}) t(b_1 b_0) \\
&= t(b_1 b_0)
\end{aligned}$$

if and only if $\mu_\emptyset$ satisfies the condition

$$\mu_\emptyset^2 = (-i)^{|\Omega|/2} \chi(b_0^4 a_{\emptyset,\Omega}). \tag{11.4.13}$$

In this case,

$$\sigma_1^2 = 1 \tag{11.4.14}$$

since σ_1^2 commutes with $\hat{\mathfrak{g}}[\theta_{(3)}]$ by (11.4.6). We therefore assume that $\mu_\emptyset$ satisfies (11.4.13). Notice that (11.4.11) now gives us a formula for $\sigma_1(t(b b_0^{-1}))$ for b as in (11.4.12).

Continuing to argue as in Section 11.3, we begin proving (11.4.8) by setting

$$Y'(b, z) = Y(\sigma_1(\iota(b)), z), \tag{11.4.15}$$

$$X'(b, z) = X(\sigma_1(\iota(b)), z) = Y'(b, z)z^{\langle \bar{b}, \bar{b}\rangle/2} \tag{11.4.16}$$

for $b \in \hat{L}_0$, acting on V'_{L_1}. As we have commented in Section 11.1, the results of Chapter 9 apply in the generality of the remarks at the ends of Sections 9.1–9.4; recall from (11.1.48) that

$$\theta a_k\big|_{T_{L_1}} = -a_k\big|_{T_{L_1}} \quad \text{for} \quad k \in \Omega. \tag{11.4.17}$$

Thus (11.2.30) and Theorem 9.3.1 as modified by Remark 9.3.11 imply that on V'_{L_1},

$$\begin{aligned}[Y(\iota(a_k)^+, z_1), Y'(b, z_2)] &= [Y(\sigma_1(\alpha_k(-1)), z_1), Y(\sigma_1(\iota(b)), z_2)] \\ &= \operatorname{Res}_{z_0} z_2^{-1} Y(\sigma_1 Y_{\mathbb{Z}}(\alpha_k(-1), z_0)\iota(b), z_2) \\ &\quad \cdot e^{-z_0(\partial/\partial z_1)}((z_1/z_2)^{1/2}\delta(z_1/z_2)) \\ &= \langle \alpha_k, \bar{b}\rangle Y'(b, z_2) z_2^{-1}(z_1/z_2)^{1/2}\delta(z_1/z_2).\end{aligned} \tag{11.4.18}$$

For $\beta \in \mathfrak{h}$ set

$$\begin{aligned}E'^{\pm}(\beta, z) &= \exp\left(\sum_{n \in \pm(\mathbb{N}+1/2)} \sum_{k \in \Omega} \frac{\langle \beta, \alpha_k\rangle}{\langle \alpha_k, \alpha_k\rangle} \frac{x^+_{a_k}(n)}{n} z^{-n}\right) \\ &= \sigma_1 E^{\pm}(\beta, z)\sigma_1^{-1},\end{aligned} \tag{11.4.19}$$

with $E^{\pm}(\beta, z)$ as in (9.1.1), and for $b \in \hat{L}_0$ define

$$Z'(b, z) = E'^{-}(\bar{b}, z)X'(b, z)E'^{+}(\bar{b}, z). \tag{11.4.20}$$

The Heisenberg algebra

$$\mathfrak{m} = \operatorname{span}\{x^+_{a_k}(n), 1_{V'_{L_1}} \mid k \in \Omega, n \in \mathbb{Z} + \tfrac{1}{2}\} = \sigma_1 \hat{\mathfrak{h}}_{\mathbb{Z}+1/2}\sigma_1^{-1} \tag{11.4.21}$$

has vacuum space T_{L_1}, and for $k \in \Omega$, $n \in \mathbb{Z} + \frac{1}{2}$ and $b \in \hat{L}_0$,

$$[x^+_{a_k}(n), X'(b, z)] = \langle \alpha_k, \bar{b}\rangle z^n X'(b, z), \tag{11.4.22}$$

$$[x^+_{a_k}(n), Z'(b, z)] = 0. \tag{11.4.23}$$

Hence

$$[\mathfrak{m}, Z'(b, z)] = 0, \tag{11.4.24}$$

and since

$$[d, Z'(b, z)] = -DZ'(b, z), \tag{11.4.25}$$

we have

$$Z'(b, z) = z'_b(0)z^0 = z'_b(0) \tag{11.4.26}$$

where $z'_b(0)$ is an operator of degree 0 on V'_{L_1}, preserving T_{L_1} and commuting with $\mathfrak{m}$.

Let $b \in \hat{L}_0$. Then (11.4.8) for $v = \iota(b)$ is equivalent to the assertion that

$$\sigma_1 X(b, z)\sigma_1^{-1} = X'(b, z) \tag{11.4.27}$$

or that

$$\sigma_1 \circ b \circ \sigma_1^{-1} = 2^{\langle \bar{b}, \bar{b} \rangle} z'_b(0), \tag{11.4.28}$$

b viewed as an operator of degree 0 on V'_{L_1}, preserving T_{L_1} and commuting with $\hat{\mathfrak{h}}_{\mathbb{Z}+1/2}$. This would follow from the identity

$$\sigma_1 \circ b = 2^{\langle \bar{b}, \bar{b} \rangle} z'_b(0) \circ \sigma_1 \tag{11.4.29}$$

on T_{L_1}.

We now prove (11.4.29) for

$$b \in \hat{L}_0, \quad \bar{b} = \tfrac{1}{2}\alpha_C \quad \text{for} \quad C \in \mathfrak{C}. \tag{11.4.30}$$

Let

$$b_1 \in \hat{L}_0, \quad \bar{b}_1 = -\tfrac{1}{2}\alpha_{C_1}, \quad C_1 \in \mathfrak{C}. \tag{11.4.31}$$

By (11.4.11) we see that

$$\begin{aligned}\sigma_1(b(t(b_1 b_0))) &= \sigma_1(t(bb_1 b_0)) \\ &= \mu_\emptyset (-1)^{|C_1 \setminus C|} i^{|C+C_1|/2} t(bb_1 b_0^{-1}),\end{aligned}$$

and using (11.2.34) we also have

$$\begin{aligned}&2^{\langle \bar{b}, \bar{b} \rangle} z'_b(0)\sigma_1(t(b_1 b_0)) \\ &\quad = 2^{|C|/2} z'_b(0)\sigma_1(t(b_1 b_0)) \\ &\quad = \mu_\emptyset i^{|C_1|/2} 2^{|C|/2} z'_b(0) t(b_1 b_0^{-1}) \\ &\quad = \mu_\emptyset i^{|C_1|/2} (-1)^{|C|/2} \sum_{T \subset C} x_{a_\emptyset, T} {}_b(0) t(b_1 b_0^{-1}) \\ &\quad = \mu_\emptyset i^{|C_1|/2} (-2)^{-|C|/2} \sum_{T \subset C} t(a_{\emptyset, T} bb_1 b_0^{-1}) \\ &\quad = \mu_\emptyset i^{|C_1|/2} (-2)^{-|C|/2} \sum_{T \subset C} (-1)^{|T \cap (C+C_1)|} i^{|T|} t(bb_1 b_0^{-1}).\end{aligned}$$

But

$$\sum_{T\subset C}(-1)^{|T\cap(C+C_1)|}i^{|T|} = \sum_{T\subset C}(-1)^{|T\cap(C\setminus C_1)|}i^{|T|}$$

$$= (1+i)^{|C\cap C_1|}(1-i)^{|C\setminus C_1|}$$

$$= i^{|C\cap C_1|}(1-i)^{|C|}$$

$$= i^{|C\cap C_1|}(-2i)^{|C|/2},$$

so that

$$2^{\langle \bar{b},\bar{b}\rangle}z_b'(0)\sigma_1(t(b_1b_0))$$

$$= \mu_\emptyset i^{|C_1|/2+|C\cap C_1|+|C|/2}(-2)^{-|C|/2}(-2)^{|C|/2}t(bb_1b_0^{-1})$$

$$= \mu_\emptyset i^{|C+C_1|/2+2|C\cap C_1|}t(bb_1b_0^{-1})$$

$$= \sigma_1(b(t(b_1b_0))),$$

proving (11.4.29) and therefore (11.4.27) for b as in (11.4.30). In particular,

$$\sigma_1 Y(b,z)\sigma_1^{-1} = Y(\sigma_1(\iota(b)),z) \quad \text{for such} \quad b. \tag{11.4.32}$$

As in Section 11.3, let $v \in V_{L_0}$, $k \in \Omega$ and $u = \iota(a_k)^-$, which is negated by θ. Then $\sigma_1 u = -\iota(a_k)^-$, and by (11.1.14), (11.1.48) and Theorem 9.5.3,

$$z_0^{-1}\delta\left(\frac{z_1-z_2}{z_0}\right)Y(\sigma_1 u,z_1)Y(\sigma_1 v,z_2) - z_0^{-1}\delta\left(\frac{z_2-z_1}{-z_0}\right)Y(\sigma_1 v,z_2)Y(\sigma_1 u,z_1)$$

$$= z_2^{-1}\delta\left(\frac{z_1-z_0}{z_2}\right)Y(Y_{\mathbb{Z}}(\sigma_1 u,z_0)\sigma_1 v,z_2) \tag{11.4.33}$$

and

$$z_0^{-1}\delta\left(\frac{z_1-z_2}{z_0}\right)Y(u,z_1)Y(v,z_2) - z_0^{-1}\delta\left(\frac{z_2-z_1}{-z_0}\right)Y(v,z_2)Y(u,z_1)$$

$$= z_2^{-1}\delta\left(\frac{z_1-z_0}{z_2}\right)Y(Y_{\mathbb{Z}}(u,z_0)v,z_2). \tag{11.4.34}$$

Thus if (11.4.8) holds for the element v, then

$$z_2^{-1}\delta\left(\frac{z_1-z_0}{z_2}\right)\sigma_1 Y(Y_{\mathbb{Z}}(u,z_0)v,z_2)\sigma_1^{-1}$$

$$= z_2^{-1}\delta\left(\frac{z_1-z_0}{z_2}\right)Y(\sigma_1(Y_{\mathbb{Z}}(u,z_0)v),z_2), \tag{11.4.35}$$

using (11.2.30) and (11.4.6), and so

$$\sigma_1 Y(Y_{\mathbb{Z}}(u, z_0)v, z_2)\sigma_1^{-1} = Y(\sigma_1(Y_{\mathbb{Z}}(u, z_0)v), z_2), \tag{11.4.36}$$

as in (11.3.36). In particular, (11.4.8) holds for (each expansion coefficient of) $Y_{\mathbb{Z}}(u, z_0)v$.

Now let u be either $\alpha_k(-1)$ or $\iota(a_k)^+$, which is negated or fixed, respectively, by θ. Then $\sigma_1 u$ is $\iota(a_k)^+$ or $\alpha_k(-1)$, which is θ-fixed or negated, respectively. By (11.1.14) and (11.1.48), Theorem 9.5.3 now gives

$$z_0^{-1}\delta\left(\frac{z_1 - z_2}{z_0}\right)Y(\sigma_1 u, z_1)Y(\sigma_1 v, z_2) - z_0^{-1}\delta\left(\frac{z_2 - z_1}{-z_0}\right)Y(\sigma_1 v, z_2)Y(\sigma_1 u, z_1)$$
$$= z_2^{-1}\left(\frac{z_1 - z_0}{z_2}\right)^{1/2}\delta\left(\frac{z_1 - z_0}{z_2}\right)Y(Y_{\mathbb{Z}}(\sigma_1 u, z_0)\sigma_1 v, z_2) \tag{11.4.37}$$

and

$$z_0^{-1}\delta\left(\frac{z_1 - z_2}{z_0}\right)Y(u, z_1)Y(v, z_2) - z_0^{-1}\delta\left(\frac{z_2 - z_1}{-z_0}\right)Y(v, z_2)Y(u, z_1)$$
$$= z_2^{-1}\left(\frac{z_1 - z_0}{z_2}\right)^{1/2}\delta\left(\frac{z_1 - z_0}{z_2}\right)Y(Y_{\mathbb{Z}}(u, z_0)v, z_2). \tag{11.4.38}$$

If (11.4.8) holds for v, then

$$z_2^{-1}\left(\frac{z_1 - z_0}{z_2}\right)^{1/2}\delta\left(\frac{z_1 - z_0}{z_2}\right)\sigma_1 Y(Y_{\mathbb{Z}}(u, z_0)v, z_2)\sigma_1^{-1}$$
$$= z_2^{-1}\left(\frac{z_1 - z_0}{z_2}\right)^{1/2}\delta\left(\frac{z_1 - z_0}{z_2}\right)Y(\sigma_1(Y_{\mathbb{Z}}(u, z_0)v), z_2) \tag{11.4.39}$$

and

$$\sigma_1 Y(Y_{\mathbb{Z}}(u, z_0)v, z_2)\sigma_1^{-1} = Y(\sigma_1(Y_{\mathbb{Z}}(u, z_0)v), z_2) \tag{11.4.40}$$

by the argument of (11.3.40), so that (11.4.8) again holds for (each expansion coefficient of) $Y_{\mathbb{Z}}(u, z_0)v$.

Hence as in (11.3.41),

$$\{v \in V_{L_0} \mid (11.4.8) \text{ holds}\} \quad \text{is a } \hat{\mathfrak{g}}\text{-submodule of } V_{L_0}, \tag{11.4.41}$$

and it then follows from (11.4.32) that (11.4.8) is valid for all $v \in V_{L_0}$. As in Section 11.3, V'_{L_1} is absolutely irreducible under the set (11.3.50), which can be used to construct the operator $L(0)$. Writing the scalar $\mu_\emptyset$ as μ, we see that we have proved:

Theorem 11.4.1: *There exists a linear automorphism*

$$\sigma_1: V'_{L_1} \to V'_{L_1} \tag{11.4.42}$$

such that

$$\sigma_1 Y(v, z)\sigma_1^{-1} = Y(\sigma_1 v, z) \quad \text{for} \quad v \in V_{L_0}, \tag{11.4.43}$$

where on the right-hand side, σ_1 is as in Theorem 11.2.1,

$$\sigma_1 \text{ is grading-preserving} \tag{11.4.44}$$

and

$$\sigma_1(t(b_0)) = \mu t(b_0^{-1}), \tag{11.4.45}$$

μ a fixed element of $\mathbb{F}^\times$ and b_0 a fixed element of $\hat{L}$ such that

$$\bar{b}_0 = \tfrac{1}{4}\alpha_\Omega. \tag{11.4.46}$$

Then in particular, the assertions of (11.2.36)–(11.2.39) and (11.2.49)–(11.2.51) hold for the present map σ_1, and

$$\sigma_1(t(bb_0)) = \mu(-1)^{|S|} i^{|C|/2} t(bb_0^{-1}) \tag{11.4.47}$$

where

$$b \in \hat{L}_0 \quad \text{with} \quad \bar{b} = \tfrac{1}{2}\varepsilon_S \alpha_C \quad \text{for} \quad C \in \mathcal{C}, \quad S \subset C. \tag{11.4.48}$$

If (and only if)

$$\mu^2 = (-i)^{|\Omega|/2}\chi(b_0^4 a_{\emptyset,\Omega}), \tag{11.4.49}$$

using the notation of (11.1.16) and (11.2.20), we have

$$\sigma_1^2 = 1. \tag{11.4.50}$$

The space V'_{L_1} is absolutely irreducible under

$$\hat{\mathfrak{h}}_{\mathbb{Z}+1/2} \cup B \tag{11.4.51}$$

where B is any subset of $\hat{L}_0$ such that $\bar{B}$ generates the lattice L_0, and any operator commuting with (11.4.51) is a scalar. In particular, the linear automorphism σ_1 is uniquely determined up to a normalizing factor by (11.2.36) and (11.2.39), and is determined precisely by the further condition (11.4.45).

Remark 11.4.2 (cf. Remark 11.3.3): From (11.4.28), the automorphism σ_1 transports the action of $\hat{L}_0$ on V'_{L_1} to an equivalent action of $\hat{L}_0$ on V'_{L_1}:

$$b \mapsto 2^{\langle \bar{b}, \bar{b} \rangle} z'_b(0) \quad \text{for} \quad b \in \hat{L}_0, \tag{11.4.52}$$

where the (necessarily invertible) operators $z'_b(0)$ are given by (11.4.20) and (11.4.26).

11.5. Summary

Here we collect the main results of the last three sections.

Theorem 11.5.1: *In the setting of Section 11.1, there exists a linear automorphism*

$$\sigma_1 \colon W_L \to W_L \tag{11.5.1}$$

such that

$$\sigma_1 V_{L_0} = V_{L_0}, \quad \sigma_1 \colon V_{L_1} \to V'_{L_0}, \quad \sigma_1 \colon V'_{L_0} \to V_{L_1}, \quad \sigma_1 \colon V'_{L_1} \to V'_{L_1}, \tag{11.5.2}$$

$$\sigma_1 Y(v, z)\sigma_1^{-1} = Y(\sigma_1 v, z) \quad \text{for} \quad v \in V_{L_0}, \tag{11.5.3}$$

$$\sigma_1^2 = 1, \tag{11.5.4}$$

$$\sigma_1 \text{ is grading-preserving}, \tag{11.5.5}$$

$$\sigma_1 = \sigma_0 \quad \text{on} \quad \mathfrak{g} \subset V_{L_0}, \tag{11.5.6}$$

$$\sigma_1(\iota(b)) = (-2)^{-|C|/2} \sum_{T \subset C} (-1)^{|S \cap T|} \iota(a_{S,T} b) \tag{11.5.7}$$

for

$$b \in \hat{L}_0 \quad \text{with} \quad \bar{b} = \tfrac{1}{2}\varepsilon_S \alpha_C, \quad \text{for} \quad C \in \mathcal{C}, \quad S \subset C, \tag{11.5.8}$$

where $a_{S,T}$ is given by (11.2.20),

$$\sigma_1(\iota(1)) = \iota(1), \tag{11.5.9}$$

$$\sigma_1(t(1)) = \iota(b_0), \tag{11.5.10}$$

$$\sigma_1(t(b_0)) = \mu t(b_0^{-1}), \tag{11.5.11}$$

where b_0 is a fixed element of $\hat{L}$ such that

$$\bar{b}_0 = \tfrac{1}{4}\alpha_\Omega \tag{11.5.12}$$

and provided that $\mu \in \mathbb{F}^\times$ exists and is chosen so that

$$\mu^2 = (-i)^{|\Omega|/2} \chi(b_0^4 a_{\emptyset,\Omega}), \tag{11.5.13}$$

using the notation χ of (11.1.16). Then the assertions of (11.2.36)–(11.2.39) and (11.2.46)–(11.2.51) hold, $\sigma_1(t(b_1))$ and $\sigma_1(\iota(b_1 b_0))$ are given by (11.3.47) for

$$b_1 \in \hat{L}_0 \quad \text{with} \quad \bar{b}_1 = -\tfrac{1}{2}\alpha_C \quad \text{for} \quad C \in \mathcal{C} \tag{11.5.14}$$

and $\sigma_1(t(bb_0))$ and $\sigma_1(t(bb_0^{-1}))$ are given by (11.4.47) for b as in (11.5.8). Moreover, σ_1 is uniquely determined by (11.2.36), (11.2.39), (11.5.2), (11.5.4), (11.5.9), (11.5.10) and a choice of μ in (11.5.11).

Remark 11.5.2: Our involution σ_1 of W_L is based (canonically, except for the normalization constants) on the Lie algebra involution σ_0 of $\mathfrak{g}$, given by (11.1.72), and hence on the Lie algebra involution σ of $\mathfrak{sl}(2, \mathbb{F})$ given by (3.1.44) (recall Remark 11.2.2). This last involution should be viewed as follows: Consider the basis

$$y_1 = i\alpha_1, \quad y_2 = x_{\alpha_1}^-, \quad y_3 = ix_{\alpha_1}^+ \tag{11.5.15}$$

of $\mathfrak{sl}(2, \mathbb{F})$, in the notation of Section 3.1. In this basis, the commutation relations are

$$[y_1, y_2] = 2y_3 \tag{11.5.16}$$

and its cyclic permutations, so that there are "manifest" Lie algebra automorphisms permuting the three pairs $\{\pm y_k\}$. The involution σ is the one given by:

$$\sigma: y_1 \leftrightarrow y_3, \quad y_2 \leftrightarrow -y_2. \tag{11.5.17}$$

One could repeat the considerations of this chapter for the two analogous involutions which permute the other pairs of indices. The symmetric group $\mathcal{S}_3$ on the three indices lifts to an extension of $\mathcal{S}_3$ by a four-element group of sign changes acting on $\mathfrak{sl}(2, \mathbb{F})$. The lifting of $\mathcal{S}_3$ to corresponding automorphisms of V_{L_0} and isomorphisms among V_{L_1}, V'_{L_0} and V'_{L_1} is what we mean by the term "triality." We have chosen to study σ instead of the involution permuting the indices 1 and 2 because σ is defined even if $\mathbb{F}$ does not contain i. Later we shall enlarge (a modification of) σ_1 to an action of $\mathcal{S}_3$ on our modules in the process of constructing the action of the Monster on the moonshine module. The additional involution needed will turn out to be much easier to describe than σ_1.

Remark 11.5.3: In case our type I code $\mathcal{C}$ is also of type II [recall (10.1.3)], the constants entering into the description of σ_1 simplify; see (11.5.7), (11.3.47) and (11.4.47). For instance, (11.3.47) reduces to:

$$\sigma_1(t(b_1)) = \imath(b_1 b_0). \tag{11.5.18}$$

12 The Main Theorem

This chapter is the focal point of the book. Here we apply the results of Chapters 8, 9, 10 and 11 and collect all the ingredients of our main construction of the moonshine module $V^\natural$, the vertex operator algebra associated with $V^\natural$ and the action of the Monster on $V^\natural$. The main theorem states that the graded dimension of the moonshine module $V^\natural$ is the modular function $J(q)$, that $V^\natural$ is a vertex operator algebra and that the Monster acts as automorphisms of this structure, in the terminology of Section 8.10. In addition, the theorem describes the Griess algebra $\mathfrak{B}$ and its commutative affinization $\hat{\mathfrak{B}}$ in terms of weight-two general vertex operators and cross-brackets, and it asserts the irreducibility of $V^\natural$ under $\hat{\mathfrak{B}}$ and the faithfulness of the action of the Monster on $\mathfrak{B}$ and on $\hat{\mathfrak{B}}$. The identity element of (our presentation of) the Griess algebra gives rise to the Virasoro algebra, and the 196883-dimensional orthogonal complement of the identity element consists of lowest weight vectors for the Virasoro algebra, giving rise to primary fields, in physics terminology. As we remark in the Introduction, our previously announced results [FLM2] included everything but the vertex operator algebra structure, which is in fact generated by $\hat{\mathfrak{B}}$, and this structure was announced in [Borcherds 3]. In particular, we recover here (as announced in [FLM2]) Griess's construction [Griess 2, 3] of the Monster as a group of automorphisms of $\mathfrak{B}$; one may then quote results of [Griess 2, 3] as simplified in [Conway 9, 10] and [Tits 4–6], to show that the group is indeed finite and simple, with the group C as the centralizer of an involution. Combining the main theorem with Tits's result that the Monster is the *full* automorphism group of $\mathfrak{B}$, we see immediately that the Monster is the full automorphism group of the vertex operator algebra. We also show that the Monster preserves a rational structure and in the case of the complex field, a positive definite hermitian form. In the course of formulating the main theorem and preparing for the completion of the proof in Chapter 13, we must present some technical constructions and results, which we suggest that the reader skim on first reading, so that he or she can quickly approach the statement of the main theorem (Theorem 12.3.1), and then gain an overview of the proof, described at the beginning of the next chapter in Section 13.1.

In Section 12.1 we introduce the lattices, central extensions and other data appropriate for the application of the results of the earlier chapters. Then in Section 12.2 we modify the involution σ_1 from Chapter 11 to an involution σ which will be an element of the Monster. In Section 12.3 we define the Monster M as the group of linear automorphisms of $V^\natural$ generated by C and σ, we define the vertex operators associated with all the elements of $V^\natural$, and we state the main theorem and begin the proof. We also give explicit formulas describing the action of σ on $\mathfrak{B}$, complementing the explicit description of $\mathfrak{B}$ itself and of the action of C on it presented earlier. We show the existence of an M-invariant rational form and an M-invariant positive definite hermitian form (if $\mathbb{F} = \mathbb{C}$) in the last two sections, 12.4 and 12.5, respectively. These structures will enter into the proof in Chapter 13. Our main results in this chapter were announced in [FLM2]; the assertion that M preserves a vertex operator algebra structure on $V^\natural$ was made in [Borcherds 3]. See also [FLM3–5] and [Tits 8].

12.1. The Main Setting

Having constructed the moonshine module $V^\natural$ in Chapter 10, we shall construct the Monster by adjoining to the group C of Section 10.4 an extra automorphism σ of $V^\natural$. This automorphism will be a modification of the involution σ_1 constructed in Chapter 11 in the context of the Golay code as the code $\mathcal{C}$ in Section 11.1, the Leech lattice as the lattice L_0' in (11.1.9) [recall (10.2.12)] and the Niemeier lattice $N(A_1^{24})$ of type A_1^{24} (recall Remark 10.2.8) as the lattice L_0 in (11.1.6). In the course of proving the required properties of our construction, we shall have to extend these structures so as to include the Niemeier lattice $N(D_4^6)$, which we shall denote $\mathfrak{L}_{00}$, and it is convenient to build up all the necessary codes, lattices and central extensions now. But the reader should be aware that for the statement, as opposed to the proof, of our main theorem, only the structures associated with L and Λ (and not $\mathcal{L}$) are needed.

Let Ω be a 24-element set. Let $\mathcal{C} \subset \mathcal{P}(\Omega)$ be a binary Golay code (see Section 10.1). Fix a 4-element subset T_0 of Ω. There are then exactly six 4-element subsets T_j, $0 \leq j \leq 5$, congruent to T_0 modulo $\mathcal{C}$, forming a partition of Ω (a sextet), as we have seen in Proposition 10.1.8. Set

$$\mathcal{C}_0 = \{C \in \mathcal{C} \mid |C \cap T_0| \in 2\mathbb{Z}\}, \tag{12.1.1}$$

$$\mathcal{C}' = \mathcal{C}_0 \oplus \mathbb{F}_2 T_0, \tag{12.1.2}$$

$$\mathcal{C}_1 = \mathcal{C}' + \mathcal{C} = \mathcal{C} \oplus \mathbb{F}_2 T_0. \tag{12.1.3}$$

We then have the inclusions and indices

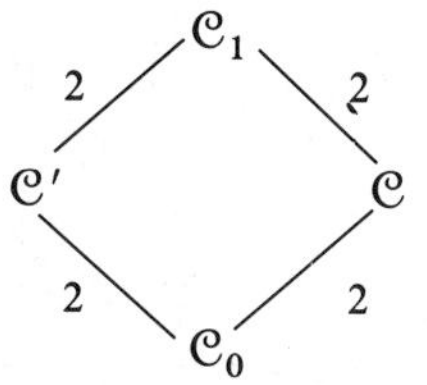

(12.1.4)

Note that $\mathcal{C}$ and $\mathcal{C}'$ are of type II, while $\mathcal{C}_1$ is of type I (recall Section 10.1).

As in Chapter 11, let

$$\mathfrak{h} = \coprod_{k \in \Omega} \mathbb{F}\alpha_k \tag{12.1.5}$$

be a vector space with basis $\{\alpha_k \mid k \in \Omega\}$, and consider the symmetric bilinear form $\langle \cdot, \cdot \rangle$ on $\mathfrak{h}$ determined by

$$\langle \alpha_k, \alpha_l \rangle = 2\delta_{k,l} \tag{12.1.6}$$

for $k, l \in \Omega$. For $S \subset \Omega$ set

$$\alpha_S = \sum_{k \in S} \alpha_k, \tag{12.1.7}$$

and let ε_S be the involution of $\mathfrak{h}$ such that

$$\varepsilon_S \colon \alpha_k \mapsto \begin{cases} -\alpha_k & \text{for} \quad k \in S \\ \alpha_k & \text{for} \quad k \notin S. \end{cases} \tag{12.1.8}$$

Set

$$Q = \coprod_{k \in \Omega} \mathbb{Z}\alpha_k, \tag{12.1.9}$$

and for a code $\mathcal{D} \subset \mathcal{P}(\Omega)$ let

$$L(\mathcal{D}) = \sum_{C \in \mathcal{D}} \mathbb{Z}\tfrac{1}{2}\alpha_C + Q. \tag{12.1.10}$$

Note that $2L(\mathcal{D}) \subset Q$. Set

$$L_0 = L(\mathcal{C}), \tag{12.1.11}$$

$$L_{00} = L(\mathcal{C}_0), \tag{12.1.12}$$

$$\mathcal{L}_{00} = L(\mathcal{C}'), \tag{12.1.13}$$

$$\mathcal{L}_0 = L(\mathcal{C}_1). \tag{12.1.14}$$

We also consider the Leech lattice Λ given by

$$\Lambda = \sum_{C \in \mathcal{C}} \mathbb{Z}\tfrac{1}{2}\alpha_C + \sum_{k,l \in \Omega} \mathbb{Z}(\alpha_k + \alpha_l) + \sum_{k \in \Omega} \mathbb{Z}(\tfrac{1}{4}\alpha_\Omega - \alpha_k) \tag{12.1.15}$$

as in (10.2.12) and set

$$L_1 = L_0 + \tfrac{1}{4}\alpha_\Omega, \quad \mathcal{L}_1 = \mathcal{L}_0 + \tfrac{1}{4}\alpha_\Omega, \tag{12.1.16}$$

$$L = L_0 \cup L_1 = L_0 + \Lambda, \tag{12.1.17}$$

$$\mathcal{L} = \mathcal{L}_0 \cup \mathcal{L}_1 = \mathcal{L}_{00} + \Lambda \tag{12.1.18}$$

$$\Lambda_0 = \Lambda \cap L_0, \tag{12.1.19}$$

$$\Lambda_1 = \Lambda \cap L_1, \tag{12.1.20}$$

$$\Lambda_{00} = \Lambda \cap \mathcal{L}_{00} \tag{12.1.21}$$

[cf. (11.1.6)–(11.1.9)]. We then have the inclusions

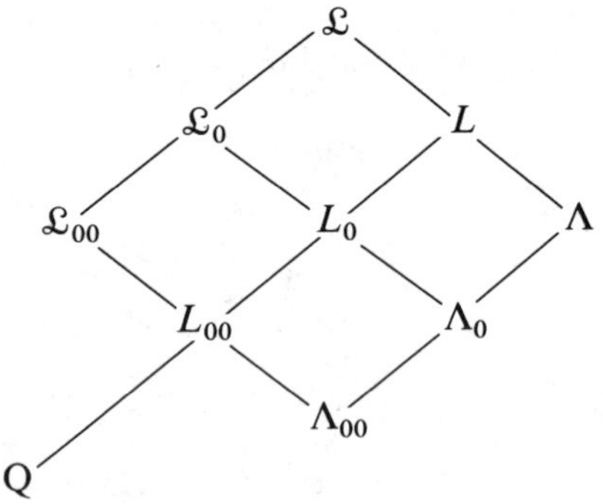

(12.1.22)

where all the indices are 2 except that

$$|L_{00} : Q| = |\mathcal{C}_0| = 2^{11}. \tag{12.1.23}$$

Note that $2\mathcal{L} \subset \Lambda_{00}$.

Remark 12.1.1: The only part of the diagram (12.1.22) needed for the statement of our theorem is the part including Λ, L, L_0, Λ_0 and Q.

Remark 12.1.2: The lattice L_0 is the (even unimodular) Niemeier lattice $N(A_1^{24})$, as we have seen in Remark 10.2.8. Analogously, $\mathfrak{L}_{00}$ is the Niemeier lattice $N(D_4^6)$. The fact that it is even unimodular follows from Proposition 10.2.1. The elements $\alpha \in \mathfrak{L}_{00}$ such that $\langle \alpha, \alpha \rangle = 2$ form a root system of type D_4^6, where the D_4 root system arises not as in Remark 6.3.4, but rather in a form analogous to the version Φ of the root system of type E_8 (see Remark 10.2.7).

Using Proposition 5.2.3, we specify a central extension

$$1 \to \langle \kappa \mid \kappa^4 = 1 \rangle \hookrightarrow \hat{\mathfrak{L}} \xrightarrow{-} \mathfrak{L} \to 1 \tag{12.1.24}$$

by giving its commutator map

$$c_0 \colon \mathfrak{L} \times \mathfrak{L} \to \mathbb{Z}/4\mathbb{Z}$$

as follows:

$$c_0(\alpha + \beta, \gamma + \delta) = 2(\langle \alpha, \gamma \rangle + \langle \alpha, \delta \rangle - \langle \beta, \gamma \rangle + \langle \beta, \delta \rangle) + 4\mathbb{Z} \tag{12.1.25}$$

for $\alpha, \gamma \in \mathfrak{L}_{00}$, $\beta, \delta \in \Lambda$. This is well defined since $\Lambda_{00} = \mathfrak{L}_{00} \cap \Lambda$ and $\mathfrak{L}$ are dual lattices [recall (6.1.12)], and is indeed alternating.

Since

$$c_0(\alpha, \beta) = 2\langle \alpha, \beta \rangle + 4\mathbb{Z} \tag{12.1.26}$$

for $\alpha, \beta \in \Lambda$ or $\alpha, \beta \in L_0$ or $\alpha, \beta \in \mathfrak{L}_{00}$, we know that commutators of vertex operators can be calculated as usual. Moreover,

$$c_0(\alpha_k, \beta) = 2\langle \alpha_k, \beta \rangle + 4\mathbb{Z} \tag{12.1.27}$$

for all $k \in \Omega$, $\beta \in \mathfrak{L}$, as in (11.1.13), so that the results of triality will hold for the constructions associated with both $\mathfrak{L}$ and L.

Remark 12.1.3: It is easy to see that the restriction of c_0 to $L \times L$ is again given by formula (12.1.25), but for $\alpha, \gamma \in L_0$ and $\beta, \delta \in \Lambda$.

We also see that

$$\mathfrak{L}_{00} = \{\alpha \in \mathfrak{L} \mid c_0(\alpha, \beta) = 2\langle \alpha, \beta \rangle + 4\mathbb{Z} \quad \text{for all} \quad \beta \in \mathfrak{L}\} \tag{12.1.28}$$

and that

$$\Lambda = \{\beta \in \mathfrak{L} \mid c_0(\alpha, \beta) = 2\langle \alpha, \beta \rangle + 4\mathbb{Z} \quad \text{for all} \quad \alpha \in \mathfrak{L}\}; \tag{12.1.29}$$

analogous formulas hold for the restriction of c_0 to $L \times L$.

Remark 12.1.4: The group of isometries of $\mathfrak{L}$ (with respect to $\langle \cdot , \cdot \rangle$) which preserve c_0 is exactly the group of those isometries which stabilize both $\mathfrak{L}_{00}$ and Λ, and similarly for L, L_0 and Λ.

As in Section 11.1, *we assume that* $\mathbb{F}$ *contains* i, and we form the untwisted spaces

$$V_L = S(\hat{\mathfrak{h}}_{\mathbb{Z}}^{-}) \otimes \mathbb{F}\{L\}, \tag{12.1.30}$$

$$V_{\mathfrak{L}} = S(\hat{\mathfrak{h}}_{\mathbb{Z}}^{-}) \otimes \mathbb{F}\{\mathfrak{L}\} \tag{12.1.31}$$

[see (11.1.16)–(11.1.21)]. Any subset M of $\mathfrak{L}$ leads to a space V_M as in (11.1.40); in particular,

$$V_\Lambda = S(\hat{\mathfrak{h}}_{\mathbb{Z}}^{-}) \otimes \mathbb{F}\{\Lambda\}, \tag{12.1.32}$$

just as in (10.3.1). Then

$$V_\Lambda \subset V_L \subset V_{\mathfrak{L}}. \tag{12.1.33}$$

For the twisted construction, we shall later fix elements $a_k \in \hat{L}$ for $k \in \Omega$ such that

$$\bar{a}_k = \alpha_k$$

as in (11.1.22), and given this, we form the character ψ and the corresponding induced modules

$$T_L = \mathrm{Ind}_{\hat{Q}}^{\hat{L}} \mathbb{F}_\psi, \tag{12.1.34}$$

$$T_{\mathfrak{L}} = \mathrm{Ind}_{\hat{Q}}^{\hat{\mathfrak{L}}} \mathbb{F}_\psi \tag{12.1.35}$$

[see (11.1.22)–(11.1.26)]. Then

$$T_L \subset T_{\mathfrak{L}}. \tag{12.1.36}$$

We must postpone the definition of T_Λ. Keeping in mind (11.1.27)–(11.1.31), we form

$$V_L' = V_L^{T_L} = S(\hat{\mathfrak{h}}_{\mathbb{Z}+1/2}^{-}) \otimes T_L, \tag{12.1.37}$$

$$V_{\mathfrak{L}}' = V_{\mathfrak{L}}^{T_{\mathfrak{L}}} = S(\hat{\mathfrak{h}}_{\mathbb{Z}+1/2}^{-}) \otimes T_{\mathfrak{L}}, \tag{12.1.38}$$

so that

$$V_L' \subset V_{\mathfrak{L}}'. \tag{12.1.39}$$

Soon we shall fix an involution z_0 of our central extensions:

$$z_0 \in \mathrm{Aut}(\hat{\mathfrak{L}}; \kappa, \langle \cdot , \cdot \rangle) \tag{12.1.40}$$

such that

$$z_0 \in \mathrm{Aut}(\hat{L}; \kappa, \langle \cdot , \cdot \rangle) \tag{12.1.41}$$

as well,

$$\bar{z}_0 = -1 \quad \text{on} \quad \mathfrak{L} \quad \text{and} \quad z_0^2 = 1. \tag{12.1.42}$$

[Involutions of this sort have typically been designated θ, as in (11.1.32)–(11.1.35).] We shall have

$$z_0 a_k = a_k^{-1} \quad \text{for} \quad k \in \Omega, \tag{12.1.43}$$

and z_0 will extend canonically to involutions

$$z_0 \colon V_{\mathfrak{L}} \to V_{\mathfrak{L}}, \quad z_0 \colon V_L \to V_L. \tag{12.1.44}$$

Then Chapter 9 will apply in the generality occurring in Chapter 11.

Now we form

$$W_{\mathfrak{L}} = V_{\mathfrak{L}} \oplus V'_{\mathfrak{L}}, \tag{12.1.45}$$

$$W_L = V_L \oplus V'_L, \tag{12.1.46}$$

so that

$$W_L \subset W_{\mathfrak{L}}. \tag{12.1.47}$$

Now using the fact that

$$\langle \tfrac{1}{4}\alpha_\Omega, \tfrac{1}{4}\alpha_\Omega \rangle = 3 \in \mathbb{Z} \tag{12.1.48}$$

[cf. (10.2.11)], we see that

$$\langle \mathfrak{L}, \mathfrak{L} \rangle \subset \tfrac{1}{2}\mathbb{Z} \tag{12.1.49}$$

and that

$$\langle \alpha, \alpha \rangle \in \mathbb{Z} \quad \text{for} \quad \alpha \in \mathfrak{L}. \tag{12.1.50}$$

Thus the vertex operators $Y_{\mathbb{Z}+1/2}(v, z)$ are defined for all $v \in V_{\mathfrak{L}}$. We shall use the notation of (11.1.37)–(11.1.39), for instance,

$$Y(v, z) = Y_{\mathbb{Z}}(v, z) \oplus Y_{\mathbb{Z}+1/2}(v, z). \tag{12.1.51}$$

Now it is time to fix our choices of z_0 and the a_k. In doing so, we shall also fix a double covering, which we shall call $\hat{\Lambda}^{(2)}$, of the Leech lattice, equivalent to the covering designated $\hat{\Lambda}$ in Section 10.3. This group $\hat{\Lambda}^{(2)}$ will be a subgroup of the fourfold covering $\hat{\Lambda} \subset \hat{L}$ of Λ, and z_0 will restrict to the canonical involution θ_0 of $\hat{\Lambda}^{(2)}$ [recall (10.3.12)].

Remark 12.1.5: As usual, we shall write $\hat{M}$ for the inverse image in $\hat{\mathfrak{L}}$ of any subset M of $\mathfrak{L}$. This is why we use the special notation $\hat{\Lambda}^{(2)}$ for our *double* covering of Λ. We shall later use the corresponding notation $\hat{M}^{(2)}$ to designate the inverse image in $\hat{\Lambda}^{(2)}$ of any subset M of Λ.

It is convenient to start with the choice of a group K_Λ which will be identified with the group K of (10.3.17). This will be the kernel of our 2^{12}-dimensional irreducible $\hat{\Lambda}^{(2)}$-module, which we shall call T_Λ (and which was called T in Proposition 10.3.1). The group K_Λ will map isomorphically onto 2Λ under $^{-}$ [recall (10.3.21)].

Proposition 12.1.6: *There exists a subgroup K_Λ of $\hat{\mathfrak{L}}$ satisfying the conditions*

$$K_\Lambda \cap \langle \kappa \rangle = 1, \quad \bar{K}_\Lambda = 2\Lambda, \tag{12.1.52}$$

$$\langle \alpha^4 \kappa^{2\langle \bar{a}, \bar{a} \rangle} \mid a \in \hat{\mathfrak{L}} \rangle \subset K_\Lambda \subset \langle a^2 \mid a \in \hat{\Lambda} \rangle, \tag{12.1.53}$$

and in particular,

$$\langle a^4 \kappa^{2\langle \bar{a}, \bar{a} \rangle} \mid a \in \hat{L} \rangle \subset K_\Lambda \subset \langle a^2 \mid a \in \hat{\Lambda} \rangle. \tag{12.1.54}$$

Proof: Set

$$\begin{aligned} A &= \{a^4 \kappa^{2\langle \bar{a}, \bar{a} \rangle} \mid a \in \hat{\mathfrak{L}}\}, \\ B &= \{a^2 \mid a \in \hat{\Lambda}\}. \end{aligned} \tag{12.1.55}$$

Then $A \subset B$ by (12.1.50) and the fact that $2\mathfrak{L} \subset \Lambda$. But A is a group since

$$a^4 \kappa^{2\langle \bar{a}, \bar{a} \rangle} b^4 \kappa^{2\langle \bar{b}, \bar{b} \rangle} = (ab)^4 \kappa^{2\langle \bar{a}, \bar{a} \rangle + 2\langle \bar{b}, \bar{b} \rangle + 6c_0(\bar{a}, \bar{b})}$$

and

$$6c_0(\bar{a}, \bar{b}) = 4\langle \bar{a}, \bar{b} \rangle + 4\mathbb{Z}.$$

Also,

$$A \cap \langle \kappa \rangle = 1, \quad \bar{A} = 4\mathfrak{L}. \tag{12.1.56}$$

Since

$$a^2 b^2 = (ab)^2 \kappa^{c_0(\bar{a}, \bar{b})}$$

and

$$c_0(\Lambda, \Lambda) = 2\mathbb{Z}/4\mathbb{Z},$$

it follows that B is a group. Moreover,

$$B \cap \langle \kappa \rangle = \langle \kappa^2 \rangle, \quad \bar{B} = 2\Lambda. \tag{12.1.57}$$

Since

$$c_0(2\Lambda, 2\Lambda) = 0,$$

we also have that B is abelian and that

$$B \simeq \langle \kappa^2 \rangle \times 2\Lambda. \tag{12.1.58}$$

We can therefore choose K_Λ/A to be any complement of $(\langle \kappa^2 \rangle \times A)/A$ in the elementary abelian 2-group B/A, and we take K_Λ to be the inverse image of K_Λ/A under the map $B \to B/A$. ∎

Remark 12.1.7: For the statement, as opposed to the proof, of our main theorem, we do not need (12.1.53), but it has been no extra effort to prove it.

Remark 12.1.8: As in (10.3.21), we have

$$\widehat{2\Lambda} = \langle \kappa \rangle \times K_\Lambda = \text{Cent}\,\hat{\Lambda}. \tag{12.1.59}$$

Also,

$$\langle \kappa^2 \rangle \times K_\Lambda = \langle a^2 \mid a \in \hat{\Lambda} \rangle. \tag{12.1.60}$$

We fix a subgroup K_Λ with the properties specified in Proposition 12.1.6. Next we define

$$\hat{\Lambda}^{(2)} = \{a \in \hat{\Lambda} \mid a^2 \kappa^{\langle \bar{a}, \bar{a} \rangle} \in K_\Lambda\}. \tag{12.1.61}$$

Proposition 12.1.9: *The set* $\hat{\Lambda}^{(2)}$ *is a group and*

$$\hat{\Lambda}^{(2)} \cap \langle \kappa \rangle = \langle \kappa^2 \rangle, \quad \overline{\hat{\Lambda}^{(2)}} = \Lambda. \tag{12.1.62}$$

In particular, the sequence

$$1 \to \langle \kappa^2 \rangle \to \hat{\Lambda}^{(2)} \xrightarrow{-} \Lambda \to 1 \tag{12.1.63}$$

is exact.

Proof: Let $a, b \in \hat{\Lambda}^{(2)}$. Then

$$c_0(\bar{a}, \bar{b}) = 2\langle \bar{a}, \bar{b} \rangle + 4\mathbb{Z},$$

so that

$$a^2 \kappa^{\langle \bar{a}, \bar{a} \rangle} b^2 \kappa^{\langle \bar{b}, \bar{b} \rangle} = (ab)^2 \kappa^{\langle \bar{a}+\bar{b}, \bar{a}+\bar{b} \rangle}$$

and thus $ab \in \hat{\Lambda}^{(2)}$. Also, $a^{-1} \in \hat{\Lambda}^{(2)}$ since $\langle \bar{a}, \bar{a} \rangle \in 2\mathbb{Z}$, and $\hat{\Lambda}^{(2)}$ is a group. It is clear that $\hat{\Lambda}^{(2)} \cap \langle \kappa \rangle = \langle \kappa^2 \rangle$. Let $\alpha \in \Lambda$, $a \in \hat{\Lambda}$ with $\bar{a} = \alpha$. Then $a^2 \kappa^{\langle \alpha, \alpha \rangle} \in \langle \kappa^2 \rangle \times K_\Lambda$ by (12.1.60), so that either $a \in \hat{\Lambda}^{(2)}$ or $a\kappa \in \hat{\Lambda}^{(2)}$. Hence $\overline{\hat{\Lambda}^{(2)}} = \Lambda$. ∎

Remark 12.1.10: As desired, the group $\hat{\Lambda}^{(2)}$ is a double covering of Λ, and the commutator map of the extension (12.1.63) is given by:

$$c_0(\alpha, \beta) = \langle \alpha, \beta \rangle + 2\mathbb{Z} \quad \text{for} \quad \alpha, \beta \in \Lambda. \tag{12.1.64}$$

Thus by Proposition 5.2.3, the extension (12.1.63) is equivalent to the extension (10.3.2), which was fundamental in Chapter 10, and we have an identification of the old and new spaces $\mathbb{F}\{\Lambda\}$ and hence V_Λ. Moreover, under the identification, K_Λ agrees with the group K of (10.3.17).

We shall next fix an automorphism $z_0 \in \mathrm{Aut}(\hat{\mathfrak{L}}; \kappa)$ with $\bar{z}_0 = -1$.

Proposition 12.1.11: *There exists an automorphism $z_0 \in \mathrm{Aut}(\hat{\mathfrak{L}} : \kappa)$ such that*

$$\bar{z}_0 = -1, \tag{12.1.65}$$

$$K_\Lambda = \{z_0(a)a^{-1} \mid a \in \hat{\Lambda}\}. \tag{12.1.66}$$

All such automorphisms satisfy the conditions

$$z_0(a) \in a^{-1}\kappa^{\langle \bar{a}, \bar{a} \rangle}\langle \kappa^2 \rangle \quad \textit{for} \quad a \in \hat{\mathfrak{L}} \tag{12.1.67}$$

and

$$z_0^2 = 1. \tag{12.1.68}$$

Proof: Using Proposition 5.4.1, we choose an automorphism $z' \in \mathrm{Aut}(\hat{\mathfrak{L}}; \kappa)$ with $\overline{z'} = -1$. Let $\{\beta_j \mid 1 \leq j \leq 2\} \cup \{\gamma_k \mid 1 \leq k \leq 22\}$ be a base of $\mathfrak{L}$ with

$$\Lambda = \coprod_{j=1}^{2} 2\mathbb{Z}\beta_j \oplus \coprod_{k=1}^{22} \mathbb{Z}\gamma_k.$$

Choose $b_j, c_k \in \hat{\mathfrak{L}}$ with

$$\bar{b}_j = \beta_j, \quad \bar{c}_k = \gamma_k$$

for $1 \leq j \leq 2$, $1 \leq k \leq 22$. Let

$$z'(b_j^2)b_j^{-2} \in K_\Lambda \kappa^{t_j},$$

$$z'(c_k)c_k^{-1} \in K_\Lambda \kappa^{u_k}$$

for $t_j, u_k \in \mathbb{Z}/4\mathbb{Z}$. Since by (12.1.60)

$$z'(b_j^2)b_j^{-2} = (z'(b_j)b_j^{-1})^2 \in \langle \kappa^2 \rangle \times K_\Lambda,$$

we have

$$t_j \in 2\mathbb{Z}/4\mathbb{Z}.$$

Choose $v_j \in \mathbb{Z}/4\mathbb{Z}$ so that $2v_j = t_j$ and define

$$\lambda \in \mathrm{Hom}(\mathfrak{L}, \mathbb{Z}/4\mathbb{Z})$$

by:

$$\lambda(\beta_j) = -v_j$$

$$\lambda(\gamma_k) = -u_k$$

for $1 \leq j \leq 2$, $1 \leq k \leq 22$. Then $z_0 = z'\lambda^*$ [using the notation (5.4.4)] satisfies the conditions

$$z_0(b_j^2)b_j^{-2} \in K_\Lambda,$$

$$z_0(c_k)c_k^{-1} \in K_\Lambda.$$

Since for $a, b \in \hat{\Lambda}$ we have

$$z_0(a)a^{-1}z_0(b)b^{-1} = z_0(a)z_0(b)a^{-1}b^{-1}\kappa^{c_0(\bar{a},\bar{b})} = z_0(ab)(ab)^{-1},$$

$c_0(\bar{a}, \bar{b})$ being even, we then have $z_0(a)a^{-1} \in K_\Lambda$ for all

$$a \in \langle b_j^2, c_k, \kappa \mid 1 \le j \le 2,\ 1 \le k \le 22\rangle = \hat{\Lambda},$$

and (12.1.66) follows.

For (12.1.67), note that by (12.1.53) we have

$$a^4\kappa^{2\langle\bar{a},\bar{a}\rangle} \in K_\Lambda$$

for $a \in \hat{\mathfrak{L}}$. But

$$z_0(a^2)a^{-2} = a^{-4}\kappa^{-2\langle\bar{a},\bar{a}\rangle} \Leftrightarrow$$

$$z_0(a^2) = a^{-2}\kappa^{-2\langle\bar{a},\bar{a}\rangle} \Leftrightarrow$$

$$z_0(a) \in a^{-1}\kappa^{\langle\bar{a},\bar{a}\rangle}\langle\kappa^2\rangle,$$

and these assertions are true if z_0 satisfies (12.1.66).

Finally, (12.1.68) follows from (5.4.5). ∎

Remark 12.1.12: Of course, $\hat{\mathfrak{L}}$ can be replaced by $\hat{L}$ in this result; the proof for $\hat{L}$ is no easier.

Remark 12.1.13: As promised, z_0 restricts to the involution θ_0 of Chapter 10 [recall (10.3.17)], under our identification. In particular,

$$z_0(a) = a^{-1}\kappa^{\langle\bar{a},\bar{a}\rangle}$$

on $\hat{\Lambda}^{(2)}$ [see (10.3.12)].

We fix an automorphism z_0 as in Proposition 12.1.11.

We shall finally fix suitable elements $a_k \in \hat{\mathfrak{L}}$ such that $\bar{a}_k = \alpha_k$. We shall use the group

$$K_L = \{a \in K_\Lambda \mid \bar{a} \in 2\Lambda_0\} \qquad (12.1.69)$$

and soon we shall also use

$$K_{\mathfrak{L}} = \{a \in K_\Lambda \mid \bar{a} \in 2\Lambda_{00}\}. \qquad (12.1.70)$$

Remark 12.1.14: In the notation of Proposition 5.4.8, (12.1.53) and (12.1.70) amount to the choice of

$$f_{\mathfrak{L}}(\alpha) = 2\langle\alpha, \alpha\rangle + 4\mathbb{Z}, \quad \alpha \in \mathfrak{L}$$

$$s_{\mathfrak{L}} = 0$$

with respect to the pair $(\hat{\mathfrak{L}}, K_{\mathfrak{L}})$. These maps are of course invariant under isometries of $\mathfrak{L}$. Later we shall apply Proposition 5.4.8 in situations of this type.

Proposition 12.1.15: *There exist $a_k \in \hat{\mathfrak{L}}$ for $k \in \Omega$ such that*

$$\bar{a}_k = \alpha_k, \tag{12.1.71}$$

$$\langle a_k \mid k \in \Omega\rangle \supset K_L, \quad i.e., \quad K_L = \widehat{2\Lambda_0} \cap \langle a_k \mid k \in \Omega\rangle, \tag{12.1.72}$$

$$z_0(a_k) = a_k^{-1}, \tag{12.1.73}$$

$$\langle a_k \mid k \in \Omega\rangle \supset \{a^{-1}z_0(a) \mid a \in \hat{\mathfrak{L}}_{00}\}. \tag{12.1.74}$$

Proof: Set

$$R = \{a \in \hat{Q} \mid z_0(a) = a^{-1}\},$$

$$S_1 = \{a^{-1}z_0(a) \mid a \in \hat{L}_0\},$$

$$S_2 = \{a^{-1}z_0(a) \mid a \in \hat{\mathfrak{L}}_{00}\}.$$

Since $\langle\alpha, \alpha\rangle \in 2\mathbb{Z}$ for $\alpha \in Q$, it follows from (12.1.67) that

$$\bar{R} = Q.$$

Since $c_0(Q, Q) = 0$, R is a group and we have

$$R \cap \langle\kappa\rangle = \langle\kappa^2\rangle, \quad R \simeq Q \times \langle\kappa^2\rangle.$$

Let $a, b \in \hat{L}_0$ or $\hat{\mathfrak{L}}_{00}$. Since

$$a^{-1}z_0(a)b^{-1}z_0(b) = b^{-1}a^{-1}z_0(a)z_0(b)\kappa^{c_0(2\bar{a}, \bar{b})}$$

$$= (ab)^{-1}z_0(ab)\kappa^{4\langle\bar{a}, \bar{b}\rangle} = (ab)^{-1}z_0(ab),$$

S_1 and S_2 are groups, and

$$z_0(a^{-1}z_0(a)) = z_0(a)^{-1}a = (a^{-1}z_0(a))^{-1},$$

so that S_1 and S_2 are contained in R. We have

$$S_1 \cap \langle\kappa\rangle = 1, \quad S_1 \simeq 2L_0,$$

$$S_2 \cap \langle\kappa\rangle = 1, \quad S_2 \simeq 2\mathfrak{L}_{00},$$

$$S_1 \cap S_2 = \{a^{-1}z_0(a) \mid a \in \hat{L}_{00}\} \simeq 2L_{00},$$

and thus

$$S_1S_2 \cap \langle\kappa\rangle = 1, \quad S_1S_2 \simeq 2\mathfrak{L}_0.$$

In addition,

$$R^2 = \{a^2 \mid a \in R\} = \{a^{-1}z_0(a) \mid a \in R\} \subset S_1 \cap S_2,$$

so that we can choose $\tilde{Q}/S_1S_2$ to be any complement of $(\langle\kappa^2\rangle \times S_1S_2)/S_1S_2$ in R/S_1S_2, and we define $\tilde{Q}$ to be the inverse image of $\tilde{Q}/S_1S_2$ under the map $R \to R/S_1S_2$. Finally, let a_k be the element in $\tilde{Q}$ with $\bar{a}_k = \alpha_k$. The a_k then have the required properties since

$$K_L = \{a^{-1}z_0(a) \mid a \in \hat{\Lambda}_0\rangle \subset S_1. \qquad \blacksquare$$

Remark 12.1.16: Formula (12.1.74) (concerning $\mathfrak{L}$) will not be used until Chapter 13.

Remark 12.1.17: Formulas (12.1.71) and (12.1.73) show that the a_k have the properties needed to apply the results of Chapter 11.

We fix $\{a_k \mid k \in \Omega\} \subset \hat{\mathfrak{L}}$ as indicated.

Having in this way fixed K_Λ, $\hat{\Lambda}^{(2)}$, z_0 and $\{a_k\}$, we are ready to complete the description of our main setting. *For M a union of cosets of Q in $\mathfrak{L}$, set*

$$T_M = \sum_{\bar{a} \in M} \mathbb{F}t(a) \tag{12.1.75}$$

as in (11.1.41) [cf. (12.1.34)–(12.1.36)]. Also define

$$\begin{aligned} T_\Lambda &= \{v \in T_L \mid bv = v \quad \text{for all} \quad b \in K_\Lambda\}, \\ T_{\Lambda_0} &= T_\Lambda \cap T_{L_0}, \quad T_{\Lambda_1} = T_\Lambda \cap T_{L_1}. \end{aligned} \tag{12.1.76}$$

When T_M is defined in one of these ways, set

$$\begin{aligned} V'_M &= S(\hat{\mathfrak{h}}^-_{\mathbb{Z}+1/2}) \otimes T_M, \\ W_M &= V_M \oplus V'_M \end{aligned} \tag{12.1.77}$$

[cf. (12.1.45)–(12.1.47)]. In particular, we have the embeddings

$$W_\Lambda \subset W_L \subset W_{\mathfrak{L}}. \tag{12.1.78}$$

We shall now show that under the identifications we have made with the structures in Section 10.3, T_Λ agrees with the 2^{12}-dimensional irreducible $\hat{\Lambda}$-module T described in Proposition 10.3.1.

Proposition 12.1.18: *For $M = \Lambda$, L or $\mathfrak{L}$, T_M is the unique (up to equivalence) irreducible $\hat{M}$-module with kernel K_M and such that*

$$\kappa \mapsto i \quad on \quad T_M. \tag{12.1.79}$$

We have

$$\dim T_\Lambda = 2^{12}, \quad \dim T_L = 2^{13}, \quad \dim T_{\mathfrak{L}} = 2^{14}. \tag{12.1.80}$$

Proof: By (12.1.72) we have

$$K_{\mathfrak{L}} \subset K_L \subset \langle a_k \mid k \in \Omega \rangle = \operatorname{Ker} \psi.$$

Since $\Lambda_{00} = \mathfrak{L}^\circ$ [using the notation (6.1.12)], the definition of c_0 gives

$$2\Lambda_{00} = \text{radical}(c_0),$$

so that

$$K_{\mathfrak{L}} \times \langle \kappa \rangle = \operatorname{Cent} \hat{\mathfrak{L}}.$$

Hence $K_{\mathfrak{L}}$ acts trivially on $T_{\mathfrak{L}}$. Similarly, since

$$2\Lambda_0 = \text{radical}(c_0|_{L \times L}),$$

K_L acts trivially on T_L. The finite groups $\hat{\mathfrak{L}}/K_{\mathfrak{L}}$ and $\hat{L}/K_L$ satisfy the hypotheses of Theorem 5.5.1 (see also Remark 5.5.2); since $\hat{Q}$ is a maximal abelian subgroup of $\hat{\mathfrak{L}}$ (and also of $\hat{L}$), that theorem shows that $T_{\mathfrak{L}}$ and T_L have the desired properties.

Let $b \in K_\Lambda \backslash K_L$, $k \in \Omega$. Then

$$c_0(\alpha_k, \bar{b}) = 2 + 4\mathbb{Z},$$

so that

$$a_k b = b a_k \kappa^2.$$

Thus a_k interchanges the ± 1-eigenspaces of b on T_L. Since $b^2 \in K_L$ and $K_\Lambda = \langle b \rangle K_L$ it follows that

$$\dim T_\Lambda = 2^{12}.$$

Since

$$2\Lambda = \text{radical}(c_0|_{\Lambda \times \Lambda})$$

we have

$$K_\Lambda \times \langle \kappa \rangle = \operatorname{Cent} \hat{\Lambda},$$

and so T_Λ is invariant under $\hat{\Lambda}$. Applying Theorem 5.5.1 to $\hat{\Lambda}/K_\Lambda$ shows that T_Λ has the required properties. ∎

Thus we have the identifications

$$T_\Lambda = T \tag{12.1.81}$$

(recall Proposition 10.3.1) and

$$V'_\Lambda = V_\Lambda^T \tag{12.1.82}$$

[see (10.3.16)].

Remark 12.1.19: Recall from Section 10.4 that we have constructed groups C_0, C_T, $\hat{C}$ and C which act on appropriate subspaces of W_Λ. Here we note that C_0, originally defined as a group of automorphisms of $\hat{\Lambda}^{(2)}$ (earlier denoted $\hat{\Lambda}$), extends naturally to a group of automorphisms of our fourfold cover $\hat{\Lambda}$ fixing κ, and hence C_0 acts on $\mathbb{F}\{\Lambda\}$ by the same formula $\iota(a) \mapsto \iota(ga)$ as in (10.4.55). Thus the results of Section 10.4 remain valid in the present setting.

12.2. The Extra Automorphism σ

We now apply the results of triality, Theorem 11.5.1, to define an automorphism σ of $V^\natural$, which together with the group C will generate the Monster. We shall freely use the notation introduced in Section 12.1.

As we have explained, Theorem 11.5.1 applies to both the type II code $\mathcal{C}$ and the type I code $\mathcal{C}_1$. Postponing our choices of $b_0 \in \hat{L}$ and $\mu \in \mathbb{F}^\times$ [recall (11.5.12), (11.5.13)], we have an operator

$$\sigma_1 \colon W_\mathfrak{L} \to W_\mathfrak{L} \tag{12.2.1}$$

preserving W_L, as in Theorem 11.5.1.

As in (11.2.41)–(11.2.43), we set

$$v^\pm = v \pm z_0(v) \tag{12.2.2}$$

$$X^\pm(v, z) = X(v^\pm, z) = \sum_{n \in \mathbb{Q}} x_v^\pm(n) z^{-n} \tag{12.2.3}$$

$$Y^\pm(v, z) = Y(v^\pm, z) \tag{12.2.4}$$

for $v \in V_\mathfrak{L}$, and similarly for $X^\pm(a, z)$, etc., for $a \in \hat{\mathfrak{L}}$.

Proposition 12.2.1: *Let $b \in \hat{\Lambda}_0$, $\bar{b} = \frac{1}{2}\varepsilon_S \alpha_C$, $C \in \mathcal{C}$, $S \subset C$, $|S| \in 2\mathbb{Z}$. Then*

$$\sigma_1(\iota(b)^+) = 2^{-|C|/2} \sum_{\substack{T \subset C \\ |T| \in 2\mathbb{Z}}} (-1)^{|S \cap T|} \iota(a_{S,T} b)^+. \tag{12.2.5}$$

In particular,

$$\sigma_1 X^+(b,z)\sigma_1^{-1} = 2^{-|C|/2} \sum_{\substack{T\subset C\\ |T|\in 2\mathbb{Z}}} (-1)^{|S\cap T|} X^+(a_{S,T}b, z). \tag{12.2.6}$$

Proof: Since $|C| \in 4\mathbb{Z}$ for $C \in \mathfrak{C}$ and

$$a_{C\setminus S, C\setminus T} = a_{T,S} = z_0(a_{S,T}),$$

(11.5.7) gives

$$\begin{aligned}\sigma_1 \iota(b)^+ &= 2^{-|C|/2} \sum_{T\subset C} ((-1)^{|S\cap T|}\iota(a_{S,T}b)\\ &\quad + (-1)^{|(C\setminus S)\cap(C\setminus T)|}\iota(a_{C\setminus S, C\setminus T} z_0(b)))\\ &= 2^{-|C|/2}\Bigg(\sum_{\substack{T\subset C\\ |T|\in 2\mathbb{Z}}} (-1)^{|S\cap T|}\iota(a_{S,T}b)^+\\ &\quad + \sum_{\substack{T\subset C\\ |T|\in 2\mathbb{Z}+1}} (-1)^{|S\cap T|}\iota(a_{S,T}b)^-\Bigg).\end{aligned}$$

But since

$$z_0(b)b^{-1} \in K_L \subset \langle a_k \mid k \in \Omega\rangle,$$

we see that

$$z_0(b)b^{-1} = a_{S,T}a_{S,C\setminus T},$$

so that

$$z_0(a_{S,T}b) = a_{T,S}z_0(b) = a_{S,C\setminus T}b.$$

Hence the terms corresponding to T and C/T in the second sum cancel each other. ∎

Now we fix an element

$$k_0 \in \Omega \tag{12.2.7}$$

and we choose $b_{k_0} \in \hat{\Lambda}^{(2)}$ such that

$$\bar{b}_{k_0} = \tfrac{1}{4}\alpha_\Omega - \alpha_{k_0}. \tag{12.2.8}$$

Set

$$\begin{aligned} b_0 &= a_{k_0} b_{k_0},\\ b_k &= a_k^{-1} b_0 \end{aligned} \tag{12.2.9}$$

for all $k \in \Omega$. The element b_0 will be our b_0 of (11.5.12). Using (12.1.72) we see that

$$a_k^2 a_l^{-2} \in K_L \subset K_\Lambda$$

for $k, l \in \Omega$, and so we have $a_k a_l^{-1} \in \hat{\Lambda}^{(2)}$ for $k, l \in \Omega$ by (12.1.61). Hence $b_k \in \hat{\Lambda}^{(2)}$ for $k \in \Omega$.

Set

$$\begin{aligned} \Lambda_2 &= \Lambda_0 + \alpha_{k_0} \\ \Lambda_3 &= \Lambda_0 + \tfrac{1}{4}\alpha_\Omega . \end{aligned} \tag{12.2.10}$$

Then the Λ_j, $0 \le j \le 3$, are the cosets of Λ_0 in L.

Keeping in mind that as operators on T_L, $a_k^4 = b_k^4 = 1$ and a_k^2 commutes with b_l^2 for $k, l \in \Omega$, we define a decomposition

$$T_L = \coprod_{j=0}^{3} T_{\Lambda_j} \tag{12.2.11}$$

by letting the T_{Λ_j} be the eigenspaces of a_k^2 and b_k^2 in T_L having the following eigenvalues:

	a_k^2	b_k^2
T_{Λ_0}	1	1
T_{Λ_1}	-1	1
T_{Λ_2}	1	-1
T_{Λ_3}	-1	-1

(12.2.12)

Note that since $a_k^2 a_l^{-2} \in K_L$ and $b_k^2 b_l^{-2} \in K_L$ for all $k, l \in \Omega$, these eigenspaces do not depend on the choice of $k \in \Omega$. We have

$$\begin{aligned} T_{L_0} &= T_{\Lambda_0} \oplus T_{\Lambda_2}, T_{L_1} = T_{\Lambda_1} \oplus T_{\Lambda_3}, \\ T_\Lambda &= T_{\Lambda_0} \oplus T_{\Lambda_1} \end{aligned} \tag{12.2.13}$$

and hence the notation T_{Λ_0}, T_{Λ_1} is consistent with that introduced in (12.1.76).

Remark 12.2.2: The T_{Λ_j}, $0 \le j \le 3$, are nonisomorphic 2^{11}-dimensional absolutely irreducible $\hat{\Lambda}_0/K_L$-modules.

Recall the automorphism z_0 of $\hat{\mathfrak{L}}$ from Proposition 12.1.11. While we have extended z_0 to V_L, we have not defined an involution on V_L'; we have only defined an action of θ_0 on $V_\Lambda^T = V_\Lambda'$ [cf. (10.3.32)]. We proceed to define an appropriate involution on V_L'.

Now z_0 leaves $\hat{L}$ and K_L invariant [recall (12.1.66)], so that z_0 induces an automorphism of $\hat{L}/K_L$ fixing κK_L. Proposition 5.5.3 thus shows that there exists an operator z_T on T_L such that $\operatorname{int}(z_T)$ and z_0 induce the same automorphism of $\hat{L}/K_L$. To determine z_T, note that by (12.1.66) and (12.1.69), z_0 induces the identity automorphism on $\hat{\Lambda}_0/K_L$. In view of Remark 12.2.2, z_T must act as a scalar on each subspace T_{Λ_j}. But

$$z_0(a_k) = a_k^{-1}, \quad z_0(b_k) = b_k^{-1}, \tag{12.2.14}$$

using (12.1.73) and Remark 12.1.13, and

$$\begin{aligned} a_k &: T_{\Lambda_0} \leftrightarrow T_{\Lambda_2}, \quad T_{\Lambda_1} \leftrightarrow T_{\Lambda_3} \\ b_k &: T_{\Lambda_0} \leftrightarrow T_{\Lambda_1}, \quad T_{\Lambda_2} \leftrightarrow T_{\Lambda_3} \end{aligned} \tag{12.2.15}$$

since

$$(a_k, b_k^2) = (a_k^2, b_k) = \kappa^2$$

[$(\cdot, \cdot)$ denoting commutator]. Thus we find that, up to a nonzero scalar multiple,

$$z_T = (-1_{T_{\Lambda_0}}) \oplus (-1_{T_{\Lambda_1}}) \oplus (-1_{T_{\Lambda_2}}) \oplus 1_{T_{\Lambda_3}}. \tag{12.2.16}$$

We define z_T to be exactly this map, which extends -1_{T_Λ}.

Remark 12.2.3: The involution z_T is analogous to, and extends, the previously fixed action -1 of θ_0 on T_Λ [recall (9.2.87)].

By analogy with the definition of the action of $\hat{C}$ on W_Λ (recall Section 10.4 and Remark 12.1.19), we let $\hat{z}_0$ be the operator on W_L determined by

$$\hat{z}_0 = (z_0, z_T) \in \operatorname{Aut}(\hat{L}; \kappa) \times \operatorname{Aut} T_L \tag{12.2.17}$$

as in (10.4.56). Then $\hat{z}_0$ extends the action of the involution $\theta_0 = \hat{\theta}_0 \hat{\theta}$ of W_Λ; see Remark 10.4.3.

We set

$$\begin{aligned} V_{\Lambda_j} &= S(\hat{\mathfrak{h}}_{\mathbb{Z}}^-) \otimes \mathbb{F}\{\Lambda_j\} \\ V'_{\Lambda_j} &= S(\hat{\mathfrak{h}}_{\mathbb{Z}+1/2}^-) \otimes T_{\Lambda_j} \end{aligned} \tag{12.2.18}$$

for $0 \le j \le 3$,

$$V_M^{\pm} = \{v \in V_M \mid \hat{z}_0 v = \pm v\} \tag{12.2.19}$$

for any subset M of L [cf. (12.2.2)–(12.2.4)] and

$$V_M'^{\pm} = \{v \in V'_M \mid \hat{z}_0 v = \pm v\} \tag{12.2.20}$$

for any subset M of L for which V'_M (and T_M) are defined. We shall consider the decomposition

$$W_L = \coprod_{0\le j\le 3} V^+_{\Lambda_j} \oplus \coprod_{0\le j\le 3} V^-_{\Lambda_j} \oplus \coprod_{0\le j\le 3} V'^+_{\Lambda_j} \oplus \coprod_{0\le j\le 3} V'^-_{\Lambda_j}. \quad (12.2.21)$$

Then we have (cf. Theorem 10.4.6):

Proposition 12.2.4: *For $v \in V_L$,*

$$\hat{z}_0 Y(v, z)\hat{z}_0^{-1} = Y(\hat{z}_0 v, z).$$

For $v \in V_{\Lambda_0}$, the components $x_v(n)$ or v_n $(n \in \mathbb{Q})$ of the vertex operators $Y(v, z)$ preserve each of the 8 spaces V_{Λ_j}, V'_{Λ_j} $(0 \le j \le 3)$, and for $v \in V_L^+$ these components preserve each of the spaces $V_L^{\pm}$, $V_L'^{\pm}$. In particular, for $v \in V^+_{\Lambda_0}$, these components preserve each of the 16 spaces $V^{\pm}_{\Lambda_j}$, $V'^{\pm}_{\Lambda_j}$ $(0 \le j \le 3)$.

Our next goal is to show that σ_1 permutes the 16 subspaces $V^{\pm}_{\Lambda_j}$, $V'^{\pm}_{\Lambda_j}$. We shall establish this by means of an irreducibility result which will also be used later.

Keeping in mind (10.2.44)–(10.2.46) and (10.3.40), we observe:

Proposition 12.2.5: *The subspace $(V^+_{\Lambda_0})_{-1}$ of $V^+_{\Lambda_0}$ of degree -1, or equivalently, weight 2, is spanned by:*

$$\begin{aligned}&\{\alpha_k(-1)\alpha_l(-1) \mid k, l \in \Omega\} \cup \{\iota(b)^+ \mid \bar{b} \in (\Lambda_0)_4\}\\ &\quad = \{\alpha_k(-1)\alpha_l(-1) \mid k, l \in \Omega\} \cup \{\iota(a_k a_l)^+, \iota(a_k a_l^{-1})^+ \mid k \neq l\}\\ &\quad\quad \cup \{\iota(b)^+ \mid b \in \hat{\Lambda}_0, \bar{b} = \tfrac{1}{2}\varepsilon_S \alpha_C, C \in \mathcal{C}, |C| = 8, S \subset C, |S| \in 2\mathbb{Z}\}.\end{aligned} \quad (12.2.22)$$

Then we have:

Proposition 12.2.6: *The set of operators*

$$\{x_v(n) \mid v \in (V^+_{\Lambda_0})_{-1}, n \in \mathbb{Q}\} \quad (12.2.23)$$

acts irreducibly, and in fact absolutely irreducibly, on each of the spaces $V^{\pm}_{\Lambda_j}$, $V'^{\pm}_{\Lambda_j}$ for $0 \le j \le 3$.

Proof: We extend the field $\mathbb{F}$ arbitrarily to handle the absolute irreducibility. Let U be a nonzero invariant subspace of one of the 16 spaces.

We first show that U is invariant under

$$\{\alpha_k(m)\alpha_l(n) \mid k \neq l \in \Omega,\ m, n \in \mathbb{Q}\}. \tag{12.2.24}$$

Using (8.6.50) or (9.3.67) we have (for fixed k) that

$$[\gamma_k, \alpha_r(p)] = \delta_{k,r} p\alpha_r(p), \tag{12.2.25}$$

where

$$\gamma_k = -\tfrac{1}{4}\alpha_k^2(0). \tag{12.2.26}$$

It follows that U is the direct sum of its eigenspaces under γ_k. Let $u \in U$ be such that

$$\gamma_k u = \lambda u, \quad \lambda \in \mathbb{Q}. \tag{12.2.27}$$

For $k \neq l$ we have

$$\alpha_k\alpha_l(m + n)u = \sum_{\substack{p,q \in \mathbb{Q} \\ p+q=m+n}} \alpha_k(p)\alpha_l(q)u$$

(finite sum). By (12.2.25) and (12.2.27),

$$\gamma_k\alpha_k(p)\alpha_l(q)u = (p + \lambda)\alpha_k(p)\alpha_l(q)u$$

and it follows that

$$a_k(m)\alpha_l(n)u \in U.$$

By linearity,

$$\alpha_k(m)\alpha_l(n)U \subset U. \tag{12.2.28}$$

Next we show more generally that U is invariant under

$$\{\alpha_k(m)\alpha_l(n) \mid k, l \in \Omega,\ m, n \in \mathbb{Q}\}. \tag{12.2.29}$$

Let $k \neq l$. Again by (8.6.50) or (9.3.67),

$$[-\tfrac{1}{2}\alpha_k\alpha_l(0), \alpha_k(m)\alpha_l(n)] = n\alpha_k(m)\alpha_k(n) + m\alpha_l(m)\alpha_l(n). \tag{12.2.30}$$

Using (12.2.25) we now see that U is invariant under $\alpha_k(m)\alpha_k(n)$ if $m + n \neq 0$. Bracketing with $\alpha_k^2(-m - n)$ and assuming that $m \neq 0$, we find that U is invariant under

$$m\alpha_k(-n)\alpha_k(n) + n\alpha_k(m)\alpha_k(-m),$$

and hence, choosing $-m$ very large, we see that any given element of U is taken into U by $\alpha_k(-n)\alpha_k(n)$.

In completing the argument, for notational convenience we restrict our attention to the case $V_{\Lambda_0}^+$, the other cases being similar.

Since

$$L(0) = -\sum_{k \in \Omega} \gamma_k, \tag{12.2.31}$$

U is graded. Let v be a nonzero homogeneous vector of maximal degree in U. Since U is invariant under the set (12.2.29), v must then be of the form

$$v = \sum_j \mu_j \iota(b_j)$$

for some finite set of $b_j \in \hat{\Lambda}_0$, $\mu_j \in \mathbb{F}^\times$. Using the operators $\alpha_k(0)\alpha_l(0)$, $k, l \in \Omega$, we may assume that

$$v = \iota(b)^+$$

for some $b \in \hat{\Lambda}_0$, since knowledge of all the quantities $\langle \alpha_k, \bar{b}\rangle\langle \alpha_l, \bar{b}\rangle$ determines $\bar{b}$ up to sign.

Using the fact that the Golay code is generated by the octads (recall Remark 10.1.9), we see that

$$(\Lambda_0)_4 \text{ generates } \Lambda_0. \tag{12.2.32}$$

It follows that

$$v \in \mathbb{F}^\times \iota(1),$$

and again by (12.2.32), we see that U contains the vectors $\iota(b)^+$ for all $b \in \hat{\Lambda}_0$. Then application of the operators $\alpha_k(m)\alpha_l(n)$ generates

$$S^+ \otimes (\iota(b)^+) \oplus S^- \otimes (\iota(b)^-),$$

where $S^{+(-)}$ denotes the space of polynomials of even (odd) degree in $S(\hat{\mathfrak{h}}_{\bar{\mathbb{Z}}})$. ∎

Proposition 12.2.7: *The operator σ_1 interchanges the following spaces:*

$$V_{\Lambda_0}^+ \to V_{\Lambda_0}^+, \quad V_{\Lambda_0}^- \leftrightarrow V_{\Lambda_2}^+, \quad V_{\Lambda_2}^- \to V_{\Lambda_2}^- \tag{12.2.33}$$

$$V_{\Lambda_0}^{\prime -} \leftrightarrow V_{\Lambda_3}^-, \quad V_{\Lambda_2}^{\prime -} \leftrightarrow V_{\Lambda_3}^+, \quad V_{\Lambda_0}^{\prime +} \leftrightarrow V_{\Lambda_1}^-, \quad V_{\Lambda_2}^{\prime +} \leftrightarrow V_{\Lambda_1}^+ \tag{12.2.34}$$

$$V_{\Lambda_1}^{\prime -} \to V_{\Lambda_1}^{\prime -}, \quad V_{\Lambda_3}^{\prime +} \to V_{\Lambda_3}^{\prime +}, \quad V_{\Lambda_1}^{\prime +} \leftrightarrow V_{\Lambda_3}^{\prime -}. \tag{12.2.35}$$

Proof: Since

$$\sigma_1(V_{\Lambda_0}^+)_{-1} = (V_{\Lambda_0}^+)_{-1} \tag{12.2.36}$$

by Theorem 11.5.1 and Proposition 12.2.1, and σ_1 normalizes the set (12.2.23), it is enough, by Proposition 12.2.6, to calculate the image under σ_1 of an arbitrary nonzero vector in each case. Now σ_1 fixes $\iota(1)$, so that

σ_1 leaves $V_{\Lambda_0}^+$ invariant. Let $k \in \Omega$. By (11.5.6) and (11.1.72),

$$\sigma_1: \alpha_k(-1) \mapsto \iota(a_k)^+,$$

so that

$$\sigma_1: V_{\Lambda_0}^- \leftrightarrow V_{\Lambda_2}^+.$$

Similarly,

$$\sigma_1: V_{\Lambda_2}^- \to V_{\Lambda_2}^-.$$

By (11.5.10),

$$\sigma_1: t(1) \mapsto \iota(b_0).$$

We have

$$t(b_k^2) = t(b_0^{-2}b_0^2b_k^2) = t(b_0^{-2})\psi(b_0^2b_k^2),$$

$$\psi(b_0^2b_k^2) = \psi(a_kb_ka_kb_kb_k^2) = \psi(b_k^4)i^{-c_0(\alpha_k,(1/4)\alpha_\Omega-\alpha_k)} = -i,$$

since $b_k^4 \in K_L \subset \operatorname{Ker}\psi$, and

$$z_0(b_0) = z_0(a_kb_k) = a_k^{-1}b_k^{-1} = b_k^{-1}a_k^{-1}\kappa = b_0^{-1}\kappa.$$

Hence

$$\sigma_1 t(b_k^2) = \sigma_1(-i)t(b_0^{-2}) = -i\iota(b_0^{-1}) = -\iota(z_0(b_0)) \tag{12.2.37}$$

by (11.3.47), so that

$$\sigma_1: (t(1) \pm t(b_k^2)) \mapsto \iota(b_0)^{\mp}. \tag{12.2.38}$$

Thus

$$\sigma_1: V_{\Lambda_0}'^- \leftrightarrow V_{\Lambda_3}^-, \quad V_{\Lambda_2}'^- \leftrightarrow V_{\Lambda_3}^+.$$

Using this calculation and (11.2.36), we find:

$$\sigma_1: \alpha_k(-\tfrac{1}{2}) \otimes (t(1) \pm t(b_k^2)) \mapsto x_{a_k}^+(-\tfrac{1}{2})\iota(b_0)^{\mp} = \iota(a_k^{-1}b_0)^{\mp}, \tag{12.2.39}$$

so that

$$\sigma_1: V_{\Lambda_0}'^+ \leftrightarrow V_{\Lambda_1}^-, \quad V_{\Lambda_2}'^+ \leftrightarrow V_{\Lambda_1}^+.$$

By (11.5.11), we have

$$\sigma_1: t(b_0) \mapsto \mu t(b_0^{-1}).$$

Moreover,

$$t(b_k^2b_0) = t(a_k^{-1}b_0a_k^{-1}b_0b_0) = t(b_0^3)i^{-5c_0(\alpha_k,(1/4)\alpha_\Omega)} = -it(b_0^3),$$

so that

$$\sigma_1: t(b_k^2b_0) \mapsto -i\mu t(b_0) \tag{12.2.40}$$

by (11.4.47). Hence

$$\begin{aligned}\sigma_1\colon (t(b_0) \pm t(b_k^2 b_0)) &\mapsto \mu(t(b_0^{-1}) \pm (-i)t(b_0))\\ &= -i\mu(it(b_0^3)\psi(b_0^{-4}) \pm t(b_0))\\ &= -i\mu(t(b_k^2 b_0) \pm t(b_0))\end{aligned} \tag{12.2.41}$$

since $\psi(b_0^{-4}) \in \psi(\kappa^2 K_L) = \{-1\}$. Thus σ_1 leaves both $V_{\Lambda_1}'^{-}$ and $V_{\Lambda_3}'^{+}$ invariant. We also get

$$\sigma_1\colon \alpha_k(-\tfrac{1}{2}) \otimes (t(b_0) + t(b_k^2 b_0)) \mapsto x_{a_k}^{+}(-\tfrac{1}{2})(-i)\mu(t(b_k^2 b_0) + t(b_0)) \in V_{\Lambda_3}'^{-}, \tag{12.2.42}$$

so that

$$\sigma_1\colon V_{\Lambda_1}'^{+} \leftrightarrow V_{\Lambda_3}'^{-}. \qquad \blacksquare$$

Recall the homomorphism ν from (10.4.51). We shall also denote by ν the homomorphism

$$\nu\colon \hat{\mathfrak{L}} \to \operatorname{Aut} W_{\mathfrak{L}}, \tag{12.2.43}$$

where $\nu(a)$ ($a \in \hat{\mathfrak{L}}$) is the operator determined by:

$$\nu(a) = (\operatorname{int}(a), \pi(a)) \in (\operatorname{Aut} \hat{\mathfrak{L}}) \times (\operatorname{Aut} T_{\mathfrak{L}}),$$

i.e.,

$$\nu(a) = 1 \otimes \operatorname{int}(a) \oplus 1 \otimes \pi(a), \tag{12.2.44}$$

where π denotes the representation on $T_{\mathfrak{L}}$. Of course, we also have

$$\nu\colon \hat{L} \to \operatorname{Aut} W_L, \tag{12.2.45}$$

given by the obvious analogue of (12.2.44), and ν induces the map given by (10.4.51).

Note that by Proposition 12.2.7, while σ_1 does not leave

$$V^{\natural} = V_{\Lambda_0}^{+} \oplus V_{\Lambda_1}^{+} \oplus V_{\Lambda_0}'^{+} \oplus V_{\Lambda_1}'^{+} \tag{12.2.46}$$

invariant, the composition

$$\nu(a_{k_0}) \circ \sigma_1 \tag{12.2.47}$$

[recall (12.2.7)] preserves $V^{\natural}$ since $\nu(a_{k_0})$ interchanges the spaces

$$V_{\Lambda_1}^{+} \leftrightarrow V_{\Lambda_1}^{-}, \quad V_{\Lambda_0}'^{+} \leftrightarrow V_{\Lambda_2}'^{+}, \quad V_{\Lambda_1}'^{+} \leftrightarrow V_{\Lambda_3}'^{-}. \tag{12.2.48}$$

Set

$$h = 1_{V_{\mathfrak{L}_0}} \oplus (-i)1_{V_{\mathfrak{L}_1}} \oplus 1_{V_{\mathfrak{L}_0}'} \oplus 1_{V_{\mathfrak{L}_1}'} \tag{12.2.49}$$

[recall (12.1.16)] and define

$$\sigma = h \circ \nu(a_{k_0}) \circ \sigma_1 \in \operatorname{Aut} W_{\mathcal{L}}. \tag{12.2.50}$$

Of course,

$$\sigma \in \operatorname{Aut} W_L \tag{12.2.51}$$

as well. The reason for this final modification by h is that we want $\sigma^2 = 1$:

Proposition 12.2.8: *We have*

$$\sigma V^{\natural} = V^{\natural} \tag{12.2.52}$$

$$\sigma Y(v, z)\sigma^{-1} = Y(\sigma v, z) \quad \textit{for} \quad v \in V_{\mathcal{L}_0} \tag{12.2.53}$$

and

$$\sigma^2 = 1 \quad \textit{on} \quad W_{\mathcal{L}}. \tag{12.2.54}$$

Proof: We have already shown that σ leaves $V^{\natural}$ invariant. Now

$$\sigma_1 Y(v, z)\sigma_1^{-1} = Y(\sigma_1 v, z) \quad \text{for} \quad v \in V_{\mathcal{L}_0} \tag{12.2.55}$$

by (11.5.3),

$$\nu(a_{k_0})Y(v, z)\nu(a_{k_0})^{-1} = Y(\nu(a_{k_0})v, z) \quad \text{for} \quad v \in V_{\mathcal{L}} \tag{12.2.56}$$

by an obvious analogue for $\mathcal{L}$ of (10.4.67) and, clearly,

$$hY(v, z)h^{-1} = Y(hv, z) \quad \text{for} \quad v \in V_{\mathcal{L}_0}. \tag{12.2.57}$$

Thus (12.2.53) holds.

For (12.2.54) we observe that

$$\nu(a_{k_0})\iota(b) = i^{c_0(\alpha_{k_0}, \bar{b})}\iota(b) \tag{12.2.58}$$

for $b \in \hat{\mathcal{L}}$, and hence by (11.5.6) and (11.5.7),

$$\nu(a_{k_0})\sigma_1(\iota(b)) = \sigma_1 \nu(a_{k_0})\iota(b) \tag{12.2.59}$$

for

$$b \in \hat{\mathcal{L}}_0 \quad \text{with} \quad \bar{b} = \tfrac{1}{2}\alpha_C \quad \text{for} \quad C \in \mathcal{C}_1 \tag{12.2.60}$$

[cf. (11.5.8)] or if

$$\bar{b} = \pm\alpha_k \quad \text{for} \quad k \in \Omega. \tag{12.2.61}$$

Thus for such b,

$$\nu(a_{k_0})^2\iota(b) = \iota(b) \tag{12.2.62}$$

and

$$\sigma^2\iota(b) = \iota(b), \tag{12.2.63}$$

using (11.5.4). Similarly,

$$\sigma^2\alpha_k(-1) = \alpha_k(-1) \quad \text{for} \quad k \in \Omega. \tag{12.2.64}$$

It follows from (12.2.53) that

$$\sigma^2 Y(v, z) = Y(v, z)\sigma^2 \tag{12.2.65}$$

for

$$v = \iota(b) \quad \text{with} \quad b \quad \text{as in (12.2.60) or (12.2.61)} \tag{12.2.66}$$

or

$$v = \alpha_k(-1). \tag{12.2.67}$$

Thus by the irreducibility assertions in Theorems 11.2.1, 11.3.1 and 11.4.1, we see that to prove (12.2.54) it is sufficient to verify that σ^2 fixes some nonzero element of each of the spaces $V_{\mathfrak{L}_0}$, $V_{\mathfrak{L}_1}$, $V'_{\mathfrak{L}_0}$, $V'_{\mathfrak{L}_1}$.

Certainly

$$\sigma^2\iota(1) = \iota(1) \tag{12.2.68}$$

(and we have already in fact checked that $\sigma^2 = 1$ on some elements of $V_{\mathfrak{L}_0}$). We also have

$$\sigma t(1) = h\nu(a_{k_0})\sigma_1 t(1) = h\nu(a_{k_0})\iota(b_0) = hi\iota(b_0) = \iota(b_0), \tag{12.2.69}$$

$$\sigma\iota(b_0) = h\nu(a_{k_0})\sigma_1\iota(b_0) = h\nu(a_{k_0})t(1) = ht(a_{k_0}) = t(1), \tag{12.2.70}$$

$$\sigma t(b_0) = h\nu(a_{k_0})\sigma_1 t(b_0) = h\nu(a_{k_0})\mu t(b_0^{-1}) = -i\mu t(b_0^{-1}), \tag{12.2.71}$$

$$\sigma(-i)\mu t(b_0^{-1}) = h\nu(a_{k_0})\sigma_1(-i)\mu t(b_0^{-1}) = h\nu(a_{k_0})(-i)t(b_0) = t(b_0), \tag{12.2.72}$$

as desired. ∎

In order to embed σ into an action of the symmetric group $\mathcal{S}_3$ on $V^\natural$ (cf. Remark 11.5.2), we define

$$\tau = \nu(b_{k_0}) \tag{12.2.73}$$

[recall (12.2.8)]. We specify σ_1 and σ completely by choosing the scalar μ in Theorem 11.5.1 to be

$$\mu = -i. \tag{12.2.74}$$

We shall show that this is the unique choice in order to have $(\tau\sigma)^3 = 1$. Note that $\mu = -i$ satisfies (11.5.13) since

$$\chi(b_0^4 a_{\emptyset,\Omega}) = \psi(b_0^4) \in \psi(\kappa^2 K_L) = \{-1\}, \quad \text{i.e.,} \quad b_0^4 = a_{\Omega,\emptyset}\kappa^2. \tag{12.2.75}$$

Proposition 12.2.9: *We have*

$$\tau V^{\natural} = V^{\natural} \tag{12.2.76}$$

$$\tau Y(v, z)\tau^{-1} = Y(\tau v, z) \quad \textit{for} \quad v \in V_{\mathfrak{L}} \tag{12.2.77}$$

$$\tau^2 = 1 \quad \textit{on} \quad W_\Lambda \tag{12.2.78}$$

$$(\tau\sigma)^3 = 1 \quad \textit{on} \quad W_{\mathfrak{L}}. \tag{12.2.79}$$

In particular, we have the group isomorphism

$$\langle \sigma, \tau \rangle \big|_{V^{\natural}} \simeq \mathcal{S}_3 \tag{12.2.80}$$

and

$$gY(v, z)g^{-1} = Y(gv, z) \tag{12.2.81}$$

on $W_{\mathfrak{L}}$ *for* $g \in \langle \sigma, \tau \rangle$, $v \in V_{\mathfrak{L}_0}$.

Proof: We need only prove (12.2.79). As in the last proof, it is sufficient to show that $(\tau\sigma)^3 = 1$ on certain elements. Using (12.1.29), we note first that

$$\begin{aligned} \tau\iota(b) &= i^{c_0((1/4)\alpha_\Omega - \alpha_{k_0}, (1/2)\epsilon_S\alpha_C)}\iota(b) \\ &= i^{-2\langle (1/2)\epsilon_S\alpha_C, (1/4)\alpha_\Omega - \alpha_{k_0}\rangle}\iota(b) \\ &= i^{-|C|/2+|S|}(-1)^{|\{k_0\}\cap C|}\iota(b) \end{aligned} \tag{12.2.82}$$

for

$$b \in \hat{\mathfrak{L}}_0 \quad \text{with} \quad \bar{b} = \tfrac{1}{2}\varepsilon_S\alpha_C \quad \text{for} \quad C \in \mathcal{C}_1, \quad S \subset C. \tag{12.2.83}$$

Also, for the same b, by (11.5.7) and (12.2.58),

$$\sigma\iota(b) = (-2)^{-|C|/2}(-1)^{|\{k_0\}\cap C|} \sum_{T\subset C} (-1)^{|S\cap T|}\iota(a_{S,T}b). \tag{12.2.84}$$

In the calculation below we shall use the following summations, for $T, U, V \subset C$:

$$\begin{aligned} \sum_{T\subset C} (-1)^{|U\cap T|} i^{|T|} &= (1-i)^{|U|}(1+i)^{|C\setminus U|} \\ &= (1+i)^{|C|}(-i)^{|U|} = (2i)^{|C|/2}(-i)^{|U|}, \end{aligned}$$

$$\sum_{U\subset C} (-1)^{|U\cap V|} = (1-1)^{|V|}(1+1)^{|C\setminus V|} = \begin{cases} 0 & \text{if} \quad V \neq \emptyset \\ 2^{|C|} & \text{if} \quad V = \emptyset. \end{cases}$$

For b as in (12.2.60), we have

$$\begin{aligned}
\iota(b) &\overset{\sigma}{\mapsto} (-2)^{-|C|/2}(-1)^{|\{k_0\}\cap C|} \sum_{T\subset C} \iota(a_{\emptyset,T}b) \\
&\overset{\tau}{\mapsto} (-2)^{-|C|/2} i^{-|C|/2} \sum_{T\subset C} i^{|T|}\iota(a_{\emptyset,T}b) \\
&\overset{\sigma}{\mapsto} (-2)^{-|C|}(-1)^{|\{k_0\}\cap C|} i^{-|C|/2} \sum_{T\subset C} i^{|T|} \sum_{U\subset C} (-1)^{|T\cap U|}\iota(a_{\emptyset,U}b) \\
&= (-2)^{-|C|/2}(-1)^{|\{k_0\}\cap C|+|C|/2} \sum_{U\subset C} (-i)^{|U|}\iota(a_{\emptyset,U}b) \\
&\overset{\tau}{\mapsto} (-2)^{-|C|/2}(-1)^{|C|/2} i^{-|C|2} \sum_{U\subset C} \iota(a_{\emptyset,U}b) \\
&\overset{\sigma}{\mapsto} (-2)^{-|C|}(-1)^{|C|/2+|\{k_0\}\cap C|} i^{-|C|/2} \sum_{U,V\subset C} (-1)^{|U\cap V|}\iota(a_{\emptyset,V}b) \\
&= (-1)^{|C|/2+|\{k_0\}\cap C|} i^{-|C|/2}\, \iota(b) \\
&\overset{\tau}{\mapsto} (-1)^{|C|/2} i^{-|C|}\iota(b) = \iota(b).
\end{aligned}$$

Similar calculations show that $(\tau\sigma)^3$ fixes $\iota(b)$ for b as in (12.2.61) and $\alpha_k(-1)$ for $k \in \Omega$. Hence, as in the previous proposition, $(\tau\sigma)^3$ acts as a scalar on each of $V_{\mathfrak{L}_0}$, $V_{\mathfrak{L}_1}$, $V'_{\mathfrak{L}_0}$, $V'_{\mathfrak{L}_1}$. Since both τ and σ fix $\iota(1)$ (say), $(\tau\sigma)^3$ fixes $V_{\mathfrak{L}_0}$. We have

$$\begin{aligned}
\sigma t(b_{k_0}) &= hv(a_{k_0})\sigma_1 t(a_{k_0}^{-1}b_0) = hv(a_{k_0})\sigma_1(-i)t(b_0) \\
&= hv(a_{k_0})(-i)\mu t(b_0^{-1}) = -i\mu t(a_{k_0}b_0^{-1}) \\
&= -\mu t(b_0^{-1}a_{k_0}) = -\mu t(b_{k_0}^{-1}), \\
\sigma t(1) &= \iota(b_0)
\end{aligned}$$

by (12.2.69) and

$$\tau\iota(b_0) = i^{c_0((1/4)\alpha_\Omega-\alpha_{k_0},(1/4)\alpha_\Omega)}\iota(b_0) = -i\iota(b_0).$$

Hence

$$\begin{aligned}
t(b_{k_0}) &\overset{\sigma}{\mapsto} -\mu t(b_{k_0}^{-1}) \overset{\tau}{\mapsto} -\mu t(1) \overset{\sigma}{\mapsto} -\mu\iota(b_0) \overset{\tau}{\mapsto} i\mu\iota(b_0) \\
&\overset{\sigma}{\mapsto} i\mu t(1) \overset{\tau}{\mapsto} i\mu t(b_{k_0}) = t(b_{k_0}),
\end{aligned}$$

which shows that $(\tau\sigma)^3$ fixes a vector in each of $V_{\mathfrak{L}_1}$, $V'_{\mathfrak{L}_0}$, $V'_{\mathfrak{L}_1}$. ∎

Remark 12.2.10: Of course, $\mathfrak{L}$ can be replaced by L throughout.

12.3. The Monster M and the Statement of the Main Theorem

All the ingredients are now before us. It is time to assemble them.

We define the *Monster* to be the group M of linear automorphisms of the moonshine module $V^\natural$ generated by C and σ (recall Sections 10.3, 10.4 and 12.2):

$$M = \langle C, \sigma \rangle \subset \operatorname{Aut} V^\natural. \tag{12.3.1}$$

Viewing the restriction to $V^\natural$ of the automorphism τ [defined in (12.2.73)] as an element of C, we note that M is also the group generated by C and a symmetric group $\mathcal{S}_3$:

$$M = \langle C, \langle \sigma, \tau \rangle \rangle \tag{12.3.2}$$

(see Proposition 12.2.9). Then M preserves each homogeneous subspace of $V^\natural$,

$$M \cdot V_n^\natural = V_n^\natural \quad \text{for} \quad n \in \mathbb{Z}, \tag{12.3.3}$$

and in particular, $V_{-1}^\natural$:

$$M \cdot \mathcal{B} = \mathcal{B} \tag{12.3.4}$$

[recall (10.3.41)]. But we do not yet know that M acts faithfully on $\mathcal{B}$.

For $v \in V^z = V_\Lambda^{\theta_0}$ [recall (10.4.59)], we have the vertex operators $Y(v, z)|_{V^z}$ defined via Chapter 8 and the operators $Y(v, z)|_{V^{-z}}$ defined via Chapter 9. (As usual, the z's should not be confused.) These operators satisfy the Jacobi identity, by Theorem 8.8.9 and Corollary 9.5.4. Also, the actions on $V^\natural$ of C and of $Y(v, z)$ for such v are compatible, as we have seen in Corollaty 10.4.8.

We now define vertex operators $Y(v, z)$ on $V^\natural$ for all $v \in V^\natural$: For $v \in V^z$ we continue to take $Y(v, z)$ as before. We know from the last section that σ and τ permute the summands in the decomposition

$$\begin{aligned} V^\natural = {} & \underbrace{V_{\Lambda_0}^+ \oplus V_{\Lambda_1}^+} \oplus \underbrace{V_{\Lambda_0}'^+ \oplus V_{\Lambda_1}'^+} \\ (= {} & \qquad V^z \qquad \oplus \qquad V^{-z}) \end{aligned} \tag{12.3.5}$$

as follows:

$$\sigma\colon V_{\Lambda_0}^+ \to V_{\Lambda_0}^+, \quad \sigma\colon V_{\Lambda_1}^+ \leftrightarrow V_{\Lambda_0}'^+, \quad \sigma\colon V_{\Lambda_1}'^+ \to V_{\Lambda_1}'^+ \tag{12.3.6}$$

$$\tau\colon V_{\Lambda_0}^+ \to V_{\Lambda_0}^+, \quad \tau\colon V_{\Lambda_1}^+ \to V_{\Lambda_1}^+, \quad \tau\colon V_{\Lambda_0}'^+ \leftrightarrow V_{\Lambda_1}'^+. \tag{12.3.7}$$

For $v \in V_{\Lambda_0}'^{+}$ we set

$$Y(v, z) = \sigma Y(\sigma^{-1}v, z)\sigma^{-1} \quad \text{on} \quad V^\natural \tag{12.3.8}$$

and for $v \in V_{\Lambda_1}'^{+}$ we take

$$Y(v, z) = \tau\sigma Y((\tau\sigma)^{-1}v, z)(\tau\sigma)^{-1} \quad \text{on} \quad V^\natural. \tag{12.3.9}$$

This gives us a well-defined linear map

$$\begin{aligned} V^\natural &\to (\operatorname{End} V^\natural)\{z\} \\ v &\mapsto Y(v, z) \end{aligned} \tag{12.3.10}$$

[using the usual $\{z\}$ notation as in (8.5.6)].

We also define the related operator

$$X(v, z) = Y(v, z)z^{\operatorname{wt} v} \tag{12.3.11}$$

for a homogeneous element $v \in V^\natural$, and hence $X(v, z)$ for arbitrary $v \in V^\natural$ by linearity, as in (8.5.27), (8.5.32). In addition, we define the component operators v_n and $x_v(n)$ for $n \in \mathbb{Z}$ by

$$\begin{aligned} Y(v, z) &= \sum_{n \in \mathbb{Z}} v_n z^{-n-1}, \\ X(v, z) &= \sum_{n \in \mathbb{Z}} x_v(n) z^{-n} \end{aligned} \tag{12.3.12}$$

as in (8.5.15), (8.5.28); note that the sums are indeed over $\mathbb{Z}$. We have

$$x_v(n)\colon V^{\pm z} \to V^{\mp z} \quad \text{for} \quad v \in V^{-z}, \quad n \in \mathbb{Z}. \tag{12.3.13}$$

We observe that formula (8.5.8) now remains valid on all of $V^\natural$:

$$\lim_{z \to 0} Y(v, z)\iota(1) = v. \tag{12.3.14}$$

Recall from Proposition 10.3.6 that the space $\mathfrak{B} = V_{-1}^\natural$ of weight 2 has the structure of a commutative nonassociative algebra with product denoted $\times$ and with a nonsingular symmetric associative form $\langle \cdot, \cdot \rangle$. Moreover, $\mathfrak{B}$ has the ± 1-eigenspace decomposition $\mathfrak{k} \oplus \mathfrak{p}$ with respect to z and identity element $\frac{1}{2}\omega \in \mathfrak{k}$. By (10.4.82) and (11.2.46),

$$M \cdot \omega = \omega. \tag{12.3.15}$$

From the definitions given in Theorem 8.9.5,

$$\begin{aligned} u \times v &= u_1 \cdot v = x_u(0)v \\ \langle u, v \rangle &= u_3 \cdot v = x_u(2)v \end{aligned} \tag{12.3.16}$$

for $u, v \in \mathfrak{k}$. For explicit formulas, refer to Proposition 10.3.6.

Strictly speaking, one piece of information has been left unspecified: The form $\langle \cdot, \cdot \rangle$ on $T = T_\Lambda$, which enters into the form $\langle \cdot, \cdot \rangle$ on $\mathfrak{p}$ given by (10.3.54) (recall Remark 10.3.2), has not been given an absolute normalization. Since we have fixed our choice of σ, which mixes $\mathfrak{k}$ and $\mathfrak{p}$, there could be only one correct normalization for the form on T_Λ. Correspondingly, we fix $\langle \cdot, \cdot \rangle$ on T_Λ so that

$$\langle t(1) + t(b_{k_0}^2), t(1) + t(b_{k_0}^2) \rangle = 2. \tag{12.3.17}$$

(To see that the left-hand side is nonzero, note that the vectors $t(b) + t(b_{k_0}^2 b)$ form a basis of T_Λ if we take b in a set of representatives of $\hat{\Lambda}^{(2)}/((Q \cap \Lambda) + 2\Lambda)^{\urcorner(2)}$, and $\hat{\Lambda}^{(2)}$ acts by permuting these basis elements up to sign. Thus the symmetric bilinear form making this basis an orthonormal basis is invariant under $\hat{\Lambda}^{(2)}$. Cf. Remark 10.3.2.)

Recall also from Section 8.9 the notions of commutative affinization and of graded representation by cross-bracket. We know from Theorem 10.4.12 that $V^\natural$ is a graded module for the commutative affinization $\hat{\mathfrak{k}}$ of $\mathfrak{k}$ and that $V^\natural$ is in fact a graded $(C, \hat{\mathfrak{k}})$-module in the sense of (10.4.79):

$$gxg^{-1} = g \cdot x \tag{12.3.18}$$

on $V^\natural$, for $g \in C$, $x \in \hat{\mathfrak{k}}$.

Since we are interested in making $V^\natural$ a graded $(M, \hat{\mathfrak{B}})$-module, we define a linear map

$$\pi: \hat{\mathfrak{B}} \to \operatorname{End} V^\natural \tag{12.3.19}$$

as follows:

$$\begin{aligned} \pi: v \otimes t^n &\mapsto x_v(n) \quad \text{for} \quad v \in \mathfrak{B}, \quad n \in \mathbb{Z} \\ \pi: e &\mapsto 1, \end{aligned} \tag{12.3.20}$$

extending the action of $\hat{\mathfrak{k}}$ given by (8.9.45)–(8.9.46). Observe that for v as in (12.3.8),

$$x_v(n) = \sigma x_{\sigma^{-1}v}(n)\sigma^{-1}$$

and analogously for v as in (12.3.9).

Recall also the discussion of the modular function $J(q)$ and of the graded dimension $\dim_* V^\natural$ in Remark 10.5.9 and (10.5.55).

We are ready to state the main theorem. The first two parts concern the moonshine module and its vertex operator algebra structure, part (c) relates the Monster to the vertex operators, and the remaining parts, (d)–(h), deal with the algebra $\mathfrak{B}$ and its relations with the other structures.

Theorem 12.3.1: *(a) The graded dimension of the moonshine module is $J(q)$:*

$$\dim_* V^\natural = J(q) = q^{-1} + 0 + 196884q + \cdots. \tag{12.3.21}$$

(b) The Jacobi identity holds on $V^\natural$:

$$z_0^{-1}\delta\left(\frac{z_1 - z_2}{z_0}\right)Y(u, z_1)Y(v, z_2) - z_0^{-1}\delta\left(\frac{z_2 - z_1}{-z_0}\right)Y(v, z_2)Y(u, z_1)$$

$$= z_2^{-1}\delta\left(\frac{z_1 - z_0}{z_2}\right)Y(Y(u, z_0)v, z_2) \tag{12.3.22}$$

for $u, v \in V^\natural$. Equivalently,

$$[Y(u, z_1) \times_n Y(v, z_2)] = \mathrm{Res}_{z_0} z_0^n z_2^{-1} Y(Y(u, z_0)v, z_2)e^{-z_0(\partial/\partial z_1)}\delta(z_1/z_2) \tag{12.3.23}$$

for $n \in \mathbb{Z}$. Moreover,

$$Y(u, z)v = e^{zL(-1)}(Y(v, -z)u). \tag{12.3.24}$$

The Virasoro algebra, determined by $L(z) = Y(\omega, z)$, acts on $V^\natural$. In the terminology of Section 8.10, $V^\natural$ is a vertex operator algebra of rank 24.
(c) The actions of the Monster and of the vertex operators on $V^\natural$ are compatible:

$$gY(v, z)g^{-1} = Y(gv, z) \quad \text{for} \quad g \in M, \quad v \in V^\natural. \tag{12.3.25}$$

In particular, in the terminology of Section 8.10, M acts as automorphisms of the vertex operator algebra $V^\natural$.
(d) The group M acts as automorphisms and isometries of $\mathfrak{B}$.
(e) The map π is a graded representation of $\hat{\mathfrak{B}}$ by cross-bracket making $V^\natural$ a graded $(M, \hat{\mathfrak{B}})$-module.
(f) For $u, v \in \mathfrak{B}$,

$$u \times v = u_1 \cdot v = x_u(0)v \tag{12.3.26}$$

$$\langle u, v\rangle = u_3 \cdot v = x_u(2)v. \tag{12.3.27}$$

(g) The action of $\hat{\mathfrak{B}}$ on $V^\natural$ is absolutely irreducible.
(h) The action of M on $\mathfrak{B}$ is faithful.

We now discuss the proof. Formula (12.3.22) and part (c) will be established in Chapter 13, and everything else will be proved here, sometimes using these.

We shall prove parts (a) and (g) below. Part (b) will be proved from (12.3.22) and Theorem 10.3.4. Using (b), we shall prove (f). Clearly, (d) follows from (c) and (f). Part (e) follows from (c), (f) and the case $n = 1$ of (12.3.23), as in Section 8.9 and Theorem 10.4.12. Part (h) follows immediately from (e) and (g). In fact, the only part of (e) needed here is that the linear map $\pi: \hat{\mathfrak{B}} \to \operatorname{End} V^\natural$ is compatible with the action of M in the usual sense.

We now prove (a). By (10.5.50),

$$J + 24 = \dim_* V_\Lambda = \dim_* V_{\Lambda_0}^+ + \dim_* V_{\Lambda_0}^- + \dim_* V_{\Lambda_1}^+ + \dim_* V_{\Lambda_1}^-$$

$$J + 72 = \dim_* V_{L_0} = \dim_* V_{\Lambda_0}^+ + \dim_* V_{\Lambda_0}^- + \dim_* V_{\Lambda_2}^+ + \dim_* V_{\Lambda_2}^-.$$

Also, σ and τ give the grading-preserving isomorphisms

$$\sigma: V_{\Lambda_2}^+ \simeq V_{\Lambda_0}^-,$$

$$\sigma: V_{\Lambda_1}^+ \simeq V_{\Lambda_0}^{\prime +},$$

$$\tau: V_{\Lambda_0}^{\prime +} \simeq V_{\Lambda_1}^{\prime +},$$

and clearly,

$$\dim_* V_{\Lambda_2}^+ = \dim_* V_{\Lambda_2}^-$$

$$\dim_* V_{\Lambda_1}^+ = \dim_* V_{\Lambda_1}^-.$$

Hence

$$\begin{aligned} J &= \tfrac{3}{2}(J + 24) - \tfrac{1}{2}(J + 72) \\ &= \dim_* V_{\Lambda_0}^+ + \dim_* V_{\Lambda_0}^- + 3 \dim_* V_{\Lambda_1}^+ - \dim_* V_{\Lambda_2}^+ \\ &= \dim_* V . \end{aligned}$$

Next we prove (g). Working over any extension field, let $v \in V^\natural$, $v \neq 0$, and decompose v into its components with respect to (12.3.5). By Proposition 12.2.6, we can move any nonzero one of these components to the appropriate one of the vectors $\iota(1)$, $\iota(b_{k_0})^+$, $\sigma\iota(b_{k_0})^+$, $\tau\sigma\iota(b_{k_0})^+$. But any of these four elements can be moved to $\iota(1)$ by the application of a suitable element of $\hat{\mathfrak{B}}$; for instance,

$$\tfrac{1}{2}(\iota(b_{k_0})^+)_3\,\iota(b_{k_0})^+ = \iota(1).$$

Now we use $L(0)$ to obtain $\iota(1)$, which generates $V^\natural$ by Proposition 12.2.6 and (12.3.14).

Next we claim that for $v \in V^{\natural}$,

$$[L(-1), Y(v, z)] = Y(L(-1)v, z) = \frac{d}{dz} Y(v, z), \tag{12.3.28}$$

$$e^{z_0 L(-1)} Y(v, z) e^{-z_0 L(-1)} = Y(e^{z_0 L(-1)} v, z) = Y(v, z + z_0), \tag{12.3.29}$$

$$Y(v, z)\iota(1) = e^{zL(-1)} v. \tag{12.3.30}$$

For $v \in V^z = V_\Lambda^{\theta_0}$, these have already been established in Chapters 8 and 9 [cf. (8.8.5)–(8.8.6), (9.5.3) and Theorem 10.3.4]. But since M fixes ω [recall (12.3.15)], M commutes with the Virasoro algebra:

$$L(z)g = gL(z) \quad \text{for} \quad g \in M \tag{12.3.31}$$

by part (c), or rather, by Corollary 10.4.9 for $g \in C$ and by Theorem 11.5.1 for $g = \sigma$. In particular, M commutes with $L(-1)$, and the claim follows by conjugation by σ or $\tau\sigma$.

Now (12.3.24) follows from (12.3.22) exactly as indicated in Remark 8.8.13 and the proof of Proposition 8.8.3. [Of course, only part of the information in (12.3.28)–(12.3.30) is needed here.]

As for (f), recall from Section 10.3 that the formulas (12.3.26) and (12.3.27) are definitions for $u \in \mathfrak{k}$ (and $v \in \mathfrak{k}$ or $\mathfrak{p}$). For $u \in \mathfrak{p}$ and $v \in \mathfrak{k}$, simply extract the coefficients of z^{-2} and z^{-4} in (12.3.24), as in Section 8.9.

Next we prove (12.3.27) for $u, v \in \mathfrak{p}$. We know that $\langle u, v\rangle$ and $u_3 \cdot v$ both define C-invariant forms on $\mathfrak{p}$, by Proposition 10.4.11 and (12.3.25). Using the absolute irreducibility of $\mathfrak{p} = \mathfrak{h} \otimes T$ under C (recall Remark 10.4.13), it suffices to verify the equality for

$$u = v = \alpha_{k_0} \otimes (t(1) + t(b_{k_0}^2)) \in \mathfrak{p}.$$

But using (10.3.54) and (12.3.17) we find that

$$\langle u, v\rangle = \tfrac{1}{2}\langle \alpha_{k_0}, \alpha_{k_0}\rangle \langle t(1) + t(b_{k_0}^2), t(1) + t(b_{k_0}^2)\rangle = 2,$$

and from (12.2.39) we obtain

$$u_3 \cdot v = (\sigma u)_3 \cdot (\sigma v) = (\iota(b_{k_0})^+)_3 \cdot \iota(b_{k_0})^+ = 2,$$

as desired.

Finally, we see just as in (8.9.36) and (8.9.37) that the product $u_1 \cdot v$ on $\mathcal{B}$ is commutative and is associative with respect to the form $u_3 \cdot v$ on $\mathcal{B}$, and keeping in mind (12.3.13) we find that (12.3.26) for $u, v \in \mathfrak{p}$ follows from (10.3.56).

This completes the proof of the Theorem modulo (12.3.22) and part (c).

Remark 12.3.2: The only restriction on our field $\mathbb{F}$ of characteristic 0 is that is should contain i. In Section 12.4 we shall remove this restriction.

Remark 12.3.3: Let us denote by $\mathcal{B}_0$ the orthogonal complement of the identity element $\frac{1}{2}\omega$ in $\mathcal{B}$:

$$\mathcal{B}_0 = \{x \in \mathcal{B} \mid \langle \tfrac{1}{2}\omega, x\rangle = 0\}. \tag{12.3.32}$$

Since M fixes $\frac{1}{2}\omega$,

$$M \cdot \mathcal{B}_0 = \mathcal{B}_0, \tag{12.3.33}$$

and clearly, M acts faithfully on the 196883-dimensional space $\mathcal{B}_0$. This action is irreducible, in view of the statements about irreducibility under C in Remark 10.4.13.

Having already described explicit formulas for the structure of the algebra $\mathcal{B}$ in Section 10.3, and the action of C on it in Section 10.4, we now do the same for the automorphism σ.

For $b \in \hat{L}$, set

$$t_\Lambda(b) = t(b) + t(b_{k_0}^2 b) \in T = T_\Lambda. \tag{12.3.34}$$

These elements span T, even if we restrict b to lie in $\hat{\Lambda}$ or $\hat{\Lambda}^{(2)}$. In what follows, this restriction may be made. The action of the involution σ on a spanning set of $\mathcal{B}$ may be described as follows:

$$\sigma\alpha_k^2 = \alpha_k^2 \quad \text{for} \quad k \in \Omega, \tag{12.3.35}$$

$$\sigma\alpha_k\alpha_l = x_{a_k a_l}^+ + x_{a_k a_l^{-1}}^+ \quad \text{for} \quad k \neq l, \tag{12.3.36}$$

$$\sigma(x_{a_k a_l}^+ - x_{a_k a_l^{-1}}^+) = x_{a_k a_l}^+ - x_{a_k a_l^{-1}}^+ \quad \text{for} \quad k \neq l, \tag{12.3.37}$$

using (11.2.46)–(11.2.48);

$$\sigma x_b^+ = \tfrac{1}{16}(-1)^{|\{k_0\}\cap C|} \sum_{\substack{T \subset C \\ |T| \in 2\mathbb{Z}}} (-1)^{|S\cap T|} x_{a_S, Tb}^+ \tag{12.3.38}$$

for

$$b \in \hat{\Lambda}_0 \quad \text{with} \quad \bar{b} = \tfrac{1}{2}\varepsilon_S \alpha_C, \quad C \in \mathcal{C}, \quad |C| = 8, \quad S \subset C, \quad |S| \in 2\mathbb{Z}, \tag{12.3.39}$$

from Proposition 12.2.1;

$$\sigma x_{bb_k}^+ = (-1)^{|\{k_0\}\cap C|}\alpha_k \otimes t_\Lambda(b) \tag{12.3.40}$$

for

$$b \in \hat{\Lambda}_0 \quad \text{with} \quad \bar{b} = -\tfrac{1}{2}\alpha_C, \quad C \in \mathcal{C}, \quad k \notin C, \tag{12.3.41}$$

using (11.3.47) to calculate $\sigma_1 t_\Lambda(b)$; and

$$\sigma(\alpha_k \otimes t_\Lambda(b)) = (-1)^{|C|/4+|(\{k_0\}+\{k\})\cap C|}\alpha_k \otimes t_\Lambda(b) \qquad (12.3.42)$$

for

$$b \in \hat{\Lambda}_1 \quad \text{with} \quad \bar{b} = \varepsilon_C(\tfrac{1}{4}\alpha_\Omega - \alpha_{k_0}), \quad C \in \mathcal{C}, \qquad (12.3.43)$$

by (11.4.47) [cf. (12.2.41), (12.2.42)].

In [Griess 3], the algebra and bilinear form that we have denoted $\mathcal{B}$ and $\langle \cdot, \cdot \rangle$, or more precisely, a "deformation" of $\mathcal{B}$ in which multiplication by ω is zero, was described, and the automorphism σ (or rather, the automorphism $\tau\sigma\tau^{-1}$, in our notation) of $\mathcal{B}$ preserving $\langle \cdot, \cdot \rangle$ was constructed. Together with the natural action of the group C on $\mathcal{B}$, Griess used this automorphism to generate the group which in our notation is the restriction of M to $\mathcal{B}$. This portion of [Griess 3] is equivalent to part (d) of Theorem 12.3.1. Griess then showed that this group is a finite simple group with C as the centralizer of an involution—the Monster. This last part of the existence proof has been simplified in [Conway 9] and [Tits 6].

Thus by Theorem 12.3.1(h) the group that we have denoted M and called the Monster is indeed the finite simple group with the same name, and hence this group acts on $V^\natural$ as described above.

Also, Tits has shown ([Tits 4], [Tits 6]) that the Monster is the full automorphism group of the algebra $\mathcal{B}$:

$$M = \operatorname{Aut} \mathcal{B}. \qquad (12.3.44)$$

Combining this with Theorem 12.3.1(c) and (f), and with obvious analogues of (e) and (h), using (g), we find:

Theorem 12.3.4: *The Monster is precisely the automorphism group of the vertex operator algebra $V^\natural$, or equivalently, the group of grading-preserving linear automorphisms g of $V^\natural$ such that*

$$gY(v, z)g^{-1} = Y(gv, z) \quad \text{for} \quad v \in V^\natural, \qquad (12.3.45)$$

or equivalently, such that

$$g(u_n v) = (gu)_n(gv) \quad \text{for} \quad u, v \in V^\natural, \quad n \in \mathbb{Z}. \qquad (12.3.46)$$

The action of the group of such automorphisms on $\mathcal{B}$ is faithful.

Borcherds associated a family of vertex operators with the space V_L for an even lattice L (in the notation of Chapter 8) and announced some properties, which he axiomatized in a notion of "vertex algebra" (see

[Borcherds 3]). Using [FLM2], which included announcements of parts (a) and (d)–(h) of Theorem 12.3.1, he stated that the M-invariant action of $\hat{\mathfrak{B}}$ on $V^\natural$ could be extended to an M-invariant vertex algebra structure on $V^\natural$.

In Chapter 8, we have supplied a proof of Borcherds' theorem about V_L, which includes the commutator formula (8.6.31) and formulas (8.8.7) and (8.8.31) (in component form). We have also presented generalizations and analogues of the theorem, including Theorems 8.6.1, 8.8.9 and 8.8.23, and in Chapter 9, Theorems 9.3.1 and 9.5.3. Parts (b) and (c) of Theorem 12.3.1 prove Borcherds' statement about $V^\natural$, using Chapter 9. This statement can also be proved (in a less explicit form) using Chapter 8, Theorem 12.3.1(e) and properties of the Monster. In Section 8.10, we have discussed the notion of "vertex operator algebra," a slight modification of Borcherds' concept of "vertex algebra."

The association of a family of vertex operators to the elements of spaces such as V_L is familiar in string theory and two-dimensional conformal quantum field theory, as we have indicated in the Introduction.

12.4. The M-Invariant $\mathbb{Q}$-Form $V^\natural_{\mathbb{Q}}$

Let $\mathbb{F}_0$ be a subfield of $\mathbb{F}$, U a vector space over $\mathbb{F}$. We recall that an $\mathbb{F}_0$-subspace U_0 of U is called an $\mathbb{F}_0$-*form* if the canonical map

$$\begin{aligned} \mathbb{F} \otimes_{\mathbb{F}_0} U_0 &\to U \\ \alpha \otimes u &\mapsto \alpha u \end{aligned} \tag{12.4.1}$$

is an $\mathbb{F}$-isomorphism. Equivalently, an $\mathbb{F}_0$-basis of U_0 is an $\mathbb{F}$-basis of U. We shall define a $\mathbb{Q}$-form $V^\natural_{\mathbb{Q}}$ in $V^\natural$, invariant under M and the components of the vertex operators $Y(v, z)$ for $v \in V^\natural_{\mathbb{Q}}$.

First define $\mathbb{Q}$-forms in $\mathfrak{h}$, $\hat{\mathfrak{h}}_Z$, $S(\hat{\mathfrak{h}}_Z^-)$ for $Z = \mathbb{Z}, \mathbb{Z} + \frac{1}{2}$ by setting

$$\begin{aligned} \mathfrak{h}_{\mathbb{Q}} &= \coprod_{k \in \Omega} \mathbb{Q}\alpha_k = \mathbb{Q} \otimes_{\mathbb{Z}} \Lambda, \\ \hat{\mathfrak{h}}_{Z,\mathbb{Q}} &= \mathfrak{h}_{\mathbb{Q}} \otimes \coprod_{n \in Z} \mathbb{Q}t^n \oplus \mathbb{Q}c, \\ \hat{\mathfrak{h}}^-_{Z,\mathbb{Q}} &= \hat{\mathfrak{h}}^-_Z \cap \hat{\mathfrak{h}}_{Z,\mathbb{Q}}, \\ S(\hat{\mathfrak{h}}^-_Z)_{\mathbb{Q}} &= S(\hat{\mathfrak{h}}^-_{Z,\mathbb{Q}}). \end{aligned} \tag{12.4.2}$$

In $\mathbb{F}\{\Lambda\}$ we take

$$\mathbb{Q}\{\Lambda\} = \sum_{a \in \hat{\Lambda}^{(2)}} \mathbb{Q}\iota(a) \tag{12.4.3}$$

as our $\mathbb{Q}$-form. Then $\mathbb{Q}\{\Lambda\}$ is invariant under the action of C_0 since $\hat{\Lambda}^{(2)}$ is invariant (recall Remark 12.1.19). In T_Λ we define the $\mathbb{Q}$-form

$$T_{\Lambda,\mathbb{Q}} = \sum_{b \in \hat{\Lambda}^{(2)}} \mathbb{Q}(t(b) + t(b_{k_0}^2 b)) \tag{12.4.4}$$

[cf. (12.3.17), (12.3.34)]. This is invariant under the action of $\hat{\Lambda}^{(2)}/K_\Lambda$ since $b_{k_0}^2 \in \operatorname{Cent} \hat{\Lambda}^{(2)}$. Formulas (10.4.30), (10.4.33) then show that $T_{\Lambda,\mathbb{Q}}$ is invariant under the action of C_T. Now set

$$\begin{aligned} V_{\Lambda,\mathbb{Q}} &= S(\hat{\mathfrak{h}}_{\mathbb{Z}}^-)_\mathbb{Q} \otimes_\mathbb{Q} \mathbb{Q}\{\Lambda\}, \\ V'_{\Lambda,\mathbb{Q}} &= S(\hat{\mathfrak{h}}_{\mathbb{Z}+1/2}^-)_\mathbb{Q} \otimes_\mathbb{Q} T_{\Lambda,\mathbb{Q}}, \\ W_{\Lambda,\mathbb{Q}} &= V_{\Lambda,\mathbb{Q}} \oplus V'_{\Lambda,\mathbb{Q}}; \end{aligned} \tag{12.4.5}$$

these are $\mathbb{Q}$-forms in V_Λ, V'_Λ, W_Λ. By the remarks above, $W_{\Lambda,\mathbb{Q}}$ is then invariant under $\hat{C}$, and in particular under $\hat{\theta}_0\hat{\theta}$ [recall (10.4.41) and Remark 10.4.3]. Hence

$$V^\natural_\mathbb{Q} = V^\natural \cap W_{\Lambda,\mathbb{Q}} \tag{12.4.6}$$

is a $\mathbb{Q}$-form in $V^\natural$, invariant under C.

Proposition 12.4.1: *The $\mathbb{Q}$-form $V^\natural_\mathbb{Q}$ in $V^\natural$ is invariant under the action of σ.*

Proof: Recall from Proposition 12.2.6 that all of the spaces $V^+_{\Lambda_j}$, $V'^+_{\Lambda_j}$, $j = 0, 1$, are irreducible under the action of the set of operators

$$\{x_v(n) \mid v \in (V^+_{\Lambda_0} \cap V^\natural_\mathbb{Q})_{-1},\ n \in \mathbb{Q}\}. \tag{12.4.7}$$

Since

$$\begin{aligned} (V^+_{\Lambda_0} \cap V^\natural_\mathbb{Q})_{-1} &= \mathbb{Q}\text{-span}\{\alpha_k(-1)\alpha_l(-1) \mid k, l \in \Omega\} \\ &\cup \{\iota(a_k a_l)^+, \iota(a_k a_l^{-1})^+ \mid k \neq l\} \\ &\cup \{\iota(b)^+ \mid b \in \hat{\Lambda}_0^{(2)}, \bar{b} = \tfrac{1}{2}\varepsilon_S \alpha_C, C \in \mathcal{C}, |C| = 8, S \subset C, |S| \in 2\mathbb{Z}\} \end{aligned}$$

is invariant under σ by (12.3.35)–(12.3.39), the set of operators (12.4.7) is normalized by σ. It is therefore enough to show that σ maps a nonzero vector in each of the $\mathbb{Q}$-forms $V^+_{\Lambda_j} \cap V^\natural_\mathbb{Q}$, $V'^+_{\Lambda_j} \cap V^\natural_\mathbb{Q}$, $j = 0, 1$, into $V^\natural_\mathbb{Q}$. Certainly

$$\sigma\iota(1) = \iota(1) \in V^\natural_\mathbb{Q}.$$

For $k \in \Omega$, consider

$$\alpha_k(-\tfrac{1}{2}) \otimes (t(1) + t(b_{k_0}^2)) \in V'^+_{\Lambda_0} \cap V^\natural_\mathbb{Q}.$$

Then (12.3.40) gives:

$$\sigma(\alpha_k(-\tfrac{1}{2}) \otimes (t(1) + t(b_{k_0}^2))) = \iota(b_k) + \iota(z_0(b_k)) \in V^+_{\Lambda_1} \cap V^\natural_\mathbb{Q}. \tag{12.4.8}$$

Finally, by (12.3.42),

$$\sigma(\alpha_k(-\tfrac{1}{2}) \otimes (t(b_{k_0}) + t(b_{k_0}^2 b_{k_0})))$$
$$= \alpha_k(-\tfrac{1}{2}) \otimes (t(b_{k_0}) + t(b_{k_0}^2 b_{k_0})) \in V_{\Lambda_1}^{\prime +} \cap V_{\mathbb{Q}}^{\natural}. \quad \blacksquare \tag{12.4.9}$$

Corollary 12.4.2: *The* $\mathbb{Q}$*-form* $V_{\mathbb{Q}}^{\natural}$ *in* $V^{\natural}$ *is invariant under the action of* M.

Remark 12.4.3: The constants in the map h [recall (12.2.49)] were choosen so that Proposition 12.4.1 would hold.

From the definitions in Chapters 8 and 9 and (12.3.8), (12.3.9), we have:

Proposition 12.4.4: *For* $v \in V_{\mathbb{Q}}^{\natural}$ *and* $n \in \mathbb{Z}$, *the components* v_n *of* $Y(v, z)$ *preserve* $V_{\mathbb{Q}}^{\natural}$:

$$v_n V_{\mathbb{Q}}^{\natural} \subset V_{\mathbb{Q}}^{\natural}. \tag{12.4.10}$$

Remark 12.4.5: These considerations justify Remark 12.3.2; when Theorem 12.3.1 is proved, it will hold over any field of characteristic zero.

12.5. The *M*-Invariant Positive Definite Hermitian Form

In this section we introduce a hermitian form on $W_{\mathfrak{L}}$ whose restriction to $V^{\natural}$ is invariant under M; throughout this section we assume that $\mathbb{F} = \mathbb{C}$.

Recall from Section 1.8 that as a consequence of Proposition 1.8.2 there is a unique hermitian form (linear in the second variable)

$$(\cdot, \cdot): S(\hat{\mathfrak{h}}_{\bar{Z}}) \times S(\hat{\mathfrak{h}}_{\bar{Z}}) \to \mathbb{C} \tag{12.5.1}$$

such that (d being the degree operator)

$$(d \cdot v, w) = (v, d \cdot w) \tag{12.5.2}$$

$$((h \otimes t^n) \cdot v, w) = (v, (h \otimes t^{-n}) \cdot w) \tag{12.5.3}$$

$$(1, 1) = 1 \tag{12.5.4}$$

for $h \in \mathfrak{h}_{\mathbb{R}} = \Lambda \otimes_{\mathbb{Z}} \mathbb{R}$, $n \in Z = \mathbb{Z}$ or $\mathbb{Z} + \frac{1}{2}$, $v, w \in S(\hat{\mathfrak{h}}_{\bar{Z}})$. Then the form $(\cdot, \cdot)$ is positive definite on $S(\hat{\mathfrak{h}}_{\bar{Z}})$, as we see by choosing an orthonormal basis $\{h_i\}$ of $\mathfrak{h}_{\mathbb{R}}$ with respect to the form $\langle \cdot, \cdot \rangle$, and by considering the basis of $S(\hat{\mathfrak{h}}_{\bar{Z}})$ consisting of the monomials in the elements $h_i \otimes t^{-n}$, $n > 0$.

Define a positive definite hermitian form

$$(\cdot\,,\cdot\,): \mathbb{C}\{L\} \times \mathbb{C}\{L\} \to \mathbb{C} \tag{12.5.5}$$

by the condition

$$(\iota(a), \iota(b)) = \begin{cases} 0 & \text{if} \quad \bar{a} \neq \bar{b} \\ 1 & \text{if} \quad a = b. \end{cases} \tag{12.5.6}$$

This is possible since

$$(\iota(\kappa a), \iota(\kappa a)) = (i\iota(a), i\iota(a)) = (\iota(a), \iota(a))$$

for $a \in \hat{\mathfrak{L}}$. We then have

$$(a \cdot v, a \cdot w) = (v, w), \tag{12.5.7}$$

$$(h(0)v, w) = (v, h(0)w) \tag{12.5.8}$$

for $a \in \hat{\mathfrak{L}}$, $h \in \mathfrak{h}_{\mathbb{R}}$, $v, w \in \mathbb{C}\{\mathfrak{L}\}$.

We also define a positive definite hermitian form

$$(\cdot\,,\cdot\,): T_{\mathfrak{L}} \times T_{\mathfrak{L}} \to \mathbb{C} \tag{12.5.9}$$

by the condition

$$(t(a), t(b)) = \begin{cases} 0 & \text{if} \quad a\hat{Q} \neq b\hat{Q} \\ 1 & \text{if} \quad a = b \end{cases} \tag{12.5.10}$$

[cf. (12.3.17)]. This is possible since

$$(t(aq), t(aq)) = (\psi(q)t(a), \psi(q)t(a)) = (t(a), t(a))$$

for $a \in \hat{\mathfrak{L}}$, $q \in \hat{Q}$. This form is also $\hat{\mathfrak{L}}$-invariant, i.e.,

$$(a \cdot v, a \cdot w) = (v, w) \tag{12.5.11}$$

for $a \in \hat{\mathfrak{L}}$, $v, w \in T_{\mathfrak{L}}$.

Using these forms we define positive definite hermitian forms $(\cdot\,,\cdot\,)$ on $V_{\mathfrak{L}}$ (respectively, $V'_{\mathfrak{L}}$) by:

$$(v_1 \otimes w_1, v_2 \otimes w_2) = (v_1, v_2)(w_1, w_2) \tag{12.5.12}$$

for $v_1, v_2 \in S(\hat{\mathfrak{h}}_{\mathbb{Z}}^-)$ [respectively, $S(\hat{\mathfrak{h}}_{\mathbb{Z}+1/2}^-)$], $w_1, w_2 \in \mathbb{C}\{\mathfrak{L}\}$ (respectively, $T_{\mathfrak{L}}$). Taking an orthogonal direct sum of these gives us a positive definite hermitian form $(\cdot\,,\cdot\,)$ on $W_{\mathfrak{L}}$, and by restriction we have such forms on W_L, W_Λ and $V^\natural$.

Note that the form $(\cdot\,,\cdot\,)$ on $V^\natural$ restricts to a rational-valued positive definite symmetric bilinear form on the $\mathbb{Q}$-form $V_{\mathbb{Q}}^\natural$ constructed in Section 12.4.

We denote by T^* the adjoint of an operator T whenever it exists, i.e.,

$$(Tv, w) = (v, T^*w) \tag{12.5.13}$$

with v, w in any of the spaces above.

Proposition 12.5.1: *On $W_{\mathfrak{L}}$, we have*

$$X(a, z)^* = X(a^{-1}, z^{-1}) \tag{12.5.14}$$

for $a \in \hat{\mathfrak{L}}$, i.e.,

$$x_a(n)^* = x_{a^{-1}}(-n) \tag{12.5.15}$$

for $a \in \hat{\mathfrak{L}}$, $n \in \mathbb{Q}$.

Proof: Consider first the untwisted construction. Denote also by $(\cdot, \cdot)$ the obvious hermitian forms

$$(\cdot, \cdot): V_{\mathfrak{L}}\{z\} \times V_{\mathfrak{L}}\{z\}_{\text{fin}} \to \mathbb{C}\{z\},$$

$$(\cdot, \cdot): V_{\mathfrak{L}}\{z\}_{\text{fin}} \times V_{\mathfrak{L}}\{z\} \to \mathbb{C}\{z\},$$

where $V_{\mathfrak{L}}\{z\}_{\text{fin}}$ denotes the space of finite sums $\sum_{n \in \mathbb{C}} v_n z^n$, $v_n \in V_{\mathfrak{L}}$. We then have, for $v, w \in V_{\mathfrak{L}}$, $a \in \hat{\mathfrak{L}}$, $\bar{a} = \alpha$,

$$\begin{aligned}
(X(a, z)v, w) &= (E^-(-\alpha, z)E^+(-\alpha, z)az^{\alpha + \langle \alpha, \alpha \rangle/2}v, w) \\
&= (E^+(-\alpha, z)az^{\alpha + \langle \alpha, \alpha \rangle/2}v, E^+(\alpha, z^{-1})w) \\
&= (az^{\alpha + \langle \alpha, \alpha \rangle/2}v, E^-(\alpha, z^{-1})E^+(\alpha, z^{-1})w) \\
&= (z^{\alpha + \langle \alpha, \alpha \rangle/2}v, a^{-1}E^-(\alpha, z^{-1})E^+(\alpha, z^{-1})w) \\
&= (v, z^{\alpha + \langle \alpha, \alpha \rangle/2}a^{-1}E^-(\alpha, z^{-1})E^+(\alpha, z^{-1})w) \\
&= (v, E^-(\alpha, z^{-1})E^+(\alpha, z^{-1})a^{-1}(z^{-1})^{-\alpha + \langle \alpha, \alpha \rangle/2}w) \\
&= (v, X(a^{-1}, z^{-1})w).
\end{aligned} \tag{12.5.16}$$

Similarly, in the twisted construction, for $v, w \in V'_{\mathfrak{L}}$, $a \in \hat{\mathfrak{L}}$, $\bar{a} = \alpha$,

$$\begin{aligned}
(X(a, z)v, w) &= (2^{-\langle \alpha, \alpha \rangle}E^-(-\alpha, z)E^+(-\alpha, z)av, w) \\
&= 2^{-\langle \alpha, \alpha \rangle}(E^+(-\alpha, z)av, E^+(\alpha, z^{-1})w) \\
&= 2^{-\langle \alpha, \alpha \rangle}(av, E^-(\alpha, z^{-1})E^+(\alpha, z^{-1})w) \\
&= (v, 2^{-\langle \alpha, \alpha \rangle}a^{-1}E^-(\alpha, z^{-1})E^+(\alpha, z^{-1})w) \\
&= (v, X(a^{-1}, z^{-1})w).
\end{aligned} \tag{12.5.17}$$

Proposition 12.5.2: *The form $(\cdot,\cdot)$ on W_Λ is invariant under the action of $\hat{C}$.*

Proof: Let $k = (g, g_T) \in \hat{C}$. We prove that

$$(kv, kw) = (v, w) \tag{12.5.18}$$

for $v, w \in (V_\Lambda)_{-n}$ by induction on n. Since

$$\begin{aligned}(kh(-m)x, kw) &= ((\bar{g}h)(-m)kx, kw)\\ &= (kx, (\bar{g}h)(m)kw)\\ &= (kx, kh(m)w)\\ &= (x, h(m)w) = (h(-m)x, w)\end{aligned} \tag{12.5.19}$$

for $h \in \mathfrak{h}_{\mathbb{R}}$, $m \in \mathbb{Z}$, $m > 0$, $x \in (V_\Lambda)_{m-n}$, we may assume that

$$v = \iota(a), \quad w = \iota(a')$$

with $a, a' \in \hat{\Lambda}$. Then (12.5.18) is clearly satisfied. Now consider the twisted space. By the same argument it is enough to show that the form on T_Λ is invariant under C_T. Since T_Λ is irreducible, a $\hat{\Lambda}$-invariant hermitian form on T_Λ is unique up to scalar multiple. Since C_* normalizes $\hat{\Lambda}$, C_* transforms $(\cdot,\cdot)$ into $\mathbb{R}^\times(\cdot,\cdot)$. Hence

$$C_T = (C_*, C_*)$$

leaves $(\cdot,\cdot)$ invariant. ∎

Proposition 12.5.3: *The form $(\cdot,\cdot)$ on $W_\mathfrak{L}$ is invariant under σ.*

Proof: Let

$$v, w \in (W_j)_{1-(n/2)},$$

$j = 0, 1, 2, 3$, $n \in \mathbb{Z}$, $n \geq 0$. We prove that

$$(\sigma v, \sigma w) = (v, w) \tag{12.5.20}$$

by induction on n. We have

$$\begin{aligned}(\sigma\alpha_k(-m)x, \sigma w) &= (x^+_{a_k}(-m)\sigma x, \sigma w)\\ &= (\sigma x, x^+_{a_k}(m)\sigma w) = (\sigma x, \sigma\alpha_k(m)w)\\ &= (x, \alpha_k(m)w) = (\alpha_k(-m)x, w)\end{aligned}$$

for $k \in \Omega$, $m \in \frac{1}{2}\mathbb{Z}$, $m > 0$, $x \in (W_j)_{1-(n/2)+m}$. Hence we may assume that either

$$v = \iota(b_1), \quad w = \iota(b_2)$$

or

$$v = t(b_1), \quad w = t(b_2)$$

with $b_1, b_2 \in \hat{\mathfrak{L}}$. In the second case, (12.5.20) follows from (11.3.47), (11.4.47). In the first case, if

$$\bar{b}_1 \notin \{0, \tfrac{1}{2}\varepsilon_S\alpha_C, \tfrac{1}{4}\varepsilon_C\alpha_\Omega \mid C \in \mathfrak{C}_1,\ S \subset C\},$$

then there exists $k \in \Omega$ such that

$$\langle \overline{a_k^{\pm 1}b_1}, \overline{a_k^{\pm 1}b_1} \rangle < \langle \bar{b}_1, \bar{b}_1 \rangle.$$

Hence with appropriate $m \in \frac{1}{2}\mathbb{Z}$, $m > 0$, we have

$$\iota(b_1) = x_{a_k^{\mp 1}}(-m)\iota(a_k^{\pm 1}b_1),$$

so that by (11.5.6),

$$\begin{aligned}
(\sigma\iota(b_1), \sigma w) &= (\sigma x_{a_k^{\mp 1}}(-m)\iota(a_k^{\pm 1}b_1), \sigma w) \\
&= \tfrac{1}{2}((\alpha_k(-m) \pm x_{a_k}^-(-m))\sigma\iota(a_k^{\pm 1}b_1), \sigma w) \\
&= \tfrac{1}{2}(\sigma\iota(a_k^{\pm 1}b_1), (\alpha_k(m) \mp x_{a_k}^-(m))\sigma w) \\
&= (\sigma\iota(a_k^{\pm 1}b_1), \sigma x_{a_k^{\pm 1}}(m)w) \\
&= (\iota(a_k^{\pm 1}b_1), x_{a_k^{\pm 1}}(m)w) \\
&= (\iota(b_1), w).
\end{aligned} \tag{12.5.21}$$

We may therefore assume that

$$\bar{b}_1, \bar{b}_2 \in \{0, \tfrac{1}{2}\varepsilon_S\alpha_C, \tfrac{1}{4}\varepsilon_C\alpha_\Omega \mid C \in \mathfrak{C}_1,\ S \subset C\},$$

in which case (12.5.20) follows from (11.5.9), (11.5.7) and (11.3.47); in case

$$\bar{b}_1 \in \{\tfrac{1}{2}\varepsilon_S\alpha_C \mid C \in \mathfrak{C}_1,\ S \subset C\},$$

we have

$$\iota(b_1) = x_{b_1}(-m)\iota(1)$$

for suitable $m \in \frac{1}{2}\mathbb{Z}$, $m > 0$, so that

$$
\begin{aligned}
(\sigma\iota(b_1), \sigma w) &= (-2)^{-|C|/2}(-1)^{|C\cap\{k_0\}|}\left(\sum_{T\subset C}(-1)^{|S\cap T|}x_{a_{S,T}b_1}(-m)\iota(1), \sigma w\right)\\
&= (-2)^{-|C|/2}(-1)^{|C\cap\{k_0\}|}\left(\iota(1), \sum_{T\subset C}(-1)^{|S\cap T|}x_{b_{\bar{1}}^{-1}a_{T,S}}(m)\sigma w\right)\\
&= (\iota(1), \sigma x_{b_{\bar{1}}^{-1}}(m)w)\\
&= (\iota(1), x_{b_{\bar{1}}^{-1}}(m)w)\\
&= (x_{b_1}(-m)\iota(1), w)\\
&= (\iota(b_1), w) \qquad (12.5.22)
\end{aligned}
$$

since

$$
\begin{aligned}
\sigma x_{b_{\bar{1}}^{-1}}(m)\sigma^{-1} &= (-2)^{-|C|/2}(-1)^{|C\cap\{k_0\}|}\sum_{T\subset C}(-1)^{|(C\setminus S)\cap(C\setminus T)|}x_{a_{T,S}b_{\bar{1}}^{-1}}(m)\\
&= (-2)^{-|C|/2}(-1)^{|C\cap\{k_0\}|}\sum_{T\subset C}(-1)^{|S\cap T|}x_{b_{\bar{1}}^{-1}a_{T,S}}(m). \qquad \blacksquare
\end{aligned}
$$

Corollary 12.5.4: *The positive definite hermitian form* $(\cdot,\cdot)$ *on* $V^{\natural}$ *is invariant under M. In particular, the* $\mathbb{Q}$*-valued positive definite symmetric bilinear form* $(\cdot,\cdot)$ *on* $V_{\mathbb{Q}}^{\natural}$ *is invariant under M.*

13 Completion of the Proof

In this final chapter we complete the proof of the main result, Theorem 12.3.1, by establishing formula (12.3.22)—the Jacobi identity for the vertex operator algebra $V^\natural$—and Theorem 12.3.1(c)—the compatibility of the actions of the Monster and of the vertex operators on $V^\natural$. In order to motivate the steps we proceed by a series of reductions to progressively more technical lemmas. Together with structures based on the Leech lattice Λ and the Niemeier lattice $N(A_1^{24})$, which play a fundamental role in the formulation of our results (recall Chapter 12), analogous larger structures based on the Niemeier lattice $N(D_4^6)$ play a fundamental role in the proof. For each of the lattices Λ, $N(A_1^{24})$ and $N(D_4^6)$, we use a corresponding group closely related to a certain maximal 2-local subgroup of the Monster (cf. [Ronan–Smith], [Tits 5, 8]); the group corresponding to Λ is the group C, and we denote the other two by $\tilde{N}$ and $\tilde{H}$. These three groups are also closely related to the automorphism groups of corresponding vertex-operator-algebraic structures associated with the lattices Λ, L and $\mathfrak{L}$, where L is the sum of Λ and a copy of $N(A_1^{24})$ and $\mathfrak{L}$ is the sum of Λ and a copy of $N(D_4^6)$. In addition to material from the earlier chapters, in particular, Chapter 5, we quote a few constructions and results from elementary group theory. We also summarize and use some basic results from group cohomology theory.

In Section 13.1 we reduce the proof of the main theorem to two lemmas. One of them, Lemma 13.1.2, is then proved in Section 13.2 by working in the direct sum W_L of the untwisted and twisted spaces associated with the lattice L. Sections 13.3 and 13.4 reduce the remaining result, Lemma 13.1.6, to Lemma 13.4.7, a special fact about a particular finite group. This reduction is accomplished by working in the space $W_{\mathfrak{L}}$, analogous to and containing W_L. The groups $\tilde{N}$ and $\tilde{H}$ mentioned above are used in Sections 13.2 and 13.4, respectively, in connection with the vertex-operator-algebraic structures defined on W_L and $W_{\mathfrak{L}}$, respectively. In Section 13.5 we survey some basic results from group cohomology theory, taken from [Cartan–Eilenberg], [Grothendieck] and [Serre 2]. Finally, we prove Lemma 13.4.7 in Section 13.6 using the group cohomology results of the preceding section.

13.1. Reduction to Two Lemmas

In this section we reduce the proof of our main result, Theorem 12.3.1, to two lemmas which will be proved in Sections 13.2 and 13.3–13.6. Recall that we need only prove (12.3.22) and Theorem 12.3.1(c). We shall freely use notations from the earlier chapters, often without explicit reference.

Let G be a group and let H be a subset of G. We use the standard group theory notations

$$\mathbf{N}_G(H) = \{g \in G \mid gHg^{-1} = H\},$$
$$\mathbf{C}_G(H) = \{g \in G \mid gh = hg \quad \text{for} \quad h \in H\} \tag{13.1.1}$$

for the normalizer, respectively, centralizer, of H in G. Also, for subsets $H_1, H_2, \ldots$ of G and elements $x_1, x_2, \ldots$ of G,

$$\langle H_1, H_2, \ldots, x_1, x_2, \ldots \rangle \tag{13.1.2}$$

denotes the subgroup of G generated by $H_1, H_2, \ldots, x_1, x_2, \ldots$, as usual. For $x, y \in G$ set

$${}^x y = xyx^{-1}, \quad y^x = x^{-1}yx, \tag{13.1.3}$$

the left and right conjugations.

Throughout this section, unless otherwise indicated, all groups and operators shall be considered restricted to our subspace $V^\natural$ of $W_{\mathfrak{L}}$. For example, σ shall denote $\sigma|_{V^\natural}$.

Recall from (12.3.1) that the Monster M is defined by:

$$M = \langle C, \sigma \rangle \subset \operatorname{Aut} V^\natural. \tag{13.1.4}$$

Recall that we have defined homomorphisms

$$\nu: \hat{\Lambda}^{(2)}/K_\Lambda \to \hat{C}, \tag{13.1.5}$$

$$\pi_0: \hat{C} \to C \tag{13.1.6}$$

in (10.4.51), (10.4.46). Using the $^{(2)}$ notation of Remark 12.1.10, set

$$F = \pi_0 \nu(\widehat{2L}^{(2)}/K_\Lambda) \subset C, \tag{13.1.7}$$

$$z_1 = \pi_0 \nu(a_k^2 K_\Lambda) \in F, \tag{13.1.8}$$

where $k \in \Omega$. Then

$$F = \langle z, z_1 \rangle \simeq (\mathbb{Z}/2\mathbb{Z})^2. \tag{13.1.9}$$

Note that since $a_k^2 K_\Lambda = a_l^2 K_\Lambda$ for $k, l \in \Omega$, z_1 does not depend on the choice of $k \in \Omega$. Since the eigenspaces of z_1 on $V^\natural$ are:

$$\begin{aligned} &+1\text{-eigenspace: } V_{\Lambda_0}^+ \oplus V_{\Lambda_0}'^+, \\ &-1\text{-eigenspace: } V_{\Lambda_1}^+ \oplus V_{\Lambda_1}'^+, \end{aligned} \tag{13.1.10}$$

we have the conjugations

$$^\sigma z = z_1, \quad ^\sigma z_1 = z, \tag{13.1.11}$$

so that σ normalizes F. Note also that

$$^\tau z = z, \quad ^\tau z_1 = z z_1.$$

Set

$$N_0 = \mathbf{C}_C(F) = \mathbf{C}_C(z_1) \subset C. \tag{13.1.12}$$

Recall from (10.4.53) that the sequence

$$1 \to \pi_0 \nu(\hat{\Lambda}^{(2)}/K_\Lambda) \hookrightarrow C \xrightarrow{-} \mathrm{Co}_1 \to 1 \tag{13.1.13}$$

is exact, where $^-$ denotes the homomorphism on C induced by

$$\begin{aligned} ^-: \hat{C} &\to \mathrm{Co}_0 \\ (g, g_T) &\mapsto \bar{g}. \end{aligned} \tag{13.1.14}$$

Lemma 13.1.1: *There is an exact sequence*

$$1 \to \pi_0 \nu(\hat{\Lambda}_0^{(2)}/K_\Lambda) \hookrightarrow N_0 \xrightarrow{-} \mathrm{Aut}(\Lambda; \Lambda_0, \langle \cdot, \cdot \rangle)/\langle \pm 1 \rangle \to 1 \tag{13.1.15}$$

where

$$\mathrm{Aut}(\Lambda; \Lambda_0, \langle \cdot, \cdot \rangle) = \{g \in \mathrm{Co}_0 \mid g\Lambda_0 = \Lambda_0\}.$$

Proof: We have

$$\mathbf{C}_{\pi_0\nu(\hat{\Lambda}^{(2)}/K_\Lambda)}(F) = \pi_0\nu(\hat{\Lambda}_0^{(2)}/K_\Lambda)$$

since Λ_0 and L are dual lattices and

$$c_0|_{\Lambda\times\Lambda} = 2\langle\cdot,\cdot\rangle + 4\mathbb{Z}.$$

Let $(g, g_T) \in \hat{C}$. Then

$$^{\pi_0(g,g_T)}z_1 = \pi_0\nu(g(a_k^2)K_\Lambda). \tag{13.1.16}$$

Hence

$$\bar{N}_0 \subset \{g \in \mathrm{Co}_0 \mid g2L = 2L\}/\langle\pm 1\rangle = \mathrm{Aut}(\Lambda; \Lambda_0, \langle\cdot,\cdot\rangle)/\langle\pm 1\rangle.$$

Conversely, let $h \in C$ be the image of $(g, g_T) \in \hat{C}$ such that

$$\bar{g} \in \mathrm{Aut}(\Lambda; \Lambda_0, \langle\cdot,\cdot\rangle).$$

Then (13.1.16) shows that

$$^{h}z_1 = z_1 \quad \text{or} \quad zz_1,$$

so that either h or $h\tau$ is in N_0. Since

$$\overline{h\tau} = \bar{h},$$

the lemma follows. ∎

Note that since $\tau \in C$ and τ normalizes F, τ normalizes N_0. The following result will be proved in Section 13.2:

Lemma 13.1.2: *The automorphism σ normalizes N_0.*

Assuming Lemma 13.1.2, we define the group

$$N = \langle N_0, \sigma, \tau\rangle = N_0 \rtimes \langle\sigma, \tau\rangle, \tag{13.1.17}$$

the semidirect product of N_0 with the group $\mathcal{S}_3$ of (12.2.80).

Recall the sextet $\{T_j \mid 0 \le j \le 5\}$ from Section 12.1. Set

$$E = \pi_0\nu(\widehat{2\mathfrak{L}}^{(2)}/K_\Lambda) \subset C, \tag{13.1.18}$$

$$z_2 = \pi_0\nu(a_{T_j,\emptyset}K_\Lambda) \in E, \tag{13.1.19}$$

where $0 \le j \le 5$, and note that z_2 does not depend on the choice of j. Then

$$E = \langle z, z_1, z_2\rangle \simeq (\mathbb{Z}/2\mathbb{Z})^3. \tag{13.1.20}$$

Lemma 13.1.3: *The automorphism σ centralizes z_2. In particular, σ normalizes E.*

Proof: In this argument we consider σ and z_2 extended to $W_{\mathfrak{L}}$ by:

$$z_2 = \nu(a_{Tj,\emptyset});$$

see (12.2.43). Let $C_0 \in \mathfrak{C}$ be such that

$$|T_0 \cap C_0| \in 1 + 2\mathbb{Z}.$$

Then $\mathfrak{L}_{00}$ together with

$$\mathfrak{L}_{01} = \mathfrak{L}_{00} + \tfrac{1}{2}\alpha_{C_0} \subset \mathfrak{L}_0,$$

$$\mathfrak{L}_{10} = \mathfrak{L}_{00} + \tfrac{1}{4}\alpha_{\Omega},$$

$$\mathfrak{L}_{11} = \mathfrak{L}_{00} + \tfrac{1}{4}\varepsilon_{C_0}\alpha_{\Omega}$$

are the cosets of $\mathfrak{L}_{00}$ in $\mathfrak{L}$. The eigenspaces of z_2 on $W_{\mathfrak{L}}$ are then given by:

$$+1\text{-eigenspace: } V_{\mathfrak{L}_{00}} \oplus V_{\mathfrak{L}_{10}} \oplus V'_{\mathfrak{L}_{00}} \oplus V'_{\mathfrak{L}_{10}}$$

$$-1\text{-eigenspace: } V_{\mathfrak{L}_{01}} \oplus V_{\mathfrak{L}_{11}} \oplus V'_{\mathfrak{L}_{01}} \oplus V'_{\mathfrak{L}_{11}}.$$

By means of (11.3.47), (11.4.47) and an argument analogous to that of Propositions 12.2.6 and 12.2.7, this time using the vertex operators

$$\alpha_k(z),\quad Y(b, z)\quad \text{for}\quad b \in \hat{\mathfrak{L}}_{00},\quad \bar{b} = \pm\alpha_k \quad\text{or}\quad \bar{b} = \tfrac{1}{2}\varepsilon_S\alpha_C,\quad C \in \mathfrak{C}',$$

we find that σ interchanges the spaces

$$V_{\mathfrak{L}_{10}} \leftrightarrow V'_{\mathfrak{L}_{00}},\quad V_{\mathfrak{L}_{11}} \leftrightarrow V'_{\mathfrak{L}_{01}} \tag{13.1.21}$$

while it leaves each of

$$V_{\mathfrak{L}_{00}},\quad V_{\mathfrak{L}_{01}},\quad V'_{\mathfrak{L}_{10}},\quad V'_{\mathfrak{L}_{11}} \tag{13.1.22}$$

invariant. Hence σ and z_2 commute. ∎

We identify E with the space $\mathrm{Mat}_{1\times 3}(\mathbb{F}_2)$ of row vectors via

$$z = [1, 0, 0],\quad z_1 = [0, 1, 0],\quad z_2 = [0, 0, 1]. \tag{13.1.23}$$

This induces an identification of $\mathrm{Aut}\, E$ with the general linear group $GL(3, \mathbb{F}_2)$. For $g \in \mathbf{N}_M(E)$ we let $\bar{g}$ denote the matrix $C \in GL(3, \mathbb{F}_2)$ such that

$$AC = B$$

whenever $A, B \in \mathrm{Mat}_{1\times 3}(\mathbb{F}_2) = E$ such that

$$A^g = B.$$

There should be no confusion with the other notations $^-$.

Set

$$N_{00} = \mathbf{C}_C(E) \subset N_0, \tag{13.1.24}$$

$$Q_1 = \mathbf{N}_C(E), \tag{13.1.25}$$

$$P_1 = \left\{ \begin{bmatrix} 1 & 0 & 0 \\ * & * & * \\ * & * & * \end{bmatrix} \in \mathrm{GL}(3, \mathbb{F}_2) \right\}$$

$$= \text{stabilizer of } z \text{ in } \mathrm{GL}(3, \mathbb{F}_2). \tag{13.1.26}$$

Lemma 13.1.4: *We have an exact sequence*

$$1 \to N_{00} \hookrightarrow Q_1 \twoheadrightarrow P_1 \to 1. \tag{13.1.27}$$

Proof: From the structure of $\hat{\Lambda}^{(2)}/K_\Lambda$ we see that

$$(\pi_0 \nu(\hat{\Lambda}^{(2)}/K_\Lambda))^- = \left\{ \begin{bmatrix} 1 & 0 & 0 \\ * & 1 & 0 \\ * & 0 & 1 \end{bmatrix} \right\}.$$

Let $C_0 \in \mathcal{C}$ be such that

$$|T_0 \cap C_0| \in 1 + 2\mathbb{Z}$$

and let $(g, g_T) \in \hat{C}$ be such that $\bar{g} = \varepsilon_{C_0} \in \mathrm{Co}_0$. Then

$$\pi_0((g, g_T))^- = \begin{bmatrix} 1 & 0 & 0 \\ \alpha & 1 & 0 \\ \beta & 1 & 1 \end{bmatrix}$$

for some $\alpha, \beta \in \mathbb{F}_2$. Recall from Remark 10.2.12(b) the automorphism $\rho \in \mathrm{Co}_0$. Let $(h, h_T) \in \hat{C}$ be such that $\bar{h} = \rho$. Then

$$\pi_0((h, h_T))^- = \begin{bmatrix} 1 & 0 & 0 \\ \gamma & 1 & 1 \\ \delta & 0 & 1 \end{bmatrix}$$

for some $\gamma, \delta \in \mathbb{F}_2$. Since the matrices above generate P_1, we obtain $\bar{Q}_1 = P_1$. ∎

Set

$$Q_2 = \mathbf{N}_N(E). \tag{13.1.28}$$

Then $\sigma \in Q_2$ by Lemma 13.1.3, and in addition $\tau \in Q_2$. Since

$$Q_2 \cap N_0 = \mathbf{N}_{N_0}(E) = Q_1 \cap N_0, \tag{13.1.29}$$

we then have

$$Q_2 = \langle Q_1 \cap N_0, \sigma, \tau\rangle = (Q_1 \cap N_0) \rtimes \langle \sigma, \tau\rangle. \tag{13.1.30}$$

Note also that

$$\mathbf{C}_{Q_2}(E) = \mathbf{C}_{Q_2 \cap N_0}(E) = N_{00}. \tag{13.1.31}$$

Set

$$P_2 = \left\{\begin{bmatrix} * & * & 0 \\ * & * & 0 \\ * & * & 1 \end{bmatrix} \in \mathrm{GL}(3, \mathbb{F}_2)\right\}$$
$$= \text{stabilizer of } F \text{ in } \mathrm{GL}(3, \mathbb{F}_2). \tag{13.1.32}$$

Lemma 13.1.5: *We have an exact sequence*

$$1 \to N_{00} \hookrightarrow Q_2 \xrightarrow{-} P_2 \to 1. \tag{13.1.33}$$

Proof: By Lemma 13.1.4,

$$(Q_1 \cap N_0)^- = \left\{\begin{bmatrix} 1 & 0 & 0 \\ 0 & 1 & 0 \\ * & * & 1 \end{bmatrix}\right\}.$$

Thus since

$$\bar{\sigma} = \begin{bmatrix} 0 & 1 & 0 \\ 1 & 0 & 0 \\ 0 & 0 & 1 \end{bmatrix},$$
$$\bar{\tau} = \begin{bmatrix} 1 & 0 & 0 \\ 1 & 1 & 0 \\ 0 & 0 & 1 \end{bmatrix}, \tag{13.1.34}$$

the proof is complete. ∎

Define

$$H = \langle Q_1, Q_2\rangle = \langle Q_1, \sigma\rangle. \tag{13.1.35}$$

The following result will be proved in Sections 13.3–13.6:

Lemma 13.1.6: *The sequence*

$$1 \to N_{00} \hookrightarrow H \xrightarrow{-} \mathrm{GL}(3, \mathbb{F}_2) \to 1 \tag{13.1.36}$$

is exact, i.e.,

$$N_{00} = \mathbf{C}_H(E). \tag{13.1.37}$$

Corollary 13.1.7: *We have*

$$H = Q_2 Q_1 Q_2. \tag{13.1.38}$$

Proof: By Lemmas 13.1.4, 13.1.5, 13.1.6, we need only the relation

$$\mathrm{GL}(3, \mathbb{F}_2) = P_2 P_1 P_2.$$

But this follows from the relation

$$P_2 P_1 \begin{bmatrix} 0 \\ 0 \\ 1 \end{bmatrix} = \mathrm{Mat}_{3\times 1}(\mathbb{F}_2)\backslash\{0\},$$

together with the fact that P_2 is the stabilizer of $\begin{bmatrix} 0 \\ 0 \\ 1 \end{bmatrix}$ in $\mathrm{GL}(3, \mathbb{F}_2)$. ∎

We shall identify the group

$$E^* = \mathrm{Hom}(E, \mathbb{F}^\times) \tag{13.1.39}$$

of characters of E with the space $\mathrm{Mat}_{3\times 1}(\mathbb{F}_2)$ of column vectors in such a way that if $\lambda \in E^*$, $x \in E$ are identified as $\lambda = B \in \mathrm{Mat}_{3\times 1}(\mathbb{F}_2)$, $x = A \in \mathrm{Mat}_{1\times 3}(\mathbb{F}_2)$ then

$$\lambda(x) = (-1)^{AB}.$$

We also identify $F = \mathrm{Mat}_{1\times 2}(\mathbb{F}_2)$ via:

$$z = [1, 0], \quad z_1 = [0, 1] \tag{13.1.40}$$

and we have an induced identification $F^* = \mathrm{Mat}_{2\times 1}(\mathbb{F}_2)$ of characters of F. For $\lambda \in E^*$ (respectively, F^*) we set

$$V_\lambda^\natural = \{v \in V^\natural \mid gv = \lambda(g)v \quad \text{for} \quad g \in E \text{ (respectively, } F)\}. \tag{13.1.41}$$

Then

$$V^\natural = \coprod_{\lambda \in E^*(\text{resp.}, F^*)} V_\lambda^\natural. \tag{13.1.42}$$

We now prove Theorem 12.3.1(c). The proof of Lemma 13.1.4 shows that there exists $\pi_0((h, h_T)) \in Q_1$ such that $\bar{h} = \rho$. Since by (10.2.68) ρ and

$\mathrm{Aut}(\Lambda; \Lambda_0, \langle \cdot, \cdot \rangle)$ generate Co_0 we have

$$C = \langle N_0, Q_1 \rangle = \langle C \cap N, C \cap H \rangle, \tag{13.1.43}$$

using Lemma 13.1.1 and the fact that $\tau \in Q_1$. Hence

$$\langle N, H \rangle = \langle C, N, H \rangle = \langle C, \sigma \rangle = M, \tag{13.1.44}$$

and so it is enough to verify (12.3.25) for $g \in N \cup H$.

Corollary 10.4.8 shows that

$$\text{(12.3.25)} \quad \text{holds for} \quad g \in C, \quad v \in V_\Lambda^{\theta_0}, \tag{13.1.45}$$

so that in particular,

$$\text{(12.3.25)} \quad \text{holds for} \quad g \in N_0\langle \tau \rangle, \quad v \in V_{\Lambda_0}^+ \cup V_{\Lambda_1}^+. \tag{13.1.46}$$

From (12.2.53), (12.2.54), (12.2.79) and the definitions (12.3.8), (12.3.9), it follows that

$$\text{(12.3.25)} \quad \text{holds for} \quad g = \sigma, \quad v \in V^\natural. \tag{13.1.47}$$

Using Lemma 13.1.2 it then follows that

$$\text{(12.3.25)} \quad \text{holds for} \quad g \in N = N_0\langle \sigma, \tau \rangle, \quad v \in V^\natural. \tag{13.1.48}$$

We turn next to the case $g \in H$. Lemma 13.1.3 and (13.1.25) imply that $H = \langle Q_1, \sigma \rangle$ leaves $V^\natural_{\left[\begin{smallmatrix}0\\0\\0\end{smallmatrix}\right]}$ invariant. From (13.1.45) and (13.1.47) we see that

$$\text{(12.3.25)} \quad \text{holds for} \quad g \in H, \quad v \in V^\natural_{\left[\begin{smallmatrix}0\\0\\0\end{smallmatrix}\right]}. \tag{13.1.49}$$

Since P_2 stabilizes $\begin{bmatrix}0\\0\\1\end{bmatrix}$ and $Q_2 \subset N$ we have

$$Q_2 V^\natural_{\left[\begin{smallmatrix}0\\0\\1\end{smallmatrix}\right]} = V^\natural_{\left[\begin{smallmatrix}0\\0\\1\end{smallmatrix}\right]}, \tag{13.1.50}$$

$$\text{(12.3.25)} \quad \text{holds for} \quad g \in Q_2, \quad v \in V^\natural_{\left[\begin{smallmatrix}0\\0\\1\end{smallmatrix}\right]}. \tag{13.1.51}$$

Lemma 13.1.4 implies that

$$Q_1 V^\natural_{\left[\begin{smallmatrix}0\\0\\1\end{smallmatrix}\right]} = V^\natural_{\left[\begin{smallmatrix}0\\0\\1\end{smallmatrix}\right]} \oplus V^\natural_{\left[\begin{smallmatrix}0\\1\\0\end{smallmatrix}\right]} \oplus V^\natural_{\left[\begin{smallmatrix}0\\1\\1\end{smallmatrix}\right]}. \tag{13.1.52}$$

Since $Q_1 \subset C$, (13.1.45) shows that

$$\text{(12.3.25)} \quad \text{holds for} \quad g \in Q_1 Q_2, \quad v \in V^\natural_{\left[\begin{smallmatrix}0\\0\\1\end{smallmatrix}\right]}. \tag{13.1.53}$$

Since $Q_2 \subset N$ it follows that

$$Q_2 Q_1 Q_2 V^{\natural}_{\left[\begin{smallmatrix}0\\0\\1\end{smallmatrix}\right]} = \coprod_{\lambda \in E^*\backslash\{1\}} V^{\natural}_{\lambda}, \tag{13.1.54}$$

$$\text{(12.3.25) holds for} \quad g \in Q_2 Q_1 Q_2, \quad v \in V^{\natural}_{\left[\begin{smallmatrix}0\\0\\1\end{smallmatrix}\right]}. \tag{13.1.55}$$

By Corollary 13.1.7, $Q_2 Q_1 Q_2 = H$ is a group, so that (13.1.49), (13.1.54), (13.1.55) imply:

$$\text{(12.3.25) holds for} \quad g \in H, \quad v \in V^{\natural}, \tag{13.1.56}$$

thus completing the proof of Theorem 12.3.1(c).

We shall now prove (12.3.22). By linearity we may assume that $u \in V^{\natural}_{\lambda}$, $v \in V^{\natural}_{\mu}$ for some $\lambda, \mu \in E^*$. Since $GL(3, \mathbb{F}_2)$ acts doubly transitively on $E^*\backslash\{1\}$, there exists $g \in GL(3, \mathbb{F}_2)$ such that

$$\{{}^g\lambda, {}^g\mu\} \subset \left\{\begin{bmatrix}0\\0\\0\end{bmatrix}, \begin{bmatrix}0\\0\\1\end{bmatrix}, \begin{bmatrix}0\\1\\0\end{bmatrix}\right\}. \tag{13.1.57}$$

Let $h \in H$ be such that $\bar{h} = g$. Then

$$\begin{aligned}
&z_0^{-1}\delta\left(\frac{z_1 - z_2}{z_0}\right)Y(u, z_1)Y(v, z_2) - z_0^{-1}\delta\left(\frac{z_2 - z_1}{-z_0}\right)Y(v, z_2)Y(u, z_1)\\
&= z_0^{-1}\delta\left(\frac{z_1 - z_2}{z_0}\right)h^{-1}Y(hu, z_1)Y(hv, z_2)h\\
&\quad - z_0^{-1}\delta\left(\frac{z_2 - z_1}{-z_0}\right)h^{-1}Y(hv, z_2)Y(hu, z_1)h\\
&= z_2^{-1}\delta\left(\frac{z_1 - z_0}{z_2}\right)h^{-1}Y(Y(hu, z_0)hv, z_2)h\\
&= z_2^{-1}\delta\left(\frac{z_1 - z_0}{z_2}\right)Y(h^{-1}Y(hu, z_0)hv, z_2)\\
&= z_2^{-1}\delta\left(\frac{z_1 - z_0}{z_2}\right)Y(Y(u, z_0)v, z_2)
\end{aligned} \tag{13.1.58}$$

by (10.3.46) and (13.1.56), proving (12.3.22).

This completes the proof of Theorem 12.3.1 modulo Lemmas 13.1.2 and 13.1.6.

13.2. Groups Acting on W_L: Proof of Lemma 13.1.2

In this section we prove Lemma 13.1.2 by constructing groups $\hat{N}_0$, $\tilde{N}$ of operators on W_L such that $\hat{N}_0$ is a normal subgroup of $\tilde{N}$, $\hat{N}_0$ extends the action of N_0 on $V^\natural$ and $\sigma \in \tilde{N}$. Most groups and operators in this section shall act on W_L.

Define

$$N_1 = \mathrm{Aut}(\hat{L}; \kappa, \langle \cdot, \cdot \rangle, K_L). \tag{13.2.1}$$

By Proposition 5.4.8, Proposition 12.1.6 and Remark 12.1.8, we have the exact sequence

$$1 \to \mathrm{Hom}(L/2\Lambda_0, \mathbb{Z}/4\mathbb{Z}) \xrightarrow{*} N_1 \xrightarrow{-} \mathrm{Aut}(L; 2\Lambda_0, \langle \cdot, \cdot \rangle, c_0|_{L\times L}, s_L, f_L) \to 1,$$

where the maps s_L, f_L, determined by

$$a^2 \in \kappa^{s_L(\bar{a})}\langle \kappa^2 \rangle \times K_L \quad \text{for} \quad a \in \hat{\Lambda}_0$$

$$b^4 \in \kappa^{f_L(\bar{b})} K_L \quad \text{for} \quad b \in \hat{L}$$

are given by:

$$s_L \equiv 0 \quad \text{on} \quad \Lambda_0$$

$$f_L(\beta) = 2\langle \beta, \beta \rangle + 4\mathbb{Z} \quad \text{for} \quad \beta \in L$$

(cf. Remark 12.1.14). Since L_0 and Λ are the only even unimodular sublattices of L, any isometry of L stabilizes L_0 and Λ. Moreover,

$$2\Lambda_0 = \mathrm{radical}(c_0|_{L\times L}). \tag{13.2.2}$$

Thus the exact sequence above becomes

$$1 \to \mathrm{Inn}\,\hat{L} \hookrightarrow N_1 \xrightarrow{-} \mathrm{Aut}(L; \langle \cdot, \cdot \rangle) \to 1 \tag{13.2.3}$$

(cf. Remark 5.4.7).

Denoting by G' the commutator subgroup of a group G and by int the map induced by conjugation, we have:

Lemma 13.2.1: *(a) The sequence*

$$1 \to \mathrm{int}(\hat{\Lambda}_0) \hookrightarrow N_1' \xrightarrow{-} \mathrm{Aut}(L; \langle \cdot, \cdot \rangle) \to 1 \tag{13.2.4}$$

is exact.

(b) We have

$$N_1' = \{g \in N_1 \mid g \quad \text{induces} \quad 1 \quad \text{on} \quad \widehat{2L}/K_L\}. \tag{13.2.5}$$

(c) We have

$$N_1'' = N_1'. \tag{13.2.6}$$

(d) The group N_1 acts faithfully on $\hat{L}/K_L$.

Proof: Set

$$N_2 = \{a \in N_1 \mid a \quad \text{induces} \quad 1 \quad \text{on} \quad \widehat{2L}/K_L\}.$$

Then

$$N_2 \cap \operatorname{Inn} \hat{L} = \operatorname{int}(\hat{\Lambda}_0).$$

Now for any $k \in \Omega$, the abelian group $\widehat{2L}/K_L$ is generated by the involutions $a_k^2 K_L$, $b_k^2 K_L$ together with κK_L. Using (12.1.60) and the fact that $\operatorname{Aut}(L; \langle \cdot, \cdot \rangle)$ stabilizes L_0 and Λ, we see that

$$g(a_k^2 K_L) \in a_k^2 \langle \kappa^2 \rangle K_L,$$

$$g(b_k^2 K_L) \in b_k^2 \langle \kappa^2 \rangle K_L$$

for any $g \in N_1$. Hence

$$\bar{N}_2 = \operatorname{Aut}(L; \langle \cdot, \cdot \rangle),$$

and

$$N_1/N_2 \simeq L/\Lambda_0$$

is abelian, so that

$$N_1' \subset N_2.$$

On the other hand, using (10.2.65) and (10.2.69) it is easy to see that

$$N_2' = N_2,$$

so that

$$N_1' = N_2$$

and parts (a), (b), (c) follow. Let $a \in N_1$ act trivially on $\hat{L}/K_L$. Then $\bar{a}$ induces 1 on $L/2\Lambda_0$. Since

$$\{\alpha \in \alpha_k + 2\Lambda_0 \mid \langle \alpha, \alpha \rangle = 2\} = \{\alpha_k\}$$

for any $k \in \Omega$, we then have $\bar{a} = 1$, so that $a \in \operatorname{Inn} \hat{L}$. But it is clear that $\operatorname{Inn} \hat{L}$ acts faithfully on $\hat{L}/K_L$. ∎

Set

$$N_* = (\operatorname{int})^{-1}(N_1) \subset \operatorname{Aut} T_L, \tag{13.2.7}$$

where int denotes the canonical homomorphism

$$\operatorname{int}\colon \mathbf{N}_{\operatorname{Aut} T_L}(\pi(\hat{L}/K_L)) \to \operatorname{Aut}(\hat{L}/K_L; \kappa), \tag{13.2.8}$$

so that we have the commutative diagram with exact rows

$$\begin{array}{ccccccc} 1 \to \mathbb{F}^\times & \hookrightarrow & N_* & \xrightarrow{\text{int}} & N_1 & \longrightarrow & 1 \\ \| & & \downarrow & & \downarrow & & \\ 1 \to \mathbb{F}^\times & \hookrightarrow & \mathbf{N}_{\operatorname{Aut} T_L}(\pi(\hat{L}/K_L)) & \xrightarrow{\text{int}} & \operatorname{Aut}(\hat{L}/K_L;\kappa) & \to & 1. \end{array} \tag{13.2.9}$$

Also set

$$N_T = N_*''. \tag{13.2.10}$$

Lemma 13.2.2: *(a) The groups N_*', N_T leave each subspace T_{Λ_j}, $0 \le j \le 3$, invariant.*

(b) We have

$$\pi(\hat{\Lambda}_0^{(2)}) \subset N_T. \tag{13.2.11}$$

(c) The sequence

$$1 \to \langle -1_{T_L} \rangle \hookrightarrow N_T \xrightarrow{\text{int}} N_1' \to 1 \tag{13.2.12}$$

is exact.

(d) We have

$$N_T' = N_T. \tag{13.2.13}$$

Proof: Since the subspaces T_{Λ_j}, $0 \le j \le 3$, are the simultaneous eigenspaces of $\widehat{2L}/K_L$, part (a) follows from Lemma 13.2.1(b). Now the commutator group $(\pi(\hat{L}), N_*)$ is the group generated by the operators $\pi(g(a)a^{-1})$ for $g \in N_1$, $a \in \hat{L}$. By Lemma 13.2.1(a) and (10.2.69),

$$\pi(\hat{\Lambda}_0) = (\pi(\hat{L}), N_*) \subset N_*'.$$

Since

$$\hat{\Lambda}_0^{(2)} = \hat{\Lambda}_0 \cap \hat{\Lambda}^{(2)} = \{a \in \hat{L} \mid a^2\kappa^{\langle \bar{a}, \bar{a} \rangle} \in K_L\},$$

$\hat{\Lambda}_0^{(2)}$ is invariant under N_1. Thus Lemma 13.2.1(a) and (10.2.69) imply that

$$\pi(\hat{\Lambda}_0^{(2)}) = (\pi(\hat{\Lambda}_0^{(2)}), N_*') = (\pi(\hat{\Lambda}_0), N_*') \subset N_T,$$

proving (b). Since the representation (π, T_L) can be defined over the field $\mathbb{Q}[i]$, we have the exact sequence

$$1 \to \langle i1_{T_L} \rangle \hookrightarrow N_*' \xrightarrow{\text{int}} N_1' \to 1,$$

as in the proof of the exactness of (10.4.25). In order to show that

$$N_T \cap \langle i1_{T_L} \rangle = \langle -1_{T_L} \rangle,$$

note that by Lemma 13.2.1(b), N_1' preserves K_Λ and hence $\hat{\Lambda}^{(2)}$, and

$$N_1'|_{\hat{\Lambda}/K_\Lambda} \subset C_1$$

[recall (10.4.12)]. Consider the exact sequence of (10.4.25):

$$1 \to \langle \pm 1 \rangle \hookrightarrow C_T \overset{\text{int}}{\to} C_1 \to 1.$$

Set

$$N_3 = (\text{int})^{-1}(N_1'|_{\hat{\Lambda}/K_\Lambda}) \subset C_T,$$

so that

$$1 \to \langle \pm 1 \rangle \hookrightarrow N_3 \overset{\text{int}}{\to} N_1'|_{\hat{\Lambda}/K_\Lambda} \to 1$$

is exact. Then

$$N_*'|_{T_\Lambda} \subset \mathbf{N}_{\text{Aut}\, T_\Lambda}(\hat{\Lambda}/K_\Lambda),$$

$$\text{int}(N_*'|_{T_\Lambda}) \subset N_1'|_{\hat{\Lambda}/K_\Lambda},$$

$$N_*'|_{T_\Lambda} \subset \mathbb{F}^\times N_3,$$

so that by taking commutators,

$$N_T|_{T_\Lambda} \subset N_3.$$

Thus

$$N_T|_{T_\Lambda} \cap \mathbb{F}^\times 1_{T_\Lambda} \subset \langle \pm 1_{T_\Lambda} \rangle,$$

and so

$$N_T \cap \mathbb{F}^\times 1_{T_L} = \langle \pm 1_{T_L} \rangle.$$

Now (c) and (d) follow from Lemma 13.2.1(c). ∎

Now by analogy with (10.4.39), define the group

$$\hat{N}_0 = \{(g, g_T) \in N_1' \times N_T \mid g = \text{int}(g_T) \quad \text{on} \quad \hat{L}/K_L\}, \tag{13.2.14}$$

and identify $\hat{N}_0$ with a group of operators on W_L via

$$(g, g_T) = \bar{g} \otimes g \oplus \bar{g} \otimes g_T \tag{13.2.15}$$

as in (10.4.56). Clearly, $\hat{N}_0 \simeq N_T$.

Lemma 13.2.3: *We have*

$$N_1'|_{\hat{\Lambda}^{(2)}} \subset C_0, \tag{13.2.16}$$

$$N_T|_{T_\Lambda} \subset C_T, \tag{13.2.17}$$

$$\hat{N}_0|_{W_\Lambda} \subset \hat{C}, \tag{13.2.18}$$

$$\hat{N}_0|_{V^\natural} \subset C, \tag{13.2.19}$$

$$\nu(\hat{\Lambda}_0^{(2)}) \subset \hat{N}_0, \tag{13.2.20}$$

$$\hat{N}_0|_{V^\natural} = N_0. \tag{13.2.21}$$

Proof: We have already observed that (13.2.16) follows from Lemma 13.2.1(b). By (13.2.16) and Lemma 13.2.2(a),

$$N_T|_{T_\Lambda} \subset C_*.$$

Now (13.2.17) follows from Lemma 13.2.2(d). The inclusions (13.2.18), (13.2.19) are consequences of (13.2.16), (13.2.17) and the definitions of $\hat{N}_0$, $\hat{C}$ and C. The inclusion (13.2.20) follows from Lemma 13.2.2(b). By (13.2.19) and Lemma 13.2.1(b),

$$\hat{N}_0|_{V^\natural} \subset N_0.$$

Now (13.2.21) follows from Lemma 13.2.1(a) and Lemma 13.1.1. ∎

Recall the following notation from (11.1.43):

$$\begin{aligned} W_0 &= V_{L_0} = S(\hat{\mathfrak{h}}_{\mathbb{Z}}^-) \otimes \mathbb{F}\{L_0\}, \\ W_1 &= V_{L_1} = S(\hat{\mathfrak{h}}_{\mathbb{Z}}^-) \otimes \mathbb{F}\{L_1\}, \\ W_2 &= V'_{L_0} = S(\hat{\mathfrak{h}}_{\mathbb{Z}+1/2}^-) \otimes T_{L_0}, \\ W_3 &= V'_{L_1} = S(\hat{\mathfrak{h}}_{\mathbb{Z}+1/2}^-) \otimes T_{L_1}. \end{aligned} \tag{13.2.22}$$

Set

$$\begin{aligned} \mathfrak{h}_1 &= \{h(-1) \otimes \iota(1) \mid h \in \mathfrak{h}\}, \\ \mathfrak{h}_2 &= \mathbb{F}\text{-span}\{\iota(a_k)^+ \mid k \in \Omega\}, \\ \mathfrak{h}_3 &= \mathbb{F}\text{-span}\{\iota(a_k)^- \mid k \in \Omega\}. \end{aligned} \tag{13.2.23}$$

These are subspaces of the weight 1 (degree 0) space $(V_{L_0})_0$. Then the components of the vertex operators $Y(u, z)$ for $u \in \mathfrak{h}_a$ in any of the cases $a = 1, 2, 3$ generate a Heisenberg algebra when restricted to any subspace W_j, $j = 0, 1, 2, 3$. Let $\tilde{N}$ denote the group of all linear automorphisms g of W_L such that

$$g \text{ permutes the subspaces } W_j, \quad j = 0, 1, 2, 3, \tag{13.2.24}$$

$$g \text{ permutes the spaces } \mathfrak{h}_a, \quad a = 1, 2, 3, \tag{13.2.25}$$

$$gY(u, z)g^{-1} = Y(gu, z) \quad \text{on} \quad W_L \quad \text{for} \quad u \in V_{L_0} \tag{13.2.26}$$

$$g \text{ is grading-preserving} \tag{13.2.27}$$

$$g\iota(1) = \iota(1). \tag{13.2.28}$$

Lemma 13.2.4: *(a) Let $g \in \hat{N}_0$. Then*

$$gY(u, z)g^{-1} = Y(gu, z) \quad on \quad W_L \tag{13.2.29}$$

for $u \in V_{L_0}$.
(b) We have

$$\hat{N}_0 \subset \tilde{N}. \tag{13.2.30}$$

Proof: Part (a) is proved in the same way as Theorem 10.4.6. It follows that the elements of $\hat{N}_0$ satisfy (13.2.24), (13.2.26)–(13.2.28). Recall the automorphism z_T from (12.2.16). By Lemma 13.2.2(a), N_T and z_T commute. Since $\text{int}(z_T)$ and z_0 induce the same automorphism of $\hat{L}/K_L$ and N_1 acts faithfully on $\hat{L}/K_L$ by Lemma 13.2.1(d), it follows that z_0 centralizes N_1'. Thus $\hat{N}_0$ leaves $\mathfrak{h}_2$ and $\mathfrak{h}_3$ invariant. ∎

We make the identification

$$(\mathbb{F}^\times)^3 = \{1_{W_0} \oplus \lambda_1 1_{W_1} \oplus \lambda_2 1_{W_2} \oplus \lambda_3 1_{W_3} \mid \lambda_1, \lambda_2, \lambda_3 \in \mathbb{F}^\times\}. \tag{13.2.31}$$

Proposition 13.2.5: *The group $\tilde{N}$ is generated by σ, $\nu(\hat{L})$, $\hat{N}_0$ and $(\mathbb{F}^\times)^3$, and in fact*

$$\tilde{N} = \langle \sigma, \tau \rangle \langle \nu(\hat{L}_0), \hat{N}_0, (\mathbb{F}^\times)^3 \rangle.$$

Proof: By Theorem 11.5.1, $\sigma \in \tilde{N}$, and one shows easily that $\nu(\hat{L})$ and $(\mathbb{F}^\times)^3$ lie in $\tilde{N}$. Let $g \in \tilde{N}$. By (13.2.24) and (13.2.28), g stabilizes W_0. Since σ and $\tau = \nu(b_{k_0})$ generate all permutations of $\{W_1, W_2, W_3\}$, we may assume that g stabilizes each W_j, $0 \le j \le 3$. On W_1, $\{u_n \mid u \in \mathfrak{h}_1, n \neq 0\}$ generates a $\mathbb{Z}$-graded Heisenberg algebra while $\{u_n \mid u \in \mathfrak{h}_2\}$ and $\{u_n \mid u \in \mathfrak{h}_3\}$ generate $\mathbb{Z} + \frac{1}{2}$-graded Heisenberg algebras. Condition (13.2.25) therefore shows that $g\mathfrak{h}_1 = \mathfrak{h}_1$ and that there exists $A \in \text{Aut}\,\mathfrak{h}$ such that

$$gh(z)g^{-1} = (Ah)(z) \tag{13.2.32}$$

for $h \in \mathfrak{h}$. The Heisenberg commutation relations then show that A is an isometry with respect to the form $\langle \cdot, \cdot \rangle$. Since the spaces

$$S(\hat{\mathfrak{h}}_{\bar{\mathbb{Z}}}) \otimes \iota(b)$$

for $b \in \hat{L}$ are the simultaneous eigenspaces for the operators $h(0)$, $h \in \mathfrak{h}$, and

$$h(0) \cdot (v \otimes \iota(b)) = \langle h, \bar{b} \rangle v \otimes \iota(b)$$

for $v \in S(\hat{\mathfrak{h}}_{\bar{\mathbb{Z}}})$, it follows that $A \in \text{Aut}(L; \langle \cdot, \cdot \rangle)$. Multiplying g by a suitable element of $\hat{N}_0$, we may assume that $A = 1$, so that $g|_{\mathfrak{h}_1} = 1$ and g commutes with $h(z)$ for $h \in \mathfrak{h}$. By (13.2.25) and the fact that

$$h(0)\iota(a_k^{\pm 1}) = \langle h, \pm\alpha_k \rangle \iota(a_k^{\pm 1})$$

for $h \in \mathfrak{h}$, $k \in \Omega$, it follows that

$$g\iota(a_k) = \lambda_k \iota(a_k),$$

$$gY(a_k, z)g^{-1} = \lambda_k Y(a_k, z)$$

for some $\lambda_k \in \mathbb{F}^\times$. We must also have

$$g|_{T_L} = g_T,$$

$$g|_{V_L'} = 1 \otimes g_T$$

for some $g_T \in \operatorname{Aut} T_L$. Hence

$$g_T \pi(a_k) g_T^{-1} = \lambda_k \pi(a_k)$$

for $k \in \Omega$, where π as usual denotes the action on T_L. Since the spaces $\mathbb{F}t(b)$ for $b \in \hat{L}$ are the simultaneous eigenspaces for the operators $\pi(a_k)$, $k \in \Omega$, it follows that

$$g_T t(1) \in \mathbb{F}^\times t(a)$$

for some $a \in \hat{L}_0$. Multiplying g by $\nu(a^{-1})$, we may assume that g_T stabilizes $\mathbb{F}t(1)$, which implies that $\lambda_k = 1$ for all $k \in \Omega$. Hence $g|_{\mathfrak{h}_2} = 1$, $g|_{\mathfrak{h}_3} = 1$ and g_T centralizes $\pi(\hat{Q})$. For $b \in \hat{L}_0$ we also have

$$g\iota(b) = \lambda(\bar{b})\iota(b)$$

for some $\lambda(\bar{b}) \in \mathbb{F}^\times$, since the multiples of $\iota(b)$ are the only elements u of V_{L_0} of weight $\langle \bar{b}, \bar{b} \rangle/2$ satisfying the condition

$$h(0)u = \langle h, \bar{b} \rangle u \quad \text{for} \quad h \in \mathfrak{h}.$$

It follows that

$$gY(b, z)g^{-1} = \lambda(\bar{b})Y(b, z),$$

$$g_T \pi(b) g_T^{-1} = \lambda(\bar{b})\pi(b).$$

Squaring and using the fact that $b^2 \in \hat{Q}$, we obtain

$$\lambda(\bar{b}) \in \{\pm 1\}.$$

Using the group structure of $\hat{L}_0$ it is now easy to see that for $C \in \mathcal{C}$ and $\beta \in Q$,

$$\lambda(\tfrac{1}{2}\alpha_C + \beta) = (-1)^{|C \cap X|}$$

for some $X \subset \Omega$. Multiplying g by $\nu(a_{X,\emptyset})$ we may assume that $\lambda(\frac{1}{2}\alpha_C + \beta) = 1$ for all C, β as above, so that $g|_{W_0} = 1$ and g commutes with the vertex operators $Y(b, z)$, $b \in \hat{L}_0$, as well as $h(z)$, $h \in \mathfrak{h}$. Since the components of these vertex operators act absolutely irreducibly on each space W_j, we finally see that $g \in (\mathbb{F}^\times)^3$. ∎

Proof of Lemma 13.1.2: Since $\hat{N}_0|_{V^\natural} = N_0$ it is enough to show that $\hat{N}_0$ is normal in $\tilde{N}$. We shall show that

$$\hat{N}_0 = \tilde{N}'''. \tag{13.2.33}$$

Since

$$N_4 = \langle \nu(\hat{L}_0), \hat{N}_0, (\mathbb{F}^\times)^3 \rangle \tag{13.2.34}$$

is the subgroup of $\tilde{N}$ stabilizing each W_j, $0 \leq j \leq 3$, and

$$\tilde{N}/N_4 \simeq \mathcal{S}_3,$$

we have

$$\tilde{N}'' \subset N_4.$$

Since $(\mathbb{F}^\times)^3$ is central in N_4, (13.2.3), (13.2.20) and the fact that

$$(N_T, \pi(a_K)) \subset (N_T, N_*) \subset N_T$$

give

$$\tilde{N}''' \subset N_4' \subset \hat{N}_0.$$

But by Lemma 13.2.2(d) we have

$$\hat{N}_0 = \hat{N}_0' = \hat{N}_0''',$$

so that

$$\tilde{N}''' = \hat{N}_0. \quad ∎$$

13.3. Action of D_4^6 on $W_\mathfrak{L}$

Recalling the diagram (12.1.22), consider the decomposition

$$W_\mathfrak{L} = \coprod_{I \in \mathfrak{L}/\mathfrak{L}_{00}} V_I \oplus \coprod_{I \in \mathfrak{L}/\mathfrak{L}_{00}} V_I'. \tag{13.3.1}$$

We shall show that on each space V_I, V_I', the components of the vertex operators

$$h(z), \quad X(a, z)$$

where $h \in \mathfrak{h}$, $a \in \hat{\mathfrak{L}}_{00}$, $\langle \bar{a}, \bar{a} \rangle = 2$ or 0, span a (possibly twisted) affine algebra associated with a semisimple Lie algebra $\mathfrak{g}$ of type D_4^6.

Recall from (12.2.17) the involution $\hat{z}_0 = (z_0, z_T)$ of W_L. We use Proposition 5.5.3 to produce an automorphism $z_{\mathfrak{L}}$ of $T_{\mathfrak{L}}$ such that $\operatorname{int}(z_{\mathfrak{L}})$ and z_0 induce the same automorphism of $\hat{\mathfrak{L}}/K_{\mathfrak{L}}$ and hence of $\hat{L}/K_{\mathfrak{L}}$ and of $\hat{L}/K_L$. Since T_L is precisely the subspace of $T_{\mathfrak{L}}$ fixed by K_L, we see that $z_{\mathfrak{L}} T_L = T_L$ and that $z_{\mathfrak{L}}$ can be normalized so as to agree with z_T on T_L, z_T being characterized up to normalization by the condition that $\operatorname{int}(z_T)$ and z_0 agree on $\hat{L}/K_L$. Since $z_{\mathfrak{L}}^2$ commutes with $\hat{\mathfrak{L}}$ on $T_{\mathfrak{L}}$, $z_{\mathfrak{L}}$ must be an involution. Rewriting $z_{\mathfrak{L}}$ as z_T, we set

$$\hat{z}_0 = (z_0, z_T) \in \operatorname{Aut}(\hat{\mathfrak{L}}; \kappa) \times \operatorname{Aut} T_{\mathfrak{L}}, \tag{13.3.2}$$

which defines an involution of $W_{\mathfrak{L}}$ extending that of (12.2.17).

Define the groups A, E, both isomorphic to $(\mathbb{Z}/2\mathbb{Z})^3$, by

$$E = \nu(\widehat{2\mathfrak{L}_{00}^{(2)}}) \subset \operatorname{Aut} W_{\mathfrak{L}} \tag{13.3.3}$$

$$A = (\nu(\widehat{2\mathfrak{L}^{(2)}}) \times \langle \hat{z}_0 \rangle)/E. \tag{13.3.4}$$

Then E restricts isomorphically to the group acting on $V^\natural$ and denoted E in (13.1.18). We make the identification

$$\begin{aligned} E &\leftrightarrow \mathbb{F}_2 \oplus \mathfrak{L}_{00}/\Lambda_{00} \\ \nu(\kappa^2) &= \hat{z} \leftrightarrow (1, 0) \\ \nu(a_k^2) &= \hat{z}_1 \leftrightarrow (0, \Lambda_{00} + \alpha_k) \\ \nu(a_{T_j,\emptyset}) &= \hat{z}_2 \leftrightarrow (0, \Lambda_{00} + \tfrac{1}{2}\alpha_{T_j}), \end{aligned} \tag{13.3.5}$$

so that the commuting involutions $\hat{z}$, $\hat{z}_1$, $\hat{z}_2$ act as z, z_1, z_2, respectively, on $V^\natural$, and we have

$$\nu(\widehat{2\mathfrak{L}_{00}^{(2)}} \cap \langle a_k \mid k \in \Omega \rangle) \leftrightarrow \mathfrak{L}_{00}/\Lambda_{00}.$$

We also identify

$$\begin{aligned} A &\leftrightarrow \mathbb{F}_2 \oplus \mathfrak{L}/\mathfrak{L}_{00} \\ \hat{z}_0 E &\leftrightarrow (1, 0) \\ \nu(\widehat{2I^{(2)}})E &\leftrightarrow (0, I) \end{aligned} \tag{13.3.6}$$

for $I \in \mathfrak{L}/\mathfrak{L}_{00}$, so that

$$\nu(\widehat{2\mathfrak{L}^{(2)}})E \leftrightarrow \mathfrak{L}/\mathfrak{L}_{00}.$$

Define the nondegenerate pairing

$$\begin{aligned}&\langle \cdot, \cdot \rangle : E \times A \to \mathbb{F}_2 \\ &\langle (\lambda_1, I_1), (\lambda_2, I_2) \rangle = \lambda_1 \lambda_2 + 2\langle I_1, I_2 \rangle + 2\mathbb{Z} \in \mathbb{Z}/2\mathbb{Z} = \mathbb{F}_2\end{aligned} \tag{13.3.7}$$

for $\lambda_1, \lambda_2 \in \mathbb{F}_2$, $I_1 \in \mathfrak{L}_{00}/\Lambda_{00}$, $I_2 \in \mathfrak{L}/\mathfrak{L}_{00}$. By means of this pairing, we make the identifications $A = E^*$, $E = A^*$.

Since E acts on $W_{\mathfrak{L}}$, we have an eigenspace decomposition

$$W_{\mathfrak{L}} = \coprod_{\lambda \in E^* = A} W_\lambda . \tag{13.3.8}$$

Lemma 13.3.1: *We have*

$$W_{(0,I)} = V_I, \quad W_{(1,I)} = V_I' \tag{13.3.9}$$

for all $I \in \mathfrak{L}/\mathfrak{L}_{00}$.

Proof: From the relations

$$\begin{aligned}\hat{z}|_{V_I} &= 1, \\ \hat{z}_1|_{V_I} &= i^{c_0(2\alpha_k, I)} = (-1)^{2\langle \alpha_k, I \rangle}, \\ \hat{z}_2|_{V_I} &= i^{c_0(\alpha_{T_j}, I)} = (-1)^{2\langle (1/2)\alpha_{T_j}, I \rangle},\end{aligned}$$

we obtain $V_I \subset W_{(0,I)}$, and from the relations

$$\begin{aligned}\hat{z}|_{V_I'} &= -1, \\ \hat{z}_1|_{V_I'} &= i^{c_0(2\alpha_k, I)} = (-1)^{2\langle \alpha_k, I \rangle}, \\ \hat{z}_2|_{V_I'} &= i^{c_0(\alpha_{T_j}, I)} = (-1)^{2\langle (1/2)\alpha_{T_j}, I \rangle},\end{aligned}$$

we obtain $V_I' \subset W_{(1,I)}$. ∎

Recall from Remark 12.1.2 that $\mathfrak{L}_{00}$ is a copy of the Niemeier lattice $N(D_4^6)$. We shall identify the subspace of $V_{\mathfrak{L}_{00}}$ of weight 1 with a Lie algebra whose (possibly twisted) affinizations will act on various subspaces of $W_{\mathfrak{L}}$; cf. Sections 6.3 and 8.9. We shall have an analogue for D_4^6 of the corresponding structures for A_1^{24} set up in (11.1.49)–(11.1.70).

Set

$$\begin{aligned}\mathfrak{A} &= (V_{\mathfrak{L}_{00}})_0 \\ &= \mathbb{F}\text{-span}\{h(-1), \iota(a) \mid h \in \mathfrak{h}, a \in \hat{\mathfrak{L}}_{00}, \langle \bar{a}, \bar{a} \rangle = 2\}. \end{aligned} \tag{13.3.10}$$

Then A acts on $\mathfrak{A}$, since E acts trivially on $\mathfrak{A}$. Let

$$\mathfrak{A} = \coprod_{\mu \in A^* = E} \mathfrak{A}_\mu \tag{13.3.11}$$

be the eigenspace decomposition.

Lemma 13.3.2: *We have*

$$\mathfrak{A}_{(1,0)} = \{h(-1) \mid h \in \mathfrak{h}\}, \tag{13.3.12}$$

$$\mathfrak{A}_{(0,I)} = \mathbb{F}\text{-span}\{\iota(a)^+ \mid a \in \hat{\mathfrak{L}}_{00}, \bar{a} \in I, \langle \bar{a}, \bar{a} \rangle = 2\}, \tag{13.3.13}$$

$$\mathfrak{A}_{(1,I)} = \mathbb{F}\text{-span}\{\iota(a)^- \mid a \in \hat{\mathfrak{L}}_{00}, \bar{a} \in I, \langle \bar{a}, \bar{a} \rangle = 2\} \tag{13.3.14}$$

for $I \in \mathfrak{L}_{00}/\Lambda_{00}$, $I \neq \Lambda_{00}$.

Proof: Since

$$\hat{z}_0 h(-1) = -h(-1),$$

$$\nu(\widehat{2\mathfrak{L}^{(2)}})h(-1) = h(-1)$$

for $h \in \mathfrak{h}$, we have $h(-1) \in \mathfrak{A}_{(1,0)}$. From the relations

$$\hat{z}_0 \iota(a)^+ = \iota(a)^+,$$

$$\nu(\widehat{2I^{(2)}})\iota(a)^+ = i^{-c_0(\bar{a},2I)}\iota(a)^+ = (-1)^{-2\langle \bar{a}, I \rangle}\iota(a)^+$$

for $I \in \mathfrak{L}/\mathfrak{L}_{00}$, $a \in \hat{\mathfrak{L}}_{00}$, $\bar{a} \in I_1 \in \mathfrak{L}_{00}/\Lambda_{00}$, $\langle \bar{a}, \bar{a} \rangle = 2$, we obtain

$$\iota(a)^+ \in \mathfrak{A}_{(0,I_1)}.$$

Similarly,

$$\iota(a)^- \in \mathfrak{A}_{(1,I_1)}$$

if $\bar{a} \in I_1 \in \mathfrak{L}_{00}/\Lambda_{00}$, $\langle \bar{a}, \bar{a} \rangle = 2$. Since $\langle \alpha, \alpha \rangle \neq 2$ for all $\alpha \in \Lambda_{00}$, the lemma is proved. ∎

We shall say that a generating function

$$f(z) = \sum_{n \in \mathbb{F}} a_n z^n \in U\{z\}, \tag{13.3.15}$$

U any vector space, is *Z-graded* ($Z = \mathbb{Z}$ or $\mathbb{Z} + \frac{1}{2}$) if

$$f(z) = \sum_{n \in Z} a_n z^n,$$

i.e., if $a_n = 0$ for all $n \in \mathbb{F}\backslash Z$.

Lemma 13.3.3: *Let $\mu \in E$, $\lambda \in A$, $u \in \mathfrak{A}_\mu$. Then*

$$Y(u, z)\big|_{W_\lambda} \quad \text{is} \quad \mathbb{Z} + \tfrac{1}{2}\langle \mu, \lambda \rangle\text{-graded.} \tag{13.3.16}$$

Proof: Let

$$\mu = (\lambda_1, I_1),$$

$$\lambda = (\lambda_2, I_2),$$

$\lambda_1, \lambda_2 \in \mathbb{F}_2$, $I_1 \in \mathfrak{L}_{00}/\Lambda_{00}$, $I_2 \in \mathfrak{L}/\mathfrak{L}_{00}$. Consider first the untwisted space, i.e., the case $\lambda_2 = 0$. Since $h(z)$ is $\mathbb{Z}$-graded on $V_\mathfrak{L}$ for $h \in \mathfrak{h}$, we may assume that $I_1 \neq 0$ and that $Y(u, z) = Y^{\pm}(a, z)$ for some $a \in \hat{\mathfrak{L}}_{00}$, $\bar{a} \in I_1$, $\langle \bar{a}, \bar{a} \rangle = 2$. Since

$$Y(a, z) = E^-(-\bar{a}, z)E^+(-\bar{a}, z) \otimes az^{\bar{a}}$$

and $E^{\pm}(-\bar{a}, z)$ are $\mathbb{Z}$-graded,

$$Y(a, z)\big|_{V_{I_2}}$$

is $\mathbb{Z} + \langle \bar{a}, I_2 \rangle$-graded. Thus $Y^{\pm}(a, z)$ are $\mathbb{Z} + \frac{1}{2}\langle \mu, \lambda \rangle$-graded on V_{I_2} as required. Consider next the twisted space, i.e., the case $\lambda_2 = 1$. Since $h(z)$ is $\mathbb{Z} + \frac{1}{2}$-graded on $V'_\mathfrak{L}$ for $h \in \mathfrak{h}$, we may again assume that $I_1 \neq 0$, and that $Y(u, z) = Y^{\pm}(a, z)$ for some $a \in \hat{\mathfrak{L}}_{00}$, $\bar{a} \in I_1$, $\langle \bar{a}, \bar{a} \rangle = 2$. Since on the twisted space

$$Y(a, z) = 2^{-\langle \bar{a}, \bar{a} \rangle} E^-(-\bar{a}, z)E^+(-\bar{a}, z) \otimes az^{-1}$$

and

$$E^{\pm}(-\bar{a}, z) = E^{\pm}(\bar{a}, z)\big|_{z^{1/2} \mapsto -z^{1/2}},$$

$Y^+(a, z)$ will be $\mathbb{Z}$- (respectively, $\mathbb{Z} + \frac{1}{2}$-) graded on V'_{I_2} if $a = z_0(a)$ (respectively, $a = -z_0(a)$) on T_{I_2}. But

$$a^{-1}z_0(a)\big|_{T_{I_2}} = i^{c_0(-2\bar{a}, I_2)}\psi(a^{-1}z_0(a)) = (-1)^{-2\langle \bar{a}, I_2 \rangle}$$

by (12.1.74). Thus $Y^{\pm}(a, z)$ is $\mathbb{Z} + \frac{1}{2}\langle \mu, \lambda \rangle$-graded. ∎

In order to interpret Lemma 13.3.3 together with the commutation relations among the components of the vertex operators $Y(u, z)$, $u \in \mathfrak{A}$, we construct a Lie algebra $\mathfrak{g}$ associated with the root system

$$\{a \in \mathfrak{L}_{00} \mid \langle \alpha, \alpha \rangle = 2\} \tag{13.3.17}$$

by a slight extension of the procedure of Section 6.2: First define the vector space

$$\mathfrak{g} = \mathfrak{h} \oplus \sum_{\substack{a \in \hat{\mathfrak{L}}_{00} \\ \langle \bar{a}, \bar{a} \rangle = 2}} \mathbb{F} x_a, \tag{13.3.18}$$

where we impose only the linear relations

$$x_{\kappa a} = i x_a \tag{13.3.19}$$

[cf. (6.2.20)]. Make $\mathfrak{g}$ into a Lie algebra by defining the brackets

$$[\mathfrak{h}, \mathfrak{h}] = 0, \tag{13.3.20}$$

$$[h, x_a] = \langle h, \bar{a} \rangle x_a, \tag{13.3.21}$$

$$[x_a, x_b] = \begin{cases} 0 & \text{if } \langle \bar{a}, \bar{b} \rangle \geq 0 \\ x_{ab} & \text{if } \langle \bar{a}, \bar{b} \rangle = -1, \\ \bar{a} & \text{if } b = a^{-1}, \end{cases} \tag{13.3.22}$$

and define on $\mathfrak{g}$ a nonsingular invariant symmetric bilinear form $\langle \cdot, \cdot \rangle$ by

$$\langle \cdot, \cdot \rangle |_{\mathfrak{h} \times \mathfrak{h}} = \langle \cdot, \cdot \rangle, \tag{13.3.23}$$

$$\langle \mathfrak{h}, x_a \rangle = 0 \tag{13.3.24}$$

$$\langle x_a, x_b \rangle = \begin{cases} 0 & \text{if } \bar{a} + \bar{b} \neq 0 \\ 1 & \text{if } b = a^{-1}. \end{cases} \tag{13.3.25}$$

Then as in Remark 8.9.1, we see that we have an identification

$$\mathfrak{g} = \mathfrak{A} \tag{13.3.26}$$

and that the Lie bracket and the invariant form are given by:

$$[u, v] = u_0 \cdot v$$
$$\langle u, v \rangle = u_1 \cdot v \tag{13.3.27}$$

for $u, v \in \mathfrak{A}$ [cf. (11.1.51)]. The decomposition (13.3.11) is an orthogonal decomposition by Lemma 13.3.2.

We have an action of A on $\mathfrak{g} = \mathfrak{A}$ by automorphisms and isometries via:

$$\nu(a)E \cdot h = h,$$
$$\nu(a)E \cdot x_b = x_{aba^{-1}}, \tag{13.3.28}$$

$$\hat{z}_0 E \cdot h = -h,$$
$$\hat{z}_0 E \cdot x_b = x_{z_0(b)} \tag{13.3.29}$$

for $a \in \widehat{2\mathfrak{L}^{(2)}}$, $h \in \mathfrak{h}$, $b \in \hat{\mathfrak{L}}_{00}$, $\langle \bar{b}, \bar{b} \rangle = 2$; cf. Proposition 6.4.2. For $\lambda \in A$, consider the possibly twisted affine algebra $\hat{\mathfrak{g}}[\lambda]$ introduced in (1.6.25), (1.6.26). As usual we set

$$x_a^{\pm} = x_a \pm x_{z_0(a)} \in \mathfrak{g} \tag{13.3.30}$$

for $a \in \hat{\mathfrak{L}}_{00}$, $\langle \bar{a}, \bar{a} \rangle = 2$.

Proposition 13.3.4: *For each $\lambda \in A$, the linear map π_λ determined by*

$$\begin{aligned} \pi_\lambda &: h \otimes t^n \mapsto h(n), \\ \pi_\lambda &: x_a^{\pm} \otimes t^n \mapsto x_a^{\pm}(n), \\ \pi_\lambda &: c \mapsto 1 \end{aligned} \tag{13.3.31}$$

for $h \in \mathfrak{h}$, $a \in \hat{\mathfrak{L}}_{00}$, $\langle \bar{a}, \bar{a} \rangle = 2$, $n \in \mathbb{Z}$ or $\mathbb{Z} + \frac{1}{2}$ as appropriate, is a representation of $\hat{\mathfrak{g}}[\lambda]$ on W_λ.

Proof: Using Lemma 13.3.3, the required commutation relations follow easily from (7.1.46), Theorem 7.2.1, (7.3.15) and Theorem 7.4.1, which are applicable since $c_0|_{\mathfrak{L}_{00} \times \mathfrak{L}_{00}} = 2\langle \cdot, \cdot \rangle + 4\mathbb{Z}$. ∎

Corollary 13.3.5: *Let g be a grading-preserving linear automorphism of $W_{\mathfrak{L}}$. Assume that*

$$g \quad \textit{permutes the spaces} \quad W_\lambda, \quad \lambda \in A, \tag{13.3.32}$$

$$g \quad \textit{permutes the spaces} \quad \mathfrak{A}_\mu, \quad \mu \in E \backslash \{1\}, \tag{13.3.33}$$

$$gY(u, z)g^{-1} = Y(gu, z) \quad \textit{for} \quad u \in \mathfrak{A}. \tag{13.3.34}$$

Then there exists $R \in \operatorname{Aut} A = \operatorname{Aut} E$ such that

$$gW_\lambda = W_{R\lambda}, \quad g\mathfrak{A}_\mu = \mathfrak{A}_{R\mu} \tag{13.3.35}$$

for all $\lambda \in A$, $\mu \in E \backslash \{1\}$. (Here Aut A *and* Aut E *are identified by requiring the pairing $E \times A \to \mathbb{F}_2$ above to be invariant.)*

Proof: For each pair $\mu_1 \neq \mu_2 \in E \backslash \{1\}$ we can find components $x_u(m)$, $x_v(n)$ of $X(u, z)$, $X(v, z)$ with $u \in \mathfrak{A}_{\mu_1}$, $v \in \mathfrak{A}_{\mu_2}$ such that

$$[x_u(m), x_v(n)] \neq 0.$$

Then $[x_u(m), x_v(n)]$ is a component of a vertex operator $X(w, z)$, $w \in \mathfrak{A}_{\mu_1 + \mu_2}$. It follows that the permutation of the $\mathfrak{A}_\mu$ must be by an element

$R \in \operatorname{Aut} E$. Also, for each $\lambda \in A$ and $\mu \in E\backslash\{1\}$, there is an element $u \in \mathfrak{A}_\mu$ with $X(u, z)|_{W_\lambda} \neq 0$. Then by Lemma 13.3.3,

$$gX(u, z)g^{-1}|_{gW_\lambda} \quad \text{is} \quad \mathbb{Z} + \tfrac{1}{2}\langle \mu, \lambda \rangle\text{-graded.}$$

Since $gX(u, z)g^{-1} = X(gu, z)$ with $gu \in \mathfrak{A}_{R\mu}$ it follows that

$$gW_\lambda = W_{R\lambda}. \quad \blacksquare$$

13.4. Groups Acting on $W_{\mathfrak{L}}$

We denote by $\tilde{H}$ the group of all linear automorphisms g of $W_{\mathfrak{L}}$ such that:

$$g \quad \text{permutes the subspaces} \quad W_\lambda, \quad \lambda \in A \tag{13.4.1}$$

$$g \quad \text{leaves} \quad V^\natural \quad \text{invariant} \tag{13.4.2}$$

$$g \quad \text{permutes by conjugation the spaces} \quad \mathfrak{A}_\mu, \quad \mu \in E\backslash\{1\} \tag{13.4.3}$$

$$gY(u, z)g^{-1} = Y(gu, z) \quad \text{on} \quad W_{\mathfrak{L}} \quad \text{for} \quad u \in V_{\mathfrak{L}_{00}} \tag{13.4.4}$$

$$g \quad \text{is grading-preserving} \tag{13.4.5}$$

$$g\iota(1) = \iota(1) \tag{13.4.6}$$

$$g \quad \text{leaves the rational form} \quad V^\natural_{\mathbb{Q}} \quad \text{invariant} \tag{13.4.7}$$

$$\begin{gathered} g \quad \text{is an isometry with respect to the} \\ \mathbb{Q}\text{-valued symmetric form} \quad \langle \cdot, \cdot \rangle \quad \text{on} \quad V^\natural_{\mathbb{Q}}. \end{gathered} \tag{13.4.8}$$

(For the notations W_λ, $\mathfrak{A}_\mu$, see Section 13.3, for $V^\natural_{\mathbb{Q}}$ see Section 12.4 and for the form on $V^\natural_{\mathbb{Q}}$ see Section 12.5. Recall from Section 12.5 that the hermitian form on $V^\natural_{\mathbb{Q}} \otimes_{\mathbb{Q}} \mathbb{C}$ does indeed restrict to a rational-valued symmetric form on $V^\natural_{\mathbb{Q}}$.)

Our next goal is to construct a subgroup $\hat{Q}_1$ of $\tilde{H}$ which restricts to the subgroup Q_1 [see (13.1.25)] on $V^\natural$.

Set

$$Q_{1,0} = \operatorname{Aut}(\hat{\mathfrak{L}}; \kappa, \langle \cdot, \cdot \rangle, K_\Lambda). \tag{13.4.9}$$

By Proposition 5.4.8, Proposition 12.1.6 and Remark 12.1.8 we have the exact sequence

$$1 \to \operatorname{Hom}(\mathfrak{L}/2\Lambda, \mathbb{Z}/4\mathbb{Z}) \xrightarrow{*} Q_{1,0} \xrightarrow{-} \operatorname{Aut}(\mathfrak{L}; 2\Lambda, \langle \cdot, \cdot \rangle, c_0, s_{\mathfrak{L}}, f_{\mathfrak{L}}) \to 1 \tag{13.4.10}$$

where the maps $s_{\mathfrak{L}}, f_{\mathfrak{L}}$, determined by

$$a^2 \in \kappa^{s_{\mathfrak{L}}(\bar{a})}\langle\kappa^2\rangle \times K_\Lambda,$$

$$b^4 \in k^{f_{\mathfrak{L}}(\bar{b})} K_\Lambda$$

for $a \in \hat{\Lambda}$, $b \in \hat{\mathfrak{L}}$, are given by:

$$s_{\mathfrak{L}} \equiv 0 \quad \text{on} \quad \Lambda,$$

$$f_{\mathfrak{L}}(\beta) = 2\langle\beta, \beta\rangle + 4\mathbb{Z}, \quad \beta \in \mathfrak{L}.$$

Since $\mathfrak{L}_{00} = \Lambda_{00} + \mathbb{Z}\text{-span}(\mathfrak{L}_2)$ and $\Lambda_{00} = \mathfrak{L}^\circ$, $\mathfrak{L}_{00}$ is the unique even unimodular sublattice of $\mathfrak{L}$ isometric to the Niemeier lattice of type D_4^6 (recall Remark 12.1.2). The sequence (13.4.10) can thus be written

$$1 \to \text{int}(\hat{\Lambda}) \hookrightarrow Q_{1,0} \twoheadrightarrow \text{Aut}(\Lambda; \langle\cdot,\cdot\rangle, \Lambda_{00}) \to 1. \tag{13.4.11}$$

By definition of C_0,

$$Q_{1,0}|_{\hat{\Lambda}} \subset C_0, \tag{13.4.12}$$

so that $Q_{1,0}$ leaves $\hat{\Lambda}^{(2)}$ invariant, and since $\text{Aut}(\Lambda; \langle\cdot,\cdot\rangle, \Lambda_{00})$ stabilizes $2\Lambda_{00}$, $Q_{1,0}$ leaves $K_{\mathfrak{L}}$ invariant. The following lemma is proved in the same way as Lemma 13.2.1(d):

Lemma 13.4.1: *The group $Q_{1,0}$ acts faithfully on $\hat{\mathfrak{L}}/K_{\mathfrak{L}}$.*

Consider the exact sequence

$$1 \to \mathbb{F}^\times \to \mathbf{N}_{\text{Aut}\,T_{\mathfrak{L}}}(\pi(\hat{\mathfrak{L}}/K_{\mathfrak{L}})) \xrightarrow{\text{int}} \text{Aut}(\hat{\mathfrak{L}}/K_{\mathfrak{L}}; \kappa) \to 1 \tag{13.4.13}$$

from Proposition 5.5.3 and set

$$Q_* = (\text{int})^{-1}(Q_{1,0}) \subset \text{Aut}\, T_{\mathfrak{L}}, \tag{13.4.14}$$

so that we have the commutative diagram with exact rows

$$\begin{array}{ccccccc} 1 \to \mathbb{F}^\times & \hookrightarrow & Q_* & \xrightarrow{\text{int}} & Q_{1,0} & \longrightarrow & 1 \\ \| & & \downarrow & & \downarrow & & \\ 1 \to \mathbb{F}^\times & \hookrightarrow & \mathbf{N}_{\text{Aut}\,T_{\mathfrak{L}}}(\pi(\hat{\mathfrak{L}}/K_{\mathfrak{L}})) & \xrightarrow{\text{int}} & \text{Aut}(\hat{\mathfrak{L}}/K_{\mathfrak{L}}; \kappa) & \to & 1. \end{array} \tag{13.4.15}$$

Then Q_* leaves T_Λ invariant since T_Λ is precisely the subspace of $T_{\mathfrak{L}}$ fixed by K_Λ and $Q_{1,0}$ stabilizes K_Λ. By definition of C_*,

$$Q_*|_{T_\Lambda} \subset C_*. \tag{13.4.16}$$

Set

$$Q_{1,T} = \{g_T \in Q_* \mid g_T|_{T_\Lambda} \in C_T\}. \tag{13.4.17}$$

Since by (10.4.25)

$$C_* = C_T \mathbb{F}^\times,$$

we then have

$$Q_* = Q_{1,T}\mathbb{F}^\times, \tag{13.4.18}$$

and we also have an exact sequence

$$1 \to \langle \pm 1_{T_{\mathfrak{L}}} \rangle \to Q_{1,T} \overset{\text{int}}{\to} Q_{1,0} \to 1. \tag{13.4.19}$$

It follows from (10.4.24) that

$$\pi(\hat{\Lambda}^{(2)}/K_{\mathfrak{L}}) \subset Q_{1,T}. \tag{13.4.20}$$

Now define the group

$$\hat{Q}_1 = \{(g, g_T) \in Q_{1,0} \times Q_{1,T} \mid g = \text{int}(g_T) \quad \text{on} \quad \hat{\mathfrak{L}}/K_{\mathfrak{L}}\}, \tag{13.4.21}$$

and identify $\hat{Q}_1$ with a group of operators on $W_{\mathfrak{L}}$ via

$$(g, g_T) = \bar{g} \otimes g \oplus \bar{g} \otimes g_T \tag{13.4.22}$$

as in (10.4.56). Clearly, $\hat{Q}_1 \simeq Q_{1,T}$.

Lemma 13.4.2: *We have*

$$\hat{Q}_1|_{W_\Lambda} \subset \hat{C}, \tag{13.4.23}$$

$$\hat{Q}_1|_{V^\natural} \subset C, \tag{13.4.24}$$

$$\nu(\hat{\Lambda}^{(2)}) \subset \hat{Q}_1, \tag{13.4.25}$$

$$\hat{Q}_1|_{V^\natural} = Q_1. \tag{13.4.26}$$

Proof: The inclusions (13.4.23), (13.4.24) follow from (13.4.12) and (13.4.17), and (13.4.25) follows from (13.4.11) and (13.4.20). Since $Q_{1,0}$ stabilizes $\widehat{2\mathfrak{L}_{00}^{(2)}}$ and since

$$E = E|_{V^\natural} = \nu(\widehat{2\mathfrak{L}_{00}^{(2)}})|_{V^\natural},$$

$\hat{Q}_1|_{V^\natural}$ normalizes E, and hence $\hat{Q}_1|_{V^\natural} \subset Q_1$. Using (13.4.25) and (13.4.11) we obtain the exact sequence

$$1 \to \pi_0\nu(\hat{\Lambda}^{(2)}/K_\Lambda) \hookrightarrow \hat{Q}_1|_{V^\natural} \twoheadrightarrow \text{Aut}(\Lambda; \langle \cdot, \cdot \rangle, \Lambda_{00})/\langle \pm 1 \rangle \to 1. \tag{13.4.27}$$

But from (13.1.13) and the fact that $\bar{Q}_1$ preserves $\mathcal{L}_{00} + \Lambda = \mathcal{L}$ it is easy to see that

$$1 \to \pi_0 \nu(\hat{\Lambda}^{(2)}/K_\Lambda) \hookrightarrow Q_1 \twoheadrightarrow \mathrm{Aut}(\Lambda; \langle\cdot,\cdot\rangle, \Lambda_{00})/\langle \pm 1\rangle \to 1 \quad (13.4.28)$$

is also an exact sequence, so that (13.4.26) follows. ∎

Lemma 13.4.3: *We have*

$$\hat{Q}_1 \subset \tilde{H}. \quad (13.4.29)$$

Proof: Since $\hat{Q}_1|_{W_\Lambda} \subset \hat{C}$, $\hat{Q}_1$ satisfies (13.4.2), (13.4.6)–(13.4.8), and (13.4.5) is clear. Since $\hat{Q}_1$ normalizes $E = \nu(\widehat{2\mathcal{L}_{00}^{(2)}})$, (13.4.1) follows. Using (13.4.11), (13.4.12), (10.4.7), (10.4.8) and Remarks 12.1.8 and 12.1.13, we obtain the exact sequence

$$1 \to (\mathrm{int}(K_\Lambda) \times \langle z_0 \rangle) \to Q_{1,0} \to \mathrm{Aut}(\Lambda^{(2)}/K_\Lambda), \quad (13.4.30)$$

where the last map is the restriction to $\Lambda^{(2)}/K_\Lambda$. This in turn implies that $\hat{Q}_1$ normalizes

$$\nu(K_\Lambda \times \langle \kappa^2 \rangle) \times \langle \hat{z}_0 \rangle, \quad (13.4.31)$$

since $\pm 1|_{T_\Lambda}$ are the only scalars in C_T. Since this last group covers A, (13.4.3) is satisfied. The compatibility (13.4.4) is proved in the same way as Theorem 10.4.6. ∎

Lemma 13.4.4: *We have that* $\sigma \in \tilde{H}$.

Proof: Condition (13.4.1) was established in (the proof of) Lemma 13.1.3, (13.4.2) and (13.4.4) in Proposition 12.2.8, (13.4.5), (13.4.6) are clear, (13.4.7) was established in Proposition 12.4.1 and (13.4.8) in Proposition 12.5.3. Using the fact that by Proposition 12.1.15

$$z_0(b)b^{-1} = a_{S,T}a_{S,T_j\setminus T}$$

if $\bar{b} = \frac{1}{2}\varepsilon_S\alpha_{T_j}$, $S, T \subset T_j$, $0 \le j \le 5$, one shows as in the proof of Proposition 12.2.1 that

$$\sigma\iota(b)^+ = \begin{cases} 2^{-|T_j|/2}(-1)^{|T_j\cap\{k_0\}|} \displaystyle\sum_{\substack{T\subset T_j \\ |T| \in 2\mathbb{Z}}} (-1)^{|S\cap T|}\iota(a_{S,T}b)^+ & \text{if } |S| \in 2\mathbb{Z} \\ 2^{-|T_j|/2}(-1)^{|T_j\cap\{k_0\}|} \displaystyle\sum_{\substack{T\subset T_j \\ |T| \in 2\mathbb{Z}}} (-1)^{|S\cap T|}\iota(a_{S,T}b)^- & \text{if } |S| \in 2\mathbb{Z}+1 \end{cases} \quad (13.4.32)$$

$$\sigma\iota(b)^- = \begin{cases} 2^{-|T_j|/2}(-1)^{|T_j\cap\{k_0\}|} \sum\limits_{\substack{T\subset T_j \\ |T|\in 2\mathbb{Z}+1}} (-1)^{|S\cap T|}\iota(a_{S,T}b)^+ & \text{if } |S| \in 2\mathbb{Z} \\ 2^{-|T_j|/2}(-1)^{|T_j\cap\{k_0\}|} \sum\limits_{\substack{T\subset T_j \\ |T|\in 2\mathbb{Z}+1}} (-1)^{|S\cap T|}\iota(a_{S,T}b)^- & \text{if } |S| \in 2\mathbb{Z}+1. \end{cases} \tag{13.4.33}$$

Now (13.4.3) follows from Theorem 11.5.1 and Lemma 13.3.2. ∎

Set

$$\hat{N}_{00} = \{g \in \hat{Q}_1 \mid g|_{V^{\natural}} \in N_{00}\}, \tag{13.4.34}$$

$$\langle \pm 1\rangle^7 = \left\{ \coprod_{\lambda\in A} s_\lambda 1_{W_\lambda} \mid s_\lambda \in \{\pm 1\},\ s_1 = 1 \right\}. \tag{13.4.35}$$

Proposition 13.4.5: *We have*

$$\mathbf{C}_{\tilde{H}}(E) = \hat{N}_{00}\langle \pm 1\rangle^7 \tag{13.4.36}$$

$$\tilde{H} = \langle \hat{Q}_1, \sigma, \langle \pm 1\rangle^7\rangle. \tag{13.4.37}$$

Proof: The groups on the right-hand sides being contained in $\mathbf{C}_{\tilde{H}}(E)$ and $\tilde{H}$, respectively, we prove the reverse inclusions. By Lemmas 13.1.4 and 13.1.5, $\langle \hat{Q}_1, \sigma\rangle$ induces the full group of automorphisms of E. Hence

$$\tilde{H}/\mathbf{C}_{\tilde{H}}(E) \simeq \operatorname{Aut} E \tag{13.4.38}$$

and

$$\tilde{H} = \langle \mathbf{C}_{\tilde{H}}(E), \hat{Q}_1, \sigma\rangle,$$

so that (13.4.37) follows from (13.4.36). Let $g \in \mathbf{C}_{\tilde{H}}(E)$. Then g leaves each of the spaces W_λ, $\lambda \in A$, invariant, and by Corollary 13.3.5, g stabilizes each of the spaces $\mathfrak{A}_\mu$, in particular

$$\mathfrak{A}_{(1,0)} = \{h(-1) \mid h \in \mathfrak{h}\},$$

so that there exists $B \in \operatorname{Aut}\mathfrak{h}$ satisfying

$$gh(z)g^{-1} = (Bh)(z)$$

for $h \in \mathfrak{h}$. By the Heisenberg commutation relations, B is an isometry with respect to the form $\langle \cdot, \cdot\rangle$ on $\mathfrak{h}$. Since the spaces

$$S(\hat{\mathfrak{h}}_{\mathbb{Z}}^-) \otimes \iota(b)$$

for $b \in \hat{\mathfrak{L}}$ are the simultaneous eigenspaces of $h(0)$ on $V_{\mathfrak{L}}$ and

$$h(0)\cdot(v \otimes \iota(b)) = \langle h, \bar{b}\rangle v \otimes \iota(b)$$

for $v \in S(\hat{\mathfrak{h}}_{\mathbb{Z}}^{-})$, it follows that $B \in \mathrm{Aut}(\mathfrak{L}; \langle\cdot,\cdot\rangle)$. Since W_Λ is the smallest subspace of $W_{\mathfrak{L}}$ containing $V^{\natural}$ and invariant under the action of the components of

$$\{h(z) \mid h \in \mathfrak{h}\},$$

g leaves W_Λ invariant, and hence by the argument above,

$$B \in \mathrm{Aut}(\mathfrak{L}; \langle\cdot,\cdot\rangle, \Lambda) = \mathrm{Aut}(\Lambda; \langle\cdot,\cdot\rangle, \Lambda_{00}).$$

In addition, since g leaves invariant each V_I, $I \in \mathfrak{L}/\mathfrak{L}_{00}$, B induces 1 on Λ/Λ_{00}.

Using (13.4.11), we show that the sequence

$$1 \to \nu(\hat{\Lambda}^{(2)}) \hookrightarrow \hat{Q}_1 \xrightarrow{-} \mathrm{Aut}(\Lambda; \langle\cdot,\cdot\rangle, \Lambda_{00}) \to 1 \qquad (13.4.39)$$

is exact. To check the exactness at $\hat{Q}_1$, let $(g, g_T) \in \hat{Q}_1$ with $\bar{g} = 1$. Then $g = \mathrm{int}(a)$ on $\hat{\mathfrak{L}}$ for some $a \in \hat{\Lambda}^{(2)}$ by (13.4.11), and $g = \mathrm{int}(g_T)$ on $\hat{\mathfrak{L}}/K_{\mathfrak{L}}$. Thus $g_T \in \mathbb{F}^\times \pi(a)$, but since $g_T|_{T_\Lambda} \in C_T$ and $\pi(a)|_{T_\Lambda} \in C_T$, we must have $g_T = \pm\pi(a)$. Multiplying a by κ^2 if necessary, we may assume that $g_T = \pi(a)$, so that $(g, g_T) = \nu(a)$, proving the exactness.

The sequence

$$1 \to \nu(\hat{\Lambda}_{00}^{(2)}) \hookrightarrow \hat{N}_{00} \xrightarrow{-} \mathrm{Aut}(\Lambda; \langle\cdot,\cdot\rangle, \Lambda_{00}, 1 \text{ on } \Lambda/\Lambda_{00}) \to 1 \quad (13.4.40)$$

(using an obvious notation) is also exact: If $(g, g_T) \in \hat{N}_{00}$ with $\bar{g} = 1$, then $(g, g_T) = \nu(a)$ for $a \in \hat{\Lambda}^{(2)}$. But since $g_T|_{T_\Lambda}$ commutes with $\pi(b)|_{T_\Lambda}$ for $b \in \widehat{2\mathfrak{L}_{00}^{(2)}}$, we see that $a \in \hat{\Lambda}_{00}^{(2)}$. Now let $(g, g_T) \in \hat{Q}_1$ with $\bar{g}$ a prescribed element of $\mathrm{Aut}(\Lambda; \langle\cdot,\cdot\rangle, \Lambda_{00}, 1 \text{ on } \Lambda/\Lambda_{00})$. Then $\bar{g} = 1$ on $\mathfrak{L}/\Lambda$ and hence on $\mathfrak{L}_{00}/\Lambda_{00}$, so that for $b \in \widehat{2\mathfrak{L}_{00}^{(2)}}$,

$$gb \in b\widehat{2\mathfrak{L}_{00}^{(2)}} = b\langle\kappa^2\rangle K_{\mathfrak{L}}.$$

Then g induces a certain automorphism of $\widehat{2\mathfrak{L}_{00}^{(2)}}/K_{\mathfrak{L}}$ fixing κ^2, and by multiplying (g, g_T) by a suitable element of $\nu(\hat{\Lambda}^{(2)})$ if necessary, we may assume that

$$g = 1 \quad \text{on} \quad \widehat{2\mathfrak{L}_{00}^{(2)}}/K_{\mathfrak{L}}.$$

But then g_T centralizes $\pi(\widehat{2\mathfrak{L}_{00}^{(2)}})$ and $(g, g_T) \in \hat{N}_{00}$, proving the exactness of (13.4.40).

Hence, by multiplying g by a suitable element of $\hat{N}_{00}$, we may assume that $B = 1$, so that $g|_{\mathfrak{a}_{(1,0)}} = 1$ and g commutes with $h(z)$ for $h \in \mathfrak{h}$. Since g stabilizes both of

$$\mathbb{F}\text{-span}\{\iota(a_k)^+ \mid k \in \Omega\},$$

$$\mathbb{F}\text{-span}\{\iota(a_k)^- \mid k \in \Omega\}$$

and

$$h(0)\iota(a_k^{\pm 1}) = \langle h, \pm\alpha_k\rangle \iota(a_k^{\pm 1})$$

for $h \in \mathfrak{h}$, $k \in \Omega$, it follows that

$$g\iota(a_k) = \lambda_k \iota(a_k),$$
$$gY(a_k, z)g^{-1} = \lambda_k Y(a_k, z)$$

for some $\lambda_k \in \mathbb{F}^\times$. We must also have

$$g|_{T_{\mathfrak{L}}} = g_T,$$
$$g|_{V'_{\mathfrak{L}}} = 1 \otimes g_T$$

for some $g_T \in \mathrm{Aut}\, T_{\mathfrak{L}}$. It follows that

$$g_T \pi(a_k) g_T^{-1} = \lambda_k \pi(a_k)$$

for $k \in \Omega$. Since the spaces $\mathbb{F}t(b)$ with $b \in \hat{\mathfrak{L}}$ are the simultaneous eigenspaces for the operators $\pi(a_k)$, $k \in \Omega$, we see that g_T permutes these spaces. In particular,

$$g_T t(1) \in \mathbb{F}^\times t(a)$$
$$g_T t(b_{k_0}^2) \in \mathbb{F}^\times t(a')$$

for some $a, a' \in \hat{\mathfrak{L}}_{00}$ [recall (12.2.8)]. Recall from (12.3.34) that the elements $t_\Lambda(b)$, $b \in \hat{\Lambda}^{(2)}$, span T_Λ. Since as we saw above W_Λ is also invariant under g, we may assume that $a \in \hat{\Lambda}_{00}^{(2)}$. Hence multiplying g by $\nu(a^{-1}) \in \hat{N}_{00}$, we may assume that g_T stabilizes $\mathbb{F}t(1)$, which gives $\lambda_k = 1$ for all $k \in \Omega$. This in turn implies that g fixes $\iota(a_k)$ for $k \in \Omega$ and that g_T centralizes $\pi(\hat{Q})$. Also, for $a \in \hat{\mathfrak{L}}_{00}$,

$$h(0)\iota(a) = \langle h, \bar{a}\rangle \iota(a) \quad \text{for} \quad h \in \mathfrak{h},$$

and the multiples of $\iota(a)$ are the only elements of $V_{\mathfrak{L}_{00}}$ of weight $\langle \bar{a}, \bar{a}\rangle/2$ satisfying this condition. Thus

$$g\iota(a) = \lambda(\bar{a})\iota(a)$$

for some $\lambda(\bar{a}) \in \mathbb{F}^\times$, and so

$$gY(a, z)g^{-1} = \lambda(\bar{a})Y(a, z),$$
$$g_T \pi(a) g_T^{-1} = \lambda(\bar{a})\pi(a).$$

Since for these a, $a^2 \in \hat{Q}$, we must have $\lambda(\bar{a}) \in \{\pm 1\}$, and using the group structure of $\hat{\mathfrak{L}}_{00}$ it follows easily that

$$\lambda(\bar{a}) = (-1)^{|C \cap X|}$$

if $\bar{a} = \frac{1}{2}\alpha_C + \beta$, $C \in \mathcal{C}'$, $\beta \in Q$, for some $X \subset \Omega$. This shows that g_T and $\pi(a_{X,\emptyset})$ induce by conjugation the same automorphism of $\hat{\mathfrak{L}}_{00}$. Since $T_{\mathfrak{L}_{00}}$ is absolutely irreducible under the action of $\hat{\mathfrak{L}}_{00}$, we must then have

$$g_T\big|_{T_{\mathfrak{L}_{00}}} \in \mathbb{F}^\times \pi(a_{X,\emptyset}).$$

We also know that g_T leaves T_Λ invariant, so that $|X| \in 2\mathbb{Z}$. Now multiplying g by $\nu(a_{X,\emptyset}) \in \hat{N}_{00}$, we may assume that $\lambda(\mathfrak{L}_{00}) = 1$ and hence that g centralizes all the vertex operators in the set

$$\{h(z) \mid h \in \mathfrak{h}\} \cup \{Y(a, z) \mid a \in \hat{\mathfrak{L}}_{00}\}.$$

Since the components of these vertex operators act absolutely irreducibly on each of the spaces W_λ, $\lambda \in A$, we now get $g \in \langle \pm 1 \rangle^7$ from (13.4.6)–(13.4.8). This completes the proof of the proposition. ∎

Corollary 13.4.6: *We have*

$$\mathbf{C}_H(E) \subset N_{00}\langle \pm 1 \rangle^7\big|_{V^\natural}. \tag{13.4.41}$$

Proof: Since

$$V^\natural \cap W_\lambda \neq \{0\} \quad \text{for all} \quad \lambda \in A,$$

we have

$$\mathbf{C}_{\tilde{H}|_{V^\natural}}(E) = \mathbf{C}_{\tilde{H}}(E)\big|_{V^\natural}. \tag{13.4.42}$$

We also have

$$H = \langle Q_1, \sigma \rangle \subset \tilde{H}\big|_{V^\natural}, \tag{13.4.43}$$

so that the corollary follows from (13.4.36). ∎

We now reduce Lemma 13.1.6 to Proposition 13.5.7 and Lemma 13.4.7 below. What remains is to eliminate certain ± signs. We set

$$Y_1 = \langle \pm 1 \rangle^7\big|_{V^\natural} = \left\{ \coprod_{\lambda \in A} s_\lambda 1_{W_\lambda \cap V^\natural} \,\Big|\, s_\lambda \in \{\pm 1\}, \quad s_1 = 1 \right\}, \tag{13.4.44}$$

$$Y_2 = \left\{ \coprod_{\lambda \in A} s_\lambda 1_{W_\lambda \cap V^\natural} \,\Big|\, s_\lambda \in \{\pm 1\}, \quad s_1 = 1, \quad \prod_{\lambda \in A} s_\lambda = 1 \right\} \tag{13.4.45}$$

$$\simeq (\mathbb{Z}/2\mathbb{Z})^6.$$

Then Y_1, Y_2 and N_{00} are normal subgroups of $HY_1 = Y_1 H$, from Lemmas

13.1.4 and 13.1.5, and it is easy to see from the definitions and Lemma 13.1.1 that

$$Y_1 \cap N_{00} = Y_2 \cap N_{00} = E. \tag{13.4.46}$$

By Corollary 13.4.6 the following commutative diagram has exact rows:

$$\begin{array}{ccccccccc} 1 & \longrightarrow & Y_1N_{00}/N_{00} & \hookrightarrow & Y_1H/N_{00} & \twoheadrightarrow & GL(3, \mathbb{F}_2) & \to & 1 \\ & & \downarrow & & \downarrow & & \| & & \\ 1 & \to & Y_1N_{00}/Y_2N_{00} & \hookrightarrow & Y_1H/Y_2N_{00} & \twoheadrightarrow & GL(3, \mathbb{F}_2) & \to & 1. \end{array} \tag{13.4.47}$$

The existence of the subgroup Q_1/N_{00} of Y_1H/N_{00} mapping isomorphically onto P_1 together with Proposition 13.5.7 below show that the first and hence also the second row splits. Thus

$$Y_1H/Y_2N_{00} \simeq \mathbb{Z}/2\mathbb{Z} \times GL(3, \mathbb{F}_2), \tag{13.4.48}$$

so that

$$(Y_1H)'N_{00} \cap Y_1N_{00} \subset Y_2N_{00} \tag{13.4.49}$$

(the prime denoting commutator subgroup as usual). Lemmas 13.1.4 and 13.1.5 and calculation in $GL(3, \mathbb{F}_2)$ show that

$$\begin{aligned} H'N_{00} \supset \langle Q_1'N_{00}, Q_2'N_{00}\rangle &= \langle Q_1'N_{00}, Q_2'N_{00}, Q_1 \cap N_0, \tau\rangle \\ &= \langle Q_1'N_{00}, Q_2'N_{00}, Q_1 \cap Q_2\rangle \\ &= \langle Q_1, Q_2\rangle = H \end{aligned} \tag{13.4.50}$$

($\tau \in Q_1'N_{00}$ and $Q_1 \cap N_0 \subset Q_2'N_{00}$), so that $H = H'N_{00}$ and hence

$$\mathbf{C}_H(E) \subset H \cap Y_1N_{00} = H'N_{00} \cap Y_1N_{00} \subset (Y_1H)'N_{00} \cap Y_1N_{00} \subset Y_2N_{00}. \tag{13.4.51}$$

Thus the sequence

$$1 \to Y_2N_{00}/N_{00} \to Y_2H/N_{00} \twoheadrightarrow GL(3, \mathbb{F}_2) \to 1 \tag{13.4.52}$$

is exact.

Now $E \subset Y_2$ defines a self-dual code of type II on the 8-element set A (the elements of E corresponding to the hyperplanes in A), as in Remark 10.1.1, and the natural actions of $GL(3, \mathbb{F}_2)$ on E and on Y_2/E are contragredient under the natural pairing. By abuse of notation, set

$$A = Y_2/E = Y_2N_{00}/N_{00} \tag{13.4.53}$$

(this being naturally equivalent as a $GL(3, \mathbb{F}_2)$-module to the group already designated A), and also set

$$R = Y_2H/N_{00}, \tag{13.4.54}$$

$$G = \mathrm{GL}(3, \mathbb{F}_2) \tag{13.4.55}$$

$$\tilde{P}_1 = Q_1/N_{00} \subset R \tag{13.4.56}$$

$$\tilde{P}_2 = Q_2/N_{00} \subset R \tag{13.4.57}$$

$$\tilde{G} = H/N_{00} = \langle \tilde{P}_1, \tilde{P}_2 \rangle \subset R. \tag{13.4.58}$$

Then G acts faithfully on A under the natural action given by

$$\bar{r} \cdot a = {}^r a \tag{13.4.59}$$

for $r \in R$, $a \in A$. Thus the following lemma shows that

$$\tilde{G} \cap A = 1, \tag{13.4.60}$$

so that

$$\mathbf{C}_H(E) \subset H \cap Y_2N_{00} = N_{00}, \tag{13.4.61}$$

completing the proof of Lemma 13.1.6.

Lemma 13.4.7: *Let $G = \mathrm{GL}(3, \mathbb{F}_2)$ and let A be the G-module $\mathbb{F}_2^3$ consisting of column vectors. Consider the subgroups*

$$P_1 = \left\{\begin{bmatrix} 1 & 0 & 0 \\ * & * & * \\ * & * & * \end{bmatrix}\right\}, \quad P_2 = \left\{\begin{bmatrix} * & * & 0 \\ * & * & 0 \\ * & * & 1 \end{bmatrix}\right\},$$
$$B = P_1 \cap P_2 = \left\{\begin{bmatrix} 1 & 0 & 0 \\ * & 1 & 0 \\ * & * & 1 \end{bmatrix}\right\} \tag{13.4.62}$$

of G. Let

$$1 \to A \to R \xrightarrow{\varphi} G \to 1 \tag{13.4.63}$$

be an exact sequence such that G acts faithfully on A, necessarily according to the given G-module structure:

$${}^r a = \varphi(r) \cdot a \tag{13.4.64}$$

for $r \in R$, $a \in A$. Let $\tilde{P}_1$, $\tilde{P}_2$, $\tilde{B}$ be subgroups of R such that

$$\tilde{P}_i \cap A = 1, \tag{13.4.65}$$

$$\varphi(\tilde{P}_i) = P_i, \quad \varphi(\tilde{B}) = B, \tag{13.4.66}$$

$$\tilde{P}_1 \cap \tilde{P}_2 = \tilde{B} \tag{13.4.67}$$

for $i = 1, 2$. *Then the group*

$$\tilde{G} = \langle \tilde{P}_1, \tilde{P}_2 \rangle \tag{13.4.68}$$

satisfies the conditions

$$\tilde{G} \cap A = 1, \quad \tilde{G} \simeq G. \tag{13.4.69}$$

Everything is now reduced to Lemma 13.4.7 and Proposition 13.5.7, both of which will follow from group cohomology theory.

13.5. Some Group Cohomology

In the proof of Lemma 13.4.7 we shall use some results from group cohomology theory. We now introduce these results, following the exposition in Chapter 7 of [Serre 2].

Let G be a group and set

$$\Lambda = \mathbb{Z}[G], \tag{13.5.1}$$

the group algebra. Let A be a G-module (equivalently, a Λ-module). Consider $\mathbb{Z}$ also as the trivial G-module $\mathbb{Z}$, i.e.,

$$g \cdot n = n \tag{13.5.2}$$

for $g \in G$, $n \in \mathbb{Z}$. The cohomology groups are then

$$H^n(G, A) = \operatorname{Ext}^n_\Lambda(\mathbb{Z}, A). \tag{13.5.3}$$

In particular,

$$H^0(G, A) = \operatorname{Hom}_\Lambda(\mathbb{Z}, A) \simeq \{a \in A \mid g \cdot a = a,\ g \in G\} = A^G, \tag{13.5.4}$$

the subgroup of invariant elements. Since the Ext^n are the right derived functors of Hom, the functors $H^n(G, \cdot)$ are the right derived functors of the left exact functor

$$A \mapsto A^G. \tag{13.5.5}$$

The sequence $(H^n(G, \cdot))_{n \geq 0}$ forms a "connected sequence" of functors [Cartan–Eilenberg] or an "exact ∂-functor" [Grothendieck], i.e., given an exact sequence

$$0 \to A \to B \to C \to 0 \tag{13.5.6}$$

of G-modules there is a homomorphism

$$\partial: H^q(G, C) \to H^{q+1}(G, A) \tag{13.5.7}$$

for each $q \geq 0$ such that the sequence

$$\cdots \to H^q(G, B) \to H^q(G, C) \overset{\partial}{\to} H^{q+1}(G, A) \to H^{q+1}(G, B) \to \cdots \tag{13.5.8}$$

is exact; in addition, ∂ is natural with respect to maps of short exact sequences. If

$$\cdots \to P_1 \to P_0 \to \mathbb{Z} \to 0 \tag{13.5.9}$$

is a projective resolution of the G-module $\mathbb{Z}$, and if we set $K^i = \operatorname{Hom}_G(P_i, A)$, then the cohomology groups $H^n(G, A)$ can be calculated as the cohomology groups of the cochain complex

$$0 \to K^0 \to K^1 \to \cdots. \tag{13.5.10}$$

Let X be an abelian group. Then $\operatorname{Hom}_{\mathbb{Z}}(\Lambda, X)$ becomes a G-module under

$$(g \cdot f)(\lambda) = f(\lambda g) \tag{13.5.11}$$

for $g \in G, f \in \operatorname{Hom}_{\mathbb{Z}}(\Lambda, X)$, $\lambda \in \Lambda$. A G-module is said to be *coinduced* if it is of the form $\operatorname{Hom}_{\mathbb{Z}}(\Lambda, X)$ for some X. Each G-module A injects into a coinduced module; in fact,

$$\begin{aligned} A &\to \operatorname{Hom}_{\mathbb{Z}}(\Lambda, A) \\ a &\mapsto (\lambda \mapsto \lambda \cdot a) \end{aligned} \tag{13.5.12}$$

gives such an injection.

Proposition 13.5.1 [Serre 2]: *For a coinduced module A,*

$$H^q(G, A) = 0 \quad \textit{for all} \quad q \geq 1.$$

Proof: Let $A = \operatorname{Hom}_{\mathbb{Z}}(\Lambda, X)$, for X an abelian group, and let

$$\cdots \to P_1 \to P_0 \to \mathbb{Z} \to 0$$

be a projective resolution of the Λ-module $\mathbb{Z}$. We then have

$$\begin{aligned} K^i &= \operatorname{Hom}_{\Lambda}(P_n, \operatorname{Hom}_{\mathbb{Z}}(\Lambda, X)) \\ &= \operatorname{Hom}_{\mathbb{Z}}(\Lambda \otimes_{\Lambda} P_n, X) \\ &= \operatorname{Hom}_{\mathbb{Z}}(P_n, X), \end{aligned}$$

so that

$$H^q(K) = \operatorname{Ext}^q_{\mathbb{Z}}(\mathbb{Z}, X) = 0$$

for $q \geq 1$. ∎

Let H be a subgroup of G. For a G-module A, let $\iota(A)$ denote A regarded only as an H-module. We shall consider the exact ∂-functor $(H^n(H, \iota(\cdot)), \partial)$. Since

$$A^G \subset A^H, \tag{13.5.13}$$

the inclusion map gives us a natural transformation

$$H^0(G, \cdot) \to H^0(H, \iota(\cdot)). \tag{13.5.14}$$

By the universal property of derived functors (see e.g. [Grothendieck], no. 2.2, 2.3), this extends in a unique way to a morphism of ∂-functors

$$\mathrm{Res}_H^G\colon H^q(G, \cdot) \to H^q(H, \iota(\cdot)) \tag{13.5.15}$$

called *restriction*.

Let $x \in G$. Then the homomorphism

$$\begin{aligned} A^H &\to A^{({}^xH)} \\ a &\mapsto x \cdot a \end{aligned} \tag{13.5.16}$$

gives us a natural transformation

$$H^0(H, \iota(\cdot)) \to H^0({}^xH, \iota(\cdot)). \tag{13.5.17}$$

This also extends to a morphism of ∂-functors

$$c_x\colon H^q(H, \iota(\cdot)) \to H^q({}^xH, \iota(\cdot)). \tag{13.5.18}$$

This is because $H^q(H, \iota(\cdot))$ is "effaceable" for $q \geq 1$ in the sense of [Grothendieck], no. 2.2 (in fact, if A is coinduced for G, then A is coinduced for H, so that $H^q(H, A) = 0$ by Proposition 13.5.1).

Assume now that H has finite index in G. We then have a homomorphism (the *norm*)

$$\begin{aligned} N_{G/H}\colon A^H &\to A^G \\ a &\mapsto \sum_{s \in G/H} sa. \end{aligned} \tag{13.5.19}$$

This gives us a natural transformation

$$H^0(H, \iota(\cdot)) \to H^0(G, \cdot) \tag{13.5.20}$$

which again extends uniquely to a morphism of ∂-functors

$$\mathrm{Cor}_H^G\colon H^q(H, \iota(\cdot)) \to H^q(G, \cdot) \tag{13.5.21}$$

called *corestriction*.

Proposition 13.5.2 ([Cartan–Eilenberg], Chapter XII, Sections 8, 9): *Let B, P, P_1, P_2 be subgroups of G, $B \subset P$, $x, y \in G$. The natural transformations above satisfy the conditions*

$$Res_B^P \operatorname{Res}_P^G = \operatorname{Res}_B^G \tag{13.5.22}$$

$$c_x c_y = c_{xy} \tag{13.5.23}$$

$$\operatorname{Cor}_P^G \operatorname{Cor}_B^P = \operatorname{Cor}_B^G \tag{13.5.24}$$

$$c_x \operatorname{Res}_P^G = \operatorname{Res}_{({}^xP)}^G \tag{13.5.25}$$

$$\operatorname{Cor}_{({}^xP)}^G c_x = \operatorname{Cor}_P^G \tag{13.5.26}$$

$$\operatorname{Cor}_P^G \operatorname{Res}_P^G = |G:P| \tag{13.5.27}$$

$$\operatorname{Res}_{P_1}^G \operatorname{Cor}_{P_2}^G = \sum_{x \in P_1 \backslash G / P_2} \operatorname{Cor}_{P_1 \cap ({}^xP_2)}^{P_1} c_x \operatorname{Res}_{({}^{x^{-1}}P_1) \cap P_2}^{P_2}. \tag{13.5.28}$$

Proof: By the universal property of the ∂-functors it is enough to verify these on H^0, where they follow immediately from the definitions. ∎

We shall be interested in the image of

$$\operatorname{Res}_H^G \colon H^q(G, A) \to H^q(H, \iota(A)). \tag{13.5.29}$$

Let $a = \operatorname{Res}_H^G b$ be in this image. Then

$$c_x \operatorname{Res}_{H \cap ({}^{x^{-1}}H)}^H a = c_x \operatorname{Res}_{H \cap ({}^{x^{-1}}H)}^G b = \operatorname{Res}_{({}^xH) \cap H}^G b = \operatorname{Res}_{({}^xH) \cap H}^H a \tag{13.5.30}$$

for $x \in G$. Conversely, an element $a \in H^q(H, \iota(A))$ is called *stable* [Cartan–Eilenberg] if for all $x \in G$,

$$c_x \operatorname{Res}_{H \cap ({}^{x^{-1}}H)}^H a = \operatorname{Res}_{({}^xH) \cap H}^H a. \tag{13.5.31}$$

Proposition 13.5.3 [Cartan–Eilenberg]: *Let $a \in H^q(H, \iota(A))$ be stable. Then*

$$Res_H^G \operatorname{Cor}_H^G a = |G:H| a. \tag{13.5.32}$$

Proof: Proposition 13.5.2 gives:

$$\begin{aligned} \operatorname{Res}_H^G \operatorname{Cor}_H^G a &= \sum_{x \in H \backslash G / H} \operatorname{Cor}_{({}^xH) \cap H}^H c_x \operatorname{Res}_{H \cap ({}^{x^{-1}}H)}^H a \\ &= \sum_{x \in H \backslash G / H} \operatorname{Cor}_{({}^xH) \cap H}^H \operatorname{Res}_{({}^xH) \cap H}^H a \\ &= \sum_{x \in H \backslash G / H} |H : ({}^xH) \cap H| a \\ &= \sum_{x \in H \backslash G / H} |HxH : H| a = |G:H| a. \quad ∎ \end{aligned} \tag{13.5.33}$$

Corollary 13.5.4 [Cartan–Eilenberg]: *Let G be a finite group, P a Sylow p-subgroup and assume that $H^q(P, \iota(A))$ is a p-group. Then*

$$\mathrm{Res}_P^G\colon H^q(G, A) \to H^q(P, \iota(A)) \tag{13.5.34}$$

maps the p-primary component of $H^q(G, A)$ isomorphically onto the subgroup of stable elements in $H^q(P, \iota(A))$.

Proof: Since $\mathrm{Cor}_P^G \mathrm{Res}_P^G = |G:P|$ is relatively prime to p, the p-primary component maps injectively. The previous proposition gives surjectivity. ∎

We next describe an explicit cochain complex, the so-called inhomogeneous standard complex. For $n \geq 0$, set

$$P_n = T^{n+1}(\Lambda), \tag{13.5.35}$$

the $(n + 1)$st tensor power of Λ, and make P_n into a Λ-module via:

$$\lambda \cdot \lambda_0 \otimes \cdots \otimes \lambda_n = \lambda\lambda_0 \otimes \cdots \otimes \lambda_n \tag{13.5.36}$$

for $\lambda, \lambda_0, \ldots, \lambda_n \in \Lambda$. Define

$$d\colon P_n \to P_{n-1} \tag{13.5.37}$$

by:

$$d(g_0 \otimes \cdots \otimes g_n) = \sum_{i=0}^{n-1} (-1)^i g_0 \otimes \cdots \otimes g_i g_{i+1} \otimes \cdots \otimes g_n + (-1)^n g_0 \otimes \cdots \otimes g_{n-1} \tag{13.5.38}$$

for $g_0, \ldots, g_n \in G$, and

$$\begin{aligned} \varepsilon\colon P_0 = \Lambda &\to \mathbb{Z} \\ \lambda &\mapsto \lambda \cdot 1. \end{aligned} \tag{13.5.39}$$

Then

$$\cdots \to P_1 \xrightarrow{d} P_0 \xrightarrow{\varepsilon} \mathbb{Z} \to 0 \tag{13.5.40}$$

is a projective resolution of the trivial G-module $\mathbb{Z}$, the standard complex for the supplemented algebra (Λ, ε) [Cartan–Eilenberg]. This resolution gives:

Proposition 13.5.5: *The cohomology groups $H^n(G, A)$ can be calculated as the cohomology groups of the complex*

$$0 \to C^0 \xrightarrow{d} C^1 \to \cdots \tag{13.5.41}$$

where

$$C^0(G, A) = A,$$
$$C^n(G, A) = \{f\colon G \times \cdots \times G \to A\} \tag{13.5.42}$$

for $n \geq 1$ *and* $d\colon C^{n-1} \to C^n$ *is given by:*

$$(da)(g) = g \cdot a - a \tag{13.5.43}$$

$$(df)(g_1, \ldots, g_n) = g_1 \cdot f(g_2, \ldots, g_n) + \sum_{i=1}^{n-1} (-1)^i f(g_1, \ldots, g_i g_{i+1}, \ldots, g_n) + (-1)^n f(g_1, \ldots, g_{n-1}) \tag{13.5.44}$$

for $a \in C^0 = A, f \in C^{n-1}$.

We show next that the natural transformation Res_H^G is induced by restriction on the complex above. Set

$$\mathrm{Res}' f = f\big|_{H \times \cdots \times H}$$

for $f \in C^n(G, A)$. These give a commutative diagram

$$\begin{array}{ccccccc} 0 \to & C^0(G, A) & \to & C^1(G, A) & \to \cdots \\ & \| & & \downarrow \mathrm{Res}' & \\ 0 \to & C^0(H, A) & \to & C^1(H, A) & \to \cdots \end{array}$$

which induces

$$\mathrm{Res}'\colon H^n(G, A) \to H^n(H, A).$$

It is easy to see that

$$\mathrm{Res}'\colon (H^n(G, \cdot)) \to (H^n(H, \iota(\cdot)))$$

form a natural transformation of ∂-functors. Since in degree 0

$$\mathrm{Res}'\colon A^G \to A^H$$

is the inclusion map we have $\mathrm{Res}' = \mathrm{Res}_H^G$ in all degrees by the universal property of $(H^n(G, \cdot))_{n \geq 0}$.

We also need the relationship between $H^1(G, A)$, $H^2(G, A)$ and group extensions. As before, let A be a G-module. Let

$$1 \to A \to R \overset{\varphi}{\to} G \to 1 \tag{13.5.45}$$

be an extension of G by A compatible with the given G-module structure on A, i.e.,

$$^{r}a = \varphi(r) \cdot a \tag{13.5.46}$$

for $a \in A, r \in R$. Let $s: G \to R$ be a section of φ. Associate to (R, s) the map

$$f: G \times G \to A \quad (\in C^2(G, A)) \tag{13.5.47}$$

defined by

$$s(g_1)s(g_2) = f(g_1, g_2)s(g_1 g_2). \tag{13.5.48}$$

Then the associative law in R is equivalent to the condition that f be a 2-cocycle. Changing s modifies f by a coboundary. One obtains:

Proposition 13.5.6 [Cartan–Eilenberg]: *The correspondence*

$$R \mapsto f + B^2(G, A) \tag{13.5.49}$$

gives a bijection between the set of equivalence classes of compatible extensions of G by A and $H^2(G, A)$, under which the split extension, the semidirect product $A \rtimes G$, corresponds to the trivial cohomology class.

Note also that if H is a subgroup of G, then $f|_{H\times H}$ is a cocycle of the extension

$$1 \to A \to \varphi^{-1}(H) \to H \to 1.$$

Proposition 13.5.7: *Let G be a finite group, p a prime and B a Sylow p-subgroup of G. Let A be an abelian p-group and let*

$$1 \to A \to R \xrightarrow{\varphi} G \to 1 \tag{13.5.50}$$

be an exact sequence. If

$$1 \to A \to \varphi^{-1}(B) \to B \to 1$$

splits, then (13.5.50) splits.

Proof: Let $b \in H^2(G, A)$ be the class associated to R. Since

$$\varphi^{-1}(B) = A \rtimes \tilde{B} \tag{13.5.51}$$

we have

$$\operatorname{Res}_B^G b = 0. \tag{13.5.52}$$

Hence

$$|G:B|b = \operatorname{Cor}_B^G \operatorname{Res}_B^G b = 0, \tag{13.5.53}$$

so that $b = 0$ and the extension splits. ∎

Now consider the split extension

$$R = A \rtimes G \tag{13.5.54}$$

and for $f: G \to A$ ($\in C^1(G, A)$) set

$$G_f = \{f(g)g \mid g \in G\} \subset R. \tag{13.5.55}$$

One verifies easily:

Proposition 13.5.8: *The subset G_f is a group if and only if f is a 1-cocyle. Two subgroups G_f, $G_{f'}$ are conjugate if and only if $f \equiv f' \bmod B^1(G, A)$, i.e., if and only if f and f' are cohomologous.*

13.6. A Splitting of a Sequence: Proof of Lemma 13.4.7

For the proof of Lemma 13.4.7 we shall use the following theorem on fusion in finite groups:

Theorem 13.6.1 ([Gorenstein 1], Theorem 7.2.7): *Let G be a finite group, P a Sylow p-subgroup, A and C subsets of P conjugate in G. Then there is a sequence of subsets $A = A_0, A_1, \ldots, A_m = C$, subgroups $Q_1, \ldots, Q_m$ of P and elements $x_1, \ldots, x_m$ of G such that*

$$\begin{aligned} A_{i-1} \subset Q_i, \quad A_i \subset Q_i, \\ A_i = {}^{x_i}A_{i-1}, \\ x_i \in \mathbf{N}_G(Q_i) \end{aligned} \tag{13.6.1}$$

for $i = 1, \ldots, m$.

A simple calculation gives:

Lemma 13.6.2: *In $G = \mathrm{GL}(3, \mathbb{F}_2)$, let Q be a nontrivial subgroup of B. Then*

$$\mathbf{N}_G(Q) \subset P_i \tag{13.6.2}$$

for $i = 1$ or 2.

Proof of Lemma 13.4.7: By Proposition 13.5.7, the extension (13.4.63) splits. We may thus assume that $R = A \rtimes G$. Let

$$\begin{aligned} \tilde{P}_i &= \{f_i(p)p \mid p \in P_i\}, \\ a_i &= f_i + B^1(G, A) \in H^1(P_i, A), \\ a &= f_i|_B + B^1(B, A) = \mathrm{Res}_B^{P_i} a_i \in H^1(B, A). \end{aligned} \tag{13.6.3}$$

We want to show that a is stable. Let $x \in G$. Apply Theorem 13.6.1 to $A = B \cap ({}^{x^{-1}}B)$, $C = ({}^{x}B) \cap B$. Since $H^1(\langle 1 \rangle, A) = 0$, we may assume that A and C are nontrivial. Lemma 13.6.2 then shows that there are $x_1, \ldots, x_m \in P_1 \cup P_2$ such that if we set

$$A_k = {}^{x_k \cdots x_1}A, \tag{13.6.4}$$

then $A_k \subset B$ and $x = x_m \cdots x_1$, so that $A_m = C$. Let $x_k \in P_i$, $i = 1$ or 2. Then

$$\begin{aligned} c_{x_k} \operatorname{Res}^B_{A_{k-1}} a &= c_{x_k} \operatorname{Res}^{P_i}_{A_{k-1}} a_i \\ &= \operatorname{Res}^{P_i}_{A_k} a_i = \operatorname{Res}^B_{A_k} a, \end{aligned} \tag{13.6.5}$$

so that by induction on m,

$$c_x \operatorname{Res}^B_{B \cap ({}^{x^{-1}}B)} a = \operatorname{Res}^B_{({}^{x}B) \cap B} a \tag{13.6.6}$$

and a is stable. Proposition 13.5.3 thus gives:

$$a = |G:B| a = \operatorname{Res}^G_B \operatorname{Cor}^G_B a. \tag{13.6.7}$$

Let f be a cocycle representing $\operatorname{Cor}^G_B a$. Then G_f contains a conjugate of $\tilde{B}$. Conjugating, we may assume that $G_f \supset \tilde{B}$. We have

$$\begin{aligned} a_i &= \operatorname{Cor}^{P_i}_B \operatorname{Res}^{P_i}_B a_i = \operatorname{Cor}^{P_i}_B a \\ &= \operatorname{Cor}^{P_i}_B \operatorname{Res}^{P_i}_B \operatorname{Res}^G_{P_i} \operatorname{Cor}^G_B a \\ &= \operatorname{Res}^G_{P_i} \operatorname{Cor}^G_B a, \end{aligned} \tag{13.6.8}$$

and so

$$(P_i)_{(f|P_i)} \tag{13.6.9}$$

is conjugate to $\tilde{P}_i$ for $i = 1, 2$. Since both $(P_i)_{(f|P_i)}$ and $\tilde{P}_i$ contain $\tilde{B}$, a conjugating element must lie in

$$\mathbf{C}_A(B) = \left\{ 0, v_3 = \begin{bmatrix} 0 \\ 0 \\ 1 \end{bmatrix} \right\}. \tag{13.6.10}$$

Since v_3 commutes with P_2, it follows that

$$\begin{aligned} \tilde{P}_1 &= (P_1)_{(f|P_1)} \quad \text{or} \quad {}^{v_3}(P_1)_{(f|P_1)} \\ \tilde{P}_2 &= (P_2)_{(f|P_2)} = {}^{v_3}(P_2)_{(f|P_2)}, \end{aligned} \tag{13.6.11}$$

so that

$$\tilde{G} = \langle \tilde{P}_1, \tilde{P}_2 \rangle = G_f \quad \text{or} \quad {}^{v_3}G_f, \tag{13.6.12}$$

as required. ∎

Appendix: Complex Realization of Vertex Operator Algebras

Many facts concerning algebras of vertex operators studied in this book, in the case when $\mathbb{F}$ is the complex field $\mathbb{C}$, admit a natural interpretation in terms of elementary complex analysis. In particular, the convergence of power series automatically provides maps between algebras of rational functions and power series algebras as discussed in Section 8.10. The operations associated with expansions of zero introduced in Chapter 8 are replaced by contour integrals and the Cauchy residue formula. In this appendix we shall give a sampling of this alternative approach. In particular, we shall reprove the Jacobi identity for general vertex operators associated with the space V_L, in the case when L is a positive definite even lattice. Though we shall assume some definitions and elementary facts from the main text of the book, we shall try to make this exposition self-contained, as far as possible. This approach can in principle be extended to non-even lattices and to the twisted case studied in Chapters 8 and 9. Our goal, however, is to build a bridge between the two techniques so that the reader, if he or she wishes, can reinterpret the results concerning general vertex operators in terms of the complex realization.

The application of complex analysis to vertex operators was first employed in the early days of string theory, in particular, contour integrals (see the Introduction for discussion and physics references). These techniques were rediscovered in the construction of the basic representations of affine Lie algebras [Frenkel–Kac]. Recently this analytic approach has been extended to non-basic standard representations [Tsuchiya–Kanie 2, 3] where nontrivial holonomy groups naturally arise and the appearance of these groups leads to further relations between vertex operator algebras and complex analysis.

In Section A.1 we provide a rigorous foundation of the complex variable approach. We use absolute convergence to define and discuss such matters as the composition of linear operators on infinite-dimensional graded spaces. We are mainly interested in applying these considerations to general vertex operators. Such an approach was developed in [Lepowsky–Wilson 2]. This section can be regarded as an alternative to the formal variable technique described in Chapters 2 and 8.

Section A.2 is the main part of the appendix. Here we reprove some of the basic results of Chapter 8, including the Jacobi identity. We start by proving "commutativity" and translation properties of general vertex operators using the specific structure of the space V_L. Afterward, we prove the "associativity" property and then skew-symmetry and two forms of the Jacobi identity using only the general properties of a vertex operator algebra, as defined in Section 8.10. The same method can also be used in other examples of vertex operator algebras.

In Section A.3 we establish an explicit relation between the complex approach and the formal variable approach.

A.1. Linear Algebra of Infinite Direct Sums

We first generalize some basic facts of linear algebra. Let V be a complex vector space and let End V be the algebra of linear operators on V. We need, however, to consider more general linear operators which map elements of V into infinite sums of such elements. One possible way to extend the algebra of operators End V is to introduce a topology on V. We prefer instead to work in an algebraic setting. The additional structure of V that we shall assume is a gradation

$$V = \coprod_{i \in I} V_i, \qquad \text{(A.1.1)}$$

where I is a countable set and each V_i, $i \in I$, is a finite-dimensional vector space. We also introduce the corresponding infinite product

$$\bar{V} = \prod_{i \in I} V_i, \qquad \text{(A.1.2)}$$

and we consider the space of linear operators from V into $\bar{V}$. The space of these linear operators is no longer an algebra, since we cannot in general define products of such operators. In order to restore the product operation at least partially it will be useful to present a linear operator $A: V \to \bar{V}$ as an "infinite matrix" $(A_i^j)_{i,j \in I}$. Let V' be the "graded dual space" of V, the direct sum of the dual spaces of the finite-dimensional spaces V_i:

$$V' = \coprod_{i \in I} V_i^*. \tag{A.1.3}$$

Then $(V')^* = \bar{V}$. We denote by $\langle \cdot, \cdot \rangle$ the natural pairing between V' and $\bar{V}$. Then $A_i^j: V_i \to V_j$, $i, j \in I$, is the linear transformation such that for all $v \in V_i$, $u \in V_j^*$,

$$\langle u, Av \rangle = \langle u, A_i^j v \rangle. \tag{A.1.4}$$

The linear operator $A: V \to \bar{V}$ completely determines its matrix $(A_i^j)_{i,j \in I}$ and vice versa.

Definition A.1.1: Let $A = (A_i^j)_{i,j \in I}$ and $B = (B_i^j)_{i,j \in I}$ be linear operators from V into $\bar{V}$. We then say that the product $AB: V \to \bar{V}$ exists if and only if the series

$$\sum_{j \in I} \langle u, A_j^k B_i^j v \rangle \tag{A.1.5}$$

are absolutely convergent for any $i, k \in I$ and $v \in V_i$, $u \in V_k^*$. In this case we define $AB = ((AB)_i^k)_{i,k \in I}$ by letting

$$\langle u, (AB)_i^k v \rangle \tag{A.1.6}$$

be the sum (A.1.5).

Clearly, AB is a well-defined linear operator from V into $\bar{V}$. In essence the multiplication of two operators is defined in terms of multiplication of their infinite matrices with finite-dimensional linear operators as their entries. We denote by $\overline{\text{End}}\, V$ the linear space of operators from V into $\bar{V}$ equipped with the above product of operators, defined only for appropriate pairs of operators.

It follows from Definition A.1.1 that if $A \in \overline{\text{End}}\, V$ and $B \in \text{End}\, V$, or $B \in \overline{\text{End}}\, V$ and A is dual to an operator in $\text{End}\, V'$, then $AB \in \overline{\text{End}}\, V$. Also, the absolute convergence of the series (A.1.5) implies the associativity of the product in $\overline{\text{End}}\, V$, namely, if $A, B, C \in \overline{\text{End}}\, V$ and if the products AB, BC, $(AB)C$ and $A(BC)$ exist then

$$(AB)C = A(BC). \tag{A.1.7}$$

The product of n operators $A_1, \ldots, A_n \in \overline{\text{End}}\, V$, with a given arrangement of parentheses, is defined by means of a repeated application of Definition A.1.1. If the product is defined for all possible arrangements of parentheses then the associativity implies that it does not depend on the arrangement, and the parentheses in the product $A_1 A_2 \cdots A_n$ can be omitted. We also introduce infinite sums and infinite products:

Definition A.1.2: Let $A_n = ((A_n)_i^j)_{i,j \in I}$, $n = 1, 2, \ldots$ be linear operators in $\overline{\text{End}}\, V$. We say that the infinite sum

$$A = \sum_{n \geq 1} A_n \tag{A.1.8}$$

exists if the series

$$\sum_{n \geq 1} \langle u, (A_n)_i^j v \rangle \tag{A.1.9}$$

are absolutely convergent for any $i, j \in I$ and $v \in V_i$, $u \in V_j^*$. In this case we define $A = (A_i^j)_{i,j \in I}$ by setting $\langle u, A_i^j v \rangle$ equal to the sum (A.1.9).

The absolute convergence of (A.1.9) implies that if the infinite sum (A.1.8) exists and if $A_n B$, $n = 1, 2, \ldots$, and AB exist then the infinite sum $\sum_{n \geq 1} A_n B$ also exists and is equal to AB. One can also define the infinite product $A = \prod_{n \geq 1} A_n$ as the limit of the partial products $A_1 \cdots A_n$ as $n \to \infty$, when all the partial products are well-defined.

Next we generalize tensor algebra to infinite direct sums of vector spaces. We consider the nth tensor power

$$T^n(V) = V \otimes \cdots \otimes V \quad (n \text{ times}). \tag{A.1.10}$$

Then the vector space $T^n(V)$ is itself an infinite direct sum indexed by $I \times \cdots \times I$ (n times); thus $\overline{T^n(V)}$ is defined. We shall consider the space of linear operators

$$A\colon T^m(V) \to \overline{T^n(V)}, \quad m, n \in \mathbb{N},$$

which we denote by $\overline{\text{Hom}}(T^m(V), T^n(V))$. In the special case when $m = n = 1$, the space of linear operators with the partial product was denoted $\overline{\text{End}}\, V$ above. The elements of $\overline{\text{Hom}}(T^m(V), T^n(V))$ are again represented by generalized matrices $A = (A_{i\cdots j}^{k\cdots l})_{i,\ldots,j,k,\ldots,l \in I}$ such that for $v^i \in V_i, \ldots, v^j \in V_j$, $u_k \in V_k^*, \ldots, u_l \in V_l^*$, we have

$$\begin{aligned} &\langle u_k \otimes \cdots \otimes u_l, A v^i \otimes \cdots \otimes v^j \rangle \\ &\quad = \langle u_k \otimes \cdots \otimes u_l, A_{i\cdots j}^{k\cdots l} v^i \otimes \cdots \otimes v^j \rangle. \end{aligned} \tag{A.1.11}$$

We can define composition of these multilinear operators as in the usual case but only for special types of operators. Here we give just a sample of such a definition. The reader can generalize this definition without any difficulty.

Definition A.1.3: Let $A = (A_{ij}^k)_{i,j,k \in I}$ and $B = (B_{ij}^k)_{i,j,k \in I}$ belong to $\overline{\text{Hom}}(V \otimes V, V)$. Then we say that the composition

$$C(\cdot, \cdot, \cdot) = A(B(\cdot, \cdot), \cdot)$$

exists if the series

$$\sum_{k \in I} \langle v_m, A_{kl}^m B_{ij}^k v^i \otimes v^j \otimes v^l \rangle \tag{A.1.12}$$

are absolutely convergent for any $i, j, l, m \in I$ and $v^i \in V_i$, $v^j \in V_j$, $v^l \in V_l$, $v_m \in V_m^*$. In this case we define

$$C = (C_{ijl}^m)_{i,j,l,m \in I} \in \overline{\text{Hom}}(T^3(V), V)$$

by setting

$$\langle v_m, C_{ijl}^m v^i \otimes v^j \otimes v^l \rangle \tag{A.1.13}$$

equal to the sum (A.1.12).

Remark A.1.4: All these considerations remain valid for a general complete normed field.

Next we consider families of operators in $\overline{\text{End}}\, V$ depending on parameters. One can define integration, differentiation and other operations on such families of operators using appropriate analytic results about differentiation or integration of infinite series, thus combining linear algebra and analysis. For the application to the vertex operator algebras studied in Section A.2 we shall only need elementary facts about rational functions and their power series expansions.

Let $A(z_1, \ldots, z_n) \in \overline{\text{End}}\, V$ be an operator-valued function defined on an open domain $U \subset \mathbb{C}^n$ such that

$$\langle u, A(z_1, \ldots, z_n) v \rangle \tag{A.1.14}$$

is a rational function for any $u \in V'$, $v \in V$. We call it a *rational operator-valued function* (*on U with values in* $\overline{\text{End}}\, V$).

We also call the function $v(z_1, \ldots, z_n) \in \bar{V}$, defined on U, a *rational vector-valued function* (*on U with values in* $\bar{V}$), if $\langle u, v(z_1, \ldots, z_n) \rangle$ is a rational function for any $u \in V'$. For the domain U we denote by $\mathcal{R}(U)$, $\mathcal{R}(U, \bar{V})$ and $\mathcal{R}(U, \overline{\text{End}}\, V)$, respectively, the space of rational complex-valued functions, rational vector-valued functions and rational operator-valued functions, everywhere defined on U. The space $\mathcal{R}(U)$ has a canonical algebra structure, while the other two spaces are modules for this algebra.

Definition A.1.5: Let T be an operator defined on a subspace of $\mathcal{R}(U)$ with values in $\mathcal{R}(U')$, where U and U' are open domains in $\mathbb{C}^m$ and $\mathbb{C}^n$, respectively. Let $v(z_1, \ldots, z_m) \in \mathcal{R}(U, \bar{V})$ and $A(z_1, \ldots, z_m) \in \mathcal{R}(U, \overline{\text{End}\, V})$. We define $(Tv)(z_1, \ldots, z_n)$ and $(TA)(z_1, \ldots, z_n)$ by the identities

$$\langle u, (Tv)(z_1, \ldots, z_n)\rangle = T\langle u, v(z_1, \ldots, z_n)\rangle \tag{A.1.15}$$

$$\langle u, (TA)(z_1, \ldots, z_n)v\rangle = T\langle u, A(z_1, \ldots, z_n)v\rangle \tag{A.1.16}$$

for all $u \in V'$, $v \in V$.

The main examples of operators T that we shall consider are differentiation $\partial/\partial z_i$, contour integration $(1/2\pi i)\int_{C_i} \cdots dz_i$, multiplication by a rational function, translation by a vector, and the operation of analytic continuation, which we denote by R. The most common domains that we shall consider are the following:

$$U_n = \{(z_1, \ldots, z_n) \in (\mathbb{C}^\times)^n \mid z_i \neq z_j \quad \text{for} \quad i \neq j\} \tag{A.1.17}$$

$$U_n^+ = \{(z_1, \ldots, z_n) \in U_n \mid |z_i| > |z_j| \quad \text{for} \quad i < j\}. \tag{A.1.18}$$

We shall also often consider the algebra

$$\mathcal{R}_n = \mathbb{C}[z_i, z_i^{-1}, (z_i - z_j)^{-1}, i \neq j, i, j = 1, \ldots, n]. \tag{A.1.19}$$

Clearly, $\mathcal{R}(U_n) = \mathcal{R}_n$.

We end this section with an example. Let $A \in \mathcal{R}(U_n^+, \overline{\text{End}\, V})$ and let $\langle u, A(z_1, \ldots, z_n)v\rangle \in \mathcal{R}_n$ for all $u \in V'$, $v \in V$. Then the analytic continuation $RA \in \mathcal{R}(U_n, \overline{\text{End}\, V})$ is well defined and

$$\langle u, (RA)(z_1, \ldots, z_n)v\rangle = \langle u, A(z_1, \ldots, z_n)v\rangle \tag{A.1.20}$$

for $(z_1, \ldots, z_n) \in U_n^+$, $u \in V'$, $v \in V$.

A.2. The Vertex Operator Algebra V_L

In this section we study the algebra of general vertex operators using the generalization of linear algebra developed in the previous section. The grading used here is that defined by weights, as in Section 8.10, not by degrees.

Let L be a positive definite even lattice of finite rank with symmetric form $\langle \cdot, \cdot \rangle$, and let $(\hat{L}, {}^-)$ be its central extension by the group $\langle \pm 1 \rangle$ with commutator map defined by

$$aba^{-1}b^{-1} = (-1)^{\langle \bar{a}, \bar{b} \rangle}, \quad a, b \in \hat{L}, \tag{A.2.1}$$

as in Section 8.8.

We define V_L as in Section 7.1 with the gradation (7.1.40):

$$V_L = \coprod_{n \geq 0} (V_L)_{(n)}. \tag{A.2.2}$$

(The grading used here is that defined by weights, as in Section 8.10, not by degrees.) For any $v \in V_L$, $z \in \mathbb{C}^\times$, we define the general vertex operator $Y(v, z)$ by the same formulas as in Sections 8.4, 8.5, which however should be interpreted in the sense of Section A.1. Namely, for any $\alpha \in \mathfrak{h}$, $z \in \mathbb{C}^\times$, $n \in \mathbb{N}$, we note that $E^+(\alpha, z)$, $(d/dz)^n \alpha(z)^+ \in \operatorname{End} V_L$ and $E^-(\alpha, z)$, $(d/dz)^n \alpha(z)^-$ are dual to operators in $\operatorname{End} V_L'$, where $E^\pm(\alpha, z)$, $\alpha(z)^\pm$ are as in (8.4.1), (8.4.10). The discussion following Definition A.1.1 shows that the product defining $Y(v, z) \in \overline{\operatorname{End}}\, V_L$ exists. Moreover, for $w \in V_L$, $w' \in V_L'$, the matrix coefficients of $Y(v, z)$ satisfy

$$\langle w', Y(v, z)w \rangle \in \mathbb{C}[z, z^{-1}] \tag{A.2.3}$$

since they are built from matrix coefficients of polynomials in az^k and $\alpha(n)z^m$, $a \in \hat{L}$, $\alpha \in \mathfrak{h}$, $k, m, n \in \mathbb{Z}$. It is easy to see that the power series expansion

$$Y(v, z) = \sum_{n \in \mathbb{Z}} v_n z^{-n-1} \tag{A.2.4}$$

yields the same component operators $v_n \in \operatorname{End} V_L$ as its formal counterpart (8.5.15). The infinite sum should now be understood, however, in terms of Definition A.1.2.

We also recall from (8.5.8) that

$$\lim_{z \to 0} Y(v, z)\iota(1) = v. \tag{A.2.5}$$

Next we consider products of vertex operators.

Proposition A.2.1: *For $v^1, \ldots, v^n$, $w \in V_L$, $w' \in V_L'$ and $z_1, \ldots, z_n \in \mathbb{C}^\times$, we have:*
(i) The product of vertex operators

$$Y(v^1, z_1) \cdots Y(v^n, z_n)$$

exists in the domain U_n^+ of (A.1.18).

(ii) The matrix coefficient

$$\langle w', Y(v^1, z_1) \cdots Y(v^n, z_n)w\rangle$$

is, in the domain U_n^+, equal to a rational function in the algebra $\mathcal{R}_n$ of (A.1.19).

(iii) The rational operator-valued function

$$R(Y(v^{\sigma 1}, z_{\sigma 1}) \cdots Y(v^{\sigma n}, z_{\sigma n})) \tag{A.2.6}$$

does not depend on the permutation $\sigma \in \mathcal{S}_n$.

Proof: We shall first consider the case $v^i = a_i \in \hat{L}$, $i = 1, \ldots, n$. The definition of normal ordered product (8.4.13)–(8.4.15) and an argument analogous to the one for $Y(v, z)$ imply that

$$\langle w', {}_{\circ}^{\circ} Y(a_1, z_1) \cdots Y(a_n, z_n) {}_{\circ}^{\circ} w\rangle \in \mathbb{C}[z_i, z_i^{-1}, i = 1, \ldots, n]. \tag{A.2.7}$$

In order to relate the product of the vertex operators with the normal ordered product we recall the identity (4.3.1), which we view in the sense of Definition A.1.1. Thus for $|z_i| > |z_j|$ we have

$$E^+(\bar{a}_i, z_i)E^-(\bar{a}_j, z_j) = E^-(\bar{a}_j, z_j)E^+(\bar{a}_i, z_i)(1 - z_j/z_i)^{\langle \bar{a}_i, \bar{a}_j\rangle}. \tag{A.2.8}$$

Clearly the right-hand side of (A.2.8) is well defined in the sense of Definition A.1.1. In order to see that the left-hand side is also well defined we write the series expansion of (A.2.8),

$$\sum_{m,n \geq 0} (E_m^+(\bar{a}_i)z_i^{-m})(E_n^-(\bar{a}_j)z_j^n) = \sum_{k,m,n \geq 0} (E_n^-(\bar{a}_j)z_j^n)(E_m^+(\bar{a}_i)z_i^{-m})\left(c_k\left(\frac{z_j}{z_i}\right)^k\right),$$

using obvious notation; in particular, the c_k are appropriate binomial coefficients. We obtain the absolute convergence of the matrix coefficients of the left-hand side using the known absolute convergence of the right-hand side. This implies that the series in $z_1, \ldots, z_n$ defining the matrix coefficient of the operator product $\langle w', Y(a_1, z_1) \cdots Y(a_n, z_n)w\rangle$ is absolutely convergent for $|z_i| > |z_j|$, proving (i) in the special case. We also recall (7.1.38):

$$z_i^{\bar{a}_i} a_j = a_j z_i^{\bar{a}_i} z_i^{\langle \bar{a}_i, \bar{a}_j\rangle}.$$

Thus we arrive at the identity [see (8.4.25)]

$$Y(a_1, z_1) \cdots Y(a_n, z_n) = \prod_{i<j} (z_i - z_j)^{\langle \bar{a}_i, \bar{a}_j\rangle} {}_{\circ}^{\circ} Y(a_1, z_1) \cdots Y(a_n, z_n) {}_{\circ}^{\circ} \tag{A.2.9}$$

valid for $|z_i| > |z_j|$, $i < j$, which in combination with (A.2.7) implies (ii) and (iii) in the special case.

To prove the general case we note that a general vertex operator can be expressed as a linear combination of vertex operators of the form

$$Y(v, z) = P\left(\frac{\partial}{\partial z_i}, i = 1, \ldots, m\right) {}^{\circ}_{\circ} Y(a_1, z_1) \cdots Y(a_m, z_m) {}^{\circ}_{\circ} \Big|_{z_i = z} \quad \text{(A.2.10)}$$

where $a_1, \ldots, a_m \in \hat{L}$ and P is a polynomial in m variables (see the proof of Theorem 8.6.1). Now we can repeat the same argument as in the special case above, replacing $Y(a_k, z_k)$ by ${}^{\circ}_{\circ} Y(a_{k1}, z_{k1}) \cdots Y(a_{km_k}, z_{km_k}) {}^{\circ}_{\circ}$. All the statements are still valid for the algebra with enlarged set of variables and the same permutation group $\mathcal{S}_n$. The statements also remain true when in addition we take partial derivatives. Finally, equating the appropriate variables yields the general statement. ∎

Remark A.2.2: The letter R in (A.2.6) refers to the term "rational", as we explained at the end of Section A.1. However, in this particular case it is consistent with the physical notion of "radial" ordering. Proposition A.2.1(iii) can be thought of as "commutativity" of general vertex operators. Note however that the domains of definition of the products in (A.2.6) do not intersect, so that the vertex operators in general do not commute. "Commutativity" in physics terminology is one of the aspects of "duality."

Remark A.2.3: The matrix coefficients of the vertex operators studied in Chapters 8 and 9 are not necessarily rational functions but rather algebraic functions. In this case analytic continuation yields a family of operators parametrized by configurations of points on a Riemann surface. In this appendix we confine ourselves to the case of rational matrix coefficients and therefore to single-valued analytic continuation.

Recall from Section 8.7 the action of the Virasoro algebra on V_L by means of the operators $L(n)$, $n \in \mathbb{Z}$, with the central element c acting as the scalar rank L. The operator $L(-1)$ plays an especially important role and provides a translation property of general vertex operators.

Proposition A.2.4: *For $v \in V_L$, $z, z_0 \in \mathbb{C}^\times$ we have*

(i)
$$[L(-1), Y(v, z)] = \frac{d}{dz} Y(v, z)$$

(ii)
$$e^{zL(-1)} v = Y(v, z)\iota(1).$$

(iii) The product $e^{z_0 L(-1)} Y(v, z) e^{-z_0 L(-1)}$ *exists for* $|z_0| < |z|$ *and is the rational operator-valued function such that*

$$R(e^{z_0 L(-1)} Y(v, z) e^{-z_0 L(-1)}) = Y(v, z + z_0).$$

Proof: (i) We first consider the case $v = \imath(a)$. Using the elementary fact

$$[L(-1), \alpha(n)] = -n\alpha(n - 1)$$

we obtain

$$[L(-1), E^{\pm}(\bar{a}, z)] = -(\bar{a}(z)^{\pm} + e^{\pm})E^{\pm}(\bar{a}, z),$$

where $e^{+} = 0$, $e^{-} = -\alpha(-1)$.

On the other hand,

$$\frac{d}{dz} E^{\pm}(\bar{a}, z) = -(\bar{a}(z)^{\pm} + f^{\pm})E^{\pm}(\bar{a}, z),$$

where $f^{+} = -\alpha(0)z^{-1}$, $f^{-} = 0$. Taking the term $az^{\bar{a}}$ of $Y(a, z)$ into account we get the result.

Thus for a product of vertex operators we obtain

$$[L(-1), Y(a_1, z_1) \cdots Y(a_n, z_n)] = \left(\sum_{i=1}^{n} \frac{\partial}{\partial z_i}\right) Y(a_1, z_1) \cdots Y(a_n, z_n),$$

which implies

$$[L(-1), \mathop{\circ}\limits^{\circ} Y(a_1, z_1) \cdots Y(a_n, z_n) \mathop{\circ}\limits^{\circ}] = \left(\sum_{i=1}^{n} \frac{\partial}{\partial z_i}\right) \mathop{\circ}\limits^{\circ} Y(a_1, z_1) \cdots Y(a_n, z_n) \mathop{\circ}\limits^{\circ}$$

since $\sum_{i=1}^{n} \partial/\partial z_i$ commutes with $\prod_{i<j}(z_i - z_j)^{\langle \bar{a}_i, \bar{a}_j \rangle}$. The presentation (A.2.10) now gives (i) for arbitrary $v \in V_L$.

(ii) This follows from (i) and (A.2.5).

(iii) Let us consider the Taylor series expansion

$$\langle w', Y(v, z + z_0)w \rangle = \sum_{n \geq 0} \frac{z_0^n}{n!} \left\langle w', \left(\frac{d}{dz}\right)^n Y(v, z)w \right\rangle.$$

Since the matrix coefficients of a general vertex operator are Laurent polynomials by (A.2.3), the Taylor series is absolutely convergent for $|z_0| < |z|$, and by (i) it is equal to the series expansion of $e^{z_0 L(-1)} Y(v, z) e^{-z_0 L(-1)}$. Thus via Definition A.1.1 the latter operator is well defined in this domain and is equal to $Y(v, z + z_0)$. ∎

Besides products one can define other types of composition of general vertex operators. Since the general vertex operator $Y(v, z)$ depends linearly on $v \in V_L$, one can consider what we shall call the *universal vertex operator*

$$Y(\cdot, z)\cdot : V_L \times V_L \to \bar{V}_L, \quad z \in \mathbb{C}^{\times}. \tag{A.2.11}$$

To study compositions of universal vertex operators we apply the generalization of tensor algebra described in Section A.1. In particular, Proposition A.2.1(i),(ii) for $n = 2$ implies that

$$Y(\cdot, z_1)Y(\cdot, z_2)\cdot : T^3(V_L) \to \bar{V}_L \tag{A.2.12}$$

is a rational operator-valued function in the domain

$$U_2^+ = \{(z_1, z_2) \mid |z_1| > |z_2| > 0\}.$$

Specializing the arguments $u, v \in V_L$ yields an operator in $\overline{\text{End}}\, V_L$. In spite of the notation of the universal vertex operator one should treat both arguments of (A.2.11) in a symmetric way. Another type of composition of two universal vertex operators is

$$Y(Y(\cdot, z_1)\cdot, z_2)\cdot : T^3(V_L) \to \bar{V}_L. \tag{A.2.13}$$

The next proposition establishes conditions under which this composition is well defined.

Proposition A.2.5: *For $u, v \in V_L$, $Y(Y(u, z_0)v, z)$ is a rational operator-valued function defined in the domain $|z| > |z_0| > 0$. We have the identity*

$$R(Y(Y(u, z_0)v, z)) = R(Y(u, z + z_0)Y(v, z)). \tag{A.2.14}$$

Proof: For $u, v, w \in V_L$, $(z, z_1, z + z_0) \in U_3$ we consider the element in $\bar{V}_L$

$$R(Y(u, z + z_0)Y(v, z)Y(w, z_1)\iota(1). \tag{A.2.15}$$

By Proposition A.2.1(iii) it is equal to

$$R(Y(w, z_1)Y(u, z + z_0)Y(v, z)\iota(1).$$

By Proposition A.2.4(ii) it is equal to

$$R(Y(w, z_1)Y(u, z + z_0)e^{zL(-1)}v).$$

By Proposition A.2.4(iii) it is equal to

$$R(Y(w, z_1)e^{zL(-1)}Y(u, z_0)v).$$

We expand the latter vector-valued function as a Laurent series in z_0. This can be done for $|z_0| < |z - z_1|$, since the Laurent series coincides with the one for (A.2.15). By Proposition A.2.4(ii) each term in this Laurent series is equal to the corresponding term of the Laurent series of

$$R(Y(w, z_1)Y(Y(u, z_0)v, z)\iota(1)$$

By Proposition A.2.1(iii) each term in this series is equal to the corresponding term in the Laurent series of

$$R(Y(Y(u, z_0)v, z)Y(w, z_1)\iota(1). \tag{A.2.16}$$

The equality of the Laurent series in z_0 defined by (A.2.15) and (A.2.16) in the limit $z_1 \to 0$ implies (A.2.14) as an equality of Laurent series in z_0. Since the right-hand side of (A.2.14) is a rational operator-valued function so is the left-hand side. The domain of the definition of the operator $Y(Y(u, z_0)v, z)$ can be deduced from the expansion of the matrix coefficient $R\langle w', Y(u, z + z_0)Y(v, z)w\rangle$ in z_0 and the form of this expansion given by Proposition A.2.1(ii). ∎

Remark A.2.6: The above Proposition A.2.5 can easily be generalized to compositions of n general vertex operators. In particular, for $u, v, w \in V_L$,

$$Y(Y(Y(u, z_1)v, z_2)w, z_3) \tag{A.2.17}$$

is a rational operator-valued function in the domain $0 < |z_1| < |z_2| < |z_3|$, $|z_1 + z_2| < |z_3|$. One can also consider more general operators by combining Propositions A.2.1, A.2.4 and A.2.5. The domain of definition is always determined by the associated ring of rational functions.

Remark A.2.7 (cf. Remark 8.6.4 and Section 8.10): The identity (A.2.14) implies in particular that for $|z_1| > |z_2| > |z_1 - z_2| > 0$,

$$Y(u, z_1)Y(v, z_2) = \sum_{n \geq N} (z_1 - z_2)^n Y(w^n, z_2)$$

for appropriate $w^n \in V_L$ [the identity (A.2.14) implies that $w^n = u_{-n-1}v$] and $N \in \mathbb{Z}$ depending on u, v. This expansion is called the "operator product expansion" in the physics literature, and the identity (A.2.14) is often referred to as the "associativity" of the operator product expansion. "Associativity" is another aspect of duality.

Although the identity (A.2.14) is not an associativity in the usual sense, it does, in combination with "commutativity" (Proposition A.2.1), imply an analogue of the Jacobi identity for a Lie algebra $\mathfrak{g}$ written in the adjoint form [see (8.8.29)]:

$$[\mathrm{ad}\, x_1, \mathrm{ad}\, x_2] = \mathrm{ad}[x_1, x_2], \quad x_1, x_2 \in \mathfrak{g}.$$

This identity will be our next objective. We shall also obtain an analogue of skew-symmetry and of the Jacobi identity in its symmetric form.

Since all the operators we consider in this section have rational matrix coefficients, one can take contour integrals of these operators. This will imply various identities among the components of general vertex operators, in particular, the component form of the Jacobi identity. One has to be careful, however, with the domains of definition of compositions of vertex operators.

The identity (A.2.4) is equivalent to

$$\frac{1}{2\pi i} \int_C Y(v, z) z^n \, dz = v_n \tag{A.2.18}$$

where C is any (counterclockwise) contour surrounding $0 \in \mathbb{C}$. Similarly,

$$\frac{1}{(2\pi i)^2} \int_{C_1} \int_{C_2} Y(u, z_1) Y(v, z_2) z_1^m z_2^n \, dz_2 \, dz_1 = u_m v_n \tag{A.2.19}$$

where C_1 and C_2 are as above and

$$C_1 \subset \{z_1 \in \mathbb{C} \,|\, |z_1| > R\}, \quad C_2 \subset \{z_2 \in \mathbb{C} \,|\, |z_2| < R\}$$

for some R since the product of two vertex operators is well defined only for $|z_1| > |z_2|$. This implies that

$$\frac{1}{(2\pi i)^2} \int_{C_1} \int_{C_2} R(Y(u, z_1) Y(v, z_2)) z_1^m z_2^n \, dz_2 \, dz_1 = u_m v_n \tag{A.2.20}$$

whenever C_1 and C_2 are any two contours surrounding $0 \in \mathbb{C}$ such that C_1 also surrounds C_2. We also note that the left-hand sides of (A.2.19) and (A.2.20) are well defined if we replace $z_1^m z_2^n$ by any rational function $f(z_1, z_2)$, defined in the domain U_2. The Jacobi identity (8.8.29) will follow from Propositions A.2.1 and A.2.5.

For a rational operator-valued function $A(z)$ defined in a punctured neighborhood U of $a \in \mathbb{C}$, we set

$$\mathrm{Res}_{z=a} A(z) = \frac{1}{2\pi i} \int_{C^\varepsilon(a)} A(z) \, dz \tag{A.2.21}$$

where $C^\varepsilon(a)$ is a contour surrounding the point a which does not contain any singularity in $U \backslash \{a\}$. In particular $\mathrm{Res}_{z=0}$ corresponds to the notation Res_z of the formal variable approach [see (8.4.33)].

Proposition A.2.8: *Let* $f(z_1, z_2) \in \mathbb{C}[z_1, z_1^{-1}, z_2, z_2^{-1}, (z_1 - z_2)^{-1}]$. *Then for* $u, v \in V_L$ *we have the Jacobi identity:*

$$\begin{aligned} &Res_{z_1=0} \mathrm{Res}_{z_2=0}(Y(u, z_1)Y(v, z_2)f(z_1, z_2)) \\ &\quad - \mathrm{Res}_{z_2=0} \mathrm{Res}_{z_1=0}(Y(v, z_2)Y(u, z_1)f(z_1, z_2)) \\ &= \mathrm{Res}_{z_2=0} \mathrm{Res}_{z_1=z_2}(Y(Y(u, z_1 - z_2)v, z_2)f(z_1, z_2)) \end{aligned}$$

or in component form, (8.8.41).

Proof: Let us denote by $C_i^r(z)$ the circular contour with counterclockwise orientation of radius r around the point $z \in \mathbb{C}$ in the variable z_i, and let $C_i^r = C_i^r(0)$. We fix radii $R > \rho > r$ and $\varepsilon < \min(R - \rho, \rho - r)$. Then the identity can be rewritten as follows:

$$\begin{aligned} &\frac{1}{(2\pi i)^2} \int_{C_2^\rho} \int_{C_1^R} (Y(u, z_1)Y(v, z_2)f(z_1, z_2))\, dz_1\, dz_2 \\ &\quad - \frac{1}{(2\pi i)^2} \int_{C_2^\rho} \int_{C_1^r} (Y(v, z_2)Y(u, z_1)f(z_1, z_2))\, dz_1\, dz_2 \\ &= \frac{1}{(2\pi i)^2} \int_{C_2^\rho} \int_{C_1^\varepsilon(z_2)} (Y(Y(u, z_1 - z_2)v, z_2)f(z_1, z_2))\, dz_1\, dz_2. \quad \text{(A.2.22)} \end{aligned}$$

By Propositions A.2.1 and A.2.5 all three integrands have the same analytic continuation to a rational operator-valued function. The identity thus follows from Cauchy's theorem and the contour-deformation shown in Fig. 1. ∎

We single out an important special case of the Jacobi identity, namely when

$$f(z_1, z_2) = f_1(z_1)f_2(z_2), \quad f_i(z_i) \in \mathbb{C}[z_i, z_i^{-1}], \quad i = 1, 2.$$

We obtain the commutation relations for the components of vertex operators:

$$\begin{aligned} &[\mathrm{Res}_{z_1=0}(Y(u, z_1)f_1(z_1)), \mathrm{Res}_{z_2=0}(Y(v, z_2)f_2(z_2))] \\ &\quad = \mathrm{Res}_{z_2=0} \mathrm{Res}_{z_0=0}(Y(Y(u, z_0)v, z_2)f_1(z_0 + z_2)f_2(z_2)). \quad \text{(A.2.23)} \end{aligned}$$

In particular for $f_1(z_1) = z_1^m$, $f_2(z_2) = z_2^n$ we deduce (8.6.31), which in turn implies the commutation relations for affine Lie algebras (Section 7.2), the Virasoro algebra (Proposition 8.7.1), the affinization of the subalgebra $\mathfrak{f}$ of the Griess algebra (Section 8.9), and so on.

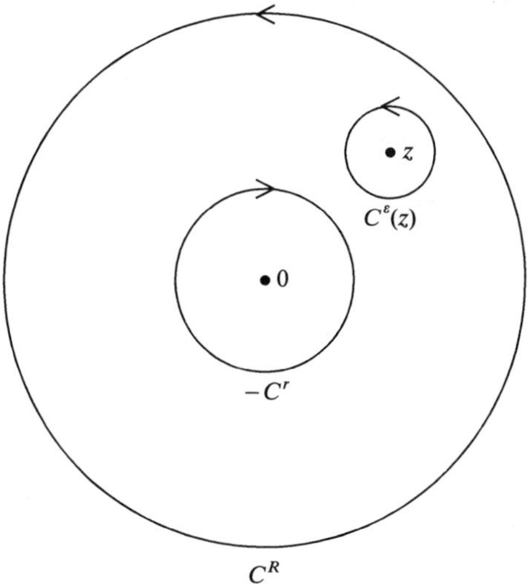

Fig. 1

The Jacobi identity immediately implies skew-symmetry:

Proposition A.2.9: *Let* $f(z_1, z_2) \in \mathbb{C}[z_1, z_1^{-1}, z_2, z_2^{-1}, (z_1 - z_2)^{-1}]$. *Then for* $u, v \in V_L$ *we have*

$$\mathrm{Res}_{z_2=0}\,\mathrm{Res}_{z_1=z_2}(Y(Y(u, z_1 - z_2)v, z_2)f(z_1, z_2))$$
$$= -\mathrm{Res}_{z_1=0}\,\mathrm{Res}_{z_2=z_1}(Y(Y(v, z_2 - z_1)u, z_1)f(z_1, z_2)).$$

We can illustrate skew-symmetry using the contour integral presentation with $R > r > 0$

$$\frac{1}{(2\pi i)^2}\int_{C_2^R}\int_{C_1^\varepsilon(z_2)}(Y(Y(u, z_1 - z_2)v, z_2)f(z_1, z_2))\,dz_1\,dz_2$$
$$= -\frac{1}{(2\pi i)^2}\int_{C_1^R}\int_{C_2^\varepsilon(z_1)}(Y(Y(v, z_2 - z_1)u, z_1)f(z_1, z_2))\,dz_2\,dz_1. \tag{A.2.24}$$

The corresponding contour picture is shown overleaf.

One can deduce from Proposition A.2.8 the following symmetric form of the Jacobi identity. We shall also give a direct proof. It again admits an interesting geometric interpretation.

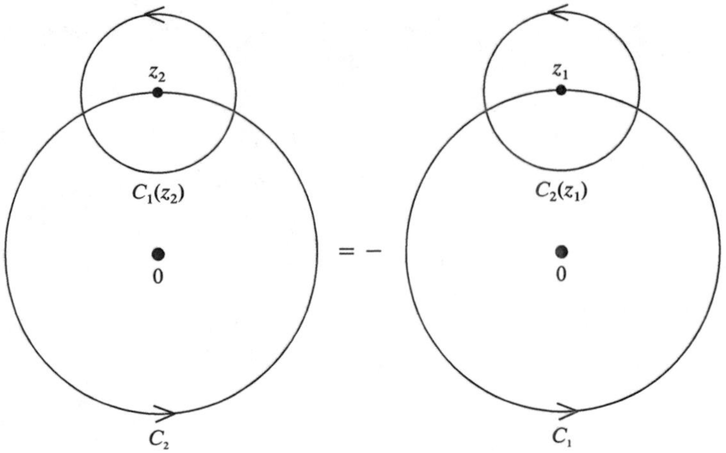

Fig. 2

Proposition A.2.10: *Let* $f(z_1, z_2, z_3) \in \mathbb{C}[z_i, z_i^{-1}, (z_i - z_j)^{-1},\ \ i \neq j = 1, 2, 3]$. *Then for* $v^1, v^2, v^3 \in V_L$ *we have*

$$\begin{aligned}
&\mathrm{Res}_{z_3=0}\,\mathrm{Res}_{z_2=z_3}\,\mathrm{Res}_{z_1=z_2}(Y(Y(Y(v^1, z_1 - z_2)v^2, z_2 - z_3)v^3, z_3)f(z_1, z_2, z_3))\\
&\quad + \mathrm{Res}_{z_1=0}\,\mathrm{Res}_{z_3=z_1}\,\mathrm{Res}_{z_2=z_3}(Y(Y(Y(v^2, z_2 - z_3)v^3, z_3 - z_1)v^1, z_1)\\
&\quad \cdot f(z_1, z_2, z_3))\\
&\quad + \mathrm{Res}_{z_2=0}\,\mathrm{Res}_{z_1=z_2}\,\mathrm{Res}_{z_3=z_1}(Y(Y(Y(v^3, z_3 - z_1)v^1, z_1 - z_2)v^2, z_2)\\
&\quad \cdot f(z_1, z_2, z_3)) = 0.
\end{aligned}$$

Proof: Let us set

$$I_{ijk} = \frac{1}{(2\pi i)^3} \int_{C_i} \int_{C_j} \int_{C_k} Y(v^i, z_i)Y(v^j, z_j)Y(v^k, z_k)f(z_1, z_2, z_3)\, dz_k\, dz_j\, dz_i \tag{A.2.25}$$

where the contour C_i contains C_j, and C_j contains C_k. Then the first term of the symmetric form of the Jacobi identity is equal to

$$I_{123} - I_{213} - I_{312} + I_{321}$$

by Proposition A.2.5, Remark A.2.6 and appropriate deformations of contours. Adding up all the cyclic permutations we get 0. ∎

The contour picture has the following form:

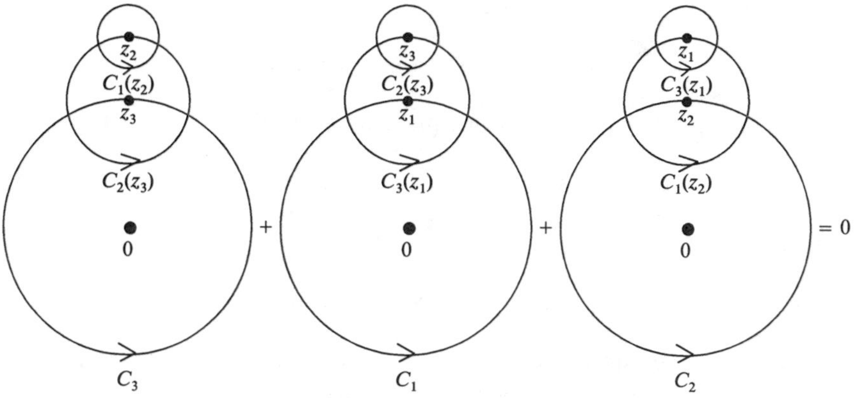

Fig. 3

Using similar techniques one can deduce further identities, for example, a generalization of the symmetric form of the Jacobi identity to n vertex operators. In fact any homologically trivial sum of configurations of contours such as in Figs. 1–3 leads to an identity for the components of general vertex operators.

We note that we have deduced the Jacobi identity using only the rationality of matrix coefficients of products of vertex operators, commutativity and the translation property. On the other hand, as was explained in Section 8.10, the Jacobi identity implies the rationality and commutativity; this fact is especially transparent in the language of formal variables.

A.3. Relation to the Formal Variable Approach

We conclude this appendix by establishing the relation of the complex variable approach with the formal variable approach used in the main body of the book (see Chapters 2 and 8).

Any element of the formal Laurent series space $\mathbb{C}[[z, z^{-1}]]$ defines a linear functional on the algebra of Laurent polynomials via the pairing

$$\langle a, b\rangle = \mathrm{Res}_z(a(z)b(z)) \tag{A.3.1}$$

where $a \in \mathbb{C}[[z, z^{-1}]]$, $b \in \mathbb{C}[z, z^{-1}]$, Res_z being the map defined by (8.4.33). An example of a formal Laurent series is the δ-function

$$\delta(z) = \sum_{n \in \mathbb{Z}} z^n \in \mathbb{C}[[z, z^{-1}]]. \tag{A.3.2}$$

For any $a \in \mathbb{C}[z, z^{-1}]$ we have

$$\langle \delta, a \rangle = \mathrm{Res}_z(\delta(z)a(z)) = a(1). \tag{A.3.3}$$

The pairing (A.3.1) can be generalized to n variables:

$$\langle a, b \rangle = \mathrm{Res}_{z_1, \ldots, z_n}(a(z_1, \ldots, z_n)b(z_1, \ldots, z_n)) \tag{A.3.4}$$

where $a \in \mathbb{C}[[z_1, z_1^{-1}, \ldots, z_n, z_n^{-1}]]$, $b \in \mathbb{C}[z_1, z_1^{-1}, \ldots, z_n, z_n^{-1}]$ and

$$\mathrm{Res}_{z_1, \ldots, z_n} = \mathrm{Res}_{z_1} \cdots \mathrm{Res}_{z_n}$$

extracts the coefficient of $z_1^{-1} \cdots z_n^{-1}$. The pairing (A.3.1) and more generally (A.3.4) can be extended to the pairing

$$(\overline{\mathrm{End}}\, V_L)[[z_1, z_1^{-1}, \ldots, z_n, z_n^{-1}]] \times \mathbb{C}[z_1, z_1^{-1}, \ldots, z_n, z_n^{-1}] \to \overline{\mathrm{End}}\, V_L \tag{A.3.5}$$

which we also denote by $\langle \cdot, \cdot \rangle$.

The matrix coefficients of compositions of vertex operators described in Propositions A.2.1, A.2.5 and Remark A.2.6 are given by rational functions in the algebra $\mathcal{R}_n$. The test functions that we used in the formulations of Propositions A.2.8, A.2.9 and A.2.10 belong to the same algebra of rational functions as the corresponding matrix coefficients. In order to pass to generating functions we introduce auxiliary variables

$$z_{ij}, \quad i, j = 1, \ldots, n, \quad i < j \tag{A.3.6}$$

and consider the algebra of Laurent polynomials

$$\mathcal{L}_n = \mathbb{C}[z_i, z_i^{-1}, z_{ij}, z_{ij}^{-1}, \quad i, j = 1, \ldots, n, \quad i < j]. \tag{A.3.7}$$

There is a natural surjection

$$s\colon \mathcal{L}_n \to \mathcal{R}_n \tag{A.3.8}$$

given by $s(z_{ij}) = z_i - z_j$, $s(z_i) = z_i$. Now if we have any identity for the components of vertex operators such as the ones given by Propositions A.2.8, A.2.9 and A.2.10 we can formulate the corresponding statement in the language of formal variables using the following δ-functions:

$$z_{12}^{-1}\delta\left(\frac{z_1 - z_2}{z_{12}}\right) = z_{12}^{-1} \sum_{n \in \mathbb{Z}} \left(\frac{z_1(1 - z_2/z_1)}{z_{12}}\right)^n \in \mathbb{C}[[z_1, z_1^{-1}, z_2, z_{12}, z_{12}^{-1}]]. \tag{A.3.9}$$

They implement the surjection in (A.3.8) via

$$\mathrm{Res}_{z_{12}}\, z_{12}^{-1}\delta\left(\frac{z_1 - z_2}{z_{12}}\right)g(z_1, z_2, z_{12}) = \iota_{12}\, g(z_1, z_2, z_1 - z_2)$$

for $g \in \mathfrak{L}_2$, using the notation ι_{12} of (8.10.37).

For any composition of n vertex operators we thus have $n(n - 1)/2$ auxiliary variables. In the complex realization we also have the same number of inequalities involving the arguments of vertex operators. To any inequality we attach a δ-function of the type (A.3.9) according to the following rule:

$$|x| > |y| \leftrightarrow z^{-1}\delta\left(\frac{x - y}{z}\right), \tag{A.3.10}$$

where z becomes an auxiliary variable. Then the given composition of n vertex operators can be multiplied by the product of $n(n - 1)/2$ δ-functions of the type (A.3.9) corresponding to the inequalities. It is crucial that the result is a well-defined formal Laurent series in $n + n(n - 1)/2$ variables. This accomplishes the translation of the contour integral identities into the corresponding formal variable identities. We illustrate this general fact by the examples below. We start with the Jacobi identity.

Proposition A.3.1: *For $u, v \in V_L$ we have*

$$z_{12}^{-1}\delta\left(\frac{z_1 - z_2}{z_{12}}\right)Y(u, z_1)Y(v, z_2) - z_{12}^{-1}\delta\left(\frac{z_2 - z_1}{-z_{12}}\right)Y(v, z_2)Y(u, z_1)$$

$$= z_1^{-1}\delta\left(\frac{z_2 + z_{12}}{z_1}\right)Y(Y(u, z_{12})v, z_2).$$

Proof: Let $w \in V_L$, $w' \in V_L'$ and set

$$f(z_1, z_2) = R\langle w', Y(u, z_1)Y(v, z_2)w\rangle$$

Then $f \in \mathcal{R}_2$ by Proposition A.2.1(ii) and

$$\langle w', Y(u, z_1)Y(v, z_2)w\rangle = \iota_{12} f(z_1, z_2)$$

by Proposition A.2.1(i) or (8.10.50). Let $g(z_1, z_2, z_{12}) \in \mathcal{L}_2$. Then

$$\operatorname{Res}_{z_1, z_2, z_{12}} z_{12}^{-1}\delta\left(\frac{z_1 - z_2}{z_{12}}\right)\langle w', Y(u, z_1)Y(v, z_2)w\rangle g(z_1, z_2, z_{12})$$

$$= \operatorname{Res}_{z_1, z_2} \iota_{12} f(z_1, z_2) \operatorname{Res}_{z_{12}} z_{12}^{-1}\delta\left(\frac{z_1 - z_2}{z_{12}}\right) g(z_1, z_2, z_{12})$$

$$= \operatorname{Res}_{z_1, z_2} \iota_{12} f(z_1, z_2) \iota_{12} g(z_1, z_2, z_1 - z_2)$$

$$= \operatorname{Res}_{z_1, z_2} \iota_{12}(f(z_1, z_2) g(z_1, z_2, z_1 - z_2))$$

$$= \operatorname{Res}_{z_1 = 0} \operatorname{Res}_{z_2 = 0} f(z_1, z_2) g(z_1, z_2, z_1 - z_2).$$

Similarly,

$$\operatorname{Res}_{z_1, z_2, z_{12}} z_{12}^{-1}\delta\left(\frac{z_2 - z_1}{-z_{12}}\right)\langle w', Y(v, z_2)Y(u, z_1)w\rangle g(z_1, z_2, z_{12})$$

$$= \operatorname{Res}_{z_2 = 0} \operatorname{Res}_{z_1 = 0} f(z_1, z_2) g(z_1, z_2, z_1 - z_2),$$

$$\operatorname{Res}_{z_1, z_2, z_{12}} z_1^{-1}\delta\left(\frac{z_2 + z_{12}}{z_1}\right)\langle w', Y(Y(u, z_{12})v, z_2)w\rangle g(z_1, z_2, z_{12})$$

$$= \operatorname{Res}_{z_2 = 0} \operatorname{Res}_{z_1 = z_2} f(z_1, z_2) g(z_1, z_2, z_1 - z_2)$$

so that the statement follows from Proposition A.2.8. ∎

The corollary about the commutators of vertex operators has an especially elegant form:

$$[Y(u, z_1), Y(v, z_2)] = \operatorname{Res}_{z_{12} = 0} z_1^{-1}\delta\left(\frac{z_2 + z_{12}}{z_1}\right) Y(Y(u, z_{12})v, z_2). \tag{A.3.11}$$

Similarly, the skew-symmetry admits the following form:

Proposition A.3.2: *For $u, v \in V_L$ we have*

$$z_1^{-1}\delta\left(\frac{z_2 + z_{12}}{z_1}\right) Y(Y(u, z_{12})v, z_2) = z_2^{-1}\delta\left(\frac{z_1 - z_{12}}{z_2}\right) Y(Y(v, -z_{12})u, z_1).$$

Finally, we obtain the following symmetric form of the Jacobi identity (cf. Remark 8.8.20):

Proposition A.3.3: *For $v^1, v^2, v^3 \in V_L$ we have*

$$z_2^{-1}\delta\left(\frac{z_3+z_{23}}{z_2}\right)z_1^{-1}\delta\left(\frac{z_3-z_{31}}{z_1}\right)z_{31}^{-1}\delta\left(\frac{z_{23}+z_{12}}{-z_{31}}\right)$$
$$\cdot\ Y(Y(Y(v^1, z_{12})v^2, z_{23})v^3, z_3)$$
$$+\ z_3^{-1}\delta\left(\frac{z_1+z_{31}}{z_3}\right)z_2^{-1}\delta\left(\frac{z_1-z_{12}}{z_2}\right)z_{12}^{-1}\delta\left(\frac{z_{31}+z_{23}}{-z_{12}}\right)$$
$$\cdot\ Y(Y(Y(v^2, z_{23})v^3, z_{31})v^1, z_1)$$
$$+\ z_1^{-1}\delta\left(\frac{z_2+z_{12}}{z_1}\right)z_3^{-1}\delta\left(\frac{z_2-z_{23}}{z_3}\right)z_{23}^{-1}\delta\left(\frac{z_{12}+z_{31}}{-z_{23}}\right)$$
$$\cdot\ Y(Y(Y(v^3, z_{31})v^1, z_{12})v^2, z_2) = 0.$$

Proof: If for $w \in V_L$, $w' \in V_L'$ we set

$$f(z_1, z_2, z_3) = R\langle w', Y(v^1, z_1)Y(v^2, z_2)Y(v^3, z_3)w\rangle$$

so that

$$\langle w', Y(Y(Y(v^1, z_{12})v^2, z_{23})v^3, z_3)w\rangle$$
$$= \iota_{3,23,12} f(z_3 + z_{23} + z_{12}, z_3 + z_{23}, z_3)$$

and similarly for the other terms. Now the argument of Proposition A.3.1 reduces the identity to Proposition A.2.10. ∎

We note that even the direct proof of Proposition A.2.10 can be translated into the language of formal variables. Thus the integrals I_{ijk} (A.2.25) give rise to the operators

$$z_{ij}^{-1}\delta\left(\frac{z_i-z_j}{z_{ij}}\right)z_{jk}^{-1}\delta\left(\frac{z_j-z_k}{z_{jk}}\right)(-z_{ki}^{-1})\delta\left(\frac{z_i-z_k}{-z_{ki}}\right)Y(v^i, z_i)Y(v^j, z_j)Y(v^k, z_k).$$

Remark A.3.4: Using the rule (A.3.10) it is not difficult to show generally that for any homologically trivial sum of configurations of contours one obtains a formal variable identity.

Specialization of the formal variable identities by setting all the elements of V_L equal to $\iota(1)$ leads to δ-function identities. Thus the identities of Propositions A.3.1–A.3.3 yield (8.8.36), (8.8.11) and an identity involving products of three δ-functions:

$$z_{12}^{-1}\delta\left(\frac{z_1 - z_2}{z_{12}}\right) - z_{12}^{-1}\delta\left(\frac{-z_2 + z_1}{z_{12}}\right) = z_2^{-1}\delta\left(\frac{z_1 - z_{12}}{z_2}\right) \quad \text{(A.3.12)}$$

$$z_1^{-1}\delta\left(\frac{z_2 + z_{12}}{z_1}\right) = z_2^{-1}\delta\left(\frac{z_1 - z_{12}}{z_2}\right) \quad \text{(A.3.13)}$$

$$\sum_{(123)} z_2^{-1}\delta\left(\frac{z_3 + z_{23}}{z_2}\right)z_1^{-1}\delta\left(\frac{z_3 - z_{31}}{z_1}\right)z_{31}^{-1}\delta\left(\frac{z_{23} + z_{12}}{-z_{31}}\right) = 0, \quad \text{(A.3.14)}$$

where the sum is taken over the cyclic permutations of the indices 1, 2 and 3. These identities are nothing but formal analogues of the homological identities presented in Figs. 1–3.

Bibliography

These references include those cited in the main text together with additional mathematical works on vertex operator algebras, the Monster or moonshine. We hope that our selection will be helpful to the reader who wishes to study these areas in more detail. There is some overlap between this list and the reference lists at the end of the Introduction, which the reader should also consult for further historical material and for information on the physics literature. We apologize to authors whose relevant work we have neglected to mention.

G. ANDREWS.
The Theory of Partitions, in: *Encyclopedia of Mathematics and Its Applications*, ed. by G.-C. Rota, Vol. 2, Addison–Wesley, Reading, Massachusetts, 1976.

A. A. BELAVIN, A. N. POLYAKOV and A. B. ZAMOLODCHIKOV.
Infinite conformal symmetries in two-dimensional quantum field theory, *Nucl. Phys.* **B241** (1984), 333–380.

D. BERNARD and J. THIERRY-MIEG.
Level one representations of the simple affine Kac–Moody algebras in their homogeneous gradations, *Commun. Math. Phys.* **111** (1987), 181–246.

R. E. BORCHERDS.
1. The Leech lattice and other lattices, Ph.D. Dissertation, Univ. of Cambridge, 1984.
2. The Leech lattice, *Proc. Royal Soc. London* **A398** (1985), 365–376.

3. Vertex algebras, Kac–Moody algebras, and the Monster, *Proc. Natl. Acad. Sci. USA* **83** (1986), 3068–3071.
4. Vertex algebras I, to appear.

R. E. Borcherds, J. H. Conway, L. Queen and N. J. A. Sloane.
A Monster Lie algebra?, *Advances in Math.* **53** (1984), 75–79.

N. Bourbaki.
1. *Groupes et algèbres de Lie*, Chaps. 4, 5, 6, Hermann, Paris, 1968.
2. *Groupes et algèbres de Lie*, Chaps. 7, 8, Hermann, Paris, 1975.

M. Broué.
Groupes finis, séries formelles et fonctions modulaires, Séminaire Groupes finis, tome I, *Publ. Math. Univ. Paris VII*, 1982, 105–127.

F. Buekenhout.
Diagram geometries for sporadic groups, in: Finite Groups—Coming of Age, Proc. 1982 Montreal Conference, ed. by J. McKay, *Contemporary Math.* **45** (1985), 1–32.

S. Capparelli.
Vertex operator relations for affine algebras and combinatorial identities, Ph.D. thesis, Rutgers Univ., 1988.

H. Cartan and S. Eilenberg.
Homological Algebra, Princeton Univ. Press, Princeton, 1956.

C. Chevalley.
The Algebraic Theory of Spinors, Colombia Univ. Press, New York, 1954.

J. H. Conway.
1. A perfect group of order 8,315,553,613,086,720,000 and the sporadic simple groups, *Proc. Natl. Acad. Sci. USA* **61** (1968), 398–400.
2. A group of order 8,315,553,613,086,720,000. *Bull. London Math. Soc.* **1** (1969), 79–88.
3. A characterization of Leech's lattice, *Invent. Math.* **7** (1969), 137–142.
4. Three lectures on exceptional groups, in: *Finite Simple Groups*, ed. by G. Higman and M. B. Powell, Chap. 7, Academic Press, London/New York, 1971, 215–247.
5. Groups, lattices, and quadratic forms, in: *Computers in Algebra and Number Theory*, SIAM-AMS Proc. IV, American Math. Soc., 1971, 135–139.
6. The miracle octad generator, in: *Topics in Group Theory and Computation*, ed. by M. P. J. Curran, Academic Press, New York, 1977, 62–68.
7. Monsters and moonshine, *Math. Intelligencer* **2** (1980), 165–172.
8. The automorphism group of the 26-dimensional even unimodular Lorentzian lattice, *J. Algebra* **80** (1983), 159–163.
9. A simple construction for the Fischer–Griess monster group, *Invent. Math.* **79** (1985), 513–540.
10. The Monster group and its 196884-dimensional space, Chap. 29 of: J. H. Conway and N. J. A. Sloane, *Sphere Packings, Lattices and Groups*, Springer-Verlag, New York, 1988, 555–567.

J. H. Conway, R. T. Curtis, S. P. Norton, R. A. Parker and R. A. Wilson.
ATLAS of Finite Groups, Oxford Univ. Press, 1985.

J. H. Conway and S. P. Norton.
Monstrous moonshine, *Bull. London Math. Soc.* **11** (1979), 308–339.

J. H. Conway, S. P. Norton and L. H. Soicher.
The Bimonster, the group Y_{555}, and the projective plane of order 3, Proc. "Computers in Algebra" Conference, Chicago, 1985, to appear.

J. H. Conway, R. A. Parker and N. J. A. Sloane.
The covering radius of the Leech lattice, *Proc. Royal Soc. London* **A380** (1982), 261–290.

J. H. Conway and A. D. Pritchard.
Hyperbolic reflections for the Bimonster and $3Fi_{24}$, *J. Algebra*, to appear.

J. H. Conway and L. Queen.
Computing the character table of a Lie group, in: Finite Groups—Coming of Age, Proc. 1982 Montreal Conference, ed. by J. McKay, *Contemporary Math.* **45** (1985), 51–87.

J. H. Conway and N. J. A. Sloane.
1. On the enumeration of lattices of determinant one, *J. Number Theory* **15** (1982), 83–94.
2. Twenty-three constructions for the Leech lattice, *Proc. Royal Soc. London* **A381** (1982), 275–283.
3. Lorentzian forms for the Leech lattice, *Bull. American Math. Soc.* **6** (1982), 215–217.
4. Leech roots and Vinberg groups, *Proc. Royal Soc. London* **A384** (1982), 233–258.
5. *Sphere Packings, Lattices and Groups*, Springer-Verlag, New York, 1988.

M. Craig.
A cyclotomic construction for Leech's lattice, *Mathematika* **25** (1978), 236–241.

R. T. Curtis.
A new combinatorial approach to M_{24}, *Math. Proc. Cambridge Philos. Soc.* **79** (1976), 25–42.

E. Date.
On a calculation of vertex operators for $E_n^{(1)}$ ($n = 6, 7, 8$), in: *Group Representations and Systems of Differential Equations*, Advanced Studies in Pure Math. **4** (1984), 225–261.

E. Date, M. Jimbo, M. Kashiwara and T. Miwa.
1. Operator approach to the Kadomtsev–Petviashvili equation—transformation groups for soliton equations III, *J. Phys. Soc. Japan* **50** (1981), 3806–3812.
2. Transformation groups for soliton equations—Euclidean Lie algebras and reduction of the KP hierarchy, *Publ. Research Inst. for Math. Sciences*, Kyoto Univ. **18** (1982), 1077–1110.

E. Date, M. Kashiwara and T. Miwa.
Vertex operators and τ functions—transformation groups for soliton equations II, *Proc. Japan Acad. Ser. A Math. Sci.* **57** (1981), 387–392.

C.-Y. Dong.
Structure of some nonstandard modules for $C_n^{(1)}$, *J. Algebra*, to appear.

C.-Y. Dong and J. Lepowsky.
A Jacobi identity for relative vertex operators and the equivalence of Z-algebras and parafermion algebras, in: *XVIIth International Colloquium on Group Theoretical Methods in Physics*, Ste-Adèle, June, 1988, Proceedings, ed. by Y. Saint-Aubin, World Scientific, to appear.

D. Dummit, H. Kisilevsky and J. McKay.
Multiplicative products of η-functions, in: Finite Groups—Coming of Age, Proc. 1982 Montreal Conference, ed. by J. McKay, *Contemporary Math.* **45** (1985), 89–98.

B. L. Feigin.
The semi-infinite cohomology of Kac–Moody and Virasoro Lie algebras, *Usp. Mat. Nauk* **39** (1984), 195–196. English transl., *Russian Math. Surveys* **39** (1984), 155–156.

B. L. Feigin and D. B. Fuks.
1. Skew-symmetric invariant differential operators on a line and Verma modules over the Virasóro algebra, *Funk. Anal. i Prilozhen.* **16** (1982), 47–63. English transl., *Funct. Anal. Appl.* **16** (1982), 114–126.
2. Representations of the Virasoro algebra, preprint.

A. Feingold.
Some applications of vertex operators to Kac–Moody algebras, in: *Vertex Operators in Mathematics and Physics, Proc. 1983 M.S.R.I. Conference*, ed. by J. Lepowsky, S. Mandelstam and I. M. Singer, Publ. Math. Sciences Res. Inst. #3, Springer-Verlag, New York, 1985, 185–206.

A. Feingold and I. B. Frenkel.
1. A hyperbolic Kac–Moody algebra and the theory of Siegel modular forms of genus 2, *Math. Ann.* **263** (1983), 87–144.

2. Classical affine algebras, *Advances in Math.* **56** (1985), 117–172.

A. Feingold and J. Lepowsky.

The Weyl–Kac character formula and power series identities, *Advances in Math.* **29** (1978), 271–309.

L. Figueiredo.

Calculus of principally twisted vertex operators, *Memoirs American Math. Soc.* **69**, 1987.

B. Fischer, D. Livingstone and M. P. Thorne.

The characters of the "Monster" simple group, Birmingham, 1978.

P. Fong.

Characters arising in the Monster-modular connection, in: The Santa Cruz Conference on Finite Groups, *Proc. Symp. Pure Math., American Math. Soc.* **37** (1980), 557–559.

I. B. Frenkel.

1. Spinor representations of affine Lie algebras, *Proc. Natl. Acad. Sci. USA* **77** (1980), 6303–6306.
2. Two constructions of affine Lie algebra representations and boson-fermion correspondence in quantum field theory, *J. Funct. Analysis* **44** (1981), 259–327.
3. Representations of affine Lie algebras, Hecke modular forms and Korteweg–deVries type equations, in: Proc. 1981 Rutgers Conf. on Lie Algebras and Related Topics, ed. by D. Winter, Springer *Lecture Notes in Math.* **933** (1982), 71–110.
4. Representations of Kac–Moody algebras and dual resonance models, in: Applications of Group Theory in Physics and Mathematical Physics, Proc. 1982 Chicago Summer Seminar, ed. by M. Flato, P. Sally and G. Zuckerman, *Lectures in Applied Math., American Math. Soc.* **21** (1985), 325–353.
5. Beyond affine Lie algebras, in: *Proc. Internat. Congr. Math.*, Berkeley, 1986, Vol. 1, American Math. Soc., 1987, 821–839.

I. B. Frenkel, H. Garland and G. Zuckerman.

Semi-infinite cohomology and string theory, *Proc. Natl. Acad. Sci. USA* **83** (1986), 8442–8446.

I. B. Frenkel, Y.-Z. Huang and J. Lepowsky.

On axiomatic approaches to vertex operator algebras and modules, to appear.

I. B. Frenkel and N. Jing.

Vertex representations of quantum affine algebras, *Proc. Natl. Acad. Sci. USA* **85** (1988).

I. B. Frenkel and V. G. Kac.

Basic representations of affine Lie algebras and dual resonance models, *Invent. Math.* **62** (1980), 23–66.

I. B. Frenkel, J. Lepowsky and A. Meurman [FLM].

1. An E_8-approach to F_1, in: Finite Groups—Coming of Age, Proc. 1982 Montreal Conference, ed. by J. McKay, *Contemporary Math.* **45** (1985), 99–120.
2. A natural representation of the Fischer–Griess Monster with the modular function J as character, *Proc. Natl. Acad. Sci. USA* **81** (1984), 3256–3260.
3. A moonshine module for the Monster, in: *Vertex Operators in Mathematics and Physics, Proc. 1983 M.S.R.I. Conference*, ed. by J. Lepowsky, S. Mandelstam and I.M. Singer, Publ. Math. Sciences Res. Inst. #3, Springer-Verlag, New York, 1985, 231–273.
4. An introduction to the Monster, in: *Unified String Theories, Proc. 1985 Inst. for Theoretical Physics Workshop*, ed. by M. Green and D. Gross, World Scientific, Singapore, 1986, 533–546.
5. Vertex operator calculus, in: *Mathematical Aspects of String Theory, Proc. 1986 Conference, San Diego*, ed. by S.-T. Yau, World Scientific, Singapore, 1987, 150–188.

H. Garland.

1. The arithmetic theory of loop algebras, *J. Algebra* **53** (1978), 480–551.
2. The arithmetic theory of loop groups, *Inst. Hautes Etudes Sci. Publ. Math.* **52** (1980), 5–136.

3. Lectures on loop algebras and the Leech lattice, Yale University, Spring 1980.

I. M. GELFAND and D. B. FUKS.
The cohomology of the Lie algebra of vector fields on a circle, *Funk. Anal. i Prilozhen.* **2** (1968), 92–93. English transl., *Funct. Anal. Appl.* **2** (1968), 342–343.

P. GODDARD, W. NAHM, D. OLIVE, H. RUEGG and A. SCHWIMMER.
Fermions and octonions, *Commun. Math. Phys.* **112** (1987), 385–408.

P. GODDARD, W. NAHM, D. OLIVE and A. SCHWIMMER.
Vertex operators for non-simply-laced algebras, *Comm. Math. Phys.* **107** (1986), 179–212.

P. GODDARD and D. OLIVE.
1. Algebras, lattices and strings, in: *Vertex Operators in Mathematics and Physics, Proc. 1983 M.S.R.I. Conference*, ed. by J. Lepowsky, S. Mandelstam and I. M. Singer, Publ. Math. Sciences Res. Inst. #3, Springer-Verlag, New York, 1985, 51–96.
2. Kac–Moody and Virasoro algebras in relation to quantum physics, *Internat. J. Modern Physics* **A1** (1986), 303–414.
3. Algebras, lattices and strings 1986, *Physica Scripta* **T15** (1987), 19–25.

R. GOODMAN and N. R. WALLACH.
1. Structure of unitary cocycle representations of loop groups and the group of diffeomorphisms of the circle, *J. Reine Angew. Math.* **347** (1984), 69–133.
2. Projective unitary positive-energy representations of Diff(S^1), *J. Funct. Anal.* **63** (1985), 299–321.

D. GORENSTEIN.
1. Finite Groups, Harper and Row, New York, 1968.
2. *Finite Simple Groups. An Introduction to their Classification*, Plenum Press, New York, 1982.
3. Classifying the finite simple groups, Colloquium Lectures, Anaheim, January 1985, Amer. Math. Soc., *Bull. Amer. Math. Soc.* (*New Series*) **14** (1986), 1–98.

R. L. GRIESS, JR.
1. The structure of the "Monster" simple group, in: *Proc. of the Conference on Finite Groups*, ed. by W. R. Scott and R. Gross, Academic Press, New York, 1976, 113–118.
2. A construction of F_1 as automorphisms of a 196,883 dimensional algebra, *Proc. Natl. Acad. Sci. USA* **78** (1981), 689–691.
3. The Friendly Giant, *Invent. Math.* **69** (1982), 1–102.
4. The sporadic simple groups and construction of the Monster, in: *Proc. Internat. Congr. Math., Warsaw, 1983*, North-Holland, Amsterdam, 1984, 369–384.
5. The Monster and its nonassociative algebra, in: Finite Groups—Coming of Age, Proc. 1982 Montreal Conference, ed. by J. McKay, *Contemporary Math.* **45** (1985), 121–157.
6. A brief introduction to the finite simple groups, in: *Vertex Operators in Mathematics and Physics, Proc. 1983 M.S.R.I. Conference*, ed. by J. Lepowsky, S. Mandelstam and I. M. Singer, Publ. Math. Sciences Res. Inst. #3, Springer-Verlag, New York, 1985, 217–229.
7. Code loops, *J. Algebra* **100** (1986), 224–234.
8. Sporadic groups, code loops and nonvanishing cohomology, *J. Pure Appl. Alg.* **44** (1978), 191–214.

R. L. GRIESS, JR., U. MEIERFRANKENFELD and Y. SEGEV.
A uniqueness proof for the Monster, to appear.

A. GROTHENDIECK.
Sur quelques points d'algèbre homologique, *Tohoku Math. J.* **9** (1957), 119–221.

K. HARADA and M. L. LANG. Some elliptic curves arising from the Leech lattice, to appear.

J. HUMPHREYS.
Introduction to Lie Algebras and Representation Theory, Third Printing, Revised, Springer-Verlag, New York, Heidelberg, Berlin, 1980.

B. HUPPERT.
Finite Groups III, Springer-Verlag, Berlin–Heidelberg–New York, 1982.

N. JACOBSON.
1. *Lie Algebras*, Wiley-Interscience, New York, 1962.
2. *Basic Algebra I*, W. H. Freeman and Company, San Francisco, 1974.
3. *Basic Algebra II*, W. H. Freeman and Company, San Francisco, 1980.

M. JIMBO and T. MIWA.
1. Soliton equations and fundamental representations of $A_{2l}^{(2)}$, *Lett. Math. Phys.* **6** (1982), 463–469.
2. Solitons and infinite dimensional Lie algebras, *Publ. Research Inst. for Math. Sciences*, Kyoto Univ. **19** (1983), 943–1001.

V. G. KAC.
1. Highest weight representations of infinite dimensional Lie algebras, in: *Proc. Internat. Congr. Math., Helsinki, 1978*, Acad. Scientiarum Fennica, Helsinki, 1980, 299–304.
2. Infinite-dimensional algebras, Dedekind's η-function, classical Möbius function and the very strange formula, *Advances in Math.* **30** (1978), 85–136.
3. An elucidation of "Infinite-dimensional . . . and the very strange formula" $E_8^{(1)}$ and the cube root of the modular invariant j, *Advances in Math.* **35** (1980), 264–273.
4. A remark on the Conway-Norton conjecture about the "Monster" simple group, *Proc. Natl. Acad. Sci. USA* **77** (1980), 5048–5049.
5. *Infinite-dimensional Lie Algebras*, 2nd ed., Cambridge Univ. Press, Cambridge, 1985.

V. G. KAC, D. A. KAZHDAN, J. LEPOWSKY and R. L. WILSON.
Realization of the basic representations of the Euclidean Lie algebras, *Advances in Math.* **42** (1981), 83–112.

V. G. KAC, R. V. MOODY and M. WAKIMOTO.
On E_{10}, to appear.

V. G. KAC and D. H. PETERSON.
1. Affine Lie algebras and Hecke modular forms, *Bull. American Math. Soc. (New Series)* **3** (1980), 1057–1061.
2. Spin and wedge representations of infinite dimensional Lie algebras and groups, *Proc. Natl. Acad. Sci. USA* **78** (1981), 3308–3312.
3. Infinite-dimensional Lie algebras, theta functions and modular forms, *Advances in Math.* **53** (1984), 125–264.
4. 112 constructions of the basic representation of the loop group of E_8, in: *Symposium on Anomalies, Geometry, Topology, 1985, Proceedings*, ed. by W. A. Bardeen and A. R. White, World Scientific, Singapore (1985), 276–298.

M. KASHIWARA and T. MIWA.
The τ function of the Kadomtsev-Petviashvili equation—transformation groups for soliton equations I, *Proc. Japan Acad. Ser. A Math. Sci.* **57** (1981), 342–347.

M. I. KNOPP.
Modular Functions in Analytic Number Theory, Markham, Chicago, 1970.

K. KOIKE.
1. Mathieu group M_{24} and modular forms, *Nagoya Math. J.* **99** (1985), 147–157.
2. Modular forms and the automorphism group of the Leech lattice, to appear.

T. KONDO.
The automorphism group of the Leech lattice and elliptic modular functions, *J. Math. Soc. Japan* **37** (1985), 337–362.

T. KONDO and T. TASAKA.
1. The theta functions of sublattices of the Leech lattice, *Nagoya Math. J.* **101** (1986), 151–179.
2. The theta functions of sublattices of the Leech lattice II, *J. Fac. Sci. Univ. Tokyo*, to appear.

M. L. LANG.
On a question raised by Conway-Norton, *J. Math. Soc. Japan*, to appear.

S. Lang.

1. *Algebra*, Benjamin/Cummings, Menlo Park, California, 1984.
2. *Introduction to Modular Forms*, Springer-Verlag, Berlin–Heidelberg–New York, 1976.

J. Leech.

1. Some sphere packings in higher space, *Canadian J. Math.* **16** (1964), 657–682.
2. Notes on sphere packings, *Canadian J. Math.* **19** (1967), 251–267.

J. Leech and N. J. A. Sloane

Sphere packings and error-correcting codes, *Canadian J. Math.* **23** (1971), 718–745.

J. Lepowsky.

1. Lie algebras and combinatorics, in: *Proc. Internat. Congr. Math., Helsinki, 1978*, Acad. Scientiarum Fennica, Helsinki, 1980, 579–584.
2. Euclidean Lie algebras and the modular function *j*, in: Santa Cruz Conference on Finite Groups, 1979, *Proc. Symp. Pure Math., American Math. Soc.* **37** (1980), 567–570.
3. Affine Lie algebras and combinatorial identities, in: Proc. 1981 Rutgers Conf. on Lie Algebras and Related Topics, ed. by D. Winter, *Lecture Notes in Math.* **933**, Springer-Verlag, Berlin–Heidelberg–New York, 1982, 130–156.
4. Calculus of twisted vertex operators, *Proc. Natl. Acad. Sci. USA* **82** (1985), 8295–8299.
5. Perspectives on vertex operators and the Monster, in: Proc. 1987 Symposium on the Mathematical Heritage of Hermann Weyl, Duke Univ., *Proc. Symp. Pure Math., American Math. Soc.* **48** (1988).
6. The algebra of general twisted vertex operators, to appear.

J. Lepowsky and A. Meurman.

An E_8-approach to the Leech lattice and the Conway group, *J. Algebra* **77** (1982), 484–504.

J. Lepowsky and M. Primc.

1. Standard modules for type one affine Lie algebras, in: Number Theory, New York, 1982, ed. by D. V. Chudnovsky, G. V. Chudnovsky, H. Cohn and M. B. Nathanson, *Lecture Notes in Math.* **1052**, Springer-Verlag, Berlin–Heidelberg–New York, 1984, 194–251.
2. Structure of the standard modules for the affine Lie algebra $A_1^{(1)}$, *Contemporary Math.* **46**, 1985.
3. Structure of the standard $A_1^{(1)}$-modules in the homogeneous picture, in: *Vertex Operators in Mathematics and Physics, Proc. 1983 M.S.R.I. Conference*, ed. by J. Lepowsky, S. Mandelstam and I. M. Singer, Publ. Math. Sciences Res. Inst. #3, Springer-Verlag, New York, 1985, 143–162.

J. Lepowsky and R. L. Wilson.

1. Construction of the affine Lie algebra $A_1^{(1)}$, *Commun. Math. Phys.* **62** (1978), 43–53.
2. A Lie theoretic interpretation and proof of the Rogers–Ramanujan identities, *Advances in Math.* **45** (1982), 21–72.
3. A new family of algebras underlying the Rogers–Ramanujan identities and generalizations, *Proc. Natl. Acad. Sci. USA* **78** (1981), 7254–7258.
4. The structure of standard modules, I: Universal algebras and the Rogers–Ramanujan identities, *Invent. Math.* **77** (1984), 199–290.
5. The structure of standard modules, II: The case $A_1^{(1)}$, principal gradation, *Invent. Math.* **79** (1985), 417–442.
6. *Z*-algebras and the Rogers–Ramanujan identities, in: *Vertex Operators in Mathematics and Physics, Proc. 1983 M.S.R.I. Conference*, ed. by J. Lepowsky, S. Mandelstam and I. M. Singer, Publ. Math. Sciences Res. Inst. #3, Springer-Verlag, New York, 1985, 97–142.

J. H. Lindsey, II

1. On the Suzuki and Conway groups, in: *Representation Theory of Finite Groups and Related Topics, Proc. Symp. Pure Math., American Math. Soc.* **21** (1971), 107–109.
2. A correlation between $PSU_4(3)$, the Suzuki group, and the Conway group, *Trans. American Math. Soc.* **157** (1971), 189–204.

I. G. MACDONALD.

1. *Symmetric Functions and Hall Polynomials*, Oxford Univ. Press, Oxford, 1979.
2. Affine Lie algebras and modular forms, Séminaire Bourbaki, exposé no. 577, 1980/81.

F. J. MACWILLIAMS and N. J. A. SLOANE.

The Theory of Error Correcting Codes, North-Holland, New York, 1977.

M. MANDIA.

Structure of the level one standard modules for the affine Lie algebras $B_l^{(1)}$, $F_4^{(1)}$ and $G_2^{(1)}$, *Memoirs American Math. Soc.* **65**, 1987.

G. MASON.

1. M_{24} and certain automorphic forms, in: Finite Groups—Coming of Age, Proc. 1982 Montreal Conference, ed. by J. McKay, *Contemporary Math.* **45** (1985), 223–244.
2. Modular forms and the theory of Thompson series, in: *Proc. Rutgers Group Theory Year, 1983–1984*, ed. by M. Aschbacher *et al.*, Cambridge Univ. Press, Cambridge, 1984, 391–407.
3. Finite groups and Hecke operators, *Math. Annalen*, to appear.
4. Finite groups and modular functions (with an appendix by S. P. Norton), In: Representations of Finite Groups, Proc. 1986 Arcata Summer Research Institute, ed. by P. Fong, Proc. Symp. Pure Math., *American Math. Soc.* **47** (1987), 181–210.

J. MCKAY.

A setting for the Leech lattice, in: *Finite Groups '72*, ed. by T. Gagen *et al.*, North-Holland, Amsterdam, 1973, 117–118.

A. MEURMAN and M. PRIMC.

Annihilating ideals of standard modules of $\mathfrak{sl}(2, \mathbb{C})^{\sim}$ and combinatorial identities, *Advances in Math.* **64** (1987), 177–240.

A. MEURMAN and A. ROCHA-CARIDI.

Highest weight representations of the Neveu–Schwarz and Ramond algebras, *Commun. Math. Phys.* **107** (1986), 263–294.

K. C. MISRA.

1. Structure of certain standard modules for $A_n^{(1)}$ and the Rogers–Ramanujan identities, *J. Algebra* **88** (1984), 196–227.
2. Structure of some standard modules for $C_n^{(1)}$, *J. Algebra* **90** (1984), 385–409.
3. Standard representations of some affine Lie algebras, in: *Vertex Operators in Mathematics and Physics, Proc. 1983 M.S.R.I. Conference*, ed. by J. Lepowsky, S. Mandelstam and I. M. Singer, Publ. Math. Sciences Res. Inst. #3, Springer-Verlag, New York, 1985, 163–183.
4. Realization of the level two standard $\mathfrak{sl}(2k + 1, \mathbb{C})^{\sim}$-modules, to appear.

D. MITZMAN.

Integral bases for affine Lie algebras and their universal enveloping algebras, *Contemporary Math.* **40,** American Math. Soc., 1985.

R. V. MOODY.

Generalized root systems and characters, in: Finite Groups—Coming of Age, Proc, 1982 Montreal Conference, ed. by J. McKay, *Contemporary Math.* **45** (1985), 245–269.

H.-V. NIEMEIER.

Definite quadratische Formen der Dimension 24 und Diskriminante 1, *J. Number Theory* **5** (1973), 142–178.

S. P. NORTON.

1. More on moonshine, in: *Computational Group Theory*, ed. by M. D. Atkinson, Academic Press, New York, 1984, 185–193.
2. The uniqueness of the Fischer–Griess Monster, in: Finite Groups—Coming of Age, Proc, 1982 Montreal Conference, ed. by J. McKay, *Contemporary Math.* **45** (1985), 271–285.

A. P. OGG.

1. Hyperelliptic modular curves, *Bull. Soc. Math. France* **102** (1974), 449–462.

2. Automorphismes des courbes modulaires, *Séminaire Delange–Pisot–Poitou, 16e année* (1974/75), no. 7.
3. Modular functions, in: Santa Cruz Conference on Finite Groups, 1979, *Proc. Symp. Pure Math., American Math. Soc.* **37** (1980), 521–532.

R. PFISTER.

Spin representations of $A_1^{(1)}$, Ph.D. thesis, Rutgers Univ., 1984.

A. PRESSLEY and G. SEGAL.

Loop Groups, Oxford Univ. Press, Oxford, 1986.

L. QUEEN.

1. Some relations between finite groups, Lie groups and modular functions, Ph.D. dissertation, Univ. of Cambridge, 1980.
2. Modular functions and finite simple groups, in: Santa Cruz Conference on Finite Groups, 1979, *Proc. Symp. Pure Math., American Math. Soc.* **37** (1980), 561–566.
3. Modular functions arising from some finite groups, *Mathematics of Computation* **37** (1981), 547–580.

A. ROCHA-CARIDI.

On highest weight and Fock space representations of the Virasoro algebra, to appear.

M. A. RONAN and S. D. SMITH.

2-local geometries for some sporadic groups, in: Santa Cruz Conference on Finite Groups, 1979, *Proc. Symp. Pure Math., American Math. Soc.* **37** (1980), 283–289.

M. SATO and Y. SATO.

1. On the bilinear equations of Hirota I (in Japanese), *Sürikaiseki Kenkyûsho Kôkyûroku* **388** (1980), 183.
2. On the bilinear equations of Hirota II (in Japanese), *Sürikaiseki Kenkyûsho Kôkyûroku* **414** (1981), 181.

G. SEGAL.

1. Unitary representations of some infinite-dimensional groups, *Commun. Math. Phys.* **80** (1981), 301–342.
2. Loop groups, in: Arbeitstagung Bonn 1984, *Lecture Notes in Math.* **1111,** Springer-Verlag, Berlin–Heidelberg–New York, 1985, 155–168.

J.-P. SERRE.

1. *A Course in Arithmetic*, Springer-Verlag, New York, Heidelberg, Berlin, 1973.
2. *Local Fields*, Springer-Verlag, New York, Heidelberg, Berlin, 1979.

N. N. SHAPOVALOV.

On a bilinear form on the universal enveloping algebra of a complex semi-simple Lie algebra, *Funct. Anal. Appl.* **6** (1972), 307–312.

N. J. A. SLOANE.

Self-dual codes and lattices, in: Relations between Combinatorics and Other Parts of Mathematics, *Proc. Symp. Pure Math., American Math. Soc.* **34** (1979), 273–308.

S. SMITH.

On the Head characters of the Monster simple group, in: Finite Groups—Coming of Age, Proc. 1982 Montreal Conference, ed. by J. McKay, *Contemporary Math.* **45** (1985), 303–313.

J. G. THOMPSON.

1. A simple subgroup of $E_8(3)$, in: *Finite Groups Symposium*, ed. by N. Iwahori, Japan Soc. Promotion Science, Tokyo, 1976, 113–116.
2. Finite groups and even lattices, *J. Algebra* **38** (1976), 523–524.
3. Uniqueness of the Fischer–Griess Monster, *Bull. London Math. Soc.* **11** (1979), 340–346.
4. Finite groups and modular functions, *Bull. London Math. Soc.* **11** (1979), 347–351.
5. Some numerology between the Fischer–Griess Monster and elliptic modular functions, *Bull. London Math. Soc.* **11** (1979), 352–353.
6. Some finite groups which appear as Gal L/K, where $K \subseteq Q(\mu_n)$, *J. Algebra* **89** (1984), 437–499.

7. Some finite groups which appear as Gal L/K, where $K \subseteq Q(\mu_n)$, in: *Group Theory, Beijing 1984*, Lecture Notes in *Math.* **1185**, Springer-Verlag, Berlin–Heidelberg–New York, 1986, 210–230.

C. B. THORN.

Computing the Kac determinant using dual model techniques and more about the no-ghost theorem, *Nucl. Phys.* **B248** (1984), 551.

J. TITS.

1. Groupes finis simples sporadiques, Séminaire Bourbaki, exposé no. 375, 1970.
2. Four presentations of Leech's lattice, in: *Finite Simple Groups II, Proc. of a London Math. Soc. Research Symposium, Durham, 1978*, ed. by M. J. Collins, Academic Press, London/New York, 1980, 303–307.
3. Quaternions over $\mathbb{Q}(\sqrt{5})$, Leech's lattice, and the sporadic group of Hall–Janko, *J. Algebra* **63** (1980), 56–75.
4. Résumé de cours, Annuaire du Collège de France, 1982–1983, 89–102.
5. Le Monstre, Séminaire Bourbaki, exposé no 620, 1983/84, *Astérisque* **121–122** (1985), 105–122.
6. On R. Griess' "Friendly Giant," *Invent. Math.* **78** (1984), 491–499.
7. Résumé de cours, Annuaire du Collège de France, 1985–1986, 101–112.
8. Le module du "moonshine," Séminaire Bourbaki, exposé no. 684, 1986/87.

A. TSUCHIYA and Y. KANIE.

1. Fock space representations of the Virasoro algebra—intertwining operators, *Publ. Research Inst. for Math. Sciences, Kyoto Univ.* **22** (1986), 259–327.
2. Vertex operators in the conformal field theory on $\mathbb{P}^1$ and monodromy representations of the braid group, *Lett. Math. Phys.* **13** (1987), 303.
3. Vertex operators in conformal field theory on $\mathbb{P}^1$ and monodromy representations of braid group, in: *Conformal Field Theory and Solvable Lattice Models, Advanced Studies in Pure Math.* **16** (1988).

H. TSUKADA.

String path integral realization of vertex operator algebras, Ph.D. thesis, Rutgers Univ., 1988.

B. B. VENKOV.

On the classification of integral even unimodular 24-dimensional quadratic forms, *Proc. Steklov Inst. Math.* **4** (1980), 63–74.

M. A. VIRASORO.

Subsidiary conditions and ghosts in dual-resonance models, *Phys. Rev.* **D1** (1970), 2933–2936.

N. R. WALLACH.

Classical invariant theory and the Virasoro algebra, in: *Vertex Operators in Mathematics and Physics, Proc. 1983 M.S.R.I. Conference*, ed. by J. Lepowsky, S. Mandelstam and I. M. Singer, Publ. Math. Sciences Res. Inst. #3, Springer-Verlag, New York, 1985, 475–482.

X.-D. WANG.

Structure of some level two standard modules for $A_{n-1}^{(1)}$, preprint, Harbin Normal University, P.R.C.

List of Frequently Used Symbols

The reader should note that some symbols have different meanings depending on the context. The Notational Conventions (page li) should also be consulted. The numbers to the right are the pages on which the symbols are explained.

Chapter 1

d	derivation, 3
$\mathfrak{g}$	Lie algebra, 3
$[\cdot,\cdot]$	Lie bracket, 3
End	algebra of endomorphisms, 4
π	representation, 4
$/$	quotient, 6, 9
Cent	center, 7
$\mathfrak{g}'$	commutator of $\mathfrak{g}$, 7
$\times$	direct product, 7
$\ltimes$, $\rtimes$	semidirect product, 8
V_α	subspace of V of degree α, 8
$\coprod$	direct sum, 8

deg	degree, 8
$\otimes$	tensor product, 10
Ind	induced module, 11, 13, 15
$\mathbb{F}[G]$	group algebra of G, 12
Aut	automorphism group, 12
e^a	basis of $\mathbb{F}[G]$, 12
$T(V)$	tensor algebra of V, 13
$T^n(V)$	nth tensor power of V, 13
$S(V)$	symmetric algebra of V, 14
$S^n(V)$	nth symmetric power of V, 14
$U(\mathfrak{g})$	universal enveloping algebra of $\mathfrak{g}$, 14
$\langle \cdot, \cdot \rangle$	symmetric bilinear form, 16, 17
$\mathbb{F}[t, t^{-1}]$	algebra of Laurent polynomials in t, 17
$d = t\frac{d}{dt}$	derivation of $\mathbb{F}[t, t^{-1}]$, 17
c	central element, 18
$\hat{\mathfrak{g}}$	affine Lie algebra, 18
$\tilde{\mathfrak{g}}$	extended affine Lie algebra, 18
$\hat{\mathfrak{g}}[\theta]$, $\tilde{\mathfrak{g}}[\theta]$	affine Lie algebra twisted by an involution θ (resp., extended twisted algebra), 20
$\mathfrak{l}^{\pm}$	positive (resp., negative) subalgebra of $\mathfrak{l}$, 21
$M(k)$, $M(\lambda)$	induced modules, 22, 26
$\mathfrak{h}$	finite-dimensional abelian Lie algebra, 24
$\hat{\mathfrak{h}}_{\mathbb{Z}}$	untwisted affinization of $\mathfrak{h}$, 25
$\hat{\mathfrak{h}}_{\mathbb{Z}+1/2}$	twisted affinization of $\mathfrak{h}$ (twisted Heisenberg algebra), 25
$Z = \mathbb{Z}$ or $\mathbb{Z} + 1/2$	grading set, 25
$(\cdot, \cdot)$	symmetric bilinear or Hermitian form, 26, 30
$T_{p(t)}$, d_n	derivations of $\mathbb{F}[t, t^{-1}]$, 31
$\mathfrak{d}$	Witt algebra, 31
$\mathfrak{v}$	Viragoro algebra, 32
L_n	generators of the Virasoro algebra, 32
$h(n)$, $L(n)$	representation of $h \otimes t^n$ (resp., of L_n), 34, 36
ω_0	distinguished element of $T^2(\mathfrak{h})$, 35
$\mathbf{1}$	vacuum vector of $S(\hat{\mathfrak{h}}_Z^-)$, 40
wt	weight, 41
$\dim_* V$	graded dimension, 42
$\eta(q)$	Dedekind η-function, 45

Chapter 2

$V\{z\}$, $V[z]$, $V[z, z^{-1}]$, $V[[z]]$, $V[[z, z^{-1}]]$	certain spaces of formal series, 48, 49
$\delta(z)$	formal δ-function, 50

$\lim_{z_1 \to z_2}$	specialization, 51
$D = D_z = z\frac{d}{dz}$	derivation, 56
D^{-1}	formal integration, 57
$x(z)$	generating function, 58
$x^{\pm} = x \pm \theta x$	θ-symmetrization (resp., antisymmetrization), 60

Chapter 3

$\theta_1, \theta_2, \sigma$	involutions of $\mathfrak{sl}(2, \mathbb{F})$, 63, 64, 66
V	twisted space, 67
$E^{\pm}(\alpha, z)$	exponential series, 69
$X(\alpha, z) = X_{Z+1/2}(\alpha, z)$	twisted vertex operator, 70
$: \ :$	normal ordering, 73
$\alpha(z)^{\pm}$	positive (resp., negative) part of generating function, 74
$X^{\pm}(\alpha, z)$	symmetrization (resp., antisymmetrization) of vertex operator, 80

Chapter 4

$E^{\pm}(\alpha, z)$	exponential series, 85
e^{α}	group algebra element, multiplication operator, 86
$X(\alpha, z) = X_Z(\alpha, z)$	untwisted vertex operator, 87
$V_{\mathfrak{h}}, V_M$	untwisted spaces, 87, 89
$\alpha(z)^{\pm}$	positive (resp., negative) part of generating function, 89
$: \ :$	normal ordering, 90
σ	isomorphism of twisted and untwisted space, 98

Chapter 5

$\langle \kappa \rangle$	group generated by κ, 102
ε_0	2-cocycle, 103
c_0	commutator map, 104
$\check{A} = A/2A$	abelian 2-group, 111
T	induced module, 118

Chapter 6

L	lattice, 122
L_m	set of elements of square length m, 123
L°	dual lattice, 123
$\theta_L(q)$	theta function, 125

$\hat{L}$	central extension, 126
Δ	root system, 127
$\mathfrak{g}$	Lie algebra associated to Δ, 127
x_a, x_α	root vector, 127, 128
$\varepsilon(\cdot,\cdot)$	multiplicative 2-cocycle, 128
Q	root lattice of $\mathfrak{g}$, 133
P	weight lattice of $\mathfrak{g}$, 133
A_n, D_n, E_n	root lattices and Lie algebra, 134
θ	lifting of -1, 138

Chapter 7

$\mathbb{F}\{L\}$	induced L-module, 146
ι	homomorphism, 146
z^h	operator in $\mathbb{F}\{L\}$, 147
$X(a, z) = X_{\mathbb{Z}}(a, z)$	untwisted vertex operator, 148
$\hat{A}_n, \hat{D}_n, \hat{E}_n$	affine Kac–Moody algebras, 156
T, T_χ	$\hat{L}$-modules, 158, 167
V_L^T	twisted space, 158
$X(a, z) = X_{\mathbb{Z}+1/2}(a, z)$	twisted vertex operator, 159
$\hat{A}_n[\theta], \hat{D}_n[\theta], \hat{E}_n[\theta]$	twisted affine Kac–Moody algebras, 169

Chapter 8

$\mathbb{F}(z)$	field of rational functions, 175
$\mathbb{F}((z))$	field of fractions of formal power series ring, 175
$\Theta, \iota_\pm$	linear maps, 176
e^{yT}	one-parameter group, 180
$a(z)$	untwisted generating function (new definition), 191
$\alpha(z)^\pm$	nonnegative (resp., negative) part of generating function (new definition), 191
$\circ\circ \ \circ\circ$	normal ordering (new definition), 191
$Y(a, z) = Y_{\mathbb{Z}}(a, z)$	untwisted vertex operator, 192
Res_{z_0}	residue, 195
$Y(v, z) = Y_{\mathbb{Z}}(v, z)$	general untwisted vertex operator, 198
v_n	component of vertex opeator, 200
ω	vector corresponding to Virasoro algebra, 217
$[\cdot \times_n \cdot], [\cdot \times_{z_0} \cdot]$	products, 229, 230
$[\cdot \times_l \cdot]_{mn}, [\cdot \times_l \cdot](m, n)$	components of products, 229
$[\cdot \times_1 \cdot]$	cross-bracket, 229
θ	involution of V_L, 239

$V_L^{\pm\theta}$	± 1-eigenspaces of θ, 239
$\mathfrak{k}$	subspace of V_L^{θ} of weight 2, 240
$\hat{\mathfrak{h}}$	commutative afinization, 241
rank V	rank of vertex operator algebra, 245

Chapter 9

$\alpha(z)$	twisted generating function (new definition), 257
$\alpha(z)^{\pm}$	positive (resp., negative) part of generating function (new definition), 257
${}^{\circ}_{\circ}\ {}^{\circ}_{\circ}$	normal ordering, 257
$Y(a, z) = Y_{\mathbb{Z}+1/2}(a, z)$	twisted vertex operator, 258
Δ_z	quadratic differential operator, 264
$Y(v, z) = Y_{\mathbb{Z}+1/2}(v, z)$	general twisted vertex operator, 266
θ	involution of V_L^T, 274
$(V_L^T)^{\pm\theta}$	± 1-eigenspace of θ, 274
$\mathfrak{p}$	subspace of V_L^T of weight 2, 295

Chapter 10

Ω	finite set, 298
$\mathfrak{C}$	binary linear code, 298
$\mathfrak{C}^{\circ}$	dual code, 299
M_{24}	Mathieu group, 302
L_0, L_0'	lattices associated with $\mathfrak{C}$, 303
Λ	Leech lattice, 304
Γ	E_8-root lattice, 305
$N(\cdot)$	Niemeier lattices, 308
Co_0, Co_1	Conway groups, 310
T	canonical $\hat{\Lambda}$-module, 313
$V^{\natural}$	moonshine module, 315
W_{Λ}	sum of untwisted and twisted modules, 315
$\mathfrak{B}$	Griess algebra, 316
$\hat{\mathfrak{B}}$	affinization of Griess algebra, 317
C	centralizer of involution in M, 324
z	central involution in C, 325
ch_*	graded character, 330
$PSL(2, \mathbb{Z})$	modular group, 336
$J(q)$	modular invariant with zero constant term, 337

Chapter 11

Q	root lattice of type A_1^{24}, 343
L_0, L_1, L	lattices associated with $\mathcal{C}$, 343
V_{L_0}, V_{L_1}, V_L	untwisted spaces, 344
V'_{L_0}, V'_{L_1}, V'_L	twisted spaces, 345
$\mathfrak{g}$	Lie algebra of type A_1^{24}, 348
σ_0	involution of $\mathfrak{g}$, 349
σ_1	involution of W_L, 349, 350, 364, 371

Chapter 12

Ω	24-element set, 375
$\mathcal{C}$	binary Golay code, 375
$\mathcal{L}, L, \mathcal{L}_0, L_0, \Lambda_0,$	
$\mathcal{L}_{00}, L_{00}, \Lambda_{00}$	lattices associated with Λ, 376
$\Lambda_0, \Lambda_1, \Lambda_2, \Lambda_3$	cosets associated with Λ, 376, 389
$K_\Lambda, K_L, K_{\mathcal{L}}$	groups associated with lattices, 380, 383
T_M, V_M, W_M	spaces associated with a union of cosets M of Q, 385
σ	automorphism of W_L and $V^\natural$, 396
τ	automorphism of $V^\natural$, 397
M	Monster, 400
$V_{\mathbb{Q}}^\natural$	$\mathbb{Q}$-form of $V^\natural$, 408
$(\cdot,\cdot)$	M-invariant positive definite Hermitian form on $V^\natural$, 411

Chapter 13

N	certain group, 420
H	certain group, 423
$\tilde{N}$	group associated with $N(A_1^{24})$, 431
$\tilde{H}$	group associated with $N(D_4^6)$, 441

Appendix

$\bar{V}$	infinite product of graded spaces, 462
$\overline{\mathrm{End}}, \overline{\mathrm{Hom}}$	spaces of linear operators from sums to products, 463, 464
R	analytic continuation, 466
$\frac{1}{2\pi i}\int_C dz$	contour integral, 473

Index

T

U

V

W

Z

PURE AND APPLIED MATHEMATICS

Vol. 1 Arnold Sommerfeld, *Partial Differential Equations in Physics*

Vol. 2 Reinhold Baer, *Linear Algebra and Projective Geometry*

Vol. 3 Herbert Busemann and Paul Kelly, *Projective Geometry and Projective Metrics*

Vol. 4 Stefan Bergman and M. Schiffer, *Kernel Functions and Elliptic Differential Equations in Mathematical Physics*

Vol. 5 Ralph Philip Boas, Jr., *Entire Functions*

Vol. 6 Herbert Busemann, *The Geometry of Geodesics*

Vol. 7 Claude Chevalley, *Fundamental Concepts of Algebra*

Vol. 8 Sze-Tsen Hu, *Homotopy Theory*

Vol. 9 A. M. Ostrowski, *Solution of Equations in Euclidean and Banach Spaces, Third Edition of Solution of Equations and Systems of Equations*

Vol. 10 J. Dieudonné, *Treatise on Analysis: Volume I, Foundations of Modern Analysis; Volume II; Volume III; Volume IV; Volume V; Volume VI; Volume VII*

Vol. 11* S. I. Goldberg, *Curvature and Homology*

Vol. 12* Sigurdur Helgason, *Differential Geometry and Symmetric Spaces*

Vol. 13 T. H. Hildebrandt, *Introduction to the Theory of Integration*

Vol. 14 Shreeram Abhyankar, *Local Analytic Geometry*

Vol. 15* Richard L. Bishop and Richard J. Crittenden, *Geometry of Manifolds*

Vol. 16* Steven A. Gaal, *Point Set Topology*

Vol. 17 Barry Mitchell, *Theory of Categories*

Vol. 18 Anthony P. Morse, *A Theory of Sets*

Vol. 19 Gustave Choquet, *Topology*

Vol. 20 Z. I. Borevich and I. R. Shafarevich, *Number Theory*

Vol. 21 José Luis Massera and Juan Jorge Schaffer, *Linear Differential Equations and Function Spaces*

Vol. 22 Richard D. Schafer, *An Introduction to Nonassociative Algebras*

Vol. 23* Martin Eichler, *Introduction to the Theory of Algebraic Numbers and Functions*

Vol. 24 Shreeram Abhyanker, *Resolution of Singularities of Embedded Algebraic Surfaces*

* Presently out of print

Vol. 25 François Treves, *Topological Vector Spaces, Distributions, and Kernels*
Vol. 26 Peter D. Lax and Ralph S. Phillips, *Scattering Theory*
Vol. 27 Oystein Ore, *The Four Color Problem*
Vol. 28* Maurice Heins, *Complex Function Theory*
Vol. 29 R. M. Blumenthal and R. K. Getoor, *Markov Processes and Potential Theory*
Vol. 30 L. J. Mordell, *Diophantine Equations*
Vol. 31 J. Barkley Rosser, *Simplified Independence Proofs: Boolean Valued Models of Set Theory*
Vol. 32 William F. Donoghue, Jr., *Distributions and Fourier Transforms*
Vol. 33 Marston Morse and Stewart S. Cairns, *Critical Point Theory in Global Analysis and Differential Topology*
Vol. 34* Edwin Weiss, *Cohomology of Groups*
Vol. 35 Hans Freudenthal and H. De Vries, *Linear Lie Groups*
Vol. 36 Laszlo Fuchs, *Infinite Abelian Groups*
Vol. 37 Keio Nagami, *Dimension Theory*
Vol. 38 Peter L. Duren, *Theory of H^p Spaces*
Vol. 39 Bodo Pareigis, *Categories and Functors*
Vol. 40* Paul L. Butzer and Rolf J. Nessel, *Fourier Analysis and Approximation: Volume I, One-Dimensional Theory*
Vol. 41* Eduard Prugovećki, *Quantum Mechanics in Hilbert Space*
Vol. 42 D. V. Widder, *An Introduction to Transform Theory*
Vol. 43 Max D. Larsen and Paul J. McCarthy, *Multiplicative Theory of Ideals*
Vol. 44 Ernst-August Behrens, *Ring Theory*
Vol. 45 Morris Newman, *Integral Matrices*
Vol. 46 Glen E. Bredon, *Introduction to Compact Transformation Groups*
Vol. 47 Werner Greub, Stephen Halperin, and Ray Vanstone, *Connections, Curvature, and Cohomology:*
Volume I, De Rham Cohomology of Manifolds and Vector Bundles
Volume II, Lie Groups, Principal Bundles, and Characteristic Classes
Volume III, Cohomology of Principal Bundles and Homogeneous Spaces
Vol. 48 Xia Dao-xing, *Measure and Integration Theory of Infinite-Dimensional Spaces: Abstract Harmonic Analysis*

Vol. 49 Ronald G. Douglas, *Banach Algebra Techniques in Operator Theory*
Vol. 50 Willard Miller, Jr., *Symmetry Groups and Theory Applications*
Vol. 51 Arthur A. Sagle and Ralph E. Walde, *Introduction to Lie Groups and Lie Algebras*
Vol. 52 T. Benny Rushing, *Topological Embeddings*
Vol. 53* James W. Vick, *Homology Theory: An Introduction to Algebraic Topology*
Vol. 54 E. R. Kolchin, *Differential Algebra and Algebraic Groups*
Vol. 55 Gerald J. Janusz, *Algebraic Number Fields*
Vol. 56 A. S. B. Holland, *Introduction to the Theory of Entire Functions*
Vol. 57 Wayne Roberts and Dale Varberg, *Convex Functions*
Vol. 58 H. M. Edwards, *Riemann's Zeta Function*
Vol. 59 Samuel Eilenberg, *Automata, Languages, and Machines: Volume A, Volume B*
Vol. 60 Morris Hirsch and Stephen Smale, *Differential Equations, Dynamical Systems, and Linear Algebra*
Vol. 61 Wilhelm Magnus, *Noneuclidean Tesselations and Their Group*
Vol. 62 François Treves, *Basic Linear Partial Differential Equations*
Vol. 63* William M. Boothby, *An Introduction to Differentiable Manifolds and Riemannian Geometry*
Vol. 64 Brayton Gray, *Homotopy Theory: An Introduction to Algebraic Topology*
Vol. 65 Robert A. Adams, *Sobolev Spaces*
Vol. 66 John J. Benedetto, *Spectral Synthesis*
Vol. 67 D. V. Wilder, *The Heat Equation*
Vol. 68 Irving Ezra Segal, *Mathematical Cosmology and Extragalactic Astronomy*
Vol. 69 I. Martin Isaacs, *Character Theory of Finite Groups*
Vol. 70 James R. Brown, *Ergodic Theory and Topological Dynamics*
Vol. 71 C. Truesdell, *A First Course in Rational Continuum Mechanics: Volume 1, General Concepts*
Vol. 72 K. D. Stroyan and W. A. J. Luxemburg, *Introduction to the Theory of Infinitesimals*
Vol. 73 B. M. Puttaswamaiah and John D. Dixon, *Modular Representations of Finite Groups*
Vol. 74 Melvyn Berger, *Nonlinearity and Functional Analysis: Lectures on Nonlinearity Problems in Mathematical Analysis*
Vol. 75 George Gratzer, *Lattice Theory*

Vol. 76 Charalambos D. Aliprantis and Owen Burkinshaw, *Locally Solid Riesz Spaces*

Vol. 77 Jan Mikusinski, *The Bochner Integral*

Vol. 78 Michiel Hazelwinkel, *Formal Groups and Applications*

Vol. 79 Thomas Jech, *Set Theory*

Vol. 80 Sigurdur Helgason, *Differential Geometry, Lie Groups, and Symmetric Spaces*

Vol. 81 Carl L. DeVito, *Functional Analysis*

Vol. 82 Robert B. Burckel, *An Introduction to Classical Complex Analysis*

Vol. 83 C. Truesdell and R. G. Muncaster, *Fundamentals of Maxwell's Kinetic Theory of a Simple Monatomic Gas: Treated as a Branch of Rational Mechanics*

Vol. 84 Louis Halle Rowen, *Polynomial Identities in Ring Theory*

Vol. 85 Joseph J. Rotman, *An Introduction to Homological Algebra*

Vol. 86 Barry Simon, *Functional Integration and Quantum Physics*

Vol. 87 Dragos M. Cvetkovic, Michael Doob, and Horst Sachs, *Spectra of Graphs*

Vol. 88 David Kinderlehrer and Guido Stampacchia, *An Introduction to Variational Inequalities and Their Applications*

Vol. 89 Herbert Seifert, W. Threlfall, *A Textbook of Topology*

Vol. 90 Grezegorz Rozenberg and Arto Salomaa, *The Mathematical Theory of L Systems*

Vol. 91 Donald W. Kahn, *Introduction to Global Analysis*

Vol. 92 Eduard Prugovečki, *Quantum Mechanics in Hilbert Space, Second Edition*

Vol. 93 Robert M. Young, *An Introduction to Nonharmonic Fourier Series*

Vol. 94 M. C. Irwin, *Smooth Dynamical Systems*

Vol. 96 John B. Garnett, *Bounded Analytic Functions*

Vol. 97 Jean Dieudonné, A Panorama of Pure Mathematics: As Seen by N. Bourbaki

Vol. 98 Joseph G. Rosenstein, *Linear Orderings*

Vol. 99 M. Scott Osborne and Garth Warner, *The Theory of Eisenstein Systems*

Vol. 100 Richard V. Kadison and John R. Ringrose, *Fundamentals of the Theory of Operator Algebras:*
Volume 1, Elementary Theory; Volume 2, Advanced Theory

Vol. 101 Howard Osborn, *Vector Bundles: Volume 1, Foundations and Stiefel-Whitney Classes*

Vol. 126 J. M. G. Fell and R. S. Doran, *Representations of *-Algebras, Locally Compact Groups, and Banach *-Algebraic Bundles: Volume 2, Induced Representations, the Imprimitivity Theorem, and the Generalized Mackey Analysis*

Vol. 127 Louis H. Rowen, *Ring Theory, Volume I*

Vol. 128 Louis H. Rowen, *Ring Theory, Volume II*

Vol. 129 Colin Bennett and Robert Sharpley, *Interpolation of Operators*

Vol. 130 Jürgen Pöschel and Eugene Trubowitz, *Inverse Spectral Theory*

Vol. 131 Jens Carsten Jantzen, *Representations of Algebraic Groups*

Vol. 132 Nolan R. Wallach, *Real Reductive Groups I*

Vol. 133 Michael Sharpe, *General Theory of Markov Processes*

Vol. 134 Igor Frenkel, James Lepowsky, and Arne Meurman, *Vertex Operator Algebras and the Monster*